TIMOTHY D. SANDERS
Occidental College
Los Angeles, California

MODERN PHYSICAL THEORY

Special Relativity and Quantum Physics

ADDISON-WESLEY PUBLISHING COMPANY
Reading, Massachusetts · Menlo Park, California · London · Don Mills, Ontario

This book is in the
ADDISON-WESLEY SERIES IN PHYSICS

Consulting Editor: DAVID LAZARUS

Published simultaneously in Canada. Library of Congress Catalog Card No. 75-100885.

PREFACE

This is a textbook on the fundamental theoretical ideas and techniques of modern physics—those embodied in special relativity and quantum theory. It is intended for seniors and beginning graduate students majoring in physics who have had good backgrounds in vector analysis, differential equations, complex variables, and intermediate electromagnetic theory. The emphasis is on mathematical formulation of fundamental ideas, together with the illustrative calculations to show how the theory is used in obtaining quantities that can be checked against experiment, thereby demonstrating that the theory is believable. Although the book is intended to stand alone as a text, it may be used effectively in a course that parallels a course on the phenomenology of modern physics, or it can be used in a course in which the material in the last third of this book is intertwined with the phenomenology.

Although previous work in linear algebra or matrix theory might be desirable for a student using this book, such background is not essential. The necessary linear algebra and tensor tools are developed in this text as the theory evolves. Most of the theory is developed in several dimensions because the student lives in three space dimensions; it is intended that he should see immediately how the theory is applicable to three- (and four-) dimensional problems.

The first chapter is a review of electromagnetic theory, beginning from the experimental laws of Coulomb and Biot–Savart, later introducing Faraday's law of electromagnetic induction, and culminating in Maxwell's equations and the wave equations for electromagnetic field quantities. This sets the stage for the failure of Galilean invariance in electromagnetic theory and the consequent fundamental disagreement between classical mechanics and electromagnetic theory. This review necessarily is brief and is not appropriate as an introduction to electromagnetic theory. Rather it is intended to set the stage for the work and fun that is to come, and to orient the reader to a point of view that the author believes to be particularly useful.

In succeeding chapters, special relativity and its tensor formulation are developed from the crucial and well-known *Gedanken Experimente*. The tensor formulation of electromagnetism is developed as an application of the theory, as are techniques for handling the kinetics of collision problems. In particular, the energy and momentum densities of the electromagnetic field are obtained from the four-dimensional formulation.

Knowing the energy and momentum densities of the electromagnetic field, one can proceed naturally to the investigation of the spectral distribution of black-body radiation. Planck's law is found in a rather conventional manner and then we consider briefly in Chapters 6 and 7 such topics as the photoelectric effect, Compton collisions, Rutherford scattering and the Bohr atom. For graduate students, the material in these chapters may be reviewed quickly; they are primarily for the benefit of undergraduates who have had only an elementary introduction to modern physics.

A chapter on the behavior of waves in general, including discussions of Fourier analysis, group velocity, and the uncertainty principles provide the basis for an introduction to matter waves. We consider the interpretation of the electromagnetic wave field as a photon probability amplitude, and this leads to the notion that material bodies also might be represented by waves, provided the wave amplitude is interpreted as a probability amplitude. Using the uncertainty principle already discussed, in conjunction with the "certainty" of the de Broglie wavelength for a particle having perfectly well-defined momentum, the wave function for a state of well-defined momentum is obtained, and the relationship between the momentum operator and the gradient vector is derived.

Mathematical questions are considered at some length at this point, and the connection between kinematic operators and group operators is mentioned; a more-or-less standard brief treatment of operators, eigenvalues, eigenfunctions, Hilbert space and normalization problems is given. Translations of the time axis are used to introduce the concept of energy, and the Klein–Gordon and Schrödinger equations follow naturally from this.

Angular momentum operators are obtained from the assumption of rotational symmetry, and the separation of coordinates in the wave equations for rotationally symmetric interactions among particles is achieved. The spherical harmonics are derived and solutions to certain radial equations are considered. This leads to the solution of the problem of a spinless one-electron nonrelativistic atom as the first realistic application of the theory, and some aspects of atomic structure are discussed subsequently. Elementary perturbation theory is introduced and is applied to semiclassical radiation theory. Some failures of the theory are noted, and as an attempt at overcoming these failures, the relativistic wave problem is reconsidered. We develop the Dirac equation, with its implications of intrinsic particle spin.

We return to the problem of atomic structure through the treatment of identical particles, together with a brief discussion on quantum statistical mechanics. The final chapter is concerned with collision theory.

The book is intended for a one-year course, but could be adapted to a one-semester course through the elimination of various sections. For example, a one-semester graduate course could be constructed by omitting Sections 4.4, 4.5, 5.3–5.7, Chapters 6 and 7, and Chapters 15–19 or 17–19 depending upon the rate at which material is covered. The decision as to what material to include or eliminate depends heavily on the preparation and intended further study of the students in the course.

The author wishes to acknowledge the thoughtful and critical comments of his students and colleagues on the manuscript and the lectures from which it evolved. The assistance of Occidental College and Addison-Wesley Publishing Company during the preparation of the manuscript also is gratefully acknowledged. The understanding and encouragement of my family during the many hours I was with the manuscript instead of with them is a gift for which I am most grateful.

T.D.S.

Pasadena, California
December 1969

CONTENTS

CHAPTER 0

A COMMENT ON PHYSICS AND PHYSICAL THEORIES

Physics is, essentially, the systematic study of the simplest aspects of the structure and interactions of matter in terms of space and time. That is not to say that physics always is simple; if it were, we would soon tire of studying it. Rather, physicists try to abstract from natural phenomena the least complicated nontrivial aspects of those phenomena, and to develop from them conceptual "models" that can be used to correlate as many facts about the phenomena as possible. In principle, the nature of the models is unimportant as long as they successfully correlate in a precise way a wide range of experimental data; as it happens, however, most useful models eventually are cast into mathematical form. Nevertheless, it must be remembered that the mathematical models of physics are not, in themselves, physics.

For physicists, mathematics is a tool and a language that is used to describe, concisely and with some precision, the underlying consistencies and order that have been discovered among the experimental data of physics. When a physical theory, in the form of a mathematical system, predicts phenomena that are not observed, or, more importantly, when a phenomenon is observed that a theory predicts to be impossible, the theory must give way to the physical facts—that is, to the palpable "truth" of the *observed* phenomenon.

Any new theory should correlate successfully all the experimental information encompassed by its predecessors, together with phenomena that the old theories failed to embrace. In short, physical theories are expedient and, as far as can be determined, accurate mathematical *models* for natural phenomena; they are man-made devices and they therefore possess the inevitable feature of obsolescence that plagues all other manufactures of man.

Theories, however imperfect the best of them may be, nevertheless are very useful. When they provide an efficient and cogent means for the presentation and discussion of physical data, they are serving their intended purpose. For this reason, there is an unusually heavy emphasis on contemporary theoretical approaches to modern physics in the material that follows. Physics has advanced very rapidly in recent years, and the transmission of accumulated data and insights to successive generations of physicists now involves an average of about ten years in the life of a prospective new physicist. There is little prospect of significantly shortening that time span in the immediate future, but perhaps through a greater utilization of existing theories—especially relatively well-established ones—in the presentation and development of physical facts and ideas in the physics curriculum, we can hope at least to continue the tradition that physics students are provided with a rather broad and modern view of their chosen subject before they take up specialized research or move into rather specialized areas of industrial work.

We here begin our studies with an investigation of the mathematical formalism that commonly is used as a model for classical electromagnetic fields. Subsequently we shall uncover a conflict between this formalism and the picture of the world that is implicit in classical mechanics. The existence of this conflict indicates that we have failed to include some physical data in one or the other, or both, of the classical formulation of physics; either Maxwellian electromagnetism or Newtonian mechanics is incomplete as a description of the real data of physics. One can think of a theory as a carrier of information—each of the mathematical symbols involved represents some bit of information about the physical system, and the juxtaposition of these symbols represents the manner in which the data are related to one another; thus the clash of the two classical theories betrays our omission of some information about the real physical world in constructing those models.

The role of modern physical theory is to find, through the analysis of more extensive experimental data, just what has been left out of classical physics, and to construct new mathematical models that are more comprehensive but that retain the "correct" features of their classical forerunners. The new formalisms that emerge are not to be considered any more sacred than the old; they are as concise and convenient representations of the facts as we are able to construct, and are *more nearly* complete than the old in that they are capable of handling a wider range of data. However, they are *not complete*, and the reader should bear in mind continually that *physics* is the phenomenon underlying *data* correlated by the mathematical symbols. Physics should not be confused with the convenient shorthand language that we happen to use in codifying that data.

READING NOTES*

Reference 16 has a very interesting introductory historical sketch in Chapter 1.

Reference 1 is a rather technical and scholarly history of nonrelativistic quantum mechanics that would provide interesting reading to accompany the present chapters on quantum theory.

Reference 2 is a collection of essays by, for, and about Einstein and his historic contributions to 20th-century physics.

---

* Works referred to in these notes will be found in the Bibliography on p. 557.

CHAPTER 1

REVIEW OF ELECTROMAGNETISM

1.1 THE DIRAC δ-FUNCTION

Let us begin by studying a mathematical expression, namely the so-called Dirac delta-function, $\delta(x)$, which is defined by the following properties:

$$\delta(x) = 0 \qquad \text{if } x \neq 0 \tag{1.1a}$$

$$\int_L \delta(x)\,dx = \begin{cases} 1 & \text{if } x = 0 \text{ is in the line segment } L \\ 0 & \text{if } x = 0 \text{ is not in the line segment } L, \end{cases} \tag{1.1b}$$

where by $\int_L$ we mean $\int_a^b$ when L is the line segment from a to b. Although $\delta(x)$ is is not strictly a function in the usual mathematical sense, it can be given a definite mathematical meaning in terms of theory of *generalized functions*, or *distributions*. Our treatment of the δ-function will be informal but plausible; for a rigorous examination of its properties, the reader should consult mathematical texts on distribution theory.

In view of the definition of $\delta(x)$, we have

$$\int_L f(x)\delta(x)\,dx = f(0) \tag{1.1c}$$

for any function $f(x)$ sufficiently well behaved (such as those one usually encounters in physics) on the line segment L containing $x = 0$.

In three dimensions we write

$$\delta^3(\mathbf{x}) = \delta(x_1)\delta(x_2)\delta(x_3), \tag{1.2}$$

where x_1, x_2 and x_3 are the Cartesian coordinates in three dimensions. This three-dimensional δ-function then has the following properties analogous to (1.1a), (1.1b) and (1.1c) above:

$$\delta^3(\mathbf{x}) = 0 \qquad \text{if } \mathbf{x} \neq 0 \tag{1.3a}$$

$$\int_V \delta^3(\mathbf{x})\,d^3x = \begin{cases} 1 & \text{if } \mathbf{x} = 0 \text{ is in volume } V \\ 0 & \text{if } \mathbf{x} = 0 \text{ is not in volume } V \end{cases} \tag{1.3b}$$

$$\int_V \delta^3(\mathbf{x}) f(\mathbf{x})\,d^3x = f(0) \quad \text{if } \mathbf{x} = 0 \text{ is in volume } V. \tag{1.3c}$$

It should be noted that (1.3c) is a consequence of the defining properties (1.3a) and (1.3b), just as property (1.1c) follows from (1.1a) and (1.1b). If the point $x = 0$ should fail to fall in the line segment L in the one-dimensional case, the integral of property (1.1c) automatically vanishes because $\delta(x)$ vanishes throughout the interval of integration; similarly, if the origin $\mathbf{x} = \mathbf{0}$ is not contained in the volume V of integration involved in property (1.3c), the integral of $\delta^3(\mathbf{x})f(\mathbf{x})$ vanishes because of the vanishing of the integrand throughout the integration volume.

Now let us demonstrate a specific representation of this three-dimensional δ-function. ($\delta(x)$ and $\delta^3(\mathbf{x})$ can be expressed in many forms, as we shall discover.) By direct differentiation, one obtains, for $\mathbf{x} \neq \mathbf{x}'$,

$$\begin{aligned}\nabla^2 \frac{1}{|\mathbf{x}-\mathbf{x}'|} &= -\nabla\cdot\frac{\mathbf{x}-\mathbf{x}'}{|\mathbf{x}-\mathbf{x}'|^3} = -\left[\frac{1}{|\mathbf{x}-\mathbf{x}'|^3}\nabla\cdot(\mathbf{x}-\mathbf{x}') + (\mathbf{x}-\mathbf{x}')\cdot\nabla\frac{1}{|\mathbf{x}-\mathbf{x}'|^3}\right] \\ &= -\left[\frac{3}{|\mathbf{x}-\mathbf{x}'|^3} - 3\frac{(\mathbf{x}-\mathbf{x}')^2}{|\mathbf{x}-\mathbf{x}'|^5}\right] = 0; \end{aligned} \tag{1.4}$$

hence $\nabla^2(1/|\mathbf{x}-\mathbf{x}'|)$ satisfies property (1.3a). Also, if $\mathbf{x}'$ lies outside a region having volume V and enclosed by a surface S, then

$$\int_V \nabla^2 \frac{1}{|\mathbf{x}-\mathbf{x}'|}\, d^3x = 0 \tag{1.5}$$

because $\mathbf{x} \neq \mathbf{x}'$ for every vector $\mathbf{x}$ contained in V, and hence $\nabla^2(1/|\mathbf{x}-\mathbf{x}'|)$ vanishes everywhere in V. On the other hand, if V is chosen so that it contains $\mathbf{x}'$, we have

$$\begin{aligned}\int_V \nabla^2 \frac{1}{|\mathbf{x}-\mathbf{x}'|}\, d^3x &= \int_V \nabla\cdot\nabla\frac{1}{|\mathbf{x}-\mathbf{x}'|}\, d^3x \\ &= \int_S \nabla\frac{1}{|\mathbf{x}-\mathbf{x}'|}\cdot d\mathbf{S} = -\int_S \frac{\mathbf{x}-\mathbf{x}'}{|\mathbf{x}-\mathbf{x}'|^3}\cdot d\mathbf{S}\end{aligned} \tag{1.6}$$

according to Gauss's theorem. Now let us choose a small sphere centered at $\mathbf{x}'$, having volume V' and surface S', and contained entirely within V. Since $\nabla^2(1/|\mathbf{x}-\mathbf{x}'|) = 0$ everywhere within the volume $V - V'$ (i.e. everywhere within V but outside of V'), we have

$$\int_V \nabla^2 \frac{1}{|\mathbf{x}-\mathbf{x}'|}\, d^3x - \int_{V'} \nabla^2 \frac{1}{|\mathbf{x}-\mathbf{x}'|}\, d^3x = 0. \tag{1.7}$$

But

$$\begin{aligned}\int_{V'} \nabla^2 \frac{1}{|\mathbf{x}-\mathbf{x}'|}\, d^3x &= -\int_{S'} \frac{\mathbf{x}-\mathbf{x}'}{|\mathbf{x}-\mathbf{x}'|^3}\cdot d\mathbf{S} \\ &= -\int_{S'} \frac{\mathbf{x}-\mathbf{x}'}{|\mathbf{x}-\mathbf{x}'|^3}\cdot(\mathbf{x}-\mathbf{x}')|\mathbf{x}-\mathbf{x}'|\, d\Omega = -\int_{S'} d\Omega \\ &= -\int_0^{2\pi} d\phi \int_0^{\pi} \sin\theta\, d\theta = -4\pi\end{aligned} \tag{1.8}$$

where $d\Omega = \sin\theta\, d\theta\, d\phi$ is an element of solid angle about the point x'. This result, combined with Eq. (1.7) leads to

$$\int_V \nabla^2 \frac{1}{|\mathbf{x} - \mathbf{x}'|} d^3\mathbf{x} = -4\pi \tag{1.9}$$

if $\mathbf{x}'$ is within V. Therefore $(-1/4\pi)\nabla^2(1/|\mathbf{x} - \mathbf{x}'|)$ possesses properties (1.3a) and (1.3b), and we may write

$$-\frac{1}{4\pi}\nabla^2 \frac{1}{|\mathbf{x} - \mathbf{x}'|} = \delta^3(\mathbf{x} - \mathbf{x}'). \tag{1.10}$$

This representation of the three-dimensional δ-function is particularly useful in the investigation of electromagnetic fields.

1.2 CHARGE AND CURRENT DENSITIES AS SOURCES OF ELECTROMAGNETIC FIELDS

Now we shall consider two of the fundamental laws of electromagnetism, Coulomb's law and the law of Biot and Savart. A mathematical generalization of the experimental facts is given in the following two expressions for the force per unit volume experienced by an electric charge density $\rho(\mathbf{x})$, and an electric current density $\mathbf{j}(\mathbf{x})$, respectively:

$$\mathbf{f}_C(\mathbf{x}) = \frac{1}{4\pi\epsilon_0}\rho(\mathbf{x}) \int_V \frac{\rho(\mathbf{x}')(\mathbf{x} - \mathbf{x}')}{|\mathbf{x} - \mathbf{x}'|^3} d^3x' \tag{1.11}$$

$$\mathbf{f}_B(\mathbf{x}) = \frac{\mu_0}{4\pi}\mathbf{j}(\mathbf{x}) \times \int_V \frac{\mathbf{j}(\mathbf{x}') \times (\mathbf{x} - \mathbf{x}')}{|\mathbf{x} - \mathbf{x}'|^3} d^3x', \tag{1.12}$$

where $\mathbf{f}_C(\mathbf{x})$ is the force density arising from the static distribution of charges, $\mathbf{f}_B(\mathbf{x})$ is the force density attributable to moving charges or currents, and ϵ_0 and μ_0 represent experimental constants. Thus the total force arising from a static system of charges or from currents is given by

$$\mathbf{F}_{C,B} = \int \mathbf{f}_{C,B}(\mathbf{x})\, d^3x, \tag{1.13}$$

where the integration is carried out over the volume containing the system of electric charges.

Since $\rho(\mathbf{x})$ is the charge per unit volume, the total charge in volume V is given by

$$Q = \int_V \rho(\mathbf{x})\, d^3x.$$

One could express the current density $\mathbf{j}(\mathbf{x})$ in terms of $\rho(\mathbf{x})$ and a velocity distribution $\mathbf{v}(\mathbf{x})$ if the velocity of each element of charge at each point in space were known; thus one would have

$$\mathbf{j}(\mathbf{x}) = \rho(\mathbf{x})\mathbf{v}(\mathbf{x}). \tag{1.14}$$

(Note that the dimensions of $\mathbf{j}(\mathbf{x})$ are those of current per unit area.) However, knowing $\mathbf{j}(\mathbf{x})$ is equivalent to knowing $\mathbf{v}(\mathbf{x})$ if $\rho(\mathbf{x})$ is given, and it is more convenient to work with the current density than with the velocity distribution. Equation (1.14) gives us a notion of the meaning of $\mathbf{j}(\mathbf{x})$; it is the amount of charge ΔQ passing across a surface ΔS perpendicular to $\mathbf{j}(\mathbf{x})$ (and therefore to $\mathbf{v}(\mathbf{x})$) at the point $\mathbf{x}$, during time interval Δt, as ΔS and Δt tend to vanish (see Fig. 1.1). Thus

$$\mathbf{j}(\mathbf{x}) = \rho(\mathbf{x})\mathbf{v}(\mathbf{x}) = \lim_{\substack{\Delta S \to 0 \\ \Delta t \to 0}} \frac{\Delta Q}{\Delta S \Delta t}\hat{\mathbf{j}}(\mathbf{x}),$$

where $\hat{\mathbf{j}}(\mathbf{x})$ is a unit vector normal to ΔS and parallel to $\mathbf{j}(\mathbf{x})$.

A familiar example of a current density is that in a circular cylindrical conductor (wire) having radius r and carrying a constant current J. Thus J is the amount of charge transferred across the conductor's cross-sectional surfaces per unit time,

$$J = \int_{S_\perp} \mathbf{j}(\mathbf{x}) \cdot d\mathbf{S} = j\pi r^2, \qquad \text{or} \qquad j = J/\pi r^2.$$

It is convenient to express the force densities given in Eqs. (1.11) and (1.12) in terms of two vector fields, the electric field $\mathbf{E}(\mathbf{x})$ and the magnetic induction field $\mathbf{B}(\mathbf{x})$. The defining relationships between the force densities and these vector fields are given by

$$\mathbf{f}_C(\mathbf{x}) = \rho(\mathbf{x})\mathbf{E}(\mathbf{x}) \tag{1.15}$$

and

$$\mathbf{f}_B(\mathbf{x}) = \mathbf{j}(\mathbf{x}) \times \mathbf{B}(\mathbf{x}), \tag{1.16}$$

where evidently

$$\mathbf{E}(\mathbf{x}) = \frac{1}{4\pi\epsilon_0}\int_V \rho(\mathbf{x}')\frac{\mathbf{x} - \mathbf{x}}{|\mathbf{x} - \mathbf{x}'|^3}\, d^3x' \tag{1.17}$$

$$\mathbf{B}(\mathbf{x}) = \frac{\mu_0}{4\pi}\int_V \hat{\mathbf{j}}(\mathbf{x}') \times \frac{\mathbf{x} - \mathbf{x}'}{|\mathbf{x} - \mathbf{x}'|^3}\, d^3x'. \tag{1.18}$$

By noting that

$$\frac{(\mathbf{x} - \mathbf{x}')}{|\mathbf{x} - \mathbf{x}'|^3} = -\nabla\frac{1}{|\mathbf{x} - \mathbf{x}'|}$$

and making use of the vector indentity $\nabla \cdot (\mathbf{U} \times \mathbf{V}) = \mathbf{V} \cdot (\nabla \times \mathbf{U}) - \mathbf{U} \cdot (\nabla \times \mathbf{V})$,

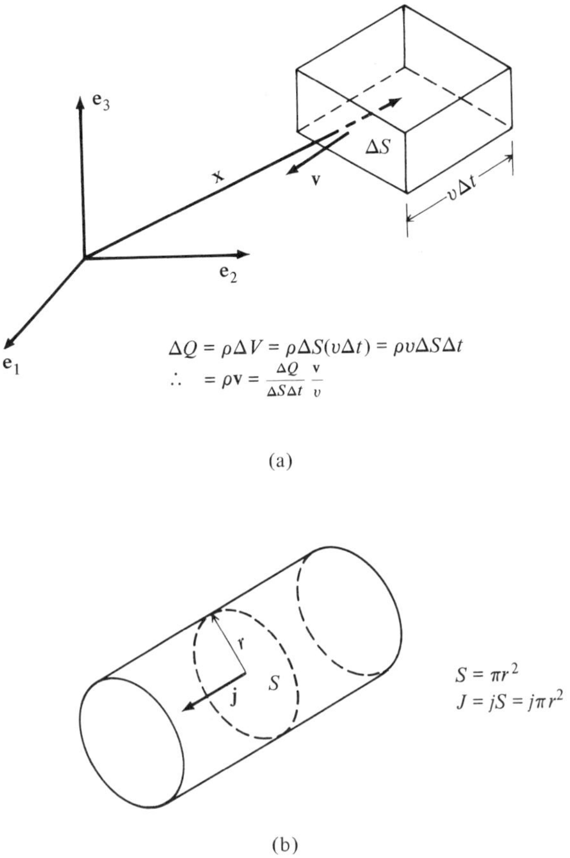

Fig. 1.1. (a) Definition of the current density **j**. (b) Relationship between the current J and the current density **j** in a cylindrical conductor.

we can find the divergences of ($\mathbf{E}(\mathbf{x})$ and $\mathbf{B}(\mathbf{x})$ from Eqs. (1.17) and (1.18):

$$\begin{aligned}\boldsymbol{\nabla}\cdot\mathbf{E}(\mathbf{x}) &= -\frac{1}{4\pi\epsilon_0}\boldsymbol{\nabla}\cdot\int_V \rho(\mathbf{x}')\boldsymbol{\nabla}\frac{1}{|\mathbf{x}-\mathbf{x}'|}d^3x' = -\frac{1}{4\pi\epsilon_0}\int_V \rho(\mathbf{x}')\boldsymbol{\nabla}^2\frac{1}{|\mathbf{x}-\mathbf{x}'|}d^3x' \\ &= -\frac{1}{\epsilon_0}\int_V \rho(\mathbf{x}')\frac{1}{4\pi}\boldsymbol{\nabla}'^2\frac{1}{|\mathbf{x}-\mathbf{x}'|}d^3x' \\ &= \frac{1}{\epsilon_0}\int_V \rho(\mathbf{x}')\delta^3(\mathbf{x}-\mathbf{x}')d^3x' = \frac{1}{\epsilon_0}\rho(\mathbf{x}),\end{aligned}$$

where we have used the fact that $\boldsymbol{\nabla}g(\mathbf{x}-\mathbf{x}') = -\boldsymbol{\nabla}'g(\mathbf{x}-\mathbf{x}')$ when $\boldsymbol{\nabla}$ means differentiation with respect to **x**,

$$\nabla = \sum_{j=1}^{3} \hat{\mathbf{i}}_j \frac{\partial}{\partial x_j},$$

and ∇' means differentiation with respect to $\mathbf{x}'$,

$$\nabla' = \sum_{j=1}^{3} \hat{\mathbf{i}}_j \frac{\partial}{\partial x'_j}.$$

Also,

$$\begin{aligned}
\nabla \cdot \mathbf{B}(\mathbf{x}) &= -\frac{\mu_0}{4\pi}\nabla \cdot \int_V \mathbf{j}(\mathbf{x}') \times \nabla \frac{1}{|\mathbf{x}-\mathbf{x}'|}\, d^3x' \\
&= -\frac{\mu_0}{4\pi}\int_V \nabla \cdot \left[\mathbf{j}(\mathbf{x}') \times \nabla \frac{1}{|\mathbf{x}-\mathbf{x}'|}\right] d^3x' \\
&= \frac{\mu_0}{4\pi}\int_V \mathbf{j}(\mathbf{x}') \cdot \left(\nabla \times \nabla \frac{1}{|\mathbf{x}-\mathbf{x}'|}\right) d^3x' = 0.
\end{aligned}$$

These are two of Maxwell's differential equations for the electromagnetic fields *in vacuo*:

$$\nabla \cdot \mathbf{E}(\mathbf{x}) = \frac{1}{\epsilon_0}\rho(x), \tag{1.19}$$

$$\nabla \cdot \mathbf{B}(\mathbf{x}) = 0. \tag{1.20}$$

We can obtain another pair of differential equations for $\mathbf{E}(\mathbf{x})$ and $\mathbf{B}(\mathbf{x})$ by calculating the curls of Eqs. (1.17) and (1.18). Taking the curl of $\mathbf{E}(\mathbf{x})$ yields

$$\nabla \times \mathbf{E}(\mathbf{x}) = -\frac{1}{4\pi\epsilon_0}\int_V \rho(\mathbf{x}')\nabla \times \nabla \frac{1}{|\mathbf{x}-\mathbf{x}'|}\, d^3x' = \mathbf{0}. \tag{1.21}$$

In calculating the curl of $\mathbf{B}(\mathbf{x})$, we must make use of the vector identity $\nabla \times (\mathbf{U} \times \mathbf{V}) = \mathbf{U}(\nabla \cdot \mathbf{V}) - \mathbf{V}(\nabla \cdot \mathbf{U}) + \mathbf{V} \cdot \nabla\mathbf{U} - \mathbf{U} \cdot \nabla\mathbf{V}$, together with the fact that $\mathbf{U} \cdot \nabla\mathbf{j}(\mathbf{x}') = 0$ and $\nabla \cdot \mathbf{j}(\mathbf{x}') = 0$ since $\mathbf{j}(\mathbf{x}')$ does not depend on the variable **x**. Thus

$$\begin{aligned}
\nabla \times \mathbf{B}(\mathbf{x}) &= -\frac{\mu_0}{4\pi}\int_V \nabla \times \left[\mathbf{j}(\mathbf{x}') \times \nabla \frac{1}{|\mathbf{x}-\mathbf{x}'|}\right] d^3x' \\
&= -\frac{\mu_0}{4\pi}\int_V \left\{\left[\left(\nabla^2 \frac{1}{|\mathbf{x}-\mathbf{x}'|}\right) + \nabla \frac{1}{|\mathbf{x}-\mathbf{x}'|} \cdot \nabla\right]\mathbf{j}(\mathbf{x}')\right. \\
&\qquad \left. - [\nabla \cdot \mathbf{j}(\mathbf{x}') + \mathbf{j}(\mathbf{x}') \cdot \nabla]\nabla \frac{1}{|\mathbf{x}-\mathbf{x}'|}\right\} d^3x'
\end{aligned}$$

$$= -\frac{\mu_0}{4\pi}\int_V \left[-4\pi\delta^3(\mathbf{x}-\mathbf{x}')\mathbf{j}(\mathbf{x}') - \mathbf{j}(\mathbf{x}')\cdot\nabla\nabla\frac{1}{|\mathbf{x}-\mathbf{x}'|}\right] d^3x'$$

$$= \mu_0\mathbf{j}(\mathbf{x}) + \frac{\mu_0}{4\pi}\nabla\left(\nabla\cdot\int_V \frac{\mathbf{j}(\mathbf{x}')}{|\mathbf{x}-\mathbf{x}'|}d^3x'\right).$$

Now the divergence appearing in the last term can be rewritten in such a way that we can make use of Gauss's theorem, as follows:

$$\nabla\cdot\int_V \frac{\mathbf{j}(\mathbf{x}')}{|\mathbf{x}-\mathbf{x}'|}d^3x' = \int_V \mathbf{j}(\mathbf{x}')\cdot\nabla\frac{1}{|\mathbf{x}-\mathbf{x}'|}d^3x' = -\int_V \mathbf{j}(\mathbf{x}')\cdot\nabla'\frac{1}{|\mathbf{x}-\mathbf{x}'|}d^3x'$$

$$= -\int_V \left\{\nabla'\cdot\left[\frac{\mathbf{j}(\mathbf{x}')}{|\mathbf{x}-\mathbf{x}'|}\right] - \frac{1}{|\mathbf{x}-\mathbf{x}'|}\nabla'\cdot\mathbf{j}(\mathbf{x}')\right\} d^3x'$$

$$= -\int_S \mathbf{j}(\mathbf{x}')\frac{1}{|\mathbf{x}-\mathbf{x}'|}\cdot d\mathbf{S} + \int_V \frac{\nabla'\cdot\mathbf{j}(\mathbf{x}')}{|\mathbf{x}-\mathbf{x}'|}d^3x',$$

where S is the surface enclosing volume V. If we are realistic in our assumptions about $\mathbf{j}(\mathbf{x}')$ for an ordinary physical problem, we shall require that $\mathbf{j}(\mathbf{x}')$ be nonvanishing only within a finite volume of space. Hence we can choose the volume of integration V to be so large that it encompasses the entire region over which the current density is nonvanishing, and then $\mathbf{j}(\mathbf{x}')$ will vanish everywhere on S. Then we are left with

$$\nabla\times\mathbf{B}(\mathbf{x}) = \mu_0\mathbf{j}(\mathbf{x}) + \frac{\mu_0}{4\pi}\nabla\int_V \frac{\nabla'\cdot\mathbf{j}(\mathbf{x}')}{|\mathbf{x}-\mathbf{x}'|}d^3x'.$$

This result can be cast into an even more useful form if we take into account another very important piece of experimental information, namely the law of conservation of electric charge. A vast amount of accumulated experimental data supports the hypothesis that electric charge cannot be created or destroyed spontaneously within a closed physical system; this circumstance can be expressed mathematically by the *continuity equation*,

$$\nabla\cdot\mathbf{j}(\mathbf{x}) + \frac{\partial\rho(\mathbf{x})}{\partial t} = 0,$$

which means that the current density emerging from a point in space ($\nabla\cdot\mathbf{j}(\mathbf{x})$) is equal to the rate of decrease of the charge density at that point ($-\partial\rho/\partial t$). (Note that this implies that $\rho(\mathbf{x})$ must depend on t as well as upon $\mathbf{x}$ wherever $\nabla\cdot\mathbf{j}(\mathbf{x})$ is nonvanishing. However, if $\mathbf{j}(\mathbf{x})$ is independent of time, $\rho(\mathbf{x})$, according to the continuity equation, will be a linear function of time. We shall say that a *steady state* exists when $\mathbf{j}(\mathbf{x})$ is time-independent, even though $\rho(\mathbf{x})$ then will contain a term proportional to time.) This additional experimental information allows us to replace $\nabla' \cdot \mathbf{j}(\mathbf{x}')$ in the above expressions for $\nabla \times \mathbf{B}(\mathbf{x})$ with $-\partial\rho(\mathbf{x}')/\partial t$; thus

$$\nabla \times \mathbf{B}(\mathbf{x}) = \mu_0 \mathbf{j}(\mathbf{x}) - \nabla \frac{\mu_0}{4\pi} \frac{\partial}{\partial t} \int \frac{\rho(\mathbf{x}')}{|\mathbf{x} - \mathbf{x}'|} d^3x'$$

$$= \mu_0 \mathbf{j}(\mathbf{x}) - \frac{\partial}{\partial t} \frac{\mu_0}{4\pi} \int_V \rho(\mathbf{x}') \nabla \frac{1}{|\mathbf{x} - \mathbf{x}'|} d^3x',$$

or, if we make use of Eq. (17),

$$\nabla \times \mathbf{B}(\mathbf{x}) = \mu_0 \mathbf{j}(\mathbf{x}) + \mu_0 \epsilon_0 \frac{\partial \mathbf{E}(\mathbf{x})}{\partial t}. \tag{1.22}$$

The second term in Eq. (1.22) represents the so-called *displacement current* discovered by Maxwell.

1.3 ELECTROMAGNETIC INDUCTION

The set of differential equations we have obtained so far, Eqs. (1.19–1.22), still does not contain all the experimental information about electromagnetic fields that is available to us; we have said nothing about Faraday's law of electromagnetic induction. Equation (1.21) is, in fact, in disagreement with experiment, as Michael Faraday found during the mid-nineteenth century. He noted first that a closed loop of conducting material placed in static external electric or magnetic fields did not carry any current; that is, static fields do not generate a net flow of current within closed conductors. This can be expressed by the equation

$$\oint_C \mathbf{j}(\mathbf{x}) \cdot d\mathbf{l} = 0, \qquad \mathbf{E}, \mathbf{B} \text{ static},$$

for any closed loop C; $d\mathbf{l}$ is an element of length along the closed loop. Now if a changing magnetic field is introduced, this equation no longer is correct, and Faraday found an experimental law that can be expressed mathematically as follows:

$$\oint_C \mathbf{j}(\mathbf{x}) \cdot d\mathbf{l} = -\sigma \int_S \frac{\partial \mathbf{B}}{\partial t} \cdot d\mathbf{s},$$

where S is a surface bounded by the loop C formed by the conductor, and σ is a constant that is characteristic of the conductor. (Stated in this form, Faraday's law is valid for most practical conductors—in fact, it is valid for the same media for which Ohm's law is valid. This is a minor technical point that need not concern us here.) It turns out that the same result holds whether or not the electric field is changing in time.

One can use Stokes's theorem to bring this equation into the form

$$\int_S \nabla \times \mathbf{j}(\mathbf{x}) \cdot d\mathbf{s} = -\sigma \int_S \frac{\partial \mathbf{B}}{\partial t} \cdot d\mathbf{s}, \qquad \text{or} \qquad \int_S \left[\nabla \times \mathbf{j}(\mathbf{x}) + \sigma \frac{\partial \mathbf{B}}{\partial t} \right] \cdot d\mathbf{s} = 0$$

or, since this equation is to hold for an arbitrary closed loop C and hence for an arbitrary surface S,

$$\nabla \times \mathbf{j}(\mathbf{x}) = -\sigma \frac{\partial \mathbf{B}}{\partial t}.$$

An additional empirical fact, known as Ohm's law, can be used to convert this formula into a relationship between the fields $\mathbf{E}(\mathbf{x})$ and $\mathbf{B}(\mathbf{x})$. Ohm's law states that the current density is proportional to the electric field; this rule is found to be an accurate description of the behavior of a vast number of conducting materials. In symbols, we have

$$\mathbf{j}(\mathbf{x}) = \sigma \mathbf{E}(\mathbf{x}),$$

where σ, called the conductivity of the conductor, turns out to be numerically equal to the constant appearing in Faraday's law. Thus we obtain, upon combining Ohm's law with Faraday's law, the relation

$$\nabla \times \mathbf{E}(\mathbf{x}) = -\frac{\partial \mathbf{B}(\mathbf{x})}{\partial t}, \tag{1.23}$$

which does not contain the constant σ. Maxwell observed that this equation should be valid regardless of the conductivity of the conductor, and that in fact it should hold even in nonconductors, including free space. Equation (1.23), then, is a corrected version of Eq. (1.21), and it is one of Maxwell's differential equations.

Evidently we now have a conflict between Eq. (1.23) and Eq. (1.17), for the latter yields $\nabla \times \mathbf{E}(\mathbf{x}) = \mathbf{0}$. For steady-state distribution of charges, and more generally for steady-state fields, Eqs. (1.17) and (1.18) do give correct descriptions for the fields $\mathbf{E}(\mathbf{x})$ and $\mathbf{B}(\mathbf{x})$, but it is clear from Eq. (1.23) that those equations are not adequate for time-varying fields in general. It is easy to see that Eq. (1.17) can be brought into line with Faraday's law if we add to the right-hand side the term

$$-\frac{\partial}{\partial t}\frac{\mu_0}{4\pi}\int_V \frac{\mathbf{j}(\mathbf{x}')}{|\mathbf{x} - \mathbf{x}'|}\, d^3x'.$$

Thus

$$\mathbf{E}(\mathbf{x}) = -\frac{1}{4\pi\epsilon_0}\left[\nabla \int_V \rho(\mathbf{x}')\frac{1}{|\mathbf{x} - \mathbf{x}'|}\, d^3x' + \mu_0\epsilon_0 \frac{\partial}{\partial t}\int_V \frac{\mathbf{j}(\mathbf{x}')}{|\mathbf{x} - \mathbf{x}'|}\, d^3x'\right] \tag{1.24}$$

is our new expression for $\mathbf{E}(\mathbf{x})$ in terms of the source charges and currents. However, this new expression for $\mathbf{E}(\mathbf{x})$ may alter Eqs. (1.19) and (1.22) which were obtained using the old expression for $\mathbf{E}(\mathbf{x})$. Therefore we must calculate $\nabla \cdot \mathbf{E}(\mathbf{x})$ and $\partial \mathbf{E}(\mathbf{x})/\partial t$ anew.

1.4 ELECTROMAGNETIC POTENTIALS AND GAUGE INVARIANCE

In order to simplify our notation in subsequent calculations, we now introduce the scalar potential

$$\phi(\mathbf{x}) = \frac{1}{4\pi\epsilon_0} \int_V \frac{\rho(\mathbf{x}')}{|\mathbf{x} - \mathbf{x}'|} d^3x' \tag{1.25}$$

and the vector potential,

$$\mathbf{A}(\mathbf{x}) = \frac{\mu_0}{4\pi} \int_V \frac{\mathbf{j}(\mathbf{x}')}{|\mathbf{x} - \mathbf{x}'|} d^3x'. \tag{1.26}$$

Using this notation, we may write Eqs. (1.18) and (1.24) in the forms

$$\mathbf{B}(\mathbf{x}) = \nabla \times \mathbf{A}(\mathbf{x}) \tag{1.27}$$

$$\mathbf{E}(\mathbf{x}) = -\nabla\phi(\mathbf{x}) - \frac{\partial}{\partial t}\mathbf{A}(\mathbf{x}). \tag{1.28}$$

It is important to recognize that the scalar and vector potentials are *auxiliary* quantities, and that we have introduced them for mathematical convenience. In fact, because of the way in which they are related to the electric field and the magnetic induction field, they must not be physically measurable quantities. In constructing the theory so far, we have assumed tacitly that electric charge and current densities are measurable—that we can, with relatively simple physical instruments, determine how much of a quantity we agree to call electric charge exists within a given small volume of space; this is one of the basic pieces of physical data we wish to investigate. We also assume that we can measure the total force exerted on the charges within an arbitrarily small volume of space. Thus, since the force density and the charge or current density are considered physically measurable quantities, the field quantities $\mathbf{E}(\mathbf{x})$ and $\mathbf{B}(\mathbf{x})$ can be considered as *defined* directly, in terms of fundamental measurable data, by Eqs. (1.15) and (1.16), and therefore we may say that we have prescribed how $\mathbf{E}(\mathbf{x})$ and $\mathbf{B}(\mathbf{x})$ are to be measured; the field vectors can be considered measurable or "physical" quantities. This cannot be said for $\phi(\mathbf{x})$ and $\mathbf{A}(\mathbf{x})$, however.

Equations (1.25) and (1.26) tell us how to *calculate* the potentials once the source densities $\rho(\mathbf{x})$ and $\mathbf{j}(\mathbf{x})$ are known; they do not, however, tell us how to *measure* the potentials, because integration is not a process of physical measurement. Also, Eqs. (1.27) and (1.28) show mathematical connections between the potentials and measurable field quantities, but again the potentials are to be obtained by integrating the physical variables $\mathbf{E}(\mathbf{x})$ and $\mathbf{B}(\mathbf{x})$. The difficulty is that the integration either of sources or of fields does not lead to a unique determination of $\phi(\mathbf{x})$ or $\mathbf{A}(\mathbf{x})$. This is demonstrated most easily by utilizing the relationships between the potentials and the fields given by Eqs. (1.27) and (1.28). Let us add to $\mathbf{A}(\mathbf{x})$ the gradient of any well-behaved function of space and time coordinates, say $\chi(\mathbf{x})$, to form a new vector potential, $\mathbf{A}'(\mathbf{x}) = \mathbf{A}(\mathbf{x}) + \nabla\chi(\mathbf{x})$; at the same time we shall form a new scalar potential $\phi'(\mathbf{x})$ by subtracting from $\phi(\mathbf{x})$ the partial derivative with respect to time of $\chi(\mathbf{x})$: $\phi'(\mathbf{x}) = \phi(\mathbf{x}) - \partial\chi(\mathbf{x})/\partial t$.* Calculating the new field vectors, $\mathbf{E}'(\mathbf{x})$ and $\mathbf{B}'(\mathbf{x})$, obtained by substituting $\phi'(\mathbf{x})$ and $\mathbf{A}'(\mathbf{x})$ for $\phi(\mathbf{x})$ and $\mathbf{A}(\mathbf{x})$ in the right-hand sides of Eqs. (1.27) and (1.28), we find

$$\mathbf{B}'(\mathbf{x}) = \nabla \times \mathbf{A}'(\mathbf{x}) = \nabla \times [\mathbf{A}(\mathbf{x}) + \nabla\chi(\mathbf{x})] = \nabla \times \mathbf{A}(\mathbf{x}) = \mathbf{B}(\mathbf{x})$$

and

$$\begin{aligned}\mathbf{E}'(\mathbf{x}) &= -\nabla\phi'(\mathbf{x}) - \frac{\partial \mathbf{A}(\mathbf{x})}{\partial t} = -\nabla\left[\phi(\mathbf{x}) - \frac{\partial}{\partial t}\chi(\mathbf{x})\right] - \frac{\partial}{\partial t}[\mathbf{A}(\mathbf{x}) + \nabla\chi(\mathbf{x})] \\ &= -\nabla\phi(\mathbf{x}) - \frac{\partial}{\partial t}\mathbf{A}(\mathbf{x}) = \mathbf{E}(\mathbf{x}).\end{aligned}$$

Thus this kind of transformation of the potentials, known as a *gauge transformation of the second kind*, leads to the *same* physically *measurable* field quantities as were obtained before the transformation was carried out. (Gauge transformations of the first kind are discussed in Chapter 9.) The observable field vectors are *invariant* under gauge transformations of the second kind.

It is clear, then, that we must distinguish carefully between variables that correspond to observable quantities and auxiliary variables that are introduced for notational convenience, and which indeed may not have physical status as measurable quantities; the values of the latter may not be uniquely defined.

1.5 MAXWELL'S EQUATIONS

Let us return now to the calculation of the divergence and of the time derivative of $\mathbf{E}(\mathbf{x})$. According to Eq. (1.28)

$$\nabla \cdot \mathbf{E}(\mathbf{x}) = -\nabla^2\phi(\mathbf{x}) - \frac{\partial}{\partial t}\nabla \cdot \mathbf{A}(\mathbf{x}).$$

Utilizing Eq. (1.26), Gauss's theorem, and the equation of continuity, we can rewrite the last term of the right-hand side in the form

$$\begin{aligned}-\frac{\partial}{\partial t}\nabla \cdot \mathbf{A}(\mathbf{x}) &= -\frac{\mu_0}{4\pi}\frac{\partial}{\partial t}\int_V \nabla \cdot \frac{\mathbf{j}(\mathbf{x}')}{|\mathbf{x} - \mathbf{x}'|}\, d^3x' = \frac{\mu_0}{4\pi}\frac{\partial}{\partial t}\int_V \mathbf{j}(\mathbf{x}') \cdot \nabla' \frac{1}{|\mathbf{x} - \mathbf{x}'|}\, d^3x' \\ &= \frac{\mu_0}{4\pi}\frac{\partial}{\partial t}\int_V \left[\nabla' \cdot \frac{\mathbf{j}(\mathbf{x}')}{|\mathbf{x} - \mathbf{x}'|} - \frac{\nabla' \cdot \mathbf{j}(\mathbf{x}')}{|\mathbf{x} - \mathbf{x}'|}\right] d^3x'\end{aligned}$$

---

* The time variable, t, is not explicitly shown here as an argument of the function $\chi(\mathbf{x})$, nor of other functions of space and time such as $\rho(\mathbf{x})$, $\mathbf{E}(\mathbf{x})$, etc. This convention will be followed throughout this discussion; every function of the spatial coordinates is assumed to be a function of time also, although it may be a constant function.

$$= \frac{\mu_0}{4\pi}\frac{\partial}{\partial t}\left[\int_S \frac{\mathbf{j}(\mathbf{x}')}{|\mathbf{x}-\mathbf{x}'|}\cdot d\mathbf{S} + \int_V \frac{\partial\rho(\mathbf{x}')/\partial t}{|\mathbf{x}-\mathbf{x}'|}d^3x'\right].$$

If the volume of integration is chosen to be so large that the current density vanishes at the surface bounding the volume, the surface integral vanishes and we obtain

$$\begin{aligned}\boldsymbol{\nabla}\cdot\mathbf{E}(\mathbf{x}) &= -\boldsymbol{\nabla}^2\phi(\mathbf{x}) + \mu_0\epsilon_0\frac{\partial^2}{\partial t^2}\frac{1}{4\pi\epsilon_0}\int_V \frac{\rho(\mathbf{x}')}{|\mathbf{x}-\mathbf{x}'|}d^3x' \\ &= -\boldsymbol{\nabla}^2\phi(\mathbf{x}) + \mu_0\epsilon_0\frac{\partial^2}{\partial t^2}\phi(\mathbf{x}).\end{aligned} \tag{1.29}$$

Now in the manipulations following Eq. (1.18) and which led to Eq. (1.19), we used a relationship equivalent to

$$\boldsymbol{\nabla}^2\phi(\mathbf{x}) = -\frac{\rho(\mathbf{x})}{\epsilon_0}; \tag{1.30}$$

however, it must be remembered that it was assumed there that we were dealing with steady-state fields, so that $\rho(\mathbf{x})$ and $\phi(\mathbf{x})$ were to be considered to vary linearly with time. Suppose we continue to accept Eq. (1.30); then, according to Eq. (1.29), $\boldsymbol{\nabla}\cdot\mathbf{E}$ would not necessarily vanish even in a volume of space V_0 where $\rho(\mathbf{x}) = 0$. Thus, since $\boldsymbol{\nabla}\cdot\mathbf{E}(\mathbf{x}) = -\boldsymbol{\nabla}^2\phi(\mathbf{x})$ for steady-state charge distributions, it would be possible to have a time-varying potential that produced the same effect on $\mathbf{E}(\mathbf{x})$ as would a nonvanishing steady-state charge distribution, even in a region of space in which there were no charges present; this follows from the presence of the last term on the right-hand side of Eq. (1.29). Such a situation is unacceptable and in fact such a result is contrary to what we mean by the notions of "electric field" and "electric charge"; the difficulty arises because of the way in which Eq. (1.30) relates the unphysical variables $\phi(\mathbf{x})$ to the physical quantity $\rho(\mathbf{x})$. This trouble is avoided if we replace Eq. (1.30) with

$$\boldsymbol{\nabla}\cdot\mathbf{E}(\mathbf{x}) = \frac{\rho(\mathbf{x})}{\epsilon_0}$$

which relates the physical variables $\mathbf{E}$ and ρ, and which is in agreement with the results obtained for the stationary case; however, it leads to a new equation for $\phi(\mathbf{x})$ in terms of the charge distribution:

$$\left(\boldsymbol{\nabla}^2 - \mu_0\epsilon_0\frac{\partial^2}{\partial t^2}\right)\phi(\mathbf{x}) = -\frac{\rho(\mathbf{x})}{\epsilon_0}. \tag{1.31}$$

Equation (1.19) has the advantage that the observable quantities $\rho(\mathbf{x})$ and $\mathbf{E}(\mathbf{x})$ are related by the same equation when the charge distribution is changing arbitrarily with time as when it is linear in the time variable. Thus Eq. (1.19) is retained as one of Maxwell's fundamental differential equations.

Our redefinition of $\mathbf{E}(\mathbf{x})$ according to Eq. (1.24) or Eq. (1.28) also may alter the relationship between $\mathbf{B}(\mathbf{x})$ and $\mathbf{E}(\mathbf{x})$ expressed in Eq. (1.22); to check this possibility, we must calculate $\partial\mathbf{E}(\mathbf{x})/\partial t$ in terms of Eq. (1.28). This yields, when the equation of continuity and Gauss's theorem are invoked once more,

$$\begin{aligned}
\frac{\partial \mathbf{E}(\mathbf{x})}{\partial t} &= -\frac{\partial}{\partial t}\nabla\phi - \frac{\partial^2}{\partial t^2}\mathbf{A} \\
&= -\nabla\frac{1}{4\pi\epsilon_0}\int_V \frac{\partial\rho(\mathbf{x}')/\partial t}{|\mathbf{x}-\mathbf{x}'|}\,d^3x' - \frac{\partial^2}{\partial t^2}\mathbf{A}(\mathbf{x}) \\
&= \nabla\frac{1}{4\pi\epsilon_0}\int_V \frac{\nabla'\cdot\mathbf{j}(\mathbf{x}')}{|\mathbf{x}-\mathbf{x}'|}\,d^3x' - \frac{\partial^2}{\partial t^2}\mathbf{A}(\mathbf{x}) \\
&= \nabla\frac{1}{4\pi\epsilon_0}\int_V \left[\nabla'\cdot\frac{\mathbf{j}(\mathbf{x}')}{|\mathbf{x}-\mathbf{x}'|} - \mathbf{j}(\mathbf{x}')\cdot\nabla'\frac{1}{|\mathbf{x}-\mathbf{x}'|}\right]d^3x' - \frac{\partial^2}{\partial t^2}\mathbf{A}(\mathbf{x}) \\
&= \nabla\frac{1}{4\pi\epsilon_0}\int_V \mathbf{j}(\mathbf{x}')\cdot\nabla\frac{1}{|\mathbf{x}-\mathbf{x}'|}\,d^3x' - \frac{\partial^2\mathbf{A}(\mathbf{x})}{\partial t^2} \\
&= \frac{1}{\epsilon_0\mu_0}\nabla[\nabla\cdot\mathbf{A}(\mathbf{x})] - \frac{\partial^2\mathbf{A}(\mathbf{x})}{\partial t^2} \\
&= \frac{1}{\mu_0\epsilon_0}\{\nabla\times[\nabla\times\mathbf{A}(\mathbf{x})] + \nabla^2\mathbf{A}(\mathbf{x})\} - \frac{\partial^2\mathbf{A}}{\partial t^2} \\
&= \frac{1}{\mu_0\epsilon_0}\nabla\times\mathbf{B}(\mathbf{x}) + \frac{1}{\mu_0\epsilon_0}\left[\nabla^2 - \mu_0\epsilon_0\frac{\partial^2}{\partial t^2}\right]\mathbf{A}(\mathbf{x}),
\end{aligned}$$

or

$$\nabla\times\mathbf{B}(\mathbf{x}) = -\left(\nabla^2 - \mu_0\epsilon_0\frac{\partial^2}{\partial t^2}\right)\mathbf{A}(\mathbf{x}) + \mu_0\epsilon_0\frac{\partial\mathbf{E}(\mathbf{x})}{\partial t}.$$

Again, our calculations with steady-state fields and sources, following Eq. (1.21) and leading to Eq. (1.22), involved a result which can be written in the form

$$\nabla^2\mathbf{A}(\mathbf{x}) = -\mu_0\mathbf{j}(\mathbf{x}), \tag{1.32}$$

and which again relates the unobservable variable $\mathbf{A}(\mathbf{x})$ to the observable current density $\mathbf{j}(\mathbf{x})$. If we accepted Eq. (1.32) as a fundamental law, it would force upon us a different relationship among the measurable quantities $\mathbf{B}(\mathbf{x})$, $\mathbf{j}(\mathbf{x})$ and $\mathbf{E}(\mathbf{x})$ for time-varying fields and currents than was found for the steady state. A situation very much like that which we just discussed concerning Eqs. (1.29) and (1.30) arises here. Rather than accept Eq. (1.32) we shall find it much more satisfactory to accept Eq. (1.22) and to adjust the equation relating to $\mathbf{A}(\mathbf{x})$ to $\mathbf{j}(\mathbf{x})$ accordingly.

Thus Eq. (1.22) is retained as one of Maxwell's equations, and we have

$$\nabla \times \mathbf{B}(\mathbf{x}) = \mu_0 \mathbf{j}(\mathbf{x}) + \mu_0 \epsilon_0 \frac{\partial \mathbf{E}(\mathbf{x})}{\partial t},$$

together with

$$\left(\nabla^2 - \mu_0 \epsilon_0 \frac{\partial^2}{\partial t^2}\right) \mathbf{A}(\mathbf{x}) = -\mu_0 \mathbf{j}(\mathbf{x}). \tag{1.33}$$

The four equations that we have found to relate the fields and source densities, and which are known as Maxwell's equations, are then

$$\nabla \cdot \mathbf{E}(\mathbf{x}) = \frac{\rho(\mathbf{x})}{\epsilon_0}, \tag{1.34}$$

$$\nabla \cdot \mathbf{B}(\mathbf{x}) = 0, \tag{1.35}$$

$$\nabla \times \mathbf{E}(\mathbf{x}) = -\frac{\partial \mathbf{B}(\mathbf{x})}{\partial t}, \tag{1.36}$$

$$\nabla \times \mathbf{B}(\mathbf{x}) = \mu_0 \mathbf{j}(\mathbf{x}) + \mu_0 \epsilon_0 \frac{\partial \mathbf{E}(\mathbf{x})}{\partial t}. \tag{1.37}$$

These are the fundamental equations of classical electromagnetic theory. Equations (1.31) and (1.33) are auxiliary differential equations in the auxiliary variables $\phi(\mathbf{x})$ and $\mathbf{A}(\mathbf{x})$; they can be altered by introducing a gauge transformation of the second kind. However, it is always possible to adjust $\mathbf{A}$ and ϕ by such a transformation so that they satisfy

$$\nabla \cdot \mathbf{A} + \mu_0 \epsilon_0 \frac{\partial \phi}{\partial t} = 0; \tag{1.38}$$

they will then satisfy Eqs. (1.31) and (1.33) also. When Eq. (1.38) is satisfied, the potentials are said to be in the *Lorentz gauge.*

The partial differential equations (1.31) and (1.33) for the potentials are recognizable as having the form of the inhomogeneous wave equation, or the inhomogeneous telegrapher's equation, in three dimensions. Similar wave equations for the field vectors **E** and **B** may be obtained readily from Maxwell's equations. Taking the curl of Eq. (1.36) and making use of the vector identity

$$\nabla \times (\nabla \times \mathbf{U}) = \nabla(\nabla \cdot \mathbf{U}) - \nabla^2 \mathbf{U},$$

one gets

$$\nabla \times (\nabla \times \mathbf{E}) = \nabla(\nabla \cdot \mathbf{E}) - \nabla^2 \mathbf{E} = \frac{1}{\epsilon_0} \nabla \rho - \nabla^2 \mathbf{E}$$

$$= -\nabla \times \frac{\partial \mathbf{B}}{\partial t} = -\mu_0 \frac{\partial \mathbf{j}}{\partial t} - \mu_0 \epsilon_0 \frac{\partial^2 \mathbf{E}}{\partial t^2},$$

or

$$\left(\nabla^2 - \mu_0\epsilon_0\frac{\partial^2}{\partial t^2}\right)\mathbf{E}(\mathbf{x}) = \frac{1}{\epsilon_0}\nabla\rho(\mathbf{x}) - \mu_0\frac{\partial\mathbf{j}(\mathbf{x})}{\partial t}. \tag{1.39}$$

A similar treatment of Eq. (1.37) gives the result

$$\begin{aligned}\nabla \times [\nabla \times \mathbf{B}(\mathbf{x})] = \nabla(\nabla\cdot\mathbf{B}) - \nabla^2\mathbf{B} &= \mu_0\nabla \times \mathbf{j} + \mu_0\epsilon_0\frac{\partial}{\partial t}\nabla \times \mathbf{E} \\ &= \mu_0\nabla \times \mathbf{j}(\mathbf{x}) - \mu_0\epsilon_0\frac{\partial^2\mathbf{B}(\mathbf{x})}{\partial t^2}\end{aligned}$$

or, since $\nabla\cdot\mathbf{B} = 0$,

$$\left(\nabla^2 - \mu_0\epsilon_0\frac{\partial^2}{\partial t^2}\right)\mathbf{B}(\mathbf{x}) = -\mu_0\nabla \times \mathbf{j}(\mathbf{x}). \tag{1.40}$$

These are the inhomogeneous wave equations for $\mathbf{E}(\mathbf{x})$ and $\mathbf{B}(\mathbf{x})$. In regions of space where $\nabla\rho(\mathbf{x})$, $\partial\mathbf{j}(\mathbf{x})/\partial t$ and $\nabla \times \mathbf{j}(\mathbf{x})$ vanish identically, the homogeneous equations

$$\left(\nabla^2 - \frac{1}{c^2}\frac{\partial^2}{\partial t^2}\right)\mathbf{E}(\mathbf{x}) = \mathbf{0} \tag{1.41}$$

$$\left(\nabla^2 - \frac{1}{c^2}\frac{\partial^2}{\partial t^2}\right)\mathbf{B}(\mathbf{x}) = \mathbf{0} \tag{1.42}$$

are valid. We have introduced the notation $c = 1/\sqrt{\mu_0\epsilon_0}$ here. The constant c plays the role of the velocity of wave propagation in this wave equation.

1.6 THE ELECTROMAGNETIC WAVE EQUATIONS, NEWTONIAN MECHANICS, AND SPECIAL RELATIVITY

As we shall see in the next chapter, the appearance of the constant c in these equations, and the very form of the equations, are in conflict with Newtonian (or Galilean) relativity, the latter being a fundamental concept of classical mechanics. The basic hypotheses of classical mechanics are consistent with the intuitively satisfying notion that a "material" medium is required to sustain any wave motion. This idea of a supporting medium led to a famous experiment, the Michelson–Morley experiment, that was designed to investigate certain properties of the "ether" which was supposed to be the supporting medium for electromagnetic waves. This experiment proved to be a crucial test for the two experimentally well-founded but mutually conflicting theories of classical physics, Maxwellian electromagnetism and Newtonian mechanics. The resolution of the paradox presented by the inconsistency of these theoretical structures is achieved through Einstein's special theory of relativity.

The widespread impression that Einstein concocted the special theory of relativity purely from abstract philosophical principles is quite erroneous. When Einstein was confronted with this great paradox, vast amounts—by the standards of the time—of experimental evidence supported each of the classical theories, and the need to resolve the conflict without coming into disagreement with one or the other sets of supporting data was the challenge that faced him and his contemporary physicists. Einstein's stroke of genius was to show the way to bringing mechanics and electromagnetism into harmony while maintaining agreement with the totality of experimental data. This was done at the cost of eliminating Galilean relativity as a strictly accurate law of physics, and replacing it with a somewhat more general law, to which Galilean relativity is a good approximation in many situations.

Since Galilean relativity involves some very plausible intuitive notions about space and time, Einstein's theory seemed to do violence to one's intuition, and it is this that gave rise to most of the debates over his theory. The special theory of relativity was founded solidly on experimental observations, whereas those who eschewed the new theory based their objections on rather more abstract and intuitive principles. The debate was healthy, however, because it forced upon physicists a thorough evaluation of the relevant data, and of the nature of physics and the role of theory in physics.

The general theory of relativity, which we shall not investigate here, was, on the other hand, Einstein's philosophical and esthetic triumph; even now its experimental basis is rather uncertain. This theory indeed *was* founded on abstract principles, and it represents one of the great adventures of the human mind. It is perhaps regrettable that its beauty as a plausible theory of space, time and matter can be appreciated only by those who have mastered the mathematical language in which it is couched. Much credit for the underlying principles of this theory must be given to Ernst Mach, who formulated some of the ideas, but not the mathematical structure, of the theory. Although much evidence supporting the general theory now exists, we must await more definitive experimental results (and also more complete mathematical studies, for the mathematics of the theory is extremely complicated and only a few approximate solutions of the basic equations are known at the present time) to determine whether the beauty of the theory will survive without succumbing to the tarnish of physical irrelevance. Considerable effort is being expended toward clarifying the status of Einstein's general theory.

READING NOTES*

There are many good texts on electromagnetic theory available. A few especially recommended are:

Reference 5 presents the fundamental theory very compactly and in a surprisingly thorough manner.

---

* Works referred to in these notes will be found in the Bibliography on p. 557.

Reference 4 is a rather comprehensive treatment with heavy emphasis on radiation.

Reference 8 is a graduate text that reviews the field quite thoroughly.

Reference 3 is another standard comprehensive graduate-level text.

Reference 7 is an old favorite among physicists; it begins, however, with special relativity.

Reference 6 provides numerous practical examples to accompany the theory in the text.

PROBLEMS

1. Use $\mathbf{A} = \hat{\mathbf{i}}A_x + \hat{\mathbf{j}}A_y + \hat{\mathbf{k}}A_z$ and similar expansions for $\mathbf{B}$ and $\mathbf{C}$ to show that $\mathbf{A} \times (\mathbf{B} \times \mathbf{C}) = (\mathbf{A} \cdot \mathbf{C})\mathbf{B} - (\mathbf{A} \cdot \mathbf{B})\mathbf{C}$.

2. Use $\nabla = \hat{\mathbf{i}}\frac{\partial}{\partial x} + \hat{\mathbf{j}}\frac{\partial}{\partial y} + \hat{\mathbf{k}}\frac{\partial}{\partial z}$ to show that
$$\nabla \times (\mathbf{A} \times \mathbf{B}) = \mathbf{B} \cdot \nabla\mathbf{A} - \mathbf{A} \cdot \nabla\mathbf{B} + (\nabla \cdot \mathbf{B})\mathbf{A} - (\nabla \cdot \mathbf{A})\mathbf{B},$$
where $\mathbf{B} \cdot \nabla\mathbf{A}$ signifies
$$\hat{\mathbf{i}}\mathbf{B} \cdot (\nabla A_x) + \hat{\mathbf{j}}\mathbf{B} \cdot (\nabla A_y) + \hat{\mathbf{k}}\mathbf{B} \cdot (\nabla A_z)$$
and a similar meaning is given to $\mathbf{A} \cdot \nabla\mathbf{B}$.

3. Show that $\nabla f(\mathbf{x} - \mathbf{x}') = -\nabla' f(\mathbf{x} - \mathbf{x}')$, where
$$\nabla = \sum_{j=1}^{3} \hat{\mathbf{i}}_j \frac{\partial}{\partial x_j}, \qquad \nabla' \equiv \sum_{j=1}^{3} \hat{\mathbf{i}}_j \frac{\partial}{\partial x'_j}$$

4. Obtain the identity
$$\nabla(\mathbf{U} \cdot \mathbf{V}) = \mathbf{U} \cdot \nabla\mathbf{V} + \mathbf{V} \cdot \nabla\mathbf{U} + \mathbf{U} \times (\nabla \times \mathbf{V}) + \mathbf{V} \times (\nabla \times \mathbf{U})$$
where $\mathbf{U} \cdot \nabla\mathbf{V}$ and $\mathbf{V} \cdot \nabla\mathbf{U}$ have the meanings indicated in Problem 2.

5. Let the "three-index totally antisymmetric symbol" ϵ_{ijk} be defined by
$$\epsilon_{ijk} = \begin{cases} +1 & \text{if } (i, j, k) \text{ is an even permutation of } (1, 2, 3) \\ -1 & \text{if } (i, j, k) \text{ is an odd permutation of } (1, 2, 3) \\ 0 & \text{otherwise.} \end{cases}$$

   a) Verify that
$$\mathbf{A} \times \mathbf{B} = \sum_{i=1}^{3} \sum_{j=1}^{3} \sum_{k=1}^{3} \epsilon_{ijk} \hat{\mathbf{i}}_i A_j B_k.$$

   b) Calculate $\sum_{k=1}^{3} \epsilon_{ijk}\epsilon_{klm}$ in terms of Kronecker δ-symbols defined by
$$\delta_{ij} = \begin{cases} 1 & \text{if } i = j \\ 0 & \text{if } i \neq j. \end{cases}$$

   c) Use this result to verify the vector identity of Problem 1.

6. Find expressions for $\delta^3(\mathbf{x})$ in terms of products of one-dimensional δ-functions for (a) cylindrical coordinates, and (b) spherical coordinates. (For Cartesian coordinates we have the product $\delta(x)\delta(y)\delta(z)$, but for spherical coordinates, $\delta^3(\mathbf{x}) \neq \delta(r)\delta(\theta)\delta(\varphi)$.)

7. Find the electric field $\mathbf{E}$, arising from a static electric dipole consisting of charges q and $-q$ located respectively at $(a/2)\hat{\mathbf{k}}$ and $(-a/2)\hat{\mathbf{k}}$, at a point $\mathbf{x}$ that is far from the dipole (i.e. $|\mathbf{x}| \gg a$). Use the expression.

$$\mathbf{E}(\mathbf{x}) = \frac{1}{4\pi\epsilon_0}\int \frac{\rho(\mathbf{x}')|\mathbf{x} - \mathbf{x}'|}{|\mathbf{x} - \mathbf{x}'|^3}\, d^3x'$$

and keep only terms up to and including first degree in $a/|\mathbf{x}|$. (Hint: What is the charge density for the dipole in terms of δ-functions? Also, use the approximation

$$\left(x^2 + \frac{a^2}{4} \pm \mathbf{a}\cdot\mathbf{x}\right)^{1/3} \simeq x^3\left[1 + \frac{3}{2}\left(\frac{a^2}{4x^2} \pm \frac{a}{x}\hat{\mathbf{x}}\cdot\hat{\mathbf{k}}\right)\right]$$

$$\simeq x^3\left(1 \pm \frac{3}{2}\frac{a}{x}\cos\theta\right),$$

where $x = |\mathbf{x}|$ and $\cos\theta = \hat{\mathbf{x}}\cdot\hat{\mathbf{k}}$.)

8. Consider a wire formed into a circular loop of radius a through which a steady current of magnitude J is flowing. Find the vector potential $\mathbf{A}$ at a point $\mathbf{x}$ such that $|\mathbf{x}| \gg a$.

9. Use the vector potential obtained in Problem 8 to obtain the magnetic induction field at $\mathbf{x}$, and show that it is approximately that of a pure magnetic dipole. Find the equivalent magnetic dipole moment in terms of a and J.

10. A classical point particle of mass m and electric charge q is moving in a circular orbit with constant angular momentum

$$\mathbf{L} = \mathbf{r} \times \mathbf{p} = \mathbf{r} \times m\mathbf{v},$$

where $\mathbf{r}$, $\mathbf{p}$ and $\mathbf{v}$ are, respectively, the location, momentum and velocity of the particle; the particle also has constant kinetic energy T. Find the average magnetic induction field $\mathbf{B}$ (neglecting radiation arising from centripetal acceleration) arising from this circulating charge, at a point $\mathbf{x}$ far from the orbiting particle ($|\mathbf{x}| \gg |r|$) when the origin of coordinates is located at the center of the particle's orbit. Compare with Problem 9.

CHAPTER 2

FOUNDATIONS OF THE SPECIAL THEORY OF RELATIVITY

The following five statements enumerate some of the properties of space and time as they were understood by physicists before 1905:

1. Space is isotropic and homogeneous; that is, the equations of motion of mechanics are unchanged by rotations and translations of the coordinate systems used to describe the positions of particles, when the rotations and translations do not depend on time as a parameter.

2. Time is independent of space; essentially, time is a parameter that measures the separation between events as seen by an observer, and the observer's location does not affect the size of the standard intervals used to measure time.

3. Time is homogeneous; the equations of motion of mechanics are unchanged by translations in the time parameter.

4. The equations of motion of mechanics are unchanged by spatial translations that involve time as a *linear* parameter, i.e. translations expressible in the form $\mathbf{x} \to \mathbf{x} - \mathbf{v}t$. Frames of reference * related by such transformations are called *inertial frames*, and the translations are called *Galilean transformations*.

5. Time is independent of inertial reference frame; that is, an observer's state of inertial motion does not affect the scale of time measurements.

Thus one has a picture of the world as a homogeneous and isotropic three-dimensional structure of space-points, with particles passing through these points according to laws formulated in terms of a parameter called time; an important property of the classical laws governing particle motion is that these laws are the same for all frames of reference moving with uniform velocity relative to one another, i.e. classical mechanical laws are the same for all frames related to one another by Galilean transformations.

The properties of Galilean transformations, including the additivity of velocities, are familiar. For example, if frame 1 is moving with velocity $\mathbf{v}_{12}$ relative to frame 2, and if frame 2 is moving relative to frame 3 with velocity $\mathbf{v}_{23}$, then a point in space given by the vector $\mathbf{x}^{(1)}$ in the coordinates of reference frame 1, is given by

* A frame of reference is simply a set of coordinate axes.

$\mathbf{x}^{(2)}$ in the coordinates of frame 2, and by $\mathbf{x}^{(3)}$ in the coordinates of frame 3, where $\mathbf{x}^{(1)}$, $\mathbf{x}^{(2)}$ and $\mathbf{x}^{(3)}$ are related by

$$\mathbf{x}^{(1)} = \mathbf{x}^{(2)} - \mathbf{v}_{12}t \quad \text{and} \quad \mathbf{x}^{(2)} = \mathbf{x}^{(3)} - \mathbf{v}_{23}t.$$

These equations imply that $\mathbf{x}^{(1)}$ is related to $\mathbf{x}^{(3)}$ by

$$\mathbf{x}^{(1)} = \mathbf{x}^{(3)} - (\mathbf{v}_{12} + \mathbf{v}_{23})t,$$

and thus that the velocity of frame 1 relative to frame 3 is just the sum of the relative velocities of frames 1 and 2 and of frames 2 and 3. In particular, if a wave (e.g. a sound wave) having velocity $\mathbf{v}_{12}$ relative to the medium through which it propagates (e.g. a mass of air) is detected by an observer who notices that the propagating medium is moving relative to him with a velocity $\mathbf{v}_{23}$, the observer will claim that the wave is moving with a speed $\mathbf{v}_{13} = (\mathbf{v}_{12} + \mathbf{v}_{23})$ relative to him. This example can be cast into *frame of reference* language if we imagine a reference frame "attached" to a phase point of the wave so that the wave is at rest relative to this frame, another frame anchored to the propagating medium, and a third frame fastened to the observer's detecting apparatus; then if these three reference frames are labeled 1, 2 and 3, respectively, their relative velocities are $\mathbf{v}_{12}$, $\mathbf{v}_{23}$ and $\mathbf{v}_{13} = (\mathbf{v}_{12} + \mathbf{v}_{23})$. (See Fig. 2.1.)

Since, in elementary mechanics, longitudinal sound waves and transverse waves along a string are familiar phenomena, one might expect that electromagnetic waves should obey the same laws that govern such mechanical waves. Such a situation would be very desirable, for then the physics of electromagnetism

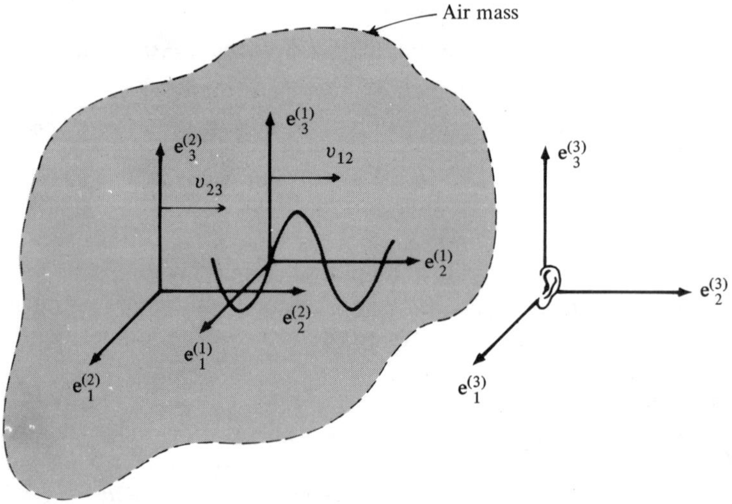

Fig. 2.1. Additivity of velocities under Galilean transformations. Here sound is used as an example.

might be unified conceptually with the physics of massive bodies. Let us therefore test the applicability of classical mechanics to electromagnetic theory by carrying out a Galilean transformation on the wave equation governing electromagnetic waves.

We are interested in the telegrapher's equation

$$\left(\nabla^2 - \frac{1}{c^2}\frac{\partial^2}{\partial t^2}\right) f(\mathbf{x}, t),$$

where $f(\mathbf{x}, t)$ is a scalar function of $\mathbf{x}, t$. Under a Galilean transformation, $\mathbf{x}_j \to \mathbf{x}_j - v_j t = \mathbf{x}'_j$, and since $f(\mathbf{x}, t)$ is a scalar function, it must have the same numerical value in both coordinate systems, although its form as a function of $\mathbf{x}', t$ may be different. Thus, let

$$f(\mathbf{x}, t) = g(\mathbf{x}', t) = g(\mathbf{x}'(\mathbf{x}, t), t),$$

where the last member demonstrates that g depends on t implicitly through the dependence of $\mathbf{x}'(\mathbf{x}, t)$ on t, as well as explicitly. Now we have

$$\frac{\partial}{\partial x_j} f(\mathbf{x}, t) = \sum_{k=1}^{3} \frac{\partial x'_k}{\partial x_j}\frac{\partial}{\partial x'_k} g(\mathbf{x}', t) = \frac{\partial g(\mathbf{x}', t)}{\partial x'_j},$$

so that

$$\nabla^2 f(\mathbf{x}, t) = \nabla'^2 g(\mathbf{x}', t).$$

However,

$$\frac{\partial}{\partial t} f(\mathbf{x}, t) = \sum_{k=1}^{3} \frac{\partial x'_k}{\partial t}\frac{\partial g(\mathbf{x}', t)}{\partial x'_k} + \frac{\partial g(\mathbf{x}', t)}{\partial t}$$

$$= v \cdot \nabla' g(\mathbf{x}', t) + \frac{\partial}{\partial t} g(\mathbf{x}', t)$$

and

$$\frac{\partial^2}{\partial t^2} f(\mathbf{x}, t) = \frac{\partial^2}{\partial t^2} g(\mathbf{x}', t) + 2\mathbf{v} \cdot \nabla' \frac{\partial}{\partial t} g(\mathbf{x}', t) + \mathbf{v} \cdot \nabla'[\mathbf{v} \cdot \nabla' g(\mathbf{x}', t)].$$

Therefore, the telegrapher's equation transforms according to the rule

$$\left(\nabla^2 - \frac{1}{c^2}\frac{\partial^2}{\partial t^2}\right) f(\mathbf{x}, t) = \left[\nabla'^2 - \left(\frac{\mathbf{v}}{c} \cdot \nabla'\right)\left(\frac{\mathbf{v}}{c} \cdot \nabla'\right) - 2\frac{\mathbf{v}}{c^2}\nabla' \frac{\partial}{\partial t} - \frac{1}{c^2}\frac{\partial^2}{\partial t^2}\right] g(\mathbf{x}', t) = 0.$$

Evidently the equation governing electromagnetic wave propagation has a different form in the primed reference frame, with coordinates x'_i, from that which it has in the unprimed reference frame; the form of the telegrapher's equation is not invariant under Galilean transformations. Since this equation represents the law of electromagnetic wave propagation, one must say that the choice of Galilean reference frame affects the laws of electromagnetism.

Classical mechanics and electromagnetism therefore cannot be welded together without some changes. But which theory should we retain, if either? As it turned out, both theories can be retained almost intact, but Newtonian mechanics has had to yield a little bit by giving up, as an exact postulate, the invariance of the equations of motion under Galilean transformations, together with the notion that time is independent of reference frame. The only way this question could be answered, however, was to turn to experiment and to try to determine physically whether the constant c that appears in the telegrapher's equation refers to the velocity of wave propagation in some *preferred* reference frame (the frame of reference in which the "ether" or propagating medium is at rest, presumably) so that Maxwell's equations are strictly correct only in that preferred frame, or whether some aspect of mechanics is a defective description of nature. Since Maxwell's equations were relative newcomers to physics, it is not surprising that the main experimental effort was aimed at detecting a preferred frame of reference to which electricity and magnetism always should be referred.

We shall briefly recount some of the experimental history of the overthrow of Galilean, or Newtonian, relativity, but the reader is referred to other textbooks and reference books for detailed accounts and for wider discussions of the relevant evidence.

2.1 THE TRAUTON AND NOBLE EXPERIMENTS

According to Eqs. (1.11) and (1.12), the total force per unit volume exerted on a charge density $\rho(\mathbf{x})$ and current density $\mathbf{j}(\mathbf{x})$ is

$$\mathbf{f}(\mathbf{x}) = \mathbf{f}_C(\mathbf{x}) + \mathbf{f}_B(\mathbf{x}) = \rho(\mathbf{x})\mathbf{E}(\mathbf{x}) + \mathbf{j}(\mathbf{x}) \times \mathbf{B}(\mathbf{x}). \tag{2.1}$$

Now if the charge density vanishes everywhere except at a single point $\mathbf{x}_0$, and is so large at $\mathbf{x}_0$ that $\int_V \rho(\mathbf{x})\, d^3x = q$ is a finite number if $\mathbf{x}_0$ is in V, then we can write

$$\rho(\mathbf{x}) = q\delta^3(\mathbf{x} - \mathbf{x}_0); \tag{2.2}$$

that is, we have a point charge of magnitude q at $\mathbf{x}_0$. Similarly, if $\mathbf{j}(\mathbf{x})$ is generated by a point charge of magnitude q moving with velocity $\mathbf{v}(\mathbf{x})$, we have

$$\mathbf{j}(\mathbf{x}) = \rho(\mathbf{x})\mathbf{v}(\mathbf{x}) = q\mathbf{v}(\mathbf{x}_0)\delta^3(\mathbf{x} - \mathbf{x}_0) \tag{2.3}$$

when the charge is at $\mathbf{x}_0$. With these expressions for the charge and current densities, the total force experienced by the charge evidently is

$$\mathbf{F} = \int_V \mathbf{f}(\mathbf{x})\, d^3x = q[\mathbf{E}(\mathbf{x}_0) + v(\mathbf{x}_0) \times \mathbf{B}(\mathbf{x}_0)]. \tag{2.4}$$

The field vectors $\mathbf{E}(\mathbf{x})$ and $\mathbf{B}(\mathbf{x})$ can be written in terms of the scalar and vector potentials, according to Eqs. (1.27) and (1.28), and for fields generated by a point charge q' at $\mathbf{x}'$, the potentials given by Eqs. (1.25) and (1.26) become

$$\phi(\mathbf{x}) = \frac{q'}{4\pi\epsilon_0}\frac{1}{|\mathbf{x}-\mathbf{x}'|}, \tag{2.5}$$

$$\mathbf{A}(\mathbf{x}) = \frac{\mu_0}{4\pi}q'\frac{\mathbf{v}'(\mathbf{x}')}{|\mathbf{x}-\mathbf{x}'|}, \tag{2.6}$$

where $\mathbf{v}'(\mathbf{x}')$ is the velocity of the source charge q'. If the velocity $\mathbf{v}'(\mathbf{x}')$ is a constant, we then have

$$-\nabla\phi(\mathbf{x}) = \frac{q'}{4\pi\epsilon_0}\frac{\mathbf{x}-\mathbf{x}'}{|\mathbf{x}-\mathbf{x}'|^3}, \qquad \frac{\partial\mathbf{A}(\mathbf{x})}{\partial t} = \mathbf{0}$$

$$\nabla\times\mathbf{A}(\mathbf{x}) = -\frac{\mu_0 q'}{4\pi}\frac{\mathbf{x}-\mathbf{x}'}{|\mathbf{x}-\mathbf{x}'|^3}\times\mathbf{v}'(\mathbf{x}').$$

Therefore the fields at point $\mathbf{x}$ arising from a uniformly moving charge q' at $\mathbf{x}'$ are

$$\mathbf{E}(\mathbf{x}) = \frac{q'}{4\pi\epsilon_0}\frac{\mathbf{x}-\mathbf{x}'}{|\mathbf{x}-\mathbf{x}'|^3}, \tag{2.7}$$

$$\mathbf{B}(\mathbf{x}) = \frac{\mu_0 q'}{4\pi}\frac{\mathbf{v}'(\mathbf{x}')\times(\mathbf{x}-\mathbf{x}')}{|\mathbf{x}-\mathbf{x}'|^3}. \tag{2.8}$$

The force on charge q at $\mathbf{x}_0$ due to these fields is

$$\mathbf{F} = \frac{qq'}{4\pi\epsilon_0}\frac{1}{|\mathbf{x}-\mathbf{x}'|^3}\{(\mathbf{x}_0-\mathbf{x}') + \mu_0\epsilon_0\mathbf{v}(\mathbf{x}_0)\times[\mathbf{v}'(\mathbf{x}')\times(\mathbf{x}_0-\mathbf{x}')]\}, \tag{2.9}$$

the first term being the electrostatic force and the second being the magnetic force. This is familiar from the elementary physics of point charges.

Now the torque on the charge at $\mathbf{x}_0$ about the point $\mathbf{x}'$ is defined to be

$$\mathbf{N} = (\mathbf{x}_0-\mathbf{x}')\times\mathbf{F} = \frac{qq'\mu_0}{4\pi|\mathbf{x}_0-\mathbf{x}'|^3}(\mathbf{x}_0-\mathbf{x}')\times\{\mathbf{v}(\mathbf{x}_0)\times[\mathbf{v}'(\mathbf{x}')\times(\mathbf{x}_0-\mathbf{x}')]\} \tag{2.10}$$

Note that the torque arises entirely from magnetic effects.

Consider the arrangement shown in Fig. 2.2, for which we take $\mathbf{v}'(\mathbf{x}') = \mathbf{v}(\mathbf{x}_0) = \mathbf{v}$. Here we have

$$\mathbf{N} = \frac{\mu_0}{4\pi}\frac{qq'}{r^3}\mathbf{r}\times[\mathbf{v}\times(\mathbf{v}\times\mathbf{r})], \tag{2.11}$$

where $\mathbf{r} = \mathbf{x}_0 - \mathbf{x}'$. If the angle between $\mathbf{v}$ and $\mathbf{r}$ is θ, the multiple cross-product

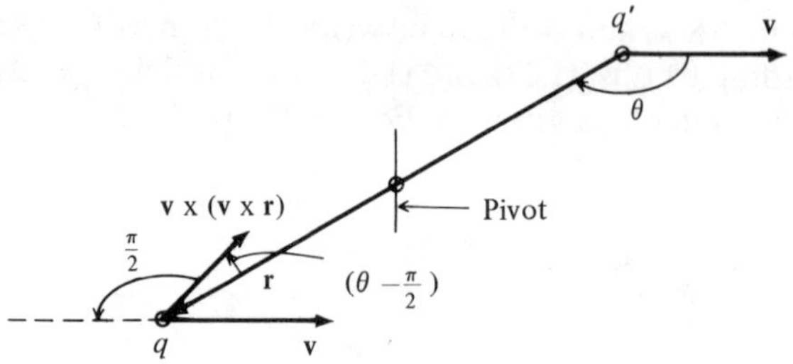

Fig. 2.2. Schematic diagram of the arrangement of the Trauton–Noble experiment.

on the right-hand side can be rewritten in terms of θ:

$$\mathbf{r} \times [\mathbf{v} \times (\mathbf{v} \times \mathbf{r})] = \mathbf{r} \times [(\mathbf{v} \cdot \mathbf{r})\mathbf{v} - \mathbf{v}^2\mathbf{r}] = (\mathbf{v} \cdot \mathbf{r})(\mathbf{r} \times \mathbf{v})$$

$$= v^2 r^2 \sin\theta \cos\theta\, (\mathbf{r} \mathbin{\hat{\times}} \mathbf{v}) = \tfrac{1}{2} v^2 r^2 \sin 2\theta\, (\mathbf{r} \mathbin{\hat{\times}} \mathbf{v}).$$

Hence the torque tending to restore the vector $\mathbf{r}$ to an orientation perpendicular to $\mathbf{v}$ is

$$\mathbf{N} = (\mathbf{r} \mathbin{\hat{\times}} \mathbf{v}) \frac{qq'}{4\pi\epsilon_0} \frac{(v/c)^2}{r} \sin 2\theta, \qquad \mathbf{r} \mathbin{\hat{\times}} \mathbf{v} \equiv \frac{\mathbf{r} \times \mathbf{v}}{|\mathbf{r} \times \mathbf{v}|}. \tag{2.12}$$

At this point we must ask, "With respect to what frame of reference are we to measure **v**?" If there were a fixed preferred frame of reference to which all electromagnetic effects should be referred, then **v** certainly would be measured relative to that frame.

In 1903, Trauton and Noble performed a series of experiments in which charged condensers were moved through space at the velocity of the earth relative to the sun; the apparatus was sufficiently sensitive to detect torques resulting from velocities somewhat less than the earth's orbital velocity, the former presumably being measured relative to some preferred reference frame. However, no torque was observed. This would seem to imply that the earth is more nearly at rest relative to the preferred frame, or the "ether," than is the sun; thus, if one accepts the notion of an absolute frame of reference for electromagnetic effects, one apparently is forced by this experiment to reconsider the possibility of a Ptolemaic or geocentric theory of the universe. A more reasonable alternative is to question the notion of a preferred reference frame, and to reason that in the absence of such a preferred frame we cannot make sense of Eq. (2.12) until the meaning of the velocity **v** is clarified—that is, until the manner in which **v** is to be measured relative to an *arbitrary* frame is specified.

Let us consider another relevant experiment.

2.2 THE MICHELSON AND MORLEY EXPERIMENT

A clever version of an optical device called an interferometer was designed by A. A. Michelson near the turn of the century; this interferometer produces circular fringes of the type discussed in Appendix A.2. Michelson and a colleague, E. W. Morley, used the properties of the fringe patterns to measure an effect predicted by the notion of an ether that sustains electromagnetic waves and that is stationary relative to the preferred reference frame.

Figure 2.3 illustrates the principle of the instrument. Optically—that is, in terms of the geometrical configuration of light rays—this apparatus is the equivalent of that shown in Appendix A.2, with a source behind the observation point, and two mirrors reflecting the source waves back to the observer. Mechanically, however, it is quite different, and it is the fact that light rays travel perpendicularly to one another in this device that makes it particularly useful in testing the ether hypothesis.

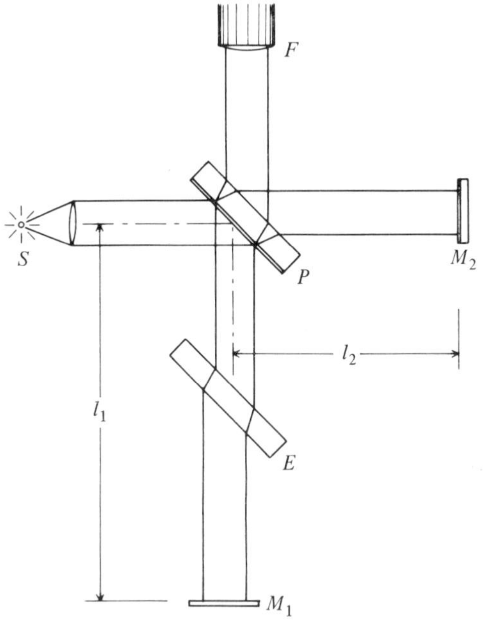

Fig. 2.3. Schematic diagram of Michelson–Morley interferometer.

Light leaving the source can follow one of two useful paths; it can be reflected in the half-silvered mirror at P, travel through the glass plate at E to mirror M_1, where it is reflected back through the same path to P and through* the lightly silvered mirror to an observer at F, or it can go from the source through the lightly silvered mirror at P to M_2, and be reflected back to P, where it is reflected† toward

* About half the light reflected back from M_1 will be returned toward the source by the lightly silvered mirror P, but this light is lost to the experiment and light following this path is not utilized.

† Again, about half the light returning from M_2 is lost by passing through P.

the observer at F. The plate E is a plate of glass exactly like that at P, except that it is unsilvered; its purpose is to render the path of light through glass the same for rays taking either of the useful paths, and thereby to nullify effects arising from having one beam pass through more glass than the other.

Assume that the interferometer is moving through the ether with velocity $\mathbf{v}$ parallel to M_1F, i.e. along the axis from mirror M_1 to the observer. The time required for light to pass from P to M_1 and back to P then is

$$t_1 = l_1\left(\frac{1}{c - v} + \frac{1}{c + v}\right) = \frac{2l_1 c}{c^2 - v^2} = \frac{2l_1}{c(1 - \beta^2)}, \qquad \beta = \frac{|\mathbf{v}|}{c} = \frac{v}{c},$$

if we make the usual classical assumptions.* The light traveling from P to M_2 and back to P must go a distance $2\sqrt{l_2^2 + (vt_2/2)^2}$ because of the motion of the instrument; hence

$$t_2 = 2\sqrt{l_2^2 + (vt_2/2)^2}/c$$

or

$$t_2 = \frac{2l_2}{c\sqrt{1 - \beta^2}}.$$

As a result, the optical path difference between the two rays is

$$\Delta = c(t_1 - t_2) = \frac{2}{\sqrt{1 - \beta^2}}\left(\frac{l_1}{\sqrt{1 - \beta^2}} - l_2\right).$$

Now if we rotate the interferometer through 90° and let the axis from M_2 to S be parallel to the velocity, we have the roles of t_1 and t_2 interchanged, i.e.

$$t_1' = \frac{2l_1}{c\sqrt{1 - \beta^2}}, \qquad t_2' = \frac{2l_2}{c(1 - \beta^2)};$$

the new optical path then has length

$$\Delta' = c(t_1' - t_2') = \frac{2}{\sqrt{1 - \beta^2}}\left(l_1 - \frac{l_2}{\sqrt{1 - \beta^2}}\right).$$

This change in optical path will cause a phase shift of

$$k(\Delta - \Delta') = \frac{2\pi}{\lambda}\frac{2(l_1 + l_2)}{\sqrt{1 - \beta^2}}\left(\frac{1}{\sqrt{1 - \beta^2}} - 1\right) = \frac{4\pi(l_1 + l_2)}{\lambda(1 - \beta^2)}(1 - \sqrt{1 - \beta^2}),$$

where k is the reduced wave number of the light waves, and λ is the wave length: $k = 2\pi/\lambda$. For $v \ll c$, i.e. $\beta \ll 1$, the phase shift is approximated by

$$k(\Delta - \Delta') \simeq \frac{4\pi(l_1 + l_2)}{\lambda(1 - \beta^2)}(1 - 1 + \tfrac{1}{2}\beta^2) \simeq 2\pi\frac{(l_1 + l_2)}{\lambda}\beta^2.$$

---

* Note the additivity of velocities in the denominators $1/(c - v)$ and $1/(c + v)$.

That is, if the bright fringe interference pattern, when **v** is parallel to M_1F, is given by

$$\cos\theta = n\frac{\lambda}{2\Delta}, \qquad n \text{ an integer,}$$

then for **v** parallel to M_2S it will be given by

$$\cos\theta = n'\frac{\lambda}{2\Delta'}, \qquad n' \text{ an integer;}$$

thus

$$2\frac{\Delta - \Delta'}{\lambda}\cos\theta = n - n'.$$

The number of fringes within angle θ will change from n to n', where

$$n' = n - 2\frac{l_1 + l_2}{\lambda}\beta^2 \cos\theta.$$

Michelson and Morley, and many other workers, carried out this experiment, and none of them (with one exception, whose data has been carefully re-analysed and finally found to be consistent with other evidence) observed any significant phase shift. Evidently, the earth must be moving very slowly, if at all, relative to any stationary ether that might exist. This result is consistent with the Trauton and Noble experiment, and with vast quantities of even more convincing data that have been collected during the last half-century.

A number of hypotheses were put forward in the attempt to resolve the contradictions between the notion of an ether or of a preferred reference frame and the null results of the Michelson and Morley, Trauton and Noble, and other related experiments. These included the Lorentz–Fitzgerald contraction hypothesis, the ether drag theories, and the emission theory for the propagation of light waves. All of the hypotheses, except one, eventually were shown by experiment to be untenable; only Einstein's work survived the turmoil intact.

2.3 THE PREMISES OF SPECIAL RELATIVITY

The reasoning behind Einstein's special theory of relativity can be summarized very nicely in terms of some *Gedanken Experimente*. Before proceeding with the *Gedanken Experimente*, however, we must state the basic hypotheses of special relativity; they are

1. The laws of physics are independent of the inertial frame of reference in which the measurements are carried out, or in terms of which the mathematical description of the laws is formulated.

2. The speed of light does not vary from one inertial frame to another; i.e. the speed of light is a fixed quantity, having the same value in all inertial frames of reference.

3. Space and time measurements are best made by comparing properties of light waves.

Actually, 3 is not independent of 1 and 2, and in fact, it is hypothesis 2 that induces us to state 3 separately for the purpose of drawing attention to the importance of the role played by electromagnetic waves in our definitions of measuring processes. Since the laws of physics are, according to hypothesis, independent of inertial reference frame, and since the speed of light also is independent of reference frame, it seems natural to use the velocity of light as a standard in comparing time and distance measurements carried out in different inertial frames. It is convenient also to use the properties of electromagnetic waves to define fundamental units of length and time; we can use this circumstance to illustrate the meaning of hypothesis 1.

Let us define the fundamental unit of time for our measurements to be the mean period of electromagnetic waves emitted by a specific decay process from a designated excited state of an atom, when measured by an apparatus *at rest relative to the decaying atom.* We don't really need to know anything about the process by which the decay takes place; all we need to know is that there *is* a well-defined mean frequency for the electromagnetic waves emitted by a certain kind of matter under specified conditions. We can imagine many ways for measuring, or, more accurately, determining the fundamental time interval τ_0. A particularly simple one, in principle, is to construct a very small wand with an electric charge on one end and a pen, or tiny flashlight, or other indicating device on the other end. The wand is balanced on a fulcrum located at a point not far from the decaying atom. As the electromagnetic wave passes it, the wand begins to oscillate in response to the changing electromagnetic field. By watching the indicator, we can count the oscillations. One complete period of oscillation then determines a time interval τ_{01}. This time interval can be recorded on some time-recording device such as an electronic timer, and a large number, say n, similar intervals can be recorded and compared with τ_{01}. The average of these time intervals then can be computed (in terms of τ_{01}, say) and used to define the standard time interval, τ_0. This standard interval now is to be recorded in time-recording devices and compared with other time intervals when measures of the latter are desired.*

Special note should be taken of the fact that the atom and measuring instrument are at rest relative to one another in this definition of a time measurement. Because of hypothesis 1, we expect that the atom will behave in the same way according to all observers at rest relative to it, no matter how the atom and its

* Note that this is not, in fact, the way time standards are determined experimentally. From a classical point of view, however, there is no reason why we could not use such a technique.

observers *together* are moving relative to any other physical systems. However, if the atom were in motion relative to the observer, the description of the time measuring process would have to be changed, for the relationship between the observer and the observed would be different. The relationship between the observer and the observed is just as important as any other part of the description of a measuring process, and without it the description is incomplete.

This suggests that a moving atom may not behave in the same way as a stationary atom *in terms of the measurements* the observer carries out on the two. However, the observer cannot properly carry out a time measurement on the moving atom's electromagnetic emissions, for the time standard we defined above provides no prescription for doing so; if he compares the period of oscillation of the moving atom's emitted electromagnetic waves with that of the stationary atom's waves, he should not expect the numerical values so obtained necessarily to agree, even if the atoms are identical except for their inertial motion. If he wants to measure the oscillation period of the radiation from the moving atom, he must find a means of accelerating himself until he is at rest with the atom before he can carry out an acceptable time measurement. *Measurements shall be considered acceptably defined only when the measuring apparatus and the object whose properties are being measured (in this case the decaying atom) are at rest relative to one another.*

One cannot emphasize too heavily that the entire physical system, including the situation of the observer and his apparatus, must be specified when a phenomenon is described in terms of physical measurements. Once the phenomenon is thus described, an observer in a different reference frame from that in which the phenomenon originally was observed then can expect to see the whole phenomenon taking place in the same way in his frame of reference, according to hypothesis 1; but he must bear the same relationship to the observed part of the phenomenon as did the original observer.

Of course it usually is not desirable nor even possible to build apparatus that is at rest relative to the observer, or to accelerate an experimental set-up so that it will be at rest relative to a particular observer. For example, decaying subnuclear particles, that are among cosmic rays, often move very rapidly relative to earth-bound observers, and it is quite out of the question to observe their decays from reference frames relative to which they are at rest. Thus it is necessary to find techniques for transforming the numerical values of physical measurements from one frame of reference to another. This, in fact, is the function performed by the mathematical formulation of relativity theory.

Having specified a method for measuring time intervals at one point in space, we can specify a standard unit of length, again in terms of electromagnetic waves. We need to know only that the velocity of light is finite, and to take into account hypothesis 2, to arrive at a satisfactory definition of a standard distance. Since the speed of light, c, is independent of reference frame, and since hypothesis 1 indicates that τ_0, if determined according to our (or some other equally well-defined) prescription, also is independent of reference frame, then the quantity $\lambda_0 = c\tau_0$ is also the same in all inertial frames. In addition, λ_0 must be finite, because both c and τ_0 are finite. Obviously λ_0 can be taken as a standard unit of length for all distance measurements made in the frame of reference for which the atom defining τ_0 is at rest. We need not have made a measurement of c before defining λ_0; in fact, we can prescribe an idealized method for measuring both simultaneously, because λ_0 can be interpreted as the wavelength of electromagnetic waves having period τ_0. We could, for example, use a pair of wands of the type discussed above, having tiny flashlights as indicators; these would be balanced on fulcrums whose relative separation is adjustable, with the flashlights aimed toward one another. If a translucent screen then were mounted so as to remain always exactly halfway between the flashlights along the direction of motion of the waves, one could adjust the relative separation of the wands until λ_{01} were the smallest separation for which the images of the flashlight beams on the translucent screen could be kept continuously coincident with one another; this separation then would represent the minimum distance between points of equal phase in the passing electromagnetic wave emitted from an atom at rest relative to the apparatus, and therefore would be the wavelength of the wave having period τ_{01}. As with the determination of the period τ_0, a large number of such measurements should be carried out and their average value should be taken as the standard distance λ_0. Knowing λ_0 and τ_0, we can calculate $c = \lambda_0/\tau_0$.

This discussion has been intended to indicate that we can, with rather simple but highly idealized apparatus, prescribe standard units of distance and time in terms of the parameters of specified electromagnetic waves. In practice this is not precisely what one does, but practical measurements always can be translated into something equivalent to the procedures we have described. Therefore we can, with these prescriptions, say that we know what is meant by time and distance; we know, in principle, how to construct clocks and measuring rods that are marked off in units of τ_0 and λ_0.

In order to proceed with the *Gedanken Experimente*, we must establish a convention for constructing reference frames that are moving uniformily relative to one another. If $\mathbf{v}$ is the velocity of relative motion as measured by an observer at rest with respect to one of the frames we wish to construct, let $\hat{\mathbf{e}}_2$ be a unit vector in the direction of $\mathbf{v}$ and at rest relative to the observer; then let $\hat{\mathbf{e}}_1$ be constructed in any direction perpendicular to $\hat{\mathbf{e}}_2$, and define $\hat{\mathbf{e}}_3 = \hat{\mathbf{e}}_1 \times \hat{\mathbf{e}}_2$. The three unit vectors $\hat{\mathbf{e}}_1$, $\hat{\mathbf{e}}_2$ and $\hat{\mathbf{e}}_3$ define a frame in which the observer is at rest; the origin of coordinates in this frame may be chosen in any convenient manner. Next, let $\hat{\mathbf{e}}'_2$ be a unit vector in the direction of $\mathbf{v}$, but attached to a point moving with velocity $\mathbf{v}$ relative to the frame we have just established; furthermore, let the point to which $\hat{\mathbf{e}}'_2$ is attached lie on the axis defined by $\hat{\mathbf{e}}_2$, so that $\hat{\mathbf{e}}'_2$ and $\hat{\mathbf{e}}_2$ are not only parallel but are collinear. Then let $\hat{\mathbf{e}}'_1$ be parallel with $\hat{\mathbf{e}}_1$, but let it be fastened to the same point as that to which $\hat{\mathbf{e}}'_2$ is attached; define $\hat{\mathbf{e}}'_3 = \hat{\mathbf{e}}'_1 \times \hat{\mathbf{e}}'_2$. The latter coordinate system, to be called the *primed* coordinate system or frame of reference, is moving with velocity $\mathbf{v}$ relative to the first, or *unprimed* frame of reference, and at some time during the course of the motion the frames will momentarily coincide, since their origins, i.e. the points to which the unit vectors are attached, are chosen to be collinear with $\mathbf{v}$. See Fig. 2.4.

We are now prepared to carry out the *Gedanken Experimente*, that is, to use the fundamental hypotheses of special relativity to compare measurements made in different frames of reference.

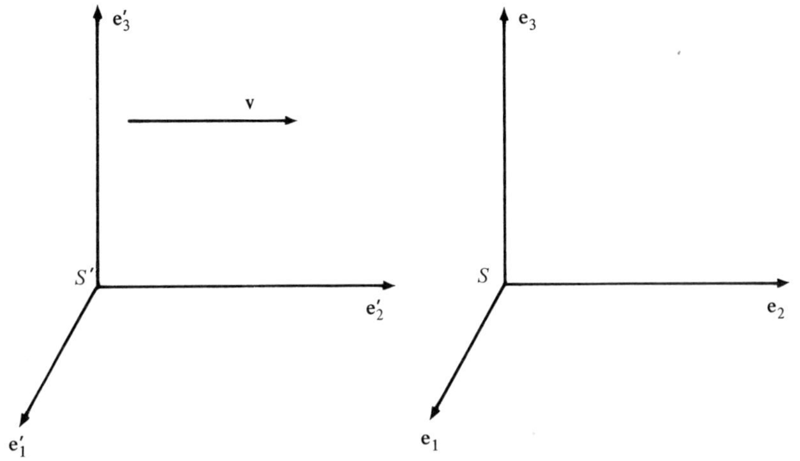

Fig. 2.4. Definition of coordinate axes.

2.4 THE COMPARISON OF LENGTHS ORIENTED PERPENDICULAR TO THE DIRECTION OF RELATIVE MOTION

We shall use S to denote the umprimed reference frame $(\hat{\mathbf{e}}_1, \hat{\mathbf{e}}_2, \hat{\mathbf{e}}_3)$ and S' to denote the primed frame $\hat{\mathbf{e}}'_1, \hat{\mathbf{e}}'_2, \hat{\mathbf{e}}'_3$; notice that since S' is moving with velocity $\mathbf{v}$ relative to S, then S is moving with velocity $-\mathbf{v}$ relative to S'.

In S a measuring rod of length l (as measured in S, of course) is placed along the $\hat{\mathbf{e}}_3$-axis, its midpoint located precisely at the origin of S. Similarly, a rod of length l (this time as measured in S') is placed along the $\hat{\mathbf{e}}'_3$-axis in S', its midpoint at the origin of S'. The endpoints of these rods are equipped with light sources focused at their respective midpoints, and these sources are somehow "tagged" (e.g. by different colors, intensities, or beam shapes) so that an observer in either frame can distinguish between the lights on his own rod, at rest relative to him, and those on the rod that is moving with respect to him. If the frames are moving toward one another initially, the $\hat{\mathbf{e}}_3$-axis and the $\hat{\mathbf{e}}'_3$-axis eventually will coincide; at the instant of coincidence, the lights are flashed and observers at the origins of both frames prepare to record the arrival of the flashes.

Since the endpoints of the rod in S' always are equidistant from the origins of both frames, the flashes originating from those endpoints will arrive together

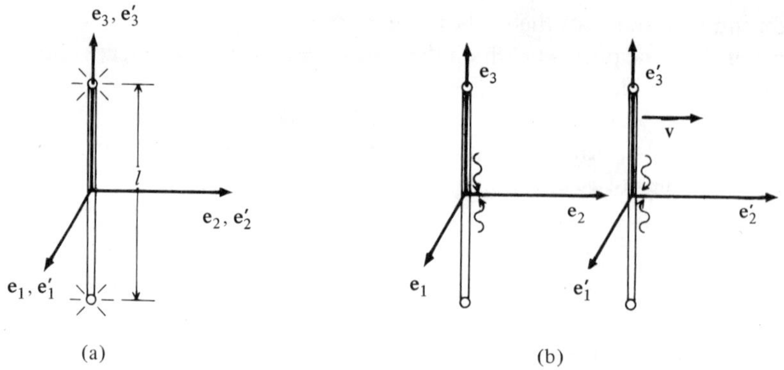

Fig. 2.5. Comparison of lengths perpendicular to the direction of motion. (a) S and S' coincident at $t = 0$. (b) Light arriving at origins at $t = l/2c$.

at the origin of either frame; a similar statement applies to the flashes from the endpoints of the rod in S. Furthermore, since the flashes all were triggered by the same event, namely the coincidence of the $\hat{\mathbf{e}}_3$- and $\hat{\mathbf{e}}'_3$-axes, the time intervals between the emission of the flashes and the arrival of the flashes at the origins are defined in exactly the same way for both frames of reference. (What bearing does hypothesis 2 above have upon this conclusion?) Therefore the flashes from the endpoints of the rod at rest in S' will arrive (at the origins of *both* frames of reference) before (or after, or simultaneously with) those from the rod in S. That is, observers in both frames will agree as to the relative lengths of the two rods—the one in S' is shorter than the one in S, they have equal lengths, or the one in S' is the longer. However, if there were any asymmetry, such as the appearance to both observers that the rod in S' were the shorter of the two, it would be possible to establish on those grounds a preferred frame of reference—e.g. of any two inertial frames of reference, that in which distances perpendicular to the direction of relative motion appear to be the shorter shall be considered "preferable" to the other, and of all inertial frames, that in which distances perpendicular to the direction of motion appear to be shortest shall be considered to be the "most preferred" frame of reference. Thus, any such asymmetry would be contrary to hypothesis 1 above, which denies the existence of any basis for the preferential treatment of any reference frames. Therefore, the rods appear to have equal lengths, according to *both* observers.

2.5 THE COMPARISON OF CLOCK RATES IN DIFFERENT REFERENCE FRAMES

In comparing the time intervals recorded by clocks in two different inertial frames, one must consider the question of simultaneity—of whether he can find a reasonable criterion for deciding that any two given events are simultaneous in all frames of reference. For example, let a pulse of light be emitted from the origins of S and

S' at the moment $t_0 = t'_0$ when they coincide. At time $t_0 + \Delta t$, according to an observer in the S frame, the wave front of the light pulse would have traveled a distance $|\Delta\mathbf{x}| = c\,\Delta t$. When an observer in S' has measured an equal length of time according to clocks at rest in his frame, his clocks will read $t' = t_0 + \Delta t$, and he too will claim that the wave front will have advanced by $|\Delta\mathbf{x}'| = |\Delta\mathbf{x}| = c\,\Delta t$ in his frame of reference. But this is absurd; it requires the wave front to be at two different places at the "same" time. For example, the wave front will be located a distance $|\mathbf{x}| = c\,\Delta t$ from the origin along the direction $\mathbf{v}$ in S' at $t' = t_0 + \Delta t$; but according to an observer in S, the origin of S' is at $\mathbf{x}_0 = \mathbf{v}\,\Delta t$ and hence the wave front must be at $\mathbf{x}_0 + \Delta\mathbf{x} = (\mathbf{v} + c\,\Delta\mathbf{x}/|\Delta\mathbf{x}|)\,\Delta t$ after the lapse of time Δt or, for $\Delta\mathbf{x}$ parallel to $\mathbf{v}$, at $\hat{\mathbf{v}}(v + c)\,\Delta t$. According to the discussion above, however, an observer in S sees the wave front at $\Delta\mathbf{x} = c\hat{\mathbf{v}}\,\Delta t$ along the direction $\mathbf{v}$ after that lapse of time, and thus, for him, the wave front must be in two places simultaneously. Because of the symmetry between reference frames implied by hypothesis 1, an observer in S' also must have a "double image" of the wave front. If this discussion is extended to other frames, one comes to the conclusion that the wave front must be everywhere at once!

What has gone wrong here? On the basis of all the experiments performed to determine the existence of an ether and to evaluate the speed of light, we feel certain that c is a fundamental constant and its value does not depend upon the reference frame in which it is measured. Therefore the wave fronts from our light pulse must be spherical in both frames, and each observer must see the wave passing points a distance $|\Delta\mathbf{x}| = c\,\Delta t$ away from the origin of his own frame at time $t_0 + \Delta t$; yet this seems to lead to an absurd conclusion.

The difficulty stems from the word "simultaneously", or from the phrase "equal length of time". Classical physics has embedded in our thinking the notion that time is a completely autonomous and independent parameter, and that everyone, given the same timekeeping equipment, will agree as to the length of a given time interval, no matter where he may be or what may be his state of motion relative to others making the same kind of measurement. That is, we assume that we already know how to compare *all* measurements of time intervals. We noted in an earlier section, however, that our definition of a time measurement in terms of a standard time interval tells us only how to compare measurements made *at rest* relative to the atom emitting the waves whose period defines the standard interval; no method is prescribed there for comparing lengths of time measured in different inertial frames. In fact, the unbelievable result we obtained when we assumed the speed of light to be the same in different frames indicates that we should look carefully at how the time intervals measured in the different frames should be compared. What we must try now is relinquishing the idea that the interval $\Delta t'$ as measured by a timepiece at rest in S' will be numerically equal to a corresponding interval Δt measured by a timepiece at rest in S and therefore moving relative to S'. Evidently this will wreak havoc with our ideas of simultaneity, but we are forced to try it.

In order to compare time intervals in different frames of reference, we shall have to use two clocks in one frame of reference, although one clock will be sufficient in the other. We have seen already that measurements of distances perpendicular to $\mathbf{v}$ are numerically the same in different frames; therefore a pulse of light traversing such a distance, since it travels with the same velocity in both frames, will provide a convenient standard for time comparison.

We imagine a rod of length l to be mounted along the $\hat{\mathbf{e}}_3'$-axis with one of its endpoints located at the origin of S'; at the end not located at the origin, is fastened a mirror facing the origin. At the instant the origins of S and S' coincide, a light is flashed at the point of common origins. A *single clock*, located at the origin of S' and at rest in S', measures the time interval required for the light pulse to travel to the mirror at the end of the rod at rest in S' to be $\Delta t' = 2(l/c)$. Two clocks are required in S to measure the corresponding time interval, because one, at the origin of S, must record the instant that the flash is emitted and another, at the point along the $\hat{\mathbf{e}}_2$-axis where the beam is returned to the origin of S', to record the time of arrival of the reflected flash at the origin of S'. (See Fig. 2.6.) The interval Δt then is obtained by taking the difference of these two times, and it is clear that it must be $\Delta t = 2\sqrt{(v\,\Delta t/2)^2 + (l^2/c)}$, since the origin of S' will have moved a distance $v\,\Delta t/2$ along the $\hat{\mathbf{e}}_3$-axis by the time the pulse reaches the mirror, and an equal distance farther by the time the pulse returns to its point of origin in S'. Hence $\Delta t = 2l/(c\sqrt{1-\beta^2})$, $\beta \equiv v/c$.

Comparison of these two measurements of the time required for light to travel to the mirror and back to the origin of S' shows that

$$\Delta t = \gamma\,\Delta t', \qquad \gamma \equiv (1-\beta^2)^{-1/2}.$$

Thus an observer in S records an interval longer by a factor γ than the interval recorded by an observer in S'. The difference between the measurements made by the two observers is that the observer in S' used a clock located at a *single point*, whereas the one in S used clocks located at *two different points* in his coordinate system. To distinguish these situations, we shall call a time interval measured by a *single clock*, kept at rest relative to the frame in which it is said to be located, a *proper time interval*. Since the required measurements in S must be made at

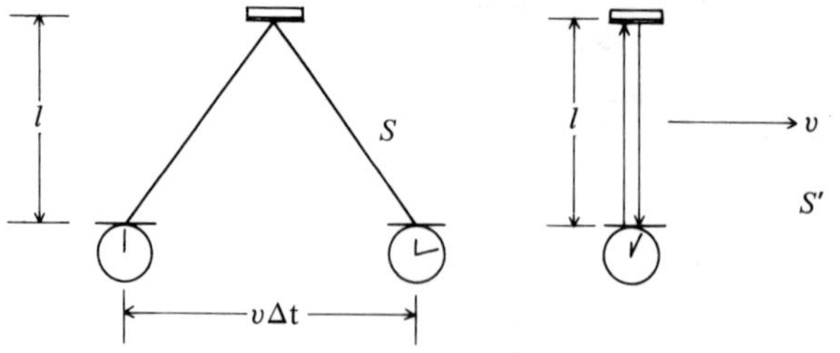

Fig. 2.6. Comparison of clock rates. (After Panofsky and Phillips, Ref. 8.)

two different points, either two clocks must be used, or a single clock must be moved from one point to another; in either case the time interval so measured could not be a proper interval. Hence we have found that improper time intervals always are dilated by a factor γ relative to proper time intervals, where in $\gamma = (1 - \beta^2)^{-1/2}$ the number β is the ratio of the velocity of the frame in which the improper interval is measured relative to the frame of the proper interval, to the velocity of light, $\beta = v/c$.

Another way of stating the meaning of a proper time interval is as follows: two events occurring at the same point in space, according to an observer in a given reference frame, are said to be separated by a proper time interval. Since a pair of events can occur at the same point in only one frame of reference, the time interval separating those events must be improper in all other inertial frames, and therefore it must be longer than the proper time when it is measured in any frame other than that in which the events coincide spatially. The proper time, then, is the *smallest* time interval separating the events.

It is because we do not have any way of comparing measurements in different frames, aside from the use of electromagnetic waves (or other signals propagating with velocities less than or equal to c), that the time intervals between events do not appear to be identical in different systems of reference. Thus, in a sense, special relativity is a theory about how we can compare measurements in different inertial frames. It would reduce to Galilean relativity if the speed of communication between observers could be made instantaneous, but disagrees with the latter because c does in fact represent an upper limit on the speed of signal propagation.

2.6 COMPARISON OF LENGTHS PARALLEL TO THE DIRECTION OF MOTION

From here forward we shall find that the comparison of time intervals as developed above will play a crucial role in determining how we compare other measurements. The comparison of lengths parallel to the direction of motion, for example, is carried out through the comparison of certain time intervals.

Consider a rod of length l, as measured in S', lying at rest along axis $\hat{e}'_2$ of S' with one of its endpoints at the origin of S'. At the end that is not at the origin of S', a mirror is mounted facing the origin of S'.

At the instant the origins of S and S' coincide, a pulse of light is emitted from the coincident origins. The pulse travels to the mirror in S' and then returns to the origin of S'. An observer in S' notes that the time required for the pulse to traverse the length of the rod and return is just $\Delta t' = 2l/c$; since the interval $\Delta t'$ is measured in terms of two events occurring at the origin of S', $\Delta t'$ is a proper time interval.

We now define a *proper distance* to be the spatial interval between two points, as determined by a measuring apparatus at rest relative to the measured interval.

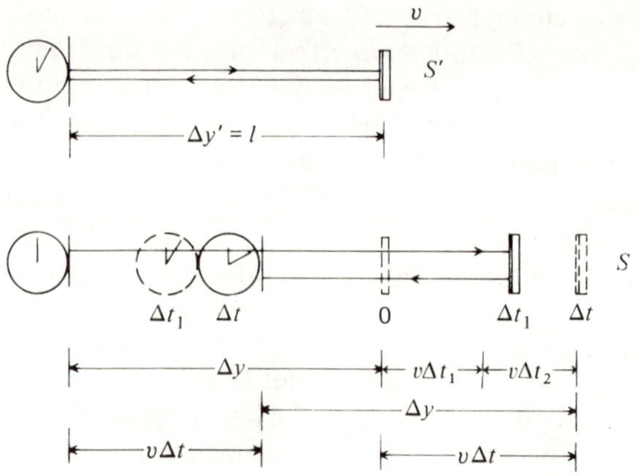

Fig. 2.7. Measurement of distances parallel to the direction of motion.

Since the mirror and the clock measuring the time separation $\Delta t'$ in S' are at rest with respect to the rod, the distance $\Delta y' = c\,\Delta t'/2 = l$ is a proper distance.

In the frame S we mean by the distance Δy the separation between the origin of S and the mirror at the end of the rod, at just the instant that the origins of S and S' coincide. From Fig. 2.7 we can conclude that the time, as measured in S, required for the light to reach the mirror is

$$\Delta t_1 = \frac{1}{c}(\Delta y + v\Delta t_1), \qquad \text{or} \qquad \Delta t_1 = \frac{\Delta y}{c - v},$$

whereas the time required for it to return to the origin of S' is

$$\Delta t_2 = \frac{1}{c}(\Delta y - v\Delta t_2), \qquad \text{or} \qquad \Delta t_2 = \frac{\Delta y}{c + v}.$$

Therefore the total time, as measured in S, between the events defining $\Delta t'$ in S' is

$$\Delta t = \Delta t_1 + \Delta t_2 = \Delta y\left(\frac{1}{c - v} + \frac{1}{c + v}\right) = \frac{2c\,\Delta y}{c^2 - v^2}$$

$$= \frac{2\,\Delta y}{c(1 - \beta^2)} = \frac{2\,\Delta y}{c}\gamma^2;$$

but since

$$\Delta t = \gamma\,\Delta t' = \gamma\frac{2l}{c} = \frac{2l}{c\sqrt{1 - \beta^2}},$$

we have

$$\Delta y = \frac{c\,\Delta t}{2\gamma^2} = \frac{2l\gamma}{2\gamma^2} = l/\gamma = l\sqrt{1-\beta^2}.$$

Thus, to an observer in S, the rod seems to have length $\Delta y = l/\gamma$, and since $\gamma > 1$ when $\mathbf{v} \neq \mathbf{0}$, the improper length Δy is less than the proper length l; the rod seems to suffer a *contraction* when observed in any other frame than the one in which it is at rest. This effect is called the Lorentz contraction because the equation relating a proper length to the corresponding length as measured in any other frame was first proposed by H. A. Lorentz as part of a prerelativistic theory he developed to explain the null results of the Michelson–Morley experiment. Note that a length appears to be *longest* when measured in the frame in which it is a *proper* length.

2.7 THE SYNCHRONIZATION OF CLOCKS

Let two clocks in S' be separated by a distance l, and let them be synchronized by arranging for a light pulse, emitted from a point midway between the clocks, to set each clock at $t' = 0$ at the instant the pulse arrives at the clock. Furthermore, imagine that the line in S' defined by the locations of the two clocks is oriented along $\hat{\mathbf{e}}'_2$ with the origin of S' at its midpoint. See Fig. 2.8.

If a light signal is generated at the origin of S and S' the instant they coincide, both clocks in S' will record the arrival of the signal a time $\Delta t' = l/2c$ after the instant when the origins coincided. However, an observer in S will note that the

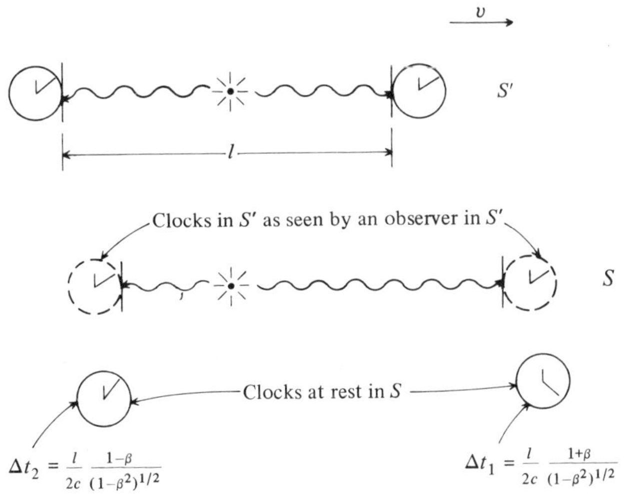

Fig. 2.8. Synchronization of clocks in different reference frames.

signal arrived at the clock on the positive $\hat{\mathbf{e}}_2'$-axis at a time

$$\Delta t_1 = \frac{l}{2c}\sqrt{1-\beta^2} + \frac{v}{c}\Delta t_1, \qquad \text{or} \qquad \Delta t_1 = \frac{l}{2c}\frac{\sqrt{1-\beta^2}}{1-\beta} = \frac{l}{2c}\frac{1+\beta}{\sqrt{1-\beta^2}}$$

after the coincidence of origins, and another observer in S will see that the signal arrived at the clock on the negative $\hat{\mathbf{e}}_2'$-axis when a time

$$\Delta t_2 = \frac{l}{2c}\sqrt{1-\beta^2} - \frac{v}{c}\Delta t_2, \qquad \text{or} \qquad \Delta t_2 = \frac{l}{2c}\frac{\sqrt{1-\beta^2}}{1+\beta} = \frac{l}{2c}\frac{1-\beta}{\sqrt{1-\beta^2}}$$

had elapsed since the emission of the signal. Therefore, since $\Delta t_1 \neq \Delta t_2$, observers in S must conclude that the clocks will not be synchronized by the emission of a light signal midway between them, even though this technique certainly *does* synchronize the clocks according to observers in S'.

Thus, clocks synchronized in their rest frame are seen to be out of synchronization by an amount

$$\Delta t_1 - \Delta t_2 = \frac{l}{2c}\gamma[(1+\beta) - (1-\beta)] = \beta\gamma\frac{l}{c}$$

in a frame moving with velocity $v = \beta c$ relative to their rest frame. In particular, the clock passing the observer last is "ahead" of the other by this amount.

READING NOTES*

Of the references given in the previous chapter, see especially Ref. 8, Chapters 15, 16. Reference 6, Chapter EI, Ref. 7, Chapter 1, Ref. 3, Chapter 11, may also be helpful.

Reference 13 is a small book that some students find quite helpful. Its Chapter 1 is relevant here.

Reference 11 gives an interesting survey of the foundations of the theory in Chapter 1.

Reference 10 is a collection of original papers by Einstein, H. A. Lorentz, H. Minkowski and others. Papers I and II are appropriate to this chapter.

Reference 12 is a remarkable book by a famous physicist, written while he still was a student. Part I of Pauli's difficult text discusses the foundations of the theory.

PROBLEMS

1. You know from your study of mechanics that if a particle has potential energy $V(|\mathbf{x} - \mathbf{x}_0|)$ when it is located at point $\mathbf{x}$, the force exerted on the particle as a result of the potential

---

* Works referred to in these notes will be found in the Bibliography on p. 577.

function is given by

$$\mathbf{F} = -\nabla V(|\mathbf{x} - \mathbf{x}_0|).$$

Show that the equations of motion for the particle in classical mechanics retain their mathematical form under Galilean transformations. Compare this with the effect of a Galilean transformation on the electromagnetic wave equations.

2. The earth's orbital speed about the sun is approximately 30 km/sec^{-1}. Estimate the magnitudes of the expected effects in the Michelson–Morley experiment for a reasonably sized interferometer.
3. A cosmic ray impinges on a molecule in the earth's atmosphere and produces a new radioactive particle that moves with velocity $0.8c$ relative to an observer, and that decays after traveling 10^4 cm according to the observer. What is the lifetime of the radioactive particle according to an observer at rest with respect to it?
4. A cosmic ray particle moves to the earth from a star four light years away from earth, with velocity $v = 0.995c$. How long does the trip take (a) according to an observer at rest relative to the particle, and (b) according to an observer at rest on the earth?
5. A spherical space probe passes by an observer on a space platform, and their relative velocity is $\frac{12}{13}c$. Draw two diagrams to scale, (a) in a plane perpendicular to the relative motion and (b) in a plane that includes the vector of relative velocity, showing the shape of the probe as measured by an observer on the platform. (c) If the radius of the probe is R, how badly out of step, according to the observer on the platform, are clocks located at the front and rear (as defined by the relative motion) of the probe? Which is ahead and which is behind? For a numerical result, let $R = 2m$.

CHAPTER 3

THE SPECIAL LORENTZ TRANSFORMATIONS

Already we have done all that is vitally necessary to derive the special Lorentz transformations; we need only put together the results of the *Gedanken Experimente* in a set of equations connecting measurements made along the coordinate axes in S with those made in S'. However, we prefer to use another approach in the interest of completeness and in order to provide the reader with a slightly different, more mathematically oriented derivation of the transformation equations.

3.1 SPACE AND TIME AS A FOUR-DIMENSIONAL CONTINUUM

Because of the impossibility of synchronizing pairs of clocks in different reference frames, the time coordinate and the space coordinates cannot be kept completely separate for observers in different reference frames; the spatial separation between clocks in one frame of reference makes these clocks seem nonsynchronous in any other frame. Similarly, if two spatially separated events occur at different times in one frame of reference, the time interval separating the events in that frame will affect the apparent spatial separation between them as measured in any other coordinate frame. Thus if x, y, z, and t are the space and time coordinates in S', we must expect that in a comparison between the coordinate systems, x may depend on x', y', z' and t', i.e.

$$x = x(x', y', z', t');$$

similarly,

$$y = y(x', y', z', t')$$

$$z = z(x', y', z', t')$$

$$t = t(x', y', z', t').$$

We therefore say that the comparison between coordinate systems may be accomplished by a coordinate transformation from (x', y', z', t') to (x, y, z, t), and the transformation is determined by the functional dependence of the unprimed coordinates upon the primed coordinates. Note that t is given the same status here as the spatial coordinates x, y, and z. Also, because of the symmetry between S and S', a transformation in the reverse direction, from the coordinates of S to those of S', must exist:

$$x' = x'(x, y, z, t)$$
$$y' = y'(x, y, z, t)$$
$$z' = z'(x, y, z, t)$$
$$t' = t'(x, y, z, t).$$

To make the notation more convenient, we introduce the *four-coordinates*

$$x^0 = ct, \qquad x^1 = x, \qquad x^2 = y, \qquad x^3 = z$$

which place space and time on the same footing. Note that x^0 has the dimensions of distance. Superscripts (which are not to be confused with exponents—coordinate symbols will be placed inside parentheses whenever they are raised to a power) are used here in anticipation of a need to distinguish between two different kinds of coordinate measure; we shall require both subscripts and superscripts to label coordinates.

Similarly, we have

$$x'^0 = ct', \qquad x'^1 = x', \qquad x'^2 = y', \qquad \text{and} \quad x'^3 = z'.$$

Hence we seek transformations that can be written in such forms as

$$x'^\mu = x'^\mu(x^0, x^1, x^2, x^3),$$

or

$$x'^\mu = x'^\mu(x^\nu), \qquad \mu, \nu = 0, 1, 2, 3.$$

3.2 THE HOMOGENEITY OF SPACE; LINEAR TRANSFORMATIONS

Now we apply the usual requirement of classical physics that space and time be homogeneous; special relativity does not affect this assumption. That is, if we change a Cartesian coordinate system by adding a constant to each coordinate axis (displace the origin by a constant vector), the coordinates in any other reference frame also will be shifted by an additive constant. Thus, if ξ^μ represents a set of four constants, and

$$x^\mu \to x^\mu + \xi^\mu, \quad \text{i.e.} \quad \begin{aligned} ct &\to ct + c\tau, \\ x &\to x + \xi, \\ y &\to y + \eta, \\ z &\to z + \zeta, \end{aligned} \quad \text{where} \quad \begin{aligned} \xi^0 &= c\tau, \\ \xi^1 &= \xi, \\ \xi^2 &= \eta, \\ \xi^3 &= \zeta, \end{aligned}$$

then the effect of this transformation on the S' coordinates must be

$$x'^\mu(x^\nu) \to x'^\mu(x^\nu + \xi^\nu) = x'^\mu(x^\nu) + \xi'^\mu(\xi^\nu)$$

since ξ^{μ} must be an *additive constant* four-coordinate vector and therefore cannot depend on x. Taking the partial derivative of x'^{μ} with respect to x^{λ}, then, we obtain

$$\frac{\partial x'^{\mu}}{\partial x^{\lambda}} = \frac{\partial x'^{\mu}}{\partial \xi^{\lambda}} = \frac{\partial \xi'^{\mu}}{\partial \xi^{\lambda}}, \qquad \mu, \lambda = 0, 1, 2, 3.$$

Since $\partial\xi'^{\mu}/\partial\xi^{\lambda}$ cannot depend on x'^{μ}, then neither can $\partial x'^{\mu}/\partial x^{\lambda}$, because they are equal. Thus

$$a^{\mu}{}_{\lambda} \equiv \frac{\partial x'^{\mu}(x^{\nu})}{\partial x^{\lambda}}$$

must be a constant, and therefore

$$\frac{\partial a^{\mu}{}_{\lambda}}{\partial x^{\nu}} = \frac{\partial^2 x'^{\mu}}{\partial x^{\nu}\partial x^{\lambda}} = 0, \qquad \text{any } \nu, \mu, \lambda,$$

identically. It follows that all higher derivatives of x'^{μ} vanish, and hence that the Taylor's series expansion of $x'^{\mu}(x^{\nu})$ in four dimensions reduces to the *linear transformation*

$$x'^{\mu}(x^{\nu}) = \sum_{\nu=0}^{3} a^{\mu}{}_{\nu}x^{\nu}, \qquad a^{\mu}{}_{\nu} = \frac{\partial x'^{\mu}(x^{\lambda})}{\partial x^{\nu}}. \tag{3.1}$$

By taking into account the symmetry between S and S', we also obtain

$$x^{\mu}(x'^{\nu}) = \sum_{\nu=0}^{3} a'^{\mu}{}_{\nu}x'^{\nu}, \qquad a'^{\mu}{}_{\nu} = \frac{\partial x^{\mu}(x'^{\lambda})}{\partial x'^{\nu}}. \tag{3.2}$$

Thus the hypothesis that space and time are homogeneous has led us to the requirement that any transformation between coordinates in different frames of reference must be a linear transformation—that is, straight lines in one frame are transformed into straight lines in another.

To make clear the meaning of Eq. (3.1) for example, let us write it out explicitly, using $(x^0, x^1, x^2, x^3) = (ct, x, y, z)$, etc. Thus

$$\begin{aligned}
x'^0 &= ct' = a^0{}_0ct + a^0{}_1x + a^0{}_2y + a^0{}_3z \\
x'^1 &= x' = a^1{}_0ct + a^1{}_1x + a^1{}_2y + a^1{}_3z \\
x'^2 &= y' = a^2{}_0ct + a^2{}_1x + a^2{}_2y + a^2{}_3z \\
x'^3 &= z' = a^3{}_0ct + a^3{}_1x + a^3{}_2y + a^3{}_3z.
\end{aligned}$$

Equations (3.1) and (3.2) therefore represent four equations each.

We can combine Eqs. (3.1) and (3.2) to get

$$x'^{\mu} = \sum_{\nu=0}^{3} a^{\mu}{}_{\nu}x^{\nu} = \sum_{\nu=0}^{3} a^{\mu}{}_{\nu} \sum_{\lambda=0}^{3} a'^{\nu}{}_{\lambda}x'^{\lambda} = \sum_{\lambda=0}^{3}\left[\sum_{\nu=0}^{3} a^{\mu}{}_{\nu}a'^{\nu}{}_{\lambda}\right]x'^{\lambda};$$

hence, since the magnitude of x'^{μ} is arbitrary, we must have

$$\sum_{\nu=0}^{3} a^{\mu}{}_{\nu} a'^{\nu}{}_{\lambda} = \delta^{\mu}{}_{\lambda}, \qquad \text{where} \qquad \delta^{\mu}{}_{\lambda} = \begin{Bmatrix} 0, \text{ if } \mu \neq \lambda \\ 1, \text{ if } \mu = \lambda \end{Bmatrix}, \tag{3.3}$$

so that

$$x^{\mu} = \sum_{\lambda=0}^{3} \delta^{\mu}{}_{\lambda} x^{\lambda} = x^{\mu}.$$

Actually, this follows immediately from differential calculus, since

$$\frac{\partial x'^{\mu}}{\partial x'^{\lambda}} = \delta^{\mu}{}_{\lambda} = \frac{\partial}{\partial x'^{\lambda}} \sum_{\nu=0}^{3} a^{\mu}{}_{\nu} x^{\nu} = \sum_{\nu=0}^{3} a^{\mu}{}_{\nu} \frac{\partial x^{\nu}}{\partial x'^{\lambda}} = \sum_{\nu=0}^{3} a^{\mu}{}_{\nu} a'^{\nu}{}_{\lambda}.$$

3.3 DERIVATION OF THE TRANSFORMATION MATRIX

Now we introduce the hypothesis that the speed of light is independent of reference frame. Let a light pulse be emitted from a source at point (ξ^1, ξ^2, ξ^3) at time ξ^0/c, i.e. at the four-point $(\xi^0, \xi^1, \xi^2, \xi^3)$. After a time $(x^0/c) - (\xi^0/c)$ the wave front will have traveled to the surface of a sphere of radius $x^0 - \xi^0$, and the equation for this surface is

$$[(x'^1 - \xi^1)^2 + (x^2 - \xi^2)^2 + (x^3 - \xi^3)^2]^{1/2} = x^0 - \xi^0, \; (x^0 > \xi^0),$$

or

$$(x^0 - \xi^0)^2 - (x^1 - \xi^1)^2 - (x^2 - \xi^2)^2 - (x^3 - \xi^3)^2 = 0. \tag{3.4}$$

Since the velocity of light is the same in S' as in S, the wave front of the pulse will have reached a sphere of radius $x'^0 - \xi'^0$ in S' after a time $(x'^0 - \xi'^0)/c$ as measured by clocks at rest in S', where ξ'^{μ} is the four-point in S' at which the pulse is emitted. Hence the equation for this surface is

$$(x'^0 - \xi'^0)^2 - (x'^1 - \xi'^1)^2 - (x'^2 - \xi'^2)^2 - (x'^3 - \xi'^3)^2 = 0. \tag{3.5}$$

The left-hand sides of Eqs. (3.4) and (3.5) have the same *form*; hence the equations for the surface representing the wave front of an electromagnetic wave are *form-invariant* (or covariant). That is, the form of the equations is the same in all frames of reference.

From Eq. (3.1) we have that

$$(x'^{\mu} - \xi'^{\mu}) = \sum_{\nu=0}^{3} a^{\mu}{}_{\nu}(x^{\nu} - \xi^{\nu})$$

or

$$(x'^{\mu} - \xi'^{\mu})^2 = \sum_{\nu,\mu=0}^{3} a^{\mu}{}_{\nu} a^{\mu}{}_{\lambda}(x^{\nu} - \xi^{\nu})(x^{\lambda} - \xi^{\lambda})$$

so that Eqs. (3.4) and (3.5) yield

$$\begin{aligned}
&(x^0 - \xi^0)^2 - (x^1 - \xi^1)^2 - (x^2 - \xi^2)^2 - (x^3 - \xi^3)^2 \\
&\quad = \sum_{\nu,\lambda=0}^{3} a^0{}_\nu a^0{}_\lambda (x-\xi)^\nu (x-\xi)^\lambda - \sum_{\nu,\lambda=0}^{3} a^1{}_\nu a^1{}_\lambda (x-\xi)^\nu (x-\xi)^\lambda \\
&\qquad - \sum_{\nu,\lambda=0}^{3} a^2{}_\nu a^2{}_\lambda (x-\xi)^\nu (x-\xi)^\lambda - \sum_{\nu,\lambda=0}^{3} a^3{}_\nu a^3{}_\lambda (x-\xi)^\nu (x-\xi)^\lambda \qquad (3.6) \\
&\quad = \sum_{\nu,\lambda=0}^{3} (a^0{}_\nu a^0{}_\lambda - a^1{}_\nu a^1{}_\lambda - a^2{}_\nu a^2{}_\lambda - a^3{}_\nu a^3{}_\lambda)(x-\xi)^\nu (x-\xi)^\lambda .
\end{aligned}$$

At this point it is convenient to introduce several innovations in our notation.

1. Let a quantity $g_{\mu\nu}$ be defined by

$$g_{\mu\nu} = \begin{Bmatrix} 0 & \text{if } \mu \neq \nu \\ +1 & \text{if } \mu = \nu = 0 \\ -1 & \text{if } \mu = \nu = 1, 2 \text{ or } 3 \end{Bmatrix}. \qquad (3.7)$$

Thus $g_{\mu\nu}$ can be written in matrix form as follows:

$$[g_{\mu\nu}] = \begin{array}{c|cccc} {}_\mu{}^\nu & 0 & 1 & 2 & 3 \\ \hline 0 & 1 & 0 & 0 & 0 \\ 1 & 0 & -1 & 0 & 0 \\ 2 & 0 & 0 & -1 & 0 \\ 3 & 0 & 0 & 0 & -1 \end{array} .$$

2. Define $x_\mu = \Sigma_{\nu=0}^{3} g_{\mu\nu} x^\nu$. Note that this introduces subscripts into our notation; also note the role played by $g_{\mu\nu}$ as an "index-lowering device". Thus, in matrix language,

$$\begin{Bmatrix} x_0 \\ x_1 \\ x_2 \\ x_3 \end{Bmatrix} = \begin{bmatrix} 1 & 0 & 0 & 0 \\ 0 & -1 & 0 & 0 \\ 0 & 0 & -1 & 0 \\ 0 & 0 & 0 & -1 \end{bmatrix} \begin{Bmatrix} x^0 \\ x^1 \\ x^2 \\ x^3 \end{Bmatrix} = \begin{Bmatrix} x^0 \\ -x^1 \\ -x^2 \\ -x^3 \end{Bmatrix} \qquad (3.8)$$

or

$$x_0 = x^0 = ct, \qquad x_1 = -x^1 = -x, \qquad x_2 = -x^2 = -y, \qquad x_3 = -x^3 = -z.$$

This leads to

$$\sum_{\mu=0}^{3} x^\mu x_\mu = x^0 x_0 + x^1 x_1 + x^2 x_2 + x^3 x_3$$
$$= (x^0)^2 - (x^1)^2 - (x^2)^2 - (x^3)^2 = c^2 t^2 - x^2 - y^2 - z^2,$$

which evidently will be a useful shorthand when we are dealing with expressions such as those in Eqs. (3.4–3.6).

3. Define $g^{\mu\nu}$ numerically equal to $g_{\mu\nu}$, but with a slightly different meaning; $g^{\mu\nu}$ will raise indices rather than lower them:

$$x^\mu = \sum_{\nu=0}^{3} g^{\mu\nu} x_\nu. \tag{3.9}$$

The quantity $g^{\mu\nu}$ is known as the *contravariant metric* and $g_{\mu\nu}$ is the *covariant metric.* In general, symbols with superscripts are called contravariants and those with subscripts are called covariant; symbols having both superscripts and subscripts are called mixed quantities. Notice that

$$\sum_{\mu=0}^{3} x^\mu x_\mu = \sum_{\mu,\nu=0}^{3} g^{\mu\nu} x_\nu x_\mu = \sum_{\mu,\nu,\lambda=0}^{3} g^{\mu\nu} g_{\mu\lambda} x_\nu x^\lambda$$
$$= \sum_{\lambda,\nu=0}^{3} \delta^\nu{}_\lambda x_\nu x^\lambda = \sum_{\lambda=0}^{3} x_\lambda x^\lambda,$$

since $\Sigma_{\mu=0}^{3} g^{\mu\nu} g_{\mu\lambda} = \delta^\nu{}_\lambda$, as is verified easily by direct computation. The mixed metric $\delta^\mu{}_\nu$ has the same values as the Kronecker δ symbol (unity when the indices are equal, zero otherwise) and its matrix representation is the unit matrix.

4. Since we frequently shall be carrying out sums over indices that are repeated in the summand (e.g. $\Sigma_{\nu=0}^{3} g^{\mu\nu} x_\nu = x^\mu$), it will be convenient to omit the summation sign and to understand that a summation is meant whenever indices are repeated and are in the proper positions (i.e. one a superscript—contravariant—and one a subscript—covariant) in an expression, *unless otherwise noted.* For example, by $g^{\mu\nu} x_\nu$ we mean $\Sigma_{\nu=0}^{3} g^{\mu\nu} x_\nu$, etc. This often is referred to as the *Einstein summation convention.*

Now let us return to Eq. (3.6); according to our new conventions this is

$$(x^\mu - \xi^\mu)(x_\mu - \xi_\mu) = g_{\rho\mu} a^\rho{}_\nu a^\mu{}_\lambda (x^\nu - \xi^\nu)(x^\lambda - \xi^\lambda)$$
$$= g_{\lambda\mu}(x^\mu - \xi^\mu)(x^\lambda - \xi^\lambda) = g_{\lambda\nu}(x^\nu - \xi^\nu)(x^\lambda - \xi^\lambda),$$

which can be rewritten in the form

$$(g_{\nu\lambda} - g_{\rho\mu} a^\rho{}_\nu a^\mu{}_\lambda)(x^\nu - \xi^\nu)(x^\lambda - \xi^\lambda) = 0. \tag{3.10}$$

Since the points x^μ are arbitrary, and hence $(x^\mu - \xi^\mu)$ also are arbitrary, we must have

$$g_{\nu\lambda} - g_{\rho\mu}a^\rho{}_\nu a^\mu{}_\lambda = 0 \tag{3.11}$$

or, after multiplication by $g^{\nu\sigma}$ and summation over ν,

$$g^{\nu\sigma}g_{\nu\lambda} - g^{\nu\sigma}g_{\rho\mu}a^\rho{}_\nu a^\mu{}_\lambda = \delta^\sigma{}_\lambda - a_\mu{}^\sigma a^\mu{}_\lambda = 0. \tag{3.12}$$

Multiplication by $a'^\lambda{}_\tau$ and summation over λ then yields

$$a'^\sigma{}_\tau = a_\mu{}^\sigma a^\mu{}_\lambda a'^\lambda{}_\tau = a_\mu{}^\sigma \delta^\mu{}_\tau = a_\tau{}^\sigma$$

where we have made use of Eq. (3.3); thus

$$a_\tau{}^\sigma = a'^\sigma{}_\tau. \tag{3.13}$$

Note that this result is a consequence of the invariance of the velocity of light.

In evaluating the $a^\mu{}_\nu$ it is easier in practice to use Eq. (3.11) than to use either of Eqs. (3.12) and (3.13). We shall evaluate the $a^\mu{}_\nu$, using Eq. (3.11), under the explicit assumption that the coordinate systems in frames S and S' are chosen here as they were in Chapter 2, so that x'^1 and x'^3, being perpendicular to the direction of motion, are equal, respectively, to x^1 and x^3. (Note the use here of the first of the *Gedanken Experimente.*) Hence we take

$$a^1{}_\nu = \delta^1{}_\nu, \qquad a^3{}_\nu = \delta^3{}_\nu, \tag{3.14}$$

and since $a^\mu{}_\nu = g^{\mu\lambda}g_{\nu\rho}a_\lambda{}^\rho$, it follows that

$$a_1{}^\nu = \delta_1{}^\nu, \qquad a_3{}^\nu = \delta_3{}^\nu \tag{3.15}$$

also. Furthermore, because of the symmetry between frames S and S', we must have

$$a'^1{}_\nu = \delta^1{}_\nu, \qquad a'^3{}_\nu = \delta^3{}_\nu \tag{3.16}$$

and

$$a'_1{}^\nu = \delta_1{}^\nu, \qquad a'_3{}^\nu = \delta_3{}^\nu. \tag{3.17}$$

Equation (3.13), used in conjunction with these last equations, then yields

$$a^\nu{}_1 = \delta^\nu{}_1, \qquad a^\nu{}_3 = \delta^\nu{}_3 \tag{3.18}$$

and

$$a_\nu{}^1 = \delta_\nu{}^1, \qquad a_\nu{}^3 = \delta_\nu{}^3. \tag{3.19}$$

Thus, if either of the indices of $a^\mu{}_\nu$ is 1 or 3, the value of $a^\mu{}_\nu$ is one or zero, depending on whether the other index is, respectively, the same or different.

Next we must evaluate $a^\mu{}_\nu$ when neither of the indices is 1 or 3. We begin by letting $\lambda = \nu = 0$ in Eq. (3.11); this gives

$$1 - g_{\rho\mu}a^\rho{}_0 a^\mu{}_0 = 0,$$

or, since $a^\mu{}_0$ vanishes if $\mu = 1$ or $\mu = 3$, according to Eq. (3.14),

$$1 - (a^0{}_0)^2 + (a^2{}_0)^2 = 0. \tag{3.20}$$

Similarly, if $\lambda = \nu = 2$ in Eq. (3.11), we obtain

$$1 - g_{\rho\mu}a^\rho{}_2 a^\mu{}_2 = 0,$$

or

$$1 - (a^0{}_2)^2 + (a^2{}_2)^2 = 0. \tag{3.21}$$

Finally, let $\lambda = 0$ and $\nu = 2$ in Eq. (3.11); this yields

$$g_{\rho\mu}a^\rho{}_0 a^\mu{}_2 = 0$$

or

$$a^0{}_0 a^0{}_2 - a^2{}_0 a^2{}_2 = 0. \tag{3.22}$$

This last equation indicates that the ratios $a^2{}_0/a^0{}_0$ and $a^0{}_2/a^2{}_2$ are equal. Let their common value be denoted $-\beta$, so that we have

$$\begin{aligned} a^2{}_0 &= -\beta a^0{}_0 \\ a^0{}_2 &= -\beta a^2{}_2. \end{aligned} \tag{3.23}$$

According to Eq. (3.20), then,

$$1 - (a^0{}_0)^2[1 - \beta^2] = 0,$$

or

$$a^0{}_0 = \pm\frac{1}{\sqrt{1 - \beta^2}}. \tag{3.24}$$

The similar result

$$a^2{}_2 = \pm\frac{1}{\sqrt{1 - \beta^2}} \tag{3.25}$$

also follows from Eq. (3.21) and Eq. (3.23).

Taken together, Eqs. (3.14), (3.23), (3.24) and (3.25) yield

$$\begin{aligned} a^0{}_\nu &= \frac{1}{\sqrt{1 - \beta^2}}(\delta^0{}_\nu - \beta\delta^2{}_\nu) \\ a^1{}_\nu &= \delta^1{}_\nu \\ a^2{}_\nu &= \frac{1}{\sqrt{1 - \beta^2}}(-\beta\delta^0{}_\nu + \delta^2{}_\nu) \\ a^3{}_\nu &= \delta^3{}_\nu. \end{aligned} \tag{3.26}$$

The choices of sign for $a^0{}_\nu$ and $a^2{}_\nu$ are dictated by the facts that the sense of time is to be the same in both frames of reference, and that the y-axes have the same sense in both S and S'; thus the positive sign is to be used in Eqs. (3.24) and (3.25). When put into matrix form, Eqs. (3.26) read

$$
[a^\mu{}_\nu] = \begin{array}{c|cccc} \mu \backslash \nu & 0 & 1 & 2 & 3 \\ \hline 0 & \frac{1}{\sqrt{1-\beta^2}} & 0 & -\frac{\beta}{\sqrt{1-\beta^2}} & 0 \\ 1 & 0 & 1 & 0 & 0 \\ 2 & -\frac{\beta}{\sqrt{1-\beta^2}} & 0 & \frac{1}{\sqrt{1-\beta^2}} & 0 \\ 3 & 0 & 0 & 0 & 1 \end{array}
$$

which is equivalent to the set of equations

$$
\begin{aligned}
x'^0 &= a^0{}_\nu x^\nu = \frac{x^0 - \beta y}{\sqrt{1-\beta^2}} = \frac{ct - \beta y}{\sqrt{1-\beta^2}} \\
x'^1 &= a^1{}_\nu x^\nu = x^1 = x \\
x'^2 &= a^2{}_\nu x^\nu = \frac{x^2 - \beta x^0}{\sqrt{1-\beta^2}} = \frac{y - \beta ct}{\sqrt{1-\beta^2}} \\
x'^3 &= a^3{}_\nu x^\nu = x^3 = z.
\end{aligned}
\tag{3.27}
$$

3.4 INTERPRETATION OF β; THE INVERSE TRANSFORMATION

Remember that β is simply a parameter, and it has yet to be given an interpretation. In order to determine the value of β, we note that if the origins of the coordinate systems of S and S' are coincident at $t = t' = 0$, the origin of S' as seen by an observer at rest in S always is given by $y_0 = vt$, whereas an observer in S' always considers this same point to be given by $y'_0 = 0$. Hence, according to the third of Eqs. (3.27),

$$
0 = \frac{vt - \beta ct}{\sqrt{1-\beta^2}}, \qquad \text{or} \qquad v = \beta c, \qquad \beta = \frac{v}{c}. \tag{3.28}
$$

Equations (3.27), when β is identified as v/c, are known as the equations for a *special Lorentz transformation*—"special" because they refer to motion along one of the coordinate axes and do not take into account any rotation of the axes in S' relative to those in S.

Notice that the special Lorentz transformation affects only the time coordinate and the space coordinate parallel to the direction of motion. If, instead of having **v** parallel to **y**, we have **v** parallel to **z**, the special Lorentz transformation would take the form

$$
\begin{aligned}
x'^0 &= \frac{x^0 - \beta x^3}{\sqrt{1-\beta^2}} \\
x'^1 &= x^1 \\
x'^2 &= x^2 \\
x'^3 &= \frac{x^3 - \beta x^0}{\sqrt{1-\beta^2}}
\end{aligned}
\tag{3.29}
$$

A similar modification of the transformation equations would be required if we had **v** parallel to **x**.

A characteristic feature of the transformation equations is the appearance of the square root $\sqrt{1-\beta^2}$. If $\beta > 1$, some of the coordinates will be represented by imaginary numbers, and since actual measurements always are represented by real numbers, we must demand that $\beta \leq 1$. Thus we require $v \leq c$ in order that the Lorentz transformations may have some physical meaning.

A simple physical argument yields the values of $a'^{\mu}{}_{\nu}$, that is, of the transformation coefficients that give the coordinates of S in terms of those of S'. Thus, since S' is assumed to move with velocity v in the positive y-direction, according to an observer in S, then S will move in the negative y'-direction relative to an observer in S'; this means that if we interchange x'^{μ} and x^{μ} and change β to $-\beta$ in Eqs. (3.27), we shall obtain equations that still will be valid and that will provide the coefficients $a'^{\mu}{}_{\nu}$ of the transformation inverse to the transformation specified by the $a^{\mu}{}_{\nu}$. Explicit values of $a'^{\mu}{}_{\nu}$ therefore can be obtained from Eqs. (3.26) simply by replacing β with $-\beta$ and appending a prime to each of the transformation coefficients on the left-hand sides.

In order to check the consistency of this procedure with the requirements imposed by Eq. (3.13), which relates certain of the $a^{\mu}{}_{\nu}$ to certain corresponding $a'^{\mu}{}_{\nu}$, we may calculate $a'^{\mu}{}_{\nu}$ directly from (3.13), using the metrics $g^{\mu\nu}$ and $g_{\mu\nu}$ to raise and lower indices when necessary. This is left as an exercise for the reader. Another check, equivalent to the one just discussed, is to solve Eqs. (3.27) directly for the x^{μ}; this also is left for the reader to do.

Let us return briefly to Eqs. (3.4) and (3.5), which were essential to our calculation of the $a^{\mu}{}_{\nu}$. These equations are mathematical expressions of the circumstance that the wave fronts emanating from a source of isotropic electromagnetic radiation are spherical in all frames of reference, and furthermore they establish a relationship between the time and space coordinates of any single phase point in an electromagnetic wave. This latter relationship is determined by the vanishing of the right-hand sides of Eqs. (3.4) and (3.5). However, in obtaining the $a^\mu{}_\nu$ we have not used the fact that the right-hand sides of these equations are zero; we have utilized only the circumstance that the left-hand of both equations must be equal. The reader is urged to retrace our derivation with this in mind, to satisfy himself that only the relationship

$$(x^\mu - \xi^\mu)(x_\mu - \xi_\mu) = (x'^\mu - \xi'^\mu)(x'_\mu - \xi'_\mu)$$

implied by Eqs. (3.4) and (3.5) has been used, and that the vanishing of both sides of this equation played no role in our work.

Thus the transformation that we have obtained is one that leaves the quantity

$$(x^\mu - \xi^\mu)(x_\mu - \xi_\mu) = s^2 \tag{3.30}$$

invariant (here s^2 is a real number, and the reason for writing it as a square will appear later); that is, s^2 has the same value in every coordinate system, so that

$$(x'^\mu - \xi'^\mu)(x'_\mu - \xi'_\mu) = s^2 \tag{3.31}$$

also. In our calculations above, the value of s^2 was never used, and its vanishing simply reflected our interpretation of x^μ as a phase point on the electromagnetic waves emitted from a source at ξ^μ.

3.5 LORENTZ TRANSFORMATIONS IN GENERAL

We have considered only one class of the transformations having the property that they leave

$$(x^\mu - \xi^\mu)(x_\mu - \xi_\mu) = s^2$$

invariant; the totality of such transformations is called the *group of Lorentz transformations.* For example, instead of letting S' move along y, we could have let it move relative to S along *any* other direction with uniform velocity, still requiring s^2 to be invariant. In fact, we even could let S and S' be at rest relative to one another and then carry out a transformation that relates the spatial coordinates of two frames at rest relative to one another; if s^2 is kept invariant under such transformations, the latter simply are rotations. Therefore, all rotations of coordinate systems also are classified as Lorentz transformations.

Let us briefly investigate rotations considered as Lorentz transformations. Since S and S' are at rest relative to one another, according to the above description of rotations, we have $\beta = v/c = 0$, and consequently $x^0 = x'^0$. Furthermore, we may choose the axis about which we shall carry out the rotation; let our axis of rotation be $x^3 = x'^3$. Thus we have $a^0{}_\nu = \delta^0{}_\nu$ and $a^3{}_\nu = \delta^3{}_\nu$, from which it follows that $a^\nu{}_0 = \delta^\nu{}_0$ and $a^\nu{}_3 = \delta^\nu{}_3$ also. The transformation equations therefore may be written as follows:

$$\begin{aligned} x'^0 &= x^0 \\ x'^1 &= a^1{}_\nu x^\nu \\ x'^2 &= a^2{}_\nu x^\nu \\ x'^3 &= x^3. \end{aligned} \tag{3.32}$$

To evaluate $a^1{}_\nu$ and $a^2{}_\nu$, we proceed as before, using Eq. (3.11), which is a direct consequence of the invariance of s^2 but is independent of its value. Thus, letting $\nu = \lambda = 1$ in Eq. (3.11), we obtain

$$-1 - g_{\rho\mu} a^\rho{}_1 a^\mu{}_1 = 0$$

or

$$(a^1{}_1)^2 + (a^2{}_1)^2 = 1. \tag{3.33}$$

Next, choosing $\nu = \lambda = 2$, we find, in the same way, that

$$(a^1{}_2)^2 + (a^2{}_2)^2 = 1. \tag{3.34}$$

Finally, letting $\nu = 1$ and $\lambda = 2$, we get

$$a^1{}_1 a^1{}_2 + a^2{}_1 a^2{}_2 = 0, \tag{3.35}$$

or we may define

$$\alpha \equiv \frac{a^1{}_2}{a^2{}_2} = -\frac{a^2{}_1}{a^1{}_1}. \tag{3.36}$$

Substituting $a^2{}_1 = -\alpha a^1{}_1$ into Eq. (3.33) yields the result

$$a^1{}_1 = \pm\frac{1}{\sqrt{1 + \alpha^2}}, \tag{3.37}$$

and using Eq. (3.36) in Eq. (3.34) gives the similar result

$$a^2{}_2 = \pm\frac{1}{\sqrt{1 + \alpha^2}} \tag{3.38}$$

where the choices of $+$ or $-$ signs in the two cases are independent of one another. For convenience, we shall choose the $+$ sign in both of Eqs. (3.37)–(3.38), but this choice is of little significance. (The reader is invited to investigate for himself the consequences of making the other three possible choices of sign, following essentially the procedure we develop below.) Combining Eqs. (3.37) and (3.38) with

(3.36), bearing in mind our choice of signs, yields

$$\begin{aligned} a^1{}_1 &= \frac{1}{\sqrt{1+\alpha^2}}, & a^1{}_2 &= \frac{\alpha}{\sqrt{1+\alpha^2}}, \\ a^2{}_1 &= -\frac{\alpha}{\sqrt{1+\alpha^2}}, & \alpha^2{}_2 &= \frac{1}{\sqrt{1+\alpha^2}}. \end{aligned} \tag{3.39}$$

The manner in which α appears throughout Eqs. (3.29) suggests that we let $\alpha = \tan\theta$, for then

$$a^1{}_1 = a^2{}_2 = \cos\theta$$

and

$$a^1{}_2 = -a^2{}_1 = \sin\theta. \tag{3.40}$$

Thus

$$\begin{aligned} x'^0 &= x^0 \\ x'^1 &= (\cos\theta)x^1 + (\sin\theta)x^2 \\ x'^2 &= (-\sin\theta)x^1 + (\cos\theta)x^2 \\ x'^3 &= x^3, \end{aligned} \tag{3.41}$$

and these are recognizable as the equations for a pure rotation through angle $\theta = \arctan\alpha$ about the z-axis.

Eqs. (3.39) in fact bear a striking resemblance to the transformation equations for a special Lorentz transformation, Eqs. (3.23)–(3.25), or their equivalents, Eqs. (3.26). The major difference is a minus sign appearing in the square root that constitutes the denominator of some of the $a^\mu{}_\nu$ for the special Lorentz transformation. The presence of this minus sign suggests letting $\beta = \tanh\zeta$, for then

$$\gamma = \frac{1}{\sqrt{1-\beta^2}} = \frac{1}{\sqrt{1-\tanh^2\zeta}} = \cosh\zeta \tag{3.42}$$

and

$$\beta\gamma = \frac{\beta}{\sqrt{1-\beta^2}} = \tanh\zeta\cosh\zeta = \sinh\zeta.$$

This leads to

$$\begin{aligned} a^0{}_0 &= \cosh\zeta, & a^0{}_2 &= -\sinh\zeta, \\ a^2{}_0 &= -\sinh\zeta, & a^2{}_2 &= \cosh\zeta, \end{aligned} \tag{3.43}$$

and finally to

$$\begin{aligned} x'^0 &= x^0 \cosh \zeta - x^2 \sinh \zeta \\ x'^1 &= x^1 \\ x'^2 &= -x^0 \sinh \zeta + x^2 \cosh \zeta \\ x'^3 &= x^3. \end{aligned} \tag{3.44}$$

The resemblance between Eqs. (3.44) and Eqs. (3.41) is conspicuous.

The special Lorentz transformation thus amounts to a *hyperbolic rotation* since the relevant transformation equations are very similar to those for ordinary rotations, except that the circular functions associated with ordinary rotations ($\sin \theta, \cos \theta$) are replaced with related hyperbolic functions ($\sinh \zeta, \cosh \zeta$). The adjective "hyperbolic", however, is essential; a special Lorentz transformation is *not a rotation* in four-dimensional space. In fact, a rotation in four-dimensional space, in which the time coordinate is not held fixed (i.e. in which $x'^0 \neq x^0$), would not leave s^2 invariant and therefore could not be considered a Lorentz transformation.

In a pure rotation about, say, the z-axis, the x- and y- axes always remain orthogonal to each other; furthermore, the curves of constant s^2 are circles lying the xy-plane and centered at the origin. For a special Lorentz transformation along the y-axis, however, the y'- and t'-axes do not remain orthogonal, and if S' moves along the positive y-axis away from the origin of S, the y'- and t-axes have an angle less than $\pi/2$ between them. Furthermore, curves of constant s^2 are hyperbolas in the yt-plane; they open toward the ends of the y-axis if $s^2 < 0$ and open toward the ends of the t-axis if $s^2 > 0$, as shown in Fig. 3.1. (It may seem rather peculiar that we allow s^2 to be negative, since the notation s^2 indicates that s^2 is the square of a number. However, we have not defined s—only s^2—and we demanded only that s^2 be a real number equal to the value of the invariant $(x^\mu - \xi^\mu)(x_\mu - \xi_\mu)$. Actually, it is clear that we must allow $s^2 < 0$, because if two events, represented by coordinates ξ^μ and η^μ, occur at the same time, i.e. $\xi^0 = \eta^0$, then the invariant quantity

$$\begin{aligned} (\xi^\mu - \eta^\mu)(\xi_\mu - \eta_\mu) &= (\xi^0 - \eta^0)^2 - \sum_{n=1}^{3} (\xi^m - \eta^m)^2 \\ &= -(\xi^1 - \eta^1)^2 - (\xi^2 - \eta^2)^2 - (\xi^3 - \eta^3)^2 \end{aligned}$$

must be less than zero, the coordinate positions ξ^μ and η^μ being real numbers. The reason that we use the symbol s^2 to represent the invariant quantity $(x^\mu - \xi^\mu)(x_\mu - \xi_\mu)$ is that the latter is a four-dimensional analog of the scalar product, or "vector squared" $(\mathbf{x} - \boldsymbol{\xi})(\mathbf{x} - \boldsymbol{\xi}) = (\mathbf{x} - \boldsymbol{\xi})^2$.) The hyperbolas all have common asymptotes, namely $x^0 = \pm x^2$, or $ct = \pm y$. If we could plot all four coordinates on a graph of this sort, we would see that these asymptotes are the four-dimensional analogs of lines generating the surface of a cone; the asymptotic surfaces are

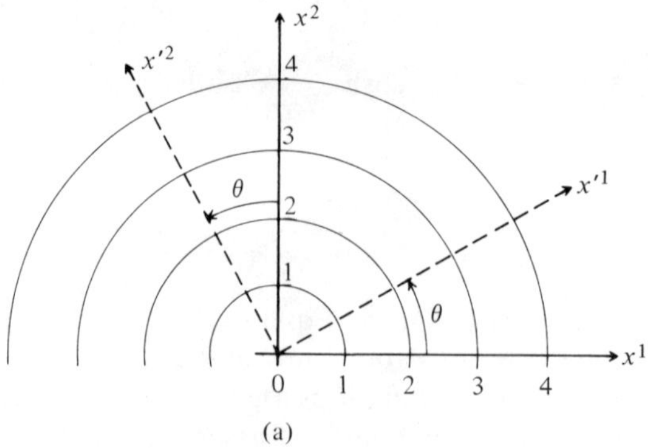

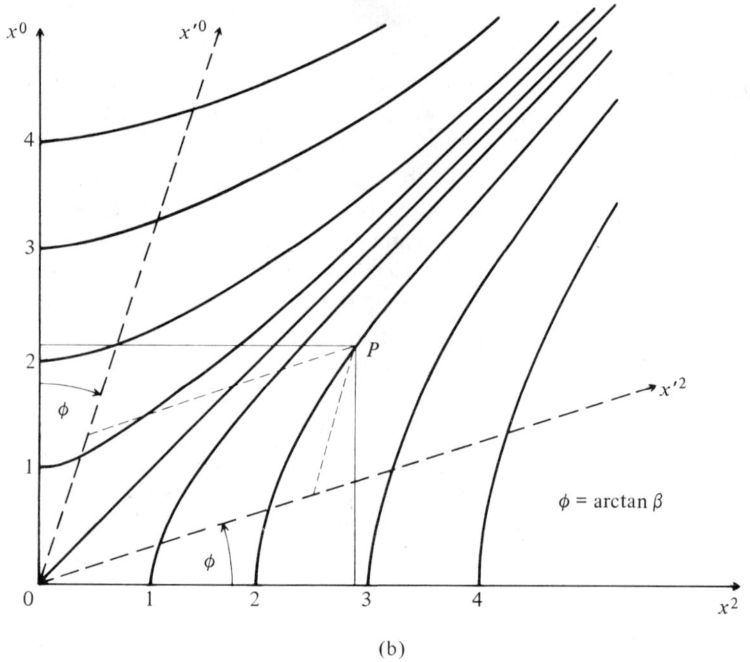

Fig. 3.1. Lorentz transformations. (a) Rotations. (b) Special Lorentz transformations. Diagrams of the kind shown in (b) are often called Minkowski diagrams.

called the "light cone" since the t'- and y'-axes converge toward the asymptotes when v approaches the speed of light, c.

Note that the scales of the x'^2 and x'^0 coordinates in Fig. 3.1 are established by the intersections of these axes with the hyperbolas corresponding to different values of s^2. It is easy to see that the scales are finest when the axes are orthogonal, i.e. when $v = 0$ is the relative velocity of S' and S. If S' moves away from the origin of S along the positive x^2-axis, the x'^0- and x'^2-axes both will lie in quadrant I, and they move toward one another, like scissors blades, as the relative velocity becomes larger and larger, until, for $v = c$, they converge to the $x^0 = x^2$ line. If S' moves along the negative x^2-axis away from the origin of S, the x'^0- and x'^2-axes both lie in quadrant II, and move toward one another in scissors-like fashion as v increases; for $v = c$ they both lie along the line $x^0 = -x^2$. This behaviour can be checked by consulting the transformation equations.

It is left to the reader to check that the angle between the x'^0-axis and the x^0-axis, and also the angle between the x'^2-axis and the x^2-axis, in a diagram such as that displayed in Fig. 3.1, is given by $\phi = \arctan \beta = \arctan(\tanh \zeta)$.

In classical mechanics (or, more accurately, in three-dimensional Euclidean geometry), the quantity $l^2 = (\mathbf{x} - \boldsymbol{\xi})^2 = (x^1 - \xi^1)^2 + (x^2 - \xi^2)^2 + (x^3 - \xi^3)^2$ represents the square of the distance between points defined by the vectors $\mathbf{x}$ and $\boldsymbol{\xi}$; this quantity is the fundamental invariant of rotation transformations in three dimensions. In the four dimensions of special relativity, the fundamental invariant, $s^2 = (x^0 - \xi^0)^2 - (\mathbf{x} - \boldsymbol{\xi})^2$ is slightly more complicated, as we have seen; in particular, s^2 can be positive, negative, or zero, in contrast with l^2, which always is non-negative.

3.6 SPACE-LIKE AND TIME-LIKE INTERVALS; THE LIGHT CONE

An interesting and very useful observation is that pairs of four-dimensional points (pairs of events), say x^μ and ξ^μ, for which $s^2 = (x^\mu - \xi^\mu)(x_\mu - \xi_\mu) < 0$, cannot communicate with each other using signals whose velocities of propagation are less than or equal to the velocity of light; such points are said to be separated by *space-like* intervals. In some frame of reference, say in S', s^2 will have the form

$$s^2 = -(\mathbf{x}' - \boldsymbol{\xi}')^2$$

so that the separation between the points x'^μ and ξ'^μ is simply a spatial interval. (To find the frame S', that is, to find its velocity relative to any other frame S in which the events are represented by the points x^μ and ξ^μ, locate the points x^μ and ξ^μ in an appropriate Minkowski diagram similar to Fig. 3.1. Connect these two points in the diagram with a straight line, and measure the angle of this line relative to the x^0-axis; the angle is equal to $\pi/2 - \zeta = \pi/2 - \operatorname{argtanh} \beta$.) In this frame, since the time interval separating the events vanishes, the points are separated by a proper space interval $l = \sqrt{(\mathbf{x}' - \boldsymbol{\xi}')^2}$. Because no signal can be transmitted instantaneously between two spatially separated points, the event represented by x'^μ in S' cannot be affected by the event represented by ξ'^μ. If x'^μ and ξ'^μ are events that cannot affect one another, then so are x^μ and ξ^μ, since the latter simply are the same events as seen from a different reference frame. An event at space-time point ξ^μ cannot be perceived by an observer at x^μ because the time interval between x^μ and ξ^μ is not sufficient to allow any signal to traverse the spatial separation represented by those two points.

Already we have observed that when x^μ is interpreted as the set of space and time coordinates for a phase point on a light wave, and when ξ^μ is taken as the event corresponding to the emission of the light wave, we have $s^2 = 0$. All points satisfying $s^2 = 0$ are said to lie on the *light cone* for the event ξ^μ, because the *hypersurface* represented by $(x^\mu - \xi^\mu)(x_\mu - \xi_\mu) = 0$ is essentially a four-dimensional analog of the three-dimensional cone $x^2 + y^2 - z^2 = 0$, as we noted earlier. Thus points, or events, on the light cone can be connected only by signals having the velocity of light.

Finally, if $s^2 > 0$ for two events x^μ and ξ^μ, the separation between them is said to be *time-like*, and the events can be connected by signals propagated with velocities less than that of light. One can find a frame S' in which the spatial separation between the events vanishes, and in that frame s^2 represents the square of the (proper) time interval between the events. (For an observer in S to find the frame S', he need only connect the points x^μ and ξ^μ on a Minkowski diagram with a straight line, and find the angle between this line and the x^0-axis. The angle found in this way is equal to $\zeta = \operatorname{argtanh} \beta$.) The "time-like world", in which events are separated by time-like intervals, is the one in which we live, for we never can learn about events whose separation from us is space-like; no signal can travel fast enough to bridge the space-like gap between us and such events.

Evidently a Lorentz transformation cannot convert a time-like separation into a space-like separation, or vice versa, since a Lorentz transformation, by definition, leaves s^2 invariant. If s^2 is positive, it must remain so in all inertial reference frames; a similar statement applies to both $s^2 = 0$ and $s^2 < 0$.

In summary, all of four-dimensional space–time is divided by the light cone into two distinct types of regions; the time-like region and the space-like region; the latter sometimes is referred to as "elsewhere" because it contains only events that cannot be communicated to any observer in the time-like region. The time-like region is divided further into an *upper cone* for $x^0 > 0$, and a *lower cone* for $x^0 < 0$; the former often is referred to as "the future" and the latter as "the past", in which case the origin must be considered the "here and now". Events occurring anywhere in the past half-cone can be known to the observer "now". An observer at a point in the future half-cone can be informed of the original observer's presence at the origin because the separation of the observers is time-like, so that there is sufficient time available for a signal to reach the observer in the future half-cone.

Because of the analogy between s^2 and $l^2 = (\mathbf{x} - \boldsymbol{\xi})^2$, we may be tempted to call $s = (s^2)^{1/2}$ an *invariant distance* or *invariant length*; this is common practice, even when the metric is as we have chosen it, so that s^2 is negative for space-like separations. However, we are interested primarily in time-like intervals, $s^2 > 0$, and it is useful to note that in the frame S' for which $(\mathbf{x}' - \boldsymbol{\xi}')^2 = 0$, the quantity $s^2 = (x'^0 - \xi'^0)^2$ is the square of a proper time interval; this is because the two time values ξ'^0/c and x'^0/c both were measured at the single space point $\boldsymbol{\xi} = \mathbf{x}$. Hence the invariant length s is the same thing as the proper time (multiplied by c)

in some frame of reference, namely S'. For this reason, s sometimes is denoted $c\tau$, and τ is called the invariant time.

3.7 THE BEHAVIOR OF SPACE AND TIME INTERVALS; THE INVARIANT FOUR-VOLUME

In order to emphasize the meaning of s as an interval, let us replace s with Δs, at the same time letting $(x^\mu - \xi^\mu) = \Delta x^\mu$. Then we have, for $(\Delta s)^2 > 0$,

$$\Delta s = \sqrt{(\Delta x^0)^2 - (\Delta \mathbf{x})^2} = c\,\Delta\tau,$$

where $\Delta\tau$ is the time interval separating the events, as measured in the frame in which the events occur at the same space point. If S' is the frame in which $\Delta\mathbf{x}' = 0$, we have $\Delta x'^0 = c\,\Delta\tau$. However, the quantitites Δx^μ in some frame S other than S' are related to the $\Delta x'^\mu$ by the Lorentz transformation equations

$$\begin{aligned} \Delta x^0 &= \gamma(\Delta x'^0 + \beta\,\Delta x'^2), \\ \Delta x^1 &= \Delta x'^1 \\ \Delta x^2 &= \gamma(\Delta x'^2 + \beta\,\Delta x'^0), \\ \Delta x^3 &= \Delta x'^3, \end{aligned} \tag{3.45}$$

which are obtained by solving Eqs. (3.29) for x^μ, and replacing x and x' with Δx and $\Delta x'$ throughout. Thus, since $\Delta\mathbf{x}' = 0$,

$$\Delta x^0 = c\,\Delta t = \gamma\,\Delta x'^0 = \gamma c\,\Delta\tau,$$

or

$$\Delta t - \gamma\,\Delta\tau. \tag{3.46}$$

That is, if $\Delta\tau$ is the proper time interval between two events, the corresponding time interval measured in an arbitrary frame of reference is $\gamma\,\Delta\tau$ where $\gamma = (1 - \beta^2)^{-1/2} = (1 - v^2/c^2)^{-1/2}$ and v is the velocity of the frame in which Δt is measured, relative to the frame in which the events are separated by the proper time interval $\Delta\tau$. This is just a restatement of the result of the second of the *Gedanken Experimente.*

Another important effect we must consider is the shrinkage of an element of three-dimensional volume, arising from the Lorentz contraction. If we have a rectangular parallelepiped with edges having lengths Δx, Δy and Δz parallel to the coordinate axes of S, and at rest relative to S, then the lengths of the edges measured in S are proper lengths and the volume element $\Delta V = \Delta x\,\Delta y\,\Delta z$ is a proper volume. In S', moving with velocity v relative to S, along the y-axis, an observer would note that his measurements of $\Delta x'$ and $\Delta z'$ were equal to Δx and Δz; however, his measurement of $\Delta y'$ would be related to Δy by the equation $\Delta y' = (1/\gamma)\cdot\Delta y$. This can be seen by noting that the observer in S' would make his measurements instantaneously so that $\Delta x'^0$ vanishes; the relations between the spatial measurements then follow from Eq. (3.45). Alternatively, one may refer back to the *Gedanken Experimente*.

Thus the volume element as measured in S', $\Delta V' = \Delta x' \, \Delta y' \, \Delta z'$, is related to $\Delta V = \Delta x \, \Delta y \, \Delta z$ by

$$\Delta V' = \frac{1}{\gamma} \Delta V. \tag{3.47}$$

The equivalent equation for infinitesimal volume elements is

$$d^3x' = dx' \, dy' \, dz' = \frac{1}{\gamma} dx \, dy \, dz = \frac{1}{\gamma} d^3x. \tag{3.48}$$

On the other hand, a four-dimensional volume element must have volume $\Delta x^0 \, \Delta x^1 \, \Delta x^2 \, \Delta x^3 = c \, \Delta t \, \Delta x \, \Delta y \, \Delta z$ in S, whereas in S' this same element of four-volume will appear to have magnitude $\Delta x'^0 \, \Delta x'^1 \, \Delta x'^2 \, \Delta x'^3 = c \, \Delta t' \, \Delta x' \, \Delta y' \, \Delta z'$. The proper time for this volume element is measured by clocks at rest relative to it, that is, by clocks at rest in S; hence $\Delta x^0 = c \, \Delta\tau$, where $\Delta\tau$ is the proper time. Therefore, according to Eq. (3.46),

$$\Delta x'^0 = \gamma \Delta x^0. \tag{3.49}$$

Also, from Eq. (3.47), we have

$$\Delta x'^1 \, \Delta x'^2 \, \Delta x'^3 = \frac{1}{\gamma} \Delta x' \, \Delta x^2 \, \Delta x^3. \tag{3.50}$$

Thus the four-dimensional volume elements are related by

$$\Delta x'^0 \, \Delta x'^1 \, \Delta x'^2 \, \Delta x'^3 = \Delta x^0 \, \Delta x' \, \Delta x^2 \, \Delta x^3 \tag{3.51}$$

or, for infinitesimal values of $\Delta x^0, \ldots, \Delta x^3$,

$$d^4x' = dx'^0 \, dx'^1 \, dx'^2 \, dx'^3 = dx^0 \, dx^1 \, dx^2 \, dx^3 = d^4x. \tag{3.52}$$

That is, for elements of four-volume, the effects of time dilation and Lorentz contraction cancel each other, and the four-volume element is invariant under Lorentz transformations. This is a very important result, and it is analogous to the invariance of three-volume elements under rotations in three-space.

READING NOTES*

Reference 10, Paper V by H. Minkowski.

Reference 9 is a brief account of the theory by its author, and although it emphasizes the general theory, it is good "cultural" reading even for special relativity. The first two chapters are appropriate here.

---

* Works referred to in these notes will be found in the Bibliography on p. 577.

Reference 3, Chapter 11.
Reference 11, Chapter 2.
Reference 8, Chapter 16.

PROBLEMS

1. Use Eq. (3.13) and the metric tensors $g_{\mu\nu}$, $g^{\mu\nu}$ to show formally that the special Lorentz transformation inverse to the one with parameters β is obtained by replacing β with $-\beta$ everywhere.
2. Solve Eqs. (3.27) directly for x^μ in terms of x'^μ and compare with the results of Problem 1.
3. Work out the consequences of choosing signs in Eqs. (3.37) and (3.38) in the three ways other than the choice adopted in the text.
4. Show that the angle ϕ between the x'^0-axis and the x^0-axis and the angle between the x'^2-axis and the x^2-axis in Fig. 3.1b is given by

$$\phi = \arctan \beta = \arctan (\tanh \zeta).$$

5. A pair of events occur simultaneously but 15×10^6 meters apart in S'. Find the relative velocities between S and S' if the events occur a time Δt apart in S, for the cases

a) $\Delta t = 1/10$ sec.
b) $\Delta t = 1$ sec.
c) $\Delta t = 10$ sec.
d) $\Delta t = 1$ min.
e) $\Delta t = 1$ hr.

6. Let $\mathbf{x}$ denote the position of an event in the coordinates of reference frame S and let $\mathbf{x}'$ denote the position of the reference frame S' moving with velocity $\mathbf{v}$ relative to S. Show that if t and t' are the corresponding times of the event,

$$\mathbf{x}' = \mathbf{x} + \boldsymbol{\beta}\left[\frac{\boldsymbol{\beta}\cdot\mathbf{x}}{\beta^2}(\gamma - 1) - \gamma ct\right],$$

$$t' = \gamma\left(t - \frac{\boldsymbol{\beta}\cdot\mathbf{x}}{c}\right) \qquad \boldsymbol{\beta} = \frac{\mathbf{v}}{c}, \gamma = \frac{1}{\sqrt{1 - \boldsymbol{\beta}^2}}.$$

7. The most general linear transformation on a two-dimensional Euclidean space has the form

$$x_i' = \sum_{j=1}^{3} a_{ij}x_j + b_i, \qquad i = 1, 2, 3,$$

where the a_{ij} and b_i are constants representing, respectively, affine transformations and translations. In matrix form this is written

$$\begin{Bmatrix} x_1' \\ x_2' \end{Bmatrix} = \begin{bmatrix} a_{11} & a_{12} \\ a_{21} & a_{22} \end{bmatrix} \begin{Bmatrix} x_1 \\ x_2 \end{Bmatrix} + \begin{Bmatrix} b_1 \\ b_2 \end{Bmatrix}$$

or, more compactly,

$$\vec{X}' = \vec{A}\cdot\vec{X} + \vec{B}$$

where the dot to the right of $\vec{A}$ means that a summation is to be carried out over the right-hand indices of the elements of $\vec{A}$; evidently the matrices $\vec{X}'$, $\vec{A}$, $\vec{X}$, and $\vec{B}$ are given by

$$\vec{X} = \begin{Bmatrix} x_1' \\ x_2' \end{Bmatrix}, \qquad \vec{A} = \begin{bmatrix} a_{11} & a_{12} \\ a_{21} & a_{22} \end{bmatrix}, \qquad \vec{X} = \begin{Bmatrix} x_1 \\ x_2 \end{Bmatrix}, \qquad \vec{B} = \begin{Bmatrix} b_1 \\ b_2 \end{Bmatrix}.$$

The "transpose" of $\vec{A}$, denoted $\vec{A}^{\mathrm{T}}$, is obtained by interchanging the indices on the elements of $\vec{A}$; thus for a 2×2 matrix,

$$\vec{A}^{\mathrm{T}} = \begin{bmatrix} a_{11} & a_{21} \\ a_{12} & a_{22} \end{bmatrix}.$$

For example, if $a_{11} = 1$, $a_{12} = 2$, $a_{21} = 3$ and $a_{22} = 4$, then

$$\vec{A} = \begin{bmatrix} 1 & 2 \\ 3 & 4 \end{bmatrix} \quad \text{and} \quad \vec{A}^{\mathrm{T}} = \begin{bmatrix} 1 & 3 \\ 2 & 4 \end{bmatrix}.$$

The *inverse* of $\vec{A}$, denoted $\vec{A}^{-1}$, has the properties

$$\vec{A}^{-1}\vec{A} = \vec{A}\vec{A}^{-1} = \vec{I}$$

where

$$\vec{I} = \begin{bmatrix} 1 & 0 \\ 0 & 1 \end{bmatrix}$$

has components $I_{ij} = \delta_{ij}$ (Kronecker δ) and is called the *identity* matrix. A matrix with the property $\vec{A}^{-1} = \vec{A}^{\mathrm{T}}$ is called an *orthogonal* matrix.

a) Find the elements a_{ij} of the matrix (transformation) that leaves

$$\alpha(x_1)^2 + \beta(x_2)^2$$

invariant (i.e. $\alpha(x_1)^2 + \beta(x_2)^2 = \alpha(x_1')^2 + \beta(x_2')^2$), where α and β are fixed constants; there should be one undetermined parameter in the result.

b) Under what conditions is the matrix of the a_{ij} an orthogonal matrix?

c) Choose the a_{ij} in such a way that they obviously represent a rotation in the x_1x_2-plane, and determine whether the matrix obtained is orthogonal.

8. Let the transformation matrix $b^\mu{}_\nu$ represent a rotation about the z-axis through angle θ in the unprimed frame of reference, so that coordinates *after the rotation* are given by

$$\bar{x}^\mu = b^\mu{}_\nu x^\nu,$$

where the x^ν are the coordinates before the rotation. A special Lorentz transformation is performed so that

$$x'^\mu = a^\mu{}_\nu x^\nu,$$

where the x^ν, again, are the coordinates before the rotation. Now the coordinates after the rotation, when viewed in the primed reference frame, are given by

$$\bar{x}'^\mu = a^\mu{}_\nu \bar{x}^\nu = a^\mu{}_\nu b^\nu{}_\sigma x^\sigma.$$

Thus, if we wish to represent the rotation in the unprimed frame as some general Lorentz transformation in the primed frame, we must write

$$\bar{x}'^{\mu} = B^{\mu}{}_{\rho} x'^{\rho} = B^{\mu}{}_{\rho} a^{\rho}{}_{\lambda} x^{\lambda}.$$

a) Write down the matrix equation for $B^{\mu}{}_{\nu}$ in terms of $a^{\mu}{}_{\nu}$ and $b^{\mu}{}_{\nu}$, and perhaps their inverses $a'^{\mu}{}_{\nu}$ and $b'^{\mu}{}_{\nu}$.

b) Let the $a^{\mu}{}_{\nu}$ be appropriate coefficients for a special Lorentz transformation along the x-axis and let $b^{\mu}{}_{\nu}$ represent the rotation described above, and calculate the coefficients $B^{\mu}{}_{\nu}$.

9. The usual mathematical formulation for the transformation of volume elements under coordinate transformations is obtained by writing

$$dx'^{1}\, dx'^{2} \cdots dx'^{n} = \left| \frac{\partial(x'^{1}, x'^{2}, \ldots, x'^{n})}{\partial(x^{1}, x^{2}, \ldots, x^{n})} \right| dx^{1}\, dx^{2} \cdots dx^{n}$$

where $x'^{i} = x'^{i}(x^{1}, x^{2}, \ldots, x^{n})$ are the transformed coordinates, the x^{j} being the original coordinates, and where

$$\left| \frac{\partial(x'^{1}, x'^{2}, \ldots, x'^{n})}{\partial(x^{1}, x^{2}, \ldots, x^{n})} \right| = \begin{vmatrix} \dfrac{\partial x'^{1}}{dx^{1}} & \dfrac{\partial x'^{1}}{\partial x^{2}} & \cdots & \dfrac{\partial x'^{1}}{\partial x^{n}} \\ \dfrac{\partial x'^{2}}{\partial x^{1}} & \dfrac{\partial x'^{2}}{\partial x^{2}} & \cdots & \dfrac{\partial x'^{2}}{\partial x^{n}} \\ \vdots & \vdots & & \vdots \\ \dfrac{\partial x'^{n}}{\partial x^{1}} & \dfrac{\partial x'^{n}}{\partial x^{2}} & \cdots & \dfrac{\partial x'^{n}}{\partial x^{n}} \end{vmatrix}$$

is the Jacobian determinant for the transformation. Using this method, show that

$$d^{4}x' = d^{4}x$$

for special Lorentz transformations and for rotations.

CHAPTER 4

SOME TECHNIQUES OF LINEAR ALGEBRA AND RELATIVISTIC KINEMATICS

4.1 THE LINEAR ALGEBRA OF FOUR-VECTORS. TENSOR TRANSFORMATIONS

We begin this chapter with a discussion of relevant mathematical techniques for handling the four-dimensional entities of special relativity. We shall avoid complete and detailed discussion of mathematical questions that do not have a direct bearing upon quickly developing an ability to manipulate relativistic quantities. Our aim shall be to provide the reader with techniques for setting up mathematical formulations of elementary problems in kinematics, and for solving these problems in terms of such formulations; that is, we wish to develop more fully the standard shorthand notation for handling relativistic problems.

First, let us consider with some care the meaning of the positions of indices in transformation coefficients such as $a^{\mu}{}_{\nu}$. These coefficients were defined by

$$a^{\mu}{}_{\nu} = \frac{\partial x'^{\mu}}{\partial x^{\nu}} = \frac{\partial x'^{\mu}(x^{\lambda})}{\partial x^{\nu}}$$

when x'^{μ} is considered as a function of the x^{ν}, and the x^{ν} are taken to be the independent variables. The importance of the fact that here the x^{ν} are considered independent variables, was not emphasized in our earlier discussion; however, the choice of independent variables has a direct bearing upon the location of indices. For example, we found that if we defined $a'^{\mu}{}_{\nu}$ as

$$a'^{\mu}{}_{\nu} = \frac{\partial x^{\mu}}{\partial x'^{\nu}} = \frac{\partial x^{\mu}(x'^{\lambda})}{\partial x'^{\nu}}$$

with the x'^{ν} considered as independent variables and the x^{μ} thought of as functions of the x'^{ν}, we obtained the result

$$a'^{\mu}{}_{\nu} = a_{\nu}{}^{\mu}.$$

Since the $a'^{\mu}{}_{\nu}$ for special Lorentz transformations can be obtained from the $a^{\mu}{}_{\nu}$ by replacing β with $-\beta$ wherever it occurs, it is quite clear that in general $a^{\mu}{}_{\nu} \neq a_{\nu}{}^{\mu} = a'^{\mu}{}_{\nu}$; the locations of indices are significant.

In general, the left-hand index on a transformation coefficient refers to the dependent variable, and the right-hand index to the independent variable. Whether an index is raised or lowered will depend upon whether the corresponding variable is covariant or contravariant. If the dependent variable is contravariant (covariant), the left-hand index *also* is contravariant (covariant). On the other hand, if the independent variable is contravariant (covariant), the right-hand index is covariant (contravariant). Thus the index on the transformation coefficient corresponding to the independent variable has "height" opposite to the index on the independent variable itself.

Table 4.1 gives the various transformation coefficients and some relationships that exist among them. The first column lists four a coefficients and the second column gives the partial derivatives defining these coefficients; in the third column are the a' coefficients obtained from the a coefficients by way of Eq. (3.13) and its analogs, and the fourth column gives the partial derivatives defining the corresponding a' coefficients. It is left as a problem for the reader to check the consistency of these relationships.

Table 4.1

$$a^{\mu}{}_{\nu} = \frac{\partial x'^{\mu}}{\partial x^{\nu}} = a'_{\nu}{}^{\mu} = \frac{\partial x_{\nu}}{\partial x'_{\mu}}$$

$$a_{\mu}{}^{\nu} = \frac{\partial x'_{\mu}}{\partial x_{\nu}} = a'^{\nu}{}_{\mu} = \frac{\partial x^{\nu}}{\partial x'^{\mu}}$$

$$a^{\mu\nu} = \frac{\partial x'^{\mu}}{\partial x_{\nu}} = a'^{\nu\mu} = \frac{\partial x^{\nu}}{\partial x'_{\mu}}$$

$$a_{\mu\nu} = \frac{\partial x'_{\mu}}{\partial x^{\nu}} = a'_{\nu\mu} = \frac{\partial x_{\nu}}{\partial x'^{\mu}}$$

Note that because of Eq. (3.3), namely $a^{\mu}{}_{\nu} \cdot a'^{\nu}{}_{\lambda} = \delta^{\mu}{}_{\lambda}$, we have $a^{\mu}{}_{\nu} \cdot a_{\lambda}{}^{\nu} = a^{\mu}{}_{\nu} \cdot a'^{\nu}{}_{\lambda} = \delta^{\mu}{}_{\lambda}$. Certainly one could not obtain this result just by looking at the defining partial derivatives,

$$a^{\mu}{}_{\nu} \cdot a_{\lambda}{}^{\nu} = \frac{\partial x'^{\mu}}{\partial x^{\nu}} \cdot \frac{\partial x'_{\lambda}}{\partial x_{\nu}},$$

because the invariance of the velocity of light is implicitly taken into account here via the relation $a'^{\nu}{}_{\lambda} = a_{\lambda}{}^{\nu}$; that is, such results do not follow directly from the calculus, but contain information about the restrictions imposed by the hypotheses of special relativity.

For the linear transformations in which we are interested, the four-component quantities x^{μ} serve as prototypes for a whole class of mathematical objects called *four-vectors* or *tensors of rank one*,* which is defined by the behavior of the objects

---

* Under the less restrictive conditions encountered in differential geometry and in general relativity, the x^{μ} do not qualify as vectors. However, in special relativity, only homogeneous spaces, and therefore only linear transformations, are considered; under this restriction, the x^{μ} indeed are tensors of rank one, or vectors.

under the transformations prescribed by the $a^\mu{}_\nu$. In general, a *contravariant four-vector*, q^μ, is a four-component quantity defined in terms of the coordinates of reference frame S such that whenever $a^\mu{}_\nu$ prescribes a linear (general Lorentz) transformation from frame S to frame S', the correct numerical description q'^μ of this four-component quantity in terms of the coordinates of S' is given by

$$q'^\mu = a^\mu{}_\nu q^\nu.$$

A *covariant four-vector* is defined similarly, with $q'_\mu = a_\mu{}^\nu q_\nu$.

An *invariant* quantity, sometimes called a *tensor of rank zero*, is a one-component quantity that has the same value in all reference frames. Examples are c, $s^2 = x^\mu x_\mu$, and the proper time interval between a pair of events, $\Delta\tau = \sqrt{(x^\mu - \xi^\mu)(x_\mu - \xi_\mu)}/c$.

Using the fact that $c\Delta\tau = \sqrt{(\Delta s)^2} = \sqrt{\Delta x^\mu \, \Delta x_\mu}$ is an invariant, we can construct a particularly useful four-vector. The velocity of one reference frame relative to another can be defined in terms of

$$\mathbf{u} = \lim_{\Delta\tau \to 0} \frac{\Delta \mathbf{x}}{\Delta\tau} = \frac{d\mathbf{x}}{d\tau}, \tag{4.1}$$

where $\Delta\mathbf{x}$ is a proper measurement of the change of position of the origin of one frame relative to another in the proper time interval $\Delta\tau$. Generalizing this to four dimensions, we have

$$u^\mu = \lim_{\Delta\tau \to 0} \frac{\Delta x^\mu}{\Delta\tau} = \frac{dx^\mu}{d\tau} \tag{4.2}$$

for the four-velocity of a point stationary in some frame Σ, as seen by an observer in S. Clearly the clock measuring $\Delta\tau$ must be at rest relative to Σ, for otherwise two clocks would be required, one at each endpoint of the spatial interval traversed by the point that is at rest in Σ, and the time interval would not be a proper one. On the other hand, Δx^μ must be measured in S, since the point is at rest in Σ; thus the proper distance traversed by the point is to be measured in S whereas the proper time interval is to be measured in Σ.

The four-velocity defined in this way is a four-vector, since Δx^μ is a four-vector and $\Delta\tau$ is an invariant. Specifically,

$$u'^\mu = \lim_{\Delta\tau \to 0} \frac{\Delta x'^\mu}{\Delta\tau} = \lim_{\Delta\tau \to 0} \frac{a^\mu{}_\nu \Delta x^\nu}{\Delta\tau} = a^\mu{}_\nu \lim_{\Delta\tau \to 0} \frac{\Delta x^\nu}{\Delta\tau} = a^\mu{}_\nu \frac{dx^\nu}{d\tau}, \tag{4.3}$$

since $a^\mu{}_\nu$ is independent of space and time intervals.

Note, however, that the usual definition of velocity is not the $\mathbf{u}$ that we defined above, but rather is

$$\mathbf{v} = \lim_{\Delta t \to 0} \frac{\Delta \mathbf{x}}{\Delta t} = \frac{d\mathbf{x}}{dt}. \tag{4.4}$$

Since $\Delta t = \gamma_v \, \Delta\tau = (1 - v^2/c^2)^{-1/2} \, \Delta\tau$, when v is the relative velocity of the frames in which Δt and $\Delta\tau$ are measured, we have

$$\mathbf{v} = \frac{d\tau}{dt}\frac{d\mathbf{x}}{d\tau} = \frac{1}{\gamma_v}\mathbf{u}. \tag{4.5}$$

Of course the velocity one actually measures is $\mathbf{v}$ and not $\mathbf{u}$, because $\Delta\mathbf{x}$ and Δt are always measured in the same frame; both of them cannot be proper measurements, therefore, unless $\Delta\mathbf{x} = \mathbf{0}$ and consequently $\mathbf{v} = \mathbf{0}$.

We now are in a position to write u^μ in terms of $\mathbf{u}$ and $\mathbf{v}$. First,

$$\frac{dx^0}{d\tau} = \frac{d(ct)}{d\tau} = c\frac{dt}{d\tau} = c\gamma_v;$$

hence

$$u^\mu = (c\gamma_v, u_x, u_y, u_z) = (c\gamma_v, \mathbf{u}). \tag{4.6}$$

Since γ_v appears in this expression, and since it is $\mathbf{v} = (1/\gamma_v)\mathbf{u}$ that one measures in practice, a more useful expression of u^μ is

$$\begin{aligned} u^\mu &= (c\gamma_v, \gamma_v v_x, \gamma_v v_y, \gamma_v v_z) = (c\gamma_v, \gamma_v \mathbf{v}) \\ &= \gamma_v(c, \mathbf{v}) \end{aligned} \tag{4.7}$$

Hence, to construct a four-vector from $\mathbf{v}$, one must multiply each component of $\mathbf{v}$ by $\gamma_v = (1 - \beta^2)^{-1/2}$, $\beta = v/c$, and append the zeroth component, $\gamma_v c$. The four-velocity u^μ thus is related to the usual three-dimensional velocity $\mathbf{v}$ in a somewhat more complicated way than x^μ is related to $\mathbf{x}$.

An important invariant in physics is the *proper mass* or *rest mass* of a particle; as the name suggests, this invariant is just the mass as measured by an observer at rest relative to the particle. Breaking slightly with tradition, we shall call this *the* mass of the particle, designating it m; no other meaning will be given the word "mass". It has been traditional to let "mass" mean the ratio of momentum to speed, so that it becomes an inertial property that varies from one reference frame to another. However, we shall follow a convention that has been widely adopted in high energy physics, and shall consider mass to mean an invariant property of a particle that can be used, together with other invariant properties, to define the nature of the particle in all reference frames. The adoption of this convention is strictly a matter of taste, for both uses of the word "mass" are now widespread; it should be borne firmly in mind that throughout this text the word "mass" has a meaning identical with the phrase "rest mass" used in many other texts.

Given the invariant mass, m, of a particle, we can define another important four-vector, the four-momentum

$$p^\mu = mu^\mu = m\frac{dx^\mu}{d\tau}, \tag{4.8}$$

for which we have

$$p'^{\mu} = mu'^{\mu} = ma^{\mu}{}_{\nu}u^{\nu} = a^{\mu}{}_{\nu}(mu^{\nu}) = a^{\mu}{}_{\nu}p^{\nu}. \tag{4.9}$$

In view of Eq. (4.7), we can write

$$p^{\mu} = (m\gamma_{v}c, m\gamma_{v}\mathbf{v}). \tag{4.10}$$

These examples illustrate the meanings of the words "four-vector" and "invariant", and demonstrate how such mathematical objects may be constructed. We must consider tensors of higher rank also. In general, a 4^n-component quantity $T^{\lambda\mu\nu\cdots\tau}$, where n is the number of indices $\lambda, \mu, \nu, \ldots, \tau$, is called a *contravariant tensor of rank n* if (and only if) its components are given correctly in the frame S' by

$$T'^{\lambda\mu\nu\cdots\tau} = a^{\lambda}{}_{\alpha}a^{\mu}{}_{\beta}a^{\nu}{}_{\gamma}\cdots a^{\tau}{}_{\kappa}T^{\alpha\beta\gamma\cdots\kappa} \tag{4.11}$$

whenever the $T^{\alpha\beta\gamma\cdots\kappa}$ are the components in frame S. Similarly, $T_{\lambda\mu\nu\cdots\tau}$ is a *covariant tensor of rank n* if (and only if)

$$T'_{\lambda\mu\nu\cdots\tau} = a_{\lambda}{}^{\alpha}a_{\mu}{}^{\beta}a_{\nu}{}^{\gamma}\cdots a_{\tau}{}^{\kappa}T_{\alpha\beta\gamma\cdots\kappa} \tag{4.12}$$

correctly gives the transformation of the components of $T_{\alpha\beta\gamma\cdots\kappa}$ from their values in S to their values in S'. Finally, a *mixed tensor of rank* $m + n$ is a 4^{m+n}-component object whose transformation from S to S' is given correctly by

$$T'^{\mu\cdots\sigma}{}_{\tau\cdots\omega} = a^{\mu}{}_{\alpha}\cdots a^{\sigma}{}_{\zeta}a_{\tau}{}^{\eta}\cdots a_{\omega}{}^{\kappa}T^{\alpha\cdots\zeta}{}_{\eta\cdots\kappa}. \tag{4.13}$$

We can check that the metrics $g^{\mu\nu}$ and $g_{\mu\nu}$ are respectively contravariant and covariant tensors of rank two, and that $\delta^{\mu}{}_{\nu}$ is a mixed tensor, contravariant of rank 1 and covariant of rank 1. These quantities were defined without reference to any frame, and hence they have identical components in all frames; i.e. we have $g'^{\mu\nu} = g^{\mu\nu}$, $g'_{\mu\nu} = g_{\mu\nu}$ and $\delta'^{\mu}{}_{\nu} = \delta^{\mu}{}_{\nu}$. Let us check that $g'^{\mu\nu} = a^{\mu}{}_{\alpha}a^{\nu}{}_{\beta}g^{\alpha\beta}$, etc.:

$$\begin{aligned} a^{\mu}{}_{\alpha}a^{\nu}{}_{\beta}g^{\alpha\beta} &= a^{\mu}{}_{\alpha}a^{\nu\alpha} = a^{\mu}{}_{\alpha}a'^{\alpha\nu} = a^{\mu}{}_{\alpha}a'^{\alpha}{}_{\gamma}g^{\gamma\nu} \\ &= \delta^{\mu}{}_{\gamma}g^{\gamma\nu} = g^{\mu\nu} = g'^{\mu\nu}; \end{aligned}$$

a similar procedure establishes the tensor property of $g_{\mu\nu}$, and we have the immediate demonstration

$$a^{\mu}{}_{\alpha}a_{\nu}{}^{\beta}\delta^{\alpha}{}_{\beta} = a^{\mu}{}_{\alpha}a_{\nu}{}^{\alpha} = a^{\mu}{}_{\alpha}a'^{\alpha}{}_{\nu} = \delta^{\mu}{}_{\nu} = \delta'^{\mu}{}_{\nu}$$

for $\delta^{\mu}{}_{\nu}$.

Two simple theorems are readily established:

1. For every tensor $T^{\mu\nu\cdots\omega}$, any one of the contravariant indices can be converted to a covariant index, by constructing $T^{\mu\nu\cdots\rho\cdots\omega}g_{\rho\sigma}$, and the resulting tensor also will be a tensor.

2. Any tensor, when multiplied by an invariant, remains a tensor.

The proof of 2 follows immediately from the definitions of invariant and tensor. The proof of 1 goes as follows: let $T^{\mu\cdots}{}_{\sigma}{}^{\cdots\omega} = T^{\mu\cdots\rho\cdots\omega}g_{\rho\sigma}$; then

$$\begin{aligned} a^{\mu}{}_{\alpha}\cdot\ldots\cdot a_{\sigma}{}^{\epsilon}\cdot\ldots\cdot a^{\omega}{}_{\zeta}T^{\alpha\cdots}{}_{\epsilon}{}^{\cdots\zeta} &= a^{\mu}{}_{\alpha}\cdot\ldots\cdot a_{\sigma}{}^{\epsilon}\cdot\ldots\cdot a^{\omega}{}_{\zeta}g_{\rho\epsilon}T^{\alpha\cdots\rho\cdots\zeta} \\ &= a^{\mu}{}_{\alpha}\cdot\ldots\cdot a_{\sigma\rho}\cdot\ldots\cdot a^{\omega}{}_{\zeta}T^{\alpha\cdots\rho\cdots\zeta} = a^{\mu}{}_{\alpha}\cdot\ldots\cdot a^{\lambda}{}_{\rho}\cdot\ldots\cdot a^{\omega}{}_{\zeta}T^{\alpha\cdots\rho\cdots\zeta}g_{\lambda\sigma} \\ &= T'^{\mu\cdots\lambda\cdots\omega}g_{\lambda\sigma} = T'^{\mu\cdots}{}_{\sigma}{}^{\cdots\omega}; \end{aligned} \tag{4.14}$$

that is, if

$$T'^{\mu\cdots\lambda\cdots\omega} = a^{\mu}{}_{\alpha}\cdot\ldots\cdot a^{\lambda}{}_{\gamma}\cdot\ldots\cdot a^{\omega}{}_{\zeta}T^{\alpha\cdots\gamma\cdots\zeta},$$

then also

$$T'^{\mu\cdots}{}_{\sigma}{}^{\cdots\omega} \equiv T'^{\mu\cdots\lambda\cdots\omega}g_{\lambda\sigma} = a^{\mu}{}_{\alpha}\cdot\ldots\cdot a_{\sigma}{}^{\gamma}\cdot\ldots\cdot a^{\omega}{}_{\zeta}T^{\alpha\cdots}{}_{\gamma}{}^{\cdots\zeta}$$

whenever $T^{\alpha\cdots}{}_{\gamma}{}^{\cdots\zeta} \equiv g_{\gamma\tau}T^{\alpha\cdots\tau\cdots\zeta}$.

One can show also that if $A^{\alpha\cdots\zeta}$ and $B^{\mu\cdots\omega}$ are both tensors, then $A^{\alpha\cdots\zeta}B^{\mu\cdots\omega}$ is a tensor also; the tensor thus formed has rank equal to the sum of the ranks of $A^{\alpha\cdots\zeta}$ and $B^{\mu\cdots\omega}$.

Multiplying any tensor, $T^{\mu\cdots\rho\cdots\tau\cdots\omega}$, by $g_{\rho\tau}$ and summing over ρ and τ, produces a new tensor, $S^{\mu\cdots\omega} = g_{\rho\tau}T^{\mu\cdots\rho\cdots\tau\cdots\omega}$ whose rank is two units lower than the rank of $T^{\mu\cdots\rho\cdots\tau\cdots\omega}$. This procedure is called *contracting* the indices ρ and τ.

We can add to our set of examples of four-vectors the differential operator $\partial/\partial x^{\mu}$, for if we consider the x'^{μ} as independent variables, we have

$$\frac{\partial}{\partial x'^{\mu}} = \frac{\partial x^{\nu}}{\partial x'^{\mu}}\frac{\partial}{\partial x^{\nu}} = a'^{\nu}{}_{\mu}\frac{\partial}{\partial x^{\nu}} = a_{\mu}{}^{\nu}\frac{\partial}{\partial x^{\nu}}. \tag{4.15}$$

Thus $\partial/\partial x^{\mu}$ is a *covariant* tensor of rank one. Similarly, $\partial/\partial x_{\mu}$ is a *contravariant* tensor of rank one:

$$\frac{\partial}{\partial x'_{\mu}} = \frac{\partial x_{\nu}}{\partial x'_{\mu}}\frac{\partial}{\partial x_{\nu}} = a'_{\nu}{}^{\mu}\frac{\partial}{\partial x_{\nu}} = a^{\mu}{}_{\nu}\frac{\partial}{\partial x_{\nu}}. \tag{4.16}$$

Henceforward we shall denote $\partial/\partial x^{\mu}$ with the symbol ∂_{μ} and $\partial/\partial x_{\mu}$ with ∂^{μ}; similarly, ∂'_{μ} shall mean $\partial/\partial x'^{\mu}$ and ∂'^{μ} shall stand for $\partial/\partial x'_{\mu}$.

The Lorentz coefficients $a^{\mu}{}_{\nu}$ themselves are tensors; in fact, this illustrates an important point, which, when throughly understood, can prevent confusion in working with tensor quantities. When we say, in relativity discussion, that $a^{\mu}{}_{\nu}$ is a tensor, we mean simply that *the collection of numbers* $\{a^{\mu}{}_{\nu}\!:\mu, \nu = 0, 1, 2, 3\}$ defined in some coordinate frame *behaves like a tensor when a Lorentz transformation,* say $\alpha^{\mu}{}_{\nu}$, *is applied to them.* Here we think of the $a^{\mu}{}_{\nu}$ as defining a relationship between two vectors in one frame of reference. To see how this applies to $a^{\mu}{}_{\nu}$, forget for a moment about the meaning of $a^{\mu}{}_{\nu}$ as Lorentz transformation coefficients; rather let y^{μ} be a four-vector in S, related to another four-vector x^{μ} in S by the relation $y^{\mu} = a^{\mu}{}_{\nu}x^{\nu}$. Under the transformation $\alpha^{\mu}{}_{\nu}$ from S to a new frame

Σ, we have $\eta^\mu = \alpha^\mu{}_\beta y^\beta$ and $\xi^\nu = \alpha^\nu{}_\sigma x^\sigma$; hence, if we let $\alpha'^\nu{}_\sigma$ be the transformation from Σ to S, so that $x^\nu = \alpha'^\nu{}_\sigma \xi^\sigma$ and $\alpha'^\nu{}_\sigma \alpha^\sigma{}_\beta = \delta^\nu{}_\beta$, then we have

$$\eta^\mu = \alpha^\mu{}_\beta y^\beta = \alpha^\mu{}_\beta a^\beta{}_\nu x^\nu = \alpha^\mu{}_\beta a^\beta{}_\nu \alpha'^\nu{}_\sigma \xi^\sigma$$
$$= (\alpha^\mu{}_\beta \alpha_\sigma{}^\nu a^\beta{}_\nu)\xi^\sigma.$$

If we write $\eta^\mu = A^\mu{}_\sigma \xi^\sigma$, we have

$$A^\mu{}_\sigma = \alpha^\mu{}_\beta \alpha_\sigma{}^\nu a^\beta{}_\nu;$$

that is, the coefficients $a^\mu{}_\nu$ connecting y^μ and x^μ in S transform to the $A^\mu{}_\nu$ connecting the corresponding vectors η^μ and ξ^μ in Σ, in just the way a mixed tensor of rank two transforms. Since tensors are defined by their transformation properties, $a^\mu{}_\nu$ is a mixed tensor of rank two. Note that here we have treated the $a^\mu{}_\nu$ simply as numbers, not as Lorentz transformation coefficients. (See, however, Problem 18.)

It is clear that tensor transformations can be thought of as products of sets of tensors with contractions of certain of the indices. Thus, for example, if we consider $a^{\mu\lambda}x^\nu$, we have a tensor of rank three, contravariant in all indices; contracting the indices λ and ν, we then obtain $a^{\mu\lambda}x^\nu g_{\lambda\nu} = a^\mu{}_\nu x^\nu = x'^\mu$. Another example is the formation of an invariant from a tensor of rank two, a specific case of which is

$$x^\mu x^\lambda g_{\mu\lambda} = x^\mu x_\mu = s^2.$$

An important property of Lorentz transformations is their *group* property. Any set A of objects for which an associative law of combination, $\cdot$, is defined such that

1. If $a \in A$ and $b \in A$, then $a \cdot b \in A$;
2. There exists $e \in A$ such that for any $a \in A$, $a \cdot e = e \cdot a = a$;
3. For every element $a \in A$, there exists another element of A, called a^{-1}, such that $aa^{-1} = a^{-1}a = e$;

is called a mathematical *group*. If we let A be the set of Lorentz transformations, and if $a = [a^\mu{}_\nu]$, $b = [b^\mu{}_\nu]$ are elements of that set, with $a \cdot b$ defined as

$$a \cdot b \equiv \left[\sum_\lambda a^\mu{}_\lambda b^\lambda{}_\nu\right] = [a^\mu{}_\lambda b^\lambda{}_\nu]$$

it is easy to show that A is a group.

Let $\{a^\mu{}_\nu\}$ be the set of Lorentz coefficients for a transformation from frame S to frame S': $x'^\mu = a^\mu{}_\nu x^\nu$, and let $\{b^\mu{}_\nu\}$ be the set of coefficients for a transformation from S' to S''; $x''^\mu = b^\mu{}_\nu x'^\nu$. Then, if $\{c^\mu{}_\nu\}$ is the set of coefficients for transforming from S to S'', one has

$$x''^\mu = c^\mu{}_\nu x^\nu = b^\mu{}_\lambda x'^\lambda = b^\mu{}_\lambda a^\lambda{}_\nu x^\nu$$

or, since x^ν is arbitrary,

$$c = [c^\mu{}_\nu] = [b^\mu{}_\lambda a^\lambda{}_\nu] = b \cdot a$$

is the transformation from S to S''; i.e. $c = b \cdot a \in A$. Next, let $[\delta^\mu{}_\nu] = e$; then $a \cdot e = [a^\mu{}_\lambda \delta^\lambda{}_\nu] = [a^\mu{}_\nu] = a$ and $e \cdot a = [\delta^\mu{}_\lambda a^\lambda{}_\nu] = a^\mu{}_\nu = a$. The transformation $e = [\delta^\mu{}_\nu]$ is just the identity transformation, e.g. from S to S; hence $e \in A$ is the identity that is required. Furthermore, let $a^{-1} = [a'^\mu{}_\nu]$; i.e. a^{-1} is the set of transformation coefficients for $S' \rightarrow S$. Then

$$a^{-1} \cdot a = [a'^\mu{}_\lambda a^\lambda{}_\nu] = [\delta^\mu{}_\nu] = [a^\mu{}_\lambda a'^\lambda{}_\nu] = aa^{-1} = e,$$

and requirement 3 is satisfied. Since ordinary multiplication and addition are associative, we also have, for any $a, b, c \in A$,

$$(a \cdot b) \cdot c = [(a^\mu{}_\nu b^\nu{}_\lambda) c^\lambda{}_\sigma] = [a^\mu{}_\nu (b^\nu{}_\lambda c^\lambda{}_\sigma)] = a \cdot (b \cdot c)$$

so that our law of combination indeed is associative. Hence the set A is a group, and for this reason the set of all Lorentz transformations often is referred to as the *Lorentz Group.*

It is a simple matter to show that the set of special Lorentz transformations along any given coordinate axis constitutes a *subgroup* of the Lorentz group; that is, taken by themselves, such a set of Lorentz transformations satisfy the requirements that define a group. To show this, it is necessary to demonstrate only that requirements 1 and 3 are satisfied, since the other requirement follows from the fact that a special Lorentz transformation is an element of the Lorentz group. Three-dimensional rotations also constitute a subgroup of the Lorentz group. The proofs that special Lorentz transformations and rotations are, individually, subgroups of the Lorentz group are left as exercises for the reader.

As an application of the group property among special Lorentz transformations, along the x-axis, say, we shall find the transformation that carried four-vectors as measured in S'' to four-vectors in S, by calculating the result of first transforming from S into some intermediate frame S', and then transforming from S' into S''. Thus if

$$x'^\mu = a^\mu{}_\nu x^\nu, \qquad x''^\mu = b^\mu{}_\nu x'^\nu,$$

$$x''^\mu = c^\mu{}_\nu x^\nu$$

we have

$$x''^\mu = c^\mu{}_\nu x^\nu = b^\mu{}_\lambda x'^\lambda = b^\mu{}_\lambda a^\lambda{}_\nu x^\nu,$$

or since the x^ν are arbitrary, as above,

$$c^\mu{}_\nu = b^\mu{}_\lambda a^\lambda{}_\nu . \tag{4.17}$$

If v is the velocity of S' relative to S, and if v' is the velocity of S'' relative to S', we have as the matrices for $a^{\mu}{}_{\nu}$ and $b^{\mu}{}_{\nu}$,

$$[a^{\lambda}{}_{\nu}] = \begin{bmatrix} \gamma_v & -\gamma_v\beta & 0 \\ -\gamma_v\beta & \gamma_v & 0 \\ 0 & 0 & 1 \\ 0 & 0 & 0 \end{bmatrix}, \qquad [b^{\mu}{}_{\lambda}] = \begin{bmatrix} \gamma_{v'} & -\gamma_{v'}\beta' & 0 & 0 \\ -\gamma_{v'}\beta' & \gamma_{v'} & 0 & 0 \\ 0 & 0 & 1 & 0 \\ 0 & 0 & 0 & 1 \end{bmatrix}.$$

Therefore the matrix for $c^{\mu}{}_{\nu}$ is given by

$$[c^{\mu}{}_{\nu}] = \begin{bmatrix} \gamma_{v'} & -\gamma_{v'}\beta' & 0 & 0 \\ -\gamma_{v'}\beta' & \gamma_{v'} & 0 & 0 \\ 0 & 0 & 1 & 0 \\ 0 & 0 & 0 & 1 \end{bmatrix} \begin{bmatrix} \gamma_v & -\gamma_v\beta & 0 & 0 \\ -\gamma_v\beta & \gamma_v & 0 & 0 \\ 0 & 0 & 1 & 0 \\ 0 & 0 & 0 & 1 \end{bmatrix}$$

$$= \begin{bmatrix} \gamma_v\gamma_{v'}(1 + \beta\beta') & -\gamma_v\gamma_{v'}(\beta + \beta') & 0 & 0 \\ -\gamma_v\gamma_{v'}(\beta + \beta') & \gamma_v\gamma_{v'}(1 + \beta\beta') & 0 & 0 \\ 0 & 0 & 1 & 0 \\ 0 & 0 & 0 & 1 \end{bmatrix}$$

$$= \begin{bmatrix} \gamma_V & -\gamma_V\beta_V & 0 & 0 \\ -\gamma_V\beta_V & \gamma_V & 0 & 0 \\ 0 & 0 & 1 & 0 \\ 0 & 0 & 0 & 1 \end{bmatrix}, \qquad \beta = \frac{v}{c},\ \beta' = \frac{v'}{c},\ \beta_V = \frac{V}{c},$$

where V is the relative velocity of S'' and S. This leads to

$$\gamma_V = \gamma_v\gamma_{v'}(1 + \beta\beta') \tag{4.18}$$

and

$$\gamma_V\beta_V = \gamma_v\gamma_{v'}(\beta + \beta') = \gamma_V\frac{\beta + \beta'}{1 + \beta\beta'},$$

or

$$V = c\beta_V = \frac{v + v'}{1 + vv'/c^2} = c\,\frac{\beta + \beta'}{1 + \beta\beta'}\,. \tag{4.19}$$

The reader should check the consistency of Eqs. (4.18) and (4.19).

4.2 RELATIVISTIC KINEMATICS AND DYNAMICAL VARIABLES; TRANSFORMATION PROPERTIES

The transformation of classical velocities **v** under special Lorentz transformations is not directly analogous to the transformation of space vectors such as **x**. The correct transformation properties are obtained from the tensor transformation equation $u'^\mu = a^\mu{}_\nu u^\nu$, where u^μ is the four-velocity of a particle as measured by an observer at rest in the S-frame of reference. Since $u^\mu = dx^\mu/d\tau = (\gamma_v c, \gamma_v \mathbf{v})$, a special Lorentz transformation for velocity V along the z — or x^3-axis yields

$$\begin{aligned} u'^0 &= \gamma_{v'} c = a^0{}_\nu u^\nu = \gamma_V(u^0 - \beta_V u^3) \\ &= \gamma_V(\gamma_v c - \beta_V \gamma_v v_z) = \gamma_V \gamma_v \left(c - \frac{V v_z}{c}\right) \\ &= c\gamma_V \gamma_v \left(1 - \frac{\mathbf{V}\cdot\mathbf{v}}{c^2}\right) = c\gamma_V\gamma_v(1 - \boldsymbol{\beta}_V \cdot \boldsymbol{\beta}),\ \boldsymbol{\beta} = \frac{\mathbf{v}}{c}, \end{aligned}$$

or

$$\gamma_{v'} = \gamma_V \gamma_v \left(1 - \frac{\mathbf{V}\cdot\mathbf{v}}{c^2}\right), \tag{4.20}$$

where v' is the speed in frame S' of a particle having speed v in S. Furthermore,

$$\begin{aligned} u'^1 &= \gamma_{v'} v'_x = u' = \gamma_v v_x \\ u'^2 &= \gamma_{v'} v'_y = u^2 = \gamma_v v_y \\ u'^3 &= \gamma_{v'} v'_z = \gamma_V(u^3 - \beta_V u^0) = \gamma_V(\gamma_v v_z - \beta_V \gamma_v c) \\ &= \gamma_V \gamma_v (v_z - V). \end{aligned}$$

Combining the last three equations with (4.20), one obtains

$$v'_x = \frac{1}{\gamma_V} \frac{v_x}{1 - (\mathbf{V}\cdot\mathbf{v})/c^2} = \sqrt{1 - \frac{\mathbf{V}^2}{c^2}}\, \frac{v_x}{1 - (\mathbf{V}\cdot\mathbf{v})/c^2} \tag{4.21}$$

$$v'_y = \frac{1}{\gamma_V} \frac{v_y}{1 - (\mathbf{V}\cdot\mathbf{v})/c^2} = \sqrt{1 - \frac{\mathbf{V}^2}{c^2}}\, \frac{v_y}{1 - (\mathbf{V}\cdot\mathbf{v})/c^2} \tag{4.22}$$

$$v'_z = \frac{v_z - V}{1 - (\mathbf{V}\cdot\mathbf{v})/c^2}. \tag{4.23}$$

Equations (4.21)–(4.23) are commonly referred to as the *transformation equations for velocities*. The reader should notice the relationship between Eq. (4.23) and Eq. (4.19).

Consider now the invariant quantity

$$u^\mu u_\mu = (c^2\gamma_v^2 - \mathbf{v}^2\gamma_v^2) = \frac{c^2 - v^2}{1 - v^2/c^2} = c^2. \tag{4.24}$$

At first sight it may seem strange that the "square" of the velocity four-vector should be independent of the frame of reference and, indeed, of the classical velocity as measured in any reference frame. However, since we have insisted that u^μ be a four-vector, we have built this feature into our definition of u^μ; according to our definition of a four-vector, the quantity $u^\mu u_\mu = u^\mu g_{\mu\nu} u^\nu$ must be an invariant. It is interesting to note that

$$x^\mu u_\mu = x^0 u^0 - \mathbf{x} \cdot \mathbf{u} = \gamma_v c^2 t - \gamma_v \mathbf{x} \cdot \mathbf{v} = c\gamma_v(ct - \mathbf{x} \cdot \boldsymbol{\beta})$$

necessarily is an invariant quantity, if $\mathbf{v}$ is the velocity of a particle relative to the frame in which $\mathbf{x}$ and t are measured. If the z-axis is parallel to $\mathbf{v}$, we have $\mathbf{x} \cdot \mathbf{v} = x^3 v = zv$, and

$$x^\mu u_\mu = c\gamma_v(ct - \beta z) = c\gamma_v(x^0 - \beta x^3).$$

At this point, the reader is referred to Equation (3.29), where he may note that the above quantity is just c times the time corresponding to t, as measured in the frame in which the particle is at rest; that is

$$x^\mu u_\mu = c\tau, \tag{4.25}$$

since the time in the frame of the particle must be the proper time; in any other frame but the rest frame of the particle, two clocks would be required to measure a time interval relating events occurring at the position of the particle.

The four-vector for momentum, p^μ, evidently yields the invariant

$$p^\mu p_\mu = (mu^\mu)(mu_\mu) = m^2 u^\mu u_\mu = m^2 c^2; \tag{4.26}$$

thus the value of the mass can always be calculated from the momentum four-vector:

$$m = \sqrt{\frac{p^\mu p_\mu}{c^2}}. \tag{4.27}$$

The momentum four-vector is the first genuine physical quantity we have encountered in our study of relativity thus far; we should consider it in some detail. The spatial components of p^μ are just the nonrelativistic momentum, $m\mathbf{v}$, multiplied by $\gamma_v = (1 - v^2/c^2)^{-1/2}$; however, we do not have any interpretation for the temporal component, $p^0 = \gamma_v mc$. In order to obtain an interpretation, let us recall Newton's second law of motion.

According to Newton's second law, the time rate of change of momentum of a body is equal to the imposed force. Thus, classically,

$$\frac{d\mathbf{p}_{\text{class}}}{dt} = \mathbf{F}_{\text{class}}.$$

The relativistic analog of this law is

$$\frac{d}{d\tau} p^\mu = F^\mu, \tag{4.28}$$

where F^μ is the four-vector analog of force. The derivative of the four-momentum is taken with respect to the invariant time τ in order that we may relate the momentum four-vector directly to a force four-vector.

Let us consider a particle of fixed mass, m, moving with velocity $\mathbf{v}$ relative to us. Its four-momentum then is $p^\mu = mu^\mu$, so that

$$F^\mu = \frac{d}{d\tau}p^\mu = \frac{d}{d\tau}mu^\mu = m\frac{d}{d\tau}u^\mu. \tag{4.29}$$

Therefore we must find, by calculation, the acceleration four-vector

$$a^\mu \equiv \frac{d}{d\tau}u^\mu = \frac{d}{d\tau}(\gamma_v c, \gamma_v \mathbf{v}) = \left(c\frac{d\gamma_v}{d\tau}, \mathbf{v}\frac{d\gamma_v}{d\tau} + \gamma_v\frac{d\mathbf{v}}{d\tau}\right).$$

Since

$$\frac{d}{d\tau}\gamma_v = \frac{d}{d\tau}\left(1 - \frac{v^2}{c^2}\right)^{-1/2} = \left(1 - \frac{v^2}{c^2}\right)^{-3/2}\frac{\mathbf{v}}{c^2}\cdot\frac{d\mathbf{v}}{d\tau} = \gamma_v^3\frac{\mathbf{v}}{c^2}\cdot\frac{d\mathbf{v}}{d\tau},$$

we have

$$\begin{aligned} a^\mu = \frac{du^\mu}{d\tau} &= \left(\gamma_v^3\frac{\mathbf{v}}{c}\cdot\frac{d\mathbf{v}}{d\tau}, \gamma_v^3\left[\frac{\mathbf{v}}{c}\cdot\frac{d\mathbf{v}}{d\tau}\right]\frac{\mathbf{v}}{c} + \gamma_v\frac{d\mathbf{v}}{d\tau}\right) \\ &= \left(\gamma_v^4\frac{\mathbf{v}}{c}\cdot\frac{d\mathbf{v}}{dt}, \gamma_v^4\left[\frac{\mathbf{v}}{c}\cdot\frac{d\mathbf{v}}{dt}\right]\frac{\mathbf{v}}{c} + \gamma_v^2\frac{d\mathbf{v}}{dt}\right), \end{aligned} \tag{4.30}$$

where we have used $dt/d\tau = \gamma_v$. Hence

$$F^\mu = ma^\mu = \left(m\gamma_v^4\frac{\mathbf{v}}{c}\cdot\frac{d\mathbf{v}}{dt}, m\gamma_v^4\left[\frac{\mathbf{v}}{c}\cdot\frac{d\mathbf{v}}{dt}\right]\frac{\mathbf{v}}{c} + m\gamma_v^2\frac{d\mathbf{v}}{dt}\right),$$

or, if $\mathbf{F}$ represents the spatial components of F^μ,

$$\mathbf{F} = m\gamma_v^2\left[\gamma_v^2\left(\frac{\mathbf{v}}{c}\cdot\frac{d\mathbf{v}}{dt}\right)\frac{\mathbf{v}}{c} + \frac{d\mathbf{v}}{dt}\right], \tag{4.31}$$

$$F^0 = m\gamma_v^4\frac{\mathbf{v}}{c}\cdot\frac{d\mathbf{v}}{dt}. \tag{4.32}$$

Equation (4.31) differs in two important ways from the classical equation relating force to acceleration. First, the force has a component in the direction of $\mathbf{v}$, even when $d\mathbf{v}/dt$ and $\mathbf{v}$ are not parallel; second, the term proportional to $d\mathbf{v}/dt$ contains a factor that does not appear in the classical equation.

Because of the factor $m\mathbf{v}\cdot(d\mathbf{v}/dt)$ in the right-hand side of Eq. (4.32), F^0 looks suspiciously as though it might be related to the power delivered to the mass while it is being acted on by the force. However, the spatial part of F^μ is not proportional to $d\mathbf{v}/dt$ and we must actually calculate the power to check whether it coincides with F^0. Thus,

$$\begin{aligned}\text{Power} = \mathbf{F}\cdot\mathbf{v} &= m\gamma_v^4\left[\left(\frac{\mathbf{v}}{c}\cdot\frac{d\mathbf{v}}{dt}\right)\frac{\mathbf{v}}{c} + \frac{1}{\gamma_v^2}\frac{d\mathbf{v}}{dt}\right]\cdot\mathbf{v}\\ &= m\gamma_v^4\left[\frac{v^2}{c^2}\mathbf{v}\cdot\frac{d\mathbf{v}}{dt} + \frac{1}{\gamma_v^2}\mathbf{v}\cdot\frac{d\mathbf{v}}{dt}\right]\\ &= m\gamma_v^4\left(\mathbf{v}\cdot\frac{dv}{dt}\right)\left[\frac{v^2}{c^2} + \left(1 - \frac{v^2}{c^2}\right)\right] = m\gamma_v^4\mathbf{v}\cdot\frac{d\mathbf{v}}{dt} = F^0 c.\end{aligned} \tag{4.33}$$

The temporal component F^0, when multiplied by c, does indeed represent the power delivered to the particle by the force **F**! This relationship can be expressed in the form

$$F^0 = \frac{\mathbf{v}}{c}\cdot\mathbf{F} \tag{4.34}$$

or

$$\frac{d}{d\tau}p^0 = \frac{\mathbf{v}}{c}\cdot\frac{d\mathbf{p}}{d\tau} \tag{4.35}$$

with $\mathbf{p} = m\gamma_v\mathbf{v}$; since $d\tau = (1/\gamma_v)\,dt$, this can be written as

$$\frac{d}{dt}p^0 = \frac{\mathbf{v}}{c}\cdot\frac{d\mathbf{p}}{dt} \tag{4.36}$$

also.

Power is defined to be the energy delivered per unit time; hence if we interpret this to mean energy delivered per unit invariant time, we have

$$\frac{d}{d\tau}E = cF^0 = c\frac{d}{d\tau}p^0$$

or

$$E = p^0 c + \epsilon_0$$

where ϵ_0 is constant in time. That is, the energy possessed by a particle is equal to c times the zeroth component of the momentum four-vector, plus a constant. The constant ϵ_0 is arbitrary, and we can choose it to be zero. In this way, we obtain

$$p^0 = mc\gamma_v = \frac{E}{c} \tag{4.37}$$

where E represents the total energy of the particle. The temporal component of momentum is to be interpreted as the total energy divided by c.

This equation not only tells us how to interpret the temporal component of momentum; it also gives us some important information about relativistic energy.

First, if $\mathbf{v} = \mathbf{0}$, we have $\gamma_{v=0} = 1$, and

$$E_0 = mc^2 \tag{4.38}$$

as the "rest energy" of a particle. Hence, the total energy of a particle includes a certain amount of the energy, mc^2, that arises just from the possession of mass.

Next, the kinetic energy, or energy of motion, for a free particle such as the one we have been considering, is $T = E - E_0$, or

$$\begin{aligned} T &= p^0 c - mc^2 = mc^2\gamma_v - mc^2 = mc^2(\gamma_v - 1) \\ &= mc^2\left[\left(1 - \frac{v^2}{c^2}\right)^{-1/2} - 1\right]. \end{aligned} \tag{4.39}$$

Does this result reduce to Newton's $T = \frac{1}{2}mv^2$ when $v \ll c$? To check, we assume $\beta = v/c \ll 1$, and obtain

$$\left(1 - \frac{v^2}{c^2}\right)^{-1/2} = 1 + \frac{1}{2}\frac{v^2}{c^2} + \frac{3}{8}\frac{v^4}{c^4} + \cdots \simeq 1 + \frac{1}{2}\frac{v^2}{c^2}.$$

Hence

$$T \simeq mc^2\left[1 + \frac{1}{2}\frac{v^2}{c^2} - 1\right] = \tfrac{1}{2}mv^2, \qquad v \ll c,$$

as we should expect.

Now let us quote some useful formulas that may be obtained directly from the equations we have just obtained. Since $p^0 = m\gamma_v c$, we have

$$\gamma_v = \frac{p_0}{mc}. \tag{4.40}$$

Furthermore, since $\mathbf{p} = m\gamma_v\mathbf{v} = mc\gamma_v\boldsymbol{\beta}_v$, where $\boldsymbol{\beta}_v = \mathbf{v}/c$, we get

$$\gamma_v\boldsymbol{\beta}_v = \frac{\mathbf{p}}{mc}. \tag{4.41}$$

Combining (4.40) and (4.41), we obtain

$$\boldsymbol{\beta}_v = \frac{\mathbf{p}}{p^0} = \frac{c\mathbf{p}}{E}. \tag{4.42}$$

For completeness, we also include the very important equation

$$(mc)^2 = \frac{E^2}{c^2} - \mathbf{p}^2. \tag{4.43}$$

Finally, if we let $E = T + mc^2$ in Eq. (4.43), we find that

$$p = |\mathbf{p}| = \sqrt{2mT + \left(\frac{T}{c}\right)^2},$$

where T is the kinetic energy. Here again, if $T \ll mc^2$, one obtains as an approximation, the Newtonian Equation

$$T \simeq \frac{\mathbf{p}^2}{2m} = \tfrac{1}{2}m\mathbf{v}^2.$$

We began this section with a discussion of the transformation of velocities, and then developed relativistic expressions for momentum, acceleration, force, and energy. We have not, however, considered the transformation properties of acceleration and force. These transformations are carried out in exactly the same way as the velocity transformation, and the reader is advised to investigate them for himself.

A word of caution now is in order. In the sequel we usually shall use **u**, **p**, **a**, **F**, etc. to represent the *spatial components of four-vectors.* The single consistent exception to this rule is **v**, which shall mean the time rate of change of position, and which therefore is *not* the spatial part of a four-vector (see 4.1). Thus **a** does *not* mean $d\mathbf{v}/dt$; rather, since

$$a^\mu = (a^0, \mathbf{a}) \equiv \frac{d}{d\tau}u^\mu = \left(\gamma_v c\,\frac{d\gamma_v}{dt}, \gamma_v \frac{d}{dt}\gamma_v\mathbf{v}\right),$$

we have

$$\mathbf{a} = \gamma_v \frac{d}{dt}(\gamma_v\,\mathbf{v}) = \gamma_v\frac{d}{dt}\mathbf{u}, \qquad \mathbf{u} = \gamma_v\mathbf{v}.$$

Similarly, **p** does not mean $m\mathbf{v}$; rather,

$$\mathbf{p} = m\gamma_v\mathbf{v}.$$

4.3 CENTER-OF-MASS AND BARYCENTRIC FRAMES OF REFERENCE

In classical mechanics, the frame of reference in which the center of mass of a system of particles is at rest plays a very important role. Remember that the location of the center of mass is given by

$$\mathbf{X} = \frac{\sum_i m_i\mathbf{x}_{(i)}}{\sum_i m_i}$$

where $\mathbf{x}_{(i)}$ is the location of the ith mass m_i. A direct relativistic generalization of this formula is not very pleasant because, for example, we do not know quite what to do with the zeroth component of the center-of-mass four-vector that would be a direct analog of **X**. If we ignore this problem, though, we can carry through a mathematically consistent analysis paralleling that of prerelativistic mechanics.

Let us define

$$X^\mu \equiv \frac{\sum_i m_i x_{(i)}{}^\mu}{\sum_i m_i}$$

where the m_i are the invariant masses (rest masses) of the particles. Thus if

$$M' \equiv \sum_i m_i$$

we have

$$M'X^\mu = \sum_i m_i x_{(i)}{}^\mu.$$

This leads to

$$P^\mu \equiv \frac{d}{d\tau} M'X^\mu = M'\frac{d}{d\tau} X^\mu = M'U^\mu = \sum_i m_i \frac{d}{d\tau} x_{(i)}{}^\mu$$
$$= \sum_i m_i u_{(i)}{}^\mu,$$

or, since $p_{(i)}{}^\mu \equiv m_i u_{(i)}{}^\mu$,

$$p^\mu = \sum_i p_{(i)}{}^\mu.$$

Now, however, things become unpleasant; for example, $P^\mu P_\mu \neq M'^2c^2$, as the reader is invited to demonstrate for a system consisting of two masses m_1 and m_2.

The difficulty is that P^μ represents the total four-momentum of the system, so that when its space components vanish, its "time" component will be the total energy of the system in the reference frame that has the center of mass X^μ at rest. But not all of the particles in the system will be at rest in this reference frame; hence the total energy of the system in this frame *exceeds* the sum of the masses m_i by the amount of their total kinetic energy! In terms of prerelativistic ideas, the apparent rest mass of the system as a whole is greater than the sum of the constituent rest masses. Thus the kinetic energies of the constituent particles become part of the *mass* of the total system, and we have an example of the equivalence of mass and energy.

In relativistic mechanics, then, the quantity $\Sigma_i m_i = M'$ plays a rather insignificant role, since it is not the total mass of the collection of particles. The significant quantity is

$$Mc^2 = \sqrt{P^\mu P_\mu}, \tag{4.45}$$

which often is called the *invariant mass*, or more simply, the *mass* of the overall system. Because of this and other deviations from the theory of centers of mass in prerelativistic physics, the term "center of mass" will be avoided in our relativistic discussions. Instead, we shall use the term *barycentric* to refer to the frame of reference in which the total momentum of the system vanishes. Thus, a frame of reference in which we measure the total four-momentum to be

$$P^\mu = (M, 0, 0, 0), \qquad \text{or } \mathbf{P} = 0,\ P^0 = M = \sqrt{\left(\sum_i p_{(i)}{}^\mu\right)\left(\sum_j p_{(j)}{}^\mu\right)},$$

is called the barycentric frame of reference.

The reader may have observed that in a system containing n particles, there are n independent momenta, so that there are $n - 1$ momenta to be specified once the total momentum P^μ has been determined. (Similarly, once we have specified the location of the center of mass in classical physics, we still have $n - 1$ coordinates to specify before the locations of all the particles can be determined.) If $n = 2$, there is a unique and eminently convenient way to specify the remaining momentum coordinate, namely, to let $p^\mu = p_{(2)}{}^\mu - p_{(1)}{}^\mu$. However, for $n > 2$, there are many apparently equally good ways to choose the momentum coordinates, and only some special features of the arrangement of particles or of the physics of their interactions can give us a clue as to which should be chosen. Quite generally, the study of three—and more—body problems is extremely difficult, and we shall restrict our considerations to two-body problems.

4.4 A KINEMATICAL APPLICATION: TWO-PARTICLE ELASTIC COLLISIONS

Now let us consider two particles colliding (or, more generally, interacting) with one another. Let one of them have mass m_1 and let the other have mass m_2; their momenta shall be $\mathbf{p}_1$ and $\mathbf{p}_2$, respectively. We can represent their collision by a diagram in which the two particles are represented by two lines, labeled with the masses and momenta of the particles, that converge toward a "blob" representing the collision or interaction process; two other lines emerge from the blob, labeled with the masses and momenta of the particles after the collision has taken place. Thus the interaction between particles is divided into three parts, namely, in chronological order,

1. the preinteraction part, when the particles each can be considered free particles with well-defined masses and momenta,

2. the interaction process, when the particles merge, or collide with one another, exchanging momenta and perhaps, for inelastic collisions, mass; and

3. the postinteraction part, when the interaction process has ceased and the system once more is resolved into free particles with definite masses and momenta. (See Fig. 4.1) We shall confine our discussion here to elastic collisions (in which the masses of the initial particles are the same as those of the final, or emerging, particles) between two particles. The former restriction can be removed without much difficulty, but the complexity of the problem increases enormously if we remove the restriction to two particles, as already has been indicated.

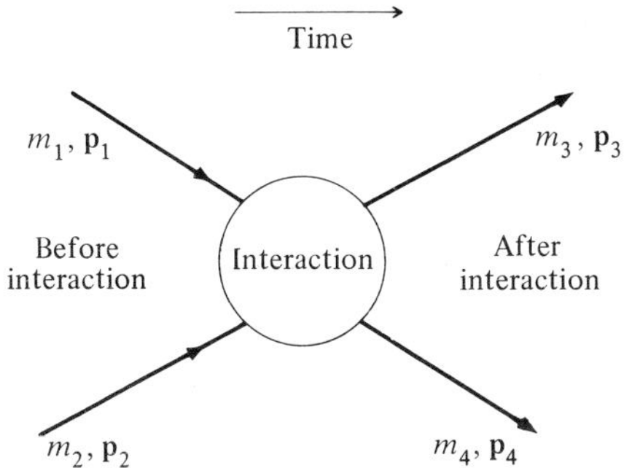

Fig. 4.1. Diagram depicting an interaction between particles 1 and 2, yielding particles 3 and 4.

The four-momenta of the initial particles are

$$p_1{}^\mu = (\sqrt{m_1^2c^2 + \mathbf{p}_1^2}, \mathbf{p}_1), \qquad p_2{}^\mu = (\sqrt{m_2^2c^2 + \mathbf{p}_2^2}, \mathbf{p}_2);$$

thus

$$p_i{}^0 = m_i\gamma_i c = \sqrt{m_i^2c^2 + \mathbf{p}_i^2}, \qquad i = 1, 2, \tag{4.46}$$

or

$$\gamma_i = \frac{p_i{}^0}{m_i c} = \sqrt{1 + \frac{\mathbf{p}_i^2}{m_i^2c^2}}, \qquad i = 1, 2. \tag{4.47}$$

The total four-momentum of the pair of particles is

$$P_1{}^\mu = p_1{}^\mu + p_2{}^\mu = (p_1{}^0 + p_2{}^0, \mathbf{p}_1 + \mathbf{p}_2) = (P^0, \mathbf{P}), \tag{4.48}$$

and the invariant mass M of the two-particle system is given by

$$\begin{aligned} M^2 = P^\mu P_\mu &= (P^0)^2 - \mathbf{P}^2 = (p_1{}^\mu + p_2{}^\mu)(p_{1\mu} + p_{2\mu}) \\ &= p_1{}^\mu p_1{}^\mu + p_2{}^\mu p_{2\mu} + 2p_1{}^\mu p_{2\mu} \\ &= m_1^2 + m_2^2 + 2(p_1{}^0 p_2{}^0 - \mathbf{p}_1 \cdot \mathbf{p}_2). \end{aligned} \tag{4.49}$$

In relativistic scattering theory, this quantity often is denoted s.

In order to specify completely the momenta involved in the initial state of our system, we must have two linearly independent momentum vectors. In addition to P^μ, we could use either $p_1{}^\mu$ or $p_2{}^\mu$, but such a choice would be rather unsymmetric and, in fact, inconvenient. An obvious choice for the second momentum is the quantity we mentioned earlier,

$$p^\mu \equiv p_2{}^\mu - p_1{}^\mu = (\sqrt{m_2^2c^2 + \mathbf{p}_2^2} - \sqrt{m_1^2c^2 + \mathbf{p}_1^2}, \mathbf{p}_2 - \mathbf{p}_1) \tag{4.50}$$

called the *relative four-momentum* for particles 1 and 2. The invariant associated with the relative four-momentum is

$$p^\mu p_\mu = (p_2{}^\mu - p_1{}^\mu)(p_2{}^\mu - p_1{}^\mu) = m_1^2 + m_2^2 - 2p_2{}^\mu p_{1\mu}. \tag{4.51}$$

In relativistic scattering theory, this invariant is often denoted by the letter t. However, s and t, or M^2 and $p^\mu p_\mu$ are related to one another, as one can see by adding Eqs. (4.49) and (4.51); thus

$$s + t = M^2 + p^\mu p_\mu = 2(m_1^2 + m_2^2), \tag{4.52}$$

or

$$t = p^\mu p_\mu = 2(m_1^2 + m_2^2) - M^2. \tag{4.53}$$

Now let us evaluate $p = |\mathbf{p}| = |\mathbf{p}_2 - \mathbf{p}_1|$, the magnitude of the relative momentum. In order to do this, we shall consider the collision as it appears to an observer in the barycentric frame. Denoting with primed symbols the variables as they are measured by an observer in this frame of reference, we have

$$\mathbf{P}' = \mathbf{p}_1' + \mathbf{p}_2' = 0, \mathbf{p}' = \mathbf{p}_2' - \mathbf{p}_1' = 2\mathbf{p}_2' = -2\mathbf{p}_1'. \tag{4.54}$$

Therefore,

$$M^2c^2 = (P'^0)^2 = (\sqrt{m_1^2c^2 + \mathbf{p}_1'^2} + \sqrt{m_2^2c^2 + \mathbf{p}_2'^2})^2$$

$$= m_1^2c^2 + m_2^2c^2 + 2\mathbf{p}_1'^2 + 2\sqrt{(m_1^2c^2 + \mathbf{p}_1'^2)(m_2^2c^2 + \mathbf{p}_1'^2)}$$

or

$$[M^2c^2 - m_1^2c^2 - m_2^2c^2 - 2\mathbf{p}_1'^2]^2 = 2[m_1^2m_2^2c^4 + 2(m_1^2 + m_2^2)c^2\mathbf{p}_1'^2 + (\mathbf{p}_1'^2)^2].$$

After some manipulation, we obtain

$$\mathbf{p}_1'^2 = \frac{1}{4M^2}\{[M^2 - (m_1 + m_2)^2][M^2 - (m_1 - m_2)^2]\}c^2;$$

that is, we get

$$|\mathbf{p}_1'| = |\mathbf{p}_2'| = \frac{1}{2M}\{[M^2 - (m_1 + m_2)^2][M^2 - (m_1 - m_2)^2]\}^{1/2}c \tag{4.55}$$

as the magnitude of the linear momentum of each of the particles, according to an observer in the barycentric frame of reference. Notice that this is an *invariant* quantity, depending *only* upon the invariants M, m_1 and m_2. According to Eq. (4.54) and (4.55), the magnitude of the *relative* linear momentum is

$$p = p' = |\mathbf{p}'| = \frac{1}{M}\{[M^2 - (m_1 + m_2)^2][M^2 - (m_1 - m_2)^2]\}^{1/2}c, \tag{4.56}$$

where, since this is an invariant quantity, it is not necessary to retain the prime on the symbol p.

We now have constructed, from the eight given independent kinematic quantities ($m_1, p_1{}^1, p_1{}^2, p_1{}^3; m_2, p_2{}^1, p_2{}^2, p_2{}^3$), a set of six independent kinematic variables, namely

$$M, P^1, P^2, P^3; m_1, m_2;$$

we do not include the invariant p in this list, because it is determined uniquely by M, m_1 and m_2. Two more "barycentric" variables are needed to specify the system completely. What have we left out in our list of six barycentric variables? We have, in fact, ignored the *direction* of the relative momentum.

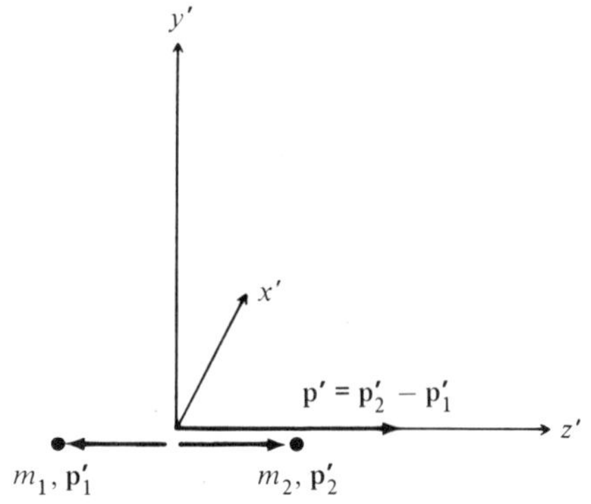

Fig. 4.2. Choice of coordinate axes in barycentric frame of reference.

It is convenient, and conventional, to let the z'-direction of a coordinate system erected in the barycentric frame be specified by the direction of the relative momentum in that frame: $\hat{\mathbf{z}}' = \mathbf{p}'/p$. Then the polar coordinates of $\mathbf{p}'$ in this coordinate system are taken to be

$$|\mathbf{p}'| = p, \qquad \theta' = 0, \qquad \varphi' = 0.$$

The remaining two coordinate axes of a Cartesian coordinate system then can be affixed to the z'-axis in a convenient way; thus we can establish a *coordinate system*, partially determined by $\mathbf{p}'$, in a *frame of reference* determined by $\mathbf{P}' = 0$. (See Fig. 4.2.) In other frames of reference also, we shall take the $\mathbf{p} = \mathbf{p}_2 - \mathbf{p}_1$ direction to be the direction of the z-axis, so that special Lorentz transformations along the direction of $\mathbf{p}$ will be special Lorentz transformations along the z-axis. The x- and y-axes then will be erected so that they are parallel to the x'- and y'-axes of the barycentric system, so that the polar coordinates of $\mathbf{p}$ in any coordinate system so arranged will be

$$|\mathbf{p}| = p, \qquad \theta = 0, \qquad \varphi = 0.$$

The eight barycentric kinematic coordinate values that are equivalent to the original eight variables m_i, $p_i^{\,1}$, $p_i^{\,2}$, $p_i^{\,3}$ $(i = 1, 2)$ are therefore

$$M, P^1, P^2, P^3; \qquad m_1, m_2; \qquad \theta = 0, \varphi = 0;$$

in the barycentric frame of reference, these become

$$M, 0, 0, 0; \qquad m_1, m_2; \qquad 0, 0,$$

since, in that frame, $\mathbf{P} = \mathbf{P}' = \mathbf{0}$.

The coordinate systems that we thus have established are particularly appropriate for the analysis of collision, or "scattering" problems. To proceed any further in the consideration of such problems, however, we must take into account the masses and momenta of the final particles—those emerging from the collision. Let the masses and momenta of the emergent particles be designated m_3, m_4, and $\mathbf{p}_3$, $\mathbf{p}_4$, respectively. We shall assume here that elastic scattering is taking place in the "blob", so that $m_1 = m_3$ and $m_2 = m_4$. The net effect of the interaction, then, is simply for the two particles to exchange some momentum, particle 1 emerging with a new momentum $\mathbf{p}_3$, and particle 2 emerging with momentum $\mathbf{p}_4$.

The first step in analysing the scattering is to apply the basic laws of physics, namely the conservation of momentum and of energy. In the relativistic formulation of mechanics, both these laws can be expressed as a single law, the conservation of four-momentum. Thus the total four-momentum, P^μ, must be constant throughout the collision process. Since

$$P^\mu = p_1{}^\mu + p_2{}^\mu$$

initially, and

$$P^\mu = p_3{}^\mu + p_4{}^\mu$$

finally, we have

$$p_1{}^\mu + p_2{}^\mu = p_3{}^\mu + p_4{}^\mu.$$

Evidently $M^2 = P^\mu P_\mu$ also is constant, and we have already specified that m_1 and m_2 are unchanged by the scattering. This means that if we define

$$p_f{}^\mu = p_4{}^\mu - p_3{}^\mu, \qquad \mathbf{p}_f = \mathbf{p}_4 - \mathbf{p}_3,$$

then, using $m_1 = m_3$, $m_2 = m_4$, we get

$$p_f = |\mathbf{p}_f| = \frac{1}{M}\{[M^2 - (m_3 + m_4)^2][M^2 - (m_3 - m_4)^2]\}^{1/2}c$$

$$= \frac{1}{M}\{[M^2 - (m_1 + m_2)^2][M^2 - (m_1 - m_2)^2]\}^{1/2}c = p.$$

Hence only the direction of the relative momentum is changed in elastic scattering.

The final relative momentum $\mathbf{p}_f$ does not necessarily lie along the z-axis, and the eight barycentric variables that describe the system in its final state are

$$M, P^1, P^2, P^3; \qquad m_1, m_2; \qquad \theta, \varphi,$$

where θ and φ are the angles giving the direction of $\mathbf{p}_f$ in terms of the coordinate system established with the z-axis along $\mathbf{p}$. Obviously θ is the angle between $\mathbf{p}$ and $\mathbf{p}_f$; the remaining angle depends upon the manner in which the x-axis is chosen.

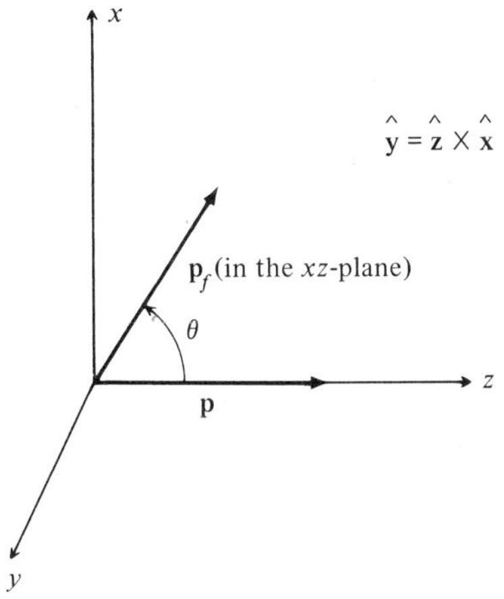

Fig. 4.3. The vectors $\mathbf{p}$ and $\mathbf{p}_f$ are used to define the xz-plane and its normal y if no asymmetry (such as spin) is present.

Unless one or both of the colliding particles has some asymmetry (such as an axis about which the particle is spinning) that makes a particular choice of coordinate orientation more convenient than any other, the usual arrangement is to let the plane of $\mathbf{p}$ and $\mathbf{p}_f$ define the xz-plane. Thus the unit vector in the x-direction is given by

$$\hat{\mathbf{x}} = \frac{\hat{\mathbf{p}}_f - (\hat{\mathbf{p}} \cdot \hat{\mathbf{p}}_f)\hat{\mathbf{p}}}{1 - (\hat{\mathbf{p}} \cdot \hat{\mathbf{p}}_f)^2}. \tag{4.57}$$

and the y-direction is taken to be

$$\hat{\mathbf{y}} = \hat{\mathbf{z}} \times \hat{\mathbf{x}}$$

so that the coordinate system will be right-handed. When this arrangement is used, the polar coordinates of $\mathbf{p}_f$ are (see Fig. 4.3)

$$p_f = p_1 \qquad \theta_f = \theta, \qquad \varphi_f = 0.$$

If one of the particles is spinning, say with spin angular momentum **s**, it may be convenient to choose the y-axis first, defining

$$\hat{\mathbf{y}} = \frac{\hat{\mathbf{s}} - (\hat{\mathbf{p}} \cdot \hat{\mathbf{s}})\hat{\mathbf{p}}}{1 - (\hat{\mathbf{p}} \cdot \hat{\mathbf{s}})^2} \,. \tag{4.58}$$

Then the x-axis is defined by the relation $\hat{\mathbf{x}} = \hat{\mathbf{y}} \times \hat{\mathbf{z}}$, so that again a right-handed coordinate system is obtained. In this coordinate system, both the angles θ_f and φ_f may be nonzero because the coordinate axes are defined without any reference to $\mathbf{p}_f$. A typical situation of this kind is illustrated in Fig. 4.4.

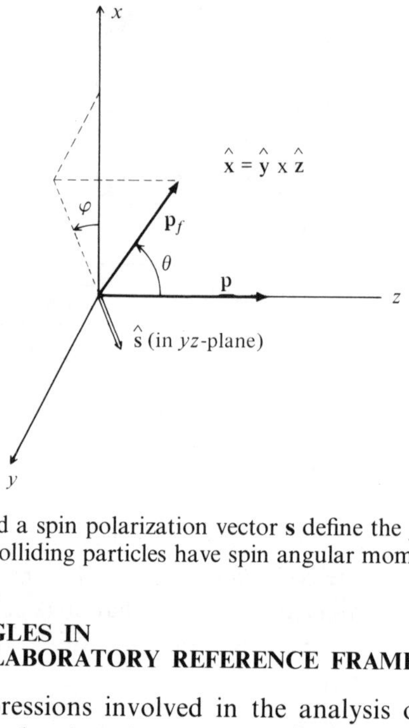

Fig. 4.4. The vector **p** and a spin polarization vector **s** define the yz-plane and its normal x when one or both of the colliding particles have spin angular momentum.

4.5 SCATTERING ANGLES IN BARYCENTRIC AND LABORATORY REFERENCE FRAMES

The mathematical expressions involved in the analysis of scattering problems generally have their simplest form in the barycentric frame of reference. However, useful collision experiments can rarely be performed in such a way that the observer is at rest in the barycentric frame. Usually a sample of particles that are at rest (except for thermal motions, which usually can be neglected) is placed in the line of fire of some other fast-moving particles (e.g. positive ions in a heavy-particle nuclear accelerator or electrons in a high-energy electron accelerator.) The experimentalist therefore usually must deal with measurements (e.g. angles, momenta, etc.) carried out in the *laboratory* frame of reference rather than in the barycentric frame. Since the theorist likes to pretend that he "lives" in the barycentric frame, so that his analyses will be as simple as possible both conceptually and computationally, we must find equations that relate measurements carried out in the laboratory frame to measurements made in the barycentric frame. These equations, which allow theorist and experimentalist to carry on meaningful dialogues, are obtained by performing a special Lorentz transformation.

In this discussion, unprimed variables will refer to quantities measured in the laboratory frame, and primed quantities will refer to the barycentric frame. In particular, we shall be interested in the angle, θ_{24}, between $\mathbf{p}_2$ and $\mathbf{p}_4$, and the corresponding angle in the barycentric system, θ, between $\mathbf{p}'$ and $\mathbf{p}'_f$. We shall restrict our attention to the elastic scattering of spinless particles, but these restrictions are removed rather easily; for example, inelastic scattering can be considered if we remove the requirement that the final masses be equal to the initial masses.

Let particle 1 be at rest in the laboratory frame, so that

$$p_1{}^\mu = (m_1 c, 0, 0, 0)$$

and let m_2 have a given momentum $\mathbf{p}_2 = |\mathbf{p}_2|\hat{\mathbf{z}}$ along the z-direction; then

$$p_2{}^\mu = (\sqrt{m_2^2c^2 + \mathbf{p}_2^2}, 0, 0, |\mathbf{p}_2|).$$

As before, the final momenta of particles 1 and 2 will be denoted $\mathbf{p}_3$ and $\mathbf{p}_4$, respectively. Conservation of 4-momentum then tells us that

$$p_1{}^\mu + p_2{}^\mu = p_3{}^\mu + p_4{}^\mu$$

or

$$m_1 c + \sqrt{m_2^2c^2 + \mathbf{p}_2^2} = \sqrt{m_1^2c^2 + \mathbf{p}_3^2} + \sqrt{m_2^2c^2 + \mathbf{p}_4^2} \tag{4.59}$$

and

$$\mathbf{p}_2 = |\mathbf{p}_2|\hat{\mathbf{z}} = \mathbf{p}_3 + \mathbf{p}_4. \tag{4.60}$$

Solution of Eq. (4.60) for $\mathbf{p}_3$, and substitution of the result into (4.59) yields

$$m_1 c + \sqrt{m_1^2c^2 + \mathbf{p}_2^2} = \sqrt{m_1^2c^2 + \mathbf{p}_2^2 + \mathbf{p}_4^2 - 2\mathbf{p}_2\cdot\mathbf{p}_4} + \sqrt{m_2^2c^2 + \mathbf{p}_4^2}. \tag{4.61}$$

This can be solved for $\cos\theta_{24} = (\mathbf{p}_2\cdot\mathbf{p}_4)/|\mathbf{p}_2|\,|\mathbf{p}_4|$; the result is

$$\cos\theta_{24} = \frac{\sqrt{m_2^2c^2 + \mathbf{p}_2^2}\sqrt{m_2^2c^2 + \mathbf{p}_4^2} - m_1 c(\sqrt{m_2^2c^2 + \mathbf{p}_4^2} - \sqrt{m_2^2c^2 + \mathbf{p}_2^2}) - m_2^2c^2}{|\mathbf{p}_2|\,|\mathbf{p}_4|}. \tag{4.62}$$

Therefore the scattering angle is determined if $\mathbf{p}_2$ is given and $|\mathbf{p}_4|$ is measured. Because of the conservation of linear momentum (Eq. 4.60), a measurement of $|\mathbf{p}_4|$ provides full knowledge of $\mathbf{p}_3$ also. Therefore, a measurement of $|\mathbf{p}_4|$ alone, or equivalently, of $p_4{}^0$, determines all the kinematic scattering parameters in elastic

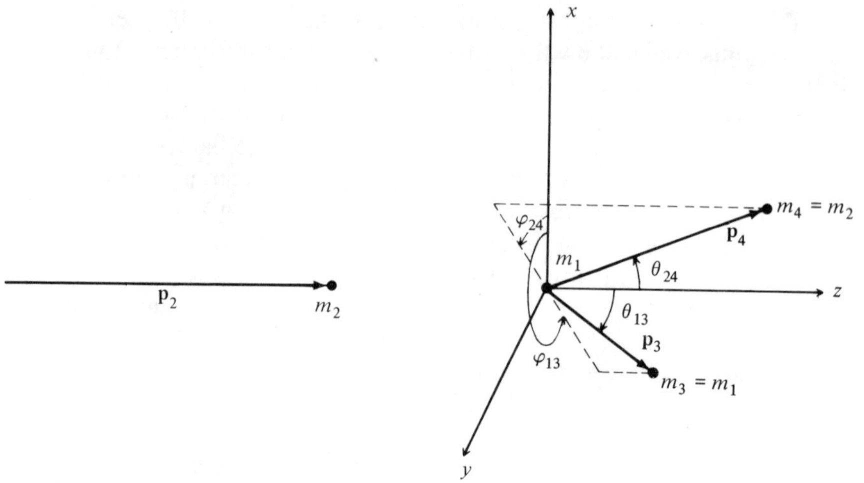

Fig. 4.5. Collision of m_2, m_1 as seen in the laboratory frame. Initially mass m_2 has momentum $\mathbf{p}_2$ and m_1 is at rest. The coordinate axes are attached to m_1 when it is at rest, and after the collision mass m_1 has momentum $\mathbf{p}_3$ with polar coordinates $|\mathbf{p}_3|, \theta_{13}, \varphi_{13}$ while mass m_2 has momentum $\mathbf{p}_4$ with polar coordinates $|\mathbf{p}_4|, \theta_{24}, \varphi_{24}$. Note that $\varphi_{24} = \varphi_{13} - \pi$; for spinless particles, $\varphi_{24} = 0, \varphi_{13} = \pi$. (a) Lab frame. $\mathbf{p}_3, \mathbf{p}_4$ in the xz-plane; $p_3^2 = p_4^2 = 0$. (b) Barycentric frame. $\mathbf{p}_3', \mathbf{p}_4'$ in the xz-plane; $p_3'^2 p_4'^2 = 0$.

collisions, except the (usually irrelevant) azimuthal angle $\varphi_{24} = \varphi_{13} - \pi$ (see Fig. 4.5). Equation (4.61) could be solved for $|\mathbf{p}_4|$ in terms of $\cos\theta_{24}$, so that a measurement of θ_{24} likewise would determine all kinematic scattering parameters for elastic collisions, except φ_{24}; in fact, the experimenter usually finds it easier to measure θ_{24} than to measure $|\mathbf{p}_4|$. The solution for $|\mathbf{p}_4|$ has a rather complicated appearance:

$$|\mathbf{p}_4| = |\mathbf{p}_2| (\cos\theta_{24}) \frac{m_2^2}{m_1^2} \frac{\left(1 + \dfrac{m_1 E_2}{m_2^2 c^2}\right)}{\left(1 + \dfrac{E_2}{m_1 c^2}\right)\left(1 - \dfrac{\mathbf{p}_2^2 c^2 \cos^2\theta_{24}}{(m_1 c^2 + E_2)^2}\right)} \cdot \left\{ 1 \pm \left[1 + \frac{\dfrac{m_1^2}{m_2^2}\left(1 - \dfrac{m_1^2}{m_2^2}\right)\left(1 + \dfrac{E_2}{m_1 c^2}\right)\left(1 - \dfrac{\mathbf{p}_2^2 c^2 \cos^2\theta_{24}}{(m_1 c^2 + E_2)^2}\right)}{\left(1 + \dfrac{m_1 E_2}{m_2^2 c^2}\cos^2\theta_{24}\right)} \right]^{1/2} \right\}, \tag{4.63}$$

where $E_2 \equiv p_2{}^0 c = c\sqrt{m_2^2 c^2 + \mathbf{p}_2^2}$ has been used for convenience in writing; the choice of signs in Eq. (4.63) is made on the basis of the requirement that $|\mathbf{p}_4|$ must have a physically meaningful value. The main point here is that $\theta_{24}, |\mathbf{p}_4|, |\mathbf{p}_3|$ and θ_{13} are not independent in elastic scattering; a measurement of one of them immediately provides a knowledge of the other three. However, in practice it is usually necessary that both $|\mathbf{p}_4|$ and θ_{24} be measured in order to insure that the scattering process was an elastic one; these measurements are equivalent to the measurement of m_4 and $|\mathbf{p}_4|$.

Ultimately, we wish to find the relationship between the laboratory variable θ_{24} and the barycentric variable θ. Therefore we seek the special Lorentz transformation that relates S, the laboratory frame, to S', the barycentric frame.

Let $\boldsymbol{\beta}$ and γ be the special Lorentz transformation parameters for $S \to S'$; then, since particle 1 is at rest in S, it will have four-momentum

$$p_1'^{\mu} = (m_1\gamma c, -m_1\gamma\boldsymbol{\beta}c) \tag{4.64}$$

in S', and particle 2 will have four-momentum

$$p_2'^{\mu} = (\gamma[p_2{}^0 - \boldsymbol{\beta}\cdot\mathbf{p}_2], \gamma[\mathbf{p}_2 - \boldsymbol{\beta}p_2{}^0]). \tag{4.65}$$

The reader should obtain these results for himself by carrying out the transformation

$$p_i'^{\mu} = a^{\mu}{}_{\nu}p_i{}^{\nu},$$

where $a^{\mu}{}_{\nu}$ are the transformation coefficients with parameters $\boldsymbol{\beta}$ and γ.

Recalling that $\mathbf{P}' = \mathbf{0}$ for the barycentric frame, we thus obtain

$$\mathbf{P}' = \mathbf{p}_1' + \mathbf{p}_2' = \gamma[\mathbf{p}_2 - \boldsymbol{\beta}p_2{}^0 - \boldsymbol{\beta}m_1c] = \mathbf{0}$$

or

$$\boldsymbol{\beta} = \frac{\mathbf{p}_2}{p_2{}^0 + m_1c} \tag{4.66}$$

and

$$\gamma = \frac{1}{\sqrt{1-\boldsymbol{\beta}^2}} = \frac{m_1c + p_2{}^0}{\sqrt{m_1^2c^2 + m_2^2c^2 + 2m_1cp_2^0}}. \tag{4.67}$$

These results will be needed below.

The angle θ_{24} in the laboratory frame is such that (see Fig. 4.6)

$$\tan\theta_{24} = \frac{p_4{}^1}{p_4{}^3}, \tag{4.68}$$

when the x-direction is as defined by Eq. (4.57). On the other hand,

$$\tan\theta = \frac{p_4'^1}{p_4'^3}. \tag{4.69}$$

In order to relate these angles, we must apply the special Lorentz transformation with parameters $-\boldsymbol{\beta}$ and γ, where $\boldsymbol{\beta}$ and γ are given in Eqs. (4.66) and (4.67), to $p_4'^{\mu}$; thus

$$p_4{}^0 = \gamma[p_4'^0 + \boldsymbol{\beta}\cdot\mathbf{p}_4'], \tag{4.70}$$

$$\mathbf{p}_4 = \mathbf{p}_4' + \boldsymbol{\beta}[(\gamma-1)\boldsymbol{\beta}\cdot\mathbf{p}_4'/\beta^2 + \gamma p_4'^0]. \tag{4.71}$$

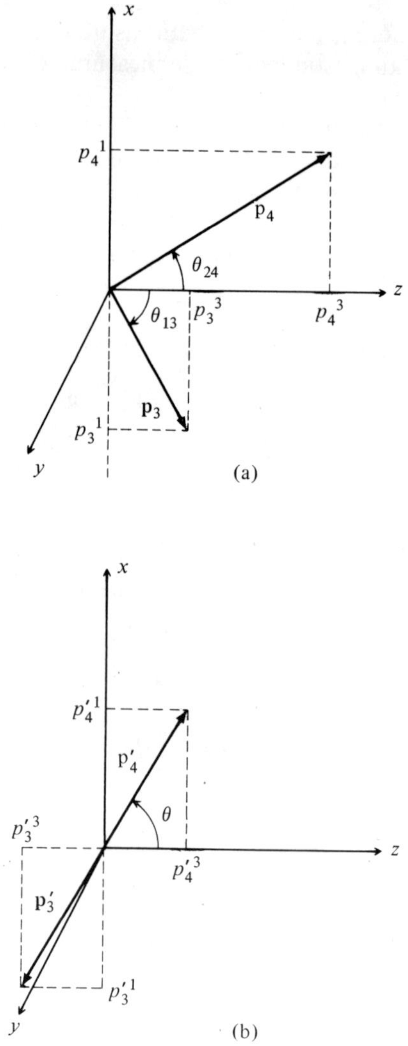

Fig. 4.6. Analysis of the components of (a) $\mathbf{p}_4$, $\mathbf{p}_3$ in the laboratory frame and (b) $\mathbf{p}_3'$, $\mathbf{p}_4'$ in the barycentric frame.

The spatial components can be written in the form

$$\begin{aligned} p_4{}^1 &= p_4'^1, \qquad p_4{}^2 = p_4'^2 \\ p_4{}^3 &= \gamma[p_4'^3 + \beta p_4'^0], \end{aligned} \tag{4.72}$$

where $\beta = |\boldsymbol{\beta}|$. Since $p_4'^1 = |\mathbf{p}_4'| \sin\theta$, and $p_4'^3 = |\mathbf{p}_4'| \cos\theta$, substitution of Eqs.

(4.72) into Eq. (4.68) yields

$$\tan\theta_{24} = \frac{p_4'^1}{\gamma[p_4'^3 + \beta p_4'^0]} = \frac{|\mathbf{p}_4'|\sin\theta}{\gamma[|\mathbf{p}_4'|\cos\theta + \beta\sqrt{m_2^2c^2 + (\mathbf{p}_4')^2}]} \tag{4.73}$$

with γ and $\beta = |\boldsymbol{\beta}|$ given by Eqs. (4.66) and (4.67).

It is a simple matter to verify that Eq. (4.73) reduces, in the nonrelativistic approximation, to

$$\tan\theta_{24} \simeq \frac{\sin\theta}{(m_2/m_1) + \cos\theta}. \tag{4.74}$$

READING NOTES*

Reference 10, Paper IV.

Reference 3, Chapters 11, 12.

Reference 7, Chapter 2.

Reference 11, Chapters 3, 4.

Reference 8, Chapter 17.

Reference 12, Parts II, III (a).

PROBLEMS

1. Show that the relationships among transformation coefficients given in Table 4.1 are consistent with one another.
2. Show that the set of all special Lorentz transformations, with multiplication here defined as matrix multiplication, forms a group that is a subgroup of the group of general Lorentz transformations.
3. Show that the set of spatial rotations is a subgroup of the general Lorentz group.
4. Consider the Lorentz-covariant definitions of force and acceleration. Show how they transform under Lorentz transformations, paying special attention to the spatial components. Show clearly the difference between the effects of Lorentz transformations and Galilean transformations on force and acceleration. Show that the Lorentz-transformed quantities have appropriate limits as $\boldsymbol{\beta} \to \mathbf{0}$.
5. Show that for a system consisting of two free particles with masses m_1 and m_2, the total momentum $P^\mu = p_1{}^\mu + p_2{}^\mu$ does *not* satisfy $P^\mu P_\mu = (m_1 + m_2)^2c^2$.
6. Derive Eq. (4.62).
7. Derive Eqs. (4.63).
8. Derive Eqs. (4.64) and (4.65).
9. Derive Eq. (4.67).
10. Derive Eq. (4.74).

---

* Works referred to in these notes will be found in the Bibliography on p. 577.

11. Show that Eq. (4.18) is consistent with the formula for the velocity V given by (4.19) (the relativistic formula for the "addition" of parallel velocities).

12. A π^+-meson, having a (rest) mass of 139 MeV (139×10^6 eV) has a total energy of 5 GeV (5×10^9 eV) in the laboratory frame of reference. *Assuming* that this meson can be thought of properly as a highly bound state of a proton (mass = 938 MeV) and an antineutron (mass = 940 MeV), state whether one might expect the 5 GeV meson to decay into a free proton and a free antineutron in the laboratory frame. Justify your answer. (The masses given here actually are the rest energies, i.e. the masses times c^2.)

13. Show that for two free particles the four-fectors $P^\mu = p_1{}^\mu + p_2{}^\mu$ and $p^\mu = p_2{}^\mu - p_1{}^\mu$ are linearly independent.

14. Let the amplitude of a light wave of frequency ν and wavelength λ be represented by $A \cos(\omega t - \mathbf{k} \cdot \mathbf{x})$, where $\omega = 2\pi\nu$ and $\mathbf{k} = (2\pi/\lambda)\hat{\mathbf{k}}$, $\hat{\mathbf{k}}$ being the direction of propagation of the wave. Since, according to special relativity, light travels with the same speed in all inertial frames of reference, $\omega t - \mathbf{k} \cdot \mathbf{x}$ must be an invariant under Lorentz transformations; $\omega t - \mathbf{k} \cdot \mathbf{x} = \omega' t' - \mathbf{k}' \cdot \mathbf{x}'$ where the primed quantities are in the S'-frame moving with velocity $\mathbf{v}$ (along the y-axis, say) relative to S. This suggests that we define $k^\mu = (k^0, \mathbf{k}) = (\omega/c, \mathbf{k})$ so that $k^\mu x_\mu = \omega t - \mathbf{k} \cdot \mathbf{x}$.

a) From the fact that $k^\mu x_\mu$ is an invariant, show that k^μ is a true four-vector.
b) Find the value of $k^\mu k_\mu$, which clearly is invariant.
c) Let ω and $\mathbf{k}$ be measured in S and find ω' and $\mathbf{k}'$, as measured in S', in terms of ω, $\mathbf{k}$ and $\mathbf{v}$.
d) Let $\hat{\mathbf{k}}$ be along the direction of $\mathbf{v}$ and write down the values of the components of k'^μ in terms of the components of k^μ and $\mathbf{v}$. These equations show the relativistic form of the longitudinal Doppler effect, "longitudinal" referring to the fact that the wave is being propagated in the same direction as $\mathbf{v}$.
e) Let $\hat{\mathbf{k}}$ be perpendicular to $\mathbf{v}$ and write down the values of the components of k'^μ as measured in S'. This gives the relativistic transverse Doppler effect. Does it have an analog in classical acoustics? If so, describe the classical effect.
f) Let $v/c = \sqrt{3/2}$ and find the angular frequency ω' in S' when
   i) $\hat{\mathbf{k}}$ is parallel to $\mathbf{v}$ and
   ii) $\hat{\mathbf{k}}$ is perpendicular to $\mathbf{v}$.

15. In a given frame of reference, a particle of mass m is moving with velocity $\mathbf{v} = dx/dt$ and experiences (three-vector) acceleration $\mathbf{a} = d^2\mathbf{x}/dt^2$. The angle between $\mathbf{v}$ and $\mathbf{a}$ is designated ϕ. An observer in this frame (S) notices that the force $\mathbf{F} = d\mathbf{p}/dt$ causing the acceleration is not parallel to $\mathbf{a}$ but is inclined at an angle θ relative to it. Find either $\cos\theta$ in terms of $\cos\phi$ and $\mathbf{v}$ or $\tan\theta$ in terms of $\tan\phi$ and $\mathbf{v}$. See Fig. 4.7.

16. Let $\mathbf{a}' = d\mathbf{v}'/dt$ be the (three-vector) acceleration experienced by a particle in its rest frame S', and let $\mathbf{a} = d\mathbf{v}/dt$ be its acceleration in frame S relative to which it moves with velocity $\mathbf{v}$. Show that

$$(\mathbf{a}')^2 = \left[\mathbf{a}^2 - \left(\frac{\mathbf{a} \times \mathbf{v}}{c}\right)^2\right]\gamma_v{}^6.$$

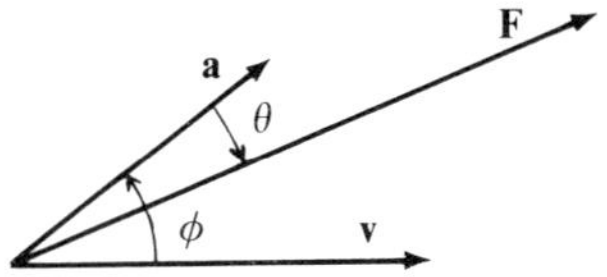

Fig. 4.7. Relationships among vectors for Problem 15.

17. The "late 1950's generation" of high-energy accelerators was intended to produce protons with at least sufficient energy that one of them impinging on a hydrogen nucleus would produce a proton–antiproton pair:

$$p + p \to p + p + (p + \bar{p}),$$

where $\bar{p}$ represents an antiproton, which has the same mass as a proton. What is the minimum kinetic energy required for this process to occur

a) in the barycentric frame and
b) in the laboratory frame?

18. In the discussion related to Eq. (4.17), we considered the $a^\mu{}_\nu$ to define a relationship between two vectors in a single frame. If, instead, we think of the $a^\mu{}_\nu$ as defining a relationship between a vector ξ'^μ in frame S' and ξ^μ in some other predetermined frame S, then when we transform from S' to a still different frame S'', we shall have

$$\xi''^\mu = b^\mu{}_\nu \xi'^\nu = b^\mu{}_\nu a^\nu{}_\lambda \xi^\lambda,$$

where $b^\mu{}_\nu$ are the coefficients for the transformation from S' to S''. Thus the relationship between a vector in S'' and one in the predetermined frame S no longer is given by the $a^\mu{}_\nu$. What rank tensor—if it is a tensor—is $a^\mu{}_\nu$ when thought of as a set of transformation coefficients between the predetermined frame S and any other frame? Discuss the genesis of the difference between this result and the one described by Eq. (4.17).

CHAPTER 5

THE FOUR-DIMENSIONAL FORMULATION OF ELECTROMAGNETIC THEORY

We began our discussion of special relativity by demonstrating that the demands of Maxwell's theory of electromagnetism were not consistent with the hypotheses of Newtonian mechanics. Einstein (with Minkowski's help) has shown us how to reformulate physics, in terms of a rather elegant four-dimensional mathematical framework, such that the two classical theories could be welded together into a unified structure. In the last chapter we illustrated how the four-dimensional formulation can be applied to mechanics, using as a special example the kinematics of elastic collisions. It is appropriate that now we should return to Maxwell's theory and work it into the same four-dimensional framework. That this reformulation of electromagnetism will yield rather elegant and apparently simple equations should be a gratifying reward for our efforts.

5.1 THE FOUR-VECTOR CURRENT DENSITY AND FOUR-VECTOR POTENTIAL

In our original expression for the current density vector we had

$$\mathbf{j} = \rho \mathbf{v}$$

where ρ is the charge density for a distribution of charge moving with velocity $\mathbf{v}$ relative to an observer. This suggests that the four-vector representation for the current density might be

$$j^\mu = \rho u^\mu = \rho \frac{dx^\mu}{d\tau}, \tag{5.1}$$

where ρ is the invariant charge density, i.e. the charge density as seen by an observer at rest relative to the charges involved. The components of this *current density four-vector* are given by

$$j^\mu = (\rho \gamma_v c, \rho \gamma_v \mathbf{v}), \tag{5.2}$$

which is obtained by letting $d\tau = (1/\gamma_v)\, dt$ and using $\mathbf{v} = d\mathbf{x}/dt$. This is an intuitively acceptable result because if we let

$$\rho = \lim_{\Delta x, \Delta y, \Delta z \to 0} \left(\frac{\Delta Q}{\Delta x\, \Delta y\, \Delta z} \right) = \lim_{\Delta V \to 0} \frac{\Delta Q}{\Delta V}$$

in the rest frame of the charge element ΔQ, then in a frame moving with velocity $\mathbf{v}$ relative to the charge element, the volume element is contracted so that we have

$$\Delta V' = \frac{1}{\gamma_v}\Delta V$$

or, in this new frame, the charge density arising from the charge element ΔQ appears to be

$$\rho' = \lim_{\Delta V' \to 0}\left(\frac{\Delta Q}{\Delta V'}\right) = \gamma_v \lim_{\Delta V \to 0}\left(\frac{\Delta Q}{\Delta V}\right) = \gamma_v \rho$$

Thus the current density in S', the frame moving with velocity $\mathbf{v}$ relative to the charges, is $\rho'\mathbf{v} = \rho\gamma_v\mathbf{v}$, which is exactly what the spatial part of j^μ gives.

This discussion immediately yields an interpretation for the temporal component of j^μ. Evidently $j^0 = \rho\gamma_v c$ is just c times the charge density as seen by an observer relative to whom the charges are moving with velocity $\mathbf{v}$; that is j^0 is c times the charge density of the charges having current density $\mathbf{j} = \rho\gamma_v\mathbf{v}$. In the limit $(v/c) \ll 1$, we have $\gamma_v \simeq 1$, and

$$j^0 \simeq \rho c, \qquad \mathbf{j} \simeq \rho\mathbf{v}$$

as the nonrelativistic approximations for the components of the current density four-vector.

A particularly pleasant result of this formulation of a four-vector current density is that the equation of continuity takes a very simple form, namely

$$\partial_\mu j^\mu = \frac{\partial}{\partial x^\mu} j^\mu = 0. \tag{5.3}$$

If we had not used four-vector language, this equation would be translated into something like

$$\frac{1}{c}\frac{\partial}{\partial t}(\rho\gamma_v c) + \nabla\cdot(\rho\gamma_v\mathbf{v}) = 0.$$

The definitions of the vector potential $\mathbf{A}$ and the scalar potential ϕ, namely

$$\mathbf{A} = \frac{\mu_0}{4\pi}\int \frac{\rho(\mathbf{x}')\mathbf{v}(\mathbf{x}')}{|\mathbf{x} - \mathbf{x}'|}\, d^3x',$$

$$\phi = \frac{1}{4\pi\epsilon_0}\int \frac{\rho(\mathbf{x}')}{|\mathbf{x} - \mathbf{x}'|}\, d^3x' = c^2\frac{\mu_0}{4\pi}\int \frac{\rho(\mathbf{x}')}{|\mathbf{x} - \mathbf{x}'|}\, d^3x',$$

suggest that we define a four-vector potential, A^μ, in terms of the current density j^μ. To generalize these formulas for $\mathbf{A}$ and ϕ directly to a four-vector potential, however, would demand that we handle the denominators and the integrals in a covariant (relativistic) manner. Instead, we simply note that in such a definition the temporal component would be expected to behave like ϕ/c, since ϕ is equal to c times the constant $\mu_0/4\pi$ times an integral over $\rho(\mathbf{x}')c$, whereas the spatial component would be expected to behave like $\mathbf{A}$, since $\mathbf{A}$ is equal to the same constant, $\mu_0/4\pi$, times a similar integral over $\rho(\mathbf{x}')\mathbf{v}(\mathbf{x}')$. To check this, we recall Eqs. (1.31) and (1.33):

$$\left[\nabla^2 - \frac{1}{c^2}\frac{\partial^2}{\partial t^2}\right]\phi = -\frac{\rho}{\epsilon_0}$$

$$\left[\nabla^2 - \frac{1}{c^2}\frac{\partial^2}{\partial t^2}\right]\mathbf{A} = -\mu_0\mathbf{j},$$

and note that the relativistic interpretation of these equations would require us to replace ρ by $\rho\gamma_v$ and to think of $\mathbf{j}$ as being equal to $\rho\gamma_v\mathbf{v}$ instead of just $\rho\mathbf{v}$. Since the d'Alembertian

$$\nabla^2 - \frac{1}{c^2}\frac{\partial^2}{\partial t^2}$$

takes on the four-dimensional form

$$-\partial^\mu\partial_\mu = \left[\nabla^2 - \frac{1}{c^2}\frac{\partial^2}{\partial t^2}\right] \tag{5.4}$$

we then have

$$\partial^\mu\partial_\mu\phi = \frac{\gamma_v\rho}{\epsilon_0} = \frac{j^0}{c\epsilon_0} = c\mu_0 j^0$$

$$\partial^\mu\partial_\mu\mathbf{A} = \mu_0\mathbf{j} = \mu_0\rho\gamma_v\mathbf{v},$$

or, if we let

$$A^\nu = \left(\frac{1}{c}\phi, \mathbf{A}\right) \tag{5.5}$$

we have the relativistically covariant equation

$$\partial^\mu\partial_\mu A^\nu = \mu_0 j^\nu. \tag{5.6}$$

Note that both sides of this equation transform like four-vectors under Lorentz transformations.

Since we know that Eq. (5.6) (which is the inhomogeneous wave equation for the four-vector potential) is relativistically covariant, its solutions will provide a relativistically covariant definition for A^ν in terms of the current density j^ν, just as Eqs. (1.25) and (1.26) (which give solutions to Eqs. (1.31) and (1.33)) gave pre-relativistic definitions for ϕ and $\mathbf{A}$.

5.2 THE ELECTROMAGNETIC FIELD TENSOR

Previously we found that the fields **E** and **B** were given in terms of **A** and ϕ by the following equations:

$$\mathbf{E} = -\nabla\phi - \frac{\partial \mathbf{A}}{\partial t}, \qquad \mathbf{B} = \nabla \times \mathbf{A}; \tag{5.7}$$

thus

$$E_x = -\frac{\partial}{\partial x^1}cA^0 - \frac{\partial}{\partial x^0}cA^1 = \frac{\partial}{\partial x_1}cA^0 - \frac{\partial}{\partial x_0}cA^1,$$

$$E_y = -\frac{\partial}{\partial x^2}cA^0 - \frac{\partial}{\partial x^0}cA^2 = \frac{\partial}{\partial x_2}cA^0 - \frac{\partial}{\partial x_0}cA^2,$$

$$E_z = -\frac{\partial}{\partial x^3}cA^0 - \frac{\partial}{\partial x^0}cA^3 = \frac{\partial}{\partial x_3}cA^0 - \frac{\partial}{\partial x_0}cA^3,$$

$$B_x = \frac{\partial}{\partial x^2}A^3 - \frac{\partial}{\partial x^3}A^2 = -\left(\frac{\partial}{\partial x_2}A^3 - \frac{\partial}{\partial x_3}A^2\right),$$

$$B_y = \frac{\partial}{\partial x^3}A^1 - \frac{\partial}{\partial x^1}A^3 = -\left(\frac{\partial}{\partial x_3}A^1 - \frac{\partial}{\partial x_1}A^3\right),$$

$$B_z = \frac{\partial}{\partial x^1}A^2 - \frac{\partial}{\partial x^2}A^1 = -\left(\frac{\partial}{\partial x_1}A^2 - \frac{\partial}{\partial x_2}A^1\right).$$

This suggests that we define a new quantity,

$$F^{\mu\nu} \equiv \frac{\partial}{\partial \mathbf{x}_\mu}A^\nu - \frac{\partial}{\partial x_\nu}A^\mu = \partial^\mu A^\nu - \partial^\nu A^\mu = -F^{\nu\mu}, \tag{5.8}$$

so that

$$\begin{aligned} F^{10} &= \frac{1}{c}E_x, & F^{20} &= \frac{1}{c}E_y, & F^{30} &= \frac{1}{c}E_z \\ F^{12} &= -B_z, & F^{23} &= -B_x, & F^{31} &= B_y \end{aligned} \tag{5.9}$$

The quantities $F^{\mu\nu}$ can be represented in the form of an antisymmetric matrix,

$$[F^{\mu\nu}] = \begin{bmatrix} 0 & -\frac{1}{c}E_x & -\frac{1}{c}E_y & -\frac{1}{c}E_z \\ \frac{1}{c}E_x & 0 & -B_z & B_y \\ \frac{1}{c}E_y & B_z & 0 & -B_x \\ \frac{1}{c}E_z & -B_y & B_x & 0 \end{bmatrix}, \tag{5.10}$$

and they transform as the components of a *tensor of rank two* under Lorentz transformations; thus

$$F'^{\mu\nu} = a^{\mu}{}_{\kappa} a^{\mu}{}_{\lambda} F^{\kappa\lambda}, \tag{5.11}$$

where the $F'^{\mu\nu}$ are components of the electromagnetic field tensor as seen in a frame S'. $F^{\mu\nu}$ are the corresponding components in frame S, and the $a^{\mu}{}_{\nu}$ are the Lorentz transformation coefficients that carry tensors in S to tensors in S'.

Evidently **E** and **B** cannot be generalized directly as four-vectors. Rather, the components of **E** and **B** actually are components of an antisymmetric tensor of rank two. This is a rather surprising and very important result; it should serve as a caution against *assuming* that a classical vector always is at least proportional to the spatial part of the four-vector. In fact **E** and **B** *do not* transform as would the spatial parts of four-vectors under Lorentz transformations.

Since **E** and **B** both are accommodated in the tensor $F^{\mu\nu}$, we might guess that Maxwell's equations take rather simple forms in terms of $F^{\mu\nu}$. This guess is correct provided we introduce a new symbol, whose meaning we have utilized before but whose explicit form we have not yet considered. Let us first discuss the three-dimensional analog of the symbol we are interested in. The completely antisymmetric tensor (under pure three-dimensional rotations) of rank three is defined by

$$\epsilon_{ijk} \equiv \begin{cases} 0 & \text{unless } (i, j, k) \text{ is a permutation of } (1, 2, 3) \\ +1 & \text{if } (i, j, k) \text{ is an even permutation of } (1, 2, 3) \\ -1 & \text{if } (i, j, k) \text{ is an odd permutation of } (1, 2, 3). \end{cases}$$

Thus

$$\epsilon_{123} = \epsilon_{231} = \epsilon_{312} = +1$$
$$\epsilon_{321} = \epsilon_{213} = \epsilon_{132} = -1$$

and all the other ϵ_{ijk}'s vanish.

An important application of this tensor in three-dimensional problems is the formation of the outer products, or cross products, in vector analysis. The ith component of the vector $\mathbf{U} \times \mathbf{V}$ is given by

$$(\mathbf{U} \times \mathbf{V})_i = \epsilon_{ijk} U^j V^k \qquad \left(= \sum_{j,k} \epsilon_{ijk} U^j V^k \right);$$

that is

$$(\mathbf{U} \times \mathbf{V})_1 = U^2 V^3 - U^3 V^2,$$
$$(\mathbf{U} \times \mathbf{V})_2 = U^3 V^1 - U^1 V^3,$$
$$(\mathbf{U} \times \mathbf{V})_3 = U^1 V^2 - U^2 V^1.$$

Since the metric tensor in Euclidean three-dimensional space is equal to the Kronecker tensor, $g_{ij} = \delta_{ij}$ (that is, each diagonal element of the tensor is unity and all off-diagonal elements are zero), contravariant and covariant vectors are indistinguishable: $U_i = U^i$. This gives us the familiar results $(\mathbf{U} \times \mathbf{V})_1 = U_2 V_3 - U_3 V_2$, etc.

It can be verified easily that ϵ_{ijk} is indeed a tensor of rank three under pure rotations, and in applications such as the one given above, it converts tensors of rank two (e.g. $U^j V^k$) into *pseudovectors* or *axial* vectors. A pseudovector, or axial vector, transforms just like an ordinary vector under rotations, but if an odd number of coordinates are inverted (e.g. $x \to -x$, $y \to -y$, $z \to -z$), an ordinary vector changes sign ($\mathbf{V} \to -\mathbf{V}$) whereas an axial vector does not change sign ($\mathbf{U} \times \mathbf{V} \to \mathbf{U} \times \mathbf{V}$).

The four-dimensional quantity that we are interested in and that corresponds to the three-dimensional ϵ_{ijk} is the *completely antisymmetric covariant tensor of rank four* defined by

$$\epsilon_{\mu\nu\rho\sigma} \equiv \begin{cases} 0 & \text{unless } (\mu, \nu, \rho, \sigma) \text{ is a permutation of } (0, 1, 2, 3) \\ +1 & \text{if } (\mu, \nu, \rho, \sigma) \text{ is an even permutation of } (0, 1, 2, 3) \\ -1 & \text{if } (\mu, \nu, \rho, \sigma) \text{ is an odd permutation of } (0, 1, 2, 3) \end{cases} \tag{5.12}$$

Clearly $\epsilon_{\mu\nu\rho\sigma} = -\epsilon_{\mu\rho\nu\sigma}$; in general, a change of sign occurs each time a pair of adjacent indices is interchanged. This tensor has $4^4 = 256$ elements, so we shall not write down all of them. The nonzero elements are

$$\begin{aligned} \epsilon_{1230} &= \epsilon_{1023} = \epsilon_{1302} = \epsilon_{2310} = \epsilon_{2031} = \epsilon_{2103} = \epsilon_{3201} \\ &= \epsilon_{3120} = \epsilon_{3012} = \epsilon_{0321} = \epsilon_{0132} = \epsilon_{0213} = -1, \\ \epsilon_{2301} &= \epsilon_{2130} = \epsilon_{2013} = \epsilon_{3210} = \epsilon_{3021} = \epsilon_{3102} = \epsilon_{0312} \\ &= \epsilon_{0231} = \epsilon_{0123} = \epsilon_{1032} = \epsilon_{1203} = \epsilon_{1320} = 1. \end{aligned}$$

Now we may return to discussing Maxwell's equations. One of them involves $\nabla \cdot \mathbf{E}$, and because of the manner in which $F^{\mu\nu}$ is constructed, this suggests that we calculate the four-divergence of $F^{\mu\nu}$, that is, calculate $\partial_\mu F^{\mu\nu}$, as a first attempt at obtaining the desired equations. Thus

$$\partial_\mu F^{\mu\nu} = \partial_\mu(\partial^\mu A^\nu - \partial^\nu A^\mu) = \partial_\mu \partial^\mu A^\nu - \partial^\nu(\partial_\mu A^\mu).$$

Recall that observable electromagnetic fields are unchanged by gauge transformations of the second kind (see the discussion following Eq. 1.28). Let us consider adding to A^μ another four-vector $\alpha^\mu(x^\nu) = \alpha^\mu$:

$$A^\mu \to A''^\mu = A^\mu + \alpha^\mu;$$

then we get

$$\begin{aligned} F''^{\mu\nu} &= \partial^\mu A''^\nu - \partial^\nu A''^\mu = (\partial^\mu A^\nu - \partial^\nu A^\mu) + (\partial^\mu \alpha^\nu - \partial^\nu \alpha^\mu) \\ &= F^{\mu\nu} + f^{\mu\nu}, \qquad f^{\mu\nu} \equiv \partial^\mu \alpha^\nu - \partial^\nu \alpha^\mu. \end{aligned}$$

If $f^{\mu\nu} = 0$, that is, if α^μ is such that $\partial^\mu \alpha^\nu - \partial^\nu \alpha^\mu = 0$, then $\alpha^\mu(x^\nu)$ is a gauge transformation of the second kind, and it leaves $F^{\mu\nu}$ unchanged. Because of this freedom we have in choosing A^μ, we may always assume $\partial_\mu A^\mu = 0$. If, for example, we let $A^\mu \rightarrow A'^\mu - A^\mu + \alpha^\mu$, with α^μ satisfying $\partial_\mu \alpha^\mu = -\partial_\mu A^\mu$ and $\partial^\mu \alpha^\nu = \partial^\nu \alpha^\mu = 0$, then $\partial_\mu A'^\mu = 0$, and we may continue the discussion, replacing A^μ everywhere with $A''^\mu = A^\mu + \alpha^\mu$. For instance, we may let $\alpha^\mu = \partial^\mu \chi$, where χ satisfies the inhomogeneous wave equation $\partial_\mu \partial^\mu \chi = -\partial_\mu A^\mu$.

Taking $\partial_\mu A^\mu = 0$, then, we get for the four-divergence of $F^{\mu\nu}$ the result

$$\partial_\mu F^{\mu\nu} = \partial_\mu \partial^\mu A^\nu = \mu_0 j^\nu; \tag{5.13}$$

here we have utilized Eq. (5.6). If we now let $\nu = 0$, this equation becomes

$$\partial_0 F^{00} + \partial_1 F^{10} + \partial_2 F^{20} + \partial_3 F^{30} = \frac{1}{c}\nabla \cdot \mathbf{E} = \mu_0 c \rho \gamma_v = \frac{1}{c}\frac{\rho\gamma_v}{\epsilon_0}$$

or

$$\nabla \cdot \mathbf{E} = \frac{\rho'}{\epsilon_0}, \qquad \rho' = \rho\gamma_v$$

which is Eq. (1.34) when the apparent increase in charge density, arising from the motion of the charges making up ρ with velocity v relative to the observer, is taken into account. Similarly, if we let $\nu = 1$, we get

$$\partial_0 F^{01} + \partial_1 F^{11} + \partial_2 F^{21} \mp \partial_3 F^{31} = -\frac{1}{c^2}\frac{\partial}{\partial t}E_x + \frac{\partial}{\partial y}B_z - \frac{\partial}{\partial z}B_y = \mu_0 j_x,$$

or

$$(\nabla \times \mathbf{B})_x = \mu_0 j_x + \frac{1}{c^2}\frac{\partial}{\partial t}E_x,$$

which is the x-component of Eq. (1.37). The y- and z-components are obtained by calculating $\partial_\mu F^{\mu 2}$ and $\partial_\mu F^{\mu 3}$, so that we obtain

$$\nabla \times \mathbf{B} = \mu_0 \mathbf{j} + \frac{1}{c^2}\frac{\partial}{\partial t}\mathbf{E}.$$

Equation (5.13), then, yields two of Maxwell's four equations. What of the other two, Eqs. (1.35) and (1.36)? To obtain them, we utilize the symbol $\epsilon_{\mu\nu\rho\sigma}$ and define

$$G_{\mu\nu} = \tfrac{1}{2}\epsilon_{\mu\nu\rho\sigma}F^{\rho\sigma}. \tag{5.14}$$

Taking the divergence of this quantity, we obtain

$$\begin{aligned}\partial^\mu G_{\mu\nu} &= \tfrac{1}{2}\epsilon_{\mu\nu\rho\sigma}\partial^\mu(\partial^\rho A^\sigma - \partial^\sigma A^\rho)\\ &= -\tfrac{1}{2}\epsilon_{\nu\mu\rho\sigma}[\partial^\mu\partial^\rho A^\sigma - \partial^\mu\partial^\sigma A^\rho]\\ &= 0.\end{aligned}$$

That this divergence vanishes follows from the fact that $\epsilon_{\mu\nu\rho\sigma}$ is antisymmetric under the interchange of indices, for

$$\epsilon_{\nu\mu\rho\sigma}\partial^\mu\partial^\rho = -\epsilon_{\nu\rho\mu\sigma}\partial^\mu\partial^\rho,$$

and upon relabeling the dummy indices on the right-hand side (certainly

$$\sum_{\mu=0}^{3}\sum_{\nu=0}^{3} a_{\mu\nu}b^{\mu\nu} = \sum_{\nu=0}^{3}\sum_{\mu=0}^{3} a_{\nu\mu}b^{\nu\mu},$$

since all that has been done here is to exchange the labels on the first and second indices), we obtain

$$\epsilon_{\nu\mu\rho\sigma}\partial^\mu\partial^\rho = -\epsilon_{\nu\mu\rho\sigma}\partial^\rho\partial^\mu = -\epsilon_{\nu\mu\rho\sigma}\partial^\mu\partial^\rho,$$

which must vanish since the expression on the far right is the negative of that on the far left. Hence we have

$$\partial^\mu G_{\mu\nu} = \tfrac{1}{2}\epsilon_{\mu\nu\rho\sigma}\partial^\mu F^{\rho\sigma} = -\tfrac{1}{2}\epsilon_{\nu\mu\rho\sigma}\partial^\mu F^{\rho\sigma} = 0. \tag{5.15}$$

Letting $\nu = 0$ in Eq. (5.15) yields

$$\begin{aligned}0 &= -\tfrac{1}{2}\epsilon_{0\mu\rho\sigma}\partial^\mu F^{\rho\sigma} = -\tfrac{1}{2}\{\partial^1F^{23} - \partial^1F^{32} + \partial^2F^{31} - \partial^2F^{13} + \partial^3F^{12} - \partial^3F^{21}\}\\ &= \partial^1F^{32} + \partial^2F^{13} + \partial^3F^{21} = -\partial_xB_x - \partial_yB_y - \partial_zB_z = -\nabla\cdot\mathbf{B}\end{aligned}$$

or

$$\nabla\cdot\mathbf{B} = 0,$$

which is identical with Eq. (1.35). For $\nu = 1$, the result

$$\begin{aligned}0 &= -\tfrac{1}{2}\epsilon_{1\mu\rho\sigma}\partial^\mu F^{\rho\sigma} = \tfrac{1}{2}\{\partial^0F^{23} - \partial^0F^{32} + \partial^2F^{30} - \partial^2F^{03} + \partial^3F^{02} - \partial^3F^{20}\}\\ &= \partial^0F^{23} + \partial^2F^{30} + \partial^3F^{02} = -\frac{1}{c}\frac{\partial}{\partial t}B_x - \frac{\partial}{\partial y}\left(\frac{1}{c}E_z\right) + \frac{\partial}{\partial z}\left(\frac{1}{c}E_y\right),\end{aligned}$$

or

$$(\nabla\times\mathbf{E})_x = -\frac{\partial}{\partial t}B_x$$

follows. This is the x-component of Eq. (1.36); we get the y- and z-components by letting $\nu = 2$ and 3, respectively, so that the last of Maxwell's four vector equations,

$$\nabla \times \mathbf{E} = -\frac{\partial \mathbf{B}}{\partial t},$$

is finally obtained from our four-dimensional formalism.

Equation (5.15) is often written in the equivalent form

$$\partial^\lambda F^{\mu\nu} + \partial^\nu F^{\lambda\mu} + \partial^\mu F^{\nu\lambda} = 0 \tag{5.16}$$

where (λ, μ, ν) is taken successively to be (0, 1, 2), (1, 2, 3), (2, 3, 0) and (3, 0, 1), thus yielding four equations equivalent to the vector equations (1.35) and (1.36).

5.3 ELECTRODYNAMICS AND THE ENERGY-MOMENTUM TENSOR

Another important relationship that is expressed economically in four-dimensional tensor form is the law of forces in electrodynamics, namely

$$f^\nu = F^{\nu\lambda} j_\lambda, \tag{5.17}$$

where f^ν is the force per unit volume exerted upon a current density $j_\lambda = g_{\lambda\mu} j^\mu$ by the field $F^{\nu\lambda}$. Thus

$$f^0 = F^{0\lambda} j_\lambda = \frac{1}{c}\mathbf{E}\cdot\mathbf{j}, \qquad \mathbf{j} = \rho\gamma_v \mathbf{v} \tag{5.18a}$$

$$\mathbf{f} = \rho\gamma_v[\mathbf{E} + (\mathbf{v} \times \mathbf{B})]. \tag{5.18b}$$

Equation (5.18b) summarizes the laws of Coulomb and of Biot–Savart, whereas the right-hand side of (5.18a) is equal to $1/c$ times the rate, per unit volume, at which the field expends energy in accelerating the charge distribution ρ. All of this is implicit in the fundamental law of electrodynamics as expressed in Eq. (5.17).

If we make use of Eq. (5.13) in Eq. (5.17), we obtain

$$f^\nu = \frac{1}{\mu_0} F^{\nu\lambda} \partial^\mu F_{\mu\lambda} = \frac{1}{\mu_0}[\partial^\mu (F^{\nu\lambda} F_{\mu\lambda}) - (\partial^\mu F^{\nu\lambda}) F_{\mu\lambda}]. \tag{5.19}$$

By interchanging indices in the last term of Eq. (5.19), we obtain

$$\begin{aligned}(\partial^\mu F^{\nu\lambda})F_{\mu\lambda} &= (\partial^\lambda F^{\nu\mu})F_{\lambda\mu} = -(\partial^\lambda F^{\nu\mu})F_{\mu\lambda} \\ &= (\partial^\lambda F^{\mu\nu})F_{\mu\lambda} = \tfrac{1}{2}(\partial^\mu F^{\nu\lambda} + \partial^\lambda F^{\mu\nu})F_{\mu\lambda}.\end{aligned} \tag{5.20}$$

Here we invoke Eq. (5.16) to convert the right-hand side of Eq. (5.20) into a single term, so that

$$(\partial^\mu F^{\nu\lambda})F_{\mu\lambda} = -\tfrac{1}{2}(\partial^\nu F^{\lambda\mu})F_{\mu\lambda} = -\tfrac{1}{4}\partial^\nu(F^{\lambda\mu}F_{\mu\lambda})$$
$$= \tfrac{1}{4}\partial^\nu(F^{\lambda\mu}F_{\lambda\mu}).$$

Substitution of this result into (5.19) gives

$$\begin{aligned} f^\nu &= \frac{1}{\mu_0}[\partial^\mu(F^{\nu\lambda}F_{\mu\lambda}) - \tfrac{1}{4}\partial^\nu(F^{\mu\lambda}F_{\mu\lambda})] \\ &= \partial^\mu\left\{\frac{1}{\mu_0}[(F^{\nu\lambda}F_{\mu\lambda}) - \tfrac{1}{4}\delta^\nu{}_\mu(F^{\kappa\lambda}F_{\kappa\lambda})]\right\} \\ &= \partial^\mu S^\nu{}_\mu, \end{aligned} \tag{5.21}$$

where

$$S^\nu{}_\mu \equiv \frac{1}{\mu_0}[F^{\nu\lambda}F_{\mu\lambda} - \tfrac{1}{4}\delta^\nu{}_\mu F^{\kappa\lambda}F_{\kappa\lambda}]. \tag{5.22}$$

The force density f^ν, then, can be expressed as the four-divergence of a mixed tensor $S^\nu{}_\mu$.

Since f^ν is the time rate of change of momentum per unit volume, the total change in momentum imparted to a set of charges and currents, having four-current density j_λ, during a time interval $\Delta t = t_2 - t_1$, must be

$$\begin{aligned} \Delta P^\mu &= \int_{t_1}^{t_2} dt \int d^3x\, f^\mu = \frac{1}{c}\int_{\Delta t} d^4x\, f^\mu \\ &= \frac{1}{c}\int_{\Delta t} d^4x\, \partial^\nu S^\mu{}_\nu, \end{aligned} \tag{5.23}$$

where the spatial integral is taken over the extent of the charge distribution and the temporal integral is taken over the time interval Δt. Now there is a divergence theorem, or Gauss's theorem, for integrals over four-dimensional intervals that is analogous to the three-dimensional theorem, and the proof of the theorem is a direct extension of the proof for three dimensions. For a three-dimensional volume V bounded by a surface Σ, the divergence theorem is

$$\int_V \nabla \cdot \mathbf{U}\, d^3x = \int_V \partial_i U^i\, d^3x = \int_\Sigma U^i\, ds_i = \int_\Sigma \mathbf{U} \cdot d\mathbf{s},$$

where $d\mathbf{s}$ is an element of two-dimensional surface area with vector direction normal to the surface, and $\mathbf{U}$ is a three-vector. The four-dimensional analog, for a four-volume V_4 "enclosed" by hypersurface Σ_4, is

$$\int_{V_4} \partial_\mu U^\mu\, d^4x = \int_{\Sigma_4} U^\mu\, d\sigma_\mu, \tag{5.24}$$

where $d\sigma_\mu$ is a four-vector normal to an element of hypersurface having *hyperarea* of magnitude $\sqrt{d\sigma_\mu\, d\sigma^\mu}$.

This divergence theorem allows us to write Eq. (5.23) in the form

$$\Delta P^\mu = \int_{\Sigma_4(\Delta t)} d\sigma^\nu S^\mu{}_\nu. \tag{5.25}$$

In the particular case that the four-dimensional "hypervolume" in our reference frame is essentially a four-dimensional cylinder, bounded at its ends by the planes $t = t_1$ and $t = t_2$ and having spheres—*hyperdisks of radius R*—as *hypercross-sections*, the radius R can be allowed to increase without limit, so that each hypercross-section is an integral over the entire spatial volume. Then the hypersurface $\Sigma_4(\Delta t)$ can be divided into three parts, namely two hyperplanes, $t = t_1$ and $t = t_2$, the normal to which is parallel to the x_0-axis, and a hypercylinder of infinite radius, the normal to which is always, in our reference frame, orthogonal to the x^0-direction. (See Fig. 5.1.) Thus, since $d\sigma^0 = d^3x$, we have

$$\Delta P^\mu = \frac{1}{c}\left(\int_{\Sigma(t_2)} d^3x S^\mu{}_0 - \int_{\Sigma(t_1)} d^3x S^\mu{}_0 + \int_{\Sigma(\text{sphere},\, \nu\neq 0)} d\sigma^\nu S^\mu{}_\nu\right), \tag{5.26}$$

in which the last term represents the four-momentum *passing out* through the surface of the (infinite) three-dimensional sphere during the time $\Delta t = t_2 - t_1$. The first two terms give the change in four-momentum of the four-current density *within* the sphere during the time period Δt. Here we shall concern ourselves only with the first two terms; that is, we shall assume that we are dealing with a closed system so that no momentum or energy is allowed to pass through the surface of the three-dimensional sphere.

Evidently the temporal component, ΔP^0, of ΔP^μ is the energy gained by the charge distribution as it is accelerated by the electromagnetic field during time Δt; it is the amount of energy removed from the "storage vault" of the electromagnetic field and converted into mechanical energy through the motions of the charges. Because we believe in the conservation of energy, we consider the increase in mechanical energy, ΔP^0, to be matched by a corresponding decrease in electromagnetic field energy; thus the charge distribution always moves in such a way as to minimize the electromagnetic field energy. The magnitude of ΔP^0 therefore yields a measure of the capacity of the electromagnetic field to store energy, and the negative of ΔP^0 will be called the change in electromagnetic field energy, as the above discussion indicates.

According to Eq. (5.26), then we should interpret $-S^0{}_0$ as the density of energy stored by the electromagnetic field $F^{\mu\nu}$, so that $(-S^0{}_0)_{t_2} - (-S^0{}_0)_{t_1}$, when $t_2 > t_1$, will be the change in the energy density of the field during the time interval $\Delta t = t_2 - t_1$.

A parallel discussion of each of the spatial components of ΔP^μ would show that the negative of $\Delta \mathbf{P}$ is the momentum stored by the electromagnetic field, and that therefore the $-(1/c)S^\mu{}_0$, for $\mu \neq 0$, are the spatial components of the momentum stored per unit of volume by the field $F^{\mu\nu}$; $-(1/c)S^\mu{}_0$, $\mu \neq 0$, are the components of the field momentum density.

In summary, we see that the electromagnetic field must "carry" with it energy and momentum because

1. the field is capable of imparting energy and momentum to mechanical systems, and

2. we impose the condition that energy and momentum must be conserved.

Since such fields generally are not well localized (the question of localization of fields will arise again in the discussion of quantum mechanics that is to follow), we think of the momentum and energy as being distributed over some three-dimensional volume with a well-defined momentum and energy density.

In order to evaluate the momentum and energy densities, let us calculate $S^\nu{}_\mu$ in terms of $\mathbf{E}$ and $\mathbf{B}$. The first term in Eq. (5.22) is proportional to

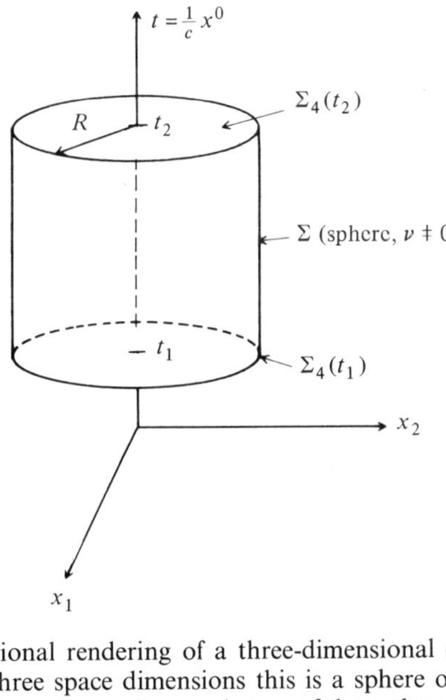

Fig. 5.1. A two-dimensional rendering of a three-dimensional cross-section of the hypersurface $\Sigma_4(\Delta t)$. In the three space dimensions this is a sphere of radius R during the time interval from t_1 to t_2. Since we can show only two of the orthogonal spatial axes, it appears here as a disk generating a cylinder along the t-axis rather than as a sphere generating a hypercylinder.

$$[F^{\nu\lambda}F_{\mu\lambda}] = [-F^{\nu\lambda}F_{\lambda\mu}] = [-F^{\nu\lambda}g_{\lambda\rho}F^{\rho\sigma}g_{\sigma\mu}]$$

$$= -\begin{bmatrix} 0 & -\frac{1}{c}E_x & -\frac{1}{c}E_y & -\frac{1}{c}E_z \\ \frac{1}{c}E_x & 0 & -B_z & B_y \\ \frac{1}{c}E_y & B_z & 0 & -B_x \\ \frac{1}{c}E_z & -B_y & B_x & 0 \end{bmatrix} \begin{bmatrix} 1 & 0 & 0 & 0 \\ 0 & -1 & 0 & 0 \\ 0 & 0 & -1 & 0 \\ 0 & 0 & 0 & -1 \end{bmatrix}$$

$$\cdot \begin{bmatrix} 0 & -\frac{1}{c}E_x & -\frac{1}{c}E_y & -\frac{1}{c}E_z \\ \frac{1}{c}E_x & 0 & -B_z & B_y \\ \frac{1}{c}E_y & B_z & 0 & -B_x \\ \frac{1}{c}E_z & -B_y & B_x & 0 \end{bmatrix} \begin{bmatrix} 1 & 0 & 0 & 0 \\ 0 & -1 & 0 & 0 \\ 0 & 0 & -1 & 0 \\ 0 & 0 & 0 & -1 \end{bmatrix} \tag{5.27}$$

$$= -\begin{bmatrix} 0 & \frac{1}{c}E_x & \frac{1}{c}E_y & \frac{1}{c}E_z \\ \frac{1}{c}E_x & 0 & B_z & -B_y \\ \frac{1}{c}E_y & -B_z & 0 & B_x \\ \frac{1}{c}E_z & B_y & -B_x & 0 \end{bmatrix}^2$$

$$= \begin{bmatrix} -\frac{1}{c}\mathbf{E}^2 & -\frac{1}{c}(\mathbf{E}\times\mathbf{B})_x & -\frac{1}{c}(\mathbf{E}\times\mathbf{B})_y & -\frac{1}{c}(\mathbf{E}\times\mathbf{B})_z \\ -\frac{1}{c}(\mathbf{E}\times\mathbf{B})_x & -\frac{1}{c^2}E_x^2 + B_y^2 + B_z^2 & -\frac{1}{c^2}E_xE_y - B_xB_y & -\frac{1}{c^2}E_xE_z - B_xB_z \\ -\frac{1}{c}(\mathbf{E}\times\mathbf{B})_y & -\frac{1}{c^2}E_xE_y - B_xB_y & -\frac{1}{c^2}E_y^2 + B_z^2 + B_x^2 & -\frac{1}{c^2}E_yE_z - B_yB_z \\ -\frac{1}{c}(\mathbf{E}\times\mathbf{B})_z & -\frac{1}{c^2}E_zE_x - B_zB_x & -\frac{1}{c^2}E_yE_z - B_yB_z & -\frac{1}{c^2}E_z^2 + B_x^2 + B_y^2 \end{bmatrix}.$$

The last term of (5.22) is proportional to the *trace*, that is, to the sum of the diagonal elements, of the first term:

$$F^{\kappa\lambda}F_{\kappa\lambda} = -F^{\kappa\lambda}F_{\lambda\kappa} = -\sum_{\rho,\sigma,\lambda} \delta^{\sigma}{}_{\rho}F^{\rho\lambda}F_{\lambda\sigma};$$

thus we have

$$-\tfrac{1}{4}\delta^{\nu}{}_{\mu}F^{\kappa\lambda}F_{\lambda\kappa} = -\tfrac{1}{4}\delta^{\nu}{}_{\mu}\left[-\frac{1}{c^2}\mathbf{E}^2 - \frac{1}{c^2}E_x^2 + B_y^2 + B_z^2 - \frac{1}{c^2}E_y^2 + B_z^2 + B_x^2 - \frac{1}{c^2}E_z^2 + B_x^2 + B_y^2\right]$$

$$= \tfrac{1}{2}\delta^{\nu}{}_{\mu}\left[\frac{1}{c^2}\mathbf{E}^2 - \mathbf{B}^2\right],$$

or

$$[-\tfrac{1}{4}\delta^{\nu}{}_{\mu}F^{\kappa\lambda}F_{\lambda\kappa}] = \begin{bmatrix} \tfrac{1}{2}\left(\frac{1}{c^2}\mathbf{E}^2 - \mathbf{B}^2\right) & 0 & 0 & 0 \\ 0 & \tfrac{1}{2}\left(\frac{1}{c^2}\mathbf{E}^2 - \mathbf{B}^2\right) & 0 & 0 \\ 0 & 0 & \tfrac{1}{2}\left(\frac{1}{c^2}\mathbf{E}^2 - \mathbf{B}^2\right) & 0 \\ 0 & 0 & 0 & \tfrac{1}{2}\left(\frac{1}{c^2}\mathbf{E}^2 - \mathbf{B}^2\right) \end{bmatrix}. \tag{5.28}$$

Using Eqs. (5.28) and (5.27) in (5.22), we get

$$[-S^{\nu}{}_{\mu}] = \frac{1}{\mu_0}\begin{bmatrix} \tfrac{1}{2}\left(\frac{1}{c^2}\mathbf{E}^2 + \mathbf{B}^2\right) & \frac{1}{c}(\mathbf{E}\times\mathbf{B})_x & \frac{1}{c}(\mathbf{E}\times\mathbf{B})_y & \frac{1}{c}(\mathbf{E}\times\mathbf{B})_z \\ \frac{1}{c}(\mathbf{E}\times\mathbf{B})_x & \frac{1}{c^2}(E_x^2 - \tfrac{1}{2}\mathbf{E}^2) + (B_x^2 - \tfrac{1}{2}\mathbf{B}^2) & \frac{1}{c^2}E_xE_y + B_xB_y & \frac{1}{c^2}E_zE_x + B_zB_x \\ \frac{1}{c}(\mathbf{E}\times\mathbf{B})_y & \frac{1}{c^2}E_xE_y + B_xB_y & \frac{1}{c^2}(E_y^2 - \tfrac{1}{2}\mathbf{E}^2) + (B_y^2 - \mathbf{B}^2) & \frac{1}{c^2}E_yE_z + B_yB_z \\ \frac{1}{c}(\mathbf{E}\times\mathbf{B})_z & \frac{1}{c^2}E_zE_x + B_zB_x & \frac{1}{c^2}E_yE_z + B_yB_z & \frac{1}{c^2}(E_z^2 - \tfrac{1}{2}\mathbf{E}^2) + (B_z - \tfrac{1}{2}\mathbf{B}^2) \end{bmatrix} \tag{5.29}$$

Hence, according to our interpretation of $-S^{0}{}_{0}$ as the energy density $\mathscr{E}$ of the field, we have

$$\mathscr{E} = \tfrac{1}{2}\left[\epsilon_0\mathbf{E}^2 + \frac{1}{\mu_0}\mathbf{B}^2\right]; \tag{5.30}$$

here we have used the relationship $1/c^2 = \epsilon_0\mu_0$. Furthermore, since $-(1/c)S^m{}_0$, for $m = 1, 2, 3$, are the components of the momentum density $\mathscr{P}$, we also have the result

$$\mathscr{P} = \epsilon_0 \mathbf{E} \times \mathbf{B}. \tag{5.31}$$

Sometimes one writes $\mathscr{P} = (1/c^2)[\mathbf{E} \times (1/\mu_0)\mathbf{B}]$, and $[\mathbf{E} \times (1/\mu_0)\mathbf{B}]$ is called the *Poynting vector.* It is important to note that the energy and momentum *densities*, $\mathscr{E}$ and $\mathscr{P}$, are not components of a four-vector, but rather are components of a tensor (the energy–momentum tensor) of rank two, just as **E** and **B** are not components of four-vectors but rather are components of rank-two tensors.

We have ignored most of the components of $S^\mu{}_\nu$. Since the matrix representing $S^\mu{}_\nu$ is symmetric in its two indices, that is, $S^\mu{}_\nu = S_\nu{}^\mu$, there are only 10, rather than 16, independent components of the tensor. We have discussed four of these components; the other six define the so-called Maxwell *stress–energy tensor*, which is a tensor under three-dimensional rotations. These do not play any important role in our discussion of the four-dimensional formulation of electromagnetic theory, and we shall not consider them further. The interested student is referred to advanced texts on electricity and magnetism for discussions of their meaning and applications.

5.4 PLANE WAVE SOLUTIONS TO THE HOMOGENEOUS WAVE EQUATION

Soon we shall turn away from relativity theory and shall investigate another paradox in physics, a paradox that arose shortly after the conflict between classical Newtonian–Maxwellian theory and the experimental data of Michelson and Morley, Trauton and Noble, etc., was discovered. The first moderately satisfactory solution of this problem was found by Max Planck in 1900. The problem involved the distribution of energy among electromagnetic waves contained within a closed cavity. It is appropriate that we should turn to this problem now, since we have just completed deriving an expression for the energy density of the electromagnetic field. However, a brief investigation of a particular type of solution to the electromagnetic wave equation is in order before we go on to consider the second paradox.

Equation (5.6) is the inhomogeneous wave equation for the four-vector potential—inhomogeneous because of the term $\mu_0 j^\nu$ on the right-hand side. We shall consider solutions to this equation in regions of space where $j^\nu = 0$; that is, we shall be concerned only with solutions to the homogeneous wave equation,

$$\partial_\mu \partial^\mu A^\nu = 0, \tag{5.32}$$

or

$$\left[\left(\frac{\partial}{\partial x^0}\right)^2 - \left(\frac{\partial}{\partial x^1}\right)^2 - \left(\frac{\partial}{\partial x^2}\right)^2 - \left(\frac{\partial}{\partial x^3}\right)^2\right] A^\nu(x^0, x^1, x^2, x^3) = 0. \tag{5.33}$$

In particular, we shall consider solutions to (5.33) that are confined to a three-dimensional parallelepiped bounded by the coordinate axes and the planes

$x^m = L_m, m = 1, 2, 3,$ and confined within the fixed time interval $0 \leq x_0/c = t \leq T$. With these restrictions, Eq. (5.33) can be solved by the method of separation of variables. In this method, we assume that the solution $A^\nu(x^\mu)$ can be factored into four functions, each involving only one of four space-time variables χ^μ; thus we assume

$$A^\nu(x^0, \ldots, x^3) = X_0{}^{(\nu)}(x^0)X_1{}^{(\nu)}(x^1)X_2{}^{(\nu)}(x_2)X_3{}^{(\nu)}(x_3). \tag{5.34}$$

Upon substituting (5.34) into (5.33) and dividing the result by $A^\nu(x^0, \ldots, x^3)$, we find

$$\frac{1}{X_0{}^{(\nu)}}\left(\frac{\partial}{\partial x^0}\right)^2 X_0{}^{(\nu)} - \frac{1}{X_1{}^{(\nu)}}\left(\frac{\partial}{\partial x^1}\right)^2 X_1{}^{(\nu)} - \frac{1}{X_2{}^{(\nu)}}\left(\frac{\partial}{\partial x^2}\right)^2 X_2{}^{(\nu)} - \frac{1}{X_3{}^{(\nu)}}\left(\frac{\partial}{\partial x^3}\right)^2 X_3{}^{(\nu)} = 0.$$

Each term on the left-hand side depends only upon one variable; in particular

$$\frac{1}{X_0{}^{(\nu)}}\left(\frac{\partial}{\partial x^0}\right)^2 X_0{}^{(\nu)} = \sum_{m=1}^{3} \frac{1}{X_m{}^{(\nu)}}\left(\frac{\partial}{\partial x^m}\right)^2 X_m{}^{(\nu)} = -(k^0)^2, \tag{5.35}$$

where $(k^0)^2$ is a constant, because the left-hand member depends only upon x^0 and the middle member depends only upon $\mathbf{x}$. We use the symbol $-(k^0)^2$ for the constant because it turns out that the desired solutions are arrived at more quickly when we take the constant to be negative definite, and because k^0 is to represent the temporal component of a four-vector, as the left-hand member of (5.35) suggests.

Now we may solve the spatial and temporal parts of Eqs. (5.35) independently. In the first place, we have

$$(1/X_0{}^{(\nu)})\left(\frac{d}{dx^0}\right)^2 X_0{}^{(\nu)} = -(k^0)^2$$

where we have replaced partial derivatives with ordinary derivatives because $X_0{}^{(\nu)} = X_0{}^{(\nu)}(x^0)$ depends upon only one variable, x^0. A more familiar form for this equation is

$$\left(\frac{d}{dx^0}\right)^2 X_0{}^{(\nu)} + (k^0)^2 X_0{}^{(\nu)} = 0, \tag{5.36}$$

and as is well-known, the most general solution for this equation can be written in the form

$$X_0{}^{(\nu)}(x^0) = \alpha_0{}^{(\nu)}e^{ik^0x^0} + \beta_0{}^{(\nu)}e^{-ik^0x^0}. \tag{5.37}$$

The spatial portion of Eq. (5.35) can be solved in a similar way. First we separate the $m = 1$ term from the others in the sum, and write the equation in the form

$$\begin{aligned}\frac{1}{X_1{}^{(\nu)}}\left(\frac{d}{dx^1}\right)^2 X_1{}^{(\nu)} &= -\left[(k^0)^2 + \sum_{m=2}^{3} \frac{1}{X_m{}^{(\nu)}}\left(\frac{d}{dx^m}\right)^2 X_m{}^{(\nu)}\right] \\ &= -(k^1)^2,\end{aligned}$$

where $(k^1)^2$ again is a constant, since the left-hand member here is dependent upon x^1 only, whereas the middle member depends only upon x^2 and x^3. The solution for $X_1{}^{(\nu)}$ then can be written in the form

$$X_1{}^{(\nu)}(x^1) = \alpha_1{}^{(\nu)}e^{ik^1x^1} + \beta_1{}^{(\nu)}e^{-ik^1x^1} \tag{5.38}$$

Similarly, by continuing these arguments, we obtain

$$X_2{}^{(\nu)} = \alpha_2{}^{(\nu)}e^{ik^2x^2} + \beta_2{}^{(\nu)}e^{-ik^2x^2}, \tag{5.39}$$

$$X_3{}^{(\nu)} = \alpha_3{}^{(\nu)}e^{ik^3x^3} + \beta_3{}^{(\nu)}e^{-ik^3x^3}, \tag{5.40}$$

where k^0, k^1, k^2 and k^3 must satisfy

$$(k^0)^2 - \sum_{m=1}^{3} (k^m)^2 = (k^0)^2 - (\mathbf{k})^2 = 0. \tag{5.41}$$

Equation (5.41) must be satisfied in all coordinate frames because the equation that we are solving, (5.32), is relativistically covariant, i.e. it is valid in every inertial frame of reference. Therefore $(k^0)^2 - (\mathbf{k})^2$ must be an invariant quantity, and since it has the form of the inner product of two four-vectors, we conclude that

$$k^\mu = (k^0, k^1, k^2, k^3) = (k^0{}_0\mathbf{k}) \tag{5.42}$$

is a contravariant four-vector. Similarly,

$$k_\mu = g_{\mu\nu}k^\nu = (k^0, -\mathbf{k}) \tag{5.43}$$

is a covariant four-vector.

Now $\partial_\mu A^\nu$ must be a mixed tensor of rank two, and for this reason, only some of the possible combinations of values of $\alpha_0{}^{(\nu)}, \ldots, \alpha_3{}^{(\nu)}$, $\beta_0{}^{(\nu)}, \ldots, \beta_3{}^{(\nu)}$ represent manifestly acceptable solutions for our system of electromagnetic waves in a box. Because of this restriction that $\partial_\mu A^\nu$ be a tensor, we must have $\alpha_1{}^{(\nu)} = \alpha_2{}^{(\nu)} = \alpha_3{}^{(\nu)} = 0$ whenever $\alpha_0{}^{(\nu)} \neq 0$ and $\beta_1{}^{(\nu)} = \beta_2{}^{(\nu)} = \beta_3{}^{(\nu)} = 0$ whenever $\beta_0{}^{(\nu)} \neq 0$, so that only the invariant form $k_\mu x^\mu$ will appear in the exponents that result when the $X_0{}^{(\nu)}, \ldots, X_3{}^{(\nu)}$ given by Eqs. (5.37)–(5.40) are multiplied together. Thus the most general solution of this kind is

$$A^\nu(x^0, \ldots, x^3) = A_+{}^\nu e^{ik_\mu x^\mu} + A_-{}^\nu e^{-ik_\mu x^\mu}, \tag{5.44}$$

where

$$A_+{}^\nu = \alpha_0{}^{(\nu)}\beta_1{}^{(\nu)}\beta_2{}^{(\nu)}\beta_3{}^{(\nu)}$$

$$A_-{}^\nu = \beta_0{}^{(\nu)}\alpha_1{}^{(\nu)}\alpha_2{}^{(\nu)}\alpha_3{}^{(\nu)}.$$

(See, however, Problem 8.)

The invariant that appears in the exponents of Eq. (5.44) can be written

$$k_\mu x^\mu = k^0x^0 - \mathbf{k}\cdot\mathbf{x} = k^0ct - \mathbf{k}\cdot\mathbf{x}$$

where if we let $k^0 = \omega/c = 2\pi\nu/c$, and let $\mathbf{k} = (2\pi/\lambda)\hat{\mathbf{n}}$, with $\hat{\mathbf{n}} = \mathbf{k}/|\mathbf{k}|$ representing the direction of the vector $\mathbf{k}$, we get

$$k_\mu x^\mu = \omega t - \mathbf{k}\cdot\mathbf{x} = 2\pi\left(\nu t - \frac{\hat{\mathbf{n}}\cdot\mathbf{x}}{\lambda}\right). \tag{5.45}$$

Already it should be evident that if the k^μ, and hence ν and λ, are real, then ν represents an oscillation frequency and λ represents a wavelength. Since

$$0 = k_\mu k^\mu = (k^0)^2 - (\mathbf{k})^2 = \left(\frac{\omega}{c}\right)^2 - |\mathbf{k}|^2 = (2\pi)^2\left[\left(\frac{\nu}{c}\right)^2 - \frac{1}{\lambda^2}\right],$$

or

$$\nu = \pm\frac{c}{\lambda}, \tag{5.46}$$

we see that the familiar relationship between wavelength, frequency and the velocity of propagation hold if we adopt the positive sign. Our interest will be confined to real values of the k^μ, since we are concerned with the existence and properties of electromagnetic waves.

The solutions given by Eq. (5.44), then, represent waves with frequency $\nu = \omega/2\pi = k^0c/2\pi$ and wavelength $\lambda = 2\pi/|\mathbf{k}|$. Equation (5.45) indicates that the waves are traveling in direction $\mathbf{k}$ with velocity $c = \lambda\nu$. These waves are called plane waves, because at every point in a plane perpendicular to $\mathbf{k}$, the value of the phase $k_\mu x^\mu$ will be the same; that is, the wave fronts of these waves are planes perpendicular to $\mathbf{k}$.

In many applications, it is more convenient to work with the *wave vector* $\mathbf{k}$ or the *reduced wave number* $|\mathbf{k}| = k = 2\pi/\lambda$ than with a wavelength λ. Similarly, it often is more convenient to deal with the *angular frequency* $\omega = 2\pi\nu = k^0/c$ than with the frequency ν. A possible point of confusion is the distinction, made by many authors, between the reduced wave number, $k = 2\pi/\lambda$, and the wave number $1/\lambda$. In modern practice the wave number is used only rarely, and as a consequence the shorter term, *wave number*, is used interchangeably with *reduced wave number*. In fact, we shall follow this latter practice throughout this book, despite the attendant dangers. Difficulties conceivably may arise when the reader is consulting literature concerning optics or atomic spectroscopy; the literature in these areas consistently uses the term "wave number" to refer to the quantity $1/\lambda$.

Note that for electromagnetic waves in vacuum, $k^0 = |\mathbf{k}|$, or $\omega = 2\pi c/\lambda$. However, this relationship does not apply to all kinds of waves, nor even to electromagnetic waves in material media; the index of refraction must be taken into account in material media, and normally it will exhibit some frequency dependence. Since our current discussion is related to electromagnetic waves in a vacuum, however, we may use the relation $|\mathbf{k}| = k^0$.

5.5 GAUGE TRANSFORMATIONS OF THE SECOND KIND

Because of the freedom allowed by the principle that physical variables are invariant under gauge transformations of the second kind, we always may impose the *Lorentz condition*

$$\partial_\mu A^\mu = 0$$

as we have seen already. Incidentally, when one imposes this condition, he is said to be working in the *Lorentz gauge.* Even when this condition is imposed, however, there is still an infinite number of four-potentials A^μ that will give the same physical effects *and* satisfy the Lorentz condition. This is because we are still free to add another four-vector to A^μ, of the form $\partial^\mu \xi$ and satisfying

$$\partial_\mu \partial^\mu \xi = 0$$

without disturbing any physics and without upsetting the Lorentz gauge. The existence of this *special* gauge transformation of the second kind allows us to impose the additional condition that

$$\nabla \cdot \mathbf{A} = 0 \tag{5.47}$$

in some convenient frame of reference. This gauge is unlike the Lorentz gauge in that it is not Lorentz invariant. Therefore, once we have chosen a frame of reference in which we wish to establish the *radiation gauge*, that is, in which we wish to take $\nabla \cdot \mathbf{A} = 0$, we must be careful to apply any conclusions drawn upon this condition only to that reference frame. Note that in the Lorentz gauge, $\nabla \cdot \mathbf{A} = 0$ is equivalent to $\partial_0 A^0 = 0$.

If we use our solution (5.44), this means that we may work in the radiation gauge only if

$$A^0 = 0 \qquad \text{and} \qquad \mathbf{k} \cdot \mathbf{A} = 0; \tag{5.48}$$

otherwise the only solutions we shall obtain will be the trivial ones with $k = |\mathbf{k}| = 0$. Since $\hat{\mathbf{k}}$ is the direction of wave propagation, Eq. (5.48) implies that the vector $\mathbf{A}$ is transverse to the wave vector $\mathbf{k}$ in the chosen reference frame.

During the remainder of this discussion of solutions A^ν to the homogeneous wave equation we shall assume that a reference frame has been chosen, and that the radiation gauge has been established in that frame; this is legitimate because one can establish such a gauge in any frame he chooses.

5.6 RADIATION ENERGY AND MOMENTUM

We now wish to study the energy and momentum densities of the fields specified by the solutions A^ν that we have found for the inhomogeneous wave equation. If we use Eq. (5.8) in Eq. (5.22), we obtain the following expression for $S^\nu{}_\mu$:

$$S^{\nu}{}_{\mu} = \frac{1}{\mu_0}[(\partial^{\nu}A^{\lambda} - \partial^{\lambda}A^{\nu})(\partial_{\mu}A_{\lambda} - \partial_{\lambda}A_{\mu}) - \tfrac{1}{4}\delta^{\nu}{}_{\mu}(\partial^{\kappa}A^{\lambda} - \partial^{\lambda}A^{\kappa})(\partial_{\kappa}A_{\lambda} - \partial_{\lambda}A_{\kappa})]$$

$$= \frac{1}{\mu_0}\{(\partial^{\nu}A^{\lambda})(\partial_{\mu}A_{\lambda}) + (\partial^{\lambda}A^{\nu})(\partial_{\lambda}A_{\mu}) - (\partial^{\lambda}A^{\nu})(\partial_{\mu}A_{\lambda}) \tag{5.49}$$

$$- (\partial^{\nu}A^{\lambda})(\partial_{\lambda}A_{\mu}) - \tfrac{1}{2}\delta^{\nu}{}_{\mu}[(\partial^{\kappa}A^{\lambda})(\partial_{\kappa}A_{\lambda}) - (\partial^{\lambda}A^{\kappa})(\partial_{\kappa}A_{\lambda})]\}.$$

Fortunately, we need not substitute our solution of A^{ν}, given by Eq. (5.44), into all the terms of (5.49); the first three terms are sufficiently general in character that the last three terms can be treated as special cases. Thus, utilizing Eq. (5.44), we get

$$(\partial^{\alpha}A^{\beta})(\partial_{\sigma}A_{\tau}) = -k^{\alpha}k_{\sigma}[A^{\beta}_{+}e^{ik_{\mu}x^{\mu}} - A^{\beta}_{-}e^{-ik_{\mu}x^{\mu}}][A_{+,\tau}e^{ik_{\nu}x^{\nu}} - A_{-,\tau}e^{-ik_{\nu}x^{\nu}}].$$

We consider first the second term of (5.49), namely

$$(\partial^{\lambda}A^{\nu})(\partial_{\lambda}A_{\mu}) = -k^{\lambda}k_{\lambda}[A^{\nu}_{+}e^{ik_{\rho}x^{\rho}} - A^{\nu}_{-}e^{-ik_{\rho}x^{\rho}}] \cdot [A_{+,\mu}e^{ik_{\sigma}x^{\sigma}} - A_{-,\mu}e^{-ik_{\sigma}x^{\sigma}}] = 0, \tag{5.50}$$

since $k^{\lambda}k_{\lambda} = (k^0)^2 - \mathbf{k}^2 = 0$, according to Eq. (5.41). Next the third term of (5.49) yields

$$(\partial^{\lambda}A^{\nu})(\partial_{\mu}A_{\lambda}) = -k^{\lambda}k_{\mu}[A^{\nu}_{+}e^{ik_{\rho}x^{\rho}} - A^{\nu}_{-}e^{-ik_{\rho}x^{\rho}}][A_{+,\lambda}e^{ik_{\sigma}x^{\sigma}} - A_{-,\lambda}e^{-ik_{\sigma}x^{\sigma}}]$$

$$= (\partial_{\mu}A^{\nu})(\partial^{\lambda}A_{\lambda}) = 0, \tag{5.51}$$

because we may impose the Lorentz condition $\partial^{\lambda}A_{\lambda} = 0$. Special cases of (5.50) and (5.51) are

$$(\partial^{\kappa}A^{\lambda})(\partial_{\kappa}A_{\lambda}) = (\partial^{\kappa}A^{\lambda})(\partial_{\lambda}A_{\kappa}) = 0. \tag{5.52}$$

We find, therefore, that only the first term on the right-hand side of Eq. (5.49) survives; that is,

$$S^{\nu}{}_{\mu} = \frac{1}{\mu_0}(\partial^{\nu}A^{\lambda})(\partial_{\mu}A_{\lambda}) \tag{5.53}$$

whenever A^{λ} represents a solution to the inhomogeneous wave equation. From this we find easily, from Eq. (5.30),

$$\mathscr{E} = -S^{0}{}_{0} = -\frac{1}{\mu_0}(\partial^0 A^{\lambda})(\partial_0 A_{\lambda}), \tag{5.54}$$

and, for $m = 1, 2, 3$,

$$\begin{aligned}
\mathscr{P}^m &= -\frac{1}{c}S^m{}_0 = -\frac{1}{c\mu_0}(\partial^m A^\lambda)(\partial_0 A_\lambda) \\
&= \frac{1}{c\mu_0}(-k^m k_0)[A^\lambda_+ e^{ik_\rho x^\rho} - A^\lambda_- e^{-ik_\rho x^\rho}][A_{+,\lambda} e^{ik_\sigma x^\sigma} - A_{-,\lambda} e^{-ik_\sigma x^\sigma}] \\
&= \frac{1}{c\mu_0}\frac{k^m}{k^0}(\partial^0 A^\lambda)(\partial_0 A_\lambda) = \frac{1}{c}\frac{k^m}{k^0}S^0{}_0 \\
&= \frac{1}{c}\frac{k^m}{|\mathbf{k}|}S^0{}_0 = \frac{1}{c}\frac{k^m}{|\mathbf{k}|}\mathscr{E}.
\end{aligned}$$

Writing this last result in three-vector form, we have

$$\mathscr{P} = \frac{1}{c}\frac{\mathbf{k}}{|\mathbf{k}|}\mathscr{E}; \tag{5.55}$$

that is, the momentum density, $\mathscr{P}$, of the wave field is in the direction of advancing wavefronts, $\mathbf{k}/|\mathbf{k}|$, and its magnitude is equal to $1/c$ times the energy density, $\mathscr{E}$.

In the next chapter, we shall have occasion to make use of Eq. (5.55). It plays an important role in the analysis of black-body radiation.

Comparing Eq. (5.31) with (5.55), we observe that $(\mathbf{E} \times \mathbf{B})$ also is in the direction of $\mathbf{k}$, and that

$$\begin{aligned}
|\mathbf{E} \times \mathbf{B}| &= \frac{1}{\epsilon_0 c}\mathscr{E} = \frac{1}{\epsilon_0 c}\cdot\tfrac{1}{2}\left[\epsilon_0 \mathbf{E}^2 + \frac{1}{\mu_0}\mathbf{B}^2\right] \\
&= \frac{1}{2c\epsilon_0\mu_0}\left[\left(\frac{1}{c}\mathbf{E}\right)^2 + \mathbf{B}^2\right] \\
&= \frac{c}{2}\left[\left(\frac{1}{c}\mathbf{E}\right)^2 + \mathbf{B}^2\right],
\end{aligned}$$

where we have made use of Eq. (5.30).

This result can be rewritten in the form

$$\left(\frac{1}{c}\mathbf{E}\right)^2 - 2\left|\frac{1}{c}\mathbf{E} \times \mathbf{B}\right| + \mathbf{B}^2 = 0,$$

which implies that $|\mathbf{E} \times \mathbf{B}| = |\mathbf{E}|\,|\mathbf{B}|$ so that $\mathbf{E}$ is perpendicular to $\mathbf{B}$, and also that $(1/c)|\mathbf{E}| = |\mathbf{B}|$.

In summary, $\mathbf{k}$, $\mathbf{E}$ and $\mathbf{B}$, in that order or in any cyclic permutation of that order, form a right-handed system of orthogonal vectors, and the magnitude of $\mathbf{E}$ is equal to c times the magnitude of $\mathbf{B}$; the energy density of the electromagnetic radiation field therefore is $\mathscr{E} = \epsilon_0\mathbf{E}^2$. These are classical results for wave fields, and may be recalled from elementary physics courses; here we have seen how they emerge from the four-dimensional formalism.

5.7 BOUNDARY CONDITIONS; RADIATION CONFINED TO A BOX

If the radiation is confined to a finite region of space, the solutions $A^\nu(x^\mu)$ must conform to restrictions imposed by conditions at the boundaries of the confining region. At the beginning of Section 5.4, it was indicated that we would consider radiation contained within a box having edges with lengths L_m, $m = 1, 2, 3$, parallel to the coordinate axes. That the sides of this box are parallel to the coordinate planes serves to justify the separation of variables, for unless the coordinate surfaces are parallel to the boundaries of the system, the boundary conditions cannot be imposed properly.

The boundary condition that we shall use here is equivalent to the assumption that the walls of the container within which the radiation is confined have perfectly reflecting surfaces. Thus we require that the components of $A^\nu(x^\mu)$ parallel to a (perfectly reflecting) wall must vanish at the surface of that wall. Since variables have been separated when we have $A^\nu(x^\mu) = X_0{}^{(\nu)}(x^0)\cdot X_1{}^{(\nu)}(x^1)\cdot X_2{}^{(\nu)}(x^2)\cdot X_3{}^{(\nu)}(x^3)$, we may impose the boundary conditions upon Eqs. (5.38)–(5.40) separately. Thus, $X_1{}^{(2)}(x^1)$ and $X_1{}^{(3)}(x^1)$ must vanish when $x^1 = 0$ and when $x^1 = L_1$; according to Eq. (5.38), then,

$$X_1{}^{(2,3)}(0) = \alpha_1{}^{(2,3)} + \beta_1{}^{(2,3)} = 0,$$

$$X_1{}^{(2,3)}(L_1) = \alpha_1{}^{(2,3)}e^{ik_1L_1} + \beta_1{}^{(2,3)}e^{-ik_1L_1} = 0.$$

Hence

$$\beta_1{}^{(2,3)} = -\alpha_1{}^{(2,3)}$$

and

$$i2\alpha_1{}^{(2,3)} \sin k_1L_1 = 0.$$

Nontrivial solutions to the last equation can occur only if k_1 satisfies

$$k^1 = -k_1 = \frac{\pi}{L_1}n_1, \qquad n_1 = 0, 1, 2, 3, \ldots ;$$

allowing negative values for n_1 does not introduce any new solutions. Subject to this condition on k_1, the solutions $X_1{}^{(2,3)}(x^1)$ are, then, $X_1{}^{(2,3)}(x^1) = a_1{}^{(2,3)} \sin k^1x^1$ where $a_1{}^{(2,3)} = 2i\alpha_1{}^{(2,3)}$. Similarly, we obtain

$$X_2{}^{(3,1)}(x^2) = a_2{}^{(3,1)} \sin k^2x^2, \qquad X_3{}^{(1,2)}(x^3) = a_3{}^{(1,2)} \sin k^3x^3$$

with

$$k^m = \frac{\pi}{L_m}n_m, \qquad n_m = 0, 1, 2, \ldots ;\ m = 1, 2, 3. \tag{5.56}$$

Note that here we have reverted to the product form for the four-vector potential, and have not used the manifestly covariant form of Eq. (5.44). In fact, as you may see by doing Problem 8 at the end of this chapter, a certain linear combination of the version of A^μ given by Eq. (5.44) is equal to the product version given in Eq. (5.34). Hence the boundary conditions are to be applied to a linear combination of the manifestly covariant expression for A^μ, which also necessarily is a solution to the homogeneous wave equation.

The boundary condition, then, restricts the values of the spatial components of k^μ to integral multiples of π divided by the corresponding dimension of the box. This gives the result

$$(k^0)^2 = (k)^2 = \mathbf{k}^2 = \pi^2\left(\frac{n_1^2}{L_1^2} + \frac{n_2^2}{L_2^2} + \frac{n_3^2}{L_3^2}\right)$$

or

$$k^0 = k = \pi\sqrt{\frac{n_1^2}{L_1^2} + \frac{n_2^2}{L_2^2} + \frac{n_3^2}{L_3^2}} \tag{5.57}$$

with the n_m all integers.

A particularly important consequence of this result is that the wavelength of a particular spectral component of the radiation will increase in proportion to the linear dimension of the enclosure to which it is confined. For simplicity, let $L \equiv L_1 = L_2 = L_3$; then

$$k = \frac{2\pi}{\lambda} = \frac{\pi}{L}\sqrt{n_1^2 + n_2^2 + n_3^2}, \tag{5.58}$$

or

$$\lambda = \frac{2L}{\sqrt{n_1^2 + n_2^2 + n_3^2}}. \tag{5.59}$$

If L is increased continuously by an amount that is small compared with λ, no sudden jump in any of the numbers n_1, n_2, or n_3 is to be expected, for this would imply that sudden jumps in the wavelength should be observed as the volume is increased; in fact such jumps are not observed. Hence a small change in L must lead to corresponding small changes in λ. A spectral component with wavelength $\lambda = 2L/\sqrt{n_1^2 + n_2^2 + n_3^2}$, then, is characterized by the numbers n_1, n_2 and n_3; for the integers n_1, n_2, n_3 specify the essential features of the spectral component, whatever the volume may be, whereas λ changes in proportion to changes in the linear dimension of the enclosure. Thus, since $L^3 = V$, we may express the proportionality of λ to $L = V^{1/3}$ in the form

$$\frac{\lambda_1}{\lambda_2} = \frac{V_1^{1/3}}{V_2^{1/3}}, \tag{5.60}$$

where λ_1 is the wavelength of a spectral component when the volume is V_1, and λ_2 is the wavelength of the same spectral component when the volume has been changed to V_2. This conclusion holds for enclosures other than cubes, also. The entire analysis that brought us to this conclusion can be confirmed by considering the reflection of radiation from a moving wall of the container as the container is expanded; however, we shall not pursue such an analysis here.

READING NOTES*

Reference 6, Chapter EII.
Reference 9, Chapter 2.
Reference 10, Paper III.
Reference 3, Chapter 12.
Reference 11, Chapter 5.
Reference 8, Chapter 18.
Reference 12, Part III(b).

PROBLEMS

1. In some frame of reference, an electromagnetic wave has electric field vector **E** polarized along the z-axis, and is being propagated in the xy-plane at angle ϕ relative to the x-axis; thus
$$\mathbf{E} = \hat{\mathbf{i}}_3 E_0 \cos(\omega t - \mathbf{k} \cdot \mathbf{x})$$
$$= \hat{\mathbf{i}}_3 E_0 \cos(\omega t - kx\cos\phi - ky\sin\phi).$$
Find, in a frame of reference S' moving with speed v along the x-direction, the components of $\mathbf{E}'$ and $\mathbf{B}'$. Also find $|\mathbf{E}'|$ and $\mathbf{k}'$. Has the polarization $\hat{\mathbf{E}}$ of the wave been changed?

2. Let a particle of mass m and electric charge q be moving in a pure magnetic induction field **B** along the z-axis. Write down the equations of motion for the particle in covariant form and show that in the frame in which the field is purely magnetic, the particle moves in a cylindrical spiral with axis along $\hat{\mathbf{z}}$. Show further that its frequency of revolution (in the particle's frame of reference) is $\omega = qB/m$. Let the radius of revolution be ρ, and show that
$$B\rho = \frac{pc}{q}, \qquad \frac{mv^2}{\rho} = qBv$$
if $v \equiv \rho\omega$. Note that the latter has exactly the same form as the nonrelativistic equation for the same phenomenon.

3. Show that the quantities $B^2 - (1/c^2)E^2$ and $\mathbf{E} \cdot \mathbf{B}$ are Lorentz-invariant, and use this to demonstrate very simply that
   a) **E** and **B** are perpendicular in all frames of reference if they are perpendicular in one (perpendicularity of **E**, **B** is a Lorentz-covariant property);
   b) $|(1/c)\mathbf{E}| < |\mathbf{B}|$ is a Lorentz-covariant property, as are
$$\left|\frac{1}{c}\mathbf{E}\right| > |\mathbf{B}|, \qquad \left|\frac{1}{c}\mathbf{E}\right| = |\mathbf{B}|;$$

---

* Works referred to in these notes will be found in the Bibliography on p. 577.

c) the obtuseness or acuteness of angle between **E** and **B** is a Lorentz-covariant property; and

d) if **E** or **B** vanishes in some frame, then $\mathbf{E} \perp \mathbf{B}$ in all frames.

4. Show that under the transformation $\mathbf{x} \to -\mathbf{x}$ (space inversion), $\mathbf{E} \to -\mathbf{E}$ and $\mathbf{B} \to \mathbf{B}$; i.e. **E** behaves in this respect like a polar vector and **B** behaves like an axial vector.

5. Let a charge q be moving uniformly relative to S with velocity $\mathbf{v}$. Find **E** and **B** in S by writing down the fields in the rest frame of the particle and transforming them to S.

6. Show that the relations between $\mathbf{E}'$, $\mathbf{B}'$ in S' and **E**, **B** in S can be summarized through the equations

$$\mathbf{E}'_{\parallel} = \mathbf{E}_{\parallel}, \qquad \mathbf{B}'_{\parallel} = \mathbf{B}_{\parallel}$$

$$\mathbf{E}'_{\perp} = \gamma(\mathbf{E}_{\perp} + \mathbf{v} \times \mathbf{B}_{\perp}), \qquad \mathbf{B}'_{\perp} = \gamma\left(\mathbf{B}_{\perp} - \frac{\mathbf{v}}{c^2} \times \mathbf{E}_{\perp}\right)$$

where $\parallel$ refers to components parallel to **v** and $\perp$ refers to components perpendicular to **v**.

7. Let a particle with mass m, charge q initially be at rest at the origin in frame S in which perpendicular fields **E** and **B** are present. Find the particle's trajectory, stating in what inertial frame your solution is valid, and discuss the particle's motion.

8. Impose the boundary conditions of Section 5.7 on Eq. (5.44) and show that then A^μ as given in Eq. (5.44) cannot represent the most general form for a vector potential for the field. Note also that the conditions on $\alpha_\mu^{(v)}$, $\beta_\mu^{(v)}$ given in Section 5.7 are inconsistent with those in the discussion preceding Eq. (5.44). However, a linear combination of the A^μ of Eq. (5.44), each with different k^μ, also is a solution to the homogeneous wave equation. Show that the product form of A^μ, Eq. (5.34), is a certain linear combination of the A^μ given in Eq. (5.44). Hence the boundary conditions can be applied in a consistent way to a *linear combination* of the (5.44) versions of A^μ.

CHAPTER 6

THE BEGINNINGS OF QUANTUM THEORY

The second of the major revolutions in modern physics arose from the attempt to analyse, on the basis of classical theory, the thermodynamic properties of electromagnetic radiation confined to the isothermal enclosure. Profound discrepancies between predictions based upon fundamental classical laws on the one hand, and experimental observations on the other, induced Max Planck to put forward a remarkable hypothesis which effectively resolved the difficulty. This hypothesis became the cornerstone of a whole new interpretation of basic physical phenomena; the interpretation is known as quantum theory, or quantum mechanics.

Here we shall analyse the radiation contained in an isothermal cavity, with a view toward finding the temperature dependence of the energy density of components of the radiation with different frequencies. This dependence can be checked by experiments, and therefore the analysis provides a valuable experimental check on the classical theory of electromagnetic radiation.

6.1 THE PROBLEM OF RADIATION IN AN ISOTHERMAL ENCLOSURE

An important property of radiation within an isothermal enclosure is that the radiation is isotropic everywhere within the enclosure; that is, there is just as much electromagnetic radiative momentum being carried in one direction as in any other direction. Since the pressure on the walls of the enclosure is just the momentum per unit area delivered by the radiation per unit time, the isotropicity of radiation means that the radiation pressure on the walls must be uniform.

Consider a train of waves moving in the direction $\mathbf{k}/k$, and let them be incident on a small element of area, ΔS, of the enclosure wall. Now generate a cylinder with sides parallel to $\mathbf{k}/k$ and having base ΔS, as shown in Fig. 6.1; the slant height of the cylinder is Δl, and it has volume $\Delta V = \Delta\mathbf{S}\cdot\Delta\mathbf{l}$, where $\Delta\mathbf{S} = \hat{\mathbf{n}}\,\Delta S$, $\hat{\mathbf{n}}$ being the normal to the surface ΔS, and $\Delta\mathbf{l} = \mathbf{k}\,\Delta l/k$. Then the amount of radiative momentum carried by the waves we are considering in this cylinder is, according to Eq. (5.55),

$$\Delta\mathbf{P} = \frac{\mathscr{E}}{c}\,\Delta V\frac{\mathbf{k}}{k} = \frac{\mathscr{E}}{c}(\Delta\mathbf{S}\cdot\Delta\mathbf{l})\frac{\mathbf{k}}{k} = \frac{\mathscr{E}}{c}\,\Delta S\,\Delta l\,\frac{\hat{\mathbf{n}}\cdot\mathbf{k}}{k^2}\,\mathbf{k}$$

$$= \frac{\mathscr{E}}{c}\,\Delta S\,\Delta l\cos\theta\,\frac{\mathbf{k}}{k}, \qquad \cos\theta = \hat{\mathbf{n}}\cdot\hat{\mathbf{k}},$$

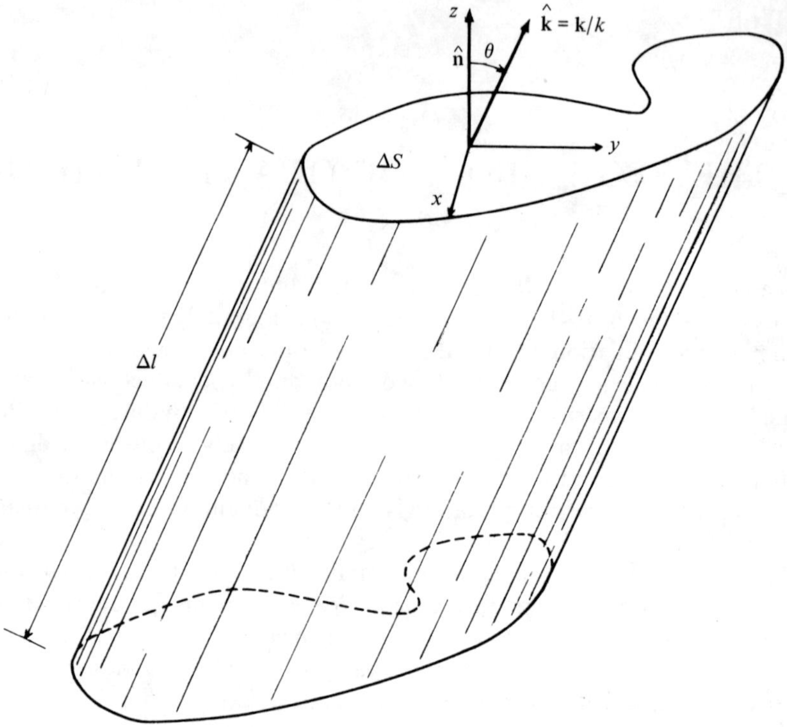

Fig. 6.1. Cylinder generated by a plane surface of area ΔS being displaced along $\hat{\mathbf{k}} = \mathbf{k}/k$ at an angle θ from the normal to the surface.

and all of it is delivered to the element of area ΔS in time $\Delta t = \Delta l/c$. If ΔS is an element of area on a perfectly smooth, reflecting wall, the change in momentum for radiation moving in direction $\mathbf{k}$ during time $\Delta t = \Delta l/c$ is

$$2\hat{\mathbf{n}} \cdot \Delta \mathbf{P} = 2 \frac{(\hat{\mathbf{n}} \cdot \mathbf{k})^2}{k^2} \frac{\mathscr{E}}{c} \Delta S \, \Delta l$$

$$= 2\mathscr{E} \, \Delta S \, \Delta t \cos^2 \theta$$

because only the normal component of momentum suffers a change, and the direction of that component is reversed, so that the change in the normal component is twice the normal component. Thus the force on ΔS arising from radiation moving along $\mathbf{k}/k$ is

$$\mathbf{F} = \frac{2\hat{\mathbf{n}} \cdot \Delta \mathbf{P}}{\Delta t} \hat{\mathbf{n}} = \hat{\mathbf{n}} 2\mathscr{E} \, \Delta S \cos^2 \theta,$$

and the resulting pressure is

$$\Pi(\hat{\mathbf{k}}) = \Pi(\theta, \varphi) = \frac{\hat{\mathbf{n}} \cdot \mathbf{F}}{\Delta S} = 2\mathscr{E} \cos^2 \theta \tag{6.1}$$

where θ and φ are the polar angles of $\mathbf{k}/k$ as measured in terms of a z-axis along $\hat{\mathbf{n}}$ and x- and y-axes in the plane of ΔS.

To find the total pressure, we must add together the pressures coming from all possible directions on one side of the surface ΔS. The quantity $\Pi(\mathbf{k}/k)$ is actually the pressure arising from waves traveling along direction $\mathbf{k}/k$ and coming from a small element of solid angle, $d\Omega$, on a hemisphere covered by ΔS, as Fig. 6.2 illustrates. We must add together the pressures arising from waves that come from all such elements of solid angle. Since, from solid analytic geometry and calculus, we know that an element of solid angle is given by $d\Omega = \sin\theta\, d\theta\, d\varphi$, where θ is the polar angle (colatitude) and φ is the azimuthal angle, we find that the total pressure is

$$\begin{aligned} \Pi &= \int_{\text{hemisphere}} \Pi(\hat{\mathbf{k}})\, d\Omega = \int_0^{2\pi} d\varphi \int_0^{\pi/2} d\theta \sin\theta\, 2\mathscr{E}\cos^2\theta \\ &= 4\pi\mathscr{E}\int_0^{\pi/2} \cos^2\theta \sin\theta\, d\theta = 4\pi\mathscr{E}\int_0^1 x^2\, dx = \frac{4\pi}{3}\mathscr{E}. \end{aligned} \tag{6.2}$$

The energy density in the same hemisphere is the sum of the energy densities arising from all waves moving along $\mathbf{k}/k$ and from all waves moving against $\mathbf{k}/k$ (i.e. along $-\mathbf{k}/k$), and these must be summed over all directions $\mathbf{k}/k$ for the hemisphere. Since the radiation is isotropic, $\mathscr{E}$ is independent of θ and φ, and we get

$$\Psi = 2\mathscr{E}\int_{\text{hemisphere}} d\Omega = 2\mathscr{E}\int_0^{2\pi} d\varphi \int_0^{\pi/2} \sin\theta\, d\theta = 4\pi\mathscr{E} \tag{6.3}$$

as the energy density of all the radiation contributing to the pressure Π; more generally, Ψ is simply the energy density at any interior point of the enclosure, since the latter is given by

$$\int_{\text{sphere}} \mathscr{E}\, d\Omega = 4\pi\mathscr{E}.$$

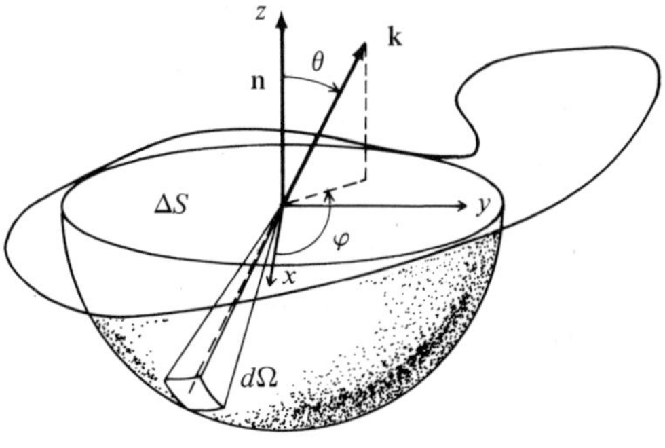

Fig. 6.2. Radiation arising from solid angle $d\Omega$ strikes the surface ΔS at an angle θ from the normal.

Equations (6.2) and (6.3) together tell us that the pressure on the walls of the enclosure is just one-third the energy density within the enclosure:

$$\Pi = \tfrac{1}{3}\Psi. \tag{6.4}$$

Now we may calculate also the flux of energy at the surface of the isothermal enclosure; that is, we may find the energy brought up to a surface element of area ΔS in time interval Δt. This energy is just

$$\begin{aligned}\Delta U &= \int_{\text{hemisphere}} \mathscr{E}\,\Delta V(\mathbf{k}/k)\,d\Omega = \mathscr{E}(\Delta S\,c\,\Delta t)\int_0^{2\pi} d\varphi \int_0^{\pi/2} d\theta \sin\theta\cos\theta \\ &= \pi c\mathscr{E}\,\Delta S\,\Delta t\end{aligned}$$

where by $\Delta V(\mathbf{k}/k)$ we mean the volume of a cylinder with base ΔS and sides having length $c\,\Delta t$ parallel to $\mathbf{k}/k$; that is, $\Delta V(\mathbf{k}/k)$ is the volume swept out by the waves passing through ΔS in direction $\mathbf{k}/k$ during the time interval Δt. Thus the desired energy flux is

$$\Phi = \frac{\Delta U}{\Delta S\,\Delta t} = \pi c\mathscr{E} = \frac{c}{4}\Psi \tag{6.5}$$

6.2 THERMODYNAMIC CONSIDERATIONS

It can be shown that a perfectly reflecting enclosure is thermodynamically equivalent to a totally absorbing body, or *black body*. Hence the emissive power, or radiative emittance of a black body is the same as that for an enclosure such as the one we are analysing. For such systems, the emissive power e is equal to the absorptive power, a; e and a represent the energy flux given off, and absorbed, respectively, by the system. A black body, by definition, would absorb all the radiation brought up to its surface in a unit of time, and would re-emit exactly the same amount of radiative energy in a unit time. Hence the energy emitted (or absorbed) per unit area per unit time is given by Eq. (6.5),

$$e = \frac{c}{4}\Psi, \tag{6.6}$$

where e is the radiative emittance of a black body having exactly the same temperature as our isothermal enclosure.

Now the confined radiation is a thermodynamic system in that it can be described in terms of the thermodynamic variables Π, V and T (pressure, volume and absolute temperature). Therefore we may utilize the well-known *energy equation* of thermodynamics,

$$\left(\frac{\partial U}{\partial V}\right)_T = T\left(\frac{\partial \Pi}{\partial T}\right)_V - \Pi, \tag{6.7}$$

where U is the internal energy of the system, $U = \Psi V$, and the subscripts on partial derivatives mean that the variable represented by the subscript is held constant when the derivative is calculated. We have found that $\Pi = \frac{1}{3}\Psi$, the energy density Ψ on the right-hand side being independent of volume; hence

$$\left(\frac{\partial U}{\partial V}\right)_T = \Psi, \qquad \left(\frac{\partial \Pi}{\partial T}\right)_V = \frac{1}{3}\left(\frac{\partial \Psi}{\partial T}\right)_V = \frac{1}{3}\frac{d\Psi}{dT}. \tag{6.8}$$

Substitution of (6.8) into (6.7) yields

$$\Psi = \frac{T}{3}\frac{d\Psi}{dT} - \frac{1}{3}\Psi,$$

or

$$4\frac{dT}{T} = \frac{d\Psi}{\Psi}. \tag{6.9}$$

Integration of Eq. (6.9) gives us

$$\Psi = bT^4, \tag{6.10}$$

where b is a constant. Insertion of this last result into Eq. (6.6) then yields the temperature dependence of the emissive power for a black body as

$$e = \frac{bc}{4}T^4.$$

Usually the constant $bc/4$ is designated σ and is called the Stephan–Boltzmann constant, so that

$$e = \sigma T^4, \tag{6.11}$$

where the experimentally determined numerical value of the constant is

$$\sigma = 5.670 \times 10^{-5} \text{ erg sec}^{-1} \text{ cm}^{-2} (\text{deg K})^{-4}.$$

This law relating e and T was discovered experimentally by Stephan in 1879, and the thermodynamic analysis was carried out later by Boltzmann; it is known as the Stephan–Boltzmann law.

Now we may use the first law of thermodynamics to obtain a relationship between temperature and volume. The differential form for the first law of thermodynamics is

$$đQ = đU + dW \tag{6.12}$$

where $đQ$ is the heat gained by a thermodynamic system, $đU$ is the increase in internal energy, and $đW$ is the amount of work done by the system. We assume that our cavity full of radiation is an isolated system and that no exchange of heat between it and the remainder of the universe is possible. Therefore we take $đQ = 0$;

the processes we may consider are adiabatic processes. Also, the internal energy of the system is $U = \Psi V$, and the work done by the system must be $dW = \Pi\, dV$, where dV is the change in volume that occurs as the process is carried out. Therefore we have

$$0 = d(\Psi V) + \Pi\, dV = V\, d\Psi + (\Psi + \Pi)\, dV, \tag{6.13}$$

or using Eq. (6.4) in (6.13)

$$\frac{d\Psi}{\Psi} = -\frac{4}{3}\frac{dV}{V}.$$

The solutions to this equation have the form

$$\Psi = (\text{const.})V^{-4/3},$$

or more explicitly

$$\Psi_1 V_1^{4/3} = \Psi_2 V_2^{4/3} \tag{6.14}$$

where Ψ_1 is the energy density of the system when the volume is V_1, and Ψ_2 is the energy density when the volume has become V_2.

Our earlier results can be used to convert Eq. (6.14) to several different forms. For example if we write (6.10) as

$$\frac{\Psi_1}{T_1^4} = \frac{\Psi_2}{T_2^4}, \tag{6.15}$$

and use this in (6.14), we have

$$(T_1 V_1^{1/3})^4 = (T_2 V_2^{1/3})^4$$

or since both T and V are real, positive quantities,

$$T_1 V_1^{1/3} = T_2 V_2^{1/3}; \tag{6.16}$$

the temperature is proportional to the inverse of the cube root of the volume. Another useful expression of Eq. (6.14) is obtained by using $U = \Psi V$, so that

$$U_1 V_1^{1/3} = U_2 V_2^{1/3}; \tag{6.17}$$

the internal energy also is proportional to the inverse of the cube root of the volume. Equations (6.15) and (6.17) taken together yield

$$\frac{U_1}{U_2} = \frac{T_1}{T_2}; \tag{6.18}$$

the internal energy is proportional to the temperature.

6.3 SPECTRAL COMPONENTS; THE WIEN DISPLACEMENT LAW

The energy contained within the enclosure is contributed by many components with different wave numbers; thus, if Ψ_k is the energy density of all waves having wave number k, then

$$\Psi = \sum_k \Psi_k .$$

If k is allowed to take on all positive real values (which, according to Eq. (5.56), is possible only if $L_1, L_2, L_3 \to \infty$), the above sum is not appropriate, and instead we must write

$$\Psi = \int_0^\infty \psi(k)\, dk$$

where $\Psi(k)$ is the energy density, per unit wave number, of waves having wave numbers between k and $k + dk$; that is, if we define

$$\Psi(k) = \int_0^k \psi(k)\, dk, \qquad \Psi = \Psi(\infty), \tag{6.19}$$

then it follows that

$$\psi(k) = \frac{d\Psi(k)}{dk}.$$

In any case, if the cavity is large, Eq. (6.19) can be used as a good approximation.

Similarly, let us write the spectral decomposition of the emissive power as

$$e = \int_0^\infty \epsilon(k)\, dk. \tag{6.20}$$

Then, according to Eq. (6.6),

$$\frac{c}{4}\Psi \quad e = \int_0^\infty \left[\frac{c}{4}\psi(k) - \epsilon(k)\right] dk = 0.$$

This suggests that

$$\epsilon(k) = \frac{c}{4}\psi(k), \tag{6.21}$$

though it does not prove that the relationship of Eq. (6.6) can be extended to spectral components. More refined arguments are required to prove Eq. (6.21), but it is a plausible result that can be shown to be true, and we shall accept it without further elaboration.

Now let us consider the temperature dependence of $\psi(k)$. In order to do this, we must recall Eq. (5.59), which can be rewritten as

$$k_1 V_1^{1/3} = k_2 V_2^{1/3}, \tag{6.22}$$

since $k = 2\pi/\lambda$. Using Eq. (6.16) in this equation, we have the result

$$k_1/k_2 = T_1/T_2. \tag{6.23}$$

This result also holds for the spectral components $k_1 + dk_1$ and $k_2 + dk_2$; thus

$$\frac{k_1}{k_2} = \frac{T_1}{T_2} = \frac{k_1 + dk_1}{k_2 + dk_2} = \frac{(k_1/k_2) + (dk_1/k_2)}{1 + (dk_2/k_2)} = \frac{(T_1/T_2) + (dk_1/k_2)}{1 + (dk_2/k_2)}$$

or

$$(T_1/T_2)[1 + (dk_2/k_2)] = (T_1/T_2) + (dk_1/k_2)$$

and hence

$$dk_1 = (T_1/T_2)\, dk_2. \tag{6.24}$$

This allows us to utilize the relation $\Psi_1/\Psi_2 = T_1^4/T_2^4$ in finding the temperature dependence of $\psi(k)$, for we have

$$\frac{\Psi_1}{\Psi_2} = \frac{\int \psi_1(k_1)\, dk_1}{\int \psi_2(k_2)\, dk_2} = \left(\frac{T_1}{T_2}\right)^4,$$

which if we use Eq. (6.24), becomes

$$\int \left[\frac{\psi_1(k_1)}{T_1^4} \cdot \frac{T_1}{T_2} - \frac{\psi_2(k_2)}{T_2^4}\right] dk_2 = 0,$$

or

$$\int \left[\frac{\psi_1(k_1)}{T_1^3} - \frac{\psi_2(k_2)}{T_2^3}\right] dk_2 = 0.$$

By the same kind of plausibility argument that led to Eq. (6.21), then, we obtain

$$\frac{\psi_1(k_1)}{\psi_2(k_2)} = \left(\frac{T_1}{T_2}\right)^3 \tag{6.25}$$

that is, $\psi(k)$ is proportional to the cube of the temperature. It follows from Eq. (6.21), then, that

$$\frac{\epsilon(k_1)}{T_1^3} = \frac{\epsilon(k_2)}{T_2^3}, \tag{6.26}$$

also. This is the form of the result that is most valuable to the experimenter because the emissivity per unit wave number, $\epsilon(k)$, is an experimentally observable quantity, as is T.

Wien observed that, because of Eq. (6.23), Eq. (6.26) could be written in the form

$$\epsilon(k_1)/T_1^3 = \epsilon(k_2)/T_2^3 = g(k/T) \tag{6.27}$$

where $g(k/T)$ does not depend independently on either k or T, but rather depends upon the ratio k/T. Wien's guess for $g(k/T)$ was that it should have the form

$$g(k/T) = a\, e^{-k/\beta T}(k/T)^3,$$

where a and β are constants. From this, Wien's law

$$\begin{aligned}\psi(k) &= (4/c)\epsilon(k) = (4a/c)\, e^{-k/\beta T}(k/T)^3 T^3 \\ &= (4a/c)\, e^{-k/\beta T} k^3\end{aligned} \tag{6.28}$$

follows. This formula has been found to fit experimental data well at high frequencies, but it fails for low frequencies, namely for $k \lesssim 3 \times 10^4\ \text{cm}^{-1}$. Rayleigh and Jeans found another form for $g(k/T)$ that works well at very low frequencies, but that fails badly at high frequencies, say $k \gtrsim 2.5 \times 10^3\ \text{cm}^{-1}$. No theory based upon classical physics has been found, that will fit experimental data in the region of frequencies between where the Wien law and the Rayleigh–Jeans law fit the data adequately.

6.4 THE RAYLEIGH–JEANS LAW

Because it serves as an introduction to methods we shall use later, and because it completes the classical picture of radiation in a cavity, the Rayleigh–Jeans analysis deserves our consideration.

A well-known result from classical statistical mechanics is that each independent dynamical coordinate that appears quadratically in the expression for the energy of the system, contributes a mean energy of $\frac{1}{2}\kappa T$ to the system, κ being Boltzmann's constant. Since for the electromagnetic field we have

$$U = \sum_{\mathbf{k}} U_{\mathbf{k}} = \sum_{\mathbf{k}} S^0{}_0(\mathbf{k})V = \frac{V}{\mu_0}\sum_{\mathbf{k}} \partial^0 A^\lambda(\mathbf{k})\partial_0 A_\lambda(\mathbf{k}), \tag{6.29}$$

where the sum is extended over all values of the wave vector $\mathbf{k}$, it is clear that the "coordinate" $\partial^0 A^\lambda(\mathbf{k})$ appears quadratically in the energy. Hence it would seem reasonable to associate with each of the $\partial^0 A^\lambda(\mathbf{k})$ a mean energy of $\frac{1}{2}\kappa T$. However, we can impose the constraints

$$\mathbf{k}\cdot\mathbf{A} = 0, \qquad A^0 = 0$$

in some frame of reference, thanks to the freedom allowed us by gauge transformations of the second kind. Therefore only two of the four components $\partial^0 A^\lambda(k)$ are independent, so that the mean energy contributed by a wave with wave vector $\mathbf{k}$ might be expected to be $2\cdot\frac{1}{2}\kappa T = \kappa T$.

This argument, however, is incorrect. We have used the rules for assigning mean energies to mechanical oscillators, when we are not in fact dealing with mechanical oscillators. To correct this, we assume that the walls of the container consist of mechanical oscillators, and that each mode of vibration in the walls is responsible for a mode of oscillation in the radiation field. The energy of each oscillator in the walls is given by

$$W_j = \frac{1}{2m_j}p_j^2 + \frac{m_j\omega_j^2}{2}x_j^2$$

where m_j is the mass of the jth oscillator, ω_j is its frequency, and p_j and x_j are, respectively, the momentum and position coordinates in the direction of the oscillation. Thus the mean energy of each mode of oscillation is $2 \cdot \frac{1}{2}\kappa T$, since two quadratic coordinates, p_j and x_j, are involved. (For the interested student, a good description of this kind of analysis is given in Lecture 41 of Volume I of *The Feynman Lectures on Physics*, Addison-Wesley Pub. Co., Reading, Mass. (1963).) Our error was to assign $\frac{1}{2}\kappa T$ to each quadratic variable of the electromagnetic field rather than to *assign* κT *to each independent mode of oscillation.* This illustrates the care that must be taken in applying the principle of equipartition of energy.

Therefore we assign to each wave vector **k** a mean energy of $2\kappa T$. This corresponds to assigning κT to each of the two polarizations of an electromagnetic wave.

Now we somehow must count the number of values of **k** that lie within the increment Δk of wave numbers between k and $k + \Delta k$. Then if this number is $\Delta N = N(k)\,\Delta k$, we shall have the following amount of energy stored in waves having wave numbers between k and $k + \Delta k$:

$$u(k)\,\Delta k = V\psi(k)\,\Delta k = 2\kappa T\,\Delta N = 2\kappa T N(k)\,\Delta k$$

where $u(k)$ is the energy per unit wave number between k and $k + \Delta k$. That is,

$$\psi(k) = \frac{2\kappa T}{V} N(k). \qquad (6.30)$$

Equation (5.56) provides us with a basis for finding $N(k)$; it says that the allowed values of **k** are those for which the components k^m are integer multiples of π/L_m. We must find out how many such allowed wave vectors correspond to wave numbers between k and $k + \Delta k$. To do this, we construct a wave-vector space, or k-space, with orthogonal coordinates corresponding to k^1, k^2 and k^3. Then to each triple of integers (n_1, n_2, n_3), there corresponds a point in this k-space, namely $(k^1, k^2, k^3) = (n_1\pi/L_1, n_2\pi/L_2, n_3\pi/L_3)$. Each such point lies at the tip of an allowed **k**-vector, and each serves as a corner for eight rectangular parallelepipeds having edges π/L_1, π/L_2 and π/L_3. This is shown in Fig. (6.3)

Although each point is a corner for eight different parallelepipeds, each parallelepiped has eight such corners; thus we may assign each point unambiguously to one parallelepiped. Hence there are as many allowed **k**-vectors per unit volume of k-space as there are parallelepipeds per unit volume of k-space, namely

$$\frac{1}{V_k} = \frac{L_1 L_2 L_3}{\pi^3} = \frac{V}{\pi^3}$$

where $V_k = (\pi/L_1)(\pi/L_2)(\pi/L_3)$ is the k-volume of one parallelepiped, and $V = L_1 L_2 L_3$ is the volume of the enclosure. The density of allowed wave vectors

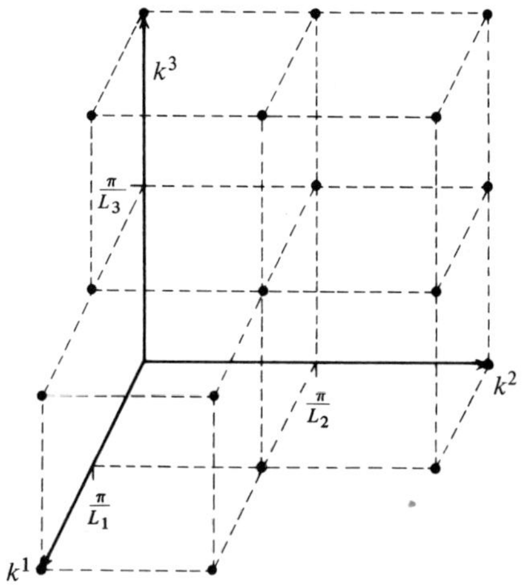

Fig. 6.3. Points in k-space corresponding to allowed values of **k**. Each such point is a vertex of a rectangular parallelepiped or "cell".

per unit volume of k-space η, then, is

$$\eta = \frac{V}{\pi^3}.$$

If k is a sufficiently large number, we can say that there are

$$N' = \eta \frac{4\pi}{3} k^3 = \frac{4\pi}{3} \frac{V}{\pi^3} k^3$$

points within a sphere having radius k in k-space. Hence, within a spherical shell of thickness Δk there are

$$\Delta N' = \frac{4\pi V}{\pi^3} k^2 \, \Delta k$$

points. Only $\frac{1}{8}$ of these points correspond to allowed wave numbers, however, because only positive values of n_m are required to give all possible solutions to the wave equations for radiation in the enclosure; therefore

$$\Delta N = \tfrac{1}{8} \Delta N' = \frac{V}{2\pi^2} k^2 \, dk = N(k) \, \Delta k,$$

or

$$N(k) = \frac{V}{2\pi^2} k^2. \tag{6.31}$$

Using Eq. (6.31) in (6.30), we find that the energy density per unit wave number is given by

$$\psi(k) = \frac{\kappa T}{\pi^2} k^2 = \frac{\kappa}{\pi^2}\left(\frac{k}{T}\right)^2 T^3. \tag{6.32}$$

This is related by Eq. (6.21) to the observable emissive power of the radiation in the enclosure. For the Rayleigh–Jeans analysis, then, one obtains

$$g(k/T) = (c\kappa/4\pi^2)(k/T)^2.$$

The unbounded increase of this function as k increases is known as the *ultraviolet catastrophe.*

In Fig. 6.4 the Rayleigh–Jeans law and Wien's hypothesis are compared with the experimental data. The failure of the classical theories is quite apparent. It was in an attempt to resolve this difficulty that Max Planck, in 1901, came upon the quantum hypothesis, which provided an expression for $\psi(k)$ that fitted the experimental data for all values of k.

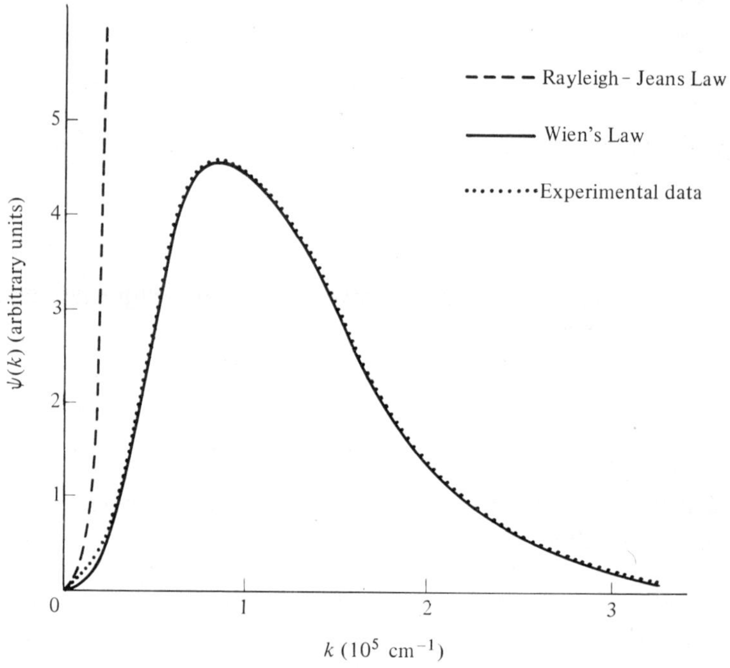

Fig. 6.4. Comparison of Wien's law and the Rayleigh–Jeans law with experiment (the vertical scale changes with temperature). The scale of the k-axis was chosen to illustrate the agreement between the Wien law and experiment for high frequencies.

6.5 PLANCK'S RESOLUTION OF THE PROBLEM

It has long been known that oscillating charges (e.g. electrons in simple harmonic motion) radiate energy in the form of electromagnetic waves, and that radiation can cause charges to oscillate. Thus there exists an interaction between matter and electromagnetic radiation, and in particular the radiation in an enclosure must be interacting continually with the walls of the container; electrons in the material of the container walls, for example, exchange energy with the radiation field. Realizing this, Planck began his analysis by investigating the *oscillators* in the walls of the enclosure and also the manner in which they are connected with the radiation field. Later he realized that similar arguments may be used to treat the electromagnetic field itself, in analogy with the manner in which the Rayleigh–Jeans theory was developed.

As a preliminary to discussing Planck's work, we must recall a fundamental result from statistical mechanics, namely that the probability for a system in thermal equilibrium to be in a particular state having energy E_s is proportional to $e^{-E_s/\kappa T}$ when there are many states available to the system. This follows from the laws of large numbers, and does not depend on any assumption of classical physics; it holds in quantum physics also.

We can indicate how this form for the probability, as a function of energy, is obtained in the case that the probability per unit energy interval varies continuously with the energy. Let $\rho_1(E_1)$ be the probability per unit energy (probability density) that system 1 is in a state with energy E_1, and let $\rho_2(E_2)$ have the corresponding meaning for system 2. Now if the systems are in equilibrium, the total system composed of systems 1 and 2 has probability per unit energy interval $\rho(E)$ of being in a state having energy $E = E_1 + E_2$. If we assume that systems 1 and 2 are *statistically independent*, that is, that the probability density $\rho(E)$ for the whole system to be in a state with energy $E_1 + E_2$ is proportional to the probability density for system 1 to be in a state with energy E_1 multiplied by the probability density for system 2 to be in a state having E_2, we have

$$\rho(E_1 + E_2) = \alpha\rho_1(E_1)\rho_2(E_2).$$

Furthermore, if systems 1 and 2 are essentially identical (and in fact even if they are not), we may assume that ρ_1, ρ_2 and ρ have the same functional dependence on energy, so that we can drop the subscripts on the ρ's. We then have

$$\frac{1}{\rho(E_1 + E_2)} \frac{d}{d(E_1 + E_2)} \rho(E_1 + E_2) = \frac{1}{\rho(E_1)} \frac{d}{dE_1} \rho(E_1) = \frac{1}{\rho(E_2)} \frac{d}{dE_2} \rho(E_2).$$

Hence

$$\frac{1}{\rho(E)} \frac{d}{dE} \rho(E) = \text{const.} = -\beta$$

and

$$\rho(E) = A\,e^{-\beta E},$$

where A is independent of energy and β is positive in order that we may get sensible results. The fact that $\beta = 1/\kappa T$ is found by investigating the possible connections between statistical quantities and thermodynamic variables.

A similar discussion leads to the analogous result when the probability does not vary continuously with energy.

Planck proposed to study the statistical behavior of oscillators when their energies were restricted to values that differed by finite amounts, and then to take the limit as the minimum difference between energy levels approached zero. He found, however, that he obtained a more satisfactory result by *not* letting the minimum energy difference vanish!

Let us take

$$P(E_s) = CN_s e^{-E_s/\kappa T}$$

as the probability that our system (an oscillator) has energy E_s. Here N_s is the number of states having energy E_s, and C is a constant to be determined by the requirement that the probability for the system to have some allowed energy is unity:

$$\sum_s P(E_s) = 1.$$

Thus

$$C = \left[\sum_s N_s e^{-E_s/\kappa T}\right]^{-1}$$

and

$$P(E_s) = \frac{N_s e^{-E_s/\kappa T}}{\sum_r N_r e^{-E_r/\kappa T}}. \tag{6.33}$$

Now consider a one-dimensional harmonic oscillator with energy

$$E(p, x) = \frac{p^2}{2m} + \frac{m\omega^2}{2} x^2.$$

The states having energy $E(p, x)$ can be represented by an ellipse in the px-plane, having semiaxes

$$p_E = \sqrt{2mE}, \qquad x_E = \sqrt{\frac{2E}{m}} \cdot \frac{1}{\omega};$$

all points on this ellipse have energy E (see Fig. 6.5). Thus all points within an elliptical ring between the ellipses defined by E and $E + \Delta E$ have energy between E and $E + \Delta E$; the area of this ring is

$$\pi(p_{E+\Delta E}x_{E+\Delta E} - p_E x_E) = \frac{2\pi}{\omega} \Delta E.$$

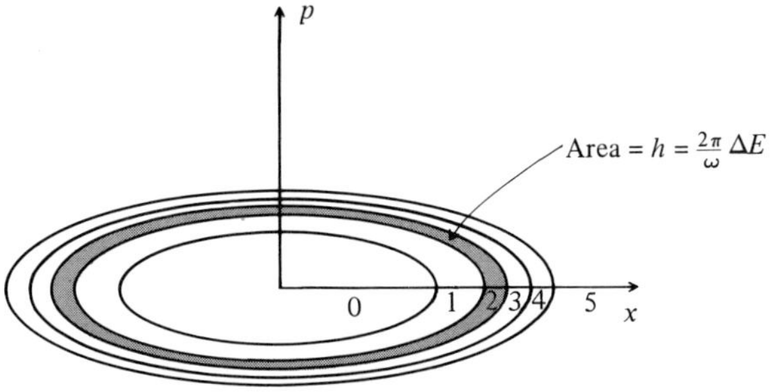

Fig. 6.5. Ellipses of constant energy in the xp-plane for a harmonic oscillator. The numbers are the indices $n = 0, 1, 2, \ldots$, and the area between ellipses is constant, namely $h = 2\pi\, \Delta E/\omega$.

The number of states, N_E, with energies between E and $E + \Delta E$ is proportional to this area:

$$N_E = \gamma \frac{2\pi}{\omega} \Delta E, \tag{6.34}$$

where γ is a constant of proportionality.

The mean energy of an oscillator in this ring is given by

$$\begin{aligned} \bar{E} &= \frac{1}{(2\pi/\omega)\, \Delta E} \int_{\text{ring}} E\, dx\, dp = \frac{\omega}{2\pi\, \Delta E} \frac{\pi}{\omega} [2E\, \Delta E + (\Delta E)^2] \\ &= E + \tfrac{1}{2}\Delta E. \end{aligned}$$

Now let us consider a set of confocal ellipses in the xp-plane, the area between each pair of ellipses being held constant at the value $2\pi\, \Delta E/\omega = h$. Then we shall assign, to all the oscillators within a given ring, an energy equal to their mean energy,

$$\bar{E}_n = E_n + \tfrac{1}{2}\Delta E, \qquad n = 0, 1, 2, \ldots,$$

where n is an index indicating the particular ring under consideration and E_n is the energy on the interior boundary of the ring. Thus, for the central ring, we have

$$\bar{E}_0 = 0 + \tfrac{1}{2}\, \Delta E = \frac{1}{2} \frac{h}{2\pi} \omega;$$

for the next ring

$$\bar{E}_1 = \Delta E + \tfrac{1}{2}\, \Delta E = \frac{3}{2} \frac{h}{2\pi} \omega;$$

for the nth ring

$$\bar{E}_n = (n + \tfrac{1}{2})\, \Delta E = (n + \tfrac{1}{2}) \frac{h}{2\pi} \omega. \tag{6.35}$$

Since we shall be dealing repeatedly with the ratio $h/2\pi$, we assign to it the symbol $\hbar = h/2\pi$.

Using (6.35) in conjunction with (6.33) and (6.34), we get

$$P(\bar{E}_n) = \frac{N_{\bar{E}_n} e^{-\bar{E}_n/\kappa T}}{\sum_{m=0}^{\infty} N_{\bar{E}_m} e^{-\bar{E}_m/\kappa T}} = \frac{e^{-n\hbar\omega/\kappa T}}{\sum_{m=0}^{\infty} e^{-m\hbar\omega/\kappa T}}$$

as the probability that a single oscillator lies in the ring with mean energy $\bar{E}_n$. The sum in the denominator is evaluated easily using the formula

$$\frac{1}{1-x} = \sum_{k=0}^{\infty} x^k$$

and letting $x = e^{-\hbar\omega/\kappa T}$. Thus we obtain

$$P(\bar{E}_n) = (1 - e^{-\hbar\omega/\kappa T})\, e^{-n\hbar\omega/\kappa T}. \tag{6.36}$$

The mean energy of an oscillator for which all the energy levels $\bar{E}_n$ are available, then, is

$$\begin{aligned}\bar{E} &= \sum_{n=0}^{\infty} \bar{E}_n P(\bar{E}_n) = \sum_{n=0}^{\infty} (n + \tfrac{1}{2})\hbar\omega(1 - e^{-\hbar\omega/\kappa T})\, e^{-n\hbar\omega/\kappa T} \\ &= \hbar\omega(1 - e^{-\hbar\omega/\kappa T})\left(\sum_{n=0}^{\infty} n\, e^{-n\hbar\omega/\kappa T}\right) + \tfrac{1}{2}\hbar\omega.\end{aligned}$$

Letting $\alpha \equiv \hbar\omega/\kappa T$ for the moment, we have

$$\begin{aligned}\sum_{n=0}^{\infty} n e^{-n\alpha} &= -\frac{d}{d\alpha}\sum_{n=0}^{\infty} e^{-n\alpha} = -\frac{d}{d\alpha}\frac{1}{1-e^{-\alpha}} \\ &= \frac{e^{-\alpha}}{(1-e^{-\alpha})^2} = e^{-\hbar\omega/\kappa T}/(1 - e^{-\hbar\omega/\kappa T})^2.\end{aligned}$$

Hence

$$\bar{E} = \frac{\hbar\omega}{e^{\hbar\omega/\kappa T} - 1} + \tfrac{1}{2}\hbar\omega$$

should be the mean energy of the oscillator.

This expression, however, again has a potential *ultraviolet catastrophe*; the first term vanishes as $\hbar\omega$ becomes arbitrarily large, whereas the second term becomes arbitrarily large. We gain nothing by taking the limit as $\hbar \to 0$, for then we get the mean energy that led to Rayleigh's formula,

$$\bar{E} \underset{\hbar\to 0}{\to} \kappa T.$$

On the other hand, if we look for the origin of the $\frac{1}{2}\hbar\omega$ term, we find that it came from taking the energy of the oscillator to be equal to the mean value of all the possible energies when p and x lay within one of the elliptical rings. This means that the lowest energy available to the oscillator would be $\frac{1}{2}\hbar\omega$. The mean level *above* the lowest level, then, is

$$\bar{E}' = \frac{\hbar\omega}{e^{\hbar\omega/\kappa T} - 1}. \tag{6.37}$$

This expression with the $\frac{1}{2}\hbar\omega$ term missing is the important one in physical applications.

Now we have analysed an oscillator that presumably exists within the wall of the enclosure, but we really are interested in the radiation field. If the radiation field interacts significantly with the oscillators in the walls, it can exchange energy, with an oscillator having frequency ω, only in "bundles" of magnitude $n\hbar\omega$, since energy must be conserved in the process of interaction. Therefore we may expect the part of the radiation field that is interacting with an oscillator having frequency ω, to behave like an oscillator with energies $n\hbar\omega$, $n = 0, 1, 2, \ldots$. It is easy to see that the mean energy of this part of the radiation field then would be

$$\bar{E}_\omega = 2\bar{E}' = \frac{2\hbar\omega}{e^{\hbar\omega/\kappa T} - 1}; \tag{6.38}$$

the factor 2 in the numerator comes from the presence of two possible polarizations for each electromagnetic wave having frequency ω.

At this point, let us return to Eq. (6.30), which gives an expression for the energy density per unit wave number when the mean energy per electromagnetic wave with frequency ω is given by $2\kappa T$. Planck's analysis suggests that we replace that $2\kappa T$ with $\bar{E}_\omega$ as given in Eq. (6.37):

$$\psi(k) = \frac{\bar{E}_\omega}{V} N(k) = \frac{2\hbar\omega}{e^{\hbar\omega/\kappa T} - 1} \frac{N(k)}{V}.$$

Using Eqs. (6.31) for $N(k)$, we get

$$\psi(k) = \frac{\hbar\omega}{e^{\hbar\omega/\kappa T} - 1} \frac{k^2}{\pi^2} = \frac{\hbar c}{\pi^2} \frac{k^3}{e^{\hbar ck/\kappa T} - 1}$$

$$= \frac{\hbar c}{\pi^2} \frac{(k/T)^3}{e^{(\hbar c/\kappa)(k/T)} - 1} T^3. \tag{6.39}$$

In the limit as $k \to 0$, we get $\bar{E}_\omega \to 2\kappa T$ and

$$\psi(k) \underset{k \to 0}{\sim} \frac{\kappa}{\pi^2} \left(\frac{k}{T}\right)^2 T^3,$$

which is Rayleigh's result, whereas in the limit as $k \to \infty$, we get $\bar{E}_\omega \sim 2\hbar\omega e^{-\hbar\omega/\kappa T}$ and

$$\psi(k) \underset{k \to \infty}{\sim} \frac{\hbar c}{\pi^2} e^{-(\hbar c/\kappa)(k/T)} \left(\frac{k}{T}\right)^3 T^3,$$

which agrees with Wien's guess as given in Eq. (6.28) if we take $a = \hbar c^2/4\pi^2$ and $\beta = \kappa/\hbar c$ in Eq. (6.28). In fact, Eq. (6.39) gives excellent agreement with experimental data for *all* wave numbers k if we choose

$$\hbar = 1.05459 \times 10^{-34} \text{ J sec.} \tag{6.40}$$

Thus, *by allowing h to remain finite*, contrary to the usual classical assumptions, and by fitting the resulting formula for the spectral distribution of energy to the experimental data, Planck was able to obtain good agreement between "theory" and experiment, and to evaluate $\hbar$ as a fundamental physical constant.

The success of this formula represents a spectacular turning point in the history of natural philosophy. However, before a clarification of the meaning of this transition in thought could be achieved, more than a quarter of a century of intensive investigation by the world's leading physicists was required.

Evidently Planck's result implies that only certain energies are allowed to an oscillator in the wall of an enclosure, namely the energies

$$\bar{E}_n = (n + \tfrac{1}{2})\hbar\omega, \qquad \omega = \text{oscillator frequency.}$$

This is a radical departure from classical ideas, which insist that the energy be allowed any value from 0 to ∞, and which do not relate the energy directly to the frequency of oscillation. If Planck's result has the significance we have given it, an oscillator has only a denumerable infinity of possible energy levels, rather than a nondenumerable infinity. Once again, the phenomenological agreement of a formula with experimental data forces us to re-examine our intuitive ideas about physics. As with Einstein's theory of special relativity, Planck's law is not based upon classical "first principles", but it *works* where the classical laws fail!

A second conclusion to be drawn from the success of Planck's formula is that the coupling that was assumed to exist between the oscillators in the walls of the enclosure and the radiation field does indeed provide a correct description of the situation. Thus the electromagnetic field oscillations have energies equal to integer multiples of the characteristic frequencies of the oscillators in the walls. This can be interpreted as meaning that the energy in the electromagnetic field corresponds to the energy emitted by one of the mechanical oscillators as it makes a transition from one state $\bar{E}_n$ to another lower state, $\bar{E}_m$:

$$\bar{E}_\omega = \bar{E}_n - \bar{E}_m = (n - m)\hbar\omega.$$

These conclusions really are more than we should swallow, however. In the first place, our analysis of the oscillators has been very naïve; they have been one-dimensional harmonic oscillators, which is hardly physically reasonable. In the second place, a study of the black-body radiation spectrum is not likely to provide us with much information concerning the interaction between the field and the oscillators. All we really know is that Planck's law works, perhaps in spite of the manner in which it was derived, and that therefore it probably gives a correct description of certain aspects of the behavior of the electromagnetic field. The oscillators, or whatever may exist in the walls of the cavity, may do what they please, but the electromagnetic field is evidently restricted to energy levels differing by integer multiples of $\hbar\omega$. This is referred to as the quantization of energy in the electromagnetic field.

This modest interpretation of Planck's result still carries the disturbing feature that the energy in the field is resolved into little bits having energies $n\hbar\omega$. In fact, this far-reaching idea was not fully incorporated into the mathematical structure of quantum theory for more than twenty-five years after Planck's discovery; the work of Bohr, de Broglie, Schrödinger, Dirac, and many others had to be digested before physicists really could come to grips with the quantum properties of electromagnetic fields.

6.6 PHOTONS AND ELECTRONS

This brings us to the phenomenon of the photoelectric effect, which, as Einstein showed in 1905, involves the quantization of energy in the electromagnetic field. Photoelectric emission was discovered by H. Hertz in 1887, and it was observed independently by J. J. Thomson (better known for his discovery of the electron in 1887) and P. Lenard in 1899.

In essence, the effect is that when high-frequency light is shone on a metal surface, electrons are emitted from the surface. The number of electrons emitted per unit time from the illuminated surface depends upon several factors, but most strikingly it is proportional to the intensity of the illumination. This is what we might naïvely expect from the classical pre-Planck's law theory of electromagnetic fields, for as the energy deposited on the surface per unit time by the radiation increases, the number of electrons gaining enough of this energy to overcome the potential that keeps them bound to the surface should increase.

However, three features of the phenomenon are at odds with what one would expect classically. First, the maximum kinetic energy of electrons emitted is independent of the intensity of illumination but increases as the frequency of the light waves is increased. Second, there is a threshold frequency; that is, if a monochromatic beam of light is used, no electrons will be emitted unless the frequency of the light is above a certain minimum value that depends on the nature of the emitting surface; this occurs no matter how intense the beam may be. Third, the emission of electrons begins almost immediately upon illumination of the surface, again no matter how intense the beam may be.

Classical theory predicts that the kinetic energy of emitted electrons should not have an upper bound and that the spectrum of electron energies should be independent of light frequency though closely related to light intensity. It also predicts that no threshold for illuminating frequency should exist, and that there should be a time lag between the onset of illumination and the onset of electron emission, this time lag becoming shorter as intensity increases. These predictions all arise from the proportionality of the radiation intensity (energy delivered per unit area per unit time) to the square of the field strength of the radiation field. Only after enough energy has been deposited in the vicinity of an electron for that electron to overcome its binding potential, should an electron escape from the metal, and as the radiation continues to impinge on the electron, the latter may continually gain energy. In the pre-Planck theory, no relation between frequency and energy existed.

Einstein solved this puzzle by applying Planck's hypothesis to the radiation, postulating that the incident radiation consisted of "energy bundles", or energy quanta, each quantum in a monochromatic beam with frequency $\nu = \omega/2\pi$ having energy $\hbar\omega = \hbar ck$. Furthermore, he assumed that the interaction mechanism between the radiation and the electron was such that the electron absorbed energy only in these discrete amounts. Thus, if the potential energy that binds the electron to the metal is W_0, and if the electron loses energy W' in working its way from a location in the interior of the metal to the surface, then the kinetic energy of the electron, upon leaving the surface, is

$$T = \hbar\omega - W_0 - W'.$$

For an electron very near the surface, W' becomes vanishingly small, and such an electron receives the maximum possible kinetic energy,

$$T_{\max} = \hbar\omega - W_0 = \hbar(\omega - \omega_0), \qquad \omega_0 = W_0/\hbar, \tag{6.41}$$

where W_0 and hence ω_0 are characteristic of the metal and the experimental conditions.

The ability of a single electron to absorb energy in a "bundle" or burst is the most startling feature of this view of the photoelectric process. This means that the energy in the radiation field can be very well *localized*, or concentrated in a very small region of space. This localization of energy is a property associated with particles rather than with fields in classical theory, and thus Einstein's analysis seems to give electromagnetic fields a duality in that they have particle-like and wave-like properties at the same time. Built into the world-view of classical physics was a dichotomy between particles, with completely localized energy content, and fields, with widely distributed energy content (and hence associated with energy *densities* rather than energy itself). Although there is no *a priori* foundation for such a view, it does agree with our normal experiences, and the dichotomy was very hard to give up. Einstein's analysis discards this dichotomy.

We might note that the simultaneous localization of energy in a field and the possession of well-defined frequency by that field really is just as incompatible with modern quantum theory as it is with the classical world view. The notion that the radiation field contains well-localized quanta (i.e. their energy densities are proportional to delta-functions $\delta^3(\mathbf{x} - \mathbf{x}_0)$) with well-defined frequencies ω and energy contents $\hbar\omega$, is not tenable in quantum theory, even though such notions gave birth to the theory! In fact, the classical theory of wave fields and quantum theory are *both* inconsistent with this notion on basically the same grounds,

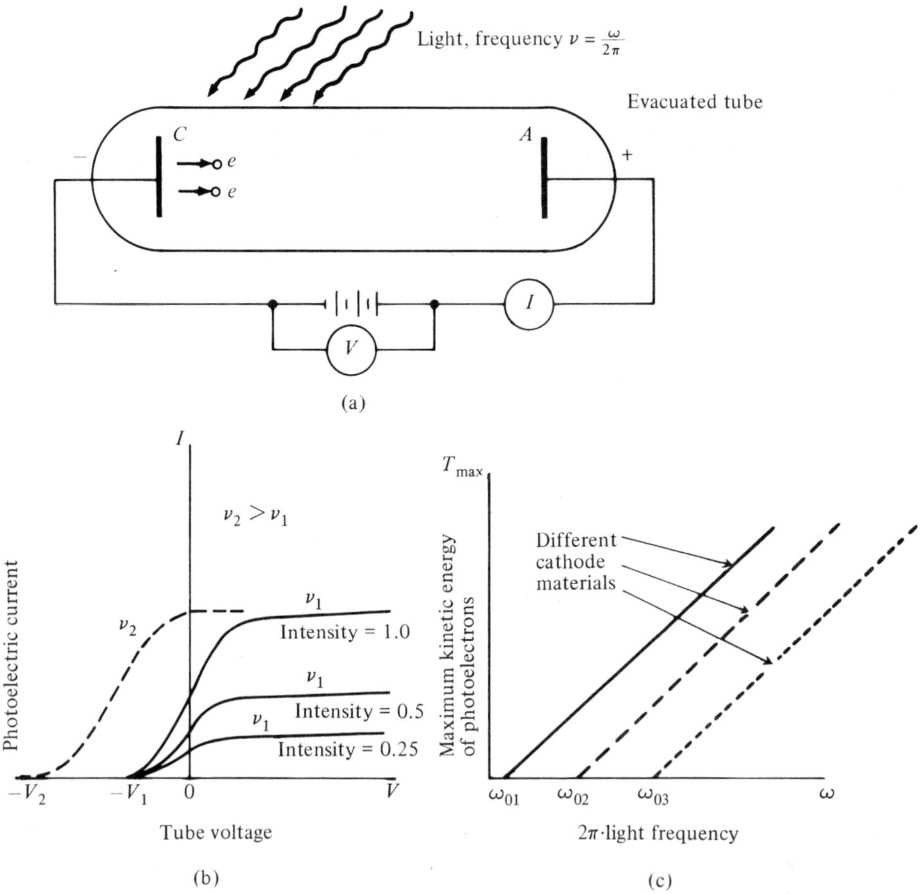

Fig. 6.6. The photoelectric effect. (a) Schematic representation of apparatus; light of frequency ν impinges on the cathode C of an evacuated tube, yielding electrons that are accelerated to anode A by voltage V, creating photoelectric current I. (b) I vs. V for two frequencies ν_2 and ν_1, $\nu_2 > \nu_1$, and for several intensities at frequency ν_1. (c) T_{max} vs. ω for three different cathode materials.

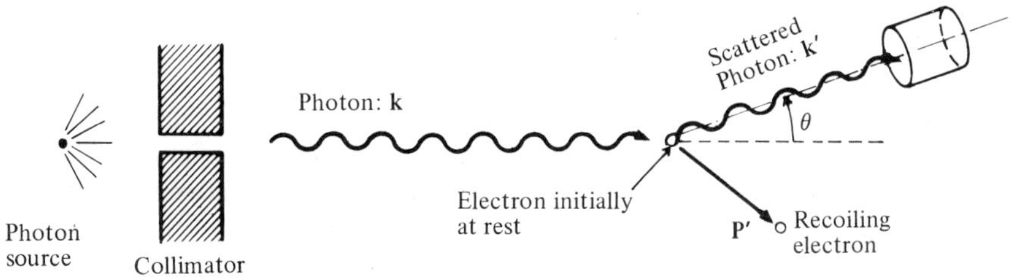

Fig. 6.7. Compton effect. A photon with initial frequency $\nu = ck/2\pi$ is scattered by a free electron and finally has frequency $\nu' = ck'/2\pi$.

as we shall see. Einstein's analysis works because in the photoelectric effect the frequency need not be *completely* well-defined, nor is the electron with which the radiation interacts *completely* localized; in each there is some leeway, or uncertainty, and by allowing a certain amount of uncertainty we can obtain a view that is consistent with the mathematical theory of wave fields. In a given experiment, says the quantum version of the theory, a phenomenon may be very *particle-like* or very *wave-like* depending upon the nature of the experiment, but no experiment can be constructed in which the phenomen is *simultaneously* as particle-like *and* as wave-like as we please; there are limits to the duality of physics.

Another quantum feature of electromagnetic radiation is the Compton effect, discovered in 1923 by Arthur H. Compton. He observed that when monochromatic x-rays are scattered from matter there is a strong component of the scattered radiation having a well-defined frequency different from that of the incident x-rays, but such that the scattered frequency is a function of the initial frequency and the angle through which the radiation is scattered. This is represented schematically in Fig. 6.7. A particularly simple case of this kind of scattering occurs when an x-ray quantum is scattered by a free electron (or other charged particle); the analysis is carried out most easily in the frame of reference in which the scattering particle is at rest initially.

We know that the electromagnetic field carries momentum, so if we are to represent it with quanta of energy, we must expect the relationship between the momenta and energies of the quanta to be the same as that between the momentum and energy of the electromagnetic field. Thus, since the energy density is related to the momentum density of the classical electromagnetic field by

$$\mathscr{P} = \hat{\mathbf{k}}\frac{\mathscr{E}}{c}$$

which, incidently, leads to $\mathscr{E}^2 - \mathscr{P}^2c^2 = 0$, we might expect that a quantum with energy $E = \hbar\omega$ would have momentum

$$\mathbf{p} = \hat{\mathbf{k}}\frac{E}{c} = \hat{\mathbf{k}}\frac{\hbar\omega}{c} = \hbar\mathbf{k} \tag{6.42}$$

for electromagnetic quanta. The relativistic equation relating mass, energy and momentum then leads to

$$m_{\text{electromagnetic quantum}} = \sqrt{E^2 - \mathbf{p}^2c^2} = 0;$$

hence the mass of a quantum of electromagnetic radiation is zero.

It is customary to refer to the particle-like, well-localized representatives of the electromagnetic field as *photons*, and to call them massless particles. (There are other massless particles also, notably neutrinos, that are not directly connected with electromagnetic radiation.) As we have noted already, it is inconsistent even in quantum theory to speak of photons with energy $\hbar\omega$; nevertheless, this terminology is widely used, and it is to be understood as referring to an approximately particle-like entity (*rather* well localized) in a beam with *rather* well-determined frequency ω. Thus a photon is thought of as a massless particle traveling at the speed of light and possessing finite energy and momentum.

Let us treat the collision of a photon and an electron as if both behaved like particles, using the conservation of four-momentum to relate the initial energy (frequency) of the photon to its final energy. If the photon momentum is $p^\mu = \hbar k^\mu$ and the electron momentum is P^μ initially, and finally they have momenta $p'^\mu = \hbar k'^\mu$ and P'^μ respectively, the conservation of momentum implies

$$p^\mu + P^\mu = p'^\mu + P'^\mu$$

or

$$(p - p')^\mu = (P' - P)^\mu. \tag{6.43}$$

However,

$$k^\mu = \left(\frac{\omega}{c}, \mathbf{k}\right), \qquad k'^\mu = \left(\frac{\omega'}{c}, \mathbf{k}'\right) \tag{6.44}$$

with

$$|\mathbf{k}| = \frac{\omega}{c}, \qquad |\mathbf{k}'| = \frac{\omega'}{c},$$

and if the electron mass is m, then

$$P^\mu = (mc, 0), \qquad P'^\mu = (m\gamma' c, \mathbf{P}'), \tag{6.45}$$

with

$$\gamma' = \sqrt{1 + (\mathbf{P}'/mc)^2}.$$

Hence Eqs. (6.43)–(6.45) yield

$$(\hbar/c)(\omega - \omega') = mc(\sqrt{1 + (\mathbf{P}'/mc)^2} - 1) \tag{6.46a}$$

and

$$\hbar(\mathbf{k} - \mathbf{k}') = \mathbf{P}'. \tag{6.46b}$$

Substituting (6.46b) into (6.46a) and using the relationship between the k's and ω's, we find

$$1 + \hbar/(mc^2)(\omega - \omega') = \sqrt{1 + (\hbar/mc_2)^2(\omega^2 + \omega'^2 - 2\omega\omega'\hat{\mathbf{k}} \cdot \hat{\mathbf{k}}')}.$$

Squaring both sides and solving for ω' yields

$$\omega' = \frac{\omega}{1 + (\hbar\omega/mc^2)(1 - \hat{\mathbf{k}} \cdot \hat{\mathbf{k}}')}. \tag{6.47}$$

A more familiar form for this relation is obtained by using $\lambda = 2\pi c/\omega$ in Eq. (6.47); thus

$$\lambda' - \lambda = \frac{2\pi\hbar}{mc}(1 - \hat{\mathbf{k}} \cdot \hat{\mathbf{k}}') = \frac{h}{mc}(1 - \hat{\mathbf{k}} \cdot \hat{\mathbf{k}}'), \tag{6.48}$$

and $\lambda_c \equiv h/mc$ is called the *Compton wavelength* for a particle of mass m. Sometimes $\lambda_c/2\pi = \hbar/mc$ is called the Compton wavelength, although "inverse Compton wave number" might be more appropriate, in order that the two quantities be distinguished. The quantity $\lambda_c/2\pi = \hbar/mc$ seems to be more useful in computation; for the electron

$$\begin{aligned} \hbar/m_e c &= 3.86159 \times 10^{-13}\text{ m} \\ &= 3.86159 \times 10^{-3}\text{ Å} \\ &= 3.86159 \times 10^{2}\text{ F} \end{aligned}$$

where $1\text{ F} = 10^{-13}\text{ cm} = 10^{-15}\text{ m}$ is called a *Fermi*, and $1\text{ Å} = 10^{-8}\text{ cm} = 10^{-10}$ is the familiar *Ångström* unit.

Evidently the shift in wavelength of a photon scattered by a particle depends only upon the mass of the scattering particle and upon the angle of scattering, $\theta = \arccos(\hat{\mathbf{k}} \cdot \hat{\mathbf{k}}')$; it is proportional to the Compton wavelength and to $(1 - \hat{\mathbf{k}} \cdot \hat{\mathbf{k}}')$. At most, the wavelength may be shifted by two Compton wavelengths:

$$(\lambda' - \lambda)_{\max} = 2(h/mc).$$

In fact, these features of the particle-like scattering of a photon agreed fully with the experiments of Compton, and this gave striking confirmation to the quantum hypothesis for electromagnetic radiation.

In these discussions, we have liberally used the term *electron*, although the essential properties of this particle have not been used in any significant way. All we really have assumed (except in giving the Compton wavelength for an electron) is that the electron is an electrically charged particle (and hence interacts with the electromagnetic field) and that it is rather easily obtained from bulk matter; the latter explains the dominant role played by electrons in studying the electromagnetic field and hence in studying photons. We summarize here some important features of electrons and of the history of their discovery.

The electron was discovered in the 1890's when experimenters observed that an electric current flowed through an evacuated tube when metal plates were placed in the tube and a high voltage applied between the plates. By measuring the deflection of the current-carrying beam by a known electric field and then measuring the magnetic field required to restore the beam to its original path, J. J. Thompson was able to show that if the beam consisted of charged particles, the ratio e/m of electric charge to mass of the particles must be $e/m = 1.76 \times 10^8\text{ C g}^{-1} = 5.27 \times 10^{17}\text{ esu g}^{-1}$. He found that the electric charge must be negative. Incidentally, this value of e/m coincided exactly with that found for the negatively charged component of the radiation from radioactive materials. The latter were originally called β-rays, and were discovered independently in 1899 by several workers, including H. Becquerel, who originally called attention to the existence of natural radioactivity; it has been shown that electrons and β-rays have the same physical properties and therefore are identical.

Later, in 1909, Millikan succeeded, in his famous oil-drop experiment, in finding the electronic charge to be about

$$e = 4.80 \times 10^{-10}\,\text{esu} = 1.60 \times 10^{-19}\,\text{C};$$

consequently the electronic mass is approximately

$$m = 9.11 \times 10^{-28}\,\text{g}.$$

That e/m should be a constant characteristic of electrons is a rather surprising result, for it means that if the electrons all have the same mass, then they all have the same charge. Millikan's experiment confirmed that this is the case. Many other experiments have been conducted (including some rather recent ones) to seek "elementary" particles with electric charges other than integer multiples of e; all these attempts have failed.

The masses of elementary particles in nature seem not to obey a simple rule, however; they are not integer multiples of a fundamental mass. This is understandable because of the mass–energy equivalence, since any interactions tending to bind particles together give rise to changes in the mass of a system. Nevertheless, the spectrum of elementary particle masses is rather peculiar, and one of the challenges of modern physics is to understand the systematics of that spectrum. Some progress seems to have been made in recent years, but most of it is based upon as-yet unconfirmed speculation.

Another deep challenge, of course, is to learn "why" elementary particles have charges differing only by integer multiples of e. It seems quite likely that the mass and charge spectra are somehow related.

READING NOTES*

Reference 23 provides an excellent introductory discussion in the first eight chapters. See especially Chapter 1.

Reference 14 is another popular modern physics text and covers parallel material in Chapters 2 and 3.

Reference 1, Chapter 1, should be consulted for a scholarly historical treatment.

Reference 15 is an excellent text on modern physics and covers the material of the present chapter in Chapter 2.

Reference 16, Chapter IV.

* Works referred to in these notes will be found in the Bibliography on p. 577.

Reference 44, Vol. I, has an excellent brief historical account in Chapter 1.

PROBLEMS

1. Let $\epsilon'(\lambda)$ be the emittivity per unit wavelength, and show that

$$\epsilon'(\lambda_2) = (T_2/T_1)^5 \epsilon'(\lambda_1)$$

gives the variation of emittivity per unit wavelength with temperature, i.e. Wien's displacement law.

2. Derive, directly from Planck's radiation law for energy density in a radiation-filled isothermal enclosure, the Wien displacement law in the form $\lambda_{\max} T = \text{constant}$, where $\lambda_{\max}$ is the wavelength at the peak of the energy density spectrum. Evaluate the constant.
3. Derive the Stephan–Boltzmann law directly from Planck's radiation law. Evaluate the Stephan–Boltzmann constant in terms of the fundamental constants in Planck's law.
4. Prove that the mean kinetic energy of a simple harmonic oscillator is half the total energy. Thus the mean total energy of such an oscillator is κT when its mean kinetic energy is $\kappa T/2$.
5. Light having frequency of about 10^5 Hz generates a photocell current in vacuum without any lag in phase. Show that the rapidity with which light energy must be converted into photoelectron energy in this case is inconsistent with classical physics, by finding the time required for yellow light ($\lambda = 6 \times 10^{-7}$ m) at an intensity of 10^4 lumen m^{-2} (621 lumen $= 1$ watt) to leave an energy of $\hbar\omega$ within an area about as big as the cross-sectional area of an atom (about 10^{-10} m in diameter) from which the photoelectron is to emerge.
6. Calculate the energy (relativistic) of the recoil electron from a Compton collision in which a photon with initial energy $\hbar\omega$ scatters from an electron at rest.

CHAPTER 7

THE ELEMENTARY STRUCTURE OF THE ATOM

The knowledge that electrons exist in matter naturally led to the speculation that they play an important role in the generation of electromagnetic radiation and in mediating the interactions between matter and the radiation field. On the other hand, most matter that we ordinarily deal with is electrically neutral; hence there must be positive charges in matter to offset the negative charges of the electrons. Evidently these positive charges are not as easily separated from matter in bulk as are electrons, since they are not readily observed in experiments such as J. J. Thomson's. What is the nature of these positive charges? What role do they play in producing radiation?* How are they and the electrons "attached" in matter? Can we visualize a model for the microscopic structure of matter such that the positive charges and electrons play roles that are consistent with the data that can be obtained from experiments?

Most students are familiar with currently accepted answers to these questions by the time they are in college. Because of the success of the atomic theory of chemistry based on Dalton's hypothesis, Gay-Lussac's law and Avagadro's rule (all of which were enunciated between 1803 and 1811) and culminating in Mendeleyev's periodic table (put together in 1869 when Maxwell was working on the final forms of his electromagnetic theory) it was mandatory that the microscopic model for matter must involve the positive and negative charges somehow stuck together in small neutral units. Thomson proposed an atomic model in which the electrons were embedded in blobs of positive charge; this was dubbed the "raisin pudding model", with the electrons serving as raisins and the positive charge representing pudding.

This model, however, was inconsistent with experiments performed by Rutherford in 1909, in which metal foils were bombarded with α-particles, the electrically positive component of emissions from naturally radioactive substances. Rutherford then proposed (in 1911) a model that did fit the data of his experiments. This model, however, seemed to be inconsistent with the laws of electromagnetism.

* It was recognized in the 1800's that any accelerated charge will produce radiation. Here we ask whether positive charges, when they are constrained as a constituent part of matter in bulk, are readily accelerated, and if so do they and the electrons share equally in the "normal" production of radiation.

Two years later, in 1913, Nils Bohr put forth a bold hypothesis that removed the essential difficulties encountered in Rutherford's model, but did so at the expense of intuitive classical notions.

Another perplexing problem was the origin of spectral lines. Atomic species have long been known to radiate strongly at certain well-defined frequencies and thereby to produce bright lines in the spectral analysis of light. Absorption of light by matter shows dark lines also. The bright lines are characteristic of the material being used to produce the radiation, and dark lines are characteristic of the absorbing material. Balmer and others in the late 1800's observed certain systematic relationships among the wave numbers at which the lines occurred for certain pure elements. It seemed rather clear that some sort of oscillator resonance was involved in the emission and absorption line phenomena, but the systematics of their frequencies did not elucidate the nature of the mechanism.

The success of Planck's quantum hypothesis for the electromagnetic field led to the suspicion that line spectra were related to quantum effects, but no successful theory for the normal modes of vibration was forthcoming. Thomson's model for the atom provided a system within which the electrons, once displaced from an equilibrium position, would exercise simple harmonic motion, and therefore would give off radiation with a frequency characteristic of their oscillations. However, the energy given up by the electrons as they radiated would have to be compensated by a corresponding loss in their potential energy relative to the "pudding" of positive charge, and hence they would tend to "sink" to the center of the positive charge distribution. Furthermore, Thomson's model failed to yield oscillation frequencies that were related according to the known systematics of atomic spectra.

The beauty of Bohr's proposal was that in unifying Rutherford's atomic model with quantum ideas, it eliminated the need for harmonic oscillators to exist within the atom. From Bohr's simple model, the basic features of the line spectrum emerged.

7.1 THE SCATTERING PROBLEM AND THE CLASSICAL NOTION OF CROSS-SECTION

Rutherford recognized that the fast charged particles emitted from radioactive substances might be used as probes to check the Thomson theory of the atom. This technique of using particles as projectiles to probe the structure of matter continues to be of utmost importance in physics. The procedure is to aim a well-collimated beam of particles, whose fundamental properties are known, at a sample of matter, and to observe the angular array of the incident particles after they have struck the *target* sample. The *scattered* array then is characteristic of the geometry of the sample, and more significantly, of its internal structure. (See Fig. 7.1.)

Let us begin with a collimated beam having cross-sectional area A and containing known particles whose velocities are all in the direction of the target.

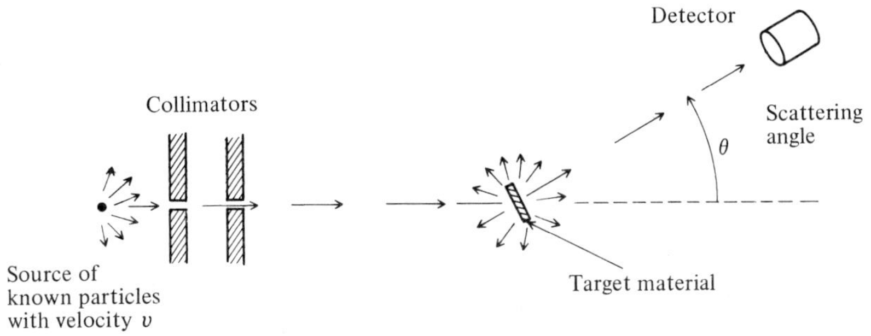

Fig. 7.1. Schematic diagram of scattering experiment.

The number of particles per unit volume in the beam will be denoted ρ_i. Thus the intensity of the beam, or the number of particles crossing unit area per unit time, is

$$I = \rho_i v,$$

and the number of particles crossing a plane perpendicular to the beam in unit time is

$$\frac{dN}{dt} = I\,A = \rho_i A v.$$

Now let us imagine an annulus, centered at an atom or scattering center in the target, with inside radius b and outside radius $b + \Delta b$; further, let us consider an angular sector at azimuthal angle φ and having angular width $\Delta\varphi$, as in Fig. 7.2. The plane of the annulus is perpendicular to the direction of the incident beam. Let us assume now that we can measure the number of particles aimed directly at the sector of the annulus just described (whose area is $\Delta\sigma(b, \varphi)$) in unit time, and denote this number $d[\Delta N(b)]/dt$. From this we could calculate the area $\Delta\sigma(b, \varphi)$, since

$$\frac{d}{dt}\Delta N(b) = \left(\frac{dN}{dt}\right)\frac{\Delta\sigma(b, \varphi)}{A};$$

thus

$$\begin{aligned} \Delta\sigma(b, \varphi) &= A\,\frac{d\,\Delta N(b)}{dt}\bigg/\frac{dN}{dt} \\ &= \frac{d\,\Delta N(b)}{dt}\bigg/ I \\ &= \frac{d\,\Delta N(b)}{dt}\bigg/ \rho_i v. \end{aligned} \tag{7.1}$$

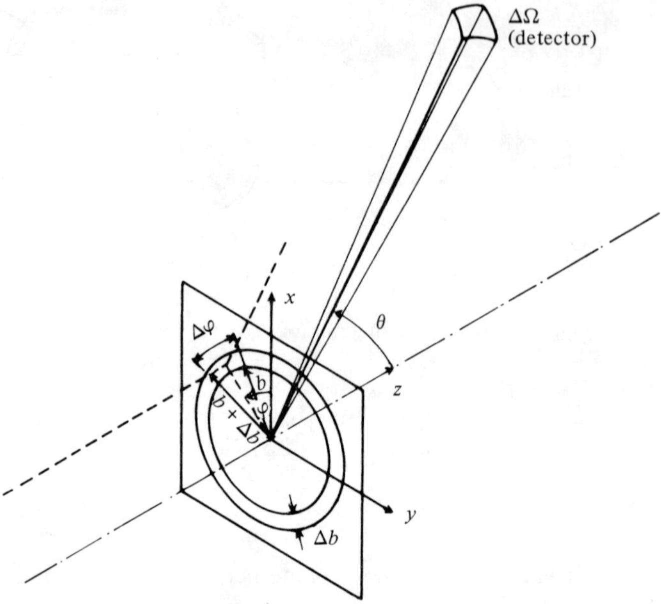

Fig. 7.2. Diagram for analysing the scattering of a classical particle incident along the dashed line from the left and aimed at a point at a distance between b and $b + db$ from the scattering center.

If we could find a way to measure $d[\Delta N(b)]/dt$, then, we would effectively measure the area $\Delta\sigma(b, \varphi)$. It turns out that $\Delta\sigma(b, \varphi)$ is, in fact, a useful quantity to use in comparing theory with experiment; it does not involve the irrelevant experimental parameter A.

One reason that $\Delta\sigma(b, \varphi)$ is a convenient parameter is that the quantity b, called the *impact parameter*, turns out to be related directly to the angle through which the particle finally is scattered when the forces responsible for the scattering are central forces. We shall see this in the Rutherford analysis. Thus $b = b(\theta)$. Now since

$$\Delta\sigma(b, \varphi) = b\,\Delta b\,\Delta\varphi,$$

we have

$$\Delta\sigma(b, \varphi) \equiv \Delta\sigma(\theta) = b\frac{db(\theta)}{d\theta}\Delta\theta\,\Delta\varphi = \frac{b(\theta)}{\sin\theta}\frac{db(\theta)}{d\theta}\sin\theta\,\Delta\theta\,\Delta\varphi.$$

Letting $\Delta\theta = d\theta$ and $\Delta\varphi = d\varphi$, and recognizing that $\sin\theta\,d\theta\,d\varphi$ is the differential of solid angle, usually denoted $d\Omega$, we get

$$\frac{d\sigma(\theta)}{d\Omega} = \frac{b(\theta)}{\sin\theta}\frac{db(\theta)}{d\theta}\cdot \tag{7.2}$$

The quantity $d\sigma(\theta)/d\Omega$ is called the *differential cross-section*, and $\sigma \equiv \int_{4\pi}(d\sigma/d\Omega)\,d\Omega$ often is called the total *cross-section* for the scattering. Note that σ has the units of area and $d\sigma/d\Omega$ has units of area per steradian. Effectively, $(d\sigma(\theta)/d\Omega)\,\Delta\Omega$ is equal to the area of the beam that gives rise to scattering into a solid angle $\Delta\Omega$ centered at angle θ away from the incident beam as a result of the presence of a single scattering center.

In most cases, the scattering experiment is cylindrically symmetric about the center of the beam, and hence the differential cross-section is independent of φ. Our derivation is not adequate when asymmetries are present, but the necessary corrections can easily be introduced.

In an actual experiment, the scattering that is observed is the result of scattering from many scattering centers, and we must somehow add together the scattering from the various centers. In the Thomson model, a sheet of metal, for example, might be thought of as being composed of closely packed spheres of positive charge with electrons stuck to the surfaces of the spheres. Thus a projectile would have to pass through many spheres of positive charge, and it might be scattered by many electrons as well. To handle this problem, one has to develop a statistical theory for the *multiple scattering* resulting from the many collisions the projectile suffers before it emerges from the metal and is observed. Rutherford did just that when he compared the model with experiment. We shall not pursue that problem here.

Rutherford's model of the atom, which unlike Thomson's was developed to give agreement with the scattering experiments, placed the positive charge within a very small volume (which turned out to have a diameter of about 10^{-13}–10^{-12} cm) at the center of the atom. Thus the interior of a piece of metal would be very sparsely populated with positive charge, and a single projectile might pass through a reasonably thin piece of material with a very small probability of being scattered more than once by positive charges. In this case, we should expect the scattered projectiles to have undergone only one "collision" when they emerge from the target.

Actually, "collision" is not an apt word in this context, for the scattering we are talking about is that arising from the electric forces between charged particles, and the particles need not come into contact to experience this force. On the other hand, the force diminishes quite rapidly with distance, and hence significant scattering effects occur only when a projectile approaches a positive nucleus of an atom fairly closely. The diminishing of the Coulomb force with increasing distance is strongly enhanced by the screening effect of the atomic electrons attached to the nucleus.

We have not discussed the role played by the atomic electrons in the scattering process. In fact, if we assume that the binding energy that holds an electron to the rest of the atom is much smaller than the kinetic energy of a projectile, we may consider the electron to be a free particle and analyse the electron–projectile collision as a free-particle collision. It is easily determined from conservation of energy and momentum alone that the momentum change in such a collision, when one of the particles is at rest, is given by

$$|\Delta \mathbf{p}_1| = \left| \frac{2m_2 E_1 p_1 \cos\theta}{(m_2 c + E_1/c)^2 - p_1^2 \cos^2\theta} \right| \leq \frac{2m_2 E_1 p_1}{m_2^2 c^2 + m_1^2 c^2 + 2m_2 E_1}$$

if $\mathbf{p}_1$ is the momentum of the incident projectile, $E_1 = \sqrt{m_1^2 c^4 + \mathbf{p}_1^2 c^2}$ is its energy, m_2 is the mass of the particle initially at rest, and θ is the angle between $\mathbf{p}_1$ and $\Delta\mathbf{p}_1$. The α-particles, which Rutherford proposed to use, have masses about equal to those of Helium atoms, or about 4×1836 electron masses; also, typical α-particle velocities for the proposed experiment would be about 2×10^9 cm sec^{-1} $\simeq \frac{1}{15} c$, so that the α-particle projectiles can be considered in nonrelativistic approximation. Then if m_1 is the α-particle mass and m_2 the electron mass, we have $m_2 \ll m_1$, $E_1 \simeq m_1 c^2$, $m_2 E_1 \ll m_1^2 c^2$, and hence

$$|\Delta \mathbf{p}_1| \leq \frac{2m_2 E_1 p_1}{2m_1^2 c^2} = \frac{m_2}{m_1} p_1 .$$

The maximum angle of deflection caused by such a collision is given by

$$\arctan \frac{\Delta \mathbf{p}_1}{\mathbf{p}_1} = \arctan \frac{m_2}{m_1} \simeq \frac{1}{4 \times 1836} \simeq 10^{-4} \text{ rad.}$$

Thus individual collisions with electrons can cause only very, very small deflections of α-particles.

Therefore we expect the scattering of α-particles to arise from individual encounters with positive nuclei in the atom of the target since, as has long been known, most of the atomic mass is carried by the positive charge. If the target is a foil of thickness τ, there will be a volume $A\tau$ of the target material lying within the beam, and the number of atoms in this volume will be

$$n = \frac{\rho_t}{\mathsf{A} N_0} A\tau \tag{7.3}$$

when ρ_t is the mass density of the target material, A is the atomic mass of the target material, and N_0 is Avagadro's number. Hence n nuclei are exposed to the beam, and each may be expected to provide a scattering center; the measured differential cross-section then should be given by

$$n \left(\frac{d\sigma}{d\Omega} \right)_{\text{single nucleus}} = n \frac{b(\theta)}{\sin\theta} \frac{db(\theta)}{d\theta} . \tag{7.4}$$

7.2 CENTER-OF-MASS CORRECTION

In the laboratory, the target is normally at rest, and aside from thermal motions, the scattering centers or atomic nuclei also are at rest. It is more convenient to carry out computations in the barycentric frame of reference than in the laboratory system, however. We shall label the particle variables measured in the barycentric frame with primes. Thus, since the barycentric frame for two particles is defined by

$$p'^{\mu} = (p_1'^{\mu} + p_2'^{\mu}) = (p_1'^0 + p_2'^0, 0),$$

or

$$\mathbf{P}' = \mathbf{p}_1' + \mathbf{p}_2' = 0,$$

where $\mathbf{p}_1'$ is the momentum of the *incident* particle and $\mathbf{p}_2'$ is the momentum of a scattering center in the target, we have

$$p_1'^0 = \sqrt{m_1^2c^2 + \mathbf{p}_1'^2}$$

$$p_2'^0 = \sqrt{m_2^2c^2 + \mathbf{p}_2'^2} = \sqrt{m_2^2c^2 + \mathbf{p}_1'^2}.$$

In the nonrelativistic approximation, that is, for $|\mathbf{p}_1'| \ll m_1c \leq p_1^0$ and $|\mathbf{p}_1'| \ll m_2c \leq p_2^0$, these become

$$p_1'^0 \simeq m_1c + \frac{\mathbf{p}_1^2}{2m_1c}, \qquad p_2'^0 = m_2c + \frac{\mathbf{p}_1'^2}{2m_2c}.$$

Hence

$$\begin{aligned} P^0 = p_1'^0 + p_2'^0 &= \sqrt{m_1c^2 + \mathbf{p}_1'^2} + \sqrt{m_2c^2 + \mathbf{p}_1'^2} \\ &\simeq (m_1 + m_2)c + \frac{1}{2c}\left(\frac{1}{m_1} + \frac{1}{m_2}\right)\mathbf{p}_1'^2 \qquad (7.5) \\ &= M'c + \frac{\mathbf{p}_1'^2}{2\mu c}, \end{aligned}$$

where $M' \equiv m_1 + m_2$ and $\mu \equiv 1/m_1 + 1/m_2 = m_1m_2/(m_1 + m_2)$ is called the reduced mass of the system. Thus, in the barycentric system (which is identical with the center-of-mass system in the nonrelativistic limit) the total energy has the form of the sum of a mass-energy term, $M'c^2$, and a kinetic-energy term, $\mathbf{p}_1'^2/2\mu$. In nonrelativistic problems, we are interested in only the kinetic-energy term; hence the total kinetic energy in the barycentric frame has the form of the kinetic energy of a single particle with mass μ and momentum $\mathbf{p}_1'$. Thus in this frame of reference, the two-body problem becomes very much like a one-body problem.

In fact, the nonrelativistic problem in which m_2 is infinite, so that $\mathbf{p}_2 = \mathbf{0}$ both before and after a collision, is identical with the nonrelativistic problem in the barycentric frame, provided the mass of the incident particle, m_1, is replaced with the reduced mass μ. In short, using the barycentric frame allows us to avoid dealing with the problems of recoiling scattering centers, or more precisely, to reduce such problems to the problem of performing a special Lorentz transformation, which nonrelativistically becomes a Galilean transformation.

7.3 THE RUTHERFORD CROSS-SECTION

Let us consider the vector diagram in which the initial momentum $\mathbf{p}'_1$ and the final momentum $\mathbf{p}'_3$ of the projectile are drawn, together with the momentum transfer $\Delta\mathbf{p}'_1 = \mathbf{p}'_3 - \mathbf{p}'_1$ (see Fig. 7.3). If θ is the scattering angle, then the angle between $\mathbf{p}'_3$ and $\Delta\mathbf{p}'_1$ is $(\pi - \theta)/2$, and

$$\Delta p'_1 = 2p'_1 \cos\frac{\pi - \theta}{2} = 2p'_1 \sin\frac{\theta}{2}. \tag{7.6}$$

Here we assume conservation of energy, so that $p'_1 = p'_3$.

Now the change in momentum is related to the force causing that change by

$$\Delta\mathbf{p}'_1 = \int_{t_i}^{t_f} \mathbf{F}(t)\, dt$$

where t is time and t_i, t_f represent the onset of the interaction causing the momentum change, and the end of the interaction, respectively. We can change the variable from time to angular position ϕ using $dt = (dt/d\phi)\, d\phi = (1/\omega(\phi))\, d\phi$, where $\omega(\phi) = d\phi/dt$ is angular velocity, together with $\phi_i = \phi(t_i)$, $\phi_f = \phi(t_f)$ (see Fig. 7.4). For the scattering problem, $\phi_i = -(\pi - \theta)/2$, $\phi_f = (\pi - \theta)/2$, so that

$$\Delta\mathbf{p}'_1 = \int_{-(\pi-\theta)/2}^{(\pi-\theta)/2} \frac{1}{\omega(\phi)} \mathbf{F}(\phi)\, d\phi.$$

In central force problems, the angular momentum is constant, and in the barycentric frame it is given by

$$\mu r^2 \omega(\phi) = p'_1 b$$

where the right-hand side is the value of the angular momentum when the incident particle is very far away. Hence

$$\frac{1}{\omega(\phi)} = \frac{\mu r^2(\phi)}{p'_1 b}$$

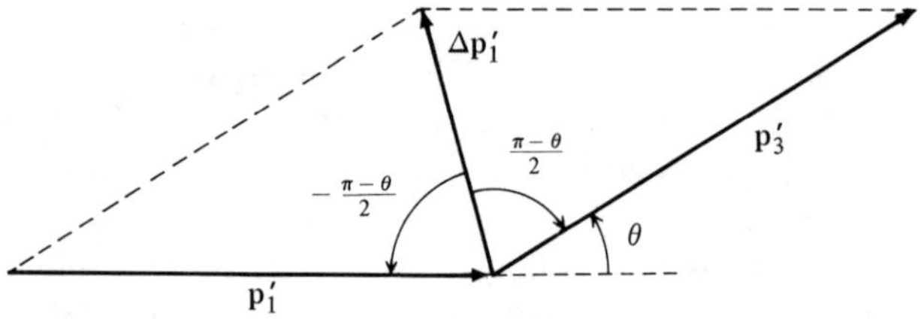

Fig. 7.3. Vector diagram for the analysis of central-force scattering.

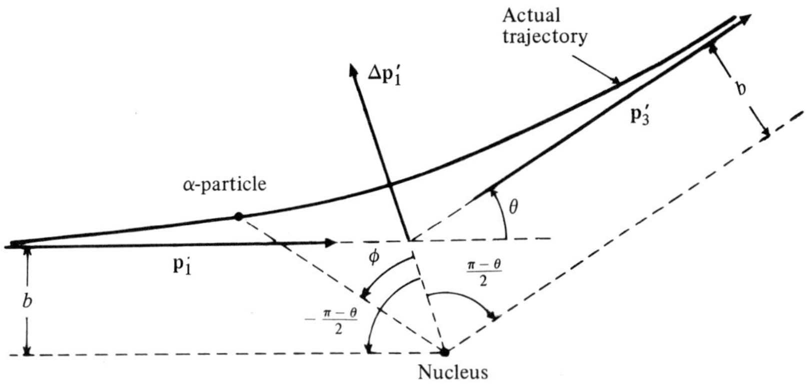

Fig. 7.4. Diagram defining variables in the Rutherford scattering analysis.

where only r depends on ϕ. Therefore, for central forces,

$$\Delta \mathbf{p}_1' = \frac{\mu}{p_1' b}\int_{-(\pi-\theta)/2}^{(\pi-\theta)/2} r^2(\phi)\mathbf{F}(\phi)\,d\phi,$$

or, if $\hat{\mathbf{n}}$ is a unit vector in the direction of $\Delta \mathbf{p}_1'$,

$$\Delta p_1' = \hat{\mathbf{n}}\cdot\Delta \mathbf{p}_1' = \frac{\mu}{p_1' b}\int_{-(\pi-\theta)/2}^{(\pi-\theta)/2} r^2(\phi)\hat{\mathbf{n}}\cdot\mathbf{F}(\phi)\,d\phi$$

$$= \frac{\mu}{p_1' b}\int_{-(\pi-\theta)/2}^{(\pi-\theta)/2} r^2(\phi)F(\phi)\cos\phi\,d\phi, \qquad |\mathbf{F}(\phi)| = F(\phi).$$

Solving for b and using Eq. (7.6), we obtain

$$b = b(\theta) = \frac{\mu}{2p_1'^2\sin(\theta/2)}\int_{-(\pi-\theta)/2}^{(\pi-\theta)/2} r^2(\phi)F(\phi)\cos\phi\,d\phi.$$

Rutherford's basic assumption was that the scattering force was essentially the Coulomb force

$$\mathbf{F}(\phi) = \mathbf{F} = \frac{Z'eZe}{4\pi\epsilon_0 r^2}\hat{\mathbf{r}},$$

where $Z'e$ is the charge on the projectile and Ze is the positive charge on the nucleus of an atom in the target. Hence, for the Rutherford problem,

$$b(\theta) = \frac{\mu Z'Ze^2}{8\pi\epsilon_0 p_1'^2}\cdot\frac{1}{\sin(\theta/2)}\int_{-(\pi-\theta)/2}^{(\pi-\theta)/2}\cos\phi\,d\phi$$
$$= \frac{\mu Z'Ze^2}{4\pi\epsilon_0 p_1'^2}\cot\frac{\theta}{2}. \qquad (7.7)$$

This yields

$$\frac{db(\theta)}{d\theta} = \frac{1}{2}\frac{\mu Z'Ze^2}{4\pi\epsilon_0 p_1'^2}\operatorname{cosec}^2\frac{\theta}{2}, \tag{7.8}$$

and hence

$$\begin{aligned} b(\theta)\frac{db(\theta)}{d\theta} &= \left(\frac{\mu Z'Ze^2}{4\pi\epsilon_0 p_1'^2}\right)^2 \frac{1}{2}\cot\frac{\theta}{2}\operatorname{cosec}^2\frac{\theta}{2} \\ &= \frac{1}{4}\left(\frac{\mu Z'Ze^2}{4\pi\epsilon_0 p_1'^2}\right)^2 \frac{\sin\theta}{\sin^4(\theta/2)}. \end{aligned} \tag{7.9}$$

Substitution of this result into Eq. (7.2) then gives the Rutherford cross-section for a single scattering center:

$$\left(\frac{d\sigma}{d\Omega}\right)_{\text{single nucleus}} = \left(\frac{Z'Ze^2}{16\pi\epsilon_0 T'}\right)^2 \frac{1}{\sin^4(\theta/2)}, \tag{7.10}$$

where we have written $T' = (1/2\mu)(p_1')^2$ as the kinetic energy in the barycentric system.

7.4 COMPARISON WITH EXPERIMENT

To compare this result with experiment we must first convert the experimental data to the form of a cross-section formula; then, recognizing that there are $n = \rho_t A\tau/\mathsf{A}N_0$ scattering centers causing the scattering, divide the experimental cross-section by n to get a value that can be compared with the Rutherford prediction.

Since the impact parameter is related in a one-to-one way with the scattering angle, we can write $\Delta N(b) = \Delta N(b(\theta)) = \Delta N[\theta]$ in Eq. (7.1). Thus, by measuring the rate at which scattered particles pass through solid angle element $\Delta\Omega$ at scattering angle θ, i.e. by measuring $(d(\Delta N[\theta])/dt)\,\Delta\Omega$, and knowing the intensity of the beam impinging on the target, we can obtain

$$\Delta\sigma = \frac{1}{I}\left(\frac{d}{dt}\Delta N[\theta]\right)\Delta\Omega$$

or

$$\frac{\Delta\sigma}{\Delta\Omega}(\theta) = \frac{d(\Delta N[\theta])/dt}{I}$$

from Eq. (7.1). Dividing this by $n = \rho_t A\tau/\mathsf{A}N_0$, we get the single atom cross-section,

$$\left(\frac{\Delta\sigma}{\Delta\Omega}(\theta)\right)_{\text{single nucleus}} = \frac{\mathsf{A}N_0}{\rho_t A\tau}\,\frac{d(\Delta N[\theta])/dt}{I}. \tag{7.11}$$

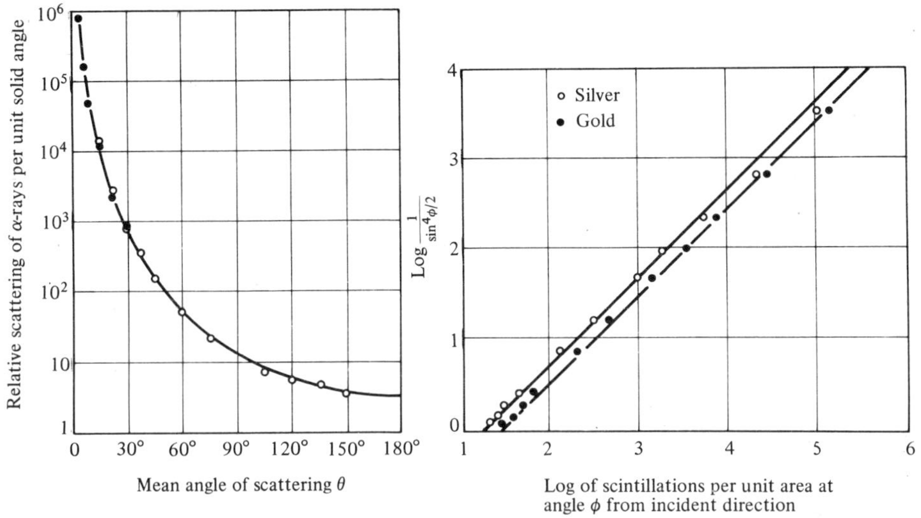

Fig. 7.5. Two different displays of the α-scattering data of Geiger and Marsden, together with theoretical curves. (a) A conventional semilog plot of the scattering by a thin film of gold. (b) A log–log plot of scattering from thin films of gold and silver, on which the theoretical curves are straight lines. (a) From R. D. Evans, *The Atomic Nucleus*, McGraw-Hill Book Co., N.Y. (1955). (b) From F. K. Richtmeyer, E. H. Kennard and T. Lauritsen, *Introduction to Modern Physics*, 5th Ed., McGraw-Hill Book Co., N.Y. (1955). Reprinted through permission of the publishers.

Thus the experimentally obtained data can be compared with Rutherford's theoretical result, Eq. (7.10). (See Fig. 7.5).

Using various materials for the targets, notably gold and silver, Rutherford and his collaborators verified that Eq. (7.10) gives essentially correct results provided the target is sufficiently thin (so that multiple scattering is unimportant) and provided the scattering angle satisfies

$$\cot\frac{\theta}{2} \gtrsim \frac{4\pi\epsilon_0 \cdot 4T'}{Z'Ze^2}.$$

(He used α-particles as projectiles, so that $Z' = 2$.) The limitation on angles corresponds to a limitation $b \gtrsim 10^{-14}$ m on the impact parameter. This limitation on angles could well result from the actual collision of the α-particles with the nuclei, or at least from a breakdown of the notion that the nucleus is infinitesimally small. In any case, the experimental results were consistent with the existence of a nucleus containing all the positive charge and having a radius of not more than 10^{-14} m. The atomic electrons then must be attached to the nucleus at distances as great as 10^{-10} m in order for the atom to have a total diameter of about 10^{-10} m. An amazing consequence of this model is that nearly all the mass of a sample of condensed matter (solid or liquid) resides in about $(10^{-4})^3 = 10^{-12}$ of its volume!

Rutherford's *notion* of the structure of the atom, in its most rudimentary aspects, has survived to the present time. Over the years, however, the model has faced severe crises and has undergone drastic modifications. The most important objection to his original model was that it provided no mechanism whereby an electron could be restrained from being accelerated toward the nucleus and merging catastrophically. The only "force" imaginable was the centrifugal effect. According to classical mechanics, an electron certainly could become a satellite, and a planetary model for the atom is quite feasible. However, the electromagnetic properties of the electron prevent this. The continual acceleration of an orbiting electron would cause it to radiate continually, and thus to move into smaller and smaller orbits in order to maintain mechanical equilibrium. This process would continue until the electron merged with the nucleus. In fact, classical physics is incapable of providing a tenable dynamically stable model for the electron–nucleus system, just as it is incapable of providing a satisfactory model for radiation in an enclosure.

7.5 BOHR'S MODEL

Aside from the instability of the Rutherford atom, the problem of predicting correctly the systematics of line spectra emitted by various species of atoms was not solved by the Rutherford atom. From the rather complicated arrays of lines within the spectrum of hydrogen, for example, several distinct series of lines had been isolated. Each of these had the form

$$k = 2\pi R\left(\frac{1}{n'^2} - \frac{1}{n^2}\right), \tag{7.12}$$

where k is the wave number of the spectral line, R is a constant, called the Rydberg after a famous spectroscopist and is equal to $109{,}678 \times 10^2\ \mathrm{m}^{-1}$, and n' and n are integers. In a given series, n' is fixed and the different lines in the series are characterized by the integer $n > n'$. For instance, $n' = 1$ characterizes the *Lyman* series, the lines of which are obtained by letting $n = 2, 3, \ldots$; similarly $n' = 2$ characterizes the *Balmer* series, $n' = 3$ the *Paschen* series, $n' = 4$ the *Brackett* series. (See Fig. 7.6.)

For atoms more complicated than the hydrogen atom with its single electron, similar spectral regularities were found, except that n' and n were not always integers in these more complicated spectra. Nevertheless, whenever a pattern among the lines was found, it had the general form

$$k = \tau_{n'} - \tau_n, \tag{7.13}$$

where $\tau_{n'}$ is fixed and $\tau_n < \tau_{n'}$ may take on a denumerably infinite number of values that decrease with increasing n. The $\tau_{n'}$ and τ_n are called *terms* of the series.

This is suspiciously similar to the situation in Planck's model for the enclosed radiation system, where the frequencies of allowed radiation are differences between the allowed energies of the oscillators in the walls, divided by $\hbar$:

$$\begin{aligned}\omega_{\text{rad}} &= [E'_{\text{osc}} - E_{\text{osc}}]/\hbar \\ &= [(n' + \tfrac{1}{2})\hbar\omega - (n + \tfrac{1}{2})\hbar\omega]/\hbar \\ &= (n' - n)\omega.\end{aligned}$$

The difficulty is that energies of the simple harmonic oscillators do not lead to radiation frequencies that are related to one another in the manner that characterizes the observed line spectra, e.g. Eq. (7.12). In particular, as $n \to \infty$ in Eq. (7.12) the values of k get closer and closer together, approaching the limit

$$k_{\infty} = \frac{2\pi R}{n'^2}.$$

The existence of such *series limits* is typical of the observed spectra, and no such limits exist for the harmonic oscillator model.

Bohr attempted to obtain, from the expression for the energy of a satellite electron, a formula for the terms τ_n such that τ_n is proportional to $1/n^2$, n being a *quantum number*, rather than proportional to n.

The total energy of an electron orbiting about its nucleus is

$$E = \frac{p^2}{2\mu} - \frac{Ze^2}{4\pi\epsilon_0 r}$$

if the charge on the nucleus is Z, and we observe the system in center-of-mass frame so that $\mu = m_e m_n/(m_e + m_n)$, m_e being the electron's mass and m_n being the mass of the nucleus. Bohr restricted his first considerations to electrons in circular orbits, so that $p^2/2\mu = \frac{1}{2}\mu r^2\omega^2$ where r is the radius of the orbit and ω is the constant angular velocity.

Since the electron here is assumed to be in a stable circular orbit, the attractive electrostatic potential must compensate exactly for the centrifugal effect of the electron's circular motion; hence

$$\mu r\omega^2 = \frac{Ze^2}{4\pi\epsilon_0 r^2} \tag{7.14}$$

or

$$\tfrac{1}{2}\mu r^2\omega^2 = \frac{Ze^2}{8\pi\epsilon_0 r}.$$

Therefore

$$E = -\frac{Ze^2}{8\pi\epsilon_0 r} = -\tfrac{1}{2}\mu r^2\omega^2. \tag{7.15}$$

It is clear from this that if r is proportional to n^2 and ω is proportional to n^{-3}, E

would be proportional to $1/n^2$, as desired. The constants of proportionality, however, must involve $\hbar$ if we are to make use of Planck's idea, and $\hbar$ has the units of energy times time, or momentum times distance.

Now whatever expression we try to *quantize*, that is restrict its values to integer multiples of a constant, should be a conserved physical quantity, so that the quantization will provide the atomic system with stability. Thus the system would remain in a state for which n is given, until disturbed by some external influence; external disturbances would be able to produce shifts from one integer n to another integer n', but not shifts involving noninteger changes in n.

The two conserved mechanical quantities involved in the satellite problem are energy and angular momentum. We do not want to quantize the energy because that would lead to the old harmonic oscillator energies, and we know already that they do not provide the proper spectral lines. The other candidate turns out to be ideal for our purposes because it has the same units as $\hbar$, namely momentum times distance, and because it is given by the formula

$$L = \mu r^2 \omega \tag{7.16}$$

for circular orbits; this latter circumstance means that if $r \propto n^2$ and $\omega \propto n^{-3}$ as we desire, then $L \propto n$. Hence we let the angular momentum be quantized according to the rule

$$L = \mu r^2 \omega = nK\hbar,$$

where n is an integer and K is a dimensionless constant.

Equation (7.14) then yields, upon multiplication by μr^3,

$$\frac{\mu Z e^2}{4\pi\epsilon_0} r = \mu^2 r^4 \omega^2 = L^2 = n^2 K^2 \hbar^2,$$

or

$$r = K^2 \frac{4\pi\epsilon_0 \hbar^2}{\mu Z e^2} n^2 \tag{7.17}$$

so that $r \propto n^2$, as we wished. It follows immediately that $\omega \propto n^{-3}$. Hence the allowed energies for the system quantized in this way are

$$E_n = -\frac{1}{K^2} \frac{\mu (Ze^2)^2}{2(4\pi\epsilon_0 \hbar)^2} \frac{1}{n^2}, \tag{7.18}$$

and the wave number of spectral lines resulting from radiative transitions between these allowed energy levels are

$$k = \frac{\omega}{c} = \frac{E_n - E_{n'}}{\hbar c} = \frac{1}{K^2} \frac{\mu}{2\hbar c} \left(\frac{Ze^2}{4\pi\epsilon_0 \hbar} \right)^2 \left(\frac{1}{n'^2} - \frac{1}{n^2} \right).$$

This provides us with a method for evaluating K, since evidently we must have

$$2\pi R = \frac{1}{K^2}\frac{\mu}{2\hbar c}\left(\frac{Ze^2}{4\pi\epsilon_0\hbar}\right)^2$$

when $Z = 1$ for hydrogen.

It turns out that $e^2/4\pi\epsilon_0\hbar c$ is a very important constant in atomic physics, and it is given a name and a symbol:

$$\alpha \equiv \frac{e^2}{4\pi\epsilon_0\hbar c} = \frac{1}{137.0360} \tag{7.19}$$

is dimensionless and is called the *fine structure constant*. For hydrogen, $m_n = 1836.1\ m_e$ so that

$$\mu c^2/\hbar c = \frac{1836.1}{1836.1 + 1}\frac{m_e c^2}{\hbar c}$$

is just 1836.1/1837.1 times the electron Compton wave number, the latter being given by

$$k_e = \frac{m_e c^2}{\hbar c} = 2.5896 \times 10^{12}\ \text{m}^{-1}.$$

Hence

$$R = \frac{1}{K^2}\frac{\mu}{m_e}k_e\alpha^2 = \frac{1}{K^2}109{,}678 \times 10^2\ \text{m}^{-1}, \tag{7.20}$$

and $K = 1$ yields excellent agreement with the experimental value for R. Thus Bohr's angular momentum quantization postulate, Eq. (7.16) becomes, with $K = 1$,

$$L = n\hbar. \tag{7.21}$$

and this provides excellent agreement with experiment.

The essential results of Bohr's circular orbit model include, then, that the electron in a one-electron atomic system is restricted to motion in orbits having radii (see Eq. (7.17) using $K = 1$)

$$a_{n,Z} = n^2\frac{1}{Z}\frac{m_e}{\mu}a_1, \qquad n = 1, 2, 3, \ldots \tag{7.22}$$

where

$$a_1 \equiv \frac{4\pi\epsilon_0\hbar c}{e^2}\frac{\hbar c}{m_e c^2} = \frac{1}{\alpha k_e} = 0.529177 \times 10^{-10}\ \text{m} \tag{7.23}$$

is the radius of the first ($n = 1$) Bohr orbit in a hydrogen atom when the mass of the nucleus is assumed to be infinite. The correction for finite nuclear mass is represented by the factor m_e/μ in (7.22) and the correction for having nuclear charge greater than unity is represented by the factor $1/Z$.

Thus the energies of the orbiting electrons are restricted to the values (use $K = 1$ in Eq. (7.18))

$$E_{n,Z} = Z^2 \frac{\mu}{m_e} E_1 \frac{1}{n^2}, \tag{7.24}$$

where

$$E_1 = -\frac{\alpha}{2} m_e c^2 = -13.6058 \text{ eV}. \tag{7.25}$$

(1 eV = 1 electron volt = 1.6022×10^{-19} J) is the energy of the first Bohr orbit in hydrogen when the nuclear mass is infinite; the mass and nuclear charge correction factors are evident in Eq. (7.24). It is interesting to note that the value of $\beta = v/c$ for an electron in the nth Bohr orbit is

$$\beta_{n,Z} = \frac{Z}{n}\alpha \tag{7.26}$$

which shows that for small Z or large n, the nonrelativistic approximation should be adequate.

The differences between allowed energy levels therefore are given by

$$E_{n,Z} - E_{n'Z} = Z^2 \frac{\mu}{m_e} E_1 \left(\frac{1}{n'^2} - \frac{1}{n^2}\right), \tag{7.27}$$

so that when an atom shifts from one energy level to another (which corresponds to an acceleration of the electron in classical theory, and hence to a radiation-producing transition of the electron from one state to another) it may be expected to give off radiation having this energy, or, according to the quantum hypothesis for the radiation, having wave number

$$k_{n'n} = Z^2 \frac{\mu}{m_e} \frac{E_1}{\hbar c} \left(\frac{1}{n'^2} - \frac{1}{n^2}\right). \tag{7.28}$$

We constructed the atomic model in just such a way that this would occur.

The electron states and the radiative transitions are frequently represented by energy level diagrams, such as Fig. 7.6a; a photograph of the Balmer lines is shown in Fig. 7.6b.

Bohr, Sommerfeld and Wilson generalized these ideas to cover elliptical as well as circular orbits and to take into account relativistic effects. Also, the theory has been applied to more complicated atoms than hydrogen, with some qualitative success; the gross features of the spectra generally agree qualitatively with the Bohr model predictions. Nevertheless, many anomolous features of spectra remain that the Bohr theory is not sufficiently rich to cope with. Further-

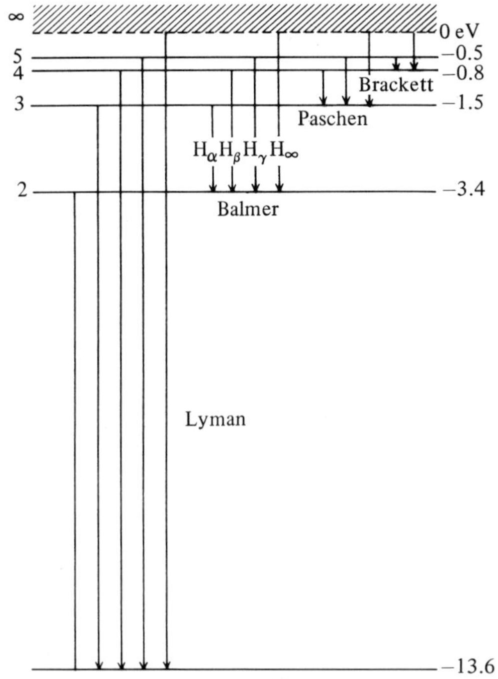

(a)

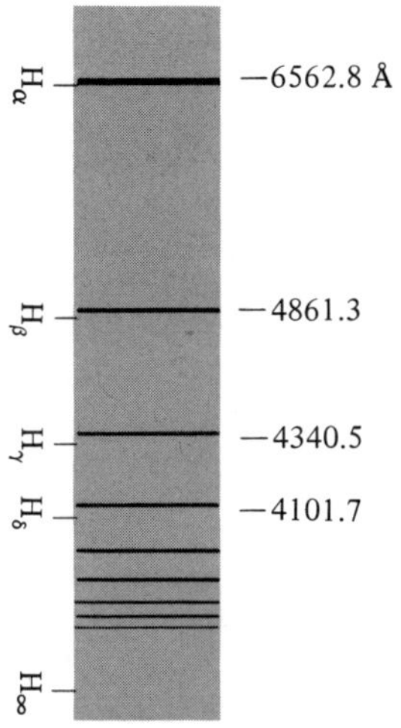

(b)

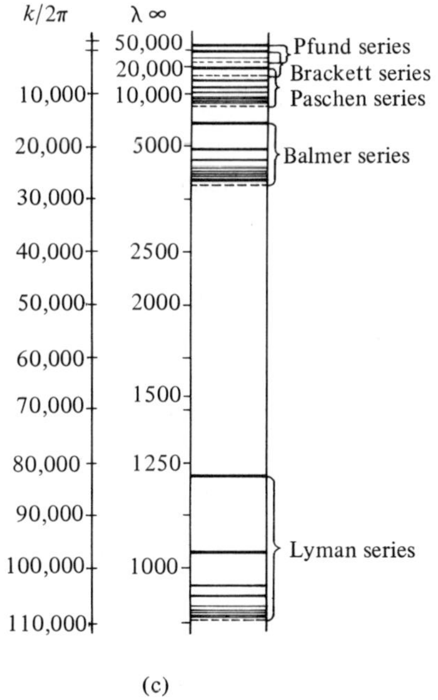

(c)

Fig. 7.6. Energy levels of the Bohr model of Hydrogen, (a), and the observed spectral series, (b) and (c); (b) is a photograph of the Balmer series. ((a) From L. KERWIN, *Atomic Physics*, Holt, Rinehart and Winston (1963), p. 182. (b) From G. HERTZBERG, *Atomic Spectra and Atomic Structure*, Dover Publications Inc., New York (1944). (c) From H. SEMAT, *Introduction to Atomic Physics*, Holt, Rinehart and Winston (p. 236 of Fourth Ed.) (1963). Reprinted through permission of the publishers.)

more, as a theory it is quite unsatisfactory because the quantization hypothesis was invoked in just such a way as to give agreement with experiment; there was no fundamental basis for quantizing angular momentum rather than energy. Wilson and Sommerfeld independently developed quantization rules that were more general and could be applied more systematically than Bohr's, but these rules still had no solid theoretical underpinnings. In fact, the rules did not always seem to give the correct results without further special modifications; the necessity of replacing the square of the angular momentum n^2h^2 with $n(n + 1)h^2$ in the analysis of the anomolous Zeeman effect provides an example.

Thus the Bohr model really did not succeed in providing us with a thoroughgoing theory for the structure of matter. Rather it provided a very important conceptual stepping stone from classical to modern physcial theory. Most importantly, through its substantial success, it made clear to physicists that Planck's constant and the quantum idea must play a central role in the physics of microscopic systems. A genuine quantum theory, rather than a series of quantization hypotheses, quite clearly was required.

READING NOTES*

Reference 23, Chapter 2.

Reference 14, Chapter 4, 5.

Reference 1, Chapter 2.

Reference 16, Chapter 5.

Reference 44, Chapter 1.

PROBLEMS

1. The Rutherford cross-section given in Eq. (7.10) was calculated nonrelativistically for the center-of-mass frame of reference. Continuing to assume that relativistic effects are not important, find $d\sigma/d\Omega$ in the laboratory frame of reference.
2. For an electron in the first Bohr orbit, find the ratio v/c. Find the ratio v/c in terms of n for higher Bohr orbits.
3. Calculate from Eq. (7.27) the energies of the least energetic photons emitted in the Lyman, Balmer, Paschen and Brackett series. Discuss whether these photons would be useful in an experiment to detect the Compton effect.
4. Consider an "atom" held together by gravitational rather than electrostatic forces.
   a) Find the "fine structure constant" as a function of the masses involved; evaluate this for the earth–sun system.
   b) Find the first "Bohr radius" as a function of the masses and evaluate it for the same system. Comment on the statement "Quantum effects apply only in the realm of small distances", on the basis of these considerations.

---

* Works referred to in these notes will be found in the Bibliography on p. 577.

CHAPTER 8

WAVES AND PARTICLES

In Chapter 6 we saw that the field energy of electromagnetic radiation seems to be "bunched up" into units of $\hbar\omega$ and that the electromagnetic field frequently behaves as if it were localizable in that energy and momentum can be transferred from the field to charged particles as if the field consisted of particle-like objects; this was the message of the photoelectric effect and of the Compton effect. On the other hand, the *quantum unit* $\hbar$ appeared in Chapter 7 when the angular momentum of an electron was quantized in the Bohr model. Thus $\hbar$ has application both in the electromagnetic field phenomena and in phenomena directly involving charged particles with mass. Could it be that both the electromagnetic wave fields, involving massless "particles", and massive particles like electrons participate in quantum phenomena in essentially similar ways? That is, can particles like electrons display wavelike properties, with $\hbar$ supplying the link between particle-like and wave-like behavior, just as electromagnetic wave fields display particle-like phenomena? Might not Planck's constant $\hbar$ help us to eliminate the wave–particle dichotomy of Classical physics and to replace it with a single unified view of microscopic phenomena? The answer, as we shall see, is "Yes".

In order to investigate this question with some depth, we shall need to know something about wave theory. Therefore we begin this chapter with a brief discussion of some general properties of waves. The important feature of this preliminary discussion, much of which will be a review for the reader, is that it deals with waves in the abstract, and the outcome of the treatment will be the discovery of certain features of waves in general, whether they be electromagnetic, acoustic, or hydrodynamic waves, or the so-called matter waves of quantum theory.

Following the discussion of abstract waves, we shall apply our knowledge about waves to the famous hypotheses of Louis de Broglie and Albert Einstein, which form the conceptual foundation of the wave theory of matter. Certain analogies between electromagnetic radiation and matter waves will be developed to help unify the quantum theory of light with the quantum theory of massive particles. The unification will not be complete, however, and matter waves will be treated from a basis that is somewhat different from that for electromagnetic waves, mainly because the latter have no mass associated with them and the signals transmitted by them always travel with a fixed—or more accurately, Lorentz-invariant—speed *in vacuo*. Nevertheless, the similarities will be quite clear, and it should be apparent that the conjectures of de Broglie and Einstein lead to a rather coherent view of radiation and matter.

The main value of the de Broglie–Einstein hypotheses is that they provide the basis for a mathematical theory that is in agreement with a far wider range of phenomena than are the theories that preceded it. Quantum theory *works*; it suggests to experimenters what kinds of apparatus will be useful, and it yields numbers that can be compared with measurements, as we shall see. This alone justifies our study of the elements of its mathematical foundations.

8.1 FUNDAMENTAL PROPERTIES OF WAVES

A physical wave is an oscillation, throughout space and time, of some physical variable. If the variations in magnitude of the physical quantity, as a function of time, are independent of the variations as a function of spatial position, then points of constant phase (i.e. maxima, minima and nodes in the waves) will be stationary in space, and such waves are called *standing waves.* If, however, the time and space oscillations are correlated in such a way that all points of constant phase move with some velocity $\mathbf{v}_p$, called the *phase velocity*, the waves are called *traveling waves.*

The mathematical study that deals most directly and appropriately with wave motion is Fourier analysis, and it can be shown, through Fourier theory, that any kind of wave motion can be represented as a linear superposition of traveling waves with different frequencies and wavelengths. The phase velocities of these traveling waves need not be equal, but each wave having fixed wavelength λ and fixed frequency ν has a constant phase velocity $\mathbf{v}_p$ of magnitude

$$|\mathbf{v}_p| = \lambda\nu = \lambdabar\omega$$

where $\lambdabar = \lambda/2\pi$ is the reduced wavelength and $\omega = 2\pi\nu$ is the angular frequency, as before. Thus we may construct any waveform, or what is the same thing, any reasonably well-behaved function of space and time, from a superposition of monochromatic waves (i.e. waves with fixed ω) having constant phase velocities. The monochromatic waves are represented mathematically by the sinusoidal functions

$$\sin(\omega t - \mathbf{k}\cdot\mathbf{x}), \qquad \cos(\omega t - \mathbf{k}\cdot\mathbf{x})$$

or equivalently

$$e^{i(\omega t - \mathbf{k}\cdot\mathbf{x})}, \qquad e^{-i(\omega t - \mathbf{k}\cdot\mathbf{x})}$$

where

$$\mathbf{k} = \frac{1}{\lambdabar}\hat{\mathbf{k}}, \qquad \text{and} \qquad \mathbf{v}_p = \lambdabar\omega\hat{\mathbf{k}},$$

$\hat{\mathbf{k}}$ being the direction of propagation of the wave.

The quantity

$$\varphi = \omega t - \mathbf{k} \cdot \mathbf{x},$$

called the *phase* of the wave, provides a concrete interpretation for $\mathbf{v}_p$. Thus a point on a wave train for which φ is constant does not remain stationary in general, for if one waits for a time Δt, the point with phase φ will have moved by $\Delta \mathbf{x}$ such that

$$\varphi = \omega(t + \Delta t) - \mathbf{k} \cdot (\mathbf{x} + \Delta x) = \omega t - \mathbf{k} \cdot \mathbf{x}$$

or

$$\omega \, \Delta t - \mathbf{k} \cdot \Delta \mathbf{x} = 0.$$

Hence

$$\mathbf{v}_p = \frac{\Delta \mathbf{x}}{\Delta t} = \frac{\omega}{k}\hat{\mathbf{k}} = \lambda\omega\hat{\mathbf{k}}$$

is the velocity with which the wave propagates in the direction $\hat{\mathbf{k}}$.

For a monochromatic wave with fixed phase velocity, points at which the magnitudes of the oscillating physical variable are equal are separated uniformly from one another in time and are equidistant from one another along the direction of propagation in space. Hence the information carried by such a wave is meager; it provides an observer with a standard for distance measurements (the wavelength), a standard for time measurements (the period), a preferred direction (the direction of propagation), and no more. It cannot even provide a uniquely defined point in space or time that might serve as a natural origin of coordinates. Thus in order that waves have any great physical significance, some mechanism must exist for either varying the amplitude of the wave as a function of space and time (amplitude modulation) or varying the frequency (frequency modulation) or phase velocity as a function of space and time.

We shall restrict our attention to amplitude modulation; this will allow as much generality as we need. Suppose, for example, that we have a wave expressible in the form

$$\begin{aligned} f(\mathbf{x}, t) &= \alpha(\mathbf{x}, t) \cos(\omega t - \mathbf{k} \cdot \mathbf{x}) + \beta(\mathbf{x}, t) \sin(\omega t - \mathbf{k} \cdot \mathbf{x}) \\ &= \gamma_+(\mathbf{x}, t)\, e^{i(\omega t - \mathbf{k}\cdot\mathbf{x})} + \gamma_-(\mathbf{x}, t)\, e^{-i(\omega t - \mathbf{k}\cdot\mathbf{x})}, \end{aligned} \tag{8.1}$$

where, since $\cos\varphi = (e^{i\varphi} + e^{-i\varphi})/2$ and $\sin\varphi = (e^{i\varphi} - e^{-i\varphi})/2i$, we have

$$\gamma_\pm(\mathbf{x}, t) = [\alpha(\mathbf{x}, t) \mp i\beta(\mathbf{x}, t)]/2.$$

Here $f(x, t)$ represents the oscillating physical variable, and the amplitudes α, β, γ_+ and γ_- are functions of $\mathbf{x}$ and t.

According to a theorem in Fourier analysis, a large class of functions, having properties very much like those we should expect physical variables to possess, can be represented as integrals over the oscillating exponential $e^{i\kappa\xi}$ in the following way: if $h(\xi)$ is in this class of functions over the real variable ξ, and if κ is a real variable, then there is a function $\eta(\kappa)$ such that

$$\begin{aligned} h(\xi) &= \frac{1}{\sqrt{2\pi}} \int_{-\infty}^{\infty} \eta(\kappa) e^{-i\kappa\xi}\, d\kappa \\ &= \frac{1}{\sqrt{2\pi}} \int_{-\infty}^{\infty} \eta(-\kappa) e^{i\kappa}\, d\kappa. \end{aligned} \tag{8.2}$$

Thus if we assume $\gamma_{\pm}(\mathbf{x}, t)$ to be members of this rather large class of functions over *each* of the variables x, y, z and t, we may write

$$\gamma_{\pm}(\mathbf{x}, t) = \frac{1}{(2\pi)^2} \int_{-\infty}^{\infty} d\omega' \int_{-\infty}^{\infty} dk_x' \int_{-\infty}^{\infty} dk_y' \int_{-\infty}^{\infty} dk_z' g_{\pm}'(\mathbf{k}', \omega')\, e^{\pm i(\omega' t - \mathbf{k}'\cdot\mathbf{x})}, \tag{8.3}$$

where the sign in the exponential is chosen, for convenience, to match the sign on $\gamma_{\pm}(\mathbf{x}, t)$.

Substituting Eq. (8.3) into (8.1) yields

$$\begin{aligned} f(\mathbf{x}, t) &= (2\pi)^{-2} \int_{-\infty}^{\infty} d\omega' \int d^3k' \{g_+'(\mathbf{k}', \omega')\, e^{i[(\omega+\omega')t - (\mathbf{k}+\mathbf{k}')\cdot\mathbf{x}]} \\ &\quad + g_-'(\mathbf{k}', \omega') e^{-i[(\omega+\omega')t - (\mathbf{k}+\mathbf{k}')\cdot\mathbf{x}]}\} \\ &= (2\pi)^{-2} \int_{-\infty}^{\infty} d\omega' \int d^3k' [g_+'(-\mathbf{k}', -\omega') + g_-'(\mathbf{k}', \omega')] \\ &\quad \cdot e^{-i[(\omega+\omega')t - (\mathbf{k}+\mathbf{k}')\cdot\mathbf{x}]} \\ &= (2\pi)^{-2} \int_{-\infty}^{\infty} d\omega'' \int d^3k'' g''(k'', \omega'')\, e^{-i(\omega'' t - \mathbf{k}''\cdot\mathbf{x})} \end{aligned} \tag{8.4}$$

in which

$$\int d^3k' \equiv \int_{-\infty}^{\infty} dk_x' \int_{-\infty}^{\infty} dk_y' \int_{-\infty}^{\infty} dk_z',$$

and we have defined $\omega'' \equiv \omega + \omega'$, and

$$g''(\mathbf{k}'', \omega'') \equiv g_+'(\mathbf{k} - \mathbf{k}'', \omega - \omega'') + g_-(\mathbf{k}'' - \mathbf{k}, \omega'' - \omega).$$

In this way, the original amplitude-modulated *monochromatic* wave, with apparent frequency ν and apparent phase velocity $\mathbf{v} = (\omega/k)\hat{\mathbf{k}}$, can be represented as a linear superposition of waves having a wide variety of frequencies and phase velocities. The amplitude of the contribution, to this superposition, by waves having frequency in the interval $\omega < \omega'' < \omega + d\omega$ and wave number in the three-dimensional interval $k_x < k_x'' < k_x + dk_x$, $k_y < k_y'' < k_y + dk_y$, $k_z < k_z'' < k_z + dk_z$, is given by $g''(\mathbf{k}, \omega)\, d\omega\, d^3k$.

Let us assume now that a dispersion relation exists between ω and $\mathbf{k}$; that is, suppose that $\omega = \omega(\mathbf{k})$ is a function of the wave number. This indeed will be the case for the physical waves in which we are interested. In order to restrict the values of ω to just those for which $\omega = \omega(\mathbf{k})$ holds, we write

$$g''(\mathbf{k}, \omega) = g(\mathbf{k}, \omega)\, \delta(\omega - \omega(\mathbf{k})). \tag{8.5}$$

Thus a choice of wave number $\mathbf{k}$ determines a value of ω, as desired.

The amplitude modulation of $f(\mathbf{x}, t)$ provides f with a signal, and this signal moves with some velocity. How shall we find the signal propagation velocity? We no longer can use the phase velocity, for there are waves with many phase velocities contributing to the signal; that is, the phase velocity is not unique. The appropriate method for finding signal velocities, or *group velocities*, was suggested in 1887 by Lord Kelvin. This method is understood best by recognizing that the representation of $f(\mathbf{x}, t)$ as a superposition of monochromatic traveling waves is simply the adding together of waves in just such a way that constructive interference occurs wherever the magnitude of $|f(\mathbf{x}, t)|$ is large, and destructive interference occurs wherever $|f(\mathbf{x}, t)|$ becomes small or vanishes. Therefore, in seeking a signal propagation velocity, we are trying to find the velocity of propagation of a portion of the waveform where constructive (or destructive) interference is maximal, and destructive (or constructive) interference is minimal, even after this portion of the wave has moved far away from its initial position or after a long period of time has elapsed. That is, we wish to find the velocity of persisting prominences in the waveform.

Space and time are the parameters used to describe interference patterns, since interference patterns yield positions in space and time where wave intensities are maximal or minimal. In the Fourier integrals representing $f(\mathbf{x}, t)$, these parameters appear only through the exponential of

$$-i\varphi_{\mathbf{x},t}(\mathbf{k}) = -i[\omega(\mathbf{k})t - \mathbf{k} \cdot \mathbf{x}]. \tag{8.6}$$

Kelvin suggested that to locate extrema in the waveform that persist in space and time, we should perform a Taylor expansion of the phase $\varphi_{\mathbf{x},t}(\mathbf{k})$ about some value $\mathbf{k}_0$ of $\mathbf{k}$. Thus, letting

$$(x, y, z) = (x^1, x^2, x^3), \qquad (k_x, k_y, k_z) = (k^1, k^2, k^3),$$

we get

$$\varphi_{\mathbf{x},t}(\mathbf{k}) = [\omega(\mathbf{k}_0)t + \mathbf{k}_0 \cdot \mathbf{x}] + \sum_{m=1}^{3} \left[\left(\frac{\partial \omega}{\partial k^m}\right)_{\mathbf{k}=\mathbf{k}_0} t - x^m\right](k^m - k_0{}^m)$$
$$+ \sum_{n=2}^{\infty} \frac{1}{n!} \sum_{m_1,\ldots,m_n=1}^{3} \left(\frac{\partial^n \omega}{\partial k^{m_1}\, \partial k^{m_2} \cdots \partial k^{m_n}}\right)_{\mathbf{k}=\mathbf{k}_0} (k^{m_1} - k_0{}^{m_1}) \cdots (k^{m_n} - k_0{}^{m_n})t.$$

For simplicity, we shall write the last term as $\mu_{\mathbf{k}_0}(\mathbf{k} - \mathbf{k}_0)t$. Hence if we write

$$\nabla_k \equiv \sum_{j=1}^{3} \hat{\mathbf{i}}_j \frac{\partial}{\partial k^j}$$

and let $\boldsymbol{\kappa} = \mathbf{k} - \mathbf{k}_0$, we have

$$\varphi_{\mathbf{x},t}(\mathbf{k}_0 + \boldsymbol{\kappa}) = [\omega(\mathbf{k}_0)t - \mathbf{k}_0 \cdot \mathbf{x}] + [\nabla_k\omega(\mathbf{k}_0)t - \mathbf{x}] \cdot \boldsymbol{\kappa} + \mu_{k_0}(\boldsymbol{\kappa})t \tag{8.7}$$

and

$$f(\mathbf{x}, t) = (2\pi)^{-2} \exp(-i[\omega(\mathbf{k}_0)t - \mathbf{k}_0 \cdot \mathbf{x}]) \cdot \int_{-\infty}^{\infty} d\omega \int d^3\boldsymbol{\kappa}\, g(\mathbf{k}_0 + \boldsymbol{\kappa}, \omega)\delta(\omega - \omega(\mathbf{k}_0 + \boldsymbol{\kappa})) \cdot \exp(-i\{[\nabla_k\omega(\mathbf{k}_0)]t - \mathbf{x}\} \cdot \boldsymbol{\kappa}) \exp(-i\mu_{\mathbf{k}_0}(\boldsymbol{\kappa})t). \tag{8.8}$$

The last factor in the integral of Eq. (8.8) is independent of space and produces interference at all nonzero times. It represents the spreading, because of this interference, of the initial prominence in the wave amplitude. We leave it as an exercise for the reader to argue that unless $\mu_{\mathbf{k}_0}(\boldsymbol{\kappa}) = 0$ the factor $\exp[-i\mu_{\mathbf{k}_0}(\boldsymbol{\kappa})t]$ leads to a reduction in the value of $|f(\mathbf{x}, t)|$ for fixed $\mathbf{x}$ as t becomes arbitrarily large. This is often referred to as the degradation of the signal contained in $f(\mathbf{x}, t)$ with increasing time. Clearly this effect is determined by the functional forms of $\omega(\mathbf{k})$, and cannot be compensated in any way. Given the dispersion relation $\omega = \omega(\mathbf{k})$, we simply must live with this signal degradation.

The next-to-last factor in the integral also represents a potentially degrading effect both as time *and* position change. In fact, the appearance of both time and position variables in this factor allow us to "follow" the signal, since this factor becomes unity for those spatial points $\mathbf{x}$ such that

$$\mathbf{x} = \mathbf{x}(t) = \nabla_k\omega(\mathbf{k}_0)t.$$

That is, an observer whose position is given by $\mathbf{x}(t) = \mathbf{v}_g t$ or whose velocity is $\mathbf{v}_g \equiv \nabla_k\omega(\mathbf{k}_0)$, will appear to be traveling at the same rate as the most persistent features of the signal.

This last comment requires some qualification. We have not yet discussed the problem of choosing a value for $\mathbf{k}_0$; clearly our choice of $\mathbf{k}_0$ has a bearing on the value of $\mathbf{v}_g$ that we shall obtain. Essentially, a choice of $\mathbf{k}_0$ is equivalent to a selection of the signal-carrying portion of the waveform, and we shall want to choose $\mathbf{k}_0$ in such a way that it somehow characterizes the signal. In practical problems the choice usually is obvious; usually one chooses the value of $\mathbf{k}_0$ that maximizes $g(\mathbf{k}_0, \omega(\mathbf{k}_0))$. Once such a choice is made, we can say with confidence that

$$\mathbf{v}_g = \nabla_k\omega(\mathbf{k}_0) \tag{8.9}$$

is the velocity of the dominant signal* carried by the wave $f(\mathbf{x}, t)$ with Fourier kernel $g(\mathbf{k}, \omega)$. In most problems $\omega(\mathbf{k}) = \omega(|\mathbf{k}|)$, from which it follows that $\hat{\mathbf{k}}_0 = \hat{\mathbf{v}}_g$, and hence $\mathbf{k}_0$ is in the direction of signal propagation.

The study of group velocity, signal degradation, and other topics suggested by the Fourier integral can be very fascinating and rewarding, but our only aims here have been to find an expression for the group velocity and to observe the presence of a signal-destroying term in the Fourier integral.

8.2 SOME ELEMENTARY PROPERTIES OF FOURIER TRANSFORMS

At this point it will be useful to introduce some elementary properties of Fourier transforms or Fourier integrals. The transforms used here are of the imaginary exponential variety. We shall not worry about such mathematical questions as the conditions under which the transforms exist; until recently many of these questions posed formidable problems even for mathematicians. Rather, we shall accept the notion that the class of functions possessing Fourier transforms is large enough to include the usual functions used in describing physical systems.

Let $f(x)$ represent a function of the single real variable x; then the function of the real variable k defined by

$$g(k) = \frac{1}{\sqrt{2\pi}} \int_{-\infty}^{\infty} f(x) e^{-ikx}\, dx \tag{8.10}$$

is called the Fourier transform of $f(x)$, or in a shorthand notation $FT_k[f(x)]$. If all the derivatives of $f(x)$ exist and if $f(x) \to 0$ as $|x| \to \infty$, then

$$\begin{aligned} FT_k[f'(x)] &= \frac{1}{\sqrt{2\pi}} \int_{-\infty}^{\infty} \left[\frac{d}{dx} f(x)\right] e^{-ikx}\, dx \\ &= \frac{1}{\sqrt{2\pi}} \left[f(x) e^{-ikx} \Big|_{-\infty}^{\infty} + ik \int_{-\infty}^{\infty} f(x) e^{-ikx}\, dx \right] \\ &= ikg(k) = ikFT_k[f(x)]. \end{aligned} \tag{8.11}$$

In addition, we have

$$\frac{1}{\sqrt{2\pi}} \int_{-\infty}^{\infty} f(ax+b) e^{-ikx}\, dx = \frac{1}{\sqrt{2\pi}} \int_{\mp\infty}^{\pm\infty} f(\xi) e^{-ik(\xi - b)/a} \frac{d\xi}{a}$$

$$= \frac{1}{\sqrt{2\pi}} e^{ik(b/a)} \frac{1}{a} \left[\pm \int_{-\infty}^{\infty} f(\xi) e^{-ik\xi/a}\, d\xi \right]$$

* The group velocity is the velocity of signal propagation in nonabsorptive media. In studying wave propagation through material media, one must often distinguish between group velocity and signal velocity, but this distinction is not necessary in our work. For further details, see L. Brillouin, *Wave Propagation and Group Velocity*, Academic Press, New York (1960).

$$= e^{ik(b/a)} \frac{1}{a}\left[\pm g\left(\frac{k}{a}\right)\right] = e^{ik(b/a)} \frac{1}{|a|} g\left(\frac{k}{a}\right),$$

where the choice of sign, when it occurs, is determined by whether $a = \pm|a|$, i.e. whether a is positive or negative; the upper sign goes with $a > 0$ and the lower with $a < 0$. The shorthand notation for this result is

$$FT_k[f(ax + b)] = \frac{e^{ik(b/a)}}{|a|} FT_{k/a}[f(x)]. \tag{8.12}$$

Now let us seek the Fourier transform of $g(k)$; that is, let us try to evaluate

$$\begin{aligned} FT_y[g(k)] &= \frac{1}{\sqrt{2\pi}} \int_{-\infty}^{\infty} g(k) e^{-iyk}\, dk \\ &= \frac{1}{2\pi} \int_{-\infty}^{\infty} dk \int_{-\infty}^{\infty} f(x) e^{-ik(x+y)}\, dx. \end{aligned}$$

Since we are assuming that $f(x)$ is "sufficiently well behaved", we take it that this assumption allows us to exchange the order of integration in the last expression and thereby to get

$$FT_y[g(k)] = \frac{1}{2\pi} \int_{-\infty}^{\infty} dx\, f(x) \int_{-\infty}^{\infty} dk\, e^{-ik(x+y)}. \tag{8.13}$$

Direct evaluation of the integral over k yields the undefined result

$$\int_{-\infty}^{\infty} e^{-ik(x+y)}\, dk = \lim_{K\to\infty} \left.\frac{e^{-ik(x+y)}}{-i(x+y)}\right|_{-K}^{K} = \lim_{K\to\infty} 2\frac{\sin K(x+y)}{x+y}.$$

However, we can obtain a result that is "almost always" (i.e. for all but a finite number of values of $x + y$) well defined, through the expedient of carefully introducing a convergence factor $e^{-|k|\epsilon}$ inside the integral and, after completing the integration, taking the limit $\epsilon \to 0$. Thus, for $\epsilon > 0$,

$$\begin{aligned} \int_{-\infty}^{\infty} e^{-ik(x+y)}\, dk &= \lim_{\epsilon\to 0} \left[\int_{-\infty}^{0} e^{-ik(x+y+i\epsilon)}\, dk + \int_{0}^{\infty} e^{-ik(x+y-i\epsilon)}\, dk\right] \\ &= \lim_{\epsilon\to 0} \left[\lim_{K\to\infty} \frac{1 - e^{-\epsilon K} e^{iK(x+y)}}{-i(x+y+i\epsilon)} + \lim_{K\to\infty} \frac{e^{-\epsilon K} e^{-iK(x+y)} - 1}{-i(x+y-i\epsilon)}\right] \\ &= \lim_{\epsilon\to 0} \left[\frac{i}{x+y+i\epsilon} - \frac{i}{x+y-i\epsilon}\right] = \lim_{\epsilon\to 0} \frac{2\epsilon}{(x+y)^2 + \epsilon^2} \\ &= \left\{\begin{matrix} 0 & \text{if} \quad x + y \neq 0 \\ \lim_{\epsilon\to 0} \frac{2}{\epsilon} & \text{if} \quad x + y = 0 \end{matrix}\right\}. \end{aligned} \tag{8.14}$$

Clearly this is a peculiar "function" of $x + y$. We can find out more about it by evaluating the iterated integral

$$\int_{-\infty}^{\infty} dx \int_{-\infty}^{\infty} e^{-ik(x+y)}\, dk,$$

bearing in mind that the function defined by Eq. (8.14) behaves rather like the δ-function $\delta(k)$. The result of evaluating this iterated integral is

$$\int_{-\infty}^{\infty} dx \int_{-\infty}^{\infty} e^{-ik(x+y)}\, dk = \lim_{\epsilon\to 0} \int_{-\infty}^{\infty} dx \frac{2\epsilon}{(x+y)^2 + \epsilon^2}$$

$$= \lim_{\epsilon\to 0} \int_{-\infty}^{\infty} d\xi \frac{2\epsilon}{\xi^2 + \epsilon^2},$$

where $\xi = x + y$. Let $\xi = \epsilon \tan \varphi$; then $\epsilon\, d\xi = \epsilon^2 \sec^2 \varphi\, d\varphi$ and the last integral becomes

$$2\int_{-\pi/2}^{\pi/2} \frac{\sec^2 \varphi}{\tan^2 \varphi + 1} d\varphi = 2\int_{-\pi/2}^{\pi/2} d\varphi = 2\pi.$$

Hence

$$\int_{-\infty}^{\infty} dx \int_{-\infty}^{\infty} dk\, e^{-ik(x+y)} = 2\pi,$$

so that by setting

$$\frac{1}{2\pi}\int_{-\infty}^{\infty} e^{-ik(x+y)}\, dk = \delta(x + y) \tag{8.15}$$

we obtain an expression for the Dirac δ-function that meets the requirements set forth at the beginning of Chapter 1. This is a convenient and frequently used representation of the δ-function.

Returning now to our search for $FT_y[g(k)]$, we find that

$$\begin{aligned} FT_y[g(k)] &= \frac{1}{\sqrt{2\pi}} \int_{-\infty}^{\infty} g(k)\, e^{-iky}\, dk \\ &= \int_{-\infty}^{\infty} dx\, f(x)\, \delta(x + y) = f(-y). \end{aligned} \tag{8.16}$$

This result is known as the *Fourier inversion formula*, and may be written in the form

$$FT_y[FT_k[f(x)]] = f(-y). \tag{8.17}$$

The three elementary properties of Fourier transforms enumerated in Eqs. (8.11), (8.12) and (8.17) may be used, for example, to verify the properties of the δ-function listed below. The same properties of the δ-function may be proved by using the definition of the δ-function given in Chapter 1. As an exercise, the student should use both methods, thereby gaining some familiarity with Fourier transforms and with the Fourier representation of the δ-function.

Before providing a list of useful δ-function identities, we note that Eq. (8.15) states that the Fourier transform of a constant is equal to that constant times $\sqrt{2\pi}$ times a δ-function. Furthermore, a constant can be represented as $\sqrt{2\pi}$ times the Fourier transform of that constant times a δ-function; this follows from Eq. (8.17):

$$FT_k[\delta(x)] = FT_k[FT_x[1/\sqrt{2\pi}]] = 1/\sqrt{2\pi}.$$

Further useful δ-function identities are

$$\delta(x) = \delta(-x) \tag{8.18a}$$

$$x\,\delta(x) = 0 \tag{8.18b}$$

$$\delta(ax) = |a|^{-1}\,\delta(x) \tag{8.18c}$$

$$\delta(x^2 - a^2) = |2a|^{-1}[\delta(x - a) + \delta(x + a)] \tag{8.18d}$$

$$f(x)\,\delta(x - a) = f(a)\,\delta(x - a) \tag{8.18e}$$

$$\int_L \delta(x - a)\,\delta(x - b)\,dx = \delta(a - b), \qquad \text{if } a, b \in L \tag{8.18f}$$

$$\delta'(x) = -\delta'(-x) \tag{8.18g}$$

$$x\,\delta'(x) = -\delta(x). \tag{8.18h}$$

Finally, it is true that if

$$g(k) = FT_k[f(x)]$$

then

$$\int_{-\infty}^{\infty} |g(k)|^2\,dk = \int_{-\infty}^{\infty} |f(x)|^2\,dx \tag{8.19}$$

under the conditions of "well-behavedness" of $f(x)$ that we have assumed. Thus

$$\int_{-\infty}^{\infty} |g(k)|^2\,dk = \int_{-\infty}^{\infty} dk\,\frac{1}{2\pi}\int_{-\infty}^{\infty} dx\,f(x)e^{-ikx}\int_{-\infty}^{\infty} dy\,f^*(y)e^{iky}$$

$$= \int_{-\infty}^{\infty} dx \int_{-\infty}^{\infty} dy\,\frac{1}{2\pi}\cdot 2\pi\delta(x - y)f(x)f^*(y)$$

$$= \int_{-\infty}^{\infty} |f(x)|^2\,dx.$$

This is a very valuable theorem in Fourier analysis.

8.3 THE UNCERTAINTY PRINCIPLE

As we have noted already, the Fourier transform

$$f(t) = \frac{1}{\sqrt{2\pi}} \int_{-\infty}^{\infty} g(\omega) e^{i\omega t} \, d\omega$$

can be interpreted as a linear superposition of monochromatic waves (in the time variable alone), the amplitude for waves having frequency ω in the interval $\omega < \omega' < \omega + d\omega$ being given by $g(\omega)\, d\omega/\sqrt{2\pi}$. The Fourier inversion formula prescribed a method for finding the amplitude functions $g(\omega)$ when the time dependence of the waveform $f(t)$ is given. Thus, according to (8.17)

$$\begin{aligned} g(\omega) &= FT_\omega[FT_{-t}[g(\omega)]] = FT_\omega[f(t)] \\ &= (1/\sqrt{2\pi}) \int_{-\infty}^{\infty} f(t) e^{-i\omega t} \, dt. \end{aligned}$$

Let us consider a particularly simple example. If a wave has the form $f(t) = f_1(t) e^{i\omega_0 t}$, where

$$f_1(t) = \begin{cases} 0 & \text{if} \quad t < T_1 \\ 1 & \text{if} \quad T_1 \leq t < T_2 \\ 0 & \text{if} \quad T_2 \leq t, \end{cases}$$

(i.e. $f(t)$ is a single "square" pulse of monochromatic waves having frequency ω_0, the pulse commencing at T_1 and lasting for a time $T_2 - T_1$), then

$$\begin{aligned} g(\omega) &= FT_\omega[f(t)] = (1/\sqrt{2\pi}) \int_{-\infty}^{\infty} f_1(t) e^{i(\omega_0 - \omega)t} \, dt \\ &= (1/\sqrt{2\pi}) \int_{T_1}^{T_2} e^{i(\omega_0 - \omega)t} \, dt \\ &= (1/\sqrt{2\pi}) [e^{i(\omega_0 - \omega)T_2} - e^{i(\omega_0 - \omega)T_1}]/i(\omega_0 - \omega). \end{aligned}$$

Letting $\Delta\omega \equiv \omega - \omega_0$ and noting that

$$\begin{aligned} \Delta\omega T_2 &= \Delta\omega \left(\frac{T_2 + T_1}{2} + \frac{T_2 - T_1}{2} \right) \\ \Delta\omega T_1 &= \Delta\omega \left(\frac{T_2 + T_1}{2} - \frac{T_2 - T_1}{2} \right) \end{aligned}$$

we obtain

$$g(\omega) = \sqrt{\frac{2}{\pi}} e^{-i\Delta\omega(T_1+T_2)/2} \frac{\sin(\Delta\omega\,\Delta T/2)}{\Delta\omega}$$
$$= e^{-i\Delta\omega(T_1+T_2)/2} g_1(\Delta\omega), \tag{8.20}$$

where $\Delta T = T_2 - T_1$ and

$$g_1(\Delta\omega) = \sqrt{\frac{2}{\pi}} \frac{\sin(\Delta\omega\,\Delta T/2)}{\Delta\omega}. \tag{8.21}$$

Aside from the oscillating exponential factor (which becomes unity if $T_1 + T_2$ is taken to be the origin on the time axis), $g(\omega)$, or better $g_1(\Delta\omega)$, is a hyperbolically damped sinusoidal function of $\Delta\omega$, as shown in Fig. 8.1. The pulse length ΔT characterizes the form of $g_1(\Delta\omega)$, and we see that when

$$\frac{\sin(\Delta\omega\,\Delta T/2)}{\Delta\omega} = \frac{\Delta T}{4} \tag{8.22}$$

$g_1(\Delta\omega)$ will have a value equal to half its maximum value. Thus the solutions to

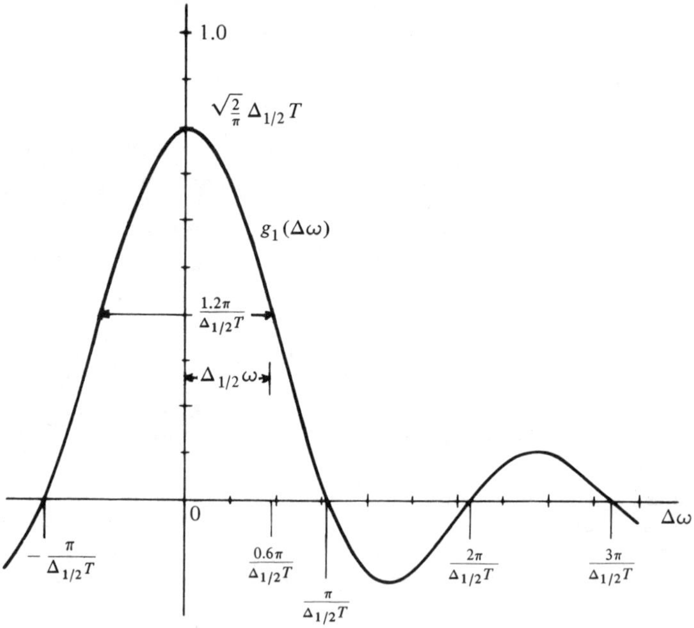

Fig. 8.1. The spectral function $g_1(\Delta\omega)$ for a square pulse. The abscissa is measured in units of $\pi/\Delta_{1/2}T = 2\pi/\Delta T$.

(8.22) will provide values of $\Delta\omega$, which we denote $\Delta_{1/2}\omega$, for which $g_1(\Delta_{1/2}\omega) = \frac{1}{2}g_1(0)$. In fact, $\Delta_{1/2}\omega$ is given by

$$\Delta_{1/2}\omega\Delta_{1/2}T = \pm 1.89 \text{ rad} = 0.6\pi \text{ rad}, \tag{8.23}$$

where we have let $\Delta_{1/2}T = \frac{1}{2}\Delta T$ represent half the width of the wave pulse, just as $\Delta_{1/2}\omega$ represents half the width of the spectral distribution of waves at half the maximum value of the spectral function $g_1(\Delta\omega)$. Thus the dominant contribution to the wave $f(t) = f_1(t)e^{i\omega_0 t}$ comes from monochromatic waves with frequencies within $\Delta_{1/2}\omega = 0.6\pi/\Delta_{1/2}T$ of ω_0.

We could be more generous and include all waves within $\pi/\Delta_{1/2}T$ of ω_0, so that the dominant contribution would be considered to come from all waves having frequencies within the central peak of $g_1(\Delta\omega)$, or we could be more miserly and include only those waves with frequencies such that $g_1^2(\Delta\omega)$ is more than half its maximum value, which would mean that only waves with frequencies within about $0.39\pi/\Delta_{1/2}T$ would be included. However, any such criterion for determining the dominant contribution to the spectral distribution yields essentially the same lesson, namely that the product of some reasonable measure of the width of $g_1(\Delta\omega)$ with some reasonable measure of the width of $f_1(t)$ always is a nonvanishing real number.

Thus if we think of the pulse of waves represented by $f(t)$ as a pulse *at* time $(T_1 + T_2)/2$ with *uncertainty* $\Delta_{1/2}T$ in the time, and having frequency ω_0 with uncertainty $\Delta_{1/2}\omega$ in the frequency, Eq. (8.23) becomes an *uncertainty relation* for the square pulse waveform, stating that the *uncertainty product* $\Delta_{1/2}\omega\ \Delta_{1/2}T$ has a nonvanishing value. Thus an increase in the certainty with which we can state that the frequency of the wave is ω_0 corresponds to a decrease in $\Delta_{1/2}\omega$ and can be achieved only at the expense of increasing $\Delta_{1/2}T$, and vice versa. The circumstance that $\Delta_{1/2}\omega$ and $\Delta_{1/2}T$ *must* satisfy such a relation is a consequence of the mathematical analysis of the wave system and is referred to as an *uncertainty principle.* The numerical value on the right-hand side of Eq. (8.23) depends on two conditions, namely the characteristics of the waveform considered and the criterion used for determining the uncertainty.

Now it is reasonable to ask, "Can the uncertainty product be made to vanish by choosing some particular waveform? If not, what is the smallest value for the uncertainty product that we can achieve?" The answer to the first query is simply "No". The answer to the second question depends upon our choice of criterion for uncertainty. The half-width at half-maximum that we used above is not always well defined, and the reasonableness of using that criterion varies according to the waveform; in fact, this is true for any criterion we might choose. A criterion that has wide applicability, though, is to let

$$\Delta\omega \equiv \sqrt{\overline{\omega^2} - \overline{\omega}^2}, \tag{8.24}$$

where

$$\overline{\omega} \equiv \int_{-\infty}^{\infty} \omega |g(\omega)|^2 \, d\omega \Big/ \int_{-\infty}^{\infty} |g(\omega)|^2 \, d\omega, \tag{8.25}$$

$$\overline{\omega^2} \equiv \int_{-\infty}^{\infty} \omega^2 |g(\omega)|^2 \, d\omega \Big/ \int_{-\infty}^{\infty} |g(\omega)|^2 \, d\omega, \tag{8.26}$$

and similarly let

$$\Delta t \equiv \sqrt{\overline{t^2} - \overline{t}^2}, \tag{8.24$'$}$$

with

$$\overline{t} \equiv \int_{-\infty}^{\infty} t |f(t)|^2 \, dt \Big/ \int_{-\infty}^{\infty} |f(t)|^2 \, dt \tag{8.25$'$}$$

$$\overline{t^2} \equiv \int_{-\infty}^{\infty} t^2 |f(t)|^2 \, dt \Big/ \int_{-\infty}^{\infty} |f(t)|^2 \, dt. \tag{8.26$'$}$$

This choice is dictated by a physical interpretation of $f(t)$ and $g(\omega)$ as being proportional to probability amplitudes.

For this definition of the uncertainties $\Delta\omega$ and Δt, it is easy to verify that $\Delta\omega \, \Delta t = \infty$ for the square pulse, so that the definition is not at all useful in that case. On the other hand, using this criterion it is possible to find a minimum value of $\Delta\omega \, \Delta t$ and, in fact, to find the waveform corresponding to that minimum uncertainty product. The actual calculations are carried out in a subsequent chapter in a different context, but we can quote the results here: For any $f(t)$ we have

$$\Delta\omega \, \Delta t \geq \tfrac{1}{2}$$

the equality holding if and only if

$$f(t) = ae^{-i\omega_0 t} e^{-bt^2}$$

where b is a positive real constant and a is a complex constant. Thus the Gaussian wave packet, for which the wave envelope is

$$f_1(t) = ae^{-bt^2}$$

yields the minimum uncertainty product.

The conclusions drawn here from the example of oscillations in the real time variable can be applied to oscillations in the position variables as well. Hence we can write, quite generally,

$$\begin{aligned} \Delta\omega \, \Delta t &\geq \tfrac{1}{2} \\ \Delta k_x \, \Delta x &\geq \tfrac{1}{2} \\ \Delta k_y \, \Delta y &\geq \tfrac{1}{2} \\ \Delta k_z \, \Delta z &\geq \tfrac{1}{2}. \end{aligned} \tag{8.27}$$

These conditions, taken together, form a general mathematical statement of an uncertainty principle, which describes fundamental limits upon the precision with which certain *conjugate* pairs of variables, such as ω and t, can be specified. This constitutes a fundamental limitation of all wave theories.

8.4 THE HYPOTHESES OF DE BROGLIE AND EINSTEIN; MATTER WAVES

In 1924 a French physics student, Louis de Broglie, asserted in his doctoral dissertation that massive particles, such as electrons and atoms, might be representable as waves. He suggested that Einstein's quantum relations between the energy and frequency and between momentum and wave number be extended to particles; thus with a particle having energy E and momentum $\mathbf{p}$ would be associated waves having frequency $\omega = E/\hbar$ and wave vector $\mathbf{k} = \mathbf{p}/\hbar$; i.e.

$$E = \hbar\omega, \qquad \mathbf{p} = \hbar\mathbf{k}. \tag{8.28}$$

These waves associated with particles having well-defined energy and momentum, and even waves associated with massive particles in general, often are referred to as de Broglie waves.

For free nonrelativistic particles, Eqs. (8.28) imply that the frequency and wave vector must be related by

$$\omega = \frac{\hbar}{2m}\mathbf{k}^2 \tag{8.29}$$

corresponding to $E = (1/2m)\mathbf{p}^2$. Free relativistic particles similarly must have their wave frequencies and wave vectors related according to the formula

$$\omega = \frac{mc^2}{\hbar}\sqrt{1 + \frac{\hbar^2\mathbf{k}^2}{m^2c^2}} \tag{8.30}$$

corresponding to

$$E = mc^2\sqrt{1 + \frac{\mathbf{p}^2}{m^2c^2}}.$$

However, since the phase velocity is $v_\mathrm{p} = \omega/k$, we have

$$v_\mathrm{p} = \frac{\hbar}{2m}k \tag{8.31}$$

for nonrelativistic particles, and for relativistic particles,

$$v_\mathrm{p} = c\frac{mc}{\hbar k}\sqrt{1 + \frac{\hbar^2k^2}{m^2c^2}} = c\sqrt{1 + \frac{m^2c^2}{\hbar^2k^2}}. \tag{8.32}$$

The relativistic phase velocity then exceeds c for all massive particles with finite momenta. This seems to be disallowed by the result from special relativity requiring that c be an upper bound on particle velocities.

It turns out, however, that Eq. (8.32) does not violate special relativity, because purely monochromatic waves with fixed phase velocity cannot carry any significant signal and hence cannot propagate any information about the location of a particle. The requirements of special relativity impose no constraint on the phase velocity, but require that the speed at which information can be propagated be limited by c. Thus $v_g < c$ for particles with nonzero mass, and $v_g = c$ for massless particles, are the required limitations.

For electromagnetic waves *in vacuo*, we have $\omega = kc$ so that $v_g = v_p = c$, whereas for relativistic particles with mass $m > 0$, Eq. (8.30) yields

$$\mathbf{v}_g = \nabla_k \omega = \frac{mc^2}{\hbar} \frac{(\hbar/mc)^2 \mathbf{k}}{\sqrt{1 + \hbar^2 \mathbf{k}^2/m^2c^2}} = c^2 \frac{\mathbf{k}}{\omega} = c^2 \frac{\mathbf{p}}{E} = c\boldsymbol{\beta} = \mathbf{v}, \tag{8.33}$$

where $\mathbf{v} = c\boldsymbol{\beta} = c(\mathbf{p}c/E)$ is the usual particle velocity. Hence the group velocity of the de Broglie waves representing a free particle is just the velocity of the classical motion of the particle, and limiting its magnitude to values less than c presents no difficulty. This seems to vindicate de Broglie's hypothesis. However, de Broglie's conjecture remained merely an interesting idea until 1927, when an experiment by Davisson and Germer confirmed the existence of wave properties for photons.

It is worth recalling at this point that an alternative to amplitude modulation of the wave is frequency modulation. If the frequency of a matter wave associated with a particle is changed as a function of time, then according to de Broglie's hypothesis, the energy of a particle will be changed in proportion. Thus frequency modulation of matter waves is equivalent to energy modulation, and the existence of frequency modulation in a wave system for a particle therefore implies the existence of external forces changing the energy of the particle. In the absence of such external forces, amplitude modulation alone effectively furnishes a wave train with a signal.

Heisenberg pointed out in 1927 that because of the fundamental limitations on what we have called the uncertainty products of conjugate pairs of wave variables, de Broglie's proposal would imply a restriction on the precision with which simultaneous measurement of conjugate pairs of mechanical variables can be carried out. Because of Eqs. (8.28), increments of energy and momentum are given in terms of frequency and wave number, respectively, by

$$\Delta E = \hbar\, \Delta\omega, \qquad \Delta p^m = \hbar\, \Delta k^m, \qquad m = 1, 2, 3.$$

Hence Eqs. (8.27) yield

$$\Delta p^\mu\, \Delta x^\mu \geq \hbar/2, \qquad \mu = 0, 1, 2, 3, \tag{8.34}$$

where we have used $p^\mu = (E/c, p^1, p^2, p^3)$ and $x^\mu = (ct, x^1, x^2, x^3)$.

No particle, then, can be localized fully in space unless its momentum is allowed to be undefined, and vice versa. In the process of localizing a particle, the localizing apparatus serves as an amplitude modulator and thereby imposes, upon the wave train representing the particle, a wave number spectrum with some finite width in the wave number variable; that is, the very process of confining a particle to a finite volume disturbs the particle's motion by giving it a spectrum of momenta rather than a single well-defined momentum.

It is curious that the uncertainty principle gives conditions upon conjugate *pairs* of variables. An uncertainty in p^1 in no way affects the accuracy with which x^2 can be measured, because p^1 and x^2 are not conjugate to each other. This is related to the duality between a coordinate direction and the notion of motion measured in terms of that direction. We shall not discuss in detail this notion of conjugacy (frequently referred to in this context as *canonical* conjugacy), but it should be noted that it extends beyond time and space variables and their conjugates, energy and momentum respectively. Also, we note that conjugate pairs appear in classical mechanics also, so that conjugacy is not strictly a feature of waves. However, in classical mechanics, the uncertainty products can be made to vanish, in principle; only the practical considerations of building sufficiently fine instruments limit the uncertainties in Newtonian mechanics and its classical successor, Hamilton–Lagrange mechanics. Thus some of the fundamental assumptions of the older theories must be regarded as valid only to a certain approximation, if de Broglie's ideas are correct.

8.5 EXPERIMENTAL VERIFICATION OF DE BROGLIE'S CONJECTURE

In several independently conducted experiments in 1927 and 1928, wave-like properties of electrons and molecules were observed. Furthermore, the data from these experiments agreed with the predictions suggested by the de Broglie–Einstein relations.

The first and most famous of these, performed by Davisson and Germer, consisted of an electron beam impinging upon a nickel crystal, together with appropriate detection apparatus with which to observe electrons scattered from the crystal. In particular, the experimenters were interested in back-scattered, or reflected electrons; earlier they had observed anomalies in the reflection pattern of electrons from a nickel target that had accidentally become crystallized. They carefully prepared a surface parallel to a convenient Bragg plane* of the crystal, and placed the surface perpendicular to the electron beam. They observed electrons scattered backward (reflected) from the surface at different scattering angles and for different energies. (See Figs. 8.2, 8.3.)

* For a discussion of Bragg planes and x-rays scattering, as well as further discussion of the experiments enumerated here, see any good text on elementary modern physics, such as those by Weidner and Sells, Semat, and Blanchard, *et al.* Especially see Richtmeyer, Kennard and Lauritsen.

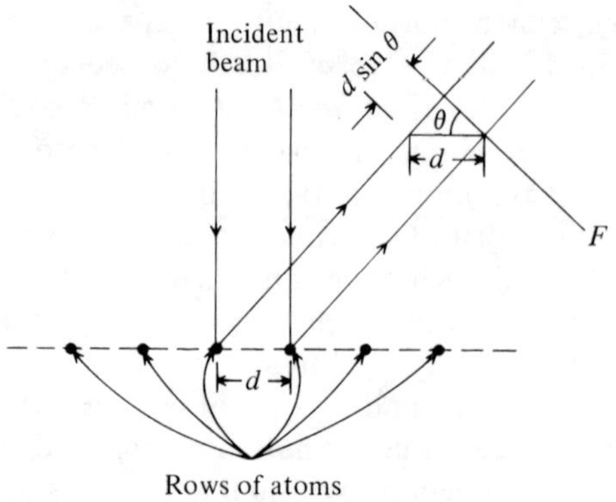

Fig. 8.2. The difference in path length for the two rays of electron waves having wave front at F is $d \sin \theta$. When this difference is $n\lambda$, constructive interference occurs.

As the energy of the incident electrons varied from 36 eV upwards, a diffraction peak in the angular distribution of electron intensities was observed, and the peak was most pronounced for 54 eV electrons; for this energy the peak reached its apex at a colatitude of 50°, as illustrated in Fig. 8.4. Assuming that the crystal acts as a diffraction grating for the electron waves, with the rows of electrons on the surface serving as rulings, we can estimate the distance d between the rows of electrons from the familiar diffraction formula of geometric optics,

$$n\lambda = d \sin \theta,$$

θ being the angle of diffraction or colatitude and n being the order of diffraction; $\lambda = 2\pi/k$ is the assumed electron wavelength, which is equal here to

$$2\pi\hbar/p = 2\pi\hbar/\sqrt{2mT},$$

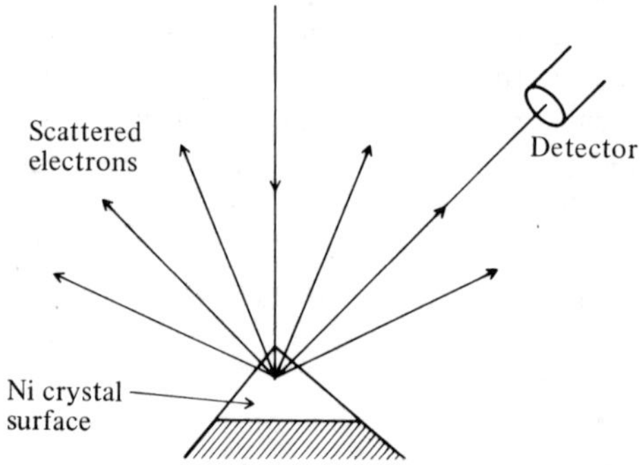

Fig. 8.3. Davisson–Germer experimental arrangement.

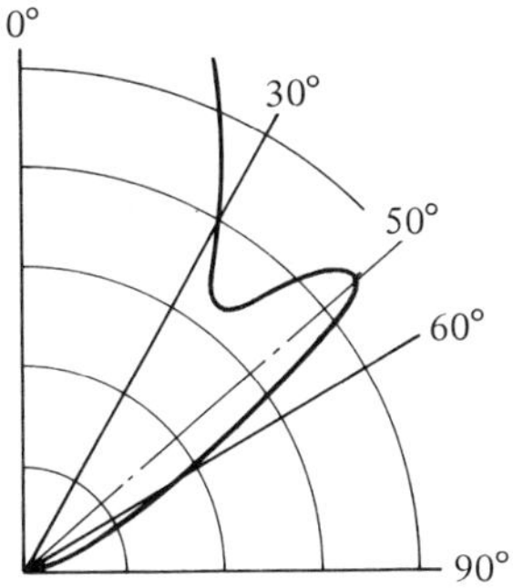

Fig. 8.4. Polar plot of scattered intensity vs. angle at 54 eV. From F. K. RICHTMEYER, E. H. KENNARD and T. LAURITSEN, *Introduction to Modern Physics*, 5th Ed., McGraw-Hill Book Co., N.Y. (1953). Reprinted through permission of the publisher.)

where m is electron mass and T is the kinetic energy of the electron. It is convenient to put the right-hand side in the form

$$2\pi\hbar c/\sqrt{2mc^2T},$$

and to use

$$\begin{aligned}\hbar c &= 197.32 \text{ MeV F},\\ mc^2 &= 0.511 \text{ MeV},\\ T &= 54 \text{ eV} = 54 \times 10^{-6} \text{ MeV},\end{aligned}$$

where as before $1 \text{ F} = 10^{-15}$ m. Thus

$$\lambda = 1.67 \times 10^5 \text{ F} = 1.67 \text{ Å}$$

Since $\sin 50° = 0.766$, and we assume this strong diffraction peak to be the first-order maximum, $n = 1$, this yields

$$d = \lambda/\sin\theta = 2.18 \text{ Å}.$$

The known spacing between the rows of atoms involved in the Davisson–Germer diffraction turns out to be 2.15 Å, as determined by x-ray studies. Thus the electron-diffraction interpretation of the Davisson–Germer experiment agrees quite well with other experimental data. More precise repetitions of the experiment have been performed and the electron-scattering and x-ray-scattering data agree to better than 1% in these experiments.

Davisson and Germer also observed evidence for electron refraction. Later, G. P. Thompson obtained diffraction rings from electron beams passed through thin metal foils; these rings are dramatically similar to x-ray diffraction rings produced by similar foils. Kikuchi independently observed diffraction patterns by passing electrons through thin sheets of mica. Three years later, Esterman,

Frisch and Stern used an elegant experimental technique to obtain Davisson–Germer-like diffraction for monoenergetic beams of α-particles incident on LiF crystals. In short, evidence for matter waves became overwhelmingly convincing by the early 1930's. Already a well-developed mathematical theory for the wave phenomena was available thanks to the work of Schrödinger, Heisenberg, Born and Dirac. Let us see how the theory is constituted.

8.6 THE MESSAGE OF QUANTUM THEORY; THE PROBABILITY INTERPRETATION

Now we must find an appropriate mathematical formulation for the wave aspect of massive objects. In addition, we must furnish a physical interpretation for the fundamental mathematical entities that will enter our formalism. Obviously the de Broglie–Einstein relations will play a basic role in our work, but they in themselves are not sufficient to provide an adequate theory. We shall need to define some function of space and time to represent the oscillating variable of the wave theory, and this variable will have to be provided with some physical meaning. To carry out this program, let us review in some detail what we know about some aspects of electromagnetic waves and the quantum behavior of electromagnetic fields.

In classical electromagnetic theory, the intensity (energy delivered to a unit area of beam cross-section per unit time) of a beam of electromagnetic radiation is given by

$$\Phi = c\mathscr{E} = c\epsilon_0 \mathbf{E}^2(\mathbf{x}) \tag{8.35}$$

where $\mathscr{E}$ is the energy density of the radiation field of the beam and $\mathbf{E}(\mathbf{x})$ is the electric field vector. (See Eq. 5.30 and the discussion following 5.55.) The very fact that this radiation is characterized in classical theory by an energy *density* indicates an important distinction between waves (or fields) and particles in classical physics. Evidently, we must think of the energy of the waves as being spread throughout the region of space where the wave field is nonvanishing: the energy is not well localized.

On the other hand, a particle, whose classical idealization is a *mass point* without geometrical extension (i.e. being of infinitesimally small size) is characterized by a perfectly localized energy. In effect, the energy density of a particle can be written in terms of a three-dimensional δ-function:

$$\mathscr{E}_{\text{particle}} = U_{\text{particle}}\, \delta^3(\mathbf{x} - \mathbf{x}_0) \tag{8.36}$$

where U_{particle} is the energy of the mass point and $\mathbf{x}_0$ is its location in space.

Correspondingly, when we make the transition from the wave theory to the particle theory of electromagnetic radiation we might think of each photon as having well-localized energy (and momentum) that it can deliver instantaneously to a well-localized particle such as an electron. Here it should be noticed that the frequency and wave number of the corresponding waves, according to the uncertainty principle, must be completely undefined. If we think of a photon as a radiation pulse with no spatial extension $\Delta x^1 = \Delta x^2 = \Delta x^3 = 0$, then the wave number spectrum must have infinite breadth, $\Delta k^1 = \Delta k^2 = \Delta k^3 = \infty$. Thus the phrase "a photon with frequency ω" seems to be self-contradictory; it bids us to think of the photon simultaneously as a wave and as a particle. More accurately we should say "a photon that gave up about $\hbar\omega$ of energy upon interacting with a massive particle", keeping in mind that the photon in this context is quite well localized, but not completely so; its energy density is sharply peaked near $\mathbf{x}_0$ but is not a δ-function. We could speak also of "a photon in a beam of radiation having dominant frequency ω".

Although the wave and particle notions, in their idealized forms, cannot be applied simultaneously, it certainly is possible to have waveforms whose spatial extension is so small that they have essentially particle-like features. This, of course, is the view that we must take in a quantum-mechanical world. The famous "wave–particle duality", then, refers to descriptions of two extreme types of behavior for a single physical phenomenon, and arises from our classical predjudices about the structure of matter rather than from any duality in nature. It has more to do with the limitations of our direct sense perceptions and the ways in which we describe our perceptions than with the fundamental properties of matter. Einstein, de Broglie and their successors have provided us with a new prejudice that gives, within a single framework, a more comprehensive representation of physical phenomena than did the old ones.

Bearing this in mind, we may think of a beam of electromagnetic radiation as a beam of photons, the mean energy per photon being $\hbar\omega$ if ω is the (angular) frequency of the radiation. Thus the energy intensity of the beam must be

$$\Phi_\omega = \Lambda_\omega E_\omega = \Lambda_\omega \hbar\omega, \qquad E_\omega \equiv \hbar\omega, \tag{8.37}$$

where Λ_ω is the photon flux—the number of photons passing through a unit area of beam cross section per unit time, that is,

$$\Lambda_\omega = \Delta N_\omega / \Delta A \, \Delta t \tag{8.38}$$

where ΔN_ω is the number of photons passing through cross-sectional area ΔA in time Δt.

Equating the intensities obtained from this particle view and from the wave view, we have

$$\Phi_\omega = \Lambda_\omega \hbar\omega = c\epsilon_0 \mathbf{E}_\omega^2(\mathbf{x})$$

or, using Eq. (8.38),

$$\Delta N_\omega = \Lambda_\omega \, \Delta A \, \Delta t = c\epsilon_0 \frac{\mathbf{E}_\omega^2(\mathbf{x})}{\hbar\omega} \Delta A \, \Delta t$$

must be the number of photons in the volume having cross-sectional area ΔA

and length $c\,\Delta t$. Since ΔA and Δt may be chosen arbitrarily, the number ΔN_ω is not necessarily an integer. Thus, either photons are not indivisible particles and fractions of photons may exist, or the interpretation of ΔN_ω as the actual number of photons in the volume $\Delta A\,c\,\Delta t$ should not be taken too literally. The former alternative would render the photon concept meaningless, so we choose the latter.

Instead of taking ΔN_ω to mean the actual number of photons in the volume $\Delta A\,c\,\Delta t$, we may take it to mean the number of particles *probably* to be found in $\Delta A\,c\,\Delta t$. Thus

$$\epsilon_0 \mathbf{E}_\omega^2(\mathbf{x})/\hbar\omega$$

may be interpreted as the *density* of the *probable number* of photons in the monochromatic beam represented by electric field vector $\mathbf{E}_\omega(\mathbf{x})$. The total number of photons in the beam then is

$$N_\omega = \int d^3x \frac{\epsilon_0 \mathbf{E}_\omega^2(\mathbf{x})}{\hbar\omega} = \frac{\epsilon_0}{\hbar\omega}\int d^3x\, \mathbf{E}_\omega^2(\mathbf{x})$$

the integral being taken over all of space, and the *probability* that a *single photon* would be found in volume element d^3x centered at $\mathbf{x}$ is

$$\mathbf{E}_\omega^2(\mathbf{x}) \Big/ \int d^3x'\, \mathbf{E}_\omega^2(\mathbf{x}');$$

this is called the *probability density* for a single photon. This indicates that the magnitude $|\mathbf{E}_\omega(\mathbf{x})|$ of the oscillating physical variable representing electromagnetic waves is proportional to the *amplitude* of the probability density in the sense that the square of

$$\left[\int d^3x' \mathbf{E}_\omega^2(\mathbf{x}')\right]^{-1/2} |\mathbf{E}_\omega(\mathbf{x})|$$

is the probability density for photons in the beam.

This suggestion that the quantum interpretation of the oscillating physical variable should be that it represents a *probability density amplitude* was contributed by Max Born in 1926, although he introduced the idea in reference to matter fields rather than to electromagnetic fields. With this view, we are able to weld together the concepts of particle and wave, replacing such statements as "a particle lies within volume element ΔV centered at $\mathbf{x}$" with, for example, "it is 90% probable that a particle will be found in volume element ΔV centered at $\mathbf{x}$", and also replacing phrases like "the wave with frequency ω" with corresponding probabilistic statements such as "the wave whose probability of having frequency between $\omega_0 - \Delta\omega$ and $\omega_0 + \Delta\omega$ is $\frac{2}{3}$." A particle-like system corresponds to a wave field whose probability amplitude is vanishingly small except within a "small" region of space–time, whereas wave-like systems correspond to wave fields with probability amplitudes that are vanishingly small except within a "small" volume of wave number–frequency space.

A succinct statement of this situation is found in Nils Bohr's *principle of complementarity*:

> At the quantum level, the most general physical properties of any system may be expressed in complementary pairs of variables, each of which can be better defined only at the expense of a corresponding loss in the degree of definition of the other.*

Let us consider this principle from a slightly different perspective from that which we adopted above. If we set up an experiment expressly designed to investigate the wave aspects of a system, we must expect that the particle aspects will not be clearly observable in the experiment: an apparatus designed to give good definition to variables that characterize waves must give correspondingly poor definition to the variables that characterize particles.

For example, if a light beam contains a large number of reasonably well-localized photons, then the addition or deletion of a few of these will not affect the behavior of the beam; the mean energy and momentum per photon is negligible compared with the energy and momentum of the entire beam. If the beam has $N \gg 1$ photons, then the fractional uncertainty in the number of photons may be expected to be on the order of $1/N \ll 1$ if the beam is not interacting strongly with its environment. Also, the fractional uncertainty in energy and momentum will be on the order of

$$\frac{\Delta E}{E} = \frac{\hbar\omega}{N\hbar\omega} = \frac{1}{N} \ll 1, \qquad \left|\frac{\Delta \mathbf{p}}{\mathbf{p}}\right| = \frac{\hbar|\mathbf{k}|}{N\hbar|\mathbf{k}|} = \frac{1}{N} \ll 1$$

when ω is the mean frequency of the beam and $\mathbf{k}$ is the corresponding wave vector. Therefore, since $|\mathbf{p}| = 2\pi\hbar/\lambda$, the uncertainty principle implies that

$$|\Delta \mathbf{x}| \geqslant \frac{\hbar}{2|\Delta \mathbf{p}|} = \frac{\hbar}{2|\mathbf{p}|}\left|\frac{\mathbf{p}}{\Delta \mathbf{p}}\right| = \frac{\lambda}{4\pi}N \gg \frac{\lambda}{4\pi};$$

the spatial resolution of the beam is poor in the sense that the beam must contain many wavelengths of radiation. A natural measure of the spatial resolution is $\lambda/4\pi|\Delta \mathbf{x}|$, which is small when the resolution is poor and is much larger than unity when the resolution is very good. Thus, in the situation described here, the photons, being many, tend to fill out a rather large region of space compared with the beam's mean wavelength, and the wavelength may be quite well defined. A large number of photons corresponds to wave-like behavior, which, for a light beam, is its classical behavior.

However, if a mechanism is available for the light beam to interact with its environment in a small region of space (e.g. if an electron or a crystal lattice, with dimensions that are small compared with a typical wavelength, lies within the spatial region covered by the beam), then single photons may be absorbed from the beam through particle-like phenomena, while the remainder of the beam per-

* D. Bohm, *Quantum Theory*, Prentice-Hall, Inc., Englewood Cliffs, N.J. (1951) p. 160.

sists in its wave-like properties. A single observation would reveal either the particle-like photon absorption, through the subsequent motion of an electron, say, *or* the wave-like behavior of the beam, but not both.

As was mentioned above, wave motion is what we expect classically for light. The condition under which we gain a *correspondence* between the newer quantum picture of the system and the older classical picture is that the number of *particle quanta* must be very large,

$$N \gg 1.$$

A similar condition yields correspondence between quantum and classical views for massive particles, except that in this case N must be the number of *wave quanta* with well-defined energy and momenta (and hence frequencies and wave vectors). Classical phenomena appear when the number of quanta with nonclassical properties becomes very large. Just as, for special relativity, the correspondence between "old" and "new" views is obtained when

$$v/c \ll 1,$$

the quantum–classical correspondence is obtained for

$$N \gg 1,$$

when N is the number of nonclassical quanta in the system.

In the last three paragraphs, the *correspondence principle* of Bohr and the *uncertainty principle* of Heisenberg have joined hands with Bohr's *principle of complementarity* to help clarify the transition from prequantum ideas in physics to new ideas that seem to be required in order that we may find unity among observable physical phenomena.*

8.7 MATTER WAVES; WAVE FUNCTIONS

We have seen that the classically observable electric field vector may be interpreted in the quantum theory as a probability-density amplitude, aside from a proportionality factor.† It was relatively simple to go from the wave theory to the quantum theory once we adopted the probability interpretation for the wave amplitude $|\mathbf{E}_\omega(\mathbf{x}, t)|$. However, the classical theory for particles with mass does not involve any such amplitude, and we must invent one in order to pass from the particle picture to the wave picture. Moreover, since all physical measurements are made, ultimately, with classical apparatus, the probability-density amplitudes for masses must not be classically observable quantities.

* For further discussion of these topics, see D. Bohm, *op. cit.*, Chapters 1–8.

† In a more sophisticated treatment we would find that it is more appropriate to consider the four-vector electromagnetic potential $A^\mu(\mathbf{x}, t)$ as the probability-density amplitude. Although A^μ is not an observable quantity in classical electromagnetic theory, it does have observable effects in the quantum theory.

It is traditional to denote the probability density amplitude for matter waves with the Greek letter ψ, and to take

$$|\psi(\mathbf{x}, t)|^2 \bigg/ \int_{\text{all space}} |\psi(\mathbf{x}', t)|^2 \, d^3x' \tag{8.39}$$

as the probability density for a massive particle. The probability of finding the particle in a specified volume element V at time t obviously is

$$P_{V,t} = \int_V |\psi(\mathbf{x}, t)|^2 \, d^3x \bigg/ \int_{\text{all space}} |\psi(\mathbf{x}', t)|^2 \, d^3x' \tag{8.40}$$

Somehow, we shall expect to endow the *wave functions* $\psi(\mathbf{x}, t)$ with the ability to carry all the relevant information about the physics of the system ψ is to represent.

8.8 AN EXAMPLE: A PARTICLE WITH WELL-DETERMINED MOMENTUM

Let us consider the wave system corresponding to a massive object having mass m and well-determined momentum $\mathbf{p} = \hbar\mathbf{k}$. Its energy is

$$E = \hbar\omega = mc^2\sqrt{1 + (\hbar\mathbf{k}/mc)^2}. \tag{8.41}$$

Since the momentum is specified precisely, the position of the object, according to the uncertainty principle, is undetermined; the object is as likely to be found at one point in space as at any other. Similarly, since the energy is well defined, the likelihood of finding the particle at a given position is unaffected by the time variable. It follows that if $\psi_{\mathbf{p},E}(\mathbf{x}, t)$ is the corresponding probability density amplitude, or wave function, then $|\psi_{\mathbf{p},E}(\mathbf{x}, t)|^2$ must be independent of $\mathbf{x}$, t; that is

$$|\psi_{\mathbf{p},E}(\mathbf{x}, t)|^2 = |\psi_{\mathbf{p},E}(\mathbf{x} + \Delta\mathbf{x}, t + \Delta t)|^2$$

for all possible $\Delta\mathbf{x}$, Δt. Thus

$$\psi_{\mathbf{p},E}(\mathbf{x}, t) = A\, e^{ih(\mathbf{x},t)}$$

where A is a constant and $h(\mathbf{x}, t)$ is a real function of $\mathbf{x}$, t.

In Appendix B it is shown that, because of the required wave periodicity in $\mathbf{x}$ and t, the only possible form for the exponent is given by

$$h(\mathbf{x}, t) = \pm(\mathbf{k} \cdot \mathbf{x} - \omega t)$$

and convention has it that the upper sign is to be chosen. Therefore the wave function for an object with fixed mass m and well-determined momentum $\mathbf{p} = \hbar\mathbf{k}$ is

$$\psi_{\mathbf{p},E}(\mathbf{x}, t) = A\, e^{i(\mathbf{k}\cdot\mathbf{x} - \omega t)} \tag{8.42}$$

with ω given by Eq. (8.41). This follows from the de Broglie–Einstein relations, the uncertainty principle, and the probability interpretation.

Noting that

$$\nabla\psi_{\mathbf{p},E}(\mathbf{x}, t) = i\mathbf{k}\psi_{\mathbf{p},E}(\mathbf{x}, t) = (i/\hbar)\mathbf{p}\psi_{\mathbf{p},E}(\mathbf{x}, t),$$

we call the differential operator

$$\mathsf{P} \equiv (\hbar/i)\nabla \tag{8.43}$$

the *momentum operator*, and we call

$$\mathsf{E} \equiv i\hbar(\partial/\partial t) \tag{8.44}$$

the *energy operator* in view of the relation

$$\frac{\partial}{\partial t}\psi_{\mathbf{p},E}(\mathbf{x}, t) = -i\omega\psi_{\mathbf{p},E}(\mathbf{x}, t) = \frac{1}{i\hbar}E\psi_{\mathbf{p},E}(\mathbf{x}, t).$$

Hence we may call the differential operators

$$\mathsf{P}^\mu \equiv i\hbar\frac{\partial}{\partial x^\mu} \equiv i\hbar\partial^\mu, \qquad \mu = 0, 1, 2, 3, \tag{8.45}$$

the *contravariant four-momentum operators.* Operating on a wave function representing an object with well-defined momentum with the momentum operator has the effect of multiplying the wave function by the momentum.

Next we observe that the effect of operating on any wave function with the operator

$$\exp(\Delta x_\mu\,\partial^\mu) = \exp\left(\frac{1}{i\hbar}\,\Delta x_\mu \mathsf{P}^\mu\right)$$

$$\equiv \sum_{n=0}^{\infty}\frac{1}{n!}\left(\frac{\Delta x_\mu}{i\hbar}\mathsf{P}^\mu\right)^n = \sum_{n=0}^{\infty}\frac{1}{n!}(\Delta x_\mu\partial^\mu)^n \tag{8.46}$$

is to translate the four-vector x^μ to $x^\mu + \Delta x^\mu$. To confirm this, note that the last member in Eq. (8.46) is simply the operational part of a four-dimensional Taylor's series expansion. We call

$$\exp(\Delta x_\mu\,\partial^\mu) = \exp(\Delta x^\mu\,\partial_\mu) = \exp((1/i\hbar)\,\Delta x_\mu \mathsf{P}^\mu) = \exp((1/i\hbar)\,\Delta x^\mu \mathsf{P}_\mu) \tag{8.47}$$

the space–time translation operator, and $(1/i\hbar)\mathsf{P}_\mu$ is called the *covariant generator of four-translations*; the contravariant generator of course is $(1/i\hbar)\mathsf{P}^\mu$.

That the four-momentum should be proportional to the generator of four-translations is not surprising, since the conservation of four-momentum is equivalent to the covariance of the physical system under translations in four-space.

Let us summarize our work. In effect, we have obtained the form for monochromatic plane waves (waves with well-determined wave vector) by requiring that $|\psi_{\mathbf{p},E}(\mathbf{x}, t)|^2$ be invariant under translations, i.e. that

$$|e^{\Delta x_\mu\partial^\mu}\psi_{\mathbf{p},E}(\mathbf{x}, t)|^2 = |\psi_{\mathbf{p},E}(\mathbf{x}, t)|^2$$

for arbitrary Δx_μ. This requirement is satisfied if and only if the effect of operating upon $\psi_{\mathbf{p},E}(\mathbf{x}, t)$ with $i\,\partial^\mu$ is to multiply the former by a real four-vector, say $R^\mu(\mathbf{p}, E)/\hbar$, that does not depend on $\mathbf{x}$, t:

$$i\,\partial^\mu \psi_{\mathbf{p},E}(\mathbf{x}, t) = \frac{1}{\hbar} R^\mu(\mathbf{p}, E)\psi_{\mathbf{p},E}(\mathbf{x}, t).$$

The further restriction that the waves represented by $\psi_{\mathbf{p},E}(\mathbf{x}, t)$ have wavelength $\lambda = 2\pi\hbar/|\mathbf{p}|$ and frequency $\omega = E/\hbar$ then leads to the conclusion that $R^\mu(\mathbf{p}, E)$ must be four-momentum,

$$R^\mu(\mathbf{p}, E) = p^\mu.$$

Thus the waves corresponding to well-defined four-momentum p^μ are

$$\psi_{\mathbf{p},E}(\mathbf{x}, t) = Ae^{-ix_\mu p^\mu} = Ae^{i(\mathbf{k}\cdot\mathbf{x} - \omega t)}.$$

These plane waves, of course, play an important role in the mathematics of quantum theory.

8.9 NORMALIZATION OF PLANE WAVES

Because of the definition of the probability density, the probability per unit volume of finding a particle with exact momentum $\mathbf{p}$ and energy E near point x at time t must be

$$|\psi_{\mathbf{p},E}(\mathbf{x}, t)|^2 \Big/ \int_{\text{all space}} |\psi_{\mathbf{p},E}(\mathbf{x}', t)|^2\, d^3x' = |A|^2 \Big/ \int_{\text{all space}} |A|^2\, d^3x$$

$$= 1 \Big/ \int_{-\infty}^{\infty} dx \int_{-\infty}^{\infty} dy \int_{-\infty}^{\infty} dz = 0.$$

This seems to indicate that monochromatic plane waves cannot exist. Such, in fact, is the case, for as we already have noted, monochromatic plane waves must have infinite extent in space and time. Nevertheless, they are a convenient approximation to waveforms whose duration is many wave periods and whose extent is many wavelengths. In general, they form a very useful mathematical device. Therefore we need a tractable way to handle them in calculations.

The problem is to obtain nonzero probability densities. Two commonly used techniques are available, and the choice between them depends upon the nature of the physical problem at hand; one chooses the approach that makes the calculations simplest.

The *box normalization* is obtained if we impose the *periodic boundary conditions*

$$\psi_{\mathbf{p},E}(\mathbf{x} + \hat{\mathbf{i}}_j L, t) = \psi_{\mathbf{p},E}(\mathbf{x}, t), \qquad j = 1, 2, 3. \tag{8.48}$$

This imposes the conditions $k_j = \pi/Ln_j$, $n_j = \ldots, -2, -1, 0, 1, 2, \ldots,$ upon the components of $\mathbf{k}$; the allowed values of the momentum then form a denumerable, or discrete, set. Integrating over a cubic box having edges of length L, we obtain

$$\int_{L^3} |\psi_{\mathbf{p},E}(\mathbf{x}', t)|^2 \, d^3x' = \int_x^{x+L} dx' \int_y^{y+L} dy' \int_z^{z+L} dz' |\psi_{\mathbf{p},E}(\mathbf{x}', t)|^2 = |A|^2 L^3. \quad (8.49)$$

Thus if we imagine that one particle is confined to each such box in all space, we can conclude that the probability density per particle is

$$|\psi_{\mathbf{p},E}(\mathbf{x}, t)|^2 \Big/ \int_{L^3} |\psi_{\mathbf{p},E}(\mathbf{x}', t)|^2 \, d^3x' = 1/L^3. \quad (8.50)$$

The periodic boundary conditions assure us that the wave functions join neatly at the boundaries of the boxes, so we do not anticipate any difficulties from discontinuities.

It is convenient to *normalize* the wave functions by insisting that

$$\int_{L^3} |\psi_{\mathbf{p},E}(\mathbf{x}, t)|^2 \, d^3x = 1 \quad (8.51)$$

so that we will not have to carry around the awkward integral in the denominator when discussing probability densities. This amounts to choosing

$$A = e^{i\alpha} L^{-3/2} \quad (8.52)$$

where α is a real number; this follows from (8.49) and (8.51). Hence we get

$$\psi_{\mathbf{p},E}(\mathbf{x}, t) = (e^{i\alpha}/L^{3/2}) e^{-ik^\mu x_\mu}.$$

Note that an arbitrary real constant, α, remains in our expression for $\psi_{\mathbf{p},E}(\mathbf{x}, t)$: normalization of the wave function did not completely remove the arbitrariness in the constant A. This is a characteristic feature of quantum mechanics; all wave functions are determined only up to a constant of modulus unity, here denoted $e^{i\alpha}$. Thus there is an infinity of distinct wave functions, all having the same absolute value, that correspond to the same physical system. This is reminiscent of the situation we encountered in prequantum electromagnetic theory, where the potential A^μ is somewhat arbitrary because *gauge transformations of the second kind* produce no physical effect. A change of the phase of a wave function by an additive real constant is a special case of what is called a *gauge transformation of the first kind* and it too produces no physical effect. These two kinds of gauge transformations can be shown to be related, but that is beyond the scope of our present interest.

We mention in passing that just as the covariance of physical laws under translations in four-space implies conservation of four-momentum, so invariance under gauge transformations implies the conservation of probability and consequently the conservation of charge.

Since α is arbitrary, we may choose it to be zero for convenience, and write

$$\psi_{\mathbf{p},E}(\mathbf{x}, t) = L^{-3/2} e^{-ik^\mu x_\mu} \tag{8.53}$$

for the box-normalized wave functions.

It must be remembered that in this form the wave function does not actually represent a probability density amplitude for a single particle, because the integral of $|\psi_{\mathbf{p},E}|^2$ over *all space* still diverges. Nevertheless, we can assume that L is very large and that the integral over the volume L^3 is a sufficient approximation to all space for most purposes. Moreover if we imagine space to be filled with adjacent boxes, we can interpret $|\psi_{\mathbf{p},E}|^2$ as the probability density per box and we can attribute the divergence in the integral over all space to the presence of infinitely many particles, one per box; the value of the integral is, effectively, the number of particles in all space. It is gratifying that in practical calculations factors involving L^3 usually cancel one another so that the final results are independent of L.

It is possible to handle the normalization of $\psi_{\mathbf{p},E}$ in a different and, for some purposes, more satisfactory way. We consider two waves having different momenta, $\psi_{\mathbf{p},E}$ and $\psi_{\mathbf{p}',E'}$, and evaluate

$$\begin{aligned} I_{\mathbf{p},\mathbf{p}'} &= \int_{-\infty}^{\infty} dx \int_{-\infty}^{\infty} dy \int_{-\infty}^{\infty} dz \ \psi^*_{\mathbf{p}',E'}(\mathbf{x}, t)\psi_{\mathbf{p},E}(\mathbf{x}, t) \\ &= A^*_{\mathbf{p}'} A_{\mathbf{p}} e^{-i(\omega-\omega')t}(2\pi)^3\, \delta^3(\mathbf{k} - \mathbf{k}') \\ &= A^*_{\mathbf{p}'} A_{\mathbf{p}} e^{-i(\omega-\omega')t}(2\pi\hbar)^3\, \delta^3(\mathbf{p} - \mathbf{p}') \\ &= |A_{\mathbf{p}}|^2(2\pi\hbar)^3\, \delta^3(\mathbf{p} - \mathbf{p}'). \end{aligned} \tag{8.54}$$

Here, recall that $f(x)\,\delta(x - a) = f(a)\,\delta(x - a)$. Letting

$$A_{\mathbf{p}} = e^{i\alpha}/(2\pi)^{3/2}$$

we get

$$\psi_{\mathbf{p},E}(\mathbf{x}, t) = \frac{e^{i\alpha}}{(2\pi)^{3/2}} e^{-ik_\mu x^\mu} \tag{8.55}$$

and

$$I_{\mathbf{p}\mathbf{p}'} = \int_{\text{all space}} d^3x \psi^*_{\mathbf{p}',E'}(\mathbf{x}, t)\psi_{\mathbf{p},E}(\mathbf{x}, t) = \delta^3(\mathbf{k} - \mathbf{k}'). \tag{8.56}$$

Alternatively,

$$A_{\mathbf{p}} = e^{i\alpha}/(2\pi\hbar)^{3/2}$$

yields

$$I_{\mathbf{p}\mathbf{p}'} = \delta^3(\mathbf{p} - \mathbf{p}'). \tag{8.57}$$

(Again we are free to let $\alpha = 0$, since it is an arbitrary real number.) The wave function so normalized is said to be *δ-function normalized.*

The δ-function normalized wave function, just as the box-normalized wave function, is not interpretable as a probability density amplitude. It is a limiting case of a wave function that can be so interpreted, however. For calculational purposes it usually is permissible to use the δ-function normalized wave amplitudes; in practice their unpleasant features usually fail to appear—they tend to cancel out.

Equations (8.56) and (8.57) demonstrate another property of the wave functions $\psi_{\mathbf{p},E}(\mathbf{x}, t)$, namely that they are mutually *orthogonal* in terms of the integral

$$\int_{\text{all space}} \psi^*_{\mathbf{p}'E'}(\mathbf{x}, t)\psi_{\mathbf{p},E}(\mathbf{x}, t)\, d^3x;$$

that is, the integral vanishes unless $\mathbf{p} = \mathbf{p}'$ (and hence $E = E'$). The same statement is valid for box-normalized wave functions. This integral is analogous to the vector dot product $\mathbf{A} \cdot \mathbf{B}$, where $\psi_{\mathbf{p}',E'}(\mathbf{x}, t)$ and $\psi_{\mathbf{p},E}(\mathbf{x}, t)$ are the analogs of $\mathbf{A}$ and $\mathbf{B}$ and the integration over d^3x is the analog of summation over the vector indices. Such integrals often are called *inner products* of wave functions. Note that the condition for orthogonality of wave functions is similar to the condition that $\mathbf{A}$ and $\mathbf{B}$ be perpendicular, namely that their respective inner products vanish. More general discussions of this kind of integral will follow in later chapters.

8.10 SCHRÖDINGER AND KLEIN–GORDON EQUATIONS

Our identification, in Section 8.8, of the differential operator $\mathsf{P}^\mu = i\hbar\, \partial^\mu$ with the four-momentum might lead us to suspect that the relativistic invariant

$$m^2c^2 = p^\mu p_\mu$$

could be replaced by a differential equation, the solution to which should be the wave function for a particle with mass m. Thus, by writing

$$m^2c^2\psi = -\hbar^2\, \partial^\mu\, \partial_\mu\psi = -\hbar^2\left(\frac{1}{c^2}\frac{\partial^2}{\partial t^2} - \nabla^2\right)\psi \tag{8.58}$$

we obtain a *wave equation* for a particle having mass m. In 1925 Erwin Schrödinger took precisely this step in an attempt to produce a genuine wave theory for massive particles. Shortly thereafter, several others, including O. Klein, W. Gordon and V. Fock, also considered Eq. (8.58), which now is known as the *Klein–Gordon equation.*

However, the solutions to this equation presented several difficulties. These difficulties could all be traced back to the existence of solutions with negative energies, which in turn arose from the presence of second-order time derivatives in Eq. (8.58). The negative energies correspond to choosing the negative square root in the solutions for E of

$$E^2 = m^2c^4 + \mathbf{p}^2c^2.$$

In classical physics we simply discard these solutions, as being meaningless physically. In the wave theory, however, it turns out to be impossible to discard negative-energy solutions and still have a physically meaningful theory. We shall not consider these circumstances here, but it is worth noting how the negative-energy solutions arise in the wave theory.

Let us solve Eq. (8.58) by separation of variables, just as we solved Eq. (5.33). The separation constants k^μ, $\mu = 0, 1, 2, 3$, here must satisfy

$$(mc/\hbar)^2 - k^\mu k_\mu = 0 \tag{8.59}$$

instead of Eq. (3.41). The elementary solutions have the forms

$$\psi_{\mathbf{k},\pm}(\mathbf{x}, t) = A_{\mathbf{k},\pm}e^{\mp ik^\mu x_\mu}, \tag{8.60}$$

where the values of k^μ are restricted only by the conditions that they are real (so that the elementary solutions will be periodic in space and time) and that they satisfy (8.59). All linear combinations

$$\psi(\mathbf{x}, t) = \int d^4k\, \delta(k^0 - \sqrt{(mc/\hbar)^2 + \mathbf{k}^2})\,(A_{\mathbf{k},+}e^{-ik^\mu x_\mu} + A_{\mathbf{k},-}e^{ik^\mu x_\mu}) \tag{8.61}$$

also are solutions: here the δ-function simply ensures conformity with Eq. (8.59).

We need to consider only the elementary solutions (8.60) here. Assume that we have a solution $\psi_{\mathbf{k},+}(\mathbf{x}, t)$ with $k^0 > 0$; then

$$\begin{aligned} \mathsf{E}\psi_{\mathbf{k},+}(\mathbf{x},t) &= i\hbar\frac{\partial}{\partial t}\psi_{\mathbf{k},+}(\mathbf{x}, t) = \hbar ck^0\psi_{\mathbf{k},+}(\mathbf{x}, t) \\ &= \hbar\omega\psi_{\mathbf{k},+}(\mathbf{x}, t) \end{aligned} \tag{8.62}$$

and hence the energy $E = \hbar ck^0 = \hbar\omega$ is positive for this solution. However, for the *same* values of the k^μ, $\psi_{\mathbf{k},-}(\mathbf{x}, t)$, corresponding to the second term in (8.61), is also a solution and appears to be just as reasonable a solution as $\psi_{\mathbf{k},+}(\mathbf{x}, t)$ except that it satisfies

$$\mathsf{E}\psi_{\mathbf{k},-}(\mathbf{x}, t) = -\hbar\omega\psi_{\mathbf{k},-}(\mathbf{x}, t); \tag{8.63}$$

that is, it has energy $E = -h\omega < 0$.

It is tempting to discard the $\psi_{\mathbf{k},-}$ solutions simply by demanding that $A_{\mathbf{k},-} = 0$. This, however, leads to insurmountable mathematical difficulties, or, more accurately, it leads to a crisis in interpreting the wave functions.

The reason the Klein–Gordon equation leads to trouble is that it contains a second-order derivative with respect to time. Because of this, two conditions on the wave functions must be given in terms of the time variable before a unique solution is determined. Classically, the evolution of a physical system is determined when its equations of motion, its spatial boundary conditions, and its state at a single time are given. Since the physics of a quantum system is somehow determined by its wave function (or vice versa), correspondence with the classical situation should be obtained when the wave equations, the spatial boundary conditions, and an initial condition on the wave function are specified. Evidently, however, the Klein–Gordon equation requires us to give a second condition involving the time variable, beyond an initial condition.

This source of the Klein–Gordon equation's unpleasantness was recognized by Schrödinger, and he retreated to the nonrelativistic relationship between energy and momentum for a free particle:

$$E = \mathbf{p}^2/2m.$$

When translated into a differential equation through the correspondences established in Eqs. (8.43) and (8.44), this becomes

$$i\hbar \frac{\partial}{\partial t} \psi(\mathbf{x}, t) = -\frac{\hbar^2}{2m} \nabla^2 \psi(\mathbf{x}, t), \tag{8.64}$$

or

$$\mathsf{E}\psi(\mathbf{x}, t) = \frac{1}{2m} \mathsf{P}^2 \psi(\mathbf{x}, t). \tag{8.65}$$

Here only the first-order time derivative appears, and only an initial condition on $\psi(\mathbf{x}, t)$ is required to give a unique solution in terms of the time variable. Equations (8.64) and (8.65) are two different ways of writing *Schrödinger's time-dependent wave equation* for a free particle.

No negative-energy solutions exist for this nonrelativistic equation and the crisis of the Klein–Gordon equation is avoided. This simplest version of Schrödinger's time-dependent wave equation can be extended to problems in which a single particle moves within a potential $V(\mathbf{x})$, simply through the conversion of the classical equation

$$E = \frac{\mathbf{p}^2}{2m} + V(\mathbf{x})$$

into the wave equation

$$i\hbar \frac{\partial}{\partial t} \psi(\mathbf{x}, t) = \left[-\frac{\hbar^2}{2m} \nabla^2 + V(\mathbf{x}) \right] \psi(\mathbf{x}, t). \tag{8.66}$$

The solution to the problem is obtained by solving the differential equation (8.66) together with appropriate boundary conditions and an initial condition. Frequently, in analogy with classical mechanics, the operator on the right-hand side of Eq. (8.65) is denoted

$$\mathsf{H}(\mathbf{P}, \mathbf{x}) \equiv \left[-\frac{\hbar^2}{2m} \nabla^2 + V(\mathbf{x}) \right] \tag{8.67}$$

and is called the Hamiltonian operator, or simply the Hamiltonian, for the system. Hence (8.66) can be written*

$$\mathsf{E}\psi(\mathbf{x}, t) = \mathsf{H}(\mathbf{P}, \mathbf{x})\psi(\mathbf{x}, t). \tag{8.68}$$

Before solving physics problems using these equations, we shall discuss the mathematical basis for the theory a bit further. The basic message that the Klein–Gordon and Schrödinger equations carry for us is that certain differential operator equations, or differential equations, replace the usual equations of classical mechanics. With this beginning, we shall develop a formalism for the formulation of quantum-mechanical problems, and later we shall apply the formalism to specific situations.

A final comment on the Klein–Gordon equation is in order. The negative-energy problems arising from the second-order time derivative actually have been overcome, and the Klein–Gordon equation plays an important role in modern particle physics. It is the basic equation for π-mesons, among other particles. The reinstatement of the Klein–Gordon equation involves some rather sophisticated developments that could occur only after the fundamental features of quantum theory had been assimilated.

READING NOTES†

Reference 22 provides an interesting introductory chapter on particles and waves.

Reference 23, Chapters 3–6.

Reference 14, Chapter 6.

Reference 16, Chapter 6.

Reference 43 discusses similar matters in Chapter 1.

Reference 44, Chapter 2.

For those who wish to pursue the mathematics further, Reference 18 can be recommended, especially Chapter 1.

Reference 17 is a compact study of the mathematical legitimacy of many of the standard manipulations in Fourier analysis.

---

* Since $\mathsf{E}\psi = \mathsf{H}\psi$ for all ψ that represent the system having Hamiltonian H, $\mathsf{E} = \mathsf{H}$ is correct for that system. There exist ψ' in the space of wave functions for which $\mathsf{E}\psi' \neq \mathsf{H}\psi'$, however, and those ψ' therefore involve a different Hamiltonian H'. Hence $\mathsf{E}\psi = \mathsf{H}\psi$ is a *dynamical equation*, and $\mathsf{E} = \mathsf{H}$ is not an identity, but an operator equality within the subspace of wave function space for which the dynamical equation holds.

† Works referred to in these notes will be found in the Bibliography on p. 577.

PROBLEMS

1. Show that unless $\mu_{\mathbf{k}}$ ($\boldsymbol{\kappa}$) $= 0$ in Eq. (8.8), the prominence in the wave amplitude is degraded as time increases.

2. Use the definition of the one-dimensional Dirac δ-function given in Chapter 1 *and* the Fourier representation of the δ-function to verify each of Eqs. (8.18a–8.18h).

3. Find the most general condition on a function $f(x)$ that ensures that its Fourier transform is real.

4. Find the Fourier transform of

$$f_1(t) = \frac{(2a)^{1/2}}{\pi^{1/4}} e^{-at^2}$$

and, assuming that this is the shape of a wave pulse, find $\Delta\omega\,\Delta t$ using at least two different criteria for defining $\Delta\omega$ and Δt.

5. Consider the function $f_1(t)$ in Problem 4 above. Demonstrate that in a certain limit (on a) it becomes the Dirac δ-function with t as its argument. Write down three other functions that in some limit become δ-functions, and demonstrate that they do so.

6. An atom with a very sharp resonance at frequency ω_0, so that it absorbs only photons with energy $\hbar\omega_0$, is exposed to a beam of electromagnetic radiation having frequency $\omega \neq \omega_0$. If the time interval during which this exposure takes place is $\tau < 1/|\omega - \omega_0|$, it is observed that the atom sometimes absorbs radiation (and re-emits it some time later). Carefully analyse this phenomenon, and explain what happens to the energy $\hbar|\omega - \omega_0|$.

7. A pencil 7 in long and weighing 0.2 oz is balanced on its point on a perfectly horizontal but nonslippery plane. Estimate the maximum time until it makes a "significant" angle with the vertical and begins to fall over.

8. Find the de Broglie wavelength of a hydrogen atom having the mean energy of an atom in hydrogen gas at room temperature.

9. What is the de Broglie wavelength of an electron with the energy imparted to it by a (a) 5 V, (b) 500 V, (c) 5×10^9 V potential difference?

10. Find the product of the de Broglie wave number with the radius of the nth Bohr orbit. Also calculate the ratio of rotational frequency to $|E_{n,Z}|/\hbar$ for the Bohr atom.

11. Compare the solutions of the Schrödinger equation

$$\nabla^2\psi = \frac{2m}{i\hbar}\frac{\partial}{\partial t}\psi$$

with those of the diffusion equation

$$\nabla^2\phi = K^2\frac{\partial}{\partial t}\phi.$$

Discuss these solutions in general terms. In particular, can a probability density be constructed for ϕ with properties similar to $\psi^*\psi$?

12. In calculus courses, you learned about limits in the mathematical sense, and the central notion involved was that one could allow the difference between two real numbers to become arbitrarily small, though remaining unequal to zero. On the other hand, you have

learned in physics that the world is made of particles, and that a characteristic diameter for some of the "most fundamental" of these is about 1 F $= 10^{-15}$ m. Does it make sense, then, to predicate the study of systems composed of these fundamental particles upon a thoroughgoing use of calculus and its underlying notion of limit? Discuss.

13. A representation for the Dirac δ-function is

$$\frac{d}{dx}\theta(x) = \delta(x),$$

where

$$\theta(x) = \begin{cases} 0 & \text{if} \quad x < 0 \\ \frac{1}{2} & \text{if} \quad x = 0 \\ 1 & \text{if} \quad x > 0. \end{cases}$$

Construct $\theta(x)$ as the limiting case of some other function and show that this does provide a legitimate representation of $\delta(x)$.

14. Show that

$$f(x)x^{\alpha}\frac{d^2}{dx^2}\theta(x) = f(x)x^{\alpha}\frac{d}{dx}\delta(x) = -\delta(x)\left[\alpha x^{\alpha-1} + x^{\alpha}\frac{d}{dx}\right]f(x),$$

where $f(x)$ is an arbitrary (well-behaved) function and $\theta(x)$ is defined as in Problem 13. (Hint: Put the expression on the left inside an integral over x.)

CHAPTER 9

MATHEMATICAL FORMULATION OF QUANTUM THEORY

Quantum theory still is in a dynamic state, and even the appropriate mathematical framework remains a topic of debate and research. Nevertheless, the early and fundamental work on the theory can be synthesized into a coherent mathematical system that serves as the foundation for continuing study.

The mathematics has been developed from several different points of view, the two earliest and most famous being the matrix approach adopted by Heisenberg and elaborated by Born and Jordan, and the wave-equation approach of Schrödinger. The matrix formulation of quantum theory relies heavily upon a knowledge of the Hamilton–Lagrange formalism of classical mechanics; for this reason it is not appropriate as an introduction even though historically it preceded the wave equation development. The Schrödinger wave picture is digested more easily in a first course, and we shall begin with it.

Both the wave-equation approach and the matrix approach can be synthesized through the notion of a Hilbert space, as the classic book by von Neumann, *Mathematical Foundations of Quantum Mechanics*, illustrates. This synthesis was initiated primarily by Dirac and Jordan. We shall not investigate the mathematical foundations of the synthesized theory in much detail, and in particular we shall avoid any rigorous discussion of Hilbert spaces. On the other hand, the use of abstract entities known as state vectors, which are elements of Hilbert Space, allows great flexibility in the treatment of quantum-physical problems. Therefore we shall introduce their fundamental properties, and shall utilize them directly in the solution of a number of problems later on.

The wave picture will be used to introduce the central ideas of the mathematics, and the state-vector approach will emerge as a generalization. Since both the matrix mechanics of Heisenberg and the wave mechanics of Schrödinger can be formulated in the language of state vectors, we shall then have a very flexible tool at our disposal.

9.1 LINEAR OPERATORS

The differential equations introduced in Section 8.10 are linear homogeneous partial differential equations of the form

$$L(\mathbf{x}, t)\psi(\mathbf{x}, t) = 0, \tag{9.1}$$

where L is a linear differential operator. For example

$$L(\mathbf{x}, t) = (mc)^2 + \hbar^2 \partial^\mu \partial_\mu \qquad \text{for Eq. (8.58)}$$

$$L(\mathbf{x}, t) = i\hbar \frac{\partial}{\partial t} + \frac{\hbar^2}{2m} \nabla^2 \qquad \text{for Eq. (8.64)}$$

$$L(\mathbf{x}, t) = i\hbar \frac{\partial}{\partial t} + \frac{\hbar^2}{2m} \nabla^2 - V(\mathbf{x}) \qquad \text{for Eq. (8.65)}$$

That $L = L(\mathbf{x}, t)$ is linear means simply that if α_1 and α_2 are arbitrary complex constants and $f_1(\mathbf{x}, t), f_2(\mathbf{x}, t)$ are arbitrary functions, L satisfies

$$L[\alpha_1 f_1(\mathbf{x}, t) + \alpha_2 f_2(\mathbf{x}, t)] = \alpha_1 L f_1(\mathbf{x}, t) + \alpha_2 L f_2(\mathbf{x}, t).$$

That the differential equation is homogeneous means simply that Eq. (9.1) holds.

We know from theorems in differential equations, that a linear combination (superposition) of solutions to a homogeneous linear differential equation is also a solution to the same equation. An analogous result from linear algebra states that a superposition of solutions to a homogeneous linear operator equation again is a solution to the same equation. In fact, in classical wave theories it is the superposition of elementary waves that leads to interference effects, as we saw in Chapter 8. The linear superposition of solutions to general homogeneous linear operator equations also leads to interference, or essentially wavelike effects. These features are not limited to homogeneous equations; in fact inhomogeneous linear equations have solutions that exhibit interference also.

Thus a fundamental assumption of quantum theory is that the *operator equations* used to define the character of a wave function *must be linear* operators. This really is a simplifying assumption since nonlinear equations of certain types may also yield interference effects. On the other hand, linear equations dominate wave theory in general, and indeed they are far simpler to handle than nonlinear ones. In order to limit the problems which we consider to ones that we are reasonably likely to be able to solve, we have traditionally clung to the assumption of linearity. Nonlinear theories have been considered, and some notable nonlinear attempts at generalized versions of quantum theory have developed in recent years. However the difficulty in obtaining theoretical results sufficiently well defined to be subjected to experimental test is formidable; no nonlinear theory as yet has both agreed with all experimental tests and been sufficiently tractable to become computationally useful.

To introduce notation and to review the assertions made above about linear equations, we shall prove these assertions. Let Q be a linear operator; it may be a differential operator, and it may involve matrices. We shall think of it primarily as a differential operator, but this is not its central feature; it is a *linear* operator:

$$\mathsf{Q}(\alpha_1 \Psi_1 + \alpha_2 \Psi_2) = \alpha_1 \mathsf{Q}_1 \Psi_1 + \alpha_2 \mathsf{Q}_2 \Psi_2 \tag{9.2}$$

whenever Ψ_1 and Ψ_2 are acceptable wave functions and α_1, α_2 are any complex numbers. If R also is a linear operator and $\{\Psi_n\}$ is a set of solutions to

$$(\mathsf{Q} - \mathsf{R})\Psi_n = 0, \; n = 1, 2, \ldots, \tag{9.3}$$

where by $(\mathsf{Q} - \mathsf{R})\Psi_n$ we mean $\mathsf{Q}\Psi_n - \mathsf{R}\Psi_n$, then for any complex constants, α_n,

$$\Psi = \sum_n \alpha_n \Psi_n$$

also satisfies

$$(\mathsf{Q} - \mathsf{R})\Psi = 0.$$

This follows directly from (9.2) and (9.3):

$$(\mathsf{Q} - \mathsf{R})\sum_n \alpha_n \Psi_n = \sum_n \alpha_n (\mathsf{Q} - \mathsf{R})\Psi_n = 0.$$

Now suppose we wish to solve the inhomogeneous equation

$$(\mathsf{Q} - \mathsf{R})\Psi = \Phi. \tag{9.4}$$

Let Ψ_0 be a solution to this equation; then

$$\Psi = \sum_n \alpha_n \Psi_n + \Psi_0$$

also is a solution, as is easily verified. Thus a solution to an inhomogeneous equation may be added to a linear combination of solutions to the corresponding homogeneous equation, and this again will be a solution.

Of course it was unnecessary to write the operator equations in terms of differences between two linear operators as in Eqs. (9.3) and (9.4). However, quantum-mechanical equations usually do have this form, as Eq. (8.64) illustrates.

It turns out that solutions to Eq. (9.3), together with appropriate boundary conditions, form a *linear vector space*. The student familiar with the axioms for vector spaces may check this easily. This vector space often is referred to as the solution space to Eq. (9.3), although it is the solution space for many other equations as well.

9.2 HERMITIAN OPERATORS

The wave function must be interpretable as a probability density amplitude in general. (Exceptions are made under circumstances such as those discussed in Section 8.9, but in such cases the "wave functions" must be thought of as limiting cases of physically acceptable functions. Plane waves are not physically acceptable because they extend uniformly throughout space, but they may be thought of as the limit of spatially restricted waves.) Therefore any wave function Ψ must satisfy

$$\int_{\text{all space}} \Psi^*(\mathbf{x}, t)\Psi(\mathbf{x}, t)\, d^3x = 1, \tag{9.5}$$

that is, it must be properly *normalized*. Any solution $\chi(\mathbf{x}, t)$ to the equation

$$(\mathsf{Q} - \mathsf{R})\chi(\mathbf{x}, t) = \Phi$$

may be normalized by letting

$$\Psi(\mathbf{x}, t) = A\chi(\mathbf{x}, t)$$

with A an arbitrary complex number, and demanding that Eq. (9.5) holds for Ψ. This yields

$$|A|^2 = \left[\int \chi^*(\mathbf{x}, t)\chi(\mathbf{x}, t)\, d^3x\right]^{-1}$$

or

$$A = e^{i\alpha}\left[\int |\chi(\mathbf{x}, t)|^2\, d^3x\right]^{-1/2}$$

with α an undetermined real number. Thus

$$\Psi(\mathbf{x}, t) = \frac{e^{i\alpha}}{\sqrt{\int |\chi(\mathbf{x}', t)|^2\, d^3x'}}\,\chi(\mathbf{x}, t) \tag{9.6}$$

is a properly normalized wave function.

As we noted in Section 8.9, α cannot be determined by any physical considerations; the phase of the wave function is unobservable. Therefore it is common simply to let $\alpha = 0$.

Henceforward, we shall assume that our wave functions are properly normalized.

Now the predictions of quantum theory are compared with experiment through the probability interpretation of the wave function. This means that we must have a way of finding the mean value, say, of an operator corresponding to a physical variable, e.g. of the momentum $\mathsf{P} = (\hbar/i)\boldsymbol{\nabla}$. For this purpose, we define the *expectation value* (or mean value) of an operator Q for a system with wave function $\Psi(\mathbf{x}, t)$ is

$$\langle \mathsf{Q} \rangle \equiv \int \Psi^*(\mathbf{x}, t)\mathsf{Q}\Psi(\mathbf{x}, t)\, d^3x. \tag{9.7}$$

For example, the mean momentum of a particle with wave function $\Psi(\mathbf{x}, t)$ is

$$\langle \mathsf{P} \rangle = \int \Psi^*(\mathbf{x}, t)\frac{\hbar}{i}\boldsymbol{\nabla}\Psi(\mathbf{x}, t)\, d^3x.$$

If $\langle \mathsf{Q} \rangle$ is to be a number subject to comparison with experiment, that is, if Q is to represent a physical variable, then $\langle \mathsf{Q} \rangle$ must be a real number; measuring apparatus give measurements in terms of real numbers. Hence we require that

$$\langle \mathsf{Q} \rangle^* = \langle \mathsf{Q} \rangle$$

or

$$\int \Psi(\mathbf{x}, t)[\mathsf{Q}\Psi(\mathbf{x}, t)]^* \, d^3x = \int \Psi^*(\mathbf{x}, t)\mathsf{Q}\Psi(\mathbf{x}, t) \, d^3x. \tag{9.8}$$

In order to translate this equation into a condition on Q, we introduce a new operator $\mathsf{Q}^\dagger$ defined by

$$\left[\int \Phi^*(\mathbf{x}, t)\mathsf{Q}\Psi(\mathbf{x}, t) \, d^3x \right]^* = \int \Psi^*(\mathbf{x}, t)\mathsf{Q}^\dagger\Phi(\mathbf{x}, t) \, d^3x$$

$$= \int [\mathsf{Q}\Psi(\mathbf{x}, t)]^*\Phi(\mathbf{x}, t) \, d^3x \tag{9.9}$$

where Φ and Ψ are arbitrary wave functions. Thus the *Hermitian conjugate* or *Hermitian adjoint*, $\mathsf{Q}^\dagger$, of Q is that operator which, when operating on $\Phi(\mathbf{x}, t)$ within the integral $\int \Psi^* \mathsf{Q}^\dagger \Phi \, d^3x$ gives the same result as if $\mathsf{Q}^\dagger$ were removed from the integral and Ψ^* were replaced by $[\mathsf{Q}\Psi]^*$.

Applying this definition to Eq. (9.8), we obtain

$$\int \Psi^*(\mathbf{x}, t)\mathsf{Q}\Psi(\mathbf{x}, t) \, d^3x = \int \Psi^*(\mathbf{x}, t)\mathsf{Q}^\dagger\Psi(\mathbf{x}, t) \, d^3x$$

or

$$\int \Psi^*(\mathbf{x}, t)[\mathsf{Q} - \mathsf{Q}^\dagger]\Psi(\mathbf{x}, t) \, d^3x = 0.$$

Since this must hold for all $\Psi(\mathbf{x}, t)$, we have

$$\mathsf{Q}^\dagger = \mathsf{Q}; \tag{9.10}$$

any *physical operator* must be equal to its Hermitian conjugate. An operator for which (9.10) holds is called a *Hermitian operator*, or sometimes a *self-conjugate* or *self-adjoint* operator.

Therefore, any physical statement, in the form of an operator equation to be satisfied by a wave function Ψ, must be given in terms of *linear Hermitian operators*.

9.3 EIGENVALUES AND EIGENFUNCTIONS

When the effect of an operator, operating on a function, is simply to multiply the function by a complex constant, that is, when

$$\mathsf{Q}\Psi(\mathbf{x}, t) = q\Psi(\mathbf{x}, t) \tag{9.11}$$

and q is a complex constant, we say that $\Psi(\mathbf{x}, t)$ is an *eigenfunction* of Q with *eigenvalue* q. Normally, we shall attach a subscript q to the function, $\Psi_q(\mathbf{x}, t)$, to indicate that it has eigenvalue q. Already we have seen several *eigenvalue equations*;

for example,

$$\mathsf{P}^{\mu}\Psi_{p^{\nu}}(\mathbf{x}, t) = i\hbar\partial^{\mu}\Psi_{p^{\nu}}(\mathbf{x}, t) = p^{\mu}\Psi_{p^{\nu}}(\mathbf{x}, t)$$

is such an equation, p_{μ} being the eigenvalues of the operator P^{μ}, and

$$\Psi_{p^{\nu}}(\mathbf{x}, t) = (2\pi)^{-3/2}e^{-(i/\hbar)p^{\mu}x_{\mu}}$$

being the corresponding eigenfunctions. Also, Eq. (8.58) is an eigenvalue equation, the solutions to which must be eigenfunctions for the operator $\mathsf{P}^{\mu}\mathsf{P}_{\mu} = -\hbar^2\partial^{\mu}\partial_{\mu}$ with eigenvalue m^2c^2.

The expectation value of an operator Q for a system whose wave function is an eigenfunction of Q, with eigenvalue q, then is

$$\begin{aligned}\langle \mathsf{Q} \rangle &= \int \Psi_q^*(\mathbf{x}, t)\mathsf{Q}\Psi_q(\mathbf{x}, t)\, d^3x = \int \Psi_q^*(\mathbf{x}, t)q\Psi_q(\mathbf{x}, t)\, d^3x \\ &= q\int |\Psi_q(\mathbf{x}, t)|^2\, d^3x = q \cdot 1 = q.\end{aligned}$$

The mean value of an operator for a system represented by an eigenfunction of the operator is the eigenvalue of that eigenfunction. In particular, since the expectation values of Hermitian operators necessarily are real, *the eigenvalues of Hermitian operators are real.*

Let us consider two *different* eigenvalues for a Hermitian operator, together with the corresponding normalized eigenfunctions; let

$$\begin{aligned}\mathsf{Q}\Psi_q(\mathbf{x}, t) &= q\Psi_q(\mathbf{x}, t), \\ \mathsf{Q}\Psi_{q'}(\mathbf{x}, t) &= q'\Psi_{q'}(\mathbf{x}, t), \qquad q' \neq q.\end{aligned}$$

Then

$$\int \Psi_{q'}^*(\mathbf{x}, t)\mathsf{Q}\Psi_q(\mathbf{x}, t)\, d^3x = q\int \Psi_{q'}^*(\mathbf{x}, t)\Psi_q(\mathbf{x}, t)\, d^3x$$

but also, since $\mathsf{Q} = \mathsf{Q}^{\dagger}$,

$$\begin{aligned}\int \Psi_{q'}^*(\mathbf{x}, t)\mathsf{Q}\Psi_q(\mathbf{x}, t)\, d^3x &= \int \Psi_{q'}^*(\mathbf{x}, t)\mathsf{Q}^{\dagger}\Psi_q(\mathbf{x}, t)\, d^3x \\ &= \int [\mathsf{Q}\Psi_{q'}(\mathbf{x}, t)]^*\Psi_q(\mathbf{x}, t)\, d^3x \\ &= \int q'^*\Psi_{q'}^*(\mathbf{x}, t)\Psi_q(\mathbf{x}, t)\, d^3x \\ &= q'\int \Psi_{q'}^*(\mathbf{x}, t)\Psi_q(\mathbf{x}, t)\, d^3x.\end{aligned}$$

Hence

$$0 = \int \Psi_{q'}^*(\mathbf{x}, t)(\mathrm{Q} - \mathrm{Q}^\dagger)\Psi_q(\mathbf{x}, t)\, d^3x$$

$$= (q - q') \int \Psi_{q'}^*(\mathbf{x}, t)\Psi_q(\mathbf{x}, t)\, d^3x.$$

Since $q \neq q'$ by hypothesis, we must have

$$\int \Psi_{q'}^*(\mathbf{x}, t)\Psi_q(\mathbf{x}, t)\, d^3x = 0. \tag{9.12}$$

Eigenfunctions corresponding to distinct eigenvalues for a Hermitian operator are therefore *orthogonal* in the sense that the integral in (9.12) vanishes.

Quite generally, for two arbitrary functions $\Psi(\mathbf{x}, t)$ and $\Phi(\mathbf{x}, t)$, the condition $(\Psi, \Phi) = 0$, where

$$(\Psi, \Phi) \equiv \int \Psi^*(\mathbf{x}, t)\Phi(\mathbf{x}, t)\, d^3x, \tag{9.13}$$

is expressed in words by saying that Ψ and Φ are orthogonal functions. From this definition of (Ψ, Φ), it is clear that

$$(\Psi, \Phi)^* = (\Phi, \Psi); \tag{9.14}$$

hence $(\Psi, \Phi) = 0$ implies that $(\Phi, \Psi) = 0$ also.

The quantity (Ψ, Φ) defined in (9.13) is called the *inner product* of the functions Ψ and Φ, and it is quite analogous to the inner product, or dot product, of two vectors

$$\mathbf{x} \cdot \mathbf{y} = \sum_m x_m y_m \tag{9.15}$$

except that instead of summing over the discrete indices m as in (9.15), we integrate over the "indices" $\mathbf{x} = (x, y, z)$.

An inner product with the property (9.14) is called a Hermitian inner product, and a vector space in which such an inner product is defined is called a *unitary* space. Hence our vector space of physically acceptable wave functions is a unitary space.

A set of functions $\{\Psi_n\}$ is said to be *linearly* independent if and only if a linear combination of them can vanish identically,

$$\alpha_1 \Psi_1(\mathbf{x}, t) + \alpha_2 \Psi_2(\mathbf{x}, t) + \cdots = \sum_n \alpha_n \Psi_n(\mathbf{x}, t) \equiv 0$$

only when all the coefficients α_n vanish. In other words, no element of a linearly independent set of wave functions can be expressed as a linear combination of the remaining elements of the set.

Now it may be that Q has two or more linearly independent eigenfunctions with the same eigenvalue:

$$\mathsf{Q}\Psi_q(\mathbf{x}, t) = q\Psi_q(\mathbf{x}, t),$$

$$\mathsf{Q}\Psi'_q(\mathbf{x}, t) = q\Psi'_q(\mathbf{x}, t);$$

$$\alpha\Psi_q(\mathbf{x}, t) + \alpha'\Psi'_q(\mathbf{x}, t) \equiv 0 \qquad \text{implies} \qquad \alpha = \alpha' = 0.$$

In such a case, the Hermiticity of $\mathsf{Q}^\dagger$ does *not* guarantee that $\Psi_q(\mathbf{x}, t)$ and $\Psi'_q(\mathbf{x}, t)$ will be orthogonal, because

$$\int \Psi_q^*(\mathbf{x}, t)[\mathsf{Q} - \mathsf{Q}^\dagger]\Psi'_q(\mathbf{x}, t)\, d^3x$$

$$= (q - q)\int \Psi_q^*(\mathbf{x}, t)\Psi'_q(\mathbf{x}, t)\, d^3x = 0$$

by virtue of the vanishing of the factor $(q - q)$; hence the value of (Ψ_q, Ψ'_q) is not determined.

We now show that *any* set of linearly independent functions, $\{\Psi^{(n)}\}$ can be transformed into a set of *orthonormal* functions $\{\Psi^{[n]}\}$, i.e. the set $\{\Psi^{[n]}\}$ satisfies

$$(\Psi^{[n]}, \Psi^{[m]}) = \delta_{n,m} \tag{9.16}$$

where

$$\delta_{n,m} = \begin{cases} 1 & \text{if} \quad n = m \\ 0 & \text{if} \quad n \neq m. \end{cases} \tag{9.17}$$

From this it follows that linearly independent eigenfunctions of a Hermitian operator either are automatically orthogonal (if they have different eigenvalues) or can be made so.

From the linearly independent set $\{\Psi^{(n)}\}$, choose one, say $\Psi^{(1)}$ and if it is not normalized already so that $(\Psi^{(1)}, \Psi^{(1)}) = 1$, form

$$\Psi^{[1]} \equiv (\Psi^{(1)}, \Psi^{(1)})^{-1/2}\Psi^{(1)}; \tag{9.18}$$

otherwise let $\Psi^{[1]} \equiv \Psi^{(1)}$ (which will yield the same result as (9.18) since $(\Psi^{(1)}, \Psi^{(1)}) = 1$ in this case). Next, construct

$$\Phi^{(2)} \equiv \Psi^{(2)} - (\Psi^{[1]}, \Psi^{(2)})\Psi^{[1]};$$

it is easy to see that $\Phi^{(2)}$ is orthogonal to $\Psi^{[1]}$, since

$$(\Psi^{[1]}, \Phi^{(2)}) = (\Psi^{[1]}, \Psi^{(2)}) - (\Psi^{[1]}, \Psi^{(2)})(\Psi^{[1]}, \Psi^{[1]})$$

$$= (\Psi^{[1]}, \Psi^{(2)}) - (\Psi^{[1]}, \Psi^{(2)}) = 0.$$

Now normalize $\Phi^{(2)}$ by forming

$$\Psi^{[2]} \equiv (\Phi^{(2)}, \Phi^{(2)})^{-1/2}\Phi^{(2)}. \tag{9.19}$$

Next, form

$$\Phi^{(3)} \equiv \Psi^{(3)} - (\Psi^{[1]}, \Psi^{(3)})\Psi^{[1]} - (\Psi^{[2]}, \Psi^{(3)})\Psi^{[2]},$$

which is orthogonal to both $\Psi^{[1]}$ and $\Psi^{[2]}$, and normalize this by writing

$$\Psi^{[3]} = (\Phi^{(3)}, \Phi^{(3)})^{-1/2}\Phi^{(3)}. \tag{9.20}$$

Proceeding this way for $m - 1$ functions, we then construct

$$\Phi^{(m)} \equiv \Psi^{(m)} - \sum_{k=1}^{m-1} (\Psi^{[k]}, \Psi^{(m)})\Psi^{[k]}$$

and the normalized function

$$\Psi^{[m]} \equiv (\Phi^{(m)}, \Phi^{(m)})^{-1/2}\Phi^{(m)}, \tag{9.21}$$

etc. Thus we obtain the orthonormal set $\{\Psi^{[n]}\}$ as promised. This technique is known as the *Schmidt orthogonalization procedure.*

The fact that the set of all linearly independent eigenfunctions of a Hermitian operator can be arranged into an orthonormal set is very useful in quantum mechanics. It means, for example, that if $\{\Psi_q{}^{(n)}\}$ is an orthonormal set of wave functions and if we have some $\Psi(\mathbf{x}, t)$ such that

$$\Psi(\mathbf{x}, t) = \sum_q \sum_n \alpha_q{}^{(n)} \Psi_q{}^{(n)}(\mathbf{x}, t), \tag{9.22}$$

where the sums range over a finite number of terms, then

$$\alpha_q{}^{(n)} = (\Psi_q{}^{(n)}, \Psi). \tag{9.23}$$

To see this, form

$$\begin{aligned}(\Psi_{q'}{}^{(n')}, \Psi) &= (\Psi_{q'}{}^{(n')}, \sum_q \sum_n \alpha_q{}^{(n)} \Psi_q{}^{(n)}) \\ &= \sum_q \sum_n \alpha_q{}^{(n)} (\Psi_{q'}{}^{(n')}, \Psi_q{}^{(n)}) \\ &= \sum_q \sum_n \alpha_q{}^{(n)} \delta_{q,q'} \delta_{n,n'} = \alpha_q{}^{(n')}.\end{aligned}$$

The restriction that the sums be finite is required to ensure that the sums can be taken outside the inner product. A mathematically rigorous treatment can be developed to remove this restriction; in general, we shall accept the existence of such a treatment and shall extend Eqs. (9.21) and (9.22) to infinite sums.

As a matter of fact, it is convenient to be able to extend the sums in Eq. (9.21) to integrals. This would mean, for example, letting the set $\{q\}$ of eigenvalues of Q become a continuous (locally connected), and therefore nondenumerable, set. But it turns out that eigenfunctions with continuous eigenvalues, like the eigenfunctions $(2\pi)^{-3/2}e^{-ik^\mu x_\mu}$ when the k^μ are allowed to have all real values, are not physically acceptable; they cannot be normalized properly. Nevertheless, they may be considered limiting cases of acceptable functions, just as an integral is a limiting case of a sum. We can include them in our calculational scheme by recognizing that the Dirac δ-function is the analog of the Kronecker δ defined by Eq. (9.17), just as an integral is analogous to a sum. We shall consider a set $\{\Psi_q^{(n)}\}$ to be orthonormal when q is a continuous eigenvalue and n is a discrete variable if

$$(\Psi_q^{(n)}, \Psi_{q'}^{(n')}) = \delta(q - q')\delta_{n,n'};$$

if n is continuous also, then $\{\Psi_q^{(n)}\}$ is an orthonormal set if

$$(\Psi_q^{(n)}, \Psi_{q'}^{(n')}) = \delta(q - q')\delta(n - n').$$

It is inconvenient to have to distinguish always between continuous and discrete eigenvalue spectra. Therefore we use

$$\underset{q}{\mathsf{S}} \equiv \begin{cases} \displaystyle\int dq & \text{if } \{q\} \text{ is continuous} \\ \displaystyle\sum_q & \text{if } \{q\} \text{ is discrete} \end{cases}$$

to represent a generalized sum, and let

$$\delta(q - q') = \left\{ \begin{array}{lll} \delta(q - q') & \text{(Dirac)} & \text{if } \{q\} \text{ is continuous} \\ \delta_{q,q'} & \text{(Kronecker)} & \text{if } \{q\} \text{ is discrete} \end{array} \right\}.$$

We then extend the discussion of Eqs. (9.21) and (9.22) to such generalized sums.

It would be very helpful if we were assured that *any* physically acceptable wave function $\Psi(\mathbf{x}, t)$ could be expanded in a linear combination of orthonormal functions, each of which is an eigenfunction of a physical operator (i.e. a Hermitian operator corresponding to a physical variable). An important assumption of quantum theory is that this indeed is so; that is, we assume that the eigenfunctions of each physical operator *span the space* of acceptable wave functions. A set of functions that span the vector space of physically acceptable functions is called a *complete* set; hence we are assuming that the eigenfunctions of each operator form a complete set.

The hypothesis of completeness for a set of eigenfunctions may be formulated mathematically. We are assuming that *any* acceptable Ψ can be expanded according to Eq. (9.21), or more generally according to

$$\Psi(\mathbf{x}, t) = \underset{q}{\mathsf{S}}\,\underset{n}{\mathsf{S}}\,\alpha_q^{(n)}\Psi_q^{(n)}(\mathbf{x}, t). \tag{9.24}$$

If we allow the generalized sums to be taken through the integral in the inner product, as we allowed the sums to do when we obtained (9.22), we get

$$\alpha_q^{(n)} = (\Psi_q^{(n)}, \Psi)$$

for the generalized case also. Hence (9.24) becomes

$$\Psi(\mathbf{x}, t) = \mathsf{S}_q \mathsf{S}_n (\Psi_q^{(n)}, \Psi)\Psi_q^{(n)}(\mathbf{x}, t)$$

$$= \mathsf{S}_q \mathsf{S}_n \Psi_q^{(n)}(\mathbf{x}, t) \int d^3x' \, \Psi_q^{(n)*}(\mathbf{x}', t)\Psi(\mathbf{x}', t)$$

$$= \int d^3x' \, [\mathsf{S}_q \mathsf{S}_n \Psi_q^{(n)*}(\mathbf{x}, t)\Psi_q^{(n)}(\mathbf{x}', t)]\Psi(\mathbf{x}', t).$$

However, since

$$\Psi(\mathbf{x}, t) = \int d^3x' \, \delta^3(\mathbf{x}' - \mathbf{x})\Psi(\mathbf{x}', t),$$

this can be written in the form

$$0 = \int d^3x' \, [\delta^3(\mathbf{x}' - \mathbf{x}) - \mathsf{S}_q \mathsf{S}_n \Psi_q^{(n)}(\mathbf{x}, t)\Psi_q^{(n)*}(\mathbf{x}', t)]\Psi(\mathbf{x}', t).$$

Since $\Psi(\mathbf{x}', t)$ is an arbitrary function, this is equivalent to

$$\mathsf{S}_q \mathsf{S}_n \Psi_q^{(n)}(\mathbf{x}, t)\Psi_q^{(n)*}(\mathbf{x}', t) = \delta^3(\mathbf{x}' - \mathbf{x}). \tag{9.25}$$

It is obvious that whenever Eq. (9.25) holds, any $\Psi(\mathbf{x}, t)$ can be expanded in the $\Psi_q^{(n)}$, and we have just shown that if (9.24) holds for any Ψ then (9.25) holds also; thus (9.24) and (9.25) are equivalent, and since (9.25) is very useful for calculation, it is called a *completeness condition* for the set $\{\Psi_q^{(n)}\}$. Note its similarity to the orthonormality condition.

Returning to the definition of expectation values, we see that the expectation value of Q for a physical system with properly normalized wave functions Ψ, $(\Psi, \Psi) = 1$, is

$$\langle \mathsf{Q} \rangle = (\Psi, \mathsf{Q}\,\Psi) = (\mathsf{S}_{q,n} \alpha_q^{(n)}\Psi_q^{(n)}, \mathsf{Q}\; \mathsf{S}_{q'n'} \alpha_{q'}^{(n')}\Psi_{q'}^{(n')})$$

$$= \mathsf{S}_{q,n} \mathsf{S}_{q',n'} \alpha_q^{(n)*}\alpha_{q'}^{(n')}(\Psi_q^{(n)}, \mathsf{Q}\,\Psi_{q'}^{(n')})$$

$$= \mathsf{S}_{q,n} \mathsf{S}_{q',n'} \alpha_q^{(n)*}\alpha_{q'}^{(n')}(\Psi_q^{(n)}, q'\Psi_{q'}^{(n')})$$

$$= \mathsf{S}_{q,n} \mathsf{S}_{q',n'} \alpha_q^{(n)*}\alpha_{q'}^{(n')}q'\,\delta(q - q')\delta(n - n') = \mathsf{S}_{q,n} |\alpha_q^{(n)}|^2 q$$

$$= \mathsf{S}_q q(\mathsf{S}_n |\alpha_q^{(n)}|^2) = \mathsf{S}_q q w_q, \tag{9.26}$$

where

$$w_q \equiv \mathsf{S}_n |\alpha_q^{(n)}|^2. \tag{9.27}$$

Evidently, since the expectation value of Q is equal to the sum over q of q times a weighting factor w_q, and since

$$\mathsf{S}_q w_q = 1, \tag{9.28}$$

(which is obtained by calculating $(\Psi, \Psi) = \langle 1 \rangle$ as above), we may interpret w_q as being the probability that the state will exhibit q as the value of the physical variable represented by Q whenever that variable is measured. Note that

$$w_q = \mathsf{S}_n |(\Psi_q^{(n)}, \Psi)|^2. \tag{9.29}$$

The importance of the completeness hypothesis can be seen from a consideration of this probability interpretation. Assume that the set $\{\Psi_q^{(n)}\}$ is *not* complete; then the wave function for a system may require a term $u(\mathbf{x}, t)$ in addition to a linear combination of the $\Psi_q^{(n)}$:

$$\Psi(\mathbf{x}, t) = \mathsf{S}_{q,n} \alpha_q^{(n)} \Psi_q^{(n)}(\mathbf{x}, t) + u(\mathbf{x}, t);$$

then

$$(\Psi, \Psi) = \mathsf{S}_{q,n} |\alpha_q^{(n)}|^2 + 2 \mathsf{S}_{q,n} \operatorname{Re} \alpha_q^{(n)}(u, \Psi_q^{(n)}) + (u, u).$$

Since u is assumed to be linearly independent of the $\Psi_q^{(n)}$, it can be made orthogonal to them through the Schmidt procedure, and then the second term vanishes. Thus if Ψ is properly normalized,

$$1 = \mathsf{S}_q w_q + (u, u) = \mathsf{S}_q w_q + \int d^3x \, |u(\mathbf{x}, t)|^2.$$

Hence

$$\mathsf{S}_q w_q < 1$$

unless $(u, u) = 0$, i.e. unless $u(\mathbf{x}, t) \equiv 0$. The probability interpretation therefore fails unless $\{\Psi_q^{(n)}\}$ constitutes a complete set.

9.4 THE PROBABILITY CURRENT AND BOUNDARY CONDITIONS

If the operator $\mathsf{V} = V(\mathbf{x})$ is simply a function of position, then

$$\mathsf{V} = \mathsf{V}^\dagger$$

implies $\langle \mathsf{V} \rangle = \langle \mathsf{V} \rangle^*$, or

$$\int d^3x \, \Psi^*(\mathbf{x}, t)[V(\mathbf{x}) - V^*(\mathbf{x})]\Psi(\mathbf{x}, t) = 0$$

for all $\Psi(\mathbf{x}, t)$. Hence, for a function of $\mathbf{x}$ to be a Hermitian operator, the function must be real valued:

$$V(\mathbf{x}) = V^*(\mathbf{x}).$$

Consider the Schrödinger equation, Eq. (8.66). Here we shall require that the operators

$$i\hbar\frac{\partial}{\partial t}, \quad -\frac{\hbar^2}{2m}\nabla^2 + V(\mathbf{x})$$

be Hermitian in order that the equation be meaningful. It is easy to see that these are Hermitian, provided $V(\mathbf{x}) = V^*(\mathbf{x})$, i.e. that the potential function is real valued, if we consider some elementary properties of Hermitian conjugates.

It is left as an exercise for the student to show that the inner product (Ψ, Φ) is antilinear–linear; that is, that it is antilinear in its first argument:

$$(\alpha_1\Psi_1 + \alpha_2\Psi_2, \Phi) = \alpha_1^*(\Psi_1, \Phi) + \alpha_2^*(\Psi_2, \Phi), \tag{9.30}$$

and linear in its second argument:

$$(\Psi, \alpha_1\Phi_1 + \alpha_2\Phi_2) = \alpha_1(\Psi, \Phi_1) + \alpha_2(\Psi, \Phi_2), \tag{9.31}$$

where α_1, α_2 are complex constants. This together with the meaning

$$(\mathsf{Q} + \mathsf{R})\Psi = \mathsf{Q}\Psi + \mathsf{R}\Psi \tag{9.32}$$

of a sum of two operators, is all that we need.

If $\mathsf{Q} = \alpha\mathsf{R} + \beta\mathsf{S}$, then we have, from the definition of the Hermitian conjugate

$$(\Psi, \mathsf{Q}^\dagger\Phi) = (\mathsf{Q}\Psi, \Phi),$$

that

$$\begin{aligned}(\Psi, \mathsf{Q}^\dagger\Phi) &= (\Psi, [\alpha\mathsf{R} + \beta\mathsf{S}]^\dagger\Phi) = ([\alpha\mathsf{R} + \beta\mathsf{S}]\Psi, \Phi)\\ &= (\alpha\mathsf{R}\Psi + \beta\mathsf{S}\Psi, \Phi) = \alpha^*(\mathsf{R}\Psi, \Phi) + \beta^*(\mathsf{S}\Psi, \Phi)\\ &= \alpha^*(\Psi, \mathsf{R}^\dagger\Phi) + \beta^*(\Psi, \mathsf{S}^\dagger\Phi) = (\Psi, [\alpha^*\mathsf{R}^\dagger + \beta^*\mathsf{S}^\dagger]\Phi).\end{aligned}$$

This can be written

$$(\Psi, \{[\alpha\mathsf{R} + \beta\mathsf{S}]^\dagger - [\alpha^*\mathsf{R}^\dagger + \beta^*\mathsf{S}^\dagger]\}\Phi) = 0,$$

or, since Ψ and Φ are arbitrary,

$$[\alpha\mathsf{R} + \beta\mathsf{S}]^\dagger = \alpha^*\mathsf{R}^\dagger + \beta^*\mathsf{S}^\dagger; \tag{9.33}$$

here, of course, α and β are complex constants. For $\beta = 0$ we obtain

$$(\alpha\mathsf{R})^\dagger = \alpha^*\mathsf{R}^\dagger \tag{9.34}$$

and for $\alpha = \beta = 1$ we have

$$[\mathsf{R} + \mathsf{S}]^\dagger = \mathsf{R}^\dagger + \mathsf{S}^\dagger. \tag{9.35}$$

To see that $i\hbar\,\partial/\partial t$ is Hermitian, we consider

$$\left(\Psi, \left\{i\hbar\frac{\partial}{\partial t} - \left[i\hbar\frac{\partial}{\partial t}\right]^\dagger\right\}\Phi\right) = \left(\Psi, \left\{i\hbar\frac{\partial}{\partial t} + i\hbar\left[\frac{\partial}{\partial t}\right]^\dagger\right\}\Phi\right)$$

$$= i\hbar\left(\Psi, \left\{\frac{\partial}{\partial t} + \left[\frac{\partial}{\partial t}\right]^\dagger\right\}\Phi\right) = i\hbar\left[\left(\Psi, \frac{\partial}{\partial t}\Phi\right) + \left(\frac{\partial}{\partial t}\Psi, \Phi\right)\right]$$

$$= i\hbar\frac{\partial}{\partial t}(\Psi, \Phi). \tag{9.36}$$

In the particular case $\Phi = \Psi$, we have $(\Psi, \Psi) = 1$ for properly normalized wave functions, and hence $\partial(\Psi, \Psi)/\partial t = \partial(1)/\partial t = 0$. Hence (9.36) leads to

$$\left(\Psi, \left\{i\hbar\frac{\partial}{\partial t} - \left[i\hbar\frac{\partial}{\partial t}\right]^\dagger\right\}\Psi\right) = 0$$

or since Ψ is arbitrary,

$$\left[i\hbar\frac{\partial}{\partial t}\right]^\dagger = i\hbar\frac{\partial}{\partial t}. \tag{9.37}$$

Now let us consider the Hamiltonian operator. Here we must evaluate

$$\left(\left[-\frac{\hbar^2}{2m}\nabla^2 + V(\mathbf{x})\right]\Psi, \Phi\right) - \left(\Psi, \left[-\frac{\hbar^2}{2m}\nabla^2 + V(\mathbf{x})\right]\Phi\right)$$

$$= -\frac{\hbar^2}{2m}[(\nabla^2\Psi, \Phi) - (\Psi, \nabla^2\Phi)] + [(V(\mathbf{x})\Psi, \Phi) - (\Psi, V(\mathbf{x})\Phi)]$$

$$= -\frac{\hbar^2}{2m}[(\nabla^2\Psi, \Phi) - (\Psi, \nabla^2\Phi)] + (\Psi, [V^*(\mathbf{x}) - V(\mathbf{x})]\Phi).$$

The first term of the last line is proportional to

$$\int \{[\nabla^2\Psi^*(\mathbf{x}, t)]\Phi(\mathbf{x}, t) - \Psi^*(\mathbf{x}, t)\nabla^2\Phi(\mathbf{x}, t)\}\, d^3x$$

$$= \int \{\nabla\cdot[(\nabla\Psi^*)\Phi - \Psi^*\nabla\Phi] - [\nabla\Psi^*\cdot\nabla\Phi - \nabla\Psi^*\cdot\nabla\Phi]\}\, d^3x$$

$$= \int \nabla\cdot[(\nabla\Psi^*)\Phi - \Psi^*\nabla\Phi]\, d^3x = \int_S [(\nabla\Psi^*)\Phi - \Psi^*\nabla\Phi]\cdot d\mathbf{S}$$

where we have used the identity

$$\nabla\cdot(\Psi^*\nabla\Phi) = \Psi^*\nabla^2\Phi + \nabla\Psi^*\cdot\nabla\Phi;$$

the last integral is to be carried out over the surface bounding the region within which the physical system is confined. We must demand the vanishing of this integral, because the kinetic energy operator

$$\mathsf{T} \equiv -(\hbar^2/2m)\nabla^2$$

must be Hermitian. Therefore we require that

$$\int_S \{[\nabla \Psi^*(\mathbf{x}, t)]\Phi(\mathbf{x}, t) - \Psi^*(\mathbf{x}, t)\nabla\Phi(\mathbf{x}, t)\} \cdot d\mathbf{S} \tag{9.38}$$

vanish; in particular, it is usual to insist that

$$\lim_{|\mathbf{x}| \to \infty} |\mathbf{x}| \Psi(\mathbf{x}, t) = 0, \tag{9.39}$$

for all wave functions. Thus on a sphere at infinity, where $d\mathbf{S} = \hat{\mathbf{n}}|\mathbf{x}|^2 \, d\Omega$, $\hat{\mathbf{n}}$ being a unit vector outward along a radius of the sphere, the integral will vanish.

Since the kinetic-energy operator must be Hermitian, the potential-energy operator must also be Hermitian. If $\mathsf{V} = V(\mathbf{x})$ is a simple function of $\mathbf{x}$, it is clear that this implies

$$V(\mathbf{x}) = V^*(\mathbf{x}), \tag{9.40}$$

as we asserted earlier. It is conceivable that the operator

$$\mathsf{V} = \mathsf{E} - \mathsf{T}$$

might be more complicated than a simple potential function—that it might, for example, be another operator W multiplying a function $V(\mathbf{x})$:

$$\mathsf{V} = \mathsf{W}V(\mathbf{x}).$$

In such a case, we have

$$\mathsf{W}^\dagger = \mathsf{W}$$

as well as (9.40) as the condition for Hermiticity of V.

Returning to Eq. (8.66) and assuming $V(\mathbf{x}) = V^*(\mathbf{x})$, let us take the complex conjugate of the Schrödinger equation:

$$-i\hbar \frac{\partial}{\partial t} \psi^*(\mathbf{x}, t) = -\frac{\hbar^2}{2m} \nabla^2 \psi^*(\mathbf{x}, t) + V(\mathbf{x})\psi^*(\mathbf{x}, t). \tag{9.41}$$

Multiplying this on the left by $\psi(\mathbf{x}, t)$, and multiplying (8.66) on the left by $\psi^*(\mathbf{x}, t)$, and subtracting the former from the latter, we obtain

$$\begin{aligned} i\hbar\left(\psi^* \frac{\partial}{\partial t}\psi + \psi \frac{\partial}{\partial t}\psi^*\right) &= i\hbar \frac{\partial}{\partial t}\psi^*\psi \\ &= -\frac{\hbar^2}{2m}[\psi^*\nabla^2\psi - (\nabla^2\psi^*)\psi] \\ &= \nabla \cdot \left\{-\frac{\hbar^2}{2m}[\psi^*\nabla\psi - (\nabla\psi^*)\psi]\right\}, \end{aligned}$$

or

$$\frac{\partial}{\partial t}\rho(\mathbf{x}, t) + \nabla \cdot \mathbf{j}(\mathbf{x}, t) = 0, \tag{9.42}$$

where

$$\rho(\mathbf{x}, t) \equiv \psi^*(\mathbf{x}, t)\psi(\mathbf{x}, t) \tag{9.43}$$

$$\begin{aligned}\mathbf{j}(\mathbf{x}, t) &\equiv \frac{\hbar}{i2m}[\psi^*\nabla\psi - (\nabla\psi^*)\psi] \\ &= \frac{1}{2m}[\psi^*\mathsf{P}\psi + (\mathsf{P}\psi)^*\psi].\end{aligned} \tag{9.44}$$

Equation (9.42) has the form of a *continuity equation*, and can be thought of as expressing the conservation of probability just as the equation preceeding Eq. (1.22) expresses the conservation of electric charge. The probability density is $\rho(\mathbf{x}, t) = |\psi(\mathbf{x}, t)|^2$ and the *probability currenty density* is given by Eq. (9.44).

It is clear that if we write

$$\mu(\mathbf{x}, t) = m\rho(\mathbf{x}, t) = m|\psi(\mathbf{x}, t)|^2$$

$$\mathbf{j}_m(\mathbf{x}, t) = m\mathbf{j}(\mathbf{x}, t) = \tfrac{1}{2}[\psi^*\mathsf{P}\psi + (\mathsf{P}\psi)^*\psi],$$

$\mu(\mathbf{x}, t)$ and $j_m(\mathbf{x}, t)$ may be interpreted as the quantum-theoretical mass density and mass current density for the particle represented by $\psi(\mathbf{x}, t)$ and having mass m. Similarly if the particle has electric charge e, then $e\rho$ and $e\mathbf{j}$ become the charge density and electric current density, respectively, for the particle.

These physical interpretations can be used to obtain conditions upon the wave function. Thus, in order that $\rho(\mathbf{x}, t)$ and $\mathbf{j}(\mathbf{x}, t)$ be well-defined and finite, we require that everywhere

$$|\psi(\mathbf{x}, t)|^2 \quad \text{be single valued,} \tag{A}$$

$$\psi(\mathbf{x}, t) \quad \text{be finite,} \tag{B}$$

$$\nabla\psi(\mathbf{x}, t) \quad \text{be finite.} \tag{C}$$

The last condition assures us that $\psi(\mathbf{x}, t)$ will be continuous everywhere. Finally, since Eq. (8.66) can be transformed into

$$\begin{aligned}\nabla^2\psi(\mathbf{x}, t) &= \frac{2m}{\hbar^2}\left[V(\mathbf{x}) - i\hbar\frac{\partial}{\partial t}\right]\psi(\mathbf{x}, t) \\ &= \frac{2m}{\hbar^2}[V(\mathbf{x}) - \mathsf{E}]\psi(\mathbf{x}, t)\end{aligned}$$

the finiteness of the potential energy and of the total energy would imply that $\nabla^2\psi(\mathbf{x}, t)$ must be finite. Actually, in some problems (rather nonphysical problems but nonetheless approximations to important physical ones), we may assume that $V(\mathbf{x})$ and E are not finite, although the kinetic energy T may be required to be finite; even in this case, $\nabla^2\psi(\mathbf{x}, t)$ will be finite, of course. A slightly weaker condition, but one more easily imposed in many cases, is that everywhere

$$\nabla\psi(\mathbf{x}, t) \qquad \text{be continuous.} \tag{D}$$

Conditions (A–D) are most important in problems involving sectionally continuous potentials (i.e. problems in which $V(\mathbf{x})$ does not vary continuously with $\mathbf{x}$ but has only "jump" discontinuities). In such problems, however, it is usually simpler to use the conditions

$$\psi(\mathbf{x}, t) \qquad \text{is continuous} \tag{C$'$}$$

and

$$\nabla \ln \psi(\mathbf{x}, t) = \frac{\nabla\psi(\mathbf{x}, t)}{\psi(\mathbf{x}, t)} \qquad \text{is continuous} \tag{D$'$}$$

in place of (C) and (D). It can be seen easily that they are equivalent to (C) and (D); the main advantage is that in condition (D′), one may multiply ψ by an arbitrary constant without changing the value of $\nabla \ln \psi$. Later we shall utilize these conditions in specific applications.

9.5 A FURTHER APPLICATION OF THE CURRENT DENSITY; INTERACTION OF A PARTICLE WITH THE ELECTROMAGNETIC FIELD; GAUGE TRANSFORMATIONS*

Unfortunately there seems to be no simple way to arrive at the proper prequantum expression for the energy (Hamiltonian) of a particle interacting with the electromagnetic field. Once the appropriate classical expression is obtained, however, the transition to quantum theory yields, very interestingly, an important relation between gauge transformations of the first and second kinds.

An elegant and extremely useful formulation of classical mechanics, which has also played a very substantial part in the development of quantum theory, is the Lagrangian formulation. Here one seeks a functional $L(x^\mu, u^\mu)$, $u^\mu = dx^\mu/dt$ such that the *action integral*

$$A(\tau_1, \tau_2) = \int_{\tau_1}^{\tau_2} L(x^\mu, u^\mu)\, d\tau$$

is a minimum. The functional $L(x^\mu, u^\mu)$ is called the *Lagrangian* and x^μ, u^μ are considered to be independent of each other, although both are functions of the invariant time: $x^\mu = x^\mu(\tau)$, $u^\mu = u^\mu(\tau)$. It may seem odd to consider $x^\mu(\tau)$ and $u^\mu(\tau)$ as independent when $u^\mu(\tau)$ is defined as $d(x^\mu/\tau)/dt$; however, we start out not knowing how $x^\mu(\tau)$ depends on τ, so of course we cannot compute $u^\mu(\tau)$ using this definition. In fact, the aim of Lagrange's method is to find the equations of motion for

* This section may be passed over without loss of continuity.

$x^\mu(\tau)$ (the solutions of which yield the functional form of $x^\mu(\tau)$ and hence of $u^\mu(\tau)$) by minimizing the action integral. We assume that $L(x^\mu, u^\mu)$ is a known function of the variables x^μ and u^μ; since x^μ and u^μ are really functions of τ, we are allowing their functional forms to vary, and by choosing the functional forms for $x^\mu(\tau)$ and $u^\mu(\tau)$ that minimize $A(\tau_1, \tau_2)$, we obtain the equations of motion (see Fig. 9.1).

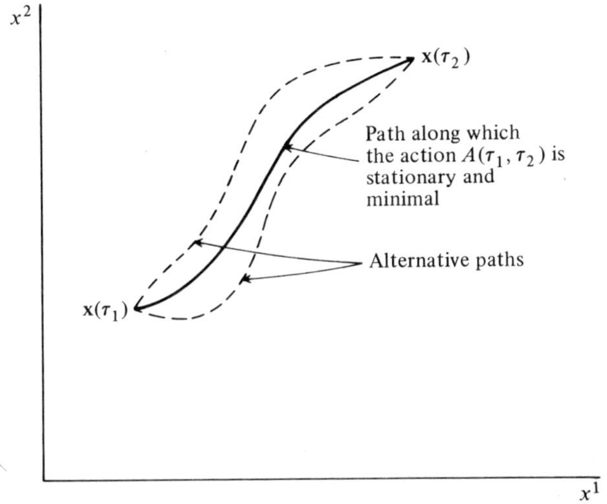

Fig. 9.1. Path of least action in the x^1x^2-plane. The actual path of a particle is that corresponding to the Lagrangian that makes the action integral $A(\tau_1, \tau_2)$ stationary and minimal.

Actually, in practice Lagrange's method is used backwards. Given the equations of motion, we find a Lagrangian such that the minimization of the action integral yields back the equations of motion. The reason for this is that the Lagrangian itself has some useful properties that may be exploited either in solving specific problems or in generalizing a theory.

We assume that a particle is known to pass through two points in four-space, namely $x^\mu(\tau_1)$ and $x^\mu(\tau_2)$; hence τ_1 and τ_2 are the invariant times associated with two fixed points on the particle's path. The true path of the particle is, according to the *principle of least action*, such that $A(\tau_1, \tau_2)$ is minimal. Thus we shall allow x^μ and u^μ to vary in such a way as to yield

$$\begin{aligned} 0 = \delta A(\tau_1, \tau_2) &= \delta \int_{\tau_1}^{\tau_2} L(x^\mu, u^\mu)\, d\tau = \int_{\tau_1}^{\tau_2} \delta L(x^\mu, u^\mu)\, d\tau \\ &= \int_{\tau_1}^{\tau_2} \left[\frac{\partial L}{\partial x^\mu} \delta x^\mu + \frac{\partial L}{\partial u^\mu} \delta u^\mu \right] d\tau, \end{aligned} \tag{9.45}$$

where the variation δ may be brought under the integral because τ_1 and τ_2 are fixed. Now we utilize $u^\mu = dx^\mu/d\tau$ and recognize that a variation in u^μ corresponds to a variation in x^μ by allowing the variation to pass through the differentiation symbol:

$$\delta u^\mu = \delta \frac{d}{d\tau} x^\mu = \frac{d}{d\tau} \delta x^\mu.$$

Thus we obtain, for the last term in the last member of (9.44),

$$\begin{aligned}\int_{\tau_1}^{\tau_2} \frac{\partial L}{\partial u^\mu} \delta u^\mu \, d\tau &= \int_{\tau_1}^{\tau_2} \frac{\partial L}{\partial u^\mu}\left(\frac{d}{d\tau}\delta x^\mu\right) d\tau \\ &= \frac{\partial L}{\partial u^\mu}\delta x^\mu \Big|_{\tau_1}^{\tau_2} - \int_{\tau_1}^{\tau_2}\left(\frac{d}{d\tau}\frac{\partial L}{\partial u^\mu}\right)\delta x^\mu \, d\tau \\ &= -\int_{\tau_1}^{\tau_2}\left(\frac{d}{dt}\frac{\partial L}{\partial u^\mu}\right)\delta x^\mu \, d\tau,\end{aligned}$$

the integrated term vanishing because x^μ is not allowed to vary at the end-points of the interval $[\tau_1, \tau_2]$; that is,

$$\delta x^\mu(\tau_1) = \delta x^\mu(\tau_2) = 0.$$

Hence Eq. (9.45) becomes

$$0 = \int_{\tau_1}^{\tau_2}\left[\frac{\partial L}{\partial x^\mu} - \frac{d}{d\tau}\frac{\partial L}{\partial u^\mu}\right]\delta x^\mu \, d\tau,$$

and since this must be true for all δx^μ in the interval (τ_1, τ_2) it follows that

$$\frac{d}{d\tau}\frac{\partial L}{\partial u^\mu} = \frac{\partial L}{\partial x^\mu}. \tag{9.46}$$

This is known as *Lagrange's equations of motion*, and the Lagrangian L must be chosen so that (9.46) corresponds to the actual equations of motion

For a free particle, the equations of motion are

$$\frac{d}{dt} p^\mu = m \frac{d}{d\tau} u^\mu = 0$$

and evidently there is a wide range of Lagrangians that will yield this equation of motion; in fact, any function of $p^\mu p_\mu$ will serve this purpose. For a nonrelativistic particle in a conservative potential $V(\mathbf{x})$, however, the equations of motion are

$$\frac{d}{dt}\mathbf{p} = -\nabla V(\mathbf{x})$$

and if we use $\tau \to t$, $x^\mu \to \mathbf{x}$ and $u^\mu \to \mathbf{v}$ to convert Eq. (9.46) into nonrelativistic form, we get

$$\tfrac{1}{2}m\mathbf{v}^2 - V(\mathbf{x}) = \frac{1}{2m}\mathbf{p}^2 - V(\mathbf{x}) = L(\mathbf{x}, \mathbf{v})$$

as the simplest appropriate Lagrangian. This suggests

$$\tfrac{1}{2}m\mathbf{v}^2$$

as the appropriate nonrelativistic free-particle Lagrangian.

Using this as a clue, we write

$$\tfrac{1}{2}mu^\mu u_\mu = L_0(x^\mu, u^\mu) \tag{9.47}$$

as the free-particle relativistic Lagrangian. For a particle in an electromagnetic field, however, the equations of motion are rather complicated looking; Eqs. (4.29), (5.17), (5.1) and (5.8) yield

$$\frac{d}{d\tau}p^\mu = m\frac{d}{d\tau}u^\mu = F^{\mu\nu}j_\nu = eF^{\mu\nu}u_\nu = e(\partial^\mu A^\nu - \partial^\nu A^\mu)u_\nu,$$

or simply

$$m\frac{d}{d\tau}u^\mu = e(\partial^\mu A^\nu - \partial^\nu A^\mu)u_\nu \tag{9.48}$$

for a particle with mass m and charge e. The simplest Lagrangian that reduces to (9.47) when $A^\mu = 0$, however, is simple indeed:

$$L(x^\mu, u^\mu) = \tfrac{1}{2}mu^\mu u_\mu + eA^\mu u_\mu. \tag{9.49}$$

To check this, we remember that $A^\mu = A^\mu(x^\nu)$, and hence

$$\frac{d}{d\tau}A^\mu(x^\nu) = \frac{\partial x^\nu}{\partial\tau}\partial_\nu A^\mu = u^\nu\partial_\nu A^\mu.$$

With this in mind, we obtain

$$\frac{d}{d\tau}\frac{\partial L}{\partial u^\mu} = \frac{d}{d\tau}(mu_\mu + eA_\mu) = m\frac{d}{d\tau}u_\mu + eu_\nu\partial^\nu A_\mu,$$

and, since $A^\mu = A^\mu(x^\nu)$ is the only part of (9.49) that depends explicitly on x^μ,

$$\frac{\partial L}{\partial x^\mu} = e(\partial_\mu A^\nu)u_\nu.$$

Substituting these into Eq. (9.46) and changing the index μ from a covariant to a contravariant position, we get Eq. (9.48) as desired.

Now Eq. (9.46) has a rather interesting form. In the simplest classical problems, $\partial L/\partial v^m$ is equal to the momentum components and $\partial L/\partial x^m$ is equal to the force components. Hence Eq. (9.46) has the usual form for equations of motion provided $\partial L/\partial u^\mu$ is interpreted as a generalized momentum (usually called the *canonical* momentum), and $\partial L/\partial x^\mu$ is thought of as a generalized force. Thus the time derivative of the canonical momentum equals the generalized force.

The canonical momentum for a particle governed by Eq. (9.48) then is, according to Eq. (9.49),

$$p_\mu = mu_\mu + eA_\mu. \tag{9.50}$$

However, since $u^\mu u_\mu = c^2$, by the definition of u^μ, we have, writing $u^\mu = (p^\mu - eA^\mu)/m$

$$m^2c^2 = (p^\mu - eA^\mu)(p_\mu - eA_\mu).$$

Solving this for the fourth component of four-momentum, we get

$$p^0 = eA^0 + \sqrt{m^2c^2 + (\mathbf{p} - e\mathbf{A})^2}. \tag{9.51}$$

Letting $p^0 = E/c$ be interpreted as the energy, we have

$$E = e\phi + \sqrt{m^2c^4 + (\mathbf{p} - e\mathbf{A})^2c^2}, \tag{9.52}$$

since $A^0 = \phi/c$, where ϕ is the usual scalar potential. Nonrelativistically, this reduces to

$$E_0 = (E - mc^2) \simeq e\phi + \frac{1}{2m}(\mathbf{p} - e\mathbf{A})^2, \tag{9.53}$$

the last term being the kinetic energy.

Thus, interactions with the electromagnetic field in general lead to a more complicated change in the energy expressions than a simple addition of potential energy. The momentum $\mathbf{p}$ must be replaced by $\mathbf{p} - e\mathbf{A}$. This, upon reflection, is quite natural, because whatever change takes place in the energy expression must take place in a Lorentz-covariant manner. The simple change $T \to T + e\phi$, if T is the kinetic energy in the form $mc^2(\gamma - 1) = \sqrt{m^2c^4 + p^2c^2} - mc^2$, clearly is not Lorentz covariant; it adds a term to the zeroth component of the four-vector, but not to any other component. The shift

$$p^\mu \to p^\mu - eA^\mu,$$

however, accomplishes the change in a covariant manner.

Since we are interested in the nonrelativistic problem for now, we shall utilize Eq. (9.53) in making the transition to quantum theory. The vector potential $\mathbf{A}(\mathbf{x})$ now may be considered an operator, $\mathsf{A} = \mathbf{A}(\mathbf{x})$, and the transition to quantum mechanics yields

$$i\hbar\frac{\partial}{\partial t}\psi(\mathbf{x}, t) = \frac{1}{2m}\left[\frac{\hbar}{i}\nabla - e\mathbf{A}(\mathbf{x})\right]^2\psi(\mathbf{x}, t) + e\phi(\mathbf{x})\psi(\mathbf{x}, t). \tag{9.54}$$

At the moment we are not interested in obtaining any particular solutions to this equation, but rather we wish to see what the current density looks like in the presence of an electromagnetic field and what effect gauge transformations may have upon this equation. Modern quantum theory is replete with problems related to these.

The only thing that is new here is the presence of $\mathbf{A}(\mathbf{x})$ on the right-hand side of (9.54). Thus, in place of the current density of Eq. (9.44), we have

$$\begin{aligned}\mathbf{j}(\mathbf{x}, t) &= \frac{1}{2m}\left\{\psi^*\left(\frac{\hbar}{i}\nabla - e\mathbf{A}\right)\psi + \left[\left(\frac{\hbar}{i}\nabla - e\mathbf{A}\right)\psi\right]^*\psi\right\} \\ &= \frac{1}{2m}\{\psi^*(\mathbf{P} - e\mathbf{A})\psi + [(\mathbf{P} - e\mathbf{A})\psi]^*\psi\},\end{aligned} \tag{9.55}$$

where $\mathbf{A}$ is assumed to take on only real values. This is what we should expect, for here again, $\mathbf{P} - e\mathbf{A}$ replaces the $\mathbf{P}$ in Eq. (9.44).

Now we may carry out a wide variety of transformations on ψ without changing $\rho(\mathbf{x}, t)$ as given in (9.43) or $\mathbf{j}(\mathbf{x}, t)$ as given in (9.55). Thus if $\psi(\mathbf{x}, t) \rightarrow \psi'(\mathbf{x}, t) = e^{if(\mathbf{x},t)}\psi(\mathbf{x}, t)$ where $f(\mathbf{x}, t)$ is a reasonably well-behaved complex function, it follows that

$$\rho(\mathbf{x}, t) \rightarrow \rho'(\mathbf{x}, t) = \rho(\mathbf{x}, t)\, e^{-2\,\mathrm{Im}\, f(\mathbf{x}, t)}$$

$$\mathbf{j}(\mathbf{x}, t) \rightarrow \mathbf{j}'(\mathbf{x}, t) = \mathbf{j}(\mathbf{x}, t)\, e^{-2\,\mathrm{Im}\, f(\mathbf{x},t)} + \rho(\mathbf{x}, t)\, e^{-2\,\mathrm{Im}\, f(\mathbf{x},t)}\frac{\hbar}{m}\,\mathrm{Re}\,\nabla f(\mathbf{x}, t).$$

Hence $\rho(\mathbf{x}, t)$ is left unchanged by any transformation of the form

$$\psi \rightarrow \psi' = \psi\, e^{if(\mathbf{x},t)}, \qquad f(\mathbf{x}, t) = f^*(\mathbf{x}, t), \tag{9.56}$$

whereas for $\mathbf{j}(\mathbf{x}, t)$ to remain unchanged requires that in addition

$$\mathbf{A} \rightarrow \mathbf{A}' = \mathbf{A} + \frac{\hbar}{e}\nabla f,$$

as the reader should verify.

Equation (9.54) is slightly less tolerant. The transformation of Eq. (9.56), substituted into (9.54), yields

$$e^{if}\left(i\hbar\frac{\partial}{\partial t} - \hbar\frac{\partial f}{\partial t}\right)\psi$$

for the left-hand side, and

$$-\frac{1}{2m}\left(\frac{\hbar}{i}\nabla - e\mathbf{A}\right)^2 e^{if}\psi + e\phi e^{if}\psi$$

for the right-hand side. The simplest way to handle the first term here is to write

$$\left(\frac{\hbar}{i}\nabla - e\mathbf{A}\right)^2 e^{if}\psi = \left(\frac{\hbar}{i}\nabla - e\mathbf{A}\right) e^{if}\, e^{-if}\left(\frac{\hbar}{i}\nabla - e\mathbf{A}\right) e^{if}\psi,$$

and use

$$\left(\frac{\hbar}{i}\nabla - e\mathbf{A}\right)e^{if}\psi = e^{if}\left(\frac{\hbar}{i}\nabla - e\mathbf{A} + \hbar\nabla f\right)\psi;$$

this gives

$$\frac{1}{2m}\left(\frac{\hbar}{i}\nabla - e\mathbf{A}\right)^2 e^{if}\psi = e^{if}\,\frac{1}{2m}\left(\frac{\hbar}{i}\nabla - e\mathbf{A} + \hbar\nabla f\right)^2\psi.$$

Hence Eq. (9.54) becomes, with the new wave function $\psi' = e^{if}\psi$,

$$\left(i\hbar\frac{\partial}{\partial t} - \hbar\frac{\partial f}{\partial t}\right)\psi = \left[\frac{1}{2m}\left(\frac{\hbar}{i}\nabla - e\mathbf{A} + \hbar\nabla f\right)^2 + e\phi\right]\psi. \tag{9.57}$$

The term involving $\partial f/\partial t$ on the left-hand side can be compensated by letting

$$\phi \to \phi' = \phi - \frac{\hbar}{e}\frac{\partial f}{\partial t}. \tag{9.58}$$

However, in order that the validity of the electric field's relation with the potentials,

$$\mathbf{E} = -\nabla\phi - \frac{\partial}{\partial t}\mathbf{A},$$

be retained, this change in ϕ must be compensated by a corresponding change in $\mathbf{A}$:

$$\mathbf{A} \to \mathbf{A}' = \mathbf{A} + \nabla\frac{\hbar}{e}f. \tag{9.59}$$

This, however, is just what is required to restore (9.57) to the form of (9.54). Therefore, the gauge transformation of the second kind given by Eqs. (9.58) and (9.59), when accompanied by the transformation of Eq. (9.56), preserves the Schrödinger equation in an electromagnetic field. In addition, the current and probability densities also are preserved.

The transformation of Eq. (9.56), with $f(\mathbf{x}, t) = f^*(\mathbf{x}, t)$, is a gauge transformation of the first kind. These considerations show the intimate connection between gauge transformations of the first kind and those of the second kind. Although gauge transformations of the second kind leave the electromagnetic field equations invariant, they do change the physics of the system if the electromagnetic field interacts with quantum-mechanical particles. Therefore they must be accompanied by related gauge transformations of the first kind applied to the particles' wave functions.

This digression is intended to illustrate the role of the current density and to elucidate the importance of gauge transformations. We shall return to the electromagnetic field and electrodynamics in later chapters.

9.6 ANOTHER LOOK AT THE KLEIN–GORDON EQUATION

At the end of Chapter 8, we mentioned that the Klein–Gordon equation is not acceptable as an ordinary quantum-mechanical wave equation for a number of reasons. We now are equipped to discern the difficulty that arises in connection with the probability interpretation.

Originally we postulated that $\psi^*\psi$ would be interpreted as the probability density. This worked very nicely for the Schrödinger equation, and we found that this interpretation yields a meaningful continuity equation, Eq. (9.42). The continuity equation arose directly from manipulation of the wave equation, and a similar continuity equation should be derivable from the Klein–Gordon equation.

To find the Klein–Gordon version of a continuity equation, we form

$$\psi^*[\partial^\mu \partial_\mu - m^2c^2]\psi - [(\partial^\mu \partial_\mu - m^2c^2)\psi^*]\psi = 0$$

or

$$\psi^* \partial^\mu \partial_\mu \psi - (\partial^\mu \partial_\mu \psi^*)\psi = 0.$$

Using the identity

$$\partial^\mu[f(x^\nu)A_\mu(x^\nu)] = [\partial^\mu f(x^\nu)]A_\mu(x^\nu) + f(x^\nu)\partial^\mu A_\mu(x^\nu),$$

this becomes

$$\partial^\mu[\psi^* \partial_\mu \psi - (\partial_\mu \psi^*)\psi] = 0, \tag{9.60}$$

which is the relativistic form for a continuity equation if the four-current is taken to be

$$j'_\mu \equiv \psi^* \partial_\mu \psi - (\partial_\mu \psi^*)\psi.$$

Thus

$$j'^0 = c\rho' \equiv \frac{1}{c}\left[\psi^* \frac{\partial}{\partial t}\psi - \left(\frac{\partial}{\partial t}\psi^*\right)\psi\right]$$

defines the density ρ' and

$$\mathbf{j}' \equiv -[\psi^* \nabla\psi - (\nabla\psi^*)\psi]$$

defines the current density. Eq. (9.60) can be written in the form

$$\frac{\partial}{\partial t}\rho' + \nabla \cdot \mathbf{j}' = 0$$

if we adopt these conventions. The units of ρ' and j' here are awkward, however; ρ' has units of volume^{-1} time^{-1}, and we should like to eliminate the time factor. This can be accomplished by multiplying ρ' by $i\hbar/2m$, the i being included so that $\partial/\partial t$ can be converted to the energy operator:

$$\begin{aligned} \rho \equiv \left(\frac{i\hbar}{2m}\right)\rho' &= \frac{1}{2mc^2}\left[\psi^* i\hbar \frac{\partial}{\partial t}\psi - \left(i\hbar \frac{\partial}{\partial t}\psi^*\right)\psi\right] \\ &= \frac{1}{2mc^2}[\psi^* \mathsf{E}\psi + (\mathsf{E}\psi)^*\psi]. \end{aligned} \tag{9.61}$$

The current density then becomes

$$\mathbf{j} \equiv \frac{i\hbar}{2m}\mathbf{j}' = \frac{1}{2m}[\psi^* \mathbf{P}\psi + (\mathbf{P}\psi)^*\psi] \tag{9.62}$$

which is identical with (9.44).

Here we see that if the probability current density is defined by Eq. (9.62), then the probability density must be given by (9.61) rather than by $\psi^*\psi$. This would be all right—we could simply revise our postulate for the probability interpretation—except that the probability density defined by (9.61) is not positive-definite. Since the Klein–Gordon equation is second order in $\partial/\partial t$, both $\psi(\mathbf{x}, t)$ and $\partial\psi(\mathbf{x}, t)/\partial t$ can be chosen arbitrarily at any time t, and obviously many such choices would yield negative values for ρ. We might try to add a postulate that such choices are ruled out, but this, it turns out, leads to still more difficulties.

These problems were overcome only when Pauli and Weisskopf proposed in 1934 that the Klein–Gordon equation be treated as an equation for a *field* much like the electromagnetic field. The reader may note that when we introduced the electromagnetic field in the preceding section, it was introduced as a *classical* entity; in the transition to quantum mechanics, the *particle* operators were replaced by their quantum-theoretical counterparts, but no change was made in $\mathbf{A}(\mathbf{x})$ or $\phi(\mathbf{x})$—they were treated as real functions of real variables just as in classical theory. In fact, incorporating electromagnetic fields into quantum theory turned out to be a formidable task and was completed in a more-or-less satisfactory way only in the late 1940's. The quantum theory of electromagnetic fields had its beginnings in 1929, and the initial formulation of the problem had been developed by the time Pauli and Weisskopf reconsidered the Klein–Gordon equation in 1934. The resulting shift in emphasis from wave equations to the study of *fields* in quantum theory brought a considerable clarification of quantum ideas as well as a broad generalization of the theory. We cannot consider fields at this point, but it is worthwhile to introduce notation that will make such generalizations easy to incorporate into the theory.

9.7 STATE VECTORS

Dirac invented a system of notation that is particularly succinct and that allows for a somewhat more general treatment of quantum-mechanical systems than does the wave-function formalism we have been using. Although the *idea* of state vectors evidently was not original with him, the currently used notation for them was, and his work certainly clarified their meaning and role.

A physical system is completely described when the values of all its simultaneously measurable physical variables are specified at once. Classically, the word "simultaneously" would not have appeared, but as we have seen, canonically conjugate variables cannot be given precise values simultaneously in quantum phenomena. If a system is given a complete description in this way, we say that it is in the physical *state* described by the values of the variables. Let a system be in a state that we arbitrarily call state 1; we shall represent this circumstance by such symbols as

$$|\text{state } 1\rangle \qquad \text{or simply} \qquad |1\rangle$$

and shall say that these symbols denote the *state vector* representing the system.

Along with the state vectors, we must have operators that may transform a given state into another state. In most circumstances the operators may be thought of as representing physical processes, but we shall not restrict ourselves to such "interpretable" operators. An operator equation will have a form such as

$$\mathfrak{R}|1\rangle = |2\rangle \qquad \text{or} \qquad (\mathfrak{Q} - \mathfrak{R})|1\rangle = \alpha|2\rangle,$$

etc., where $\mathfrak{Q}$ and $\mathfrak{R}$ here are operators and $|1\rangle$ and $|2\rangle$ represent states. By

$$(\mathfrak{Q} \pm \mathfrak{R})|1\rangle$$

we mean

$$\mathfrak{Q}|1\rangle \pm \mathfrak{R}|1\rangle,$$

and by

$$\mathfrak{Q}\mathfrak{R}|1\rangle$$

we mean

$$\mathfrak{Q}(\mathfrak{R}|1\rangle)$$

that is, if $|2\rangle \equiv \mathfrak{R}|1\rangle$, then

$$\mathfrak{Q}\mathfrak{R}|1\rangle = \mathfrak{Q}|2\rangle.$$

Now again we wish linear superpositions of states to be possible physical states; hence we shall postulate that the state vectors constitute a linear vector space. This means that a commutative and associative law of addition for state vectors is defined (as we have assumed above) such that the sum of two state vectors is again a state vector, that there is a null vector which when added to any other state vector has no effect upon it, that for every vector there is a negative vector such that the sum of these two yields the null vector, and that scalar multiplication by complex numbers is defined with the following properties:

$$\alpha|1\rangle \text{ is a state vector;}$$

$$\alpha(|1\rangle + |2\rangle) = \alpha|1\rangle + \alpha|2\rangle; \qquad (\alpha + \beta)|1\rangle = \alpha|1\rangle + \beta|1\rangle;$$

$$(\alpha\beta)|1\rangle = \alpha(\beta|1\rangle); \qquad 1|1\rangle = |1\rangle;$$

if α and β are complex numbers.

It should be noted that although $\alpha|1\rangle$ and $|1\rangle$ may be different state vectors (when $\alpha \neq 1$), they nevertheless represent the same physical state of the system. For this reason, it is sometimes said that a physical state is represented by a *ray* in the state vector space. Thus complex numbers, when considered as operators, operating on state vectors, do not change the description of the physical system. This freedom of multiplying state vectors by constants can be restricted partially by the demand that state vectors be normalized in a certain way; the freedom that then remains again will correspond to gauge transformations of the first kind.

The useful numerical content of state vectors is extracted through the definition of a Hermitian inner product. The definition of such an inner product converts the state vector space into a unitary space. To define the inner product, we proceed as follows: assuming that we have a system in state 1, we perform an experiment in which certain physical variables are measured, and we observe that the values of these variables correspond to state 2. The probability amplitude for this occurrence is denoted

$$\langle \text{state } 2|\text{state } 1\rangle \qquad \text{or simply} \qquad \langle 2|1\rangle;$$

hence this inner product $\langle 2|1\rangle$ is a complex number. Now if the variables that were measured were precisely those that define state 1, their values, by definition of a state, would have to be just those that describe state 1, and we would have $|2\rangle = |1\rangle$ and hence $|\langle 1|1\rangle|^2 = 1$. On the other hand, if some of the variables were not those used to define state 1, then state 2 may have been just one of many possible states that could be obtained from state 1, by a measurement of those variables; the probability that the measured values would correspond to state 2 rather than to one of the other possible states is given by $|\langle 2|1\rangle|^2$.

This suggests that the inner product of state vectors should have essentially the same properties as the inner product of wave functions. Hence we postulate that $\langle 2|1\rangle$ be *linear* in the right-hand label, i.e. if

$$|1\rangle = \alpha|3\rangle + \beta|4\rangle$$

then

$$\langle 2|1\rangle = \alpha\langle 2|3\rangle + \beta\langle 2|4\rangle.$$

Also, we require that the inner product be Hermitian:

$$\langle 1|2\rangle = \langle 2|1\rangle^*.$$

This implies that $\langle 2|1\rangle$ is *antilinear* in the left-hand label, since if $|1\rangle = \alpha|3\rangle + \beta|4\rangle$, then

$$\begin{aligned}\langle 1|2\rangle &= (\alpha\langle 2|3\rangle + \beta\langle 2|4\rangle)^* = \alpha^*\langle 2|3\rangle^* + \beta^*\langle 2|4\rangle^* \\ &= \alpha^*\langle 3|2\rangle + \beta^*\langle 4|2\rangle.\end{aligned}$$

A theorem of linear algebra states that a unitary vector space can be used to define another vector space, the *dual* of the original, in a manner that allows an element of the dual space to correspond in one-to-one fashion with an element of the original space. We denote the dual of $\alpha|\text{state } 1\rangle$ by $\langle \text{state } 1|\alpha^*$. Dirac calls $|\text{state } 1\rangle$ a *ket* and $\langle \text{state } 1|$ its corresponding *bra*; together, in juxtaposition, a bra-ket pair form a bracket denoting an inner product.

An operator $\mathfrak{Q}$ operating on a state $|1\rangle$ yields a new state, so that we can form the inner product

$$\langle 2|\mathfrak{Q}|1\rangle$$

of the bra corresponding to $|2\rangle$ with the state $\mathfrak{Q}|1\rangle$. This gives the probability amplitude that a state 1, after being operated on by $\mathfrak{Q}$, yields measurements corresponding to state 2. This interpretation makes calling $\langle 1|\mathfrak{Q}|1\rangle$ the expectation value of $\mathfrak{Q}$ eminently reasonable. For $\langle 1|\mathfrak{Q}|1\rangle$ to be real, we must have

$$\langle 1|\mathfrak{Q}|1\rangle^* = \langle 1|\mathfrak{Q}|1\rangle, \tag{9.63}$$

but it is not clear immediately that this is true in general; indeed it is not.

As before, we must define a Hermitian adjoint for an operator in order to get a convenient criterion to determine which operators have real expectation values in general. Let $\mathfrak{Q}^\dagger$ be the operator for which

$$\langle 2|\mathfrak{Q}^\dagger|1\rangle = \langle 1|\mathfrak{Q}|2\rangle^*; \tag{9.64}$$

then Eq. (9.63) becomes

$$\langle 1|\mathfrak{Q} - \mathfrak{Q}^\dagger|1\rangle = 0,$$

or, since state 1 is an arbitrary state,

$$\mathfrak{Q} = \mathfrak{Q}^\dagger. \tag{9.65}$$

That is, if an operator corresponds to a physical variable, it must have real expectation values and hence must equal its Hermitian adjoint; it must be Hermitian.

It is very convenient to write

$$\langle 2| = |2\rangle^\dagger$$

and call a bra the Hermitian adjoint of a ket; this is consistent with the usual practice in linear algebra. In particular, this gives

$$\langle 1|\mathfrak{Q}|2\rangle^* = [\mathfrak{Q}|2\rangle]^\dagger|1\rangle = |2\rangle^\dagger\mathfrak{Q}^\dagger|1\rangle = \langle 2|\mathfrak{Q}^\dagger|1\rangle,$$

so that we can think of the Hermitian conjugate $\mathfrak{Q}^\dagger$ as here having the same effect upon $|1\rangle$ in the inner product as $\mathfrak{Q}$ has upon $|2\rangle$ in the complex conjugate of the same inner product. We also write $\langle 2|^\dagger = (|2\rangle^\dagger)^\dagger = |2\rangle$.

9.8 EIGENSTATES; AN EXAMPLE

As indicated in Section 9.7, a state is defined by the values of the physical variables used to describe it. We might as well make this explicit in our state-vector notation whenever these values play a role in our considerations. Let $\mathfrak{Q}$, $\mathfrak{R}$, $\mathfrak{S}, \ldots$ be a set of operators corresponding to simultaneously observable variables, and let $|q, r, s, \ldots\rangle$ be a state such that

$$\mathfrak{Q}|q, r, s, \ldots\rangle = q|q, r, s, \ldots\rangle$$

$$\mathfrak{R}|q, r, s, \ldots\rangle = r|q, r, s, \ldots\rangle$$

$$\mathfrak{S}|q, r, s, \ldots\rangle = s|q, r, s, \ldots\rangle$$

where $q, r, s, \ldots$ are real numbers. Then the state $|q, r, s, \ldots\rangle$ is an eigenstate of $\mathfrak{Q}, \mathfrak{R}, \mathfrak{S}, \ldots$, with eigenvalues $q, r, s, \ldots$; obviously $\mathfrak{Q}, \mathfrak{R}, \mathfrak{S}, \ldots$, all must be Hermitian operators. In this way we shall label eigenstates with their eigenvalues.

To illustrate this and the general notion of state vectors, let us consider a simple example. We are given a simple particle with mass m and momentum $\mathbf{p}$; we denote it $|\mathbf{p}\rangle$. What is the probability that this particle may be found within a small volume element d^3x centered at the point $\mathbf{x}$? The state defined by the particle lying within this volume element we shall denote $|\mathbf{x}(d^3x)\rangle$. The desired probability amplitude then is

$$\langle \mathbf{x}(d^3x)|\mathbf{p}\rangle$$

and the desired probability is

$$|\langle \mathbf{x}(d^3x)|\mathbf{p}\rangle|^2.$$

Evidently this probability should be proportional to d^3x since a particle with fixed momentum must be thought of as having its probability density uniformly distributed in space; hence we write

$$|\langle \mathbf{x}(d^3x)|\mathbf{p}\rangle|^2 = |\langle \mathbf{x}|\mathbf{p}\rangle|^2\, d^3x$$

and $\langle \mathbf{x}|\mathbf{p}\rangle$ then is the probability density amplitude. This latter, however, may be recognized as the wave function, so that we have

$$\langle \mathbf{x}|\mathbf{p}\rangle = \psi_{\mathbf{p}}(\mathbf{x}) \propto e^{i\mathbf{k}\cdot\mathbf{x}}, \qquad \mathbf{k} = \mathbf{p}/\hbar.$$

The sum, over all volume elements d^3x, of $|\langle \mathbf{x}(d^3x)|\mathbf{p}\rangle|^2$ must be unity and hence

$$\sum_{d^3x} |\langle x(d^3x)|\mathbf{p}\rangle|^2 = \int d^3x |\langle \mathbf{x}|\mathbf{p}\rangle|^2 = 1. \tag{9.66}$$

The amplitude $\langle \mathbf{x}|\mathbf{p}\rangle$ is therefore a *normalized* wave function. As we have seen, this normalization can be achieved only if box normalization is utilized, and therefore

$$\langle \mathbf{x}|\mathbf{p}\rangle = \psi_{\mathbf{p}}(\mathbf{x}) = L^{-3/2} e^{i\mathbf{k}\cdot\mathbf{x}}. \tag{9.67}$$

If one is inclined to be a bit generous with the probability interpretation and allow δ-function normalization, the condition (9.66) becomes

$$\int d^3x \langle \mathbf{x}|\mathbf{p}'\rangle^* \langle \mathbf{x}|\mathbf{p}\rangle = \int d^3x \langle \mathbf{p}'|\mathbf{x}\rangle\langle \mathbf{x}|\mathbf{p}\rangle = \hbar^3 \delta^3(\mathbf{p}' - \mathbf{p}), \tag{9.68}$$

and

$$\langle \mathbf{x}|\mathbf{p}\rangle = \psi_{\mathbf{p}}(\mathbf{x}) = (2\pi)^{-3/2} e^{i\mathbf{k}\cdot\mathbf{x}} \tag{9.69}$$

follows.

We must take care to note that $|\mathbf{x}(d^3x)\rangle$ is *not* an eigenstate of the position operator $\mathfrak{X}$, since the former allows the particle to be spread out over some volume

d^3x around the position $\mathbf{x}$. The state $|\mathbf{x}\rangle$ however is an eigenstate of $\mathfrak{X}$, and

$$\mathfrak{X}|\mathbf{x}\rangle = \mathbf{x}|\mathbf{x}\rangle.$$

Thus the inner product $\langle \mathbf{x}|\mathbf{p}\rangle$, which we identify with a wave function, is the inner product of a position eigenstate with a momentum eigenstate. The nonnormalizability of this wave function in the usual sense, which led us to adopt the box and δ-function normalization techniques, reflects the fact that the position operator $\mathfrak{X}$ and the momentum operator $\mathfrak{P}$ are not simultaneously observable.

Quite generally, we may identify wave functions with inner products between states, e.g. $|1\rangle$, and eigenstates of position, e.g. $|\mathbf{x}, q, r, \ldots\rangle$:

$$\psi_{\text{state 1}}(\mathbf{x}; q, r, \ldots) = \langle \mathbf{x}, q, r, \ldots|1\rangle$$

is the wave function for a system in state 1. The labels $q, r, \ldots$ are included to indicate that other variables, simultaneously measurable with position, may be required to define the system completely. Furthermore

$$\psi^*_{\text{state 2}}(\mathbf{x}, q, r, \ldots) = \langle \mathbf{x}, q, r, \ldots|2\rangle^* = \langle 2|\mathbf{x}, q, r, \ldots\rangle$$

so that

$$\int \psi^*_{\text{state 2}}(\mathbf{x}, q, r, \ldots)\psi_{\text{state 1}}(\mathbf{x}, q, r, \ldots)\, d^3x$$
$$= \int \langle 2|\mathbf{x}, q, r, \ldots\rangle\langle \mathbf{x}, q, r, \ldots|1\rangle\, d^3x.$$

In the event that $\mathbf{x}$ is the only measurable quantity necessary to define the system, the above integral provides a method for calculating the inner product $\langle 2|1\rangle$, for then the labels $q, r, \ldots$ do not appear, and we have

$$(\psi_{\text{state 2}}, \psi_{\text{state 1}}) = \int d^3x\, \langle 2|\mathbf{x}\rangle\, \langle \mathbf{x}|1\rangle$$

It is easy to check that $(\psi_{\text{state 2}}, \psi_{\text{state 1}})$ and $\langle 2|1\rangle$ have precisely the same properties in this case, and therefore must be identical.

Integration over position is not a sufficiently general procedure to provide an adequate technique for calculating general inner products, however. Indeed, the above case is simply a utilization of the completeness condition for state vectors when the states can be defined completely by spatial position.

9.9 ORTHONORMALITY AND COMPLETENESS; PROJECTIONS

Two state vectors, say $|1\rangle$ and $|2\rangle$ are said to be orthogonal whenever their inner product vanishes:

$$\langle 1|2\rangle = 0.$$

Our probability amplitude interpretation for the inner product requires us to normalize the state vectors to

$$\langle 1|1\rangle = e^{i\alpha}, \qquad \alpha \text{ a real number};$$

but $\langle 1|1\rangle = \langle 1|1\rangle^*$, and hence, for any state $|1\rangle$,

$$\langle 1|1\rangle = \pm 1;$$

we choose the upper sign. Hence an orthonormal set of state vectors $\{|1\rangle, |2\rangle, \ldots\}$ is a set such that

$$\langle i|j\rangle = \delta_{ij}, \qquad i, j = 1, 2, \ldots,$$

where we interpret this to mean that $\langle i|j\rangle = 0$ unless $|i\rangle$ and $|j\rangle$ are the same state, when $\langle i|i\rangle = 1$.

As with wave functions, however, we should like to include in our formalism some state vectors that are not physically acceptable but that will be mathematically useful. This requires a slight liberalization of the mathematical requirements imposed by the probability interpretation so that we can use a more general orthonormality condition. This time let us formulate the condition in terms of eigenstates of Hermitian operators. Let $\mathfrak{Q}, \mathfrak{R}, \mathfrak{S}, \ldots$ be a set of Hermitian operators corresponding to a set of physical variables capable of completely describing a system; we call such a set a *complete set of simultaneously measurable observables*, or a complete set of *compatible* observables. Let its set of eigenstates be $\{|q, r, s, \ldots\rangle\}$ and its eigenvalues be $\{q, r, s, \ldots\}$. The set $\{q, r, s, \ldots\}$ can be arranged conveniently into an orthonormal set in the sense that

$$\langle q', r', s', \ldots|q, r, s, \ldots\rangle = \delta(q' - q)\delta(r' - r)\delta(s' - s)\cdots, \tag{9.70}$$

where $\delta(q' - q)$ is a Dirac δ-function if the set of eigenvalues $\{q\}$ is nondenumerable, or a Kronecker δ-symbol, $\delta_{q'q}$, if the set $\{q\}$ is denumerable.

If the set $\mathfrak{Q}, \mathfrak{R}, \mathfrak{S}, \ldots$ is complete, as we have assumed, then any set $\{|q, r, s, \ldots\rangle\}$ of linear independent eigenvectors is automatically an orthogonal set. This follows from the orthogonality of states corresponding to different eigenvalues of any one of the operators:

$$\begin{aligned}
0 &= \langle q, r', s, \ldots|(\mathfrak{R} - \mathfrak{R})|q, r, \ldots\rangle = \langle q, r', s, \ldots|(\mathfrak{R} - \mathfrak{R}^\dagger)|q, r, s, \ldots\rangle \\
&= \langle q, r', s, \ldots|\mathfrak{R}(q, r, s, \ldots\rangle - \langle q, r', s, \ldots|\mathfrak{R}^\dagger|q, r, s, \ldots\rangle \\
&= r\langle q, r', s, \ldots|q, r, s, \ldots\rangle - \langle q, r, s, \ldots|\mathfrak{R}|q, r', s, \ldots\rangle^* \\
&= r\langle q, r', s, \ldots|q, r, s, \ldots\rangle - r'^*\langle q, r, s, \ldots|q, r', s, \ldots\rangle^* \\
&= (r - r')\langle q, r', s, \ldots|q, r, s, \ldots\rangle,
\end{aligned}$$

with $r'^* = r'$ because $\mathfrak{R} = \mathfrak{R}^\dagger$ is assumed. Thus $r \neq r'$ implies $\langle q, r', s, \ldots|q, r, s, \ldots\rangle = 0$. Since the set of operators is complete, two eigenstates of these operators can differ meaningfully (i.e. be linearly independent) only if their eigenvalues differ for at least one of the operators. This proves our assertion and hence shows that only the normalization is specified by Eq. (9.70).

Also, as with wave functions, we want to hypothesize that complete sets of eigenstates exist, so that

$$\mathop{S}_{q,r,s,\ldots} |q, r, s, \ldots\rangle\langle q, r, s, \ldots| = 1, \tag{9.71}$$

where 1 is taken to be the identity operator.

Eq. (9.71) is equivalent to the condition that any state vector can be written as a linear superposition of the $|q, r, s, \ldots\rangle$:

$$\begin{aligned} |1\rangle &= \mathop{S}_{q,r,s,\ldots} |q, r, s, \ldots\rangle\langle q, r, s, \ldots|1\rangle; \\ \langle 1| = |1\rangle^{+} &= \mathop{S}_{q,r,s,\ldots} [|q, r, s, \ldots\rangle\langle q, r, s, \ldots|1\rangle]^{\dagger} \\ &= \mathop{S}_{q,r,s,\ldots} \langle q, r, s, \ldots|1\rangle^{*}\langle q, r, s, \ldots| \\ &= \mathop{S}_{q,r,s,\ldots} \langle 1|q, r, s, \ldots\rangle\langle q, r, s, \ldots|. \end{aligned}$$

The inner products $\langle q, r, s, \ldots|1\rangle$ are then the expansion coefficients for such superpositions, and the squares of these coefficients are interpreted as probabilities.

Note that

$$\begin{aligned} \langle 1|2\rangle = \langle 1|1|2\rangle &= \langle 1|(\mathop{S}_{q,r,s,\ldots} |q, r, s, \ldots\rangle\langle q, r, s, \ldots|)|2\rangle \\ &= \mathop{S}_{q,r,s} \langle 1|q, r, s, \ldots\rangle\langle q, r, s, \ldots|2\rangle \end{aligned} \tag{9.72}$$

justifies writing inner products as sums, over complete sets of eigenvalues, of different inner products. This is sometimes called *inserting a complete set of states* in place of the identity operator.

Since we have extended the orthonormality condition to δ-functions as well as Kronecker δ-symbols, the assumption that there exists a complete set of state vectors is equivalent to the assumption that a complete set of compatible operators exists. (By introducing the δ-function normalization convention, we gloss over a whole collection of problems related to convergence that are beyond the scope of our considerations here. Completeness is one such question; often the completeness we speak of here is subsumed under two mathematical requirements for Hilbert space which, in mathematical terms are that the space must be "complete in its metric" and that it be "separable". Thanks to Dirac's δ-function, we may avoid consideration of this question.) We must be sure that the eigenvalues of all operators of the complete set of compatible operators appear in the sum defining the completeness, Eq. (9.71); otherwise the sum will represent a *projector* and will not be equal to the identity operator. We are not interested in projectors at the moment, but this word of caution is in order.

9.10 AN APPLICATION: DOUBLE-SLIT DIFFRACTION

Let us consider now a simple but concrete example of the use of Dirac state-vector notation, namely the double-slit diffraction of waves (or particle–waves). Assume that we have a screen containing two parallel slits a distance $2d$ apart, and that behind this plane screen lies another, covered with fluorescent material or some other particle-detecting coating, a distance l away. The slits each have width w with $w \ll d \ll l$. A monoenergetic beam of particles is incident normally on the first screen; let this beam have wave number $\mathbf{k}$ and hence particle momentum $\mathbf{p} = \hbar\mathbf{k}$. See Fig. 9.2.

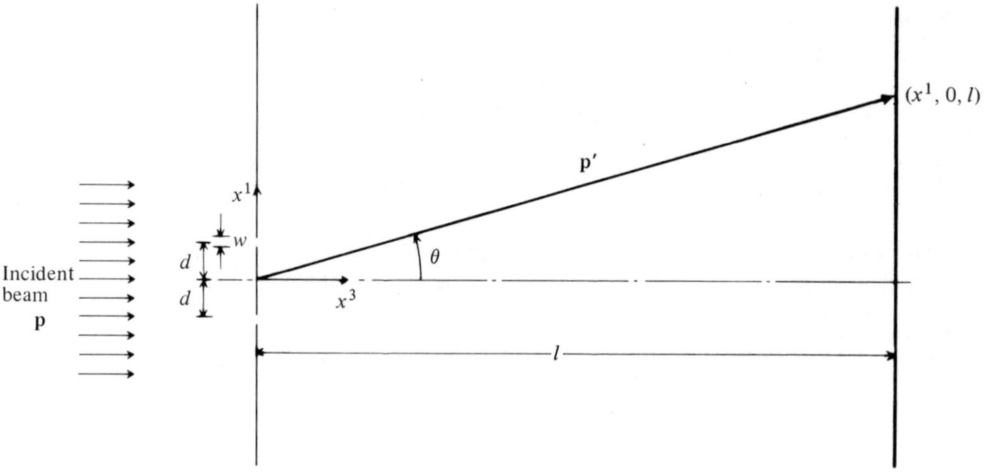

Fig. 9.2. Diagram defining the variables in the analysis of the double-slit diffraction problem. The incident beam has momentum $\mathbf{p}$ and the particle creating the fluorescent spot has momentum $\mathbf{p}'$.

Let the direction of incidence be along the x^3-axis, the plane of the first screen being $x^3 = 0$ and that of the second being $x^3 = l$. The slits are parallel to the x^2-axis and are centered on the lines $x^1 = d$, $x^3 = 0$ and $x^1 = -d$, $x^3 = 0$. We set up polar coordinates in the (x^3x^1)-plane, defining $\rho \equiv \sqrt{(x^1)^2 + (x^3)^2}$, $\theta \equiv \arctan(x^1/x^3)$. Since $d \ll l$, a particle arriving at the second screen will appear to have left the first screen along the line $x^1 = x^3 = 0$ lying halfway between the slits, and hence to a good approximation will travel in the planes $\theta = \text{const}$. If we further assume that a particle cannot be deflected parallel to the x^2-axis upon passing through the slit, the particles will travel approximately on the lines $\theta = \text{const.}$, $y = \text{const}$.

Now a fluorescent spot appears at the point (x^1, x^2, l) on the second screen. This tells us that upon leaving the first screen the particle causing the fluorescence left the first screen with momentum in the direction

$$\hat{\mathbf{p}}' = \frac{\hat{\mathbf{i}}_1 x^1 + \hat{\mathbf{i}}_3 l}{\sqrt{(x^1)^2 + l^2}}.$$

If we assume that energy was conserved as the particle passed through the first screen, its momentum upon leaving the screen was

$$\mathbf{p}' = p\,\frac{\hat{\mathbf{i}}_1 x^1 + \hat{\mathbf{i}}_3 l}{\sqrt{(x^1)^2 + l^2}},$$

which can be written in the form

$$\mathbf{p}' = p(\hat{\mathbf{i}}_1 \sin\theta + \hat{\mathbf{i}}_3 \cos\theta).$$

Now, knowing that this is the particle's momentum upon leaving the screen, we may ask what the probability is that the particle passed through the upper slit at $x^1 = d$, $x^3 = 0$. Evidently the probability amplitude corresponding to the desired probability is

$$\langle \hat{\mathbf{i}}_1 d | \mathbf{p}' \rangle = A e^{i\mathbf{k}' \cdot (\hat{\mathbf{i}}_1 d)} = A e^{ikd\sin\theta}.$$

The probability that it came through the lower slit is related to the amplitude

$$\langle -\hat{\mathbf{i}}_1 d | \mathbf{p}' \rangle = A e^{-ikd\sin\theta}.$$

If the incident beam is uniform, the state described by "the particle passes through the first screen" is given by

$$\frac{1}{\sqrt{2}}(|\hat{\mathbf{i}}_1 d\rangle + e^{i\alpha}|-\hat{\mathbf{i}}_1 d\rangle) = |1\rangle,$$

where the magnitudes of the coefficients of the two states in this linear combination are equal because uniformity of the beam implies equal likelihood of passing through either of the holes. The factor $1/\sqrt{2}$ is for convenience in normalizing the state:

$$\begin{aligned}\langle 1|1\rangle &= \tfrac{1}{2}(\langle \hat{\mathbf{i}}_1 d|\hat{\mathbf{i}}_1 d\rangle + e^{i\alpha}\langle \hat{\mathbf{i}}_1 d| - \hat{\mathbf{i}}_1 d\rangle + e^{-i\alpha}\langle -\hat{\mathbf{i}}_1 d|\hat{\mathbf{i}}_1 d\rangle \\ &\quad + \langle -\hat{\mathbf{i}}_1 d| - \hat{\mathbf{i}}_1 d\rangle) \\ &= \tfrac{1}{2}(\langle \hat{\mathbf{i}}_1 d|\hat{\mathbf{i}}_1 d\rangle + \langle -\hat{\mathbf{i}}_1 d| - \hat{\mathbf{i}}_1 d\rangle) = 1.\end{aligned}$$

This follows from the following properties of our arrangement: First, $|\hat{\mathbf{i}}_1 d\rangle$ and $|-\hat{\mathbf{i}}_1 d\rangle$ are eigenstates of the position operator $\mathfrak{X}$, with

$$\begin{aligned}\mathfrak{X}|\hat{\mathbf{i}}_1 d\rangle &= \hat{\mathbf{i}}_1 d|\hat{\mathbf{i}}_1 d\rangle \\ \mathfrak{X}| - \hat{\mathbf{i}}_1 d\rangle &= -\hat{\mathbf{i}}_1 d| - \hat{\mathbf{i}}_1 d\rangle;\end{aligned}$$

since $\mathfrak{X}$ is Hermitian and the eigenvalues are different, we have

$$\langle \hat{\mathbf{i}}_1 d | -\hat{\mathbf{i}}_1 d \rangle = 0.$$

Second, the spectrum of possible eigenvalues here is finite, $|\hat{\mathbf{i}}_1 d\rangle$ and $|-\hat{\mathbf{i}}_1 d\rangle$ representing the only two possible position states of the particle as it passes through the screen; therefore the normalization of the state is

$$\langle \hat{\mathbf{i}}_1 d | \hat{\mathbf{i}}_1 d \rangle = \langle \hat{\mathbf{i}}_1 d | -\hat{\mathbf{i}}_1 d \rangle = 1.$$

Therefore we have

$$\begin{aligned}\frac{1}{\sqrt{2}}(\langle \hat{\mathbf{i}}_1 d) + e^{-i\alpha}\langle -\hat{\mathbf{i}}_1 d|)|\mathbf{p}\rangle &= \frac{1}{\sqrt{2}}(\langle \hat{\mathbf{i}}_1 d|\mathbf{p}\rangle + e^{-i\alpha}\langle -\hat{\mathbf{i}}_1 d|\mathbf{p}\rangle) \\ &= \frac{1}{\sqrt{2}}A[e^{ikd\sin\theta} + e^{-i(\alpha + kd\sin\theta)}] \\ &= \sqrt{2}Ae^{-i\alpha/2}\cos(kd\sin\theta + \tfrac{1}{2}\alpha)\end{aligned}$$

as the probability amplitude for the observed event, namely that the particle passed through screen 1 and had momentum such that it would cause fluorescence at angle θ from the incident beam direction. The probability, proportional to $\cos^2(kd\sin\theta + \frac{1}{2}\alpha)$, therefore reaches its maxima when

$$kd\sin\theta + \alpha/2 = n\pi, \qquad n \text{ an integer}$$

and its minima when

$$kd\sin\theta + \alpha/2 = (n + \tfrac{1}{2})\pi, \qquad n \text{ an integer.}$$

If we assume the states $|\hat{\mathbf{i}}_1 d\rangle$ and $e^{i\alpha}| -\hat{\mathbf{i}}_1 d\rangle$ are in phase (i.e. $\alpha = 0$), then this leads to the standard double-slit diffraction formulas

$$\sin\theta = n\lambda/2d \qquad \text{for maxima}$$

$$\sin\theta = (n + \tfrac{1}{2})\lambda/2d \qquad \text{for minima}$$

where $2d$ is the separation between slits.

Note that in obtaining this solution we have not used all the data we had to begin with. We used the fluorescing spot on the second screen only to establish the *direction* of $\mathbf{p}'$, and nowhere did we really utilize the separation l between the two screens. Therefore the probability that we obtained is not the probability that a particle passing through the fluorescing point originated at the double slit, but rather it is the probability that a particle moving away from the double slit with momentum $\mathbf{p}'$ came through the double slit.

The main point is that, using the state vectors and the interpretation of the inner product, we have obtained interference effects that must apply to particles.

READING NOTES*

References 22, 23 and 43 are good, especially 43.

Reference 24 has long been considered the fountainhead of quantum theory, although it has also been considered very difficult for students to read. See Chapters 1–3.

Reference 37 is another standard text.

Reference 38 is a fine text on the quantum theory and is well worth consulting. See Volume I.

Reference 42, like most other quantum theory books, attacks the subject from a point of view different from the present one and may provide good supplementary reading for the student.

PROBLEMS

1. Prove that if Ψ_q is an eigenfunction of Q, with eigenvalue q, then

   $\Delta Q \equiv \sqrt{(\Psi_q, (\mathrm{Q} - q)\Psi_q)} = 0,$

   i.e. show that there is no uncertainty or dispersion in the variable Q for the state with wavefunction Ψ_q.
2. Verify Eqs. (9.30) and (9.31).
3. Calculate the current density for a free particle with wave vector $\mathbf{k} = \mathbf{p}/\hbar$.
4. For the Schrödinger equation, carry out a Galilean transformation $\mathbf{x} \to \mathbf{x}' = \mathbf{x} - \mathbf{v}t$ and show that the equation is covariant under this transformation. Also demonstrate the effect of the transformation on the current density.
5. Construct the Hamiltonian for a charged particle under the influence of a pure magnetic field along the x-axis and a pure electric field along the y-axis. Find the current density for the particle.
6. Consider the time-dependent Schrödinger equation for a particle in a real time-independent potential. Show that the real part of the wave function does not satisfy a differential equation that is of first order in the time derivative.
7. Show that the statements of completeness and orthonormality for state vectors are consistent with the corresponding statements for wave functions.
8. Show, for state vectors, that the statements of completeness and orthonormality are consistent with each other.
9. A monochromatic source of spinless particles is placed behind a large screen in which four holes have been punched, as indicated in Fig. 9.3. The holes are at the vertices of a square whose diagonal has length d, and the diameters of the holes are much less than d but much larger than $\hbar/p$, where p is the momentum of the particles. The particles are moving in a direction perpendicular to the plane of the screen. On the other side of the punctured screen is a large photographic film capable of registering the intensity of the particles impinging on any small unit of its surface area; this plane sheet of film is parallel to the punched screen and is a distance $1 \gg d$ from the latter. Using the coordinate system

* Works referred to in these notes will be found in the Bibliography on p. 577.

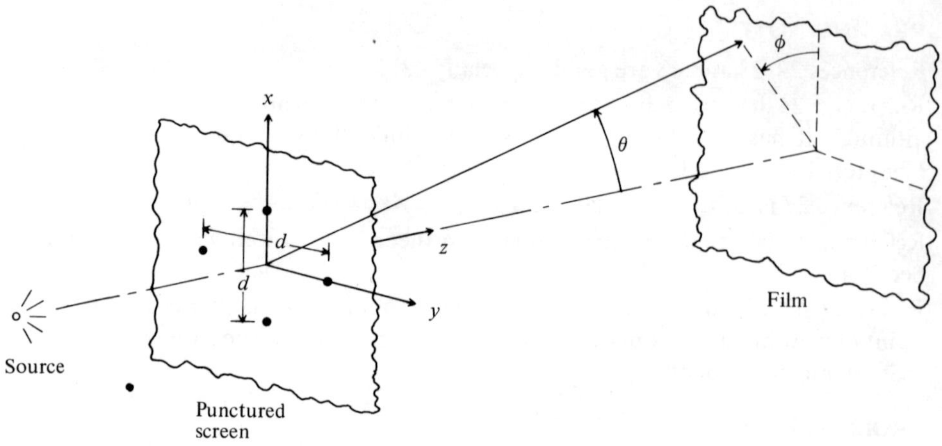

Fig. 9.3. Schematic diagram for the experiment in Problem 9.

indicated in Fig. 9.3 and utilizing the state-vector formalism, find the positions of the maxima and minima of particle intensity recorded on the film.

10. Let the Hamiltonian for a nonrelativistic system be

$$\mathsf{H} = -\frac{\hbar^2}{2m}\nabla^2 + V_1(\mathbf{r}) + iV_2(\mathbf{r})$$

where $V_1(\mathbf{r})$ and $V_2(\mathbf{r})$ are real functions of position.

a) Comment upon the reasonableness of choosing this as a physical operator.

b) Find the probability current density $\mathbf{j}$ for a system having this Hamiltonian, and find the corresponding probability density ρ, in terms of the wave function and appropriate operators. Obtain an expression for

$$\frac{\partial \rho}{\partial t} + \nabla \cdot \mathbf{j}$$

and comment upon the meaning of your result. What is happening to the system if $V_2(\mathbf{r}) > 0$? If $V_2(\mathbf{r}) < 0$?

CHAPTER 10

UNITARY AND HERMITIAN OPERATORS; COMMUTATORS

10.1 PRODUCTS OF OPERATORS; INVERSES

In what follows we shall be dealing frequently with *products* of operators. The product of two operators is defined simply by

$$(\mathfrak{Q}\mathfrak{R})|1\rangle = \mathfrak{Q}\mathfrak{R}|1\rangle = \mathfrak{Q}(\mathfrak{R}|1\rangle), \tag{10.1}$$

for all state vectors $|1\rangle$; the product means simply that the operators are applied in succession. It is not true that this is generally a commutative procedure. Indeed, that it is not generally commutative is central to some of the most interesting features of quantum theory. When two operators, say $\mathfrak{Q}$ and $\mathfrak{R}$, don't commute, $\mathfrak{Q}\mathfrak{R}|1\rangle \neq \mathfrak{R}\mathfrak{Q}|1\rangle$ for some state vector $|1\rangle$, i.e. $\mathfrak{Q}\mathfrak{R} \neq \mathfrak{R}\mathfrak{Q}$, we say they have a nonvanishing *commutator*. The commutator of $\mathfrak{Q}$ and $\mathfrak{R}$ is defined by

$$[\mathfrak{Q}, \mathfrak{R}] \equiv \mathfrak{Q}\mathfrak{R} - \mathfrak{R}\mathfrak{Q}, \tag{10.2}$$

which has the obvious property

$$[\mathfrak{Q}, \mathfrak{R}] = -[\mathfrak{R}, \mathfrak{Q}], \tag{10.3}$$

and the less obvious property

$$\big[[\mathfrak{Q}, \mathfrak{R}], \mathfrak{S}\big] + \big[[\mathfrak{S}, \mathfrak{Q}], \mathfrak{R}\big] + \big[[\mathfrak{R}, \mathfrak{S}], \mathfrak{Q}\big] = 0; \tag{10.4}$$

the latter is known as the *Jacobi identity*.

The Hermitian conjugate of a product of operators is the product of their Hermitian conjugates, in the reverse order. We show this by calculating the Hermitian conjugate of $\mathfrak{Q}\mathfrak{R}|1\rangle$ for an arbitrary state vector $|1\rangle$:

$$\begin{aligned}[(\mathfrak{Q}\mathfrak{R})|1\rangle]^\dagger &= [\mathfrak{Q}(\mathfrak{R}|1\rangle)]^\dagger = (\mathfrak{R}|1\rangle)^\dagger\mathfrak{Q}^\dagger \\ &= (\langle 1|\mathfrak{R}^\dagger)\mathfrak{Q}^\dagger = \langle 1|\mathfrak{R}^\dagger\mathfrak{Q}^\dagger.\end{aligned}$$

On the other hand,

$$[(\mathfrak{Q}\mathfrak{R})|1\rangle]^\dagger = \langle 1|(\mathfrak{Q}\mathfrak{R})^\dagger,$$

and hence

$$\langle 1|[\mathfrak{R}^\dagger\mathfrak{Q}^\dagger - (\mathfrak{Q}\mathfrak{R})^\dagger] = 0$$

or

$$(\mathfrak{Q}\mathfrak{R})^\dagger = \mathfrak{R}^\dagger\mathfrak{Q}^\dagger. \tag{10.5}$$

The inverse, $\mathfrak{Q}^{-1}$, of an operator $\mathfrak{Q}$ is the unique operator such that

$$(\mathfrak{Q}^{-1})\mathfrak{Q} = \mathfrak{Q}(\mathfrak{Q}^{-1}) = 1,$$

provided such an operator exists. From the definition it follows that when $\mathfrak{Q}^{-1}$ and $\mathfrak{R}^{-1}$ exist,

$$(\mathfrak{Q}\mathfrak{R})^{-1} = \mathfrak{R}^{-1}\mathfrak{Q}^{-1}. \tag{10.6}$$

Another proof of this arises from an equivalent definition of the inverse of an operator $\mathfrak{Q}$, namely that whenever

$$|2\rangle = \mathfrak{Q}|1\rangle,$$

$\mathfrak{Q}$ is the operator such that

$$|1\rangle = \mathfrak{Q}^{-1}|2\rangle$$

whenever such an inverse exists. Then if $|4\rangle = \mathfrak{Q}\mathfrak{R}|3\rangle$ we have

$$\mathfrak{Q}^{-1}|4\rangle = \mathfrak{R}|3\rangle$$

and

$$\mathfrak{R}^{-1}\mathfrak{Q}^{-1}|4\rangle = |3\rangle$$

i.e.

$$(\mathfrak{Q}\mathfrak{R})^{-1} = \mathfrak{R}^{-1}\mathfrak{Q}^{-1}$$

Now suppose that $\mathfrak{T}$ is an operator that possesses an inverse (i.e. a *nonsingular* operator, a *singular* one being an operator that does not have an inverse), and let its effect on a state be denoted by a prime, e.g.

$$|3'\rangle = \mathfrak{T}|3\rangle. \tag{10.7}$$

This means that if the state $\mathfrak{Q}|3\rangle$ is transformed by $\mathfrak{T}$, that is, if we let $\mathfrak{T}$ operate on $\mathfrak{Q}|3\rangle$, then

$$\mathfrak{T}(\mathfrak{Q}|3\rangle) = \mathfrak{T}\mathfrak{Q}\mathfrak{T}^{-1}\mathfrak{T}|3\rangle = \mathfrak{T}\mathfrak{Q}\mathfrak{T}^{-1}|3'\rangle. \tag{10.8}$$

Hence if we let

$$\mathfrak{Q}' \equiv \mathfrak{T}\mathfrak{Q}\mathfrak{T}^{-1} \tag{10.9}$$

we see that operating on $|3'\rangle$ with $\mathfrak{Q}'$ has the same effect as operating on $|3\rangle$ with $\mathfrak{Q}$ and then transforming the result with $\mathfrak{T}$. Another way of putting it is to say that when

$$|3\rangle \to |3'\rangle = \mathfrak{T}|3\rangle$$

we have

$$\mathfrak{Q}|3\rangle - \mathfrak{Q}'|3'\rangle = (\mathfrak{T}\mathfrak{Q}\mathfrak{T}^{-1})|3'\rangle = \mathfrak{T}\mathfrak{Q}|3\rangle.$$

Here $\mathfrak{Q}' = \mathfrak{T}\mathfrak{Q}\,\mathfrak{T}^{-1}$ is called the *similarity transformation* of $\mathfrak{Q}$ by $\mathfrak{T}$.

An important property of similarity transformations is that they preserve eigenvalues. For example, let

$$\mathfrak{Q}|q, r, \ldots\rangle = q|q, r, \ldots\rangle, \tag{10.10}$$

and then let the states all be transformed by $\mathfrak{T}$. It follows that

$$\begin{aligned} \mathfrak{T}\mathfrak{Q}|q, r, \ldots\rangle &= \mathfrak{T}\mathfrak{Q}\mathfrak{T}^{-1}(\mathfrak{T}|q, r, \ldots\rangle) \\ &= \mathfrak{Q}'(\mathfrak{T}|q, r, \ldots\rangle) \\ &= q(\mathfrak{T}|q, r, \ldots\rangle). \end{aligned} \tag{10.11}$$

Hence $\mathfrak{T}|q, r, \ldots\rangle$ is an eigenstate of $\mathfrak{Q}'$ with the same eigenvalues $\{q\}$ as those of $\mathfrak{Q}$, $\{|q, r, \ldots\rangle\}$.

10.2 REPRESENTATIONS AND UNITARY TRANSFORMATIONS

A complete set of linearly independent state vectors is often referred to as a *basis* for the space of state vectors allowed to a physical system. A complete orthonormal set is called an *orthonormal basis*; or a representation. There may be many acceptable orthonormal bases—in fact for systems that are at all interesting there generally are infinitely many—and it may be useful to shift from one basis to another. However, a change of basis is simply a shift in the mathematical mode of expression for a problem and cannot be allowed to change the physics involved; hence, whatever changes of basis we use must not be physically observable and therefore need not be represented by Hermitian operators.

For example, let $\mathfrak{Q}, \mathfrak{R}, \ldots$ be a complete set of observable operators, and let $\mathfrak{M}, \mathfrak{N}, \ldots$ be another such set, at least one operator of the latter being distinct from any one operator in the former set. The eigenstates $\{|q, r, \ldots\rangle\}$ of the first set then form a basis as do those of the second, $\{|m, n, \ldots\rangle\}$, but the two sets differ in some respect. Both sets of states are complete so that any state of the second set can be expressed as a linear combination of elements of the first set; in fact, the completeness relation for the first set yields

$$|m, n, \ldots\rangle = 1|m, n, \ldots\rangle = \underset{q,r,\ldots}{\mathsf{S}} |q, r, \ldots\rangle\langle q, r, \ldots|m, n, \ldots\rangle. \tag{10.12}$$

The completeness of both sets implies that a one-to-one correspondence between elements of the respective sets can be established. We seek an operator that will express such a one-to-one correspondence, and which at the same time will preserve inner products.

That an operator $\mathfrak{U}$ preserves inner products means that if for some arbitrary state $|3\rangle$, $\mathfrak{U}$ converts $|3\rangle$ into $|3'\rangle$, $|3'\rangle = \mathfrak{U}|3\rangle$,

then

$$\langle 2'|3'\rangle = \langle 2|\mathfrak{U}^\dagger\mathfrak{U}|3\rangle = \langle 2|3\rangle.$$

Since $|2\rangle$ and $|3\rangle$ are arbitrary, it follows that

$$\mathfrak{U}^\dagger\mathfrak{U} = 1.$$

This being so, we may write

$$\langle 2'|3'\rangle = \langle 2|\mathfrak{U}^\dagger\mathfrak{U}1|3\rangle = \langle 2|\mathfrak{U}^\dagger\mathfrak{U}\mathfrak{U}^\dagger\mathfrak{U}|3\rangle = \langle 2'|\mathfrak{U}\mathfrak{U}^\dagger|3'\rangle$$

and since $|2'\rangle$ and $|3'\rangle$ also are arbitrary,

$$\mathfrak{U}\mathfrak{U}^\dagger = 1.$$

Hence

$$\mathfrak{U}\mathfrak{U}^\dagger = \mathfrak{U}^\dagger\mathfrak{U} = 1 \tag{10.13}$$

is required, or, by the definition of $\mathfrak{U}^{-1}$,

$$\mathfrak{U}^\dagger = \mathfrak{U}^{-1}. \tag{10.14}$$

Operators that preserve the inner product, or equivalently, that satisfy (10.14) are called *unitary* operators.

Hence we want to obtain a unitary change of basis. To do this, we utilize the possibility of establishing a one-to-one correspondence between basis sets $\{|q, r, \ldots\rangle\}$ and $\{|m, n, \ldots\rangle\}$. Let $\{\mathbf{t}\}$ be a set of numbers (or ordered n-tuples of numbers, if necessary) that are in one-to-one correspondence with the elements of $\{|m, n, \ldots\rangle\}$ and with $\{|q, r, \ldots\rangle\}$, and let the state of the former set corresponding to $\mathbf{t}$ be labeled

$$m_{\mathbf{t}}, n_{\mathbf{t}}, \ldots\rangle,$$

while the state of the latter set corresponding to $\mathbf{t}$ is written

$$|q_{\mathbf{t}}, r_{\mathbf{t}}, \ldots\rangle.$$

Then the set of labels $\{\mathbf{t}\}$ can be used to represent the sets of labels $\{(q, r, \ldots)\}$ and $\{(m, n, \ldots)\}$ through the correspondence

$$\mathbf{t} \leftrightarrow (m_{\mathbf{t}}, n_{\mathbf{t}}, \ldots) \leftrightarrow (q_{\mathbf{t}}, r_{\mathbf{t}}, \ldots).$$

Now we claim that the operator

$$\mathfrak{U} = \underset{\mathbf{t}}{\mathsf{S}}\, |m_{\mathbf{t}}, n_{\mathbf{t}}, \ldots\rangle\langle q_{\mathbf{t}}, r_{\mathbf{t}}, \ldots| \tag{10.15}$$

carries each element of the set $\{|q, r, \ldots\rangle\}$ into exactly one element of $\{|m, n, \ldots\rangle\}$

and is unitary. First, note that

$$\mathfrak{U}|q, r, \ldots\rangle = \mathop{\mathsf{S}}_{\mathbf{t}} |m_{\mathbf{t}}, n_{\mathbf{t}}, \ldots\rangle\langle q_{\mathbf{t}}, r_{\mathbf{t}}, \ldots|q, r, \ldots\rangle$$
$$= \mathop{\mathsf{S}}_{\mathbf{t}} |m_{\mathbf{t}}, n_{\mathbf{t}}, \ldots\rangle\delta(q_{\mathbf{t}} - q)\delta(r_{\mathbf{t}} - r)\ldots$$

Thus only one term in the sum survives, namely that term for which $\mathbf{t}$ corresponds to the labels $(q, r, \ldots)$ designating the original state $|q, r, \ldots\rangle$; let this $\mathbf{t}$ be designated $\mathbf{t}(q, r, \ldots)$. Then

$$\mathfrak{U}|q, r, \ldots\rangle = |m_{\mathbf{t}(q,r,\ldots)}, n_{\mathbf{t}(q,r,\ldots)}, \ldots\rangle.$$

Furthermore, $\mathfrak{U}$ is unitary, since

$$\begin{aligned}\mathfrak{U}^{\dagger} &= \mathop{\mathsf{S}}_{\mathbf{t}} (|m_{\mathbf{t}}, n_{\mathbf{t}}, \ldots\rangle\langle q_{\mathbf{t}}, r_{\mathbf{t}}, \ldots|)^{\dagger} \\ &= \mathop{\mathsf{S}}_{\mathbf{t}} |q_{\mathbf{t}}, r_{\mathbf{t}}, \ldots\rangle\langle m_{\mathbf{t}}, n_{\mathbf{t}}, \ldots|\end{aligned} \tag{10.16}$$

and hence

$$\mathfrak{U}^{+}\mathfrak{U} = \mathop{\mathsf{S}}_{\mathbf{t}}\mathop{\mathsf{S}}_{\mathbf{t}'} |q_{\mathbf{t}}, r_{\mathbf{t}}, \ldots\rangle\langle m_{\mathbf{t}}, n_{\mathbf{t}}, \ldots|m_{\mathbf{t}'}, n_{\mathbf{t}'} \ldots\rangle\langle q_{\mathbf{t}'}, r_{\mathbf{t}'}, \ldots|$$
$$= \mathop{\mathsf{S}}_{\mathbf{t}}\mathop{\mathsf{S}}_{\mathbf{t}'} |q_{\mathbf{t}}, r_{\mathbf{t}}, \ldots\rangle\delta(\mathbf{t} - \mathbf{t}')\langle q_{\mathbf{t}'}, r_{\mathbf{t}'}, \ldots|$$
$$= \mathop{\mathsf{S}}_{\mathbf{t}} |q_{\mathbf{t}}, r_{\mathbf{t}}, \ldots\rangle\langle q_{\mathbf{t}}, r_{\mathbf{t}}, \ldots| = 1,$$

and similarly for $\mathfrak{U}\mathfrak{U}^{\dagger}$. Here we have used

$$\langle m_{\mathbf{t}}, n_{\mathbf{t}}, \ldots|m_{\mathbf{t}'}, n_{\mathbf{t}'}, \ldots\rangle = \delta(m_{\mathbf{t}} - m_{\mathbf{t}'})\delta(n_{\mathbf{t}} - n_{\mathbf{t}'}) = \delta(\mathbf{t} - \mathbf{t}'). \tag{10.17}$$

(Why is this equation correct?)

This operator $\mathfrak{U}$ *diagonalizes* $\mathfrak{Q}, \mathfrak{R}, \ldots$ in the basis $\{|m, n, \ldots\rangle\}$. Thus if $\mathbf{t}$ designates a state $|m_{\mathbf{t}}, n_{\mathbf{t}}, \ldots\rangle$, we have

$$|q_{\mathbf{t}}, r_{\mathbf{t}}, \ldots\rangle = \mathfrak{U}^{\dagger}|m_{\mathbf{t}}, n_{\mathbf{t}}, \ldots\rangle,$$

so that

$$\mathfrak{Q}|q_{\mathbf{t}}, r_{\mathbf{t}}, \ldots\rangle = q_{\mathbf{t}}|q_{\mathbf{t}}, r_{\mathbf{t}}, \ldots\rangle$$

becomes

$$\mathfrak{Q}\mathfrak{U}^{\dagger}|m_{\mathbf{t}}, n_{\mathbf{t}}, \ldots\rangle = q_{\mathbf{t}}\mathfrak{U}^{\dagger}|m_{\mathbf{t}}, n_{\mathbf{t}}, \ldots\rangle, \tag{10.18}$$

or

$$\mathfrak{U}\mathfrak{Q}\mathfrak{U}^{\dagger}|m_{\mathbf{t}}, n_{\mathbf{t}}, \ldots\rangle = q_{\mathbf{t}}|m_{\mathbf{t}}, n_{\mathbf{t}}, \ldots\rangle.$$

Hence the operator

$$\mathfrak{Q}' = \mathfrak{U}\mathfrak{Q}\mathfrak{U}^{\dagger} \tag{10.19}$$

is diagonal in the basis $|m, n, \ldots\rangle$ with eigenvalues $\{q\}$; the state $|m_t, n_t, \ldots\rangle$ is an eigenstate of $\mathfrak{Q}'$ with eigenvalue q_t!

Therefore, for any representation, a unitary transformation exists whereby a given Hermitian operator can be diagonalized. On the other hand, if we wish to leave the operator, say $\mathfrak{Q}$, alone, we can find a unitary operator that will transform the given representation into one in which $\mathfrak{Q}$ is diagonal, as Eq. (10.18) shows. This is an extremely important result.

It should be observed that unitary transformations preserve the Hermitian property of Hermitian operators. This must be so because inner products are preserved by such transformations, and Hermitian operators are defined in terms of inner products. We can see this directly by noting that

$$(\mathfrak{U}\mathfrak{Q}\mathfrak{U}^\dagger)^\dagger = (\mathfrak{U}^\dagger)^\dagger\mathfrak{Q}^\dagger\mathfrak{U}^\dagger = \mathfrak{U}\mathfrak{Q}^\dagger\mathfrak{U}^\dagger;$$

hence if $\mathfrak{Q}^\dagger = \mathfrak{Q}$, we have

$$(\mathfrak{U}\mathfrak{Q}\mathfrak{U}^\dagger)^\dagger = \mathfrak{U}\mathfrak{Q}^\dagger\mathfrak{U}^\dagger = \mathfrak{U}\mathfrak{Q}\mathfrak{U}^\dagger,$$

or

$$\mathfrak{Q}'^\dagger = \mathfrak{Q}'.$$

10.3 OPERATORS EXPRESSED IN TERMS OF STATE VECTORS

Since we were able to express the unitary transformation operators in terms of sums over ket–bra pairs, it may be expected that a general operator may be expressible in terms of such sums. This in fact is true, and all that is needed to show it is the completeness relation. If $\mathfrak{Q}$ is any operator and $\{|q, r, \ldots\rangle\}$ a complete set of its eigenstates, then

$$\begin{aligned} \mathfrak{Q} = \mathfrak{Q}\cdot 1 &= \mathfrak{Q} \mathop{\mathrm{S}}_{q,r,\ldots} |q, r, \ldots\rangle\langle q, r, \ldots| \\ &= \mathop{\mathrm{S}}_{q,r,\ldots} |q, r, \ldots\rangle q\langle q, r, \ldots| \end{aligned} \tag{10.20}$$

expresses $\mathfrak{Q}$ in terms of the basis $\{|q, r, \ldots\rangle\}$.

One important example is that of the *projection operators*. These are a special case of the general class of operators

$$\Pi_{12} = |1\rangle\langle 2|. \tag{10.21}$$

Applying Π_{12} to an arbitrary state vector $|3\rangle$ yields simply

$$\Pi_{12}|3\rangle = |1\rangle\langle 2|3\rangle;$$

i.e. it replaces $|3\rangle$ with $|1\rangle$ multiplied by the complex number $\langle 2|3\rangle$. In particular,

$$\Pi_{12}|2\rangle = |1\rangle.$$

The operator $\Pi_{12} = |1\rangle\langle 2|$ has two eigenvalues, namely $\langle 2|1\rangle$ and 0 and its eigenstates consist of a set containing $|1\rangle$ together with a set of linearly independent states orthogonal to $|2\rangle$. Let $\{|k'\rangle\}$, $k' = 1', 2', \ldots$ be such a set, with $|1'\rangle \equiv |1\rangle$, $\langle 2|k'\rangle = 0$ for $k' \neq 1$; then

$$\Pi_{12}|k'\rangle = |1\rangle\langle 2|k'\rangle = \begin{cases} \langle 2|1\rangle|1\rangle & \text{if } k' = 1' = 1 \\ 0 & \text{otherwise.} \end{cases} \tag{10.22}$$

Clearly, if $\langle 2|1\rangle = 0$, Π_{12} has only 0 as an eigenvalue, and its effect is to annihilate every state in its set of eigenstates:

$$\Pi_{12}|k'\rangle = 0 \qquad \text{for all states } |k'\rangle \qquad \text{if } \langle 2|1\rangle = 0.$$

Note that

$$\Pi_{12}\Pi_{34} = \langle 2|3\rangle \Pi_{14} \tag{10.23}$$

and

$$\Pi_{12}^{+} = \Pi_{21}. \tag{10.24}$$

Hence Π_{12} is Hermitian if and only if $|1\rangle = |2\rangle$. On the other hand,

$$\Pi_{12}^{+}\Pi_{12} = \Pi_{21}\Pi_{12} = |2\rangle\langle 2| = \Pi_{22}$$

is Hermitian; this shows, however, that Π_{12} is unitary if and only if $|1\rangle = |2\rangle$ and exactly one state is available to the system. (Why?)

The Hermitian operators Π_{11} are the elementary projection operators; they convert any state, say $|3\rangle$, into the state labeling the projection operator, multiplied by the inner product of the two states:

$$\Pi_{11}|3\rangle = \langle 1|3\rangle|1\rangle.$$

Thus Π_{11} has two eigenvalues, $\langle 1|1\rangle$ and 0, its eigenstates being $|1\rangle$ and all states orthogonal to $|1\rangle$, respectively. Also,

$$\Pi_{11}^{2} = \Pi_{11}\Pi_{11} = \Pi_{11}; \tag{10.25}$$

that is, Π_{11} is *idempotent.*

Since the double subscripts on the projection operators are redundant, we write

$$\Pi_{(q,r\ldots)} = |q, r, \ldots\rangle\langle q, r, \ldots| \tag{10.26}$$

for the projection operator corresponding to the eigenstate $|q, r, \ldots\rangle$ of the operators $\mathfrak{Q}, \mathfrak{R}, \ldots$, for which

$$\Pi_{(q,r,\ldots)}|q', r', \ldots\rangle = \delta(q - q')\delta(r - r')\cdots|q, r, \ldots\rangle. \tag{10.27}$$

Thus $\Pi_{(q,r,\ldots)}$ annihilates every eigenstate of these operators except the one with eigenvalues $(q, r, \ldots)$; we say that it projects any state onto the state $|q, r, \ldots\rangle$, and the eigenstates of $\mathfrak{Q}, \mathfrak{R}, \ldots$ that have any eigenvalue different from those given in the collection $(q, r, \ldots)$ have no projection along $|q, r, \ldots\rangle$—that is, they are orthogonal to it.

Next, let us form

$$\Pi^{(q)}_{(r,s,\ldots)} \equiv \mathsf{S}_{q}\, \Pi_{(q,r,s,\ldots)}. \tag{10.28}$$

This projection operator projects any state onto a subspace spanned by the eigenstates of $\mathfrak{Q}$:

$$\Pi^{(q)}_{(r,s,\ldots)}|1\rangle = \mathsf{S}_{q}\, \Pi_{(q,r,s,\ldots)}|1\rangle = \mathsf{S}_{q}\, |q, r, s, \ldots\rangle\langle q, r, s, \ldots|1\rangle.$$

Also,

$$\Pi^{(q,r)}_{(s,\ldots)}|1\rangle = \mathsf{S}_{q,r}\, \Pi_{(q,r,s,\ldots)}|1\rangle = \mathsf{S}_{q,r}\, |q, r, s, \ldots\rangle\langle q, r, s, \ldots|1\rangle.$$

Finally,

$$1 = \mathsf{S}_{q,r,s,\ldots}\, |q, r, s, \ldots\rangle\langle q, r, s, \ldots| = \mathsf{S}_{q,r,s,\ldots}\, \Pi_{(q,r,s,\ldots)} \tag{10.29}$$

projects the whole space of physical states onto the whole space and is recognizable as an expression of the completeness relation.

Note that the eigenstates of $\Pi^{(q)}_{(r,s,\ldots)}$ are all the states $\{|q, r, s, \ldots\rangle\}$ and those with eigenvalue 0 are the ones for which at least one of the eigenvalues of $\mathfrak{R}$, $\mathfrak{S}, \ldots$ is not in the ordered set $(r, s, \ldots)$; the eigenvalues of $\mathfrak{Q}$ here are irrelevant. For $\Pi^{(q,r)}_{(s,\ldots)}$, the eigenvalues of $\mathfrak{Q}$ and $\mathfrak{R}$ are irrelevant, etc. Finally, the eigenvalues of the identity operator are obviously all unity:

$$1|3\rangle = \mathsf{S}_{q,r,s,\ldots}\, \Pi_{(q,r,s,\ldots)}|3\rangle = |3\rangle.$$

This means that the expectation value of this last operator is *always* 1. On the other hand, the expectation value for a projection operator that does *not* include *all* eigenvalues for *all* the operators in a complete set of compatible observables such as $\mathfrak{Q}$, $\mathfrak{R}, \ldots$ will have expectation values less than one for some states, and for some states its expectation value will actually vanish. Thus it is *vital* that all eigenvalues of a complete set of compatible observable operators be included in the summation in the completeness relation.

10.4 COMPATIBLE OPERATORS; THE TENSOR PRODUCT

When is a set of operators compatible? It is compatible when all its elements commute with one another; thus compatible operators are mutually commuting operators. By saying that they are compatible, we mean that there is a basis, or representation, in which all the eigenvalue problems corresponding to the different operators can be solved at once; there exists a basis that consists of eigenstates of all the operators.

Consider the two eigenvalue problems

$$\mathfrak{L}|1\rangle = \lambda|1\rangle,$$
$$\mathfrak{M}|1\rangle = \mu|1\rangle; \tag{10.30}$$

we wish to find the λ and μ for which solutions to these equations exist, and the states $|\lambda(l)\rangle$ and $|\mu(m)\rangle$ (the extra indices l and m are included to take care of the cases in which several linearly independent eigenstates correspond to a single eigenvalue) that solve the equations. Can we solve them simultaneously? Are there states $|\lambda(l)\mu(m)\rangle$ that are solutions to both equations?

We assume first that simultaneous solutions exist, namely $\{|\lambda(l)\mu(m)\rangle\}$; then

$$\mathfrak{L}\mathfrak{M}|\lambda(l)\mu(m)\rangle = \mathfrak{L}\mu|\lambda(l)\mu(m)\rangle = \mu\mathfrak{L}|\lambda(l)\mu(m)\rangle = \mu\lambda|\lambda(l)\mu(m)\rangle = \lambda\mathfrak{M}|\lambda(l)\mu(m)\rangle$$
$$= \mathfrak{M}\lambda|\lambda(l)\mu(m)\rangle = \mathfrak{M}\mathfrak{L}|\lambda(l)\mu(m)\rangle. \tag{10.31}$$

Since $|\lambda(l)\mu(m)\rangle$ is an arbitrary solution, this implies that

$$\mathfrak{L}\mathfrak{M} - \mathfrak{M}\mathfrak{L} = [\mathfrak{L}, \mathfrak{M}] = 0.$$

Therefore such solutions exist only for compatible (commuting) operators.

Furthermore, whenever $[\mathfrak{L}, \mathfrak{M}] = 0$, such solutions can be found. Indeed, let $[\mathfrak{L}, \mathfrak{M}] = 0$ and solve the first eigenvalue problem, obtaining solutions $|\lambda(l, \nu)\rangle$; then

$$\mathfrak{L}\mathfrak{M}|\lambda(l, \nu)\rangle = \mathfrak{M}\mathfrak{L}|\lambda(l, \nu)\rangle$$
$$= \mathfrak{M}\lambda|\lambda(l, \nu)\rangle = \lambda\mathfrak{M}|\lambda(l, \nu)\rangle \tag{10.32}$$

Hence $\mathfrak{M}|\lambda(l, \nu)\rangle$ is a solution whenever $|\lambda(l, \nu)\rangle$ is a solution. That being the case, we now can find a unitary operator that will diagonalize $\mathfrak{M}$ (we can always do that, anyway) without undiagonalizing $\mathfrak{L}$. For a given value of λ, say λ_0, we solve

$$\mathfrak{M}|\lambda_0(l, \nu)\rangle = \mu|\lambda_0(l, \nu)\rangle,$$

here dealing only with states in the subspace defined by the projection operator $\Pi_{\lambda_0}^{(l,\nu)} = \mathsf{S}_{l,\nu}\, |\lambda_0(l, \nu)\rangle\langle\lambda_0(l, \nu)|$. The solutions to this problem we denote $|\lambda_0(l)\mu(m)\rangle$; they still satisfy $\mathfrak{L}|\lambda_0(l)\mu(m)\rangle = \lambda_0|\lambda_0(l)\mu(m)\rangle$. We can do this for each value of λ, thereby obtaining the complete set of solutions $\{|\lambda(l)\mu(m)\}$ to *both* eigenvalue problems.

Hence a set of eigenstates of two operators can be obtained if and only if those operators mutually commute. By an extension of these arguments, a set of eigenstates of three operators, say $\mathfrak{K}$, $\mathfrak{L}$ and $\mathfrak{M}$, can be obtained if and only if

$$[\mathfrak{K}, \mathfrak{L}] = 0, \qquad [\mathfrak{L}, \mathfrak{M}] = 0, \qquad \text{and} \qquad [\mathfrak{M}, \mathfrak{K}] = 0.$$

For the general case, simultaneous eigenstates can be formed if and only if all operators commute with one another Later we shall see that when two operators $\mathfrak{L}$ and $\mathfrak{M}$ commute, the uncertainty product $\Delta L\,\Delta M$ for the variables corresponding to $\mathfrak{L}$ and $\mathfrak{M}$ vanishes. Thus when operators are compatible, there is no minimum value for the uncertainty product associated with them; they can be measured simultaneously with arbitrarily good accuracy, in principle.

We discuss now a particular kind of situation in which two operators are simultaneously diagonable, or simultaneously measurable. It may happen that a physical system is composed of several quite independent subsystems. In that case one operator, say $\mathfrak{L}_1$, may be applicable to one of the subsystems but not to the others, and $\mathfrak{L}_2$ may apply only in another subsystem. Let us solve the eigenvalue problem for $\mathfrak{L}_2$, obtaining

$$\mathfrak{L}_2|\lambda_2(l_2)\rangle = \lambda_2|\lambda_2(l_2)\rangle, \tag{10.33}$$

where $\{|\lambda_2(l_2)\rangle\}$ is assumed to be a set of basis states for system 2 only, the system in which $\mathfrak{L}_2$ operates. Now we should expect $\mathfrak{L}_1|\lambda_2(l_2)\rangle = 0$.

To see this, consider a system consisting of two noninteracting particles, particle 1 and particle 2. If these two particles are distinguishable but fail to interact with one another (say an idealized neutron, with no electric charge density, and an idealized electron, with no capacity to interact with nuclear matter), then a device that will measure the energy of particle 1, but that does not recognize particle 2 (and hence operates only in the state space of particle 1) will indicate zero energy when utilized to attempt a measurement of the energy of particle 2. If $\mathfrak{L}_1$ is the energy operator for particle 1, then $\mathfrak{L}_1|\lambda_2(l_2)\rangle = 0$. Similarly, $\mathfrak{L}_2|\lambda_1(l_1)\rangle = 0$ if $\{|\lambda_1(l_1)\rangle\}$ are state vectors for particle 1.

How, then, do we construct a space for the whole system? We form the *tensor product* of the spaces of the subsystems, writing, for example,

$$|\lambda_1(l_1);\,\lambda_2(l_2)\rangle \equiv |\lambda_1(l_1)\rangle|\lambda_2(l_2)\rangle \tag{10.34}$$

as the tensor product of a pair of basis vectors. We write

$$\{|\lambda_1(l_1);\,\lambda_2(l_2)\rangle\} = \{|\lambda_1(l_1)\rangle\} \otimes \{|\lambda_2(l_2)\rangle\}$$

for the product space representing the overall system. In addition, we establish the rule that for any operator $\mathfrak{M}$ that is a linear combination of operators, each of which applies exclusively to one or the other of the two systems,

$$\mathfrak{M}|\lambda_1(l_1)\rangle|\lambda_2(l_2)\rangle = [\mathfrak{M}|\lambda_1(l_1)\rangle]|\lambda_2(l_2)\rangle + |\lambda_1(l_1)\rangle\mathfrak{M}|\lambda_2(l_2)\rangle. \tag{10.35}$$

Note that this rule is analogous to the rule for taking the first derivative of the product of two functions; for this reason, our operators are sometimes called *derivations* by mathematicians. This rule yields

$$\begin{aligned}\mathfrak{L}_1|\lambda_1(l_1);\,\lambda_2(l_2)\rangle &= [\mathfrak{L}_1|\lambda_1(l_1)\rangle]|\lambda_2(l_2)\rangle + |\lambda_1(l_1)\rangle\mathfrak{L}_1|\lambda_2(l_2)\rangle \\ &= \lambda_1|\lambda_1(l_1)\rangle|\lambda_2(l_2)\rangle = \lambda_1|\lambda_1(l_1);\,\lambda_2(l_2)\rangle, \end{aligned} \tag{10.36}$$

as we should expect.

What now, if the different independent subsystems begin to interact with one another, e.g. if the neutron becomes less "ideal" and has a magnetic moment with which the electron can interact? Then it may be that "turning on" the interaction is reflected in $\mathfrak{L}_1$ by the appearance of a term involving subsystem 2:

$$\mathfrak{L}_1 \to \mathfrak{L}_1 + \kappa_1 \mathfrak{N}_{12} = \mathfrak{L}'_1$$

where κ_1 is a parameter representing the strength of the interaction; for $\kappa_1 = 0$ the interaction is "turned off" and the systems are independent, whereas for $\kappa_1 = 1$ the interaction is "all the way on" and the systems are interacting as strongly as the operator $\mathfrak{N}_{12}$ will allow them to. Similarly we may have

$$\mathfrak{L}_2 \to \mathfrak{L}_2 + \kappa_2 \mathfrak{N}_{21} = \mathfrak{L}'_2.$$

If $[\mathfrak{L}'_1, \mathfrak{L}'_2] \neq 0$, then these operators cannot be diagonalized simultaneously and our tensor product of spaces may seem inappropriate for representing the total system.

On the other hand, the states (10.34) do form a complete set of states for the physical system, and hence any state is a linear superposition of them. Hence our tensor product space remains quite adequate, although the *basis* states may be somewhat different from the observed states of the system. All we need to do is to find a unitary transformation $\mathfrak{U}$ that diagonalizes a complete set of commuting observables (which now cannot contain both $\mathfrak{L}'_1$ and $\mathfrak{L}'_2$ but perhaps will contain one of them). This will produce a new set of states

$$\{\mathfrak{U}|\lambda_1(l_1); \lambda_2(l_2)\rangle\}$$

that are eigenstates of our new complete set of commuting observables. If we have chosen the set of observables carefully, this new set of state vectors will form a basis, the elements of which should correspond to physically observable states of the system.

10.5 UNITARY OPERATORS: IDENTIFICATION OF SOME PHYSICAL OPERATORS WITH GENERATORS OF TRANSFORMATIONS

In order to discuss more fully the compatibility of operators, we should investigate some properties of unitary operators. A rather revealing analogy exists between operators in a unitary space and complex numbers; the analogy becomes a correspondence when we realize that the eigenvalues of operators are complex numbers. Hermitian operators, with real eigenvalues, are analogs of numbers on the real line, whereas unitary operators, having eigenvalues of modulus unity, are analogs of numbers on the unit circle. An operator that is both unitary and Hermitian therefore is the analog of the numbers ± 1.

Using the symbolic notation $e^{\mathfrak{Q}}$ to refer to the infinite sum

$$e^{\mathfrak{Q}} \equiv \sum_{n=0}^{\infty} \frac{1}{n!}\mathfrak{Q}^n = 1 + \mathfrak{Q} + \frac{1}{2}\mathfrak{Q}^2 + \frac{1}{3!}\mathfrak{Q}^3 + \cdots \tag{10.37}$$

of powers of the operator $\mathfrak{Q}$, where we identify $\mathfrak{Q}^0$ with the identity operator, $\mathfrak{Q}^0 = 1$, we might expect unitary operators to be expressible in the form

$$\mathfrak{U} = e^{i\eta\mathfrak{H}} \equiv \sum_{n=0}^{\infty} \frac{1}{n!} i^n \eta^n \mathfrak{H}^n \tag{10.38}$$

with $\mathfrak{H}$ a Hermitian operator and η a real number. This is suggested by the analogy between operators and complex numbers. We shall ignore questions of convergence of such series, always assuming that they are absolutely convergent, and shall show that, with this understanding, this conjecture about the form of a unitary operator is correct; that is, we shall show that $\mathfrak{H}$ is Hermitian.

The Hermitian conjugate of $\mathfrak{U}$ is given by

$$\mathfrak{U}^\dagger = [e^{i\eta\mathfrak{H}}]^\dagger = \left[\sum_{n=0}^{\infty} \frac{1}{n!} i^n \eta^n \mathfrak{H}^n\right]^\dagger = \sum_{n=0}^{\infty} \frac{1}{n!} (-i\eta)^n (\mathfrak{H}^\dagger)^n,$$

so that the requirement $\mathfrak{U}\,\mathfrak{U}^\dagger = 1$ becomes

$$1 = \mathfrak{U}\mathfrak{U}^\dagger = \sum_{n=0}^{\infty}\sum_{m=0}^{\infty} \frac{1}{n!m!} i^{n-m} \eta^{n+m} \mathfrak{H}^n (\mathfrak{H})^m$$

$$= 1 + \sum_{k=1}^{\infty}\sum_{n=0}^{\infty} \frac{i^{2n-k}}{n!(k-n)!} \eta^k \mathfrak{H}^n (\mathfrak{H}^\dagger)^{k-n}.$$

Hence each coefficient of η^k must vanish for $k \geq 1$. The coefficient of η is simply

$$-i(\mathfrak{H}^\dagger - \mathfrak{H}),$$

so that we require

$$\mathfrak{H} = \mathfrak{H}^\dagger \tag{10.39}$$

i.e. $\mathfrak{H}$ must be Hermitian. In fact, this is sufficient to guarantee that the coefficients of all η^k vanish provided $k \geq 1$, as the reader may demonstrate for himself.

The Hermitian operator $\mathfrak{H}$ appearing in the exponent is often called the *generator* of the unitary transformation $\mathfrak{U}$, and the real number η is called the transformation *parameter*. We have obtained the result that the generators of unitary transformations are Hermitian operators. In a sense, then, every observable is a generator of a unitary transformation, and the transformation that it generates therefore leaves inner products invariant whenever the transformation parameter is real.

We have seen examples of this already for wave functions; let us generalize these examples to state-vector language. First consider a wave function for the eigenstate $|q, r, s, \ldots\rangle$, namely

$$\psi_{q,r,s\ldots}(\mathbf{x}, t, m, n, \ldots) = \langle \mathbf{x}, t, m, n, \ldots | q, r, s, \ldots\rangle,$$

where t represents time and the indices $m, n, \ldots$ are eigenvalues of whatever operators $\mathfrak{M}, \mathfrak{N}, \ldots$ are required to provide, together with $\mathfrak{X}, \mathfrak{T}$, a complete set of commuting operators. Now insert a complete set of energy eigenstates $|E, k, l, \ldots\rangle$:

$$\psi_{q,r,s,\ldots}(\mathbf{x}, t, m, n, \ldots) = \underset{E,k,l}{\mathsf{S}} \langle \mathbf{x}, t, m, n, \ldots | E, k, l, \ldots\rangle \langle E, k, l, \ldots | q, r, s, \ldots\rangle.$$

Now the energy eigenfunctions $\langle \mathbf{x}, t, m, n, \ldots | E, k, l, \ldots\rangle = \psi_{E,k,l,\ldots}(\mathbf{x}, t, m, n, \ldots)$ satisfy

$$i\hbar\frac{\partial}{\partial t}\psi_{E,k,l,\ldots}(\mathbf{x}, t, m, n, \ldots) = E\psi_{E,k,l,\ldots}(\mathbf{x}, t, m, n, \ldots)$$

which has the solutions

$$\begin{aligned}
&\psi_{E,k,l,\ldots}(\mathbf{x}, t, m, \ldots) \\
&\quad = \psi_{E,k,l,\ldots}(\mathbf{x}, 0, m, n, \ldots)e^{-iEt/\hbar} \\
&\quad = \langle \mathbf{x}, 0, m, n, \ldots | E, k, l, \ldots\rangle e^{-iEt/\hbar} \\
&\quad = \langle \mathbf{x}, 0, m, n, \ldots | e^{-iEt/\hbar} | E, k, l, \ldots\rangle \qquad (10.40) \\
&\quad = \langle \mathbf{x}, 0, m, n, \ldots | e^{-it\mathfrak{E}/\hbar} | E, k, l, \ldots\rangle \\
&\quad = \langle \mathbf{x}, t, m, n, \ldots | E, k, l, \ldots\rangle,
\end{aligned}$$

where $\mathfrak{E}$ is the energy operator for the system.

Equations (10.40) imply that

$$\begin{aligned}
\psi_{q,r,s,\ldots}(\mathbf{x}, t, m, n, \ldots) &= \underset{E,k,l,\ldots}{\mathsf{S}} \langle \mathbf{x}, 0, m, n, \ldots | E, k, l, \ldots\rangle e^{-itE/\hbar} \langle E, k, l, \ldots | q, r, s, \ldots\rangle \\
&= \underset{E,k,l,\ldots}{\mathsf{S}} \langle \mathbf{x}, 0, m, n, \ldots | e^{-it\mathfrak{E}/\hbar} | E, k, l, \ldots\rangle \langle E, k, l, \ldots | q, r, s, \ldots\rangle \\
&= \langle \mathbf{x}, 0, m, n, \ldots | e^{-it\mathfrak{E}/\hbar} | q, r, s, \ldots\rangle.
\end{aligned}$$

Since $|q, r, s, \ldots\rangle$ is arbitrary, wc havc

$$\langle \mathbf{x}, t, m, n, \ldots | = \langle \mathbf{x}, 0, m, n, \ldots | e^{-it\mathfrak{E}/\hbar},$$

or, since $\mathfrak{E} = \mathfrak{E}^\dagger$,

$$|\mathbf{x}, t, m, n, \ldots\rangle = e^{it\mathfrak{E}/\hbar} |\mathbf{x}, 0, m, n, \ldots\rangle. \qquad (10.41)$$

The effect of $e^{it\mathfrak{E}/\hbar}$ as an operator is thus to *translate* the time value of a state vector from t' to $t' + t$ (above we let $t' = 0$ for convenience). The energy operator is the generator of time translations, aside from a factor $1/\hbar$.

A similar development for space translations shows that if $\mathfrak{P}$ is the momentum operator, then $e^{-i\mathbf{x}\cdot\mathfrak{P}/\hbar}$ is the position translation operator with transformation parameters $\mathbf{x}$ and generators $-\mathfrak{P}/\hbar$. Hence, aside from the factor $(-1/\hbar)$, the momentum operator generates translations in position. Clearly, translations in four-space are accomplished through the unitary operator $e^{ix_\mu\mathfrak{P}^\mu/\hbar}$. (Note the difference in sign of the exponent from the corresponding translation operators for wave functions. Where does it come from?)

It is important to be aware that these translation operators are the so-called passive transformation, or coordinate transformation operators. They change the coordinates in terms of which a state is described, without actually changing any other feature of the state. Thus, given a state $|1\rangle$, the state $e^{-i\mathbf{x}_0 \cdot \mathfrak{P}/\hbar}|1\rangle$ is the same state except that the origin of the coordinate system in terms of which the latter is described is located at $\mathbf{x}_0$ in the coordinate system of the former (see Fig. 10.1a). Alternatively, one could think of retaining the same coordinate system for both states; then the latter state would be obtained from the former by physically moving the system from its original position $\mathbf{x}$ to a new mean position $\mathbf{x} - \mathbf{x}_0$ (see Fig. 10.1b). A coordinate (passive) translation by $\mathbf{x}_0$, then, corresponds to a shift in the physical system (active translation) by $-\mathbf{x}_0$.

In this light, we may look upon $|\mathbf{x}, t, m, \ldots\rangle$ as a representation of the coordinate system in terms of which the physical system is described. Translating the coordinate system by $\mathbf{x}_0$,

$$|\mathbf{x}, t, m, \ldots\rangle \to e^{-i\mathbf{x}_0 \cdot \mathfrak{P}/\hbar}|\mathbf{x}, t, m, \ldots\rangle = |\mathbf{x}_0 + \mathbf{x}, t, m, \ldots\rangle$$

is equivalent to moving the physical system $|1\rangle$ by $-\mathbf{x}_0$,

$$|1\rangle \to e^{i\mathbf{x}_0 \cdot \mathfrak{P}/\hbar}|1\rangle,$$

since

$$\langle \mathbf{x}_0 + \mathbf{x}, t, m, \ldots|1\rangle = \langle \mathbf{x}, t, m, \ldots|e^{i\mathbf{x}_0 \cdot \mathfrak{P}/\hbar}|1\rangle.$$

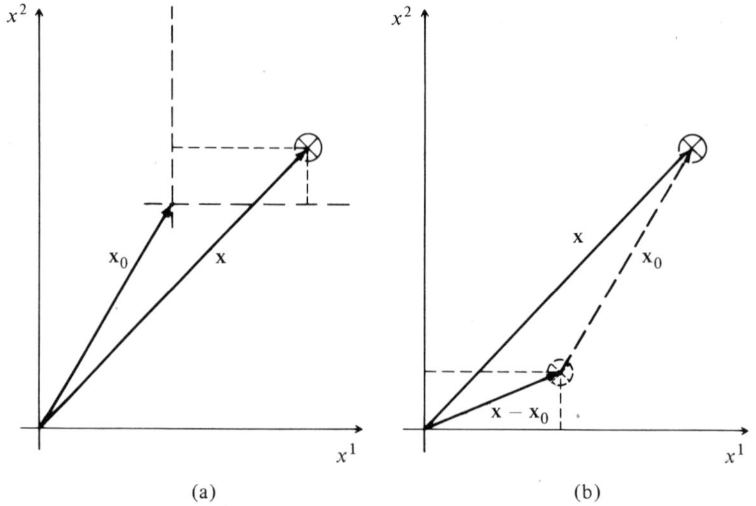

Fig. 10.1. The principle of active–passive equivalence. Changing to a new coordinate system with origin at x_0 (a) gives a description for an object at $\mathbf{x}$ equivalent to the description obtained by moving the object from $\mathbf{x}$ to $\mathbf{x} - \mathbf{x}_0$ (b).

If we wish to translate the physical system rather than the coordinate system by the amount $\mathbf{x}_0$, then, we must change the sign in the exponential. Thus $e^{i\mathbf{x}_0 \cdot \mathfrak{P}/\hbar}$ moves the whole physical system by $+\mathbf{x}_0$, while retaining the same coordinate system. This active translation by $\mathbf{x}_0$ is clearly equivalent to a passive translation in which the physical system is left alone and the coordinates are shifted instead, by $-\mathbf{x}_0$. This situation is sometimes referred to as the principle of active–passive equivalence.

10.6 CONSERVATION LAWS

The time-translation law can be used to make explicit the notion of *conservation* of a physical quantity. An operator, say $\mathfrak{Q}$, is said to represent a *conserved quantity* whenever it is unaffected by the time-translation operator:

$$e^{it\mathfrak{E}/\hbar}\mathfrak{Q}e^{-it\mathfrak{E}/\hbar} = \mathfrak{Q} \qquad \text{for all } t.$$

Equivalently,

$$e^{it\mathfrak{E}/\hbar}\mathfrak{Q} - \mathfrak{Q}e^{it\mathfrak{E}/\hbar} = [e^{it\mathfrak{E}/\hbar}, \mathfrak{Q}] = 0. \tag{10.42}$$

Put in terms of state vectors, this means that $\mathfrak{Q}$ has the same effect on a state at one time as at any other time; $\mathfrak{Q}$ operating on the time-translated state $e^{it\mathfrak{E}/\hbar}|1\rangle$ has the same effect as $\mathfrak{Q}$ operating on $|1\rangle$ followed by a time translation:

$$\mathfrak{Q}e^{it\mathfrak{E}/\hbar}|1\rangle = e^{it\mathfrak{E}/\hbar}\mathfrak{Q}|1\rangle.$$

This condition should be reducible to a simpler relation between $\mathfrak{E}$ and $\mathfrak{Q}$. We have

$$0 - \left[\sum_{n=0}^{\infty} \frac{1}{n!}\left(\frac{it}{\hbar}\right)^n \mathfrak{E}^n, \mathfrak{Q}\right] - \sum_{n=0}^{\infty} \frac{1}{n!}\left(\frac{it}{\hbar}\right)^n [\mathfrak{E}^n, \mathfrak{Q}],$$

as is verified easily since $it/\hbar$ is simply a number and therefore commutes with all operators, and since for any sum of operators, $\mathfrak{L} + \mathfrak{M}$, we have*

$$\begin{aligned}[\mathfrak{L} + \mathfrak{M}, \mathfrak{Q}] &= (\mathfrak{L} + \mathfrak{M})\mathfrak{Q} - \mathfrak{Q}(\mathfrak{L} + \mathfrak{M}) \\ &= \mathfrak{L}\mathfrak{Q} + \mathfrak{M}\mathfrak{Q} - \mathfrak{Q}\mathfrak{L} - \mathfrak{Q}\mathfrak{M} \\ &= [\mathfrak{L}, \mathfrak{Q}] + [\mathfrak{M}, \mathfrak{Q}].\end{aligned}$$

---

* This identity follows from the distributivity of multiplication of operators over addition and from the commutativity of addition of operators. The reader is invited to demonstrate that the latter follow from our definitions of operator addition through

$$(\mathfrak{L} + \mathfrak{M})|1\rangle = \mathfrak{L}|1\rangle + \mathfrak{M}|1\rangle$$

and of operator multiplication through

$$\mathfrak{L}\mathfrak{M}|1\rangle = \mathfrak{L}(\mathfrak{M}|1\rangle).$$

In order for Eq. (10.42) to be valid for all values of t, we must have

$$[\mathfrak{E}^n, \mathfrak{Q}] = 0 \qquad \text{for all } n = 0, 1, 2, \ldots .$$

However,

$$[\mathfrak{E}, \mathfrak{Q}] = 0$$

is sufficient, for this implies that

$$\begin{aligned}[\mathfrak{E}^2, \mathfrak{Q}] &= \mathfrak{E}^2\mathfrak{Q} - \mathfrak{Q}\mathfrak{E}^2 \\ &= \mathfrak{E}(\mathfrak{E}\mathfrak{Q} - \mathfrak{Q}\mathfrak{E}) - (\mathfrak{Q}\mathfrak{E} - \mathfrak{E}\mathfrak{Q})\mathfrak{E} \\ &= \mathfrak{E}[\mathfrak{E}, \mathfrak{Q}] + [\mathfrak{E}, \mathfrak{Q}]\mathfrak{E} = 0,\end{aligned}$$

etc.

Therefore an operator $\mathfrak{Q}$ represents a conserved quantity if and only if

$$[\mathfrak{E}, \mathfrak{Q}] = 0 \tag{10.43}$$

(we shall have to modify this slightly in Section 10.8 to allow dynamics to play a role in quantum systems), where $\mathfrak{E}$ is the energy operator. (Obviously the energy itself must be a conserved quantity; if it is not, we must be dealing with a system that is not isolated and hence for which we do not have complete information. Generally our discussion is restricted to isolated systems in which energy is conserved along with probability, charge, etc.) All operators that are compatible with the energy operator conserve energy. (This statement is modified slightly in Section 10.8.)

One operator that is obviously *not* conserved is the time operator $\mathfrak{T}$. Let $|t', q, \ldots\rangle$ be an eigenstate of $\mathfrak{T}$ with eigenvalue t', and consider the state $e^{it\mathfrak{E}/\hbar}|t', q, \ldots\rangle = |t' + t, q, \ldots\rangle$. We have

$$\begin{aligned}\mathfrak{T}e^{it\mathfrak{E}/\hbar}|t', q, \ldots\rangle &= \mathfrak{T}|t' + t, q, \ldots\rangle = (t' + t)|t' + t, q, \ldots\rangle \\ &= e^{it\mathfrak{E}/\hbar}(t' + t)|t', q, \ldots\rangle \\ &= e^{it\mathfrak{E}/\hbar}(t + \mathfrak{T})|t', q, \ldots\rangle.\end{aligned}$$

Hence

$$[e^{it\mathfrak{E}/\hbar}, \mathfrak{T}] = -te^{it\mathfrak{E}/\hbar} \tag{10.44}$$

or, if $\mathfrak{U}(t) \equiv e^{it\mathfrak{E}/\hbar}$,

$$[\mathfrak{U}(t), \mathfrak{T}] = -t\mathfrak{U}(t).$$

Expanding the time-translation operator in powers of t, we obtain, from Eq. (10.44),

$$\sum_{k=0}^{\infty} \frac{1}{k!}\left(\frac{it}{\hbar}\right)^k [\mathfrak{E}^k, \mathfrak{T}] = -\sum_{m=0}^{\infty} \frac{1}{m!}\left(\frac{it}{\hbar}\right)^m t\mathfrak{E}^m.$$

Equating powers of t yields

$$[\mathfrak{E}^k, \mathfrak{T}] = i\hbar k \mathfrak{E}^{k-1}, \tag{10.45}$$

or, for $k = 1$,

$$[\mathfrak{E}, \mathfrak{T}] = i\hbar. \tag{10.46}$$

For an operator $\mathfrak{Q}$ to be unaffected by translations in space, we must have

$$e^{-i\mathbf{x}\cdot\mathfrak{P}/\hbar}\mathfrak{Q}e^{i\mathbf{x}\cdot\mathfrak{P}/\hbar} = \mathfrak{Q}$$

or

$$[e^{-i\mathbf{x}\cdot\mathfrak{P}/\hbar}, \mathfrak{Q}] = 0, \qquad \text{for all } \mathbf{x}.$$

This in turn is equivalent to

$$[\mathfrak{P}, \mathfrak{Q}] = 0.$$

On the other hand, the position operator $\mathfrak{X}$ certainly is not such an operator. An analysis like that developed above for the time translation yields, for $\mathfrak{U}(\mathbf{x}) \equiv e^{-i\mathbf{x}\cdot\mathfrak{P}/\hbar}$,

$$[\mathfrak{U}(\mathbf{x}), \mathfrak{X}] = -\mathbf{x}\mathfrak{U}(\mathbf{x}).$$

In the limit of small $\mathbf{x}$, this becomes

$$[\mathfrak{P}_k, \mathfrak{X}_l] = (\hbar/i)\delta_{k,l}, \qquad k, l = 1, 2, 3. \tag{10.47}$$

For the four-position operator $\mathfrak{X}^\mu = (c\mathfrak{T}, -\mathfrak{X})$ and four-momentum operator $\mathfrak{P}^\mu = (\mathfrak{E}/c, -\mathfrak{P})$, then, we have

$$[\mathfrak{P}^\mu, \mathfrak{X}^\nu] = i\hbar g^{\mu\nu} \tag{10.48}$$

where $g^{\mu\nu}$ is the metric for special relativity's four-space.

10.7 OPERATORS FOR WAVE FUNCTIONS

It is interesting to see how the differential operators $\mathsf{E} = i\hbar\,\partial/\partial t$ and $\mathbf{P} = (\hbar/i)\nabla$ as well as the numerical operators for time, $\mathsf{T} = t$, and position $\mathbf{X} = \mathbf{x}$, can be obtained formally from the state-vector formalism. In developing these relationships, we shall see how energy conservation is built into the connection between inner products of state vectors and inner products of wave functions.

Consider the inner product between $|1\rangle$ and $\mathfrak{Q}|2\rangle$; to convert this to a statement on wave functions, we insert complete sets of states $|\mathbf{x}, t, m, \ldots\rangle$ that are eigenstates of $\mathfrak{X}$:

$$\langle 1|\mathfrak{Q}|2\rangle = \underset{\mathbf{x},t,m,\ldots}{\mathsf{S}} \langle 1|\mathbf{x}, t, m, \ldots\rangle \cdot \underset{\mathbf{x}',t',m',\ldots}{\mathsf{S}} \langle \mathbf{x}, t, m, \ldots|\mathfrak{Q}|\mathbf{x}', t', m', \ldots\rangle\langle \mathbf{x}', t', m', \ldots|2\rangle.$$

Here we have used states that also are eigenstates of time t as well as of position $\mathbf{x}$; the indices $m, \ldots$ simply represent whatever additional indices (or eigenvalues) are required to specify a complete set of states for the system under consideration. This can be rewritten

$$\langle 1|\mathfrak{Q}|2\rangle = \underset{\mathbf{x},t,m,\ldots}{\mathsf{S}} \; \underset{\mathbf{x}',t',m',\ldots}{\mathsf{S}} \; \psi_1'^*(\mathbf{x}, t, m, \ldots) \cdot \langle \mathbf{x}, t, m, \ldots|\mathfrak{Q}|\mathbf{x}', t', m', \ldots\rangle \psi_2'(\mathbf{x}', t', m', \ldots). \quad (10.49)$$

This is beginning to look like the inner product for wave functions; in fact, it can be considered a generalized version of the latter. However, there are *two* integrations over position here, and there are integrations over time, which we have not encountered previously in dealing with wave functions. Evidently $\langle \mathbf{x}, t, m, \ldots|\mathfrak{Q}|\mathbf{x}', t', m', \ldots\rangle$ plays the role of the wave-function operator corresponding to $\mathfrak{Q}$.

Usually the operators encountered in quantum theory are local in space and time, i.e. *diagonal* in the parameters $\mathbf{x}$ and t. This means that we can write

$$\langle \mathbf{x}, t, m, \ldots|\mathfrak{Q}|\mathbf{x}', t', m', \ldots\rangle = \delta^3(\mathbf{x} - \mathbf{x}')\delta(t - t')\mathsf{Q}_{m,\ldots;m',\ldots}. \quad (10.50)$$

This will eliminate one of the integrals over position and one of the integrals over time, to yield

$$\langle 1|\mathfrak{Q}|2\rangle = \int d^3x \int dt \underset{m,\ldots}{\mathsf{S}} \; \underset{m',\ldots}{\mathsf{S}} \; \psi_1'^*(\mathbf{x}, t, m, \ldots)\mathsf{Q}_{m,\ldots;m',\ldots}\psi_2(\mathbf{x}, t, m', \ldots).$$

This looks more like the usual form for inner product of wave functions, except that an integration over the time parameter is still involved.

To keep the discussion of the time integral reasonably simple, let us assume that state $|1\rangle$ is an eigenstate of energy with $\mathfrak{E}|1\rangle = E_1|1\rangle$ and that similarly $|2\rangle$ is an eigenstate of energy with $\mathfrak{E}|2\rangle = E_2|2\rangle$. Then we have

$$\begin{aligned}\psi_1'^*(\mathbf{x}, t, m, \ldots) &= \langle 1|\mathbf{x}, t, m, \ldots\rangle = \langle 1|e^{it\mathfrak{E}/\hbar}|\mathbf{x}, 0, m, \ldots\rangle \\ &= e^{itE_1/\hbar}\langle 1|\mathbf{x}, 0, m, \ldots\rangle = e^{itE_1/\hbar}\psi_1'^*(\mathbf{x}, 0, m, \ldots),\end{aligned}$$

and similarly

$$\psi_2'(\mathbf{x}, t, m', \ldots) = e^{-itE_2/\hbar}\psi_2'(\mathbf{x}, 0, m', \ldots).$$

Hence we have

$$\langle 1|\mathfrak{Q}|2\rangle = \underset{m,\ldots;m',\ldots}{\mathsf{S}} \int d^3x \int_{-\infty}^{\infty} dt\, e^{it(E_1 - E_2)/\hbar}\psi_1'^*(\mathbf{x}, 0, m, \ldots) \cdot \mathsf{Q}_{m,\ldots;m',\ldots}\psi_2'(\mathbf{x}, 0, m', \ldots).$$

But

$$\int_{-\infty}^{\infty} dt\, e^{it(E_1 - E_2)/\hbar} = 2\pi\hbar\delta(E_1 - E_2),$$

and hence

$$\langle 1|\mathfrak{Q}|2\rangle = \left[\underset{m,\ldots;m',\ldots}{\mathsf{S}} \int d^3x\, \psi_1'^*(\mathbf{x}, 0, m, \ldots)\mathrm{Q}_{m,\ldots;m',\ldots}\psi_2'(\mathbf{x}, 0, m', \ldots)\right] \cdot 2\pi\delta(E_1 - E_2).$$

Thus the energies of states 1 and 2 must be equal; the inner products of state with different energy eigenvalues are orthogonal. We knew this already, but we note here that this expression of energy conservation (why do we call this energy conservation?) arises from the integral over the time parameter.

Now we recognize the coefficient of the energy δ-function as the inner product of wave functions:

$$\begin{aligned}\langle 1|\mathfrak{Q}|2\rangle &= \underset{m,\ldots;m',\ldots}{\mathsf{S}} \int d^3x\, \psi_1^*(\mathbf{x}, m, \ldots)\mathrm{Q}_{m,\ldots;m',\ldots}\psi_2(\mathbf{x}, m', \ldots)\delta(E_1 - E_2) \\ &= (\psi_1, \mathrm{Q}\psi_2)\delta(E_1 - E_2), \end{aligned} \tag{10.51}$$

where we have defined

$$\psi_1(\mathbf{x}, m, \ldots) \equiv \sqrt{2\pi\hbar}\, \psi_1'(\mathbf{x}, 0, m, \ldots)$$

$$\psi_2(\mathbf{x}, m', \ldots) \equiv \sqrt{2\pi\hbar}\, \psi_2'(\mathbf{x}, 0, m', \ldots).$$

Here we have included in our definition of the inner product for wave functions,

$$(\psi_1, \mathrm{Q}\psi_2) = \underset{m,\ldots;m',\ldots}{\mathsf{S}} \int d^3x\, \psi_1^*(\mathbf{x}, m, \ldots)\mathrm{Q}_{m,\ldots;m',\ldots}\psi_2(\mathbf{x}, m', \ldots), \tag{10.52}$$

some sums over parameters $m, \ldots; m', \ldots$ that did not appear in our original definition of the inner product; this is to allow for the description of physical systems that involve variables other than space–time variables, and comes from the insertion of a *complete* set of states $|\mathbf{x}, t, m, \ldots\rangle$. Aside from this, our definition of $(\psi_1, \mathrm{Q}\psi_2)$ is just like the definition given in Section 9.3.

On the other hand, one may detect a conflict between what we have here and our discussion in Section 9.8. The latter discussion assumed, however, that $\mathbf{x}$ was the only measurable physical quantity. In such a case, the parameter t would not appear in a complete set of states, the δ-function of energy would not appear and we would have $\langle 1|\mathfrak{Q}|2\rangle = (\psi_1, \mathrm{Q}\psi_2)$. Therefore there is no genuine conflict.

Let us summarize our results. First, the usual (local) operators for wave functions are related to the operators for state vectors through the relations

$$\langle \mathbf{x}, t, m, \ldots|\mathfrak{Q}|\mathbf{x}', t', m', \ldots\rangle = \delta^3(\mathbf{x} - \mathbf{x}')\delta(t - t')\mathrm{Q}_{m,\ldots;m',\ldots},$$

$\mathfrak{Q}$ being the state-vector operator and $\mathrm{Q}_{m,\ldots;m',\ldots}$ being the wave-function operator. The eigenstates of energy, say $|1\rangle \equiv |E_1, 1\rangle$, are related to their wave functions

$\psi_1(\mathbf{x}, m, \ldots)$ through

$$\langle \mathbf{x}, t, m, \ldots | E_1, 1\rangle = e^{-iE_1t/\hbar}(2\pi\hbar)^{-1}\psi_1(\mathbf{x}, m, \ldots).$$

Finally, inner products for wave functions are defined by (10.52) and are related to the corresponding inner products for state vectors through (10.51).

To make this more concrete, and to illustrate the techniques involved, let us evaluate $\langle \mathbf{x}, t, m, \ldots |\mathfrak{P}| \mathbf{x}', t', m', \ldots\rangle$ in order to find the wave-function operator **P**. We insert a complete set of *momentum eigenstates*:

$$\underset{\mathbf{p},q,\ldots}{\mathsf{S}} \langle \mathbf{x}, t, m, \ldots | \mathbf{p}, q, \ldots\rangle\langle \mathbf{p}, q, \ldots |\mathfrak{P}| \mathbf{x}', t', m', \ldots\rangle$$

$$= \underset{\mathbf{p},q,\ldots}{\mathsf{S}} \langle \mathbf{x}, t, m, \ldots | \mathbf{p}, q, \ldots\rangle \mathbf{p} \langle \mathbf{p}, q, \ldots | \mathbf{x}', t', m', \ldots\rangle$$

$$= \underset{q,\ldots}{\mathsf{S}} \int d^3p\, \mathbf{p} e^{i\mathbf{p}\cdot(\mathbf{x}-\mathbf{x}')/\hbar} \cdot \langle 0, t, m, \ldots | \mathbf{p}, q, \ldots\rangle\langle \mathbf{p}, q, \ldots | 0, t', m', \ldots\rangle$$

$$= \underset{q,\ldots}{\mathsf{S}} \int d^3p\, (-\hbar/i)\nabla_{\mathbf{x}'} e^{i\mathbf{p}\cdot(\mathbf{x}-\mathbf{x}')/\hbar}\langle 0, t, m, \ldots | \mathbf{p}, q, \ldots\rangle\langle \mathbf{p}, q, \ldots | 0, t', m', \ldots\rangle$$

$$= (-\hbar/i)\nabla_{\mathbf{x}'} \underset{q,\ldots}{\mathsf{S}} \int d^3p\, e^{i\mathbf{p}\cdot(\mathbf{x}-\mathbf{x}')/\hbar}\langle 0, t, m, \ldots | \mathbf{p}, q, \ldots\rangle\langle \mathbf{p}, q, \ldots | 0, t', m', \ldots)$$

$$= (-\hbar/i)\nabla_{\mathbf{x}'} \underset{\mathbf{p},q,\ldots}{\mathsf{S}} \langle \mathbf{x}, t, m, \ldots | \mathbf{p}, q, \ldots\rangle\langle \mathbf{p}, q, \ldots | \mathbf{x}', t', m', \ldots\rangle$$

$$= (-\hbar/i)\nabla_{\mathbf{x}'}\delta^3(\mathbf{x} - \mathbf{x}')\delta(t - t')\delta(m - m')\cdots,$$

Here the $\nabla_{\mathbf{x}'}$ is operating on $\delta^3(\mathbf{x} - \mathbf{x}')$, which is undesirable. To remedy this, we consider

$$\langle 1|\mathfrak{P}|2\rangle = \underset{\mathbf{x},t,m,\ldots}{\mathsf{S}}\ \underset{\mathbf{x}',t',m',\ldots}{\mathsf{S}} \langle 1|\mathbf{x}, t, m, \ldots\rangle$$
$$\cdot[-(\hbar/i)\nabla_{\mathbf{x}'}\delta^3(\mathbf{x} - \mathbf{x}')\delta(t - t')\delta(m - m')\ldots]\langle \mathbf{x}', t', m', \ldots|2\rangle$$

$$= \underset{t,m,\ldots}{\mathsf{S}} \int d^3x' \langle 1|\mathbf{x}, t, m, \ldots\rangle[-(\hbar/i)\nabla_{\mathbf{x}'}\delta^3(\mathbf{x} - \mathbf{x}')]\langle \mathbf{x}', t', m', \ldots|2\rangle.$$

The integral over d^3x' has the form

$$\int d^3x'[\nabla_{\mathbf{x}'}\delta^3(\mathbf{x} - \mathbf{x}')]g(\mathbf{x}') = \sum_j \hat{\mathbf{i}}_j \int d^3x' \left[\frac{\partial}{\partial x_j'}\delta^3(\mathbf{x} - \mathbf{x}')\right] g(\mathbf{x}'),$$

and integration by parts yields, for the jth component of this vector,

$$\int dS^j\, \delta(\mathbf{x} - \mathbf{x}')g(\mathbf{x}') - \int d^3x'\, \delta^3(\mathbf{x} - \mathbf{x}')\frac{\partial}{\partial x_j'}g(\mathbf{x}');$$

here dS^j is an element of area normal to $\hat{\mathbf{i}}_j$ on the boundary of the region enclosing the physical system. Either $g(\mathbf{x}')$ is supposed to vanish on the boundary of the system, or periodic boundary conditions ensure the vanishing of the surface integral. In either case, the surface integral vanishes, and therefore

$$\int d^3x'[\nabla_{\mathbf{x}'}\, \delta^3(\mathbf{x} - \mathbf{x}')]\, g(\mathbf{x}') = -\int d^3x'\, \delta^3(\mathbf{x} - \mathbf{x}')\nabla_{\mathbf{x}'}\, g(\mathbf{x}').$$

That is, we may replace $\nabla_{\mathbf{x}'}\, \delta^3(\mathbf{x} - \mathbf{x}')$ with $-\delta^3(\mathbf{x} - \mathbf{x}')\, \nabla_{\mathbf{x}'}$ by virtue of boundary conditions on the wave functions for a system.

Hence we have

$$\begin{aligned}\langle \mathbf{x}, t, m, \ldots |\mathfrak{P}| \mathbf{x}', t', m', \ldots\rangle &= \delta^3(\mathbf{x} - \mathbf{x}')\, \delta(t - t')\, \delta(m - m') \cdots (\hbar/i)\nabla \\ &= \delta^3(\mathbf{x} - \mathbf{x}')\, \delta(t - t')\mathsf{P},\end{aligned}$$

or

$$\mathsf{P} = \delta(m - m') \cdots (\hbar/i)\nabla. \tag{10.53}$$

Similarly we can obtain $\mathsf{E} = \delta(m - m') \cdots i\hbar\, \partial/\partial t$. It is even simpler to get $X = \delta(m - m') \cdots \mathbf{x}$ and $T = \delta(m - m') \cdots t$. Usually the string of δ-functions is neglected, it being understood that they are present whenever other physical quantities besides time and position are required to give a complete description to a physical system.

10.8 DYNAMICS AND THE HAMILTONIAN OPERATOR

In classical physics the problem of dynamics is to find the positions $\mathbf{x}_j(t)$ of particles composing a system, as functions of time. This means being able to solve the equations of motion

$$m_j \frac{d^2}{d\tau^2} x_j^{\mu}(\tau) = F_j^{\mu}(x^{\nu}, \tau)$$

or its nonrelativistic counterpart

$$m_j \frac{d^2}{dt^2} \mathbf{x}_j(t) = \mathbf{F}_j(\mathbf{x}, t)$$

where $F_j(x^{\nu}, \tau)$ or $\mathbf{F}_j(\mathbf{x}, t)$ is the force exerted on the jth particle. For many problems it is simpler to describe the motion in terms of potential energy than directly in terms of forces, and to solve somewhat different but equivalent equations for $\mathbf{x}(t)$. In any case, the solution to the dynamical problem consists of providing a numerical description of the evolution of the system in terms of a parameter such as τ or t.

In this section we discuss how this is done in the nonrelativistic case, where the parameter is time variable. This presents a special problem because the time variable cannot appear as an operator in nonrelativistic physics; it is simply a parameter in terms of which the dynamical evolution of the physical system is to be described. Thus energy no longer appears as an operator canonically conjugate to the time operator, since the latter does not exist nonrelativistically. However, a new operator which effectively serves as a time-translation operator, and hence as an effective energy operator, is introduced. It does not, however, satisfy the canonical commutation relation between energy and time.

In effect, the act of specifying a nonrelativistic dynamical time-translation operator is tantamount to imposing a constraint among the expectation values of dynamical operators such as momentum and position. The form, in terms of dynamical variables, of the dynamical generator of time translations is dictated by the physical features of the system and constitutes part of a mathematical model for the system. The constraint imposed by the model involves a parameter, namely time. Since time is a parameter here, it must commute with all operators, including the dynamical time-translation generator. However, we still have

$$[i\hbar\, \partial/\partial t, t]f(t) = i\hbar f(t);$$

therefore if we consider a state vector $|1(t)\rangle$ to depend on the parameter t, it follows that

$$[i\hbar\, \partial/\partial t, t]|1(t)\rangle = i\hbar|1(t)\rangle$$

and this will guarantee that $i\hbar\, \partial/\partial t$ has the *same effect* on $|1(t)\rangle$ as does the new dynamical time-translation generator. Hence, although $i\hbar\, \partial/\partial t$ will not be considered a dynamical operator in our discussion of the nonrelativistic case, it behaves like an energy operator.

In the relativistic case, position and time must be handled on the same footing. However, the dynamical evolution of the system may be described in terms of the proper time of an observer. If we work in the frame of reference in which the observer is at rest, the time variable in that frame coincides with his proper time. In this frame of reference, the energy operator $\mathfrak{E}$ coincides with $i\hbar\, \partial/\partial \tau$ where τ is the proper time, and hence by writing all operators and state vectors in this frame of reference and letting $\mathfrak{E} \to i\hbar\, \partial/\partial \tau$, $\mathfrak{T} \to \tau$ and prescribing the dynamical τ-translation generator we can obtain a quantum-theoretical model for the system. Relativistic problems, however, involve several complications in quantum theory which are described briefly elsewhere in this book; we confine our attention here to nonrelativistic problems.

Let us consider a state vector $|1\rangle$ that describes the physical state of a nonrelativistic particle in terms of its spatial behavior and perhaps some internal properties (e.g. electric charge, nuclear parameters, etc.). We know that the classical energy of this particle might be given in terms of its kinetic energy and its potential energy; transcribed into quantum-mechanical operator language, this would be

$$\mathfrak{H} = \frac{1}{2M}\mathfrak{P}^2 + \mathfrak{V}(\mathfrak{X}), \tag{10.54}$$

where the second term represents the potential energy as a function of position alone. That this is the operator transcription of the classical expression for the energy of the particle suggests that we consider it the generator of dynamical time translations; that is, we may hypothesize that after a time $t - t_0$, the particle will be in a new state

$$|1(t)\rangle = e^{-i(t-t_0)\mathfrak{H}/\hbar}|1\rangle \tag{10.55}$$

if it was in state $|1\rangle$ at time t_0.

If we now define the time derivative of $|1(t)\rangle$ by

$$\begin{aligned}\frac{\partial}{\partial t}|1(t)\rangle &= \lim_{\Delta t \to 0} \frac{1}{\Delta t}[|1(t + \Delta t)\rangle - |1(t)\rangle] \\ &= \lim_{\Delta t \to 0} \frac{1}{\Delta t}(e^{-i\Delta t\mathfrak{H}/\hbar} - 1)|1(t)\rangle,\end{aligned}$$

we find that

$$i\hbar\frac{\partial}{\partial t}|1(t)\rangle = \mathfrak{H}|1(t)\rangle \tag{10.56}$$

when the operator $\mathfrak{H}$ is independent of the parameter t. This is a state-vector version of Schrödinger's equation, and the latter lies at the foundation of modern quantum theory. Hence our hypothesis that $\mathfrak{H}$ (which is called the *Hamiltonian* for the particle) generates dynamical time translations is equivalent to Schrödinger's equation together with the initial condition $|1(t_0)\rangle = |1\rangle$. Note that the state vector $|1\rangle$ does not depend on time nor explicitly upon energy; energy is introduced into the picture when we make the *dynamical hypothesis* that the Hamiltonian represents the energy of the system and hence generates time translations that are effected by the dynamical evolution of the physical system.

Any operator that in our previous discussion involved the time operator simply has $\mathfrak{T}$ replaced by the parameter t wherever it occurs. Thus a general operator $\mathfrak{Q}(\mathfrak{T})$ that depended on the time operator $\mathfrak{T}$ now becomes a function of the time parameter t: $\mathfrak{Q}(\mathfrak{T}) \to \mathfrak{Q}(t)$. We can relate $\mathfrak{Q}(t')$ at time t' to the operator $\mathfrak{Q}(t)$ at time t through the Taylor-series expansion

$$\mathfrak{Q}(t) = e^{(t-t')\,\partial/\partial t'}\mathfrak{Q}(t').$$

Let $\{|1\rangle, |2\rangle, \ldots\}$ be a complete set of states for a nonrelativistic physical system and let $\mathfrak{Q}(t)$ be some time-dependent operator of interest. We let $\{|1\rangle, |2\rangle, \ldots\}$ represent the system at time t_0, so that the matrix elements of $\mathfrak{Q}$ at that time are, for example,

$$\langle 1|\mathfrak{Q}(t_0)|2\rangle.$$

At a later time t, however, this matrix element will have become

$$\begin{aligned}\langle 1(t)|\mathfrak{Q}(t)|2(t)\rangle &= \langle 1|e^{i(t-t_0)\mathfrak{H}/\hbar}\mathfrak{Q}(t)e^{-i(t-t_0)\mathfrak{H}/\hbar}|2\rangle \\ &= \langle 1|\mathfrak{Q}_{\mathrm{H}}(t)|2\rangle,\end{aligned} \tag{10.57}$$

where

$$\mathfrak{Q}_{\mathrm{H}}(t) \equiv e^{i(t-t_0)\mathfrak{H}/\hbar}\mathfrak{Q}(t)e^{-i(t-t_0)\mathfrak{H}/\hbar}. \tag{10.58}$$

This suggests that instead of describing the dynamical evolution of the system through the time-dependent state vectors given by Eq. (10.55), we may deal with the time-independent state vectors $\{|1\rangle, |2\rangle, \ldots\}$ and place the temporal development entirely in the operators (10.58). These operators are known as the *Heisenberg* operators and the corresponding time-independent state vectors are known as Heisenberg state vectors; they form a basis for the *Heisenberg representation* (or Heisenberg picture) of the system.

Because Heisenberg originally dealt with this formulation of quantum theory through the calculation of matrix elements of the Heisenberg operators $\mathfrak{Q}_H(t)$, $\mathfrak{R}_H(t), \ldots$ and by manipulating the resulting matrices, the Heisenberg representation (or, more accurately, Heisenberg picture) is sometimes referred to as *matrix mechanics.* It is clearly equivalent to the *Schrödinger representation* (Schrödinger picture), in which one deals with the time-dependent Schrödinger state vectors $|1(t)\rangle, |2(t)\rangle, \ldots$, which contain all the dynamical information about the system, and with the Schrödinger operators $\mathfrak{Q}(t), \mathfrak{R}(t), \ldots$, which are devoid of information about the physical evolution of the system. The latter was first analysed through the solution of differential equations, and hence it gained the name *wave mechanics.* The two descriptions are related to one another through the unitary transformation

$$\mathfrak{U}(t) = e^{-i(t-t_0)\mathfrak{H}/\hbar}: \tag{10.59}$$

$$|1\rangle_H = \mathfrak{U}^\dagger(t)|1(t)\rangle_S; \qquad \mathfrak{Q}_H(t) = \mathfrak{U}^\dagger(t)\mathfrak{Q}_S(t)\mathfrak{U}(t), \tag{10.60}$$

where

$$|1(t)\rangle_S \equiv |1(t)\rangle,$$

$$\mathfrak{Q}_S(t) \equiv \mathfrak{Q}(t)$$

are the Schrödinger state vectors and operators

From Eq. (10.57) we get the useful result

$$\begin{aligned}
\frac{d}{dt}\langle 1(t)|\mathfrak{Q}(t)|2(t)\rangle &= \frac{d}{dt}\langle 1|\mathfrak{Q}_H(t)|2\rangle \\
&= \left\langle 1\left|\frac{d}{dt}\mathfrak{Q}_H(t)\right|2\right\rangle \\
&= \langle 1|e^{i(t-t_0)\mathfrak{H}/\hbar}\left\{\frac{\partial}{\partial t}\mathfrak{Q}(t) - \frac{1}{i\hbar}[\mathfrak{H}, \mathfrak{Q}(t)]\right\}e^{-i(t-t_0)\mathfrak{H}/\hbar}|2\rangle \\
&= \left\langle 1(t)\left|\frac{\partial}{\partial t}\mathfrak{Q}(t) - \frac{1}{i\hbar}[\mathfrak{H}, \mathfrak{Q}(t)]\right|2(t)\right\rangle.
\end{aligned} \tag{10.61}$$

This can be written in the form

$$\frac{d}{dt}\langle 1|\mathfrak{Q}_{\mathrm{H}}(t)|2\rangle = \left\langle 1\left|\frac{d}{dt}\mathfrak{Q}_{\mathrm{H}}(t)\right|2\right\rangle$$

$$= \left\langle 1\left|\left[\frac{\partial}{\partial t}\mathfrak{Q}(t)\right]_{\mathrm{H}}\right|2\right\rangle - \frac{1}{i\hbar}\langle 1|[\mathfrak{H}, \mathfrak{Q}_{\mathrm{H}}(t)]|2\rangle \tag{10.62}$$

or, since $|1\rangle$ and $|2\rangle$ are arbitrary states,

$$\frac{d}{dt}\mathfrak{Q}_{\mathrm{H}}(t) = \left[\frac{\partial}{\partial t}\mathfrak{Q}(t)\right]_{\mathrm{H}} - \frac{1}{i\hbar}[\mathfrak{H}, \mathfrak{Q}_{\mathrm{H}}(t)], \tag{10.63}$$

where

$$\left[\frac{\partial}{\partial t}\mathfrak{Q}_{\mathrm{H}}(t)\right]_{\mathrm{H}} \equiv e^{i(t-t_0)\mathfrak{H}/\hbar}\left[\frac{\partial}{\partial t}\mathfrak{Q}(t)\right]e^{-i(t-t_0)\mathfrak{H}/\hbar}.$$

Equations (10.61)–(10.63) are different expressions of what are called Heisenberg's equations of motion. On the right-hand sides, the first terms express the rate of change arising from the explicit time-dependence of the operator $\mathfrak{Q}(t)$ and the second terms express the rate of change in $\mathfrak{Q}_{\mathrm{H}}(t)$ arising from the dynamical evolution of the system. In (10.63) we have this expressed with no reference to state vectors.

As an example of the application of this equation, consider $\langle 1(t)|\mathfrak{X}|2(t)\rangle$, where $\mathfrak{X}$ is the (time-independent) position operator. We have

$$\frac{d}{dt}\langle 1(t)|\mathfrak{X}|2(t)\rangle = \left\langle 1\left|\frac{d}{dt}\mathfrak{X}_{\mathrm{H}}(t)\right|2\right\rangle$$

$$= \left\langle 1(t)\left| -\frac{1}{i\hbar}[\mathfrak{H}, \mathfrak{X}]\right|2(t)\right\rangle.$$

Now if $\mathfrak{H}$ is given by Eq. (10.54), then $[\mathfrak{V}(\mathfrak{X}), \mathfrak{X}] = 0$, since $\mathfrak{V}$ is assumed to depend only on the position operator, and

$$\left[\frac{1}{2M}\mathfrak{P}^2, \mathfrak{X}\right] = \frac{\hbar}{iM}\mathfrak{P};$$

hence

$$\frac{d}{dt}\langle 1(t)|\mathfrak{X}|2(t)\rangle = \left\langle 1(t)\left|\frac{1}{M}\mathfrak{P}\right|2(t)\right\rangle \tag{10.64}$$

or

$$\frac{d}{dt}\mathfrak{X}_{\mathrm{H}}(t) = -\frac{1}{i\hbar}[\mathfrak{H}, \mathfrak{X}_{\mathrm{H}}] = \frac{1}{M}\mathfrak{P}_{\mathrm{H}}. \tag{10.65}$$

The time derivative of position, then, is the velocity.

An interesting result of abandoning the notion that time can be represented by an operator, at least in nonrelativistic quantum theory, is that we lose a pair of canonically conjugate variables and their commutation relations. Since t is a parameter, we clearly have

$$[\mathfrak{H}, t] = 0.$$

On the other hand, we have Schrödinger's equation

$$i\hbar\frac{\partial}{\partial t}|1(t)\rangle = \mathfrak{H}|1(t)\rangle$$

for any arbitrary state $|1\rangle$, from which it would seem to follow that

$$\left[i\hbar\frac{\partial}{\partial t}, t\right]|1(t)\rangle = i\hbar|1(t)\rangle = [\mathfrak{H}, t]|1(t)\rangle,$$

or $[\mathfrak{H}, t] = i\hbar$. The latter, however, is incorrect since Schrödinger's equation is not an identity between $i\hbar\,\partial/\partial t$ and $\mathfrak{H}$; the former simply is a derivative with respect to a parameter whereas the latter is an operator in the Hilbert space of state vectors for the system. The equation $[i\hbar\,\partial/\partial t, t] = i\hbar$ is not an operator equation in the Hilbert space of state vectors, but rather is an identity for the differential operator $i\hbar\,\partial/\partial t$; this commutator contains no physical operators, and hence is not a canonical commutation relation of the theory. Neither is $[\mathfrak{H}, t] = 0$, since this is a simple consequence of our decision to treat time as a pure numerical parameter.

Nonrelativistically, then, the energy–time uncertainty principle, which cannot be based on the noncommutativity of the energy and time operators, does not exist on the same footing as the momentum–position uncertainty principle. However, $\mathfrak{H}$ relates energy to position and momentum operators, and Eq. (10.64) relates position and momentum operators through the time parameter. These interrelationships can be used to resurrect the energy–time uncertainty principle, as the following example illustrates: A particle is observed to have momentum $\mathbf{p}$ at time t and to have momentum $\mathbf{p} + \Delta\mathbf{p}$ at time $t + \Delta t$; then

$$\begin{aligned}\Delta E &= \frac{1}{2M}(\mathbf{p} + \Delta\mathbf{p})^2 - \frac{1}{2M}\mathbf{p}^2 \\ &= \frac{1}{M}\mathbf{p}\cdot\Delta\mathbf{p} + \frac{1}{2M}(\Delta\mathbf{p})^2 \simeq \frac{1}{M}\mathbf{p}\cdot\Delta\mathbf{p},\end{aligned}$$

if $\Delta\mathbf{p}$ is parallel to $\mathbf{p}$ and $|\Delta\mathbf{p}|/|\mathbf{p}|$ is small. Since $\mathbf{p}/M = \dot{\mathbf{x}}$, or $\Delta\mathbf{x} \simeq \mathbf{p}\,\Delta t/M$, we have

$$\Delta E\,\Delta t = \Delta\mathbf{p}\cdot\frac{\mathbf{p}}{M}\Delta t = \Delta\mathbf{p}\cdot\Delta\mathbf{x} \gtrsim \frac{\hbar}{2}, \tag{10.66}$$

where $\Delta\mathbf{p}\cdot\Delta\mathbf{x} \gtrsim \hbar/2$ is a consequence of the canonical momentum–position commutation relations, as we shall see later.

In relativistic physics this asymmetry between the energy–time and momentum–position commutation relations does not exist, and we may escape detailed consideration of the implications of the asymmetry simply by noting that nonrelativistic quantum physics is an approximation to the thoroughgoing relativistic theory, and the difficulties we have encountered here are a consequence of the approximation.

Our earlier discussion of conserved quantities is also modified by the role of time as a parameter rather than as an operator. A conserved quantity in nonrelativistic theory is one represented by an operator $\mathfrak{Q}$ such that

$$\frac{\partial}{\partial t}\mathfrak{Q}(t) = 0 \qquad \text{and} \qquad [\mathfrak{H}, \mathfrak{Q}(t)] = 0,$$

since $\mathfrak{H}$ is the generator of time translations. Normally, $\mathfrak{H}$ is such a conserved quantity, since certainly $[\mathfrak{H}, \mathfrak{H}] = 0$, and if $\mathfrak{H}$ represents the energy of an isolated physical system (no external forces), $\partial\mathfrak{H}/\partial t = 0$. Most of the cases with which we shall deal involve operators that do not depend explicitly on time, and hence $[\mathfrak{H}, \mathfrak{Q}] = 0$ is the crucial equation.

10.9 EHRENFEST'S THEOREMS

We already have proved one of Ehrenfest's equations in operator form, namely Eq. (10.65). The more common form for this equation is obtained by taking expectation values of the left- and right-hand sides:

$$\frac{d}{dt}\langle\mathfrak{X}\rangle = \left\langle\frac{d}{dt}\mathbf{x}_{\mathrm{H}}\right\rangle = \frac{1}{M}\langle\mathfrak{P}\rangle$$

$$\frac{d}{dt}\bar{\mathbf{x}} = \frac{1}{M}\bar{\mathbf{p}}. \tag{10.67}$$

Similarly, we can calculate the first total derivative of the momentum operator:

$$\frac{d}{dt}\mathfrak{P} = \frac{\partial}{\partial t}\mathfrak{P} - \frac{1}{i\hbar}[\mathfrak{H}, \mathfrak{P}] = \frac{i}{\hbar}\left[\left\{\frac{1}{2M}\mathfrak{P}^2 + \mathfrak{V}(\mathfrak{X})\right\}, \mathfrak{P}\right]$$

$$= \frac{i}{\hbar}[\mathfrak{V}(\mathfrak{X}), \mathfrak{P}].$$

Now if $\mathfrak{V}(\mathfrak{X})$ can be written as a power series, one can exploit the commutator relations for $\mathfrak{P}$ and $\mathfrak{X}$ to show that

$$[\mathfrak{P}, \mathfrak{V}(\mathfrak{X})] = \frac{\hbar}{i}\nabla_{\mathfrak{X}}\mathfrak{V}(\mathfrak{X})$$

if by $\nabla_{\mathfrak{X}}\mathfrak{V}(\mathfrak{X})$ we mean the result obtained by treating $\mathfrak{X}$ as a real vector variable.

Alternatively we can use the completeness relation to get, for any states $|1\rangle, |2\rangle$, and any pair of operators $\mathfrak{Q}$, $\mathfrak{R}$,

$$\langle 1|\mathfrak{Q}\mathfrak{R}|2\rangle = \underset{\mathbf{x},m,\ldots}{\mathsf{S}} \ \underset{\mathbf{x}',m',\ldots}{\mathsf{S}} \langle 1|\mathbf{x}', m', \ldots\rangle\langle \mathbf{x}', m', \ldots|\mathfrak{Q}\mathfrak{R}|\mathbf{x}, m, \ldots\rangle\langle \mathbf{x}, m, \ldots|2\rangle.$$

Now, letting $\mathfrak{Q} = \mathfrak{V}(\mathfrak{X})$ and $\mathfrak{R} = \mathfrak{P}$,

$$(\mathbf{x}', m', \ldots|\mathfrak{V}(\mathfrak{X})\mathfrak{P}|\mathbf{x}, m, \ldots\rangle = \delta^3(\mathbf{x} - \mathbf{x}')\,\delta(m - m')\cdots V(\mathbf{x})\mathbf{P},$$

and similarly

$$\langle \mathbf{x}', m', \ldots|\mathfrak{P}\mathfrak{V}(\mathfrak{X})|\mathbf{x}, m, \ldots\rangle = \delta^3(\mathbf{x} - \mathbf{x}')\,\delta(m - m')\cdots \mathbf{P}V(\mathbf{x}).$$

Thus

$$\begin{aligned}
\langle 1|[\mathfrak{V}(\mathfrak{X}), \mathfrak{P}]|2\rangle &= \underset{\mathbf{x},m,\ldots}{\mathsf{S}} \langle 1|\mathbf{x}, m, \ldots\rangle\left[V(\mathbf{x}), \frac{\hbar}{i}\boldsymbol{\nabla}\right]\langle \mathbf{x}, m, \ldots|2\rangle \\
&= \underset{\mathbf{x},m,\ldots}{\mathsf{S}} \langle 1|\mathbf{x}, m, \ldots\rangle(i\hbar\boldsymbol{\nabla} V(\mathbf{x}))\langle \mathbf{x}, m, \ldots|2\rangle \\
&= \underset{\mathbf{x},m,\ldots}{\mathsf{S}} \ \underset{\mathbf{x}',m',\ldots}{\mathsf{S}} \langle 1|\mathbf{x}', m', \ldots\rangle \\
&\qquad\qquad \cdot\langle \mathbf{x}', m', \ldots|i\hbar\boldsymbol{\nabla}_{\mathfrak{x}}\mathfrak{V}(\mathfrak{X})|\mathbf{x}, m, \ldots\rangle\langle \mathbf{x}, m, \ldots|2\rangle \\
&= \langle 1|i\hbar\boldsymbol{\nabla}_{\mathfrak{x}}\mathfrak{V}(\mathfrak{X})|2\rangle,
\end{aligned}$$

where by $\boldsymbol{\nabla}_{\mathfrak{x}}\mathfrak{V}(\mathfrak{X})$ we mean the result that would be obtained by taking the gradient as if $\mathfrak{X}$ were a real vector variable.

Hence we have

$$\frac{d}{dt}\mathfrak{P} = -\boldsymbol{\nabla}_{\mathfrak{x}}\mathfrak{V}(\mathfrak{X}) \tag{10.68}$$

or

$$\frac{d}{dt}\bar{\mathbf{p}} = -\overline{\boldsymbol{\nabla} V(\mathbf{x})}; \tag{10.69}$$

this clearly corresponds to the classical equation of motion for a particle under the influence of a potential.

The business of taking gradients, or in general, derivatives of operators is a bit awkward. Usually it is avoided in quantum mechanics by working in the wave function representation or some other convenient representation exclusively, so that derivatives are taken with respect to real parameters. We have avoided that alternative here in order to emphasize that the results are independent of the representation; this was done at the cost of introducing the conceptually awkward, but nevertheless useful, notion of taking a derivative with respect to an operator. It should be clear from the discussion above and that of the previous section, however, that taking a derivative is equivalent to evaluating a certain commutator, and here, in fact, we have

$$\nabla_{\mathfrak{X}}\mathfrak{Q} \equiv (i/\hbar)[\mathfrak{P}, \mathfrak{Q}]$$

as a perfectly reasonable *definition* of the gradient with respect to the operator $\mathfrak{X}$.

10.10 COMMUTATORS AND THE UNCERTAINTY PRINCIPLE

An intimate connection exists between the uncertainty principle and the commutation relations between operators. Our original statement of the uncertainty principle was based upon wave considerations, but now we can give a more general statement derived from general state-vector considerations.

The expectation value $\langle\mathfrak{Q}\rangle$ of an operator gives the mean value of an operator for a given state. Anyone who has dealt with statements about mean values or averages no doubt has observed a need for another parameter to give a measure of the *spread* of possible values about the mean value. Such a parameter is the standard deviation. Our state-vector version of the standard deviation is

$$\Delta Q \equiv \langle(\mathfrak{Q} - \langle\mathfrak{Q}\rangle)^\dagger(\mathfrak{Q} - \langle\mathfrak{Q}\rangle)\rangle^{1/2} = \langle\Delta\mathfrak{Q}^\dagger\,\Delta\mathfrak{Q}\rangle^{1/2} \geq 0, \tag{10.70}$$

where

$$\Delta\mathfrak{Q} \equiv \mathfrak{Q} - \langle\mathfrak{Q}\rangle.$$

The reader can verify easily that

$$\Delta Q = [\langle\mathfrak{Q}^\dagger\mathfrak{Q}\rangle - |\langle\mathfrak{Q}\rangle|^2]^{1/2}. \tag{10.71}$$

We call ΔQ the uncertainty in the variable Q represented by the operator $\mathfrak{Q}$. The uncertainty product for two variables, Q and R, is

$$\Delta Q\,\Delta R = \sqrt{\langle\Delta\mathfrak{Q}^\dagger\,\Delta\mathfrak{Q}\rangle\langle\Delta\mathfrak{R}^\dagger\,\Delta\mathfrak{R}\rangle}. \tag{10.72}$$

The Schwarz inequality can be expressed in the form

$$\langle\Delta\mathfrak{Q}^\dagger\,\Delta\mathfrak{Q}\rangle\langle\Delta\mathfrak{R}^\dagger\,\Delta\mathfrak{R}\rangle \geq |\langle\Delta\mathfrak{Q}^\dagger\,\Delta\mathfrak{R}\rangle|^2 \tag{10.73}$$

to yield a minimum value for $\Delta Q\,\Delta R$:

$$\Delta Q\,\Delta R \geq |\langle\Delta\mathfrak{Q}^\dagger\,\Delta\mathfrak{R}\rangle|. \tag{10.74}$$

This can be refined further to yield a minimum value involving the commutator, provided $\mathfrak{Q}$ and $\mathfrak{R}$ are Hermitian:

$$\Delta Q\,\Delta R \geq \tfrac{1}{2}|\langle[\mathfrak{Q}, \mathfrak{R}]\rangle|. \tag{10.75}$$

Below, we prove the Schwarz inequality in the form stated above, and develop the commutator version of the minimum value for the uncertainty product.

The proof of the first inequality is rather easy. We simply observe that

$$\langle 1|(\Delta\mathfrak{Q} - \alpha\,\Delta\mathfrak{R})^\dagger(\Delta\mathfrak{Q} - \alpha\,\Delta\mathfrak{R})|1\rangle = \langle 2|2\rangle \geq 0, \tag{10.76}$$

where $|2\rangle \equiv (\Delta\mathfrak{Q} - \alpha\,\Delta\mathfrak{R})|1\rangle$; then we choose a value of α that minimizes the left-hand side so that we can make the most efficient possible use of the inequality. Thus

$$0 \le \langle 1|(\Delta\mathfrak{Q} - \alpha\,\Delta\mathfrak{R})^\dagger(\Delta\mathfrak{Q} - \alpha\,\Delta\mathfrak{R})|1\rangle$$
$$= \langle 1|\Delta\mathfrak{Q}^\dagger\,\Delta\mathfrak{Q}|1\rangle + |\alpha|^2\langle 1|\Delta\mathfrak{R}^\dagger\,\Delta\mathfrak{R}|1\rangle - 2\,\mathrm{Re}\,\alpha\langle 1|\Delta\mathfrak{Q}^\dagger\,\Delta\mathfrak{R}|1\rangle. \tag{10.77}$$

The last member is minimized with respect to $|\alpha|$ if

$$|\alpha| = \mathrm{Re}\,\frac{\alpha}{|\alpha|}\langle 1|\Delta\mathfrak{Q}^\dagger\,\Delta\mathfrak{R}|1\rangle\big/\langle 1|\Delta\mathfrak{R}^\dagger\,\Delta\mathfrak{R}|1\rangle,$$

and substitution of this value of $|\alpha|$ into the inequality yields

$$0 \le \langle 1|\Delta\mathfrak{Q}^\dagger\,\Delta\mathfrak{Q}|1\rangle - \left(\mathrm{Re}\,\frac{\alpha}{|\alpha|}\langle 1|\Delta\mathfrak{Q}^\dagger\,\Delta\mathfrak{R}|1\rangle\right)^2\Big/\langle 1|\Delta\mathfrak{R}^\dagger\,\Delta\mathfrak{R}|1\rangle \tag{10.78}$$

Furthermore, if we let $\alpha/|\alpha| = e^{i\varphi}$ and $\langle 1|\Delta\mathfrak{Q}^\dagger\,\Delta\mathfrak{R}|1\rangle = \beta + i\gamma$, with β, γ real, we get

$$\mathrm{Re}\,e^{i\varphi}(\beta + i\gamma) = \beta\cos\varphi - \gamma\sin\varphi$$

which is maximized by choosing

$$\tan\varphi = -\gamma/\beta,$$

or

$$\cos\varphi = \beta/\sqrt{\gamma^2 + \beta^2}, \qquad \sin\varphi = -\gamma/\sqrt{\gamma^2 + \beta^2}.$$

Hence the maximum value of $\mathrm{Re}\,\alpha/|\alpha|\,\langle 1|\Delta\mathfrak{Q}^\dagger\,\Delta\mathfrak{R}|1\rangle$ is $(\beta^2 + \gamma^2)/\sqrt{\beta^2 + \gamma^2} = \sqrt{\beta^2 + \gamma^2}$, or

$$\mathrm{Re}\,\frac{\alpha}{|\alpha|}\langle 1|\Delta\mathfrak{Q}^\dagger\,\Delta\mathfrak{R}|1\rangle \le |\langle 1|\Delta\mathfrak{Q}^\dagger\,\Delta\mathfrak{R}|1\rangle|. \tag{10.79}$$

The minimum value of the right-hand side of (10.78) is therefore

$$\langle 1|\Delta\mathfrak{Q}^\dagger\,\Delta\mathfrak{Q}|1\rangle - |\langle 1|\Delta\mathfrak{Q}^\dagger\,\Delta\mathfrak{R}|1\rangle|^2\big/\langle 1|\Delta\mathfrak{R}^\dagger\,\Delta\mathfrak{R}|1\rangle,$$

and the most stringent form of the inequality becomes

$$|\langle 1|\Delta\mathfrak{Q}^\dagger\,\Delta\mathfrak{R}|1\rangle|^2 \le \langle 1|\Delta\mathfrak{Q}^\dagger\,\Delta\mathfrak{Q}|1\rangle\langle 1|\Delta\mathfrak{R}^\dagger\,\Delta\mathfrak{R}|1\rangle$$
$$= (\Delta Q)^2(\Delta R)^2.$$

This is the result we desired.

We have, then, Eq. (10.74),

$$\Delta Q\,\Delta R \ge |\langle 1|\Delta\mathfrak{Q}^\dagger\,\Delta\mathfrak{R}|1\rangle|$$

which, for Hermitian operators $\mathfrak{Q}$ and $\mathfrak{R}$, becomes

$$\begin{aligned}\Delta Q\,\Delta R &\geq |\langle 1|\Delta\mathfrak{Q}\,\Delta\mathfrak{R}|1\rangle| \\ &= |\langle 1|\tfrac{1}{2}[\Delta\mathfrak{Q},\Delta\mathfrak{R}] + \tfrac{1}{2}\{\Delta\mathfrak{Q},\Delta\mathfrak{R}\}|1\rangle| \\ &= \tfrac{1}{2}[|\langle 1|[\Delta\mathfrak{Q},\Delta\mathfrak{R}]|1\rangle|^2 + |\langle 1|\{\Delta\mathfrak{Q},\Delta\mathfrak{R}\}|1\rangle|^2 \\ &\qquad + 2\,\mathrm{Re}\,\langle 1|[\Delta\mathfrak{Q},\Delta\mathfrak{R}]|1\rangle\langle 1|\{\Delta\mathfrak{Q},\Delta\mathfrak{R}\}|1\rangle^*]^{1/2},\end{aligned}$$

where $\{\Delta\mathfrak{Q},\Delta\mathfrak{R}\} \equiv \Delta\mathfrak{Q},\Delta\mathfrak{R} + \Delta\mathfrak{R}\,\Delta\mathfrak{Q}$ is called the *anticommutator* of $\Delta\mathfrak{Q},\Delta\mathfrak{R}$. The last term in the square root vanishes, as the reader may verify. Furthermore,

$$[\Delta\mathfrak{Q},\Delta\mathfrak{R}] = [\mathfrak{Q},\mathfrak{R}] \tag{10.80}$$

and

$$\langle\{\Delta\mathfrak{Q},\Delta\mathfrak{R}\}\rangle = \langle\{\mathfrak{Q},\mathfrak{R}\}\rangle - 2\langle\mathfrak{Q}\rangle\langle\mathfrak{R}\rangle. \tag{10.81}$$

Putting all this together, we get

$$\Delta Q\,\Delta R \geq \tfrac{1}{2}(1\langle[\mathfrak{Q},\mathfrak{R}]\rangle|^2 + |\langle\{\mathfrak{Q},\mathfrak{R}\}\rangle - 2\langle\mathfrak{Q}\rangle\langle\mathfrak{R}\rangle|^2)^{1/2}. \tag{10.82}$$

If $[\mathfrak{Q},\mathfrak{R}] = 0$, we have the result

$$\Delta Q\,\Delta R \geq |\langle\mathfrak{Q}\mathfrak{R}\rangle - \langle\mathfrak{Q}\rangle\langle\mathfrak{R}\rangle| \geq 0;$$

but for certain states (e.g. if $|1\rangle$ is an eigenstate of $\mathfrak{R}$) we shall have $\langle\mathfrak{Q}\mathfrak{R}\rangle = \langle\mathfrak{Q}\rangle\cdot\langle\mathfrak{R}\rangle$ and the inequality becomes trivial. On the other hand, if $[\mathfrak{Q},\mathfrak{R}] \neq 0$, we obtain

$$\Delta Q\,\Delta R \geq \tfrac{1}{2}|\langle[\mathfrak{Q},\mathfrak{R}]\rangle|, \tag{10.83}$$

the equality holding only if

$$\tfrac{1}{2}\langle\{\mathfrak{Q},\mathfrak{R}\}\rangle = \langle\mathfrak{Q}\rangle\langle\mathfrak{R}\rangle, \tag{10.84}$$

i.e.

$$\langle\{\Delta\mathfrak{Q},\Delta\mathfrak{R}\}\rangle = 0 \tag{10.84$'$}$$

and if also

$$(\Delta\mathfrak{Q} - \alpha\,\Delta\mathfrak{R})|1\rangle = 0 \tag{10.85}$$

for some value of α. That (10.85) is required for the equality to hold in (10.83) follows from (10.77).

The inequality (10.83) indicates the theoretical limit of simultaneous measurability for two physical variables Q and R. From Eq. (10.48), for example, we get

$$\Delta p^\mu\,\Delta x^\nu \geq (\hbar/2)|g^{\mu\nu}| = (\hbar/2)\,\delta^{\mu\nu},$$

which justifies our earlier statement that the minimum uncertainty product for canonically conjugate four-momenta and four-position operators is $\hbar/2$. Later we shall see that the inequality (10.75) has important consequences even for operators that are not canonically conjugate, such as the components of angular momentum.

The connection between the uncertainty principle for two variables Q and R, and the simultaneous diagonability of the corresponding operators, $\mathfrak{Q}$ and $\mathfrak{R}$, should now be apparent. Two variables can be arbitrarily well defined simultaneously if and only if $\Delta Q\, \Delta R$ has 0 as a minimum value, i.e. if and only if $[\mathfrak{Q}, \mathfrak{R}] = 0$. The condition $[\mathfrak{Q}, \mathfrak{R}] = 0$, however, is the necessary and sufficient condition that $\mathfrak{Q}$ and $\mathfrak{R}$ be simultaneously diagonable, as we saw in Section 10.4.

If $[\mathfrak{Q}, \mathfrak{R}] \neq 0$, it may occur that $\langle 1|[\mathfrak{Q}, \mathfrak{R}]|1\rangle = 0$ for *some* states $|1\rangle$, and hence for such a state $\Delta Q\, \Delta R = 0$; however, there must be other states for which $\langle[\mathfrak{Q}, \mathfrak{R}]\rangle \neq 0$. On the other hand, if $[\mathfrak{Q}, \mathfrak{R}] = \text{const} \cdot 1$, that is, if $[\mathfrak{Q}, \mathfrak{R}]$ is proportional to the identity operator with a nonvanishing constant of proportionality, then $\langle[\mathfrak{Q}, \mathfrak{R}]\rangle \neq 0$ for all states. In such cases, i.e. when $\mathfrak{Q}$ and $\mathfrak{R}$ are canonically conjugate to within a constant factor, R must be completely undefined ($\Delta R = \infty$) whenever Q is well-determined ($\Delta Q = 0$). Then eigenstates of $\mathfrak{Q}$ are unchanged by shifts in the value of R, the uncertainty in R being infinite. Conversely, if $\mathfrak{Q}$ and $\mathfrak{R}$ are canonically conjugate, and if a state $|1\rangle$ is unchanged by a shift in the value of R,

$$e^{\pm iR\mathfrak{Q}/\hbar}|1\rangle = \phi|1\rangle, \qquad |\phi| = 1,$$

that is, if the effect of a unitary transformation with R as the transformation parameter is to multiply the state by a phase factor, then $|1\rangle$ must be an eigenstate of the generator of the transformation, $\mathfrak{Q}$. Hence every eigenstate is left unchanged by some set of unitary transformations and therefore is related to a symmetry operation (transformation). Every eigenvalue problem corresponds to finding the states that are essentially left alone by some set of transformations, i.e. finding some symmetry of the system represented by the states.

READING NOTES*

Reference 24, Chapters 4, 5.

Reference 38, Chapters 5, 7.

Reference 43, Chapter 6.

PROBLEMS

1. Derive Eq. (10.4) for arbitrary operators $\mathfrak{Q}$, $\mathfrak{R}$ and $\mathfrak{S}$.
2. Show that

$$\mathsf{E} = \delta(m - m') \cdots i\hbar\, \partial/\partial t,$$

$$\mathsf{X} = \delta(m - m') \cdots \mathbf{x},$$

$$\mathsf{T} = \delta(m - m') \cdots t,$$

as indicated at the end of Section 10.7.

---

* Works referred to in these notes will be found in the Bibliography on p. 577.

3. Note that $i\hbar(\partial/\partial t)t|1(t)\rangle \neq \mathfrak{H}t|1(t)\rangle$. Why not?

4. Consider a potential that can be written as a power series in $\mathbf{x}$:

$$V(\mathbf{x}) = \sum_{n_1,n_2,n_3=0}^{3} v_{n_1,n_2,n_3} x_1^{n_1} x_2^{n_2} x_3^{n_3}.$$

Transcribe this to operator notation and thereby show that for potentials

$$[\mathfrak{P}, \mathfrak{V}(\mathfrak{X})] = \frac{\hbar}{i} \nabla_{\mathfrak{x}} \mathfrak{V}(\mathfrak{X}).$$

5. Show that $\langle 1|[\mathfrak{Q}, \mathfrak{R}]|1\rangle$ is imaginary and $\langle 1|\{\mathfrak{Q}, \mathfrak{R}\}|1\rangle$ is real for any Hermitian operators $\mathfrak{Q}$ and $\mathfrak{R}$ and thus that

$$\operatorname{Re} \langle 1|[\Delta\mathfrak{Q}, \Delta\mathfrak{R}]|1\rangle \langle 1|\{\Delta\mathfrak{Q}, \Delta\mathfrak{R}\}|1\rangle^* = 0$$

in the equation preceding (10.80).

6. Note that Eqs. (10.84′) and (10.85) can be construed as conditions on the state vector for which the left-hand side of (10.83) is minimal. For example, consider motion of a particle in one dimension, the momentum and position operators in that dimension being $\mathfrak{P}$ and $\mathfrak{X}$. Let $|1\rangle$ represent a state for that particle in the frame of reference such that $\langle 1|\mathfrak{P}|1\rangle = 0$ and in terms of a coordinate system chosen so that $\langle 1|\mathfrak{X}|1\rangle = 0$. Then, assuming that state $|1\rangle$ satisfies

$$\langle 1|\{\Delta\mathfrak{P}, \Delta\mathfrak{X}\}|1\rangle = \langle 1|\{\mathfrak{P}, \mathfrak{X}\}|1\rangle = 0$$

and

$$(\Delta\mathfrak{P} - \alpha\, \Delta\mathfrak{X})|1\rangle = (\mathfrak{P} - \alpha\mathfrak{X})|1\rangle = 0$$

use *only* these equations, together with $[\mathfrak{P}, \mathfrak{X}] = \hbar/i$ to show that we must have

$$\alpha = i\hbar/(\Delta X)^2 = i4(\Delta P)^2/\hbar, \qquad \Delta X\, \Delta P = \hbar/2.$$

7. Show that the spatial translation operator is $e^{-i\mathbf{x}\cdot\mathfrak{P}/\hbar}$.

8. Find $\langle \mathbf{p}, E, m, \ldots|\mathfrak{X}|\mathbf{p}', E', m', \ldots\rangle$, $\langle \mathbf{p}, E, m, \ldots|\mathfrak{T}|\mathbf{p}', E', m', \ldots\rangle$, $\langle \mathbf{p}, E, m, \ldots|\mathfrak{E}|\mathbf{p}', E', m', \ldots\rangle$ $\langle \mathbf{p}, E, m, \ldots|\mathfrak{P}|\mathbf{p}', E', m', \ldots\rangle$ in terms of $\mathbf{p}$, E, $\partial/\partial E$ and $\nabla_{\mathbf{p}} = \sum_{j=1}^{3} \hat{\mathbf{i}}_j\, \partial/\partial p_j$.

9. Verify Eq. (10.79).

10. If $\mathfrak{A}$ is an operator, it is called antihermitian if $\mathfrak{A}^+ = -\mathfrak{A}$. Show that an arbitrary operator can be written as the sum of a Hermitian operator and an antihermitian operator; write down the Hermitian and antihermitian parts in terms of the original operator.

11. Prove the commutator relations

$$[\mathfrak{A}, \alpha_1\mathfrak{B} + \alpha_2\mathfrak{C}] = \alpha_1[\mathfrak{A}, \mathfrak{B}] + \alpha_2[\mathfrak{A}, \mathfrak{C}]$$

$$[\mathfrak{A}, \mathfrak{B}\mathfrak{C}] = [\mathfrak{A}, \mathfrak{B}]\mathfrak{C} + \mathfrak{B}[\mathfrak{A}, \mathfrak{C}]$$

$$[\mathfrak{A}, [\mathfrak{B}, \mathfrak{C}]] = [[\mathfrak{A}, \mathfrak{B}], \mathfrak{C}] + [\mathfrak{B}, [\mathfrak{A}, \mathfrak{C}]],$$

and the relation $[\mathfrak{A}, \mathfrak{B}^2] = \{\mathfrak{B}, [\mathfrak{A}, \mathfrak{B}]\}$.

12. Use the operator transcription of the potential in Problem 4 in the Hamiltonian $\mathfrak{H} = \frac{1}{2}m\mathfrak{P}^2 + \mathfrak{V}(\mathfrak{X})$.
a) Show that $\langle \mathbf{x}'|\mathfrak{V}(\mathfrak{X})|x\rangle = V(x)\,\delta^3(\mathbf{x}' - \mathbf{x})$.
b) Evaluate $[\mathfrak{H}, \mathfrak{P}_j\mathfrak{X}_k]$.
c) Consider the expectation value of this expression for some energy eigenstate $|E\rangle$ ($\mathfrak{H}|E\rangle = E|E\rangle$) and thereby justify the *virial theorem*, $\langle\mathfrak{K}\rangle = -\frac{1}{2}\langle\boldsymbol{\mathfrak{F}}\cdot\boldsymbol{\mathfrak{X}}\rangle$, where $\mathfrak{K}$ is the kinetic energy and $\boldsymbol{\mathfrak{F}}$ is the force. (Remember in classical mechanics that for conservative potentials $\mathbf{F} = -\boldsymbol{\nabla}V(\mathbf{x})$.)

13. Let γ, ε be numbers and let $\mathfrak{A}$, $\mathfrak{B}$ be operators such that $[\mathfrak{A}, \mathfrak{B}] = \varepsilon$; then show that $e^{\alpha(\mathfrak{A}+\mathfrak{B})} = e^{\alpha\mathfrak{A}}\, e^{\alpha\mathfrak{B}}\, e^{-\varepsilon\alpha^2/2}$. (Hint: take $d/d\alpha$ of the left-hand side and integrate.)

14. Show that in an eigenstate of energy for which the energy eigenvalue is a member of a discrete set of eigenvalues, a system has a vanishing expectation value for the linear momentum. Use $\mathfrak{H} = (1/2m)\boldsymbol{\mathfrak{P}}^2 + \mathfrak{V}(\boldsymbol{\mathfrak{X}})$. (Hint: consider $[\mathfrak{H}, \boldsymbol{\mathfrak{X}}]$.)

CHAPTER 11

ELABORATION AND EXAMPLES

The mathematical formalism developed in the two preceding chapters has been presented with little reference to calculations of the sort required to solve actual physics problems. We now turn to two specific problems that illustrate some of the important features of quantum theory as well as providing some insights into problem-solving techniques.

11.1 THE MINIMUM WAVE PACKET

An interesting question always arises when a minimum uncertainty product is stated; namely for what state does the minimum product hold? The answer of course will depend upon the particular variables whose uncertainties are involved. Especially interesting is the uncertainty product $\Delta p\,\Delta x$ between momentum and position in a given direction, and we now make good our earlier promise to demonstrate that the wave function yielding the minimum product is a Gaussian-shaped wave packet.

The essential features defining a *minimum uncertainty state* are expressed in Eqs. (10.84) or (10.84′), and (10.85). Equation (10.85) provides an eigenvalue problem,

$$(\mathfrak{Q} - \alpha\mathfrak{R})|1\rangle = (\langle\mathfrak{Q}\rangle - \alpha\langle\mathfrak{R}\rangle)|1\rangle \tag{11.1}$$

to be solved.

Now let $\mathfrak{Q} = \mathfrak{P}$, the momentum operator, and let $\mathfrak{R} = \mathfrak{X}$, the position operator, and construct the inner products of the right- and left-hand sides of (11.1) with $\langle x, m, \ldots|$, an eigenstate of $\mathfrak{X}$:

$$\langle x, m, \ldots|\mathfrak{P}|1\rangle - \alpha x\langle x, m, \ldots|1\rangle = (\langle\mathfrak{P}\rangle - \alpha\langle\mathfrak{X}\rangle)\langle x, m, \ldots|1\rangle$$

or, using the completeness condition and (10.53),

$$\frac{\hbar}{i}\frac{\partial}{\partial x}\langle x, m, \ldots|1\rangle = [\langle\mathfrak{P}\rangle + \alpha(x - \langle\mathfrak{X}\rangle)]\langle x, m, \ldots|1\rangle. \tag{11.2}$$

This gives us a differential equation for $\langle x, m, \ldots|1\rangle = \psi_1(x, m, \ldots)$. We are interested only in the variable x, so we can replace the partial derivative with a total derivative, and integrate the equation to get

$$\langle x, m, \ldots | 1 \rangle = \psi_1(x, m, \ldots)$$
$$= N_{m,\ldots} \exp\left\{\frac{i}{\hbar}\left[\bar{p}x + \frac{\alpha}{2}(x - \bar{x})^2\right]\right\}, \tag{11.3}$$

where we have written $\bar{p}$ for $\langle \mathfrak{P} \rangle = \langle 1|\mathfrak{P})1\rangle$ and $\bar{x}$ for $\langle \mathfrak{X} \rangle = \langle 1|\mathfrak{X}|1\rangle$ to emphasize that they merely are numbers.

Equations (10.84) and (10.85) can be used together to help normalize $\psi_1(x, m, \ldots)$. Combining those two equations yields

$$\begin{aligned}\langle 1|\Delta\mathfrak{Q}\,\Delta\mathfrak{R} + \Delta\mathfrak{R}\,\Delta\mathfrak{Q}|1\rangle &= \langle 1|\alpha^*(\Delta R)^2 + \alpha(\Delta R)^2|1\rangle \\ &= 2(\mathrm{Re}\,\alpha)\langle 1|(\Delta\mathfrak{R})^2|1\rangle\end{aligned} \tag{11.4}$$

Since $\langle 1|(\Delta\mathfrak{R})^2|1\rangle > 0$ unless $\Delta\mathfrak{R}|1\rangle = 0$, we must have

$$\mathrm{Re}\,\alpha = 0, \qquad \text{or} \qquad \alpha = i\beta, \qquad \beta \text{ real.}$$

Therefore (11.3) becomes

$$\psi_1(x, m, \ldots) = N_{m}, \ldots \exp\left\{\frac{1}{\hbar}\left[i\bar{p}x - \frac{\beta}{2}(x - \bar{x})^2\right]\right\}, \tag{11.5}$$

and the normalization condition

$$1 = \langle 1|1\rangle = \underset{x,m,\ldots}{\mathsf{S}} \langle 1|x, m, \ldots\rangle\langle x, m, \ldots|1\rangle$$
$$= \underset{m,\ldots}{\mathsf{S}} |N_{m}, \ldots|^2 \int_{-\infty}^{\infty} dx\, e^{-(\beta/\hbar)(x-\bar{x})^2}$$

yields

$$\underset{m,\ldots}{\mathsf{S}} |N_{m}, \ldots|^2 = \sqrt{\beta/\pi\hbar}. \tag{11.6}$$

If we consider a particle having no degrees of freedom besides position, then the parameters $m, \ldots$ and the summation sign in (11.6) disappear, and we get

$$N = (\beta/\pi\hbar)^{1/4}, \tag{11.6$'$}$$

where the arbitrary phase of N is chosen so that N is real.

Using the definition $\Delta x \equiv \sqrt{\langle 1|(\Delta\mathfrak{X})^2|1)}$, we can evaluate β through the condition

$$(\Delta x)^2 = \langle 1|(\Delta\mathfrak{X})^2|1\rangle$$
$$= \int_{-\infty}^{\infty} dx\,(x - \bar{x})^2|\psi_1(x)|^2 = \hbar/2\beta.$$

Hence $\beta = \hbar/[2(\Delta x)^2]$, and we have

$$\psi_1(x) = [2\pi(\Delta x)^2]^{-1/4} \exp\left[-\frac{(x-\bar{x})^2}{4(\Delta x)^2}\right] \exp(i\bar{p}x/\hbar). \tag{11.7}$$

We can reintroduce the parameters $m, \ldots$ simply by writing $N_m, \ldots$ in place of $[2\pi(\Delta x)^2]^{-1/4}$ and subjecting the $N_m, \ldots$ to (11.6) with β replaced by $\hbar/[2(\Delta x)^2]$. Compare this result with the discussion following Eq. (8.26′) in Chapter 8.

Obtaining a wave function $\psi_1(x, \ldots) = \langle x, \ldots | 1\rangle$ is equivalent to obtaining the state vector $|1\rangle$, and indeed $\langle x, \ldots | 1\rangle$ can be used as a representation for the state vector. Thus, in deriving (11.7) we have solved the problem of defining the state with minimum uncertainty.

Because of the symmetry between the roles played by $\mathfrak{P}$ and $\mathfrak{X}$, we expect $\langle p, \ldots | 1\rangle$ to have the form of (11.7) with p and x interchanged. To verify this, we simply calculate the Fourier transform of (11.7) by inserting a complete set of states in $\langle p, q, \ldots | 1\rangle$:

$$\begin{aligned}
\langle p, q, \ldots | 1\rangle &= \underset{x,m,\ldots}{\mathsf{S}} \langle p, q, \ldots | x, m, \ldots\rangle\langle x, m, \ldots | 1\rangle \\
&= \underset{m,\ldots}{\mathsf{S}} \int_{-\infty}^{\infty} dx \exp(-ipx/\hbar)\langle p, q, \ldots | 0, m, \ldots\rangle\langle x, m, \ldots | 1\rangle \\
&= \underset{m,\ldots}{\mathsf{S}} \langle p, q, \ldots | 0, m, \ldots\rangle N_{m,\ldots} \int_{-\infty}^{\infty} dx \exp\left[-\frac{(x-\bar{x})^2}{4(\Delta x)^2}\right] \\
&\qquad \times \exp\left[-\frac{i}{\hbar}(p-\bar{p})x\right] \\
&= M_{q,\ldots} \exp\left[-\frac{i}{\hbar}(p-\bar{p})\bar{x}\right] \int_{-\infty}^{\infty} dx \exp\left[\frac{(x-\bar{x})^2}{4(\Delta x)^2}\right] \\
&\qquad \times \exp\left[-\frac{i}{\hbar}(p-\bar{p})(x-\bar{x})\right] \\
&= M_{q,\ldots} \exp\left[-\frac{i}{\hbar}(p-\bar{p})\bar{x}\right] 2\sqrt{\pi}\, \Delta x \exp\left[-\frac{(p-\bar{p})^2(\Delta x)^2}{\hbar^2}\right],
\end{aligned}$$

where

$$M_{q,\ldots} = \underset{m,\ldots}{\mathsf{S}} \langle p, q, \ldots | 0, m, \ldots\rangle N_{m,\ldots}. \tag{11.8}$$

In the absence of the parameters $q, \ldots$, and $m, \ldots$, this last equation becomes

$$M = \langle p|0\rangle N = \langle p|0\rangle [2\pi(\Delta x)^2]^{-1/4} = (2\pi\hbar)^{-1/2}[2\pi(\Delta x)^2]^{-1/4}$$

and we have

$$\langle p|1\rangle = \left[\frac{2(\Delta x)^2}{\pi\hbar^2}\right]^{1/4} \exp\left(\frac{i}{\hbar}\bar{p}\bar{x}\right) \exp\left[-\frac{(p-\bar{p})^2(\Delta x)^2}{\hbar^2}\right] \exp\left[-\frac{i}{\hbar}\bar{x}p\right].$$

However, since $\Delta x\, \Delta p = \hbar/2$ for this state, we may replace Δx everywhere with $\hbar/(2\,\Delta p)$, to get

$$\langle p|1\rangle = [2\pi(\Delta p)^2]^{-1/4} \exp\left[-\frac{(p-\bar{p})^2}{4(\Delta p)^2}\right] \exp\left[-\frac{i}{\hbar}\bar{x}p\right]; \tag{11.9}$$

we have neglected the constant $\exp(i\bar{p}\bar{x}/\hbar)$ as an irrelevant phase factor. This has exactly the form of (11.7), with x and $-p$ interchanged; the sign change for p has no vital significance and results from the canonical conjugacy of x and p.

11.2 THE SPREADING OF A WAVE PACKET IN TIME

We have discussed the phenomenon of the spreading, in time, of a signal, and it would be nice to see just how the minimum wave packet, given in (11.7), spreads in time. The spreading phenomenon is dependent only upon the dispersion relation between frequency and wave number (or energy and momentum) and is not a consequence of the shape of the envelope of the wave packet. On the other hand, the *shape* of the spreading packet is determined by the shape of the initial packet.

Let us consider the time variation of the variance, $(\Delta x)^2$, of the position of a particle that is in state $|1\rangle = |1(t_0)\rangle$ at time t_0. The variance is given by

$$\begin{aligned}(\Delta x)^2_{t_0} &= \langle 1|(\mathfrak{X} - \bar{x})^2|1\rangle \\ &= \langle 1|(\mathfrak{X})^2|1\rangle - \langle 1|\mathfrak{X}|1\rangle^2\end{aligned} \tag{11.10}$$

at time t_0, whereas at time t,

$$\begin{aligned}(\Delta x)^2_t &= \langle 1(t)|(\mathfrak{X})^2|1(t)\rangle - \langle 1(t)|\mathfrak{X}|1(t)\rangle^2 \\ &= \langle 1|e^{i(t-t_0)\mathfrak{H}/\hbar}\mathfrak{X}^2 e^{-i(t-t_0)\mathfrak{H}/\hbar}|1\rangle - \langle 1|e^{-i(t-t_0)\mathfrak{H}/\hbar}\mathfrak{X}e^{i(t-t_0)\mathfrak{H}/\hbar}|1\rangle^2\end{aligned} \tag{11.11}$$

since $\partial\mathfrak{X}/\partial t = 0$. However,

$$\begin{aligned}e^{i(t-t_0)\mathfrak{H}/\hbar}\mathfrak{X}^2 e^{-i(t-t_0)\mathfrak{H}/\hbar} &= \mathfrak{X}^2 + \left[\frac{i}{\hbar}\mathfrak{H}, \mathfrak{X}^2\right](t-t_0) \\ &\quad + \frac{1}{2}\left[\frac{i}{\hbar}\mathfrak{H}, \left[\frac{i}{\hbar}\mathfrak{H}, \mathfrak{X}^2\right]\right](t-t_0)^2 + \cdots,\end{aligned} \tag{11.12}$$

and

$$[\mathfrak{H}, \mathfrak{X}^2] = \frac{1}{2M}[\mathfrak{P}^2, \mathfrak{X}^2] = \frac{\hbar}{iM}(\mathfrak{P}\cdot\mathfrak{X} + \mathfrak{X}\cdot\mathfrak{P})$$
$$= \frac{2\hbar}{iM}\mathfrak{X}\cdot\mathfrak{P} - 3\frac{\hbar^2}{M}, \tag{11.13}$$

$$[\mathfrak{H}, [\mathfrak{H}, \mathfrak{X}^2]] = \frac{2\hbar}{iM}[\mathfrak{H}, \mathfrak{X}\cdot\mathfrak{P}] = \frac{\hbar}{iM^2}[\mathfrak{P}^2, \mathfrak{X}\cdot\mathfrak{P}] + \frac{2\hbar}{iM}[\mathfrak{V}(\mathfrak{X}), \mathfrak{X}\cdot\mathfrak{P}]$$
$$= -2\left(\frac{\hbar}{M}\right)^2\mathfrak{P}^2 + \frac{2\hbar}{iM}[\mathfrak{V}(\mathfrak{X}), \mathfrak{X}\cdot\mathfrak{P}]. \tag{11.14}$$

From Eq. (11.14) it is clear that we can go no further unless the potential $\mathfrak{V}$ is specified.

Assume that $\mathfrak{V}$ vanishes: then

$$e^{i(t-t_0)\mathfrak{H}/\hbar}\mathfrak{X}^2 e^{-i(t-t_0)\mathfrak{H}/\hbar} = \mathfrak{X}^2 + \frac{1}{M}(\mathfrak{P}\cdot\mathfrak{X} + \mathfrak{X}\cdot\mathfrak{P})(t - t_0)$$
$$+ \frac{1}{M^2}\mathfrak{P}^2(t - t_0)^2,$$

and

$$e^{i(t-t_0)\mathfrak{H}/\hbar}\mathfrak{X}e^{-i(t-t_0)\mathfrak{H}/\hbar} = \mathfrak{X} + \frac{1}{M}\mathfrak{P}(t - t_0).$$

Substitution of these into Eq. (11.11) yields

$$(\Delta\mathbf{x})_t^2 = (\Delta\mathbf{x})_{t_0}^2 + \frac{1}{M}[\langle 1|\mathfrak{P}\cdot\mathfrak{X} + \mathfrak{X}\cdot\mathfrak{P}|1\rangle - 2\langle 1|\mathfrak{X}|1\rangle\cdot\langle 1|\mathfrak{P}|1\rangle](t - t_0)$$
$$+ \frac{1}{M^2}[\langle 1|\mathfrak{P}^2|1\rangle - \langle 1|\mathfrak{P}|1\rangle^2](t - t_0)^2. \tag{11.15}$$

Hence, for a free particle, that is, for $\mathfrak{V} = 0$,

$$\left.\frac{d}{dt}(\Delta\mathbf{x})_t^2\right|_{t=t_0} = \frac{1}{M}[\langle\mathfrak{P}\cdot\mathfrak{X} + \mathfrak{X}\cdot\mathfrak{P}\rangle - 2\langle\mathfrak{X}\rangle\cdot\langle\mathfrak{P}\rangle], \tag{11.16}$$

and

$$\left.\frac{d^2}{dt^2}(\Delta\mathbf{x})_t^2\right|_{t=t_0} = \frac{1}{M^2}[\langle\mathfrak{P}^2\rangle - \langle\mathfrak{P}\rangle^2] = \frac{(\Delta\mathbf{p})_{t_0}^2}{M^2}. \tag{11.17}$$

Since $(\Delta\mathbf{p})_{t_0}^2 > 0$, it follows from Eqs. (11.17) and (11.15) that $(\Delta\mathbf{x})_t^2$ eventually increases quadratically with $(t - t_0)$, and hence the spatial spread of the wave system continually increases and the "signal" therefore continually is "degraded".

For the one-dimensional Gaussian wave packet of the preceding section, we have $(\Delta p)_{t_0} = \hbar/[2(\Delta x)_{t_0}]$, so that

$$\left.\frac{d^2}{dt^2}(\Delta x)_t^2\right|_{t=t_0} = \frac{\hbar^2}{(2M)^2(\Delta x)_{t_0}^2}.$$

Also, since one of the conditions for the wave packet to be minimal was that

$$\langle\{\Delta\mathfrak{P}, \Delta\mathfrak{X}\}\rangle = \langle\mathfrak{P}\mathfrak{X}\rangle + \langle\mathfrak{X}\mathfrak{P}\rangle - 2\langle\mathfrak{X}\rangle\langle\mathfrak{P}\rangle = 0,$$

we have

$$\left.\frac{d}{dt}(\Delta x)_t^2\right|_{t=t_0} = 0$$

for this wave packet. Hence

$$(\Delta x)_t^2 = (\Delta x)_{t_0}^2 + \left(\frac{\hbar}{2M}\right)^2 \frac{(t - t_0)^2}{(\Delta x)_{t_0}^2}. \tag{11.18}$$

Therefore, by letting $\Delta x \equiv (\Delta x)_{t_0}$, $t_0 = 0$, we obtain, for $|\psi_1(x, t)|$, upon replacing Δx in Eq. (11.7) with $(\Delta x)_t$,

$$|\psi_1(x, t)| = \left\{2\pi\left[(\Delta x)^2 + \left(\frac{\hbar}{2M}\frac{t}{\Delta x}\right)^2\right]\right\}^{-1/4}$$

$$\cdot \exp\left\{-(x - \bar{x})^2/4\left[(\Delta x)^2 + \left(\frac{\hbar}{2M}\frac{t}{\Delta x}\right)^2\right]\right\}. \tag{11.19}$$

Another approach to this problem is to consider the state $|1\rangle$ evolving in time according to

$$|1(t)\rangle = e^{-i(t-t_0)\mathfrak{H}/\hbar}|1(t_0)\rangle.$$

For a free particle, $\mathfrak{H} = (1/2M)\mathfrak{P}^2$, so that inserting a complete set of states $|p, q, \ldots\rangle$ can convert the operator in the exponential to a number, for free particles:

$$|1(t)\rangle = \mathop{\mathsf{S}}_{p,q,\ldots} |p, q, \ldots\rangle\langle p, q, \ldots|e^{-i(t-t_0)p^2/2M\hbar}|1(t_0)\rangle.$$

Hence

$$\langle x, m, \ldots|1(t)\rangle = \mathop{\mathsf{S}}_{p,q,\ldots} \langle x, m, \ldots|p, q, \ldots\rangle e^{-i(t-t_0)p^2/2M\hbar}\langle p, q, \ldots|1(t_0)\rangle$$

$$= \mathop{\mathsf{S}}_{q,\ldots} \int_{-\infty}^{\infty} dp\, e^{ipx/\hbar} e^{-i(t-t_0)p^2/2M\hbar} e^{-ip\bar{x}/\hbar}$$

$$\cdot e^{-(p-\bar{p})^2(\Delta x)^2/\hbar^2}(2/\hbar)^{1/2}\,\Delta x\, M_{q,\ldots}.$$

The calculation from here is straightforward, and it is left for the reader to obtain $\psi_1(x, t) = \langle x|1(t)\rangle$ and to show that the result of this calculation leads to Eq. (11.19).

11.3 THE LINEAR HARMONIC OSCILLATOR

The linear harmonic oscillator appears throughout classical physics because it represents a first approximation to motion through small displacements from an equilibrium position. It is important in quantum physics for similar reasons. Furthermore, it provides a simple but elegant example of the application of the Heisenberg state-vector approach as opposed to the differential-equation approach of Schrödinger. It also serves as an introduction to the techniques used in studying angular momentum.

The Hamiltonian for the classical linear harmonic oscillator is given by

$$H = \frac{1}{2m} p^2 + \frac{m\omega_0^2}{2} x^2, \tag{11.20}$$

where $m\omega_0^2 = k$ is the *spring constant* or *force constant* of the oscillator, m is the mass of the particle subjected to the oscillator potential $V(x) = \frac{1}{2}kx^2$, and $\omega_0 = \sqrt{k/m}$ is the natural frequency of oscillations. Note that H can be factored conveniently into two complex-conjugate factors:

$$H = \left(\frac{p}{\sqrt{2m}} + i\sqrt{\frac{m}{2}}\,\omega_0 x\right)\left(\frac{p}{\sqrt{2m}} - i\sqrt{\frac{m}{2}}\,\omega_0 x\right). \tag{11.21}$$

In making the transition to quantum theory, we must replace p with $\mathfrak{P}$ and x with $\mathfrak{X}$; then the complex factors in Eq. (11.21) become

$$\frac{1}{\sqrt{2m}}\,\mathfrak{P} \pm i\sqrt{\frac{m}{2}}\,\omega_0 \mathfrak{X}. \tag{11.22}$$

Unfortunately, neither of these factors is Hermitian, nor is their product equal to

$$\frac{1}{2m}\,\mathfrak{P}^2 + \frac{m}{2}\,\omega_0^2 \mathfrak{X}^2. \tag{11.23}$$

On the other hand,

$$\begin{aligned}\mathfrak{H} = \tfrac{1}{2}\Bigg\{&\left(\frac{1}{\sqrt{2m}}\,\mathfrak{P} + i\sqrt{\frac{m}{2}}\,\omega_0 \mathfrak{X}\right)\left(\frac{1}{\sqrt{2m}}\,\mathfrak{P} - i\sqrt{\frac{m}{2}}\,\omega_0 \mathfrak{X}\right)\\ &+ \left(\frac{1}{\sqrt{2m}}\,\mathfrak{P} - i\sqrt{\frac{m}{2}}\,\omega_0 \mathfrak{X}\right)\left(\frac{1}{\sqrt{2m}}\,\mathfrak{P} + i\sqrt{\frac{m}{2}}\,\omega_0 \mathfrak{X}\right)\Bigg\}\end{aligned} \tag{11.24}$$

is equal to (11.23) and is also Hermitian. Let us demonstrate the validity of this assertion.

It is convenient to replace $\mathfrak{H}$, which has the dimensions of energy, with the dimensionless quantity $\mathfrak{H}/\hbar\omega_0$. With this in mind, we define

$$\mathfrak{a} \equiv \left(\frac{1}{\sqrt{2m\hbar\omega_0}}\,\mathfrak{P} - i\sqrt{\frac{m\omega_0}{2\hbar}}\,\mathfrak{X}\right) \tag{11.25}$$

and from this we obtain

$$\mathfrak{a}^\dagger = \left(\frac{1}{\sqrt{2m\hbar\omega_0}}\,\mathfrak{P} + i\sqrt{\frac{m\omega_0}{2\hbar}}\,\mathfrak{X}\right), \tag{11.26}$$

since $\mathfrak{P}$ and $\mathfrak{X}$ are Hermitian: $\mathfrak{a}$ and $\mathfrak{a}^+$ are just $(\hbar\omega_0)^{-1/2}$ times the factors given in (11.24). Hence Eq. (11.24) becomes

$$\mathfrak{H} = \tfrac{1}{2}\hbar\omega_0(\mathfrak{a}^\dagger\mathfrak{a} + \mathfrak{a}\mathfrak{a}^\dagger) = \tfrac{1}{2}\hbar\omega_0\{\mathfrak{a}^\dagger, \mathfrak{a}\}$$

and in this form it is obviously Hermitian; here $\{\mathfrak{a}^\dagger, \mathfrak{a}\} = \mathfrak{a}^\dagger\mathfrak{a} + \mathfrak{a}\mathfrak{a}^\dagger$ is the anti-commutator of $\mathfrak{a}^\dagger$ and $\mathfrak{a}$.

Knowing the commutation properties of $\mathfrak{P}$ and $\mathfrak{X}$, we can calculate $[\mathfrak{a}^\dagger, \mathfrak{a}]$ directly:

$$\begin{aligned}
[\mathfrak{a}^\dagger, \mathfrak{a}] &= \left[\left(\frac{1}{\sqrt{2m\hbar\omega_0}}\,\mathfrak{P} + i\sqrt{\frac{m\omega_0}{2\hbar}}\,\mathfrak{X}\right), \left(\frac{1}{\sqrt{2m\hbar\omega_0}}\,\mathfrak{P} - i\sqrt{\frac{m\omega_0}{2\hbar}}\,\mathfrak{X}\right)\right] \\
&= \frac{1}{2m\hbar\omega_0}[\mathfrak{P}, \mathfrak{P}] - i\frac{1}{2\hbar}[\mathfrak{P}, \mathfrak{X}] + i\frac{1}{2\hbar}[\mathfrak{X}, \mathfrak{P}] + \frac{m\omega_0}{2\hbar}[\mathfrak{X}, \mathfrak{X}] \\
&= -\frac{i}{\hbar}[\mathfrak{P}, \mathfrak{X}] = -1,
\end{aligned}$$

or

$$[\mathfrak{a}, \mathfrak{a}^\dagger] = 1. \tag{11.27}$$

Therefore $\{\mathfrak{a}^\dagger, \mathfrak{a}\} = 2\mathfrak{a}^\dagger\mathfrak{a} + 1$ and the Hamiltonian can be written

$$\mathfrak{H} = \hbar\omega_0(\mathfrak{a}^\dagger\mathfrak{a} + \tfrac{1}{2}). \tag{11.28}$$

Now let us assume that we have an eigenstate $|E\rangle$ of $\mathfrak{H}$ with eigenvalue E:

$$\mathfrak{H}|E\rangle = E|E\rangle$$

Is $\mathfrak{a}|E\rangle$ or $\mathfrak{a}^\dagger|E\rangle$ again an eigenstate of $\mathfrak{H}$? To find out, we calculate

$$\begin{aligned}
\mathfrak{H}\mathfrak{a}|E\rangle &= \hbar\omega_0(\mathfrak{a}^\dagger\mathfrak{a}\mathfrak{a} + \tfrac{1}{2}\mathfrak{a})|E\rangle = \hbar\omega_0(\mathfrak{a}\mathfrak{a}^\dagger\mathfrak{a} + [\mathfrak{a}^\dagger, \mathfrak{a}]\mathfrak{a} + \tfrac{1}{2}\mathfrak{a})|E\rangle \\
&= \hbar\omega_0\mathfrak{a}(\mathfrak{a}^\dagger a - \tfrac{1}{2})|E\rangle = \mathfrak{a}(\mathfrak{H} - \hbar\omega_0)|E\rangle \\
&= \mathfrak{a}(E - \hbar\omega_0)|E\rangle = (E - \hbar\omega_0)\mathfrak{a}|E\rangle.
\end{aligned}$$

Therefore $\mathfrak{a}|E\rangle$ is an eigenstate of $\mathfrak{H}$ with eigenvalue $E - \hbar\omega_0$. Similarly,

$$\mathfrak{H}\mathfrak{a}^\dagger|E\rangle = (E + \hbar\omega_0)\mathfrak{a}^\dagger|E\rangle$$

so that $\mathfrak{a}^\dagger|E\rangle$ is also an eigenstate of $\mathfrak{H}$ and has eigenvalue $(E + \hbar\omega_0)$.

The same argument may be applied to $\mathfrak{a}(\mathfrak{a}|E\rangle$ or $\mathfrak{a}^\dagger(\mathfrak{a}^\dagger|E\rangle)$ to show that

$$\mathfrak{H}\mathfrak{a}^2|E\rangle = (E - 2\hbar\omega_0)\mathfrak{a}^2|E\rangle$$

or

$$\mathfrak{H}(\mathfrak{a}^\dagger)^2|E\rangle = (E + 2\hbar\omega_0)(\mathfrak{a}^\dagger)^2|E\rangle. \tag{11.29}$$

Indeed it is clear that the argument may be applied as often as desired to yield

$$\begin{aligned} \mathfrak{H}\mathfrak{a}^n|E\rangle &= (E - n\hbar\omega_0)\mathfrak{a}^n|E\rangle, \\ \mathfrak{H}(\mathfrak{a}^\dagger)^n|E\rangle &= (E + n\hbar\omega_0)(\mathfrak{a}^\dagger)^n|E\rangle. \end{aligned} \tag{11.30}$$

Thus $\mathfrak{a}$ is a *step-down* or *destruction* or *annihilation* operator in that it lowers the energy of a state by $\hbar\omega_0$, whereas $\mathfrak{a}^\dagger$ is a *step-up* or *creation* operator in that it raises the energy by $\hbar\omega_0$.

If the eigenvalue E is to correspond to what we normally call energy, it must have a lower bound, E_0. If $|E_0\rangle$ is the eigenstate having this lowest energy, it follows that

$$\mathfrak{a}|E_0\rangle = 0$$

for if this were not so, the state $\mathfrak{a}|E_0\rangle$ would have energy $E_0 - \hbar\omega_0$, contradicting the assumption that E_0 is the lowest energy. On the other hand, operating on $|E_0\rangle$ with $(\mathfrak{a}^\dagger)^n$ yields

$$(\mathfrak{a}^\dagger)^n|E_0\rangle = N_n^{-1}|E_0 + n\hbar\omega_0\rangle,$$

that is, $(\mathfrak{a}^\dagger)^n|E_0\rangle$ has an energy $n\hbar\omega_0$ units above the lowest energy level. The quantity N_n^{-1} is simply a normalization constant. In picturesque language, we say that the lowest state, or ground state, $|E_0\rangle$, has no quanta—it is the vacuum state—whereas the state $(\mathfrak{a}^\dagger)^n|E_0\rangle$ has n quanta, each contributing $\hbar\omega_0$ to the energy of the state.

With this terminology in mind, we define a *number operator*

$$\mathfrak{n} \equiv \mathfrak{a}^\dagger\mathfrak{a};$$

clearly

$$\mathfrak{H} = \hbar\omega_0(\mathfrak{n} + \tfrac{1}{2}). \tag{11.31}$$

Since $\mathfrak{a}|E_0\rangle = 0$, we have

$$\mathfrak{n}|E\rangle = \mathfrak{a}^\dagger\mathfrak{a}|E_0\rangle = 0,$$

whereas

$$\begin{aligned} \mathfrak{n}\mathfrak{a}^\dagger|E_0\rangle &= \mathfrak{a}^\dagger\mathfrak{a}\mathfrak{a}^\dagger|E_0\rangle \\ &= (\mathfrak{a}^\dagger[\mathfrak{a}, \mathfrak{a}^\dagger] + \mathfrak{a}^\dagger\mathfrak{a}^\dagger\mathfrak{a})|E_0\rangle = \mathfrak{a}^\dagger|E_0\rangle. \end{aligned}$$

Hence $|E_0\rangle$ is an eigenstate of $\mathfrak{n}$ with eigenvalue 0 and $\mathfrak{a}^\dagger|E_0\rangle$ is an eigenstate of $\mathfrak{n}$ with eigenvalue 1.

By mathematical induction it is easy to show that

$$\mathfrak{n}(\mathfrak{a}^\dagger)^n = \mathfrak{n}(\mathfrak{a}^\dagger)^n + (\mathfrak{a}^\dagger)^n\mathfrak{n},$$

or

$$[\mathfrak{n}, (\mathfrak{a}^\dagger)^n] = n(\mathfrak{a}^\dagger)^n.$$

This in turn leads to the result

$$\mathfrak{n}(\mathfrak{a}^\dagger)^n|E_0\rangle = n(\mathfrak{a}^\dagger)^n|E_0\rangle$$

since $\mathfrak{n}|E_0\rangle = 0$; hence $(\mathfrak{a}^\dagger)^n|E_0\rangle$ is an eigenstate of $\mathfrak{n}$ with eigenvalue n. Since $\mathfrak{H} = \hbar\omega_0(\mathfrak{n} + \frac{1}{2})$, we have

$$\mathfrak{H}(\mathfrak{a}^\dagger)^n|E_0\rangle = \hbar\omega_0(n + \tfrac{1}{2})(\mathfrak{a}^\dagger)^n|E_0\rangle = (E_0 + n\hbar\omega_0)(\mathfrak{a}^\dagger)^n|E_0\rangle.$$

In particular, for $n = 0$, this gives

$$E_0 = \tfrac{1}{2}\hbar\omega_0 > 0;$$

the ground state has nonvanishing energy!

Because of the relation (11.31) between $\mathfrak{H}$ and $\mathfrak{n}$, it is common to label the states of a harmonic oscillator with the number of quanta rather than with the energy. States so labeled are said to be states in the *number representation*, and they are denoted $|n\rangle$; thus

$$\mathfrak{n}|n\rangle = n|n\rangle,$$

and

$$\mathfrak{H}|n\rangle = \hbar\omega_0(n + \tfrac{1}{2})|n\rangle.$$

The operator $\mathfrak{n} = \mathfrak{a}^\dagger\mathfrak{a}$ is obviously Hermitian.

Let the ground state $|0\rangle$ be given and let it be properly normalized:

$$\langle 0|0\rangle = 1.$$

Then the state $|n\rangle$ is determined in terms of $|0\rangle$;

$$|n\rangle = N_n(\mathfrak{a}^\dagger)^n|0\rangle,$$

where N_n is a normalization constant to be determined. If we demand that $\langle n|n\rangle = 1$, we have

$$1 = \langle 0|\mathfrak{a}^n|N_n|^2(\mathfrak{a}^\dagger)^n|0\rangle = |N_n|^2\langle 0|\mathfrak{a}^n(\mathfrak{a}^\dagger)^n|0\rangle. \tag{11.32}$$

Since $[\mathfrak{a}, \mathfrak{a}^\dagger] = 1$, we can calculate

$$\begin{aligned}[\mathfrak{a}, (\mathfrak{a}^\dagger)^n] &= [\mathfrak{a}, \mathfrak{a}^\dagger](\mathfrak{a}^\dagger)^{n-1} + \mathfrak{a}^\dagger[\mathfrak{a}, (\mathfrak{a}^\dagger)^{n-1}] \\ &= (\mathfrak{a}^\dagger)^{n-1} + \mathfrak{a}^\dagger[\mathfrak{a}, \mathfrak{a}^\dagger](\mathfrak{a}^\dagger)^{n-2} + \mathfrak{a}^\dagger\mathfrak{a}^\dagger[\mathfrak{a}, (\mathfrak{a}^\dagger)^{n-2}] \\ &= (\mathfrak{a}^\dagger)^{n-1} + \mathfrak{a}^\dagger(\mathfrak{a}^\dagger)^{n-2} + \mathfrak{a}^\dagger\mathfrak{a}^\dagger[\mathfrak{a}, (\mathfrak{a}^\dagger)^{n-2}] \\ &= \cdots = n(\mathfrak{a}^\dagger)^{n-1}.\end{aligned}$$

Therefore

$$\mathfrak{a}^n(\mathfrak{a}^\dagger)^n = \mathfrak{a}^{n-1}[n(\mathfrak{a}^\dagger)^{n-1} + (\mathfrak{a}^\dagger)^n\mathfrak{a}]$$

or

$$\begin{aligned}\mathfrak{a}^n(\mathfrak{a}^\dagger)^n|0\rangle &= \mathfrak{a}^{n-1}[n(\mathfrak{a}^\dagger)^{n-1} + (\mathfrak{a}^\dagger)^n\mathfrak{a}]|0\rangle \\ &= \mathfrak{a}^{n-1}n(\mathfrak{a}^\dagger)^{n-1}|0\rangle = n\mathfrak{a}^{n-1}(\mathfrak{a}^\dagger)^{n-1}|0\rangle \\ &= \cdots = n(n-1)\cdots 1|0\rangle \\ &= n!|0\rangle.\end{aligned}$$

Therefore Eq. (11.32) can be written

$$1 = |N_n|^2 n!\langle 0|0\rangle = |N_n|^2 n!,$$

or, since the phase of a state is arbitrary, we may let

$$N_n = 1/\sqrt{n!}$$

and hence the state $|n\rangle$ may be defined as

$$|n\rangle = (1/\sqrt{n!})(\mathfrak{a}^+)^n|0\rangle. \tag{11.33}$$

The states $|n\rangle$ are therefore normalized, and since $\mathfrak{n}$ is Hermitian they must be orthogonal for different eigenvalues n of $\mathfrak{n}$; hence

$$\langle n|m\rangle = \delta_{n,m}.$$

Finally, let us calculate the inner products $\langle m|\mathfrak{a}|n\rangle$ and $\langle m|\mathfrak{a}^\dagger|n\rangle$. For the former we obtain

$$\begin{aligned}\langle m|\mathfrak{a}|n\rangle &= \left\langle m\left|\mathfrak{a}\frac{1}{\sqrt{n!}}(\mathfrak{a}^\dagger)^n\right|0\right\rangle = \frac{1}{\sqrt{n!}}\langle m|n(\mathfrak{a}^\dagger)^{n-1}|0\rangle \\ &= \frac{n\sqrt{(n-1)!}}{\sqrt{n!}}\langle m|n-1\rangle = \sqrt{n}\langle m|n-1\rangle \\ &= \sqrt{n}\,\delta_{m,n-1} = \sqrt{m+1}\,\delta_{n,m+1},\end{aligned} \tag{11.34}$$

where we have made use of Eq. (11.33). Since

$$\langle m|\mathfrak{a}^\dagger|n\rangle = \langle n|\mathfrak{a}|m\rangle^*, \tag{11.35}$$

Eq. (11.34) yields

$$\langle m|\mathfrak{a}^\dagger|n\rangle = \sqrt{m}\,\delta_{n,m-1} = \sqrt{n+1}\,\delta_{m,n+1}\,.$$

It is interesting to verify that $\langle m|n\rangle = \delta_{m,n}$ directly. Thus, for $m \geq n$,

$$\langle m|n\rangle = \left\langle 0\left|\frac{\mathfrak{a}^m}{\sqrt{m!}}\frac{(\mathfrak{a}^\dagger)^n}{\sqrt{n!}}\right|0\right\rangle = \frac{1}{\sqrt{n!m!}}\langle 0|\mathfrak{a}^{m-n}\mathfrak{a}^n(\mathfrak{a}^\dagger)^n|0\rangle$$

$$= \frac{1}{\sqrt{m!n!}}\langle 0|\mathfrak{a}^{m-n}n!|0\rangle = \sqrt{\frac{n!}{m!}}\langle 0|\mathfrak{a}^{m-n}|0\rangle$$

$$= \begin{cases} 0 & \text{if } m > n \\ 1 & \text{if } m = n. \end{cases}$$

Because $\langle n|m\rangle = \langle m|n\rangle^*$, the above result holds for $n \geq m$ also, and the verification is complete.

Another interesting check on our calculations is to evaluate $\langle m|\mathfrak{n}|n\rangle = \langle m|\mathfrak{a}^\dagger\mathfrak{a}|n\rangle$ by inserting a complete set of states $|l\rangle$, and using (11.34) and (11.35).

$$\begin{aligned}\langle m|\mathfrak{n}|n\rangle &= \sum_{l=0}^{\infty} \langle m|\mathfrak{a}^\dagger|l\rangle\,\langle l|\mathfrak{a}|n\rangle \\ &= \sum_{l=0}^{\infty} \sqrt{l+1}\,\delta_{m,l+1}\sqrt{n}\,\delta_{l,n-1} \qquad (11.36) \\ &= \sqrt{mn}\,\delta_{n,m} = n\,\delta_{n,m},\end{aligned}$$

just as we expect.

11.4 MATRIX ELEMENTS

Inner products in quantum mechanics are frequently called *matrix elements.* This terminology is justified by the correspondence between the sums over a complete set of states and the definition of matrix multiplication. For example, the sum in Eqs. (11.36) is just the multiplication of the (infinite-dimensional) matrices $\langle m|\mathfrak{a}^\dagger|l\rangle$ and $\langle l|\mathfrak{a}|n\rangle$. The inner product of two states can always be written, by virtue of the completeness assumption, as the inner product of two column vectors:

$$\langle \text{state 1}|\text{state 2}\rangle = \underset{q,r,\ldots}{\mathsf{S}}\ \langle \text{state 1}|q,r,\ldots\rangle\langle q,r,\ldots|\text{state 2}\rangle$$

where $\langle q,r,\ldots|\text{state 2}\rangle$ is one column vector and $\langle q,r,\ldots|\text{state 1}\rangle$ is the other.

In this language, the matrix element of $\mathfrak{Q}$ between state 1 and state 2 is $\langle\text{state 1}|\mathfrak{Q}|\text{state 2}\rangle$. Of course it is most useful to specify states 1 and 2 as eigenstates of some complete set of mutually commuting operators; then, if $|\text{state 1}\rangle = |m, n,\ldots\rangle$ and $|\text{state 2}\rangle = |m', n',\ldots\rangle$, we have $\langle m,n,\ldots|\mathfrak{Q}|m',n',\ldots\rangle$ as the matrix element. In any case $|\text{state } 1\rangle$ and $|\text{state } 2\rangle$ can be written as linear combinations of such eigenstates and we can write

$$\langle \text{state } 1|\mathfrak{Q}|\text{state } 2\rangle$$

$$= \underset{m,n,\ldots}{\mathsf{S}} \; \underset{m',n',\ldots}{\mathsf{S}} \langle \text{state } 1|m, n, \ldots\rangle\langle m, n, \ldots|\mathfrak{Q}|m', n', \ldots\rangle\langle m', n', \ldots|\text{state } 2\rangle.$$

In particular, we might as well utilize eigenstates of $\mathfrak{Q}$, say $\{|q, r, \ldots\rangle\}$, so that

$$\langle q, r, \ldots|\mathfrak{Q}|q', r', \ldots\rangle = q\,\delta(q - q')\,\delta(r - r')\cdots$$

and

$$\langle \text{state } 1|\mathfrak{Q}|\text{state } 2\rangle = \underset{q,r,\ldots}{\mathsf{S}} \langle \text{state } 1|q, r, \ldots\rangle q\langle q, r, \ldots|\text{state } 2\rangle.$$

Inner products, then, are matrix elements of the identity operator.

The abstract operator formalism outlined in Chapters 9 and 10 can be made concrete in terms of matrix elements, and in fact it is the role of wave functions to give concrete space–time expression to the state vectors.

As an example, we write out the matrices composed of matrix elements for the harmonic oscillator operators $\mathfrak{a}^\dagger$, $\mathfrak{a}$ and $\mathfrak{n}$ in the number representation:

$$[\langle m|\mathfrak{a}^\dagger|n\rangle] = \begin{array}{c|ccccc} & n & & & & \\ m & 0 & 1 & 2 & 3 & \cdots \\ \hline 0 & 0 & \sqrt{1} & 0 & 0 & \cdots \\ 1 & 0 & 0 & \sqrt{2} & 0 & \cdots \\ 2 & 0 & 0 & 0 & \sqrt{3} & \cdots \\ 3 & 0 & 0 & 0 & 0 & \cdots \\ \vdots & \vdots & \vdots & \vdots & \vdots & \end{array}$$

$$[\langle m|\mathfrak{a}^\dagger|\mathfrak{a}\rangle] = \begin{array}{c|ccccc} & n & & & & \\ m & 0 & 1 & 2 & 3 & \cdots \\ \hline 0 & 0 & 0 & 0 & 0 & \cdots \\ 1 & \sqrt{1} & 0 & 0 & 0 & \cdots \\ 2 & 0 & \sqrt{2} & 0 & 0 & \cdots \\ 3 & 0 & 0 & \sqrt{3} & 0 & \cdots \\ \vdots & \vdots & \vdots & \vdots & \vdots & \end{array}$$

$$[\langle m|\mathfrak{n}|n\rangle] = \begin{array}{c|ccccc} & n & & & & \\ m & 0 & 1 & 2 & 3 & \cdots \\ \hline 0 & 0 & 0 & 0 & 0 & \cdots \\ 1 & 0 & 1 & 0 & 0 & \cdots \\ 2 & 0 & 0 & 2 & 0 & \cdots \\ 3 & 0 & 0 & 0 & 3 & \cdots \\ \vdots & \vdots & \vdots & \vdots & \vdots & \end{array}$$

Let us calculate $\mathfrak{a}^\dagger|n\rangle$ in terms of the states $\{|m\rangle\}$, using the matrix elements $\langle m|\mathfrak{a}^\dagger|n\rangle$; we have

$$\mathfrak{a}^\dagger|n\rangle = \sum_{m=0}^{\infty} |m\rangle\langle m|\mathfrak{a}^\dagger|n\rangle,$$

or

$$(\mathfrak{a}^\dagger|0\rangle, \mathfrak{a}^\dagger|1\rangle, \ldots) = (|0\rangle, |1\rangle, \ldots)\begin{bmatrix} 0 & 0 & 0 & \cdots \\ \sqrt{1} & 0 & 0 & \cdots \\ 0 & \sqrt{2} & 0 & \cdots \\ 0 & 0 & \sqrt{3} & \cdots \\ \vdots & \vdots & \vdots & \end{bmatrix}$$

$$= (|1\rangle, \sqrt{2}|2\rangle, \sqrt{3}|3\rangle, \ldots) \tag{11.37}$$

i.e.,

$$\mathfrak{a}^\dagger|n\rangle = \sqrt{n+1}\,|n+1\rangle. \tag{11.38}$$

Note that in terms of the matrix calculations the kets $|\rangle$ are treated as row vectors, whereas the bras $\langle|$ behave as column vectors.

We may pass to wave functions simply by forming the inner products $\langle x|n\rangle$, $\langle x|a^+|n\rangle$, etc. As an example, consider the wave-function version of the above calculation. We have

$$\langle x|\mathfrak{a}^\dagger|m\rangle = \int dx'\,\langle x|\mathfrak{a}^\dagger|x'\rangle\,\langle x'|m\rangle$$

$$= \int dx' \left\langle x\left|\frac{1}{\sqrt{2m\hbar\omega_0}}\mathfrak{P} + i\sqrt{\frac{m\omega_0}{2\hbar}}\,\mathfrak{X}\right|x'\right\rangle \langle x'|m\rangle$$

$$= \left[\frac{1}{\sqrt{2m\hbar\omega_0}}\frac{\hbar}{i}\frac{d}{dx} + i\sqrt{\frac{m\omega_0}{2\hbar}}\,x\right]\psi_m(x),$$

where $\psi_m(x) \equiv \langle x|m\rangle$. Hence Eq. (11.38) yields

$$\left[\frac{1}{\sqrt{2m\hbar\omega_0}}\frac{\hbar}{i}\frac{d}{dx} + i\sqrt{\frac{m\omega_0}{2\hbar}}\,x\right]\psi_m(x) = \sqrt{m+1}\,\psi_{m+1}(x). \tag{11.39}$$

In a later chapter we shall deal more completely with the harmonic oscillator wave functions, and it will become apparent there that the wave-function analysis of the oscillator problem is far more complex than the state-vector analysis used in this chapter. The present analysis leaves only the vacuum state undefined and gives the energies and number of quanta for all states. In addition it gives a prescription for obtaining all states from the vacuum state. In effect, all we need the wave function analysis for is to obtain a wave function for the vacuum state $\psi_0(x)$; all other wave functions can be obtained from it through Eq. (11.39). Unless we are concerned with finding the spatial probability density, there is no need to go any further than the analysis of the preceding section.

READING NOTES*

Reference 43, Chapters 3, 4 and 6.

PROBLEMS

1. Derive Eq. (11.12) and display the next two terms on the right-hand side of (11.12).
2. Complete the calculation of $\psi_1(x, t)$ according to the method of the last paragraph in Section 11.2.
3. Using $[\mathfrak{n}, (\mathfrak{a}^\dagger)^n] = n(\mathfrak{a}^\dagger)^n$, show that

$$[\mathfrak{a}, (\mathfrak{a}^\dagger)^n] = n(\mathfrak{a}^\dagger)^{n-1},$$

   which is the result obtained below Eq. (11.32).
4. Show how the nonvanishing energy of the ground state of the harmonic oscillator is connected with the uncertainty principle.
5. The wave function for a one-dimensional particle whose Hamiltonian is $\mathsf{H} = \mathsf{T} + V(x)$, where $\mathsf{T} = \mathsf{P}^2/2m$ is the kinetic energy, happens to be

$$\psi_E(x) = N_E e^{-x^2/4\alpha^2}$$

   where N_E is a constant. We happen to know that ψ_E is an eigenfunction of H with eigenvalue E, and we assume that there is no constant term in the potential energy $V(x)$. The particle, as indicated, has mass m.
   a) Find $V(x)$.
   b) Normalize the wave function so that $\int_{-\infty}^{\infty} dx|\psi_E(x)|^2 = 1$; i.e. find an appropriate value for N_E.
   c) Find the expectation values of the kinetic energy, potential energy, and total energy.
   d) Evaluate α in terms of E.
   e) Find the r.m.s. values of momentum and position, i.e. of p and of x.
   f) Determine the probability per unit momentum interval that the particle has momentum $\hbar k = p$.
   g) Using the operator $\mathsf{B} = (\alpha d/dx + x/2\alpha)$ and its Hermitian conjugate, form the Hamiltonian for the particle.
   h) Find the energies of the wave function $\mathsf{B}\psi_E(x)$ and $\mathsf{B}^\dagger\psi_E(x)$.
6. Remembering Heisenbergs's equations of motion, find the time-dependences of the operators $\mathfrak{a}$ and $\mathfrak{a}^\dagger$.

* Works referred to in these notes will be found in the Bibliography on p. 577.

CHAPTER 12

ROTATIONAL SYMMETRY AND ANGULAR MOMENTUM

In three-dimensional problems, the possibility of carrying out rotations independently about three separate axes leads to a conceptual and mathematical richness that is absent in lower-dimensional problems. Two-dimensional problems allow rotations about only one axis, and hence angular momentum is a one-dimensional quantity, whereas angular momentum in three dimensions is actually a three-dimensional (vector) quantity. This chapter is devoted to the investigation of some of the consequences for quantum theory of this richness, and to the study of the connection between rotational symmetry and angular momentum. The state-vector methods will be utilized extensively and we shall find that by exploiting state vectors we arrive at results that could not have been obtained through the exclusive use of wave functions and an appeal to classical mechanics.

As a matter of fact, we shall find that states with angular momentum equal to $\frac{1}{2}n\hbar$, n an integer, exist in a mathematical sense. Using wave functions and differential operators, we would have arrived at the conclusion that only states having angular momentum $n\hbar$, n an integer, are produced by the mathematical formalism. The latter approach was the one actually used to study most atomic systems in the early days of quantum physics, and the discovery by Uhlenbeck and Goudsmit that particles having spin $\frac{1}{2}n\hbar$ exist *physically*, led to a minor crisis for the theory. The crisis was resolved by tacking onto the theory, special two-element wave functions, or *spinors*, specifically for the purpose of handling such "peculiar" particles. By exploiting the abstract state-vector notation, we shall see that spinors occur naturally within the context of the theory of rotational transformations, without any special assumptions.

12.1 ROTATIONS IN THREE DIMENSIONS

Although we have used active transformations in our previous discussions, we use passive transformations here in order that our presentation may be fairly consistent with the existing literature on rotations. This means simply that instead of rotating thc physical system, we shall rotate the coordinate system used to describe it; the two are equivalent provided the angles of the active rotations are the negatives of the passive rotation angles, as shown in Fig. 12.1.

Our task then is to obtain unitary transformations that reorient the orthogonal coordinate axes of a Euclidean three-dimensional space. This is the analog of

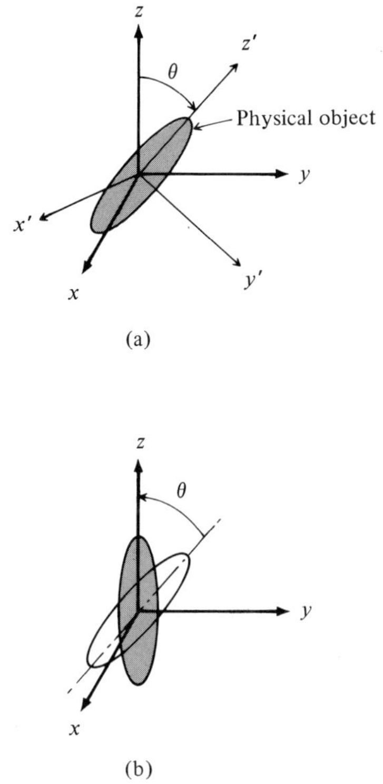

Fig. 12.1. Passive (a) and active (b) rotations. The descriptions of the object in the rotated coordinate system (x', y', z') is the same as the description of the rotated object in the original coordinate system (x, y, z).

finding the passive spatial translation operators $e^{-i\Delta\mathbf{x}\cdot\mathfrak{P}/\hbar}$ that shift the origin of coordinates (i.e. *translate* the orthogonal coordinate axes) by an amount $\Delta\mathbf{x}$. Indeed we shall use this analogy to define the generators of rotations. For a rotation through angle θ about an axis, say the z-axis, denoted $\hat{\mathbf{k}}$, we write the unitary transformation in the form $e^{-i\theta\mathfrak{J}_z/\hbar}$ and call $\mathfrak{J}_z$ the generator of rotations about the z-axis.

It is rather easy to determine that three parameters are required to specify completely the reorientation of three orthogonal (or linearly independent) vectors in space. Two parameters, e.g. the longitude and colatitude on a unit sphere, are needed to specify the change in orientation of one of the vectors; a third parameter is required to give the new direction and sense of one of the other vectors in the plane perpendicular to the new orientation of the first vector. The orientation of the third axis is determined by the fact that it remains orthogonal to the first two.

The most commonly used choice of parameters for rotations corresponds to the Euler parametrization, which gives rotations in terms of *Euler angles*; we shall adopt this parametrization. In a given orthogonal three-dimensional basis, we arbitrarily call one axis the z-axis, then label the remaining two axes according to the right-hand rule, so that we have a right-handed coordinate system. The angles defined in the following transformation are then called the Euler angles: Rotate the x- and y-axes counterclockwise about the z-axis (according to an observer looking down at the xy-plane from the tip of the z-axis) through an angle α; then label the new coordinates x', y' and z'. Thus the new coordinate unit vectors, $\hat{\mathbf{i}}'$, $\hat{\mathbf{j}}'$ and $\hat{\mathbf{k}}'$, are related to the old ones, $\hat{\mathbf{i}}$, $\hat{\mathbf{j}}$ and $\hat{\mathbf{k}}$, in such a way that $\hat{\mathbf{k}}' = \hat{\mathbf{k}}$, $\hat{\mathbf{i}}'$ is nearer than $\hat{\mathbf{i}}$ to $\hat{\mathbf{j}}$ by angle α, and $\hat{\mathbf{j}}'$ is nearer to $-\hat{\mathbf{i}}$ than was $\hat{\mathbf{j}}$, again by angle α; this is illustrated in Fig. 12.2a. Next, rotate the z'- and x'-axes about y' counterclockwise through an angle β (this rotation is in the $z'x'$-plane) and label the new unit vectors $\hat{\mathbf{i}}''$, $\hat{\mathbf{j}}''$, $\hat{\mathbf{k}}''$ with coordinates x'', y'', z''. Finally, rotate the x''- and y''- ($= y'$-) axes counterclockwise through angle γ about z'', and label the final coordinates x''', y''', and z''' ($= z''$) with unit vectors $\hat{\mathbf{i}}'''$, $\hat{\mathbf{j}}'''$ and $\hat{\mathbf{k}}'''$ ($= \hat{\mathbf{k}}''$). Figures 12.2(b) and (c) show these rotations.

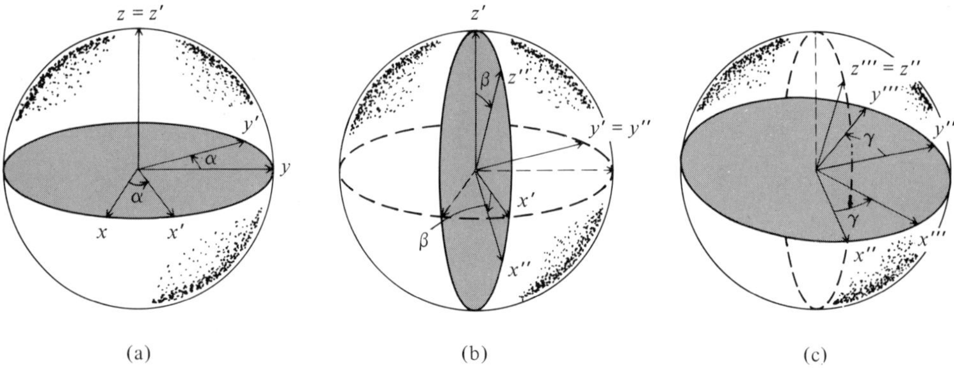

Fig. 12.2. The Euler angles. (a) shows the rotation through α about the z-axis, (b) shows the rotation about y' through angle β, and in (c) the rotation about z'' through γ is shown.

This procedure can be thought of in the following way: First the x-axis is moved in the xy-plane in such a way that the ultimate position of the z-axis (i.e. the $z'' = z'''$-axis) lies in the new $z'x'$-plane; then the z-axis is moved, by way of a rotation about the new y'-axis, into its ultimate (z''') direction. Note that the second rotation is in the $z'x'$-plane. Finaliy, now that the z'-axis, now called z'', is in its ultimate position, we need only move the x-axis (now x'') to its ultimate position in the plane perpendicular to z''. This completes the reorientation of the coordinate system.

Denoting the generator of rotations about the x, y and z-axes respectively by $\mathfrak{J}_x$, $\mathfrak{J}_y$, and $\mathfrak{J}_z$ (note that these must be Hermitian operators—why?), aside from a factor $-1/h$, we may write our overall rotation operator in the form

$$\mathfrak{R}(\alpha, \beta, \gamma) = e^{-i\gamma\mathfrak{J}_{z''}/\hbar}e^{-i\beta J_{y'}/\hbar}e^{-i\alpha J_z/\hbar} = \mathfrak{R}_{z''}(\gamma)\mathfrak{R}_{y'}(\beta)\mathfrak{R}_z(\alpha).$$

The order of these rotations must be preserved, for we do not yet know the commutation properties of the rotation operators.

Now it is not difficult to see, upon consulting Fig. 12.2, that the rotation about $\hat{\mathbf{k}}$ through α, followed by the rotation through β about $\hat{\mathbf{j}}'$, is equivalent to a rotation through β about $\hat{\mathbf{j}}$ followed by a rotation about $\hat{\mathbf{k}}$ through α. Thus

$$\mathfrak{R}_{y'}(\beta)\mathfrak{R}_z(\alpha) = \mathfrak{R}_z(\alpha)\mathfrak{R}_y(\beta),$$

or

$$\mathfrak{R}_{y'}(\beta) = \mathfrak{R}_z(\alpha)\mathfrak{R}_y(\beta)\mathfrak{R}_z^{-1}(\alpha) = \mathfrak{R}_z(\alpha)\mathfrak{R}_y(\beta)\mathfrak{R}_z^{\dagger}(\alpha), \tag{12.1}$$

since the rotation operators are unitary. Similarly,

$$\mathfrak{R}_{z''}(\gamma)\mathfrak{R}_{y'}(\beta) = \mathfrak{R}_{y'}(\beta)\mathfrak{R}_{z'}(\gamma); \tag{12.2}$$

but, just as in Eq. (12.1), we have

$$\mathfrak{R}_{z'}(\gamma) = \mathfrak{R}_z(\alpha)\mathfrak{R}_z(\gamma)\mathfrak{R}_z^{-1}(\alpha), \tag{12.3}$$

and hence, combining Eqs. (12.1), (12.2) and (12.3)

$$\begin{aligned}\mathfrak{R}_{z''}(\gamma)\mathfrak{R}_{y'}(\beta) &= \mathfrak{R}_z(\alpha)\mathfrak{R}_y(\beta)\mathfrak{R}_z^{-1}(\alpha)\mathfrak{R}_z(\alpha)\mathfrak{R}_z(\gamma)\mathfrak{R}_z^{-1}(\alpha)\\ &= \mathfrak{R}_z(\alpha)\mathfrak{R}_y(\beta)\mathfrak{R}_z(\gamma)\mathfrak{R}_z^{-1}(\alpha).\end{aligned}$$

From this it follows that the overall rotation operator can be written in any of the forms given in the equations

$$\begin{aligned}\mathfrak{R}(\alpha, \beta, \gamma) &= \mathfrak{R}_{z''}(\gamma)\mathfrak{R}_{y'}(\beta)\mathfrak{R}_z(\alpha) = \mathfrak{R}_z(\alpha)\mathfrak{R}_y(\beta)\mathfrak{R}_z(\gamma)\\ &= e^{-i\gamma\mathfrak{J}_{z''}/\hbar}e^{-i\beta\mathfrak{J}_{y'}/\hbar}e^{-i\alpha\mathfrak{J}_z/\hbar}\\ &= e^{-i\alpha\mathfrak{J}_z/\hbar}e^{-i\beta\mathfrak{J}_y/\hbar}e^{-i\gamma\mathfrak{J}_z/\hbar},\end{aligned} \tag{12.4}$$

Therefore, rotations as described in our definition of the Euler angles are equivalent to successive rotations of the coordinate axes about $\hat{\mathbf{k}}$, $\hat{\mathbf{j}}$, and $\hat{\mathbf{k}}$ by γ, β and α, respectively, rather than about $\hat{\mathbf{k}}$, $\hat{\mathbf{j}}'$ and $\hat{\mathbf{k}}''$ by α, β and γ, respectively.

By analogy with the passive spatial translation operators, whose generators are $-\mathfrak{P}/\hbar$, we *define* the operator

$$\mathfrak{J} \equiv \hat{\mathbf{i}}\mathfrak{J}_x + \hat{\mathbf{j}}\mathfrak{J}_y + \hat{\mathbf{k}}\mathfrak{J}_z \tag{12.5}$$

as the angular momentum operator; that is, the angular momentum is. apart from the factor $-1/\hbar$, the generator of rotations. For example, $\boldsymbol{\theta} = \hat{\mathbf{i}}\theta_x + \hat{\mathbf{j}}\theta_y + \hat{\mathbf{k}}\theta_z$ represents the angle $\theta = |\boldsymbol{\theta}|$ about the $\hat{\boldsymbol{\theta}} = \boldsymbol{\theta}/\theta$-axis, and

$$\mathfrak{R}(\boldsymbol{\theta}) \equiv e^{-i\boldsymbol{\theta}\cdot\mathfrak{J}/\hbar} \tag{12.6}$$

rotates the coordinate system through θ about $\hat{\boldsymbol{\theta}}$. Since, as we have mentioned, $\mathfrak{J}$ must be Hermitian, it has real expectation values and therefore may represent a physical quantity; thus our defining angular momentum in this way will not lead to obviously embarassing mathematical results.

12.2 COMMUTATORS FOR ANGULAR MOMENTUM OPERATORS

Let us consider the effect of carrying out rotations $\mathfrak{R}_z(\alpha)$ and $\mathfrak{R}_y(\beta)$ in different orders. In particular, let α and β be infinitesimal; then

$$\begin{aligned}\mathfrak{R}_z(\alpha)\mathfrak{R}_y(\beta) &\simeq \left(1 - \frac{i}{\hbar}\alpha\mathfrak{J}_z\right)\left(1 - \frac{i}{\hbar}\beta\mathfrak{J}_y\right)\\ &\simeq 1 - \frac{i}{\hbar}(\alpha\mathfrak{J}_z + \beta\mathfrak{J}_y)\\ &\simeq \left(1 - \frac{i}{\hbar}\beta\mathfrak{J}_y\right)\left(1 - \frac{i}{\hbar}\alpha\mathfrak{J}_z\right) = \mathfrak{R}_y(\beta)\mathfrak{R}_z(\alpha),\end{aligned}$$

to first-order α and β. To be quite sure that the order of carrying out the rotations is immaterial, we must consider terms of higher order; hence we consider second-order terms:

$$\begin{aligned}&\left(1 - \frac{i}{\hbar}\alpha\mathfrak{J}_z - \frac{\alpha^2}{2\hbar^2}\mathfrak{J}_z^2\right)\left(1 - \frac{i}{\hbar}\beta\mathfrak{J}_y - \frac{\beta^2}{2\hbar^2}\mathfrak{J}_y^2\right)\\ &\quad\simeq 1 - \frac{i}{\hbar}(\alpha\mathfrak{J}_z + \beta\mathfrak{J}_y) - \frac{\alpha^2}{2\hbar^2}\mathfrak{J}_z^2 - \frac{\beta^2}{2\hbar^2}\mathfrak{J}_y^2 - \frac{\alpha\beta}{\hbar}\mathfrak{J}_z\mathfrak{J}_y\\ &\quad\simeq \left(1 - \frac{i}{\hbar}\beta\mathfrak{J}_y - \frac{\beta^2}{2\hbar^2}\mathfrak{J}_y^2\right)\left(1 - \frac{i}{\hbar}\alpha\mathfrak{J}_z - \frac{\alpha^2}{2\hbar^2}\mathfrak{J}_z^2\right) + \frac{\alpha\beta}{\hbar}(\mathfrak{J}_y\mathfrak{J}_z - \mathfrak{J}_z\mathfrak{J}_y).\end{aligned}$$

Hence, to second order in α, β,

$$\mathfrak{R}_z(\alpha)\mathfrak{R}_y(\beta) = \mathfrak{R}_y(\beta)\mathfrak{R}_z(\alpha) + \frac{\alpha\beta}{\hbar^2}[\mathfrak{J}_y, \mathfrak{J}_z] \tag{12.7}$$

and the answer to the question whether the order of transformation matters is thus seen to hang on the value of $[\mathfrak{J}_y, \mathfrak{J}_z]$. Indeed, we shall use this circumstance to evaluate the commutator.

First, consider what happens to the z-axis under $\mathfrak{R}_z(\alpha)\mathfrak{R}_y(\beta)$; it becomes a unit vector given by the spherical coordinates (colatitude, aximuthal angle) = $(\theta_{\mathbf{k}'}, \varphi_{\mathbf{k}'}) = (\beta, \alpha \sin \beta)$. If the transformations about the z- and y-axes are effected in the reverse order, $\mathfrak{R}_y(\beta)\mathfrak{R}_z(\alpha)$, the new z-axis is given by $(\theta_{\mathbf{k}''}, \varphi_{\mathbf{k}''}) = (\beta, 0)$.

Next, consider the motion of the y-axis under these transformations; after $\mathfrak{R}_z(\alpha)\mathfrak{R}_y(\beta)$, the spherical coordinates of $\hat{\mathbf{j}} \to \hat{\mathbf{j}}'$ become $(\theta_{\mathbf{j}'}, \varphi_{\mathbf{j}'}) = (\pi/2, \alpha + \pi/2)$, whereas $\mathfrak{R}_y(\beta)\mathfrak{R}_z(\alpha)$ yields $(\theta_{\mathbf{j}''}, \varphi_{\mathbf{j}''}) = (\pi/2 - \alpha \sin \beta, \alpha + \pi/2)$. The x-axis directions are given by $(\theta_{\mathbf{i}'}, \varphi_{\mathbf{i}'}) = (\pi/2 + \beta, \alpha)$ under $\mathfrak{R}_z(\alpha)\mathfrak{R}_y(\beta)$ and by $(\theta_{\mathbf{i}''}, \varphi_{\mathbf{i}''}) = (\pi/2 + \beta, \alpha)$ under $\mathfrak{R}_y(\beta)\mathfrak{R}_z(\alpha)$ also. These results are illustrated in Fig. 12.3.

Thus $\mathfrak{R}_z(\alpha)\mathfrak{R}_y(\beta)$ carries the tip of $\hat{\mathbf{k}}$ a distance β along the surface of the unit sphere toward the x-axis, and a distance $\alpha \sin \beta \simeq \alpha\beta$ parallel to the xy-plane toward the y-axis, whereas $\mathfrak{R}_y(\beta)\mathfrak{R}_z(\alpha)$ accomplishes only the first of these movements, carrying $\hat{\mathbf{k}}$ a distance β on the unit sphere toward the x-axis. Similarly,

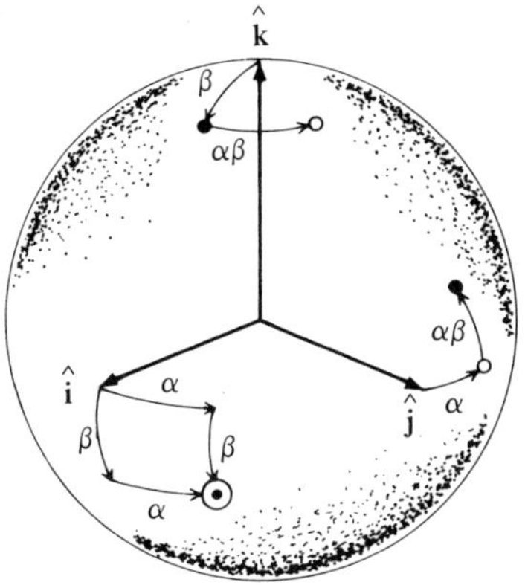

Fig. 12.3. Paths traversed on the unit sphere for infinitesimal rotations about z- and y-axes by angles α, β respectively. The angles are exaggerated for clarity. The end positions of the axes $(\hat{\mathbf{i}}, \hat{\mathbf{j}}, \hat{\mathbf{k}})$ are represented by ○ $(\hat{\mathbf{i}}', \hat{\mathbf{j}}', \hat{\mathbf{k}}')$ for $\mathfrak{R}_z(\alpha)\mathfrak{R}_y(\beta)$ and by ● $(\hat{\mathbf{i}}'', \hat{\mathbf{j}}'', \hat{\mathbf{k}}'')$ for $\mathfrak{R}_y(\beta)\mathfrak{R}_z(\alpha)$.

$\mathfrak{R}_z(\alpha)\mathfrak{R}_y(\beta)$ moves $\hat{\mathbf{j}}$ a distance α toward $-\hat{\mathbf{i}}$, whereas $\mathfrak{R}_y(\beta)\mathfrak{R}_z(\alpha)$ does the same thing but adds a movement of length $\alpha \sin \beta \simeq \alpha\beta$ toward the z-axis. On the other hand, $\mathfrak{R}_z(\alpha)\mathfrak{R}_y(\beta)$ has exactly the same effect upon $\hat{\mathbf{i}}$ as does $\mathfrak{R}_y(\beta)\mathfrak{R}_z(\alpha)$ for infinitesimal angles. The result of both is to move $\hat{\mathbf{i}}$ a distance α toward $\hat{\mathbf{j}}$ and a distance β toward $-\hat{\mathbf{k}}$ on the unit sphere.

Careful examination of these results for infinitesimal α, β reveals that $\mathfrak{R}_z(\alpha)\mathfrak{R}_y(\beta)$, followed by a second-order infinitesimal rotation about the x-axis, through $\alpha\beta$, i.e. followed by $\mathfrak{R}_x(\alpha\beta)$, is equal to $\mathfrak{R}_y(\beta)\mathfrak{R}_z(\alpha)$. Thus, to second order in α, β

$$\mathfrak{R}_x(\alpha\beta)\mathfrak{R}_z(\alpha)\mathfrak{R}_y(\beta) = \mathfrak{R}_y(\beta)\mathfrak{R}_z(\alpha).$$

However, we also have Eq. (12.7) to second order in α, β. Therefore, writing

$$\mathfrak{R}_x(\alpha\beta) \simeq \left(1 - \frac{i}{\hbar}\alpha\beta\mathfrak{J}_x\right)$$

yields

$$\mathfrak{R}_x(\alpha\beta)\mathfrak{R}_z(\alpha)\mathfrak{R}_y(\beta) = \mathfrak{R}_z(\alpha)\mathfrak{R}_y(\beta) - \frac{i}{\hbar}\alpha\beta\mathfrak{J}_x$$

to second order, and by comparing this with Eq. (12.7), we get

$$[\mathfrak{J}_y, \mathfrak{J}_z] = i\hbar\mathfrak{J}_x \tag{12.8a}$$

as a necessary condition upon the components of $\mathfrak{J}$, which is proportional to the generator of rotations. It turns out that (12.8a) is sufficient to *define* the generator of rotations, also.

Since there is nothing sacred about our labeling of axes—we could cyclically permute the labels of the axes and still retain an orthogonal right-handed coordinate basis—the same result must hold when the indices x, y and z, on the generators $\mathfrak{J}$, are cyclically permuted; thus

$$[\mathfrak{J}_z, \mathfrak{J}_x] = i\hbar\mathfrak{J}_y \tag{12.8b}$$

$$[\mathfrak{J}_x, \mathfrak{J}_y] = i\hbar\mathfrak{J}_z \tag{12.8c}$$

must hold also.

Equations (12.8) are the fundamental commutation relations for the angular-momentum operators, and in fact can be used to define angular momentum. A convenient way of codifying these three equations is to return to the number indices (1, 2, 3) in place of (x, y, z) and to utilize the completely antisymmetric symbol ϵ_{ijk}, which we recall is defined by

$$\epsilon_{ijk} = \begin{cases} +1 & \text{if } (i, j, k) \text{ is an even permutation of } (1, 2, 3), \\ -1 & \text{if } (i, j, k) \text{ is an odd permutation of } (1, 2, 3), \\ 0 & \text{otherwise.} \end{cases}$$

Indeed, Eqs. (12.8) may be written

$$[\mathfrak{J}_i, \mathfrak{J}_j] = i\hbar \sum_{k=1}^{3} \epsilon_{ijk}\mathfrak{J}_k = i\hbar\epsilon_{ijk}\mathfrak{J}_k. \tag{12.9}$$

It is useful to note the commutation relations between the $\mathfrak{J}_i$ and the squared angular-momentum operator $\mathfrak{J}^2$; we have

$$\begin{aligned} [\mathfrak{J}_i, \mathfrak{J}^2] &= \sum_{j=1}^{3} [\mathfrak{J}_i, \mathfrak{J}_j^2] = \sum_{j=1}^{3} \{[\mathfrak{J}_i, \mathfrak{J}_j]\mathfrak{J}_j + \mathfrak{J}_j[\mathfrak{J}_i, \mathfrak{J}_j]\} \\ &= \sum_{j=1}^{3}\sum_{k=1}^{3} i\hbar\epsilon_{ijk}(\mathfrak{J}_k\mathfrak{J}_j + \mathfrak{J}_j\mathfrak{J}_k) \\ &= \sum_{j=1}^{3}\sum_{k=1}^{3} i\hbar\epsilon_{ikj}(\mathfrak{J}_j\mathfrak{J}_k + \mathfrak{J}_k\mathfrak{J}_j) \\ &= -\sum_{j=1}^{3}\sum_{k=1}^{3} i\hbar\epsilon_{ijk}(\mathfrak{J}_k\mathfrak{J}_j + \mathfrak{J}_j\mathfrak{J}_k) \\ &= 0, \qquad i = 1,2,3, \end{aligned} \tag{12.10}$$

where in the third-from-last line we interchanged dummy indices, $j \leftrightarrow k$, and in the next-to-last line we utilized the antisymmetry of ϵ_{ijk} to show that $[\mathfrak{J}_i, \mathfrak{J}^2] = -[\mathfrak{J}_i, \mathfrak{J}^2] = 0$.

Let us check that these commutation relations are consistent with what we should expect on the basis of the correspondence principle. In classical physics, the angular momentum is defined by

$$\mathbf{L} = \mathbf{r} \times \mathbf{p},$$

and the transcription of this to operator notation yields

$$\mathfrak{L} = \mathfrak{X} \times \mathfrak{P} = \sum_{i=1}^{3} \sum_{j=1}^{3} \sum_{k=1}^{3} \epsilon_{ijk} \mathfrak{X}_i \mathfrak{P}_j \hat{\mathbf{i}}_k$$

if we utilize the completely antisymmetric symbol. Hence

$$\begin{aligned}
[\mathfrak{L}_i, \mathfrak{L}_j] &= \sum_{l,m} \sum_{r,s} \epsilon_{ilm} \epsilon_{jrs} [\mathfrak{X}_l \mathfrak{P}_m, \mathfrak{X}_r \mathfrak{P}_s] \\
&= \sum_{l,m,r,s} \epsilon_{ilm} \epsilon_{jrs} \{\mathfrak{X}_l [\mathfrak{P}_m, \mathfrak{X}_r] \mathfrak{P}_s + \mathfrak{X}_l \mathfrak{X}_r [\mathfrak{P}_m, \mathfrak{P}_s] \\
&\qquad + \mathfrak{X}_r [\mathfrak{X}_l, \mathfrak{P}_s] \mathfrak{P}_m + [\mathfrak{X}_l, \mathfrak{X}_r] \mathfrak{P}_m \mathfrak{P}_s\} \\
&= \sum_{l,m,r,s} \epsilon_{ilm} \epsilon_{jrs} \frac{\hbar}{i} \{\mathfrak{X}_l \mathfrak{P}_s \delta_{mr} - \mathfrak{X}_r \mathfrak{P}_m \delta_{ls}\} \\
&= \frac{\hbar}{i} \sum_{l,m,s} \epsilon_{ilm} \epsilon_{jms} \{\mathfrak{X}_l \mathfrak{P}_s - \mathfrak{X}_s \mathfrak{P}_l\}.
\end{aligned}$$

Here either $i = j$ and $l = s$, in which case $\mathfrak{X}_l \mathfrak{P}_s - \mathfrak{X}_s \mathfrak{P}_l = 0$, or $i = s$ and $j = l$, in which case $\epsilon_{ilm} = \epsilon_{ijm}$ and $\epsilon_{jms} = \epsilon_{jmi} = \epsilon_{ijm}$. Since the last member of the above equation fails to vanish only for the latter case, we have

$$\begin{aligned}
[\mathfrak{L}_i, \mathfrak{L}_j] &= \frac{\hbar}{i} \sum_m \epsilon_{ijm} \epsilon_{ijm} (\mathfrak{X}_j \mathfrak{P}_i - \mathfrak{X}_i \mathfrak{P}_j) \\
&= \frac{\hbar}{i} \sum_m \epsilon_{ijm} (-\mathfrak{L}_m) = i\hbar \epsilon_{ijm} \mathfrak{L}_m,
\end{aligned} \tag{12.11}$$

where we have the fact that

$$\epsilon_{ijm}(\mathfrak{X}_j \mathfrak{P}_i - \mathfrak{X}_i \mathfrak{P}_j) = -\epsilon_{ijm}(\mathfrak{X}_i \mathfrak{P}_j - \mathfrak{X}_j \mathfrak{P}_i) = \begin{cases} -\mathfrak{L}_m & \text{if } i \neq j \neq m \neq i \\ 0 & \text{otherwise.} \end{cases}$$

which is verified by noting that $\mathfrak{X}_1 \mathfrak{P}_2 - \mathfrak{X}_2 \mathfrak{P}_1 = \sum_{m,n} \epsilon_{m,n,3} \mathfrak{X}_m \mathfrak{P}_n = \mathfrak{L}_3$, etc.

Therefore the operators transcribed from the classical angular momentum components do indeed satisfy the commutation relations (12.9).

12.3 EIGENSTATES OF ANGULAR MOMENTUM

The commutation relations for angular momentum operators,

$$[\mathfrak{J}_i, \mathfrak{J}^2] = 0$$

$$[\mathfrak{J}_j, \mathfrak{J}_k] = i\hbar \epsilon_{jkl} \mathfrak{J}_l$$

for $j, k = 1, 2, 3,$

demonstrate that

1. Any component of angular momentum may be diagonalized simultaneously with $\mathfrak{J}^2$, and
2. Only one component of $\mathfrak{J}$ may be diagonalized at a time.

The latter is because no two of the $\mathfrak{J}_i$ commute with each other.

Therefore we shall deal with states that are eigenstates of $\mathfrak{J}^2$ and $\mathfrak{J}_3$. Let these be labeled $|\lambda m\rangle$ and let them satisfy

$$\mathfrak{J}^2|\lambda m\rangle = \hbar^2\lambda|\lambda m\rangle \tag{12.12}$$

$$\mathfrak{J}_3|\lambda m\rangle = \hbar m|\lambda m\rangle; \tag{12.13}$$

these will be called the angular momentum eigenstates.

Now we may determine the effects of $\mathfrak{J}_1$ and $\mathfrak{J}_2$ upon these states by using Eqs. (12.8). For example, we might ask whether $\mathfrak{J}_1|\lambda m\rangle$ is an eigenstate of $\mathfrak{J}_3$ (surely it *is* an eigenstate of $\mathfrak{J}^2$). To find the answer, we calculate

$$\begin{aligned}\mathfrak{J}_3\mathfrak{J}_1|\lambda m\rangle &= (\mathfrak{J}_1\mathfrak{J}_3 + [\mathfrak{J}_3, \mathfrak{J}_1])|\lambda m\rangle \\ &= (\mathfrak{J}_1\hbar m + i\hbar\epsilon_{312}\mathfrak{J}_2)|\lambda m\rangle = \hbar(m\mathfrak{J}_1 + i\mathfrak{J}_2)|\lambda m\rangle.\end{aligned} \tag{12.14}$$

Evidently $\mathfrak{J}_1|\lambda m\rangle$ is an eigenstate of $\mathfrak{J}_3$ only if $\mathfrak{J}_2|\lambda m\rangle$ also is such an eigenstate. To find out whether $\mathfrak{J}_2|\lambda m\rangle$ is an eigenstate of $\mathfrak{J}_3$, we proceed as before:

$$\begin{aligned}\mathfrak{J}_3\mathfrak{J}_2|\lambda m\rangle &= (\mathfrak{J}_2\mathfrak{J}_3 + [\mathfrak{J}_3, \mathfrak{J}_2])|\lambda m\rangle \\ &= \hbar(m\mathfrak{J}_2 - i\mathfrak{J}_1)|\lambda m\rangle.\end{aligned} \tag{12.15}$$

Here $\mathfrak{J}_1$ re-enters the problem and we seem to be going in circles.

However, by forming the linear combinations $\mathfrak{J}_1 + i\mathfrak{J}_2$, we obtain, from Eqs. (12.14) and (12.15),

$$\mathfrak{J}_3(\mathfrak{J}_1 \pm i\mathfrak{J}_2)|\lambda m\rangle = \hbar(m \pm 1)(\mathfrak{J}_1 \pm i\mathfrak{J}_2)|\lambda m\rangle;$$

the state vectors $(\mathfrak{J}_1 \pm i\mathfrak{J}_2)|\lambda m\rangle$ are eigenstates of $\mathfrak{J}_3$ with eigenvalues $(m \pm 1)$ respectively. It is convenient to define

$$\mathfrak{J}_\pm \equiv (\mathfrak{J}_1 \pm i\mathfrak{J}_2); \tag{12.16}$$

then we may write

$$\mathfrak{J}_\pm|\lambda m\rangle = \hbar N_m^\pm|\lambda, m \pm 1\rangle \tag{12.17}$$

where $N_m^\pm$ are normalization constants. The operators $\mathfrak{J}_\pm$ are called raising and lowering operators, in analogy with the harmonic-oscillator raising and lowering operators.

Our original query, whether $\mathfrak{J}_1|\lambda m\rangle$ or $\mathfrak{J}_2|\lambda m\rangle$ were eigenstates for $\mathfrak{J}_3$, now may be answered in the negative. Since

$$\mathfrak{J}_1 = \tfrac{1}{2}(\mathfrak{J}_+ + \mathfrak{J}_-) \tag{12.18a}$$

and

$$\mathfrak{J}_2 = \frac{1}{2i}(\mathfrak{J}_+ - \mathfrak{J}_-), \tag{12.18b}$$

we have

$$\mathfrak{J}_1|\lambda m\rangle = \frac{\hbar}{2}N_m^+|\lambda, m+1\rangle + \frac{\hbar}{2}N_m^-|\lambda, m-1\rangle$$

and

$$\mathfrak{J}_2|\lambda m\rangle = \frac{\hbar}{2i}N_m^+|\lambda, m+1\rangle - \frac{\hbar}{2i}N_m^-|\lambda, m-1\rangle,$$

so that $\mathfrak{J}_1|\lambda m\rangle$ and $\mathfrak{J}_2|\lambda m\rangle$ are not, in general, eigenstates of $\mathfrak{J}_3$.

The normalization constants of $N_m^{\pm}$ may be calculated through the use of the orthonormality condition

$$\langle \lambda m|\lambda' m'\rangle = \delta(\lambda - \lambda')\,\delta(m - m'). \tag{12.19}$$

Thus

$$\begin{aligned}\langle \lambda, m \pm 1|\hbar^2 N_m^{\pm *} N_{m'}^{\pm}|\lambda', m' \pm 1\rangle &= \langle \lambda m|(\mathfrak{J}_1 \pm i\mathfrak{J}_2)^\dagger(\mathfrak{J}_1 \pm i\mathfrak{J}_2)|\lambda' m'\rangle \\ &= \langle \lambda m|\mathfrak{J}_1^2 + \mathfrak{J}_2^2 \mp i[\mathfrak{J}_2, \mathfrak{J}_1]|\lambda' m'\rangle = \langle \lambda m|\mathfrak{J}^2 - \mathfrak{J}_3^2 - \hbar\mathfrak{J}_3|\lambda' m'\rangle \\ &= \langle \lambda m|\hbar^2\lambda' - \hbar^2 m'^2 \mp \hbar^2 m'|\lambda' m'\rangle = \hbar^2[\lambda' - m'(m' \pm 1)]\langle \lambda m|\lambda' m'\rangle \\ &= \hbar^2[\lambda - m(m \pm 1)]\,\delta(\lambda - \lambda')\,\delta(m - m').\end{aligned}$$

However, the first element of these equations is just

$$\hbar^2 N_m^{\pm} N_{m'}^{\pm *}\langle \lambda, m \pm 1|\lambda', m' \pm 1\rangle = \hbar^2|N_m^{\pm}|^2\,\delta(\lambda - \lambda')\,\delta(m - m').$$

Therefore, neglecting any phase factors,

$$N_m^{\pm} = [\lambda - m(m \pm 1)]^{1/2} \tag{12.20}$$

or

$$\mathfrak{J}_{\pm}|\lambda, m\rangle = \hbar\sqrt{\lambda - m(m \pm 1)}\,|\lambda, m \pm 1\rangle \tag{12.21}$$

To learn more about the eigenvalues m, we utilize some symmetry properties based on the assumption that space is isotropic. Presumably a rotation of the coordinate axes will not affect the physics of a system in any way. In particular, a rotation of coordinate axes by angle π about the y-axis has no physical effect; it just provides a new coordinate basis in terms of which to describe the physical system. Such a rotation is equivalent to the transformation $z \to z' = -z, x \to x' = -x, y \to y' = y$. If, in the original coordinates, the physical system was in an eigenstate of $\mathfrak{J}_3$ with eigenvalue $\hbar m$, the eigenvalue in the new coordinate framework would have to be $-\hbar m$, because the new z- (or 3-)axis is the negative of the old and the basis is still right-handed; the z-component of angular momentum is along the negative z-axis of the new coordinates, which coincides with the positive z-axis of the old coordinates.

According to the principle of active–passive equivalence, we could have left the coordinate axes fixed and rotated the physical system through $-\pi$ about $\hat{\mathbf{j}}$; this converts $|\lambda m\rangle$ to $|\lambda, -m\rangle$:

$$e^{-i\pi\mathfrak{J}_2/\hbar}|\lambda m\rangle = N|\lambda, -m\rangle, \tag{12.22}$$

where N is a normalization constant (with absolute value 1). In fact, it can be shown directly that

$$\mathfrak{J}_3 e^{-i\pi\mathfrak{J}_2/\hbar}|\lambda, m\rangle = -\hbar m e^{-i\pi\mathfrak{J}_2/\hbar}|\lambda, m\rangle, \tag{12.23}$$

but this involves a long and tedious calculation in which the commutator relations play a vital role.

Equation (12.22) may be written in the form

$$\begin{aligned} N|\lambda, -m\rangle &= e^{-\pi(\mathfrak{J}_+ - \mathfrak{J}_-)/2\hbar}|\lambda m\rangle \\ &= \sum_{n=0}^{\infty} \frac{(\pi/2\hbar)^n}{n!}(\mathfrak{J}_+ - \mathfrak{J}_-)^n|\lambda m\rangle. \end{aligned}$$

Although actually computing the sum in the last member is a long and arduous task, we can see that through the use of commutator relations for $\mathfrak{J}_+$, $\mathfrak{J}_-$ and $\mathfrak{J}_3$, namely

$$\begin{aligned} [\mathfrak{J}_+, \mathfrak{J}_-] &= [\mathfrak{J}_1 + i\mathfrak{J}_2, \mathfrak{J}_1 - i\mathfrak{J}_2] \\ &= i2[\mathfrak{J}_2, \mathfrak{J}_1] = 2\hbar\mathfrak{J}_3, \end{aligned} \tag{12.24a}$$

$$[\mathfrak{J}_3, \mathfrak{J}_\pm] = [\mathfrak{J}_3, \mathfrak{J}_1 \pm i\mathfrak{J}_2] = i\hbar(\mathfrak{J}_2 \mp i\mathfrak{J}_1) = \pm\hbar\mathfrak{J}_\pm, \tag{12.24b}$$

this equation may be cast into the form

$$N|\lambda, -m\rangle = \sum_{p,q,r} \alpha_{p,q,r}\mathfrak{J}_+^p\mathfrak{J}_3^q\mathfrak{J}_-^r|\lambda, m\rangle;$$

the $\mathfrak{J}_-$ all can be brought to the right, the $\mathfrak{J}_+$ all to the left, and the $\mathfrak{J}_3$ sandwiched between them.

Now we recall that for different m values, the $|\lambda m\rangle$ must be linearly independent (in fact orthogonal—why?). This means that only those terms in the sum for which

$$\mathfrak{J}_+^p\mathfrak{J}_3^q\mathfrak{J}_-^r|\lambda m\rangle = N_{p,q,r}|\lambda, -m\rangle,$$

where $N_{p,q,r}$ is a normalization constant, survive. In turn, this means that m must differ from $-m$ by an integer, since $\mathfrak{J}_\pm$ and $\mathfrak{J}_3$ can change eigenvalues of $\mathfrak{J}_3$ only by ± 1, 0 respectively. Hence $m - (-m) = n$, n an integer, or $m = \frac{1}{2}n$ for some integer n. That is, *the eigenvalues of* $\mathfrak{J}_3$ *must be integers divided by* 2, i.e. half-integers.

We can learn still more by considering the operators $\mathfrak{J}_i^2$. Let $|1\rangle$ be any state and let $\mathfrak{U}$ be a unitary operator that diagonalizes $\mathfrak{J}_i$; such a $\mathfrak{U}$ exists because $\mathfrak{J}_i$ is a Hermitian operator. Thus, we obtain, when μ is the eigenvalue of $\mathfrak{U}^\dagger\mathfrak{J}_i\mathfrak{U}$ for $|1\rangle$,

$$\langle 1|\mathfrak{U}^+\mathfrak{J}_i^2\mathfrak{U}|1\rangle = \mu^2\langle 1|\mathfrak{U}^+\mathfrak{U}|1\rangle = \mu^2\langle 1|1\rangle \geq 0,$$

since μ is real. On the other hand, if we replace $|1\rangle$ with $\mathfrak{U}^\dagger|1\rangle$, we get

$$\langle 1|\mathfrak{U}\mathfrak{U}^\dagger\mathfrak{J}_i^2\mathfrak{U}\mathfrak{U}^\dagger|1\rangle = \langle 1|\mathfrak{J}_i^2|1\rangle;$$

hence, in general

$$\langle 1|\mathfrak{J}_i^2|1\rangle \geq 0, \qquad \text{for all states } |1\rangle. \tag{12.25}$$

In fact, this is true for all Hermitian operators, since all we used in obtaining this result was the Hermiticity of $\mathfrak{J}_i$ and the consequent reality of its eigenvalues.

Equation (12.25) guarantees that the eigenvalues $\hbar^2\lambda$ of $\mathfrak{J}^2$ are positive definite, as are the eigenvalues of $\hbar^2(\lambda - m^2)$ of $\mathfrak{J}_1^2 + \mathfrak{J}_2^2 = \mathfrak{J}^2 - \mathfrak{J}_3^2$. Therefore

$$\lambda - m^2 \geq 0, \qquad \text{or} \qquad \lambda \geq m^2.$$

It follows that for some value of n,

$$(\mathfrak{J}_+)^n|\lambda, -j\rangle = 0,$$

where j is the maximum value of m, $m \leq j \leq \sqrt{\lambda}$; $-j$ is therefore the minimum value of m. Because of Eq. (12.21), we have

$$(\mathfrak{J}_+)^n|\lambda, -j\rangle = \left\{\prod_{k=0}^{n-1}[\lambda - (-j + k)(-j + k + 1)]^{1/2}\right\}|\lambda, -j + n\rangle = 0.$$

Since j must be half an integer, the number of integer steps required to get from $-j$ to $+j$ is $2j$, and hence $(\mathfrak{J}_+)^n|\lambda, -j\rangle = 0$ holds for $n \geq 2j + 1$. This requires that the $k = 2j$ factor of the product vanish, i.e. that

$$0 = \lambda - (-j + 2j)(-j + 2j + 1) = \lambda - j(j + 1),$$

or

$$\lambda = j(j + 1). \tag{12.26}$$

Because of Eq. (12.26), it is convenient to make a slight change in our notation; we let $|\lambda m\rangle = |j(j + 1), m\rangle$ be denoted $|j, m\rangle$, so that

$$\mathfrak{J}^2|jm\rangle = \hbar^2 j(j + 1)|jm\rangle \tag{12.27}$$

$$\mathfrak{J}_3|jm\rangle = \hbar m|jm\rangle \tag{12.28}$$

$$\begin{aligned}\mathfrak{J}_\pm|jm\rangle &= \hbar[j(j + 1) - m(m \pm 1)]^{1/2}|j, m \pm 1\rangle \\ &= \hbar[(j \mp m)(j \pm m + 1)]^{1/2}|j, m \pm 1\rangle.\end{aligned} \tag{12.29}$$

The eigenvalues m are just the half-integer numbers $-j, -j + 1, -j + 2, \ldots, j - 2, j - 1, j$; clearly there are $2j + 1$ of them.

We have seen that the total angular momentum squared, $\mathfrak{J}^2$, and one of the components of $\mathfrak{J}$ (we chose $\mathfrak{J}_3$) can be well defined simultaneously, but only at the expense of correspondingly poor definition of the other two components of $\mathfrak{J}$. Therefore in quantum theory we can specify the size of the total angular momentum, which has value $\hbar\sqrt{j(j + 1)}$, where j is a half-integer, and the size of the angular momentum along a single direction and sense in space, which has value $\hbar m \leq \hbar j$. The raising and lowering operators $\mathfrak{J}_+$ and $\mathfrak{J}_-$ are convenient linearly independent operators, in terms of which $\mathfrak{J}_1$ and $\mathfrak{J}_2$ may be expressed.

A common and very useful alternative set of operators to the set $\mathfrak{J}_+, \mathfrak{J}_-, \mathfrak{J}_3$ is defined by

$$\mathfrak{J}_{+1} \equiv -\frac{1}{\sqrt{2}}\mathfrak{J}_+ \tag{12.30a}$$

$$\mathfrak{J}_0 \equiv \mathfrak{J}_3 \tag{12.30b}$$

$$\mathfrak{J}_{-1} \equiv \frac{1}{\sqrt{2}}\mathfrak{J}_- . \tag{12.30c}$$

In terms of these *spherical vector components*, the square of the angular momentum has the form

$$\mathfrak{J}^2 = \sum_{\mu=-1}^{+1} (-1)^\mu \mathfrak{J}_\mu \mathfrak{J}_{-\mu} . \tag{12.31}$$

Also, we have the commutation relations

$$[\mathfrak{J}_0, \mathfrak{J}_\mu] = \mu\hbar\mathfrak{J}_\mu \tag{12.32}$$

$$[\mathfrak{J}_{-1}, \mathfrak{J}_{+1}] = \hbar\mathfrak{J}_0 , \tag{12.33}$$

and the effects of $\mathfrak{J}_\mu$ on $|jm\rangle$ are given by

$$\mathfrak{J}_0|jm\rangle = \hbar m|jm\rangle \tag{12.34a}$$

$$\mathfrak{J}_{\pm 1}|jm\rangle = \mp\hbar\sqrt{\frac{(j \mp m)(j \pm m + 1)}{2}}|jm \pm 1\rangle \tag{12.34b}$$

The chief virtue of these operators lies in their commutation relations. In a more thorough study of angular momentum than we can take up here, the commutation relations are found to play a central role, and the above form for them leads to many simplifications in the handling of operators and matrix elements. For example, Eq. (2.32) looks rather like an eigenvalue problem; under the influence of $\mathfrak{J}_0$, $\mathfrak{J}_\mu$ has eigenvalue $\mu\hbar$. This kind of behavior is highly desirable for the efficient study of operator algebras.

12.4 MORE COMMUTATORS

We have defined $\mathfrak{J}$ through its role in generating rotations rather than through the correspondence principle; hence we cannot legitimately use the commutators $[\mathfrak{X}_i, \mathfrak{P}_j] = i\hbar$ together with $\mathfrak{L} = \mathfrak{X} \times \mathfrak{P}$ to obtain the commutators of $\mathfrak{X}_i$ and $\mathfrak{P}_i$ with angular-momentum components. Instead, we consider, for example,

$$\begin{aligned}
\mathfrak{X}e^{-i\theta\mathfrak{J}_3/\hbar}|\mathbf{x}\rangle &= \mathfrak{X}|\hat{\mathbf{i}}(x\cos\theta - y\sin\theta) + \hat{\mathbf{j}}(y\cos\theta + x\sin\theta) + \hat{\mathbf{k}}z\rangle \\
&= \{\mathbf{x} + [\hat{\mathbf{i}}(\cos\theta - 1) + \hat{\mathbf{j}}\sin\theta]x + [\hat{\mathbf{i}}(-\sin\theta) + \hat{\mathbf{j}}(\cos\theta - 1)]y\} \\
&\qquad \cdot |\hat{\mathbf{i}}(x\cos\theta - y\sin\theta) + \hat{\mathbf{j}}(y\cos\theta + x\sin\theta) + \hat{\mathbf{k}}z\rangle \\
&= \{\mathbf{x} + [\hat{\mathbf{i}}(\cos\theta - 1) + \hat{\mathbf{j}}\sin\theta]x + [\hat{\mathbf{i}}(-\sin\theta) + \hat{\mathbf{j}}(\cos\theta - 1)]y\} \\
&\qquad \cdot e^{-i\theta\mathfrak{J}_3/\hbar}|\mathbf{x}\rangle \\
&= e^{-i\theta\mathfrak{J}_3/\hbar}\{\mathbf{x} + [\hat{\mathbf{i}}(\cos\theta - 1) + \hat{\mathbf{j}}\sin\theta]x \\
&\qquad + [\hat{\mathbf{i}}(-\sin\theta) + \hat{\mathbf{j}}(\cos\theta - 1)]y\}|\mathbf{x}\rangle \\
&= e^{-i\theta\mathfrak{J}_3/\hbar}\{\mathfrak{X} + [\hat{\mathbf{i}}(\cos\theta - 1) + \hat{\mathbf{j}}\sin\theta]\mathfrak{X} \\
&\qquad + [\hat{\mathbf{i}}(-\sin\theta) + \hat{\mathbf{j}}(\cos\theta - 1)]\mathfrak{Y}\}|\mathbf{x}\rangle. \qquad (12.35)
\end{aligned}$$

Since this holds for arbitrary $|\mathbf{x}\rangle$, we may equate the coefficients of $|\mathbf{x}\rangle$ and thereby obtain

$$[\mathfrak{X}, e^{-i\theta\mathfrak{J}_3/\hbar}] = e^{-i\theta\mathfrak{J}_3/\hbar}\{[\hat{\mathbf{i}}(\cos\theta - 1) + \hat{\mathbf{j}}\sin\theta]\mathfrak{X} + [\hat{\mathbf{i}}(-\sin\theta) + \hat{\mathbf{j}}(\cos\theta - 1)]\mathfrak{Y}\}.$$

Both sides here may be expanded in powers of θ, yielding

$$\begin{aligned}
&\sum_{k=1}^{\infty}\frac{1}{k!}\left(-i\frac{\theta}{\hbar}\right)^k[\mathfrak{X}, \mathfrak{J}_3^k] \\
&\quad = \left[\sum_{n=0}^{\infty}\frac{1}{n!}\left(-i\frac{\theta}{\hbar}\right)^n\right]\left[\sum_{l=1}^{\infty}\frac{(-1)^l\theta^{2l}}{(2l)!}(\hat{\mathbf{i}}\mathfrak{J}_3^n\mathfrak{X} + \hat{\mathbf{j}}\mathfrak{J}_3^n\mathfrak{Y})\right. \\
&\qquad\qquad \left. + \sum_{l=0}^{\infty}\frac{(-1)^l\theta^{2l+1}}{(2l+1)!}(\hat{\mathbf{j}}\mathfrak{J}_3^n\mathfrak{X} - \hat{\mathbf{i}}\mathfrak{J}_3^n\mathfrak{Y})\right]. \qquad (12.36)
\end{aligned}$$

This must hold for all powers of θ; in particular, coefficients of θ must be equal on both sides of the equation, and this yields

$$-\frac{i\theta}{\hbar}[\mathfrak{X}, \mathfrak{J}_3] = \theta(\hat{\mathbf{j}}\mathfrak{X} - \hat{\mathbf{i}}\mathfrak{Y}),$$

or

$$[\mathfrak{X}, \mathfrak{J}_3] = \frac{\hbar}{i}\mathfrak{Y}, \qquad [\mathfrak{Y}, \mathfrak{J}_3] = -\frac{\hbar}{i}\mathfrak{X}, \qquad [\mathfrak{Z}, \mathfrak{J}_3] = 0.$$

These relationships are easily generalized if we write $\mathfrak{X}_1$, $\mathfrak{X}_2$ and $\mathfrak{X}_3$ for $\mathfrak{X}$, $\mathfrak{Y}$ and $\mathfrak{Z}$; thus

$$[\mathfrak{J}_j, \mathfrak{X}_k] = i\hbar\epsilon_{jkl}\mathfrak{X}_l \tag{12.37}$$

in which the invariance of operator relations under cyclic permutations of coordinate axes has been utilized.

From these commutator relations we can find all the other commutators involved in Eq. (12.36), and it is not difficult to show that (12.37) ensures that (12.36) is satisfied for all powers of θ.

To obtain the commutators of $\mathfrak{J}_j$ with $\mathfrak{P}_k$, we use a simple stratagem that reduces this problem to the one we just solved. Consider

$$\langle\mathbf{p}|\exp(i\theta\mathfrak{J}_3/\hbar)|\mathbf{x}\rangle = \langle\mathbf{p}|(\hat{\mathbf{i}}\cos\theta + \hat{\mathbf{j}}\sin\theta)x + (\hat{\mathbf{j}}\cos\theta - \hat{\mathbf{i}}\sin\theta)y + \hat{\mathbf{k}}z\rangle$$

$$= (2\pi)^{-3/2}\exp\left\{-\frac{i}{\hbar}[(p_x\cos\theta + p_y\sin\theta)x + (p_y\cos\theta - p_x\sin\theta)y + p_z z]\right\}$$

$$= \langle p_x(\hat{\mathbf{i}}\cos\theta - \hat{\mathbf{j}}\sin\theta) + p_y(\hat{\mathbf{j}}\cos\theta + \hat{\mathbf{i}}\sin\theta) + p_z\hat{\mathbf{k}}|\mathbf{x}\rangle,$$

i.e.

$$e^{i\theta\mathfrak{J}_3/\hbar}|\mathbf{p}\rangle = |(\hat{\mathbf{i}}\cos\theta - \hat{\mathbf{j}}\sin\theta)p_x + (\hat{\mathbf{j}}\cos\theta + \hat{\mathbf{i}}\sin\theta)p_y + \hat{\mathbf{k}}p_z\rangle.$$

Upon changing the sign of θ and operating from the left with the momentum operator $\mathfrak{P}$, we obtain

$$\mathfrak{P}e^{-i\theta\mathfrak{J}_3/\hbar}|\mathbf{p}\rangle = \mathfrak{P}|\hat{\mathbf{i}}(p_x\cos\theta - p_y\sin\theta) + \hat{\mathbf{j}}(p_y\cos\theta + p_x\sin\theta) + \hat{\mathbf{k}}p_z\rangle.$$

This has precisely the form of the original equation, (12.35), which we used to find the commutator $[\mathfrak{X}, \mathfrak{J}_3]$ if $\mathfrak{X}$ and $\mathbf{x}$ in that equation are replaced by $\mathfrak{P}$ and $\mathbf{p}$; hence the commutators between $\mathfrak{P}$ and $\mathfrak{J}$ must have exactly the same form as those between $\mathfrak{X}$ and $\mathfrak{J}$:

$$[\mathfrak{J}_j, \mathfrak{P}_k] = i\hbar\epsilon_{jkl}\mathfrak{P}_l. \tag{12.38}$$

The student should check to see that Eqs. (12.38) agree with the commutators obtained by replacing $\mathfrak{J}_i$ with $\mathfrak{L}_i = (\mathfrak{X} \times \mathfrak{P})_i$ and using the commutators between $\mathfrak{X}_i$ and $\mathfrak{P}_j$.

12.5 ANGULAR-MOMENTUM OPERATORS FOR WAVE FUNCTIONS; ORBITAL ANGULAR MOMENTUM

In many important nonrelativistic problems, the Hamiltonian contains a potential that depends only upon the distance from the origin in a given coordinate basis, that is, the force is a central force and the potential is a central potential. For such problems the classical Hamiltonian is

$$H = \frac{1}{2m}\,\mathbf{p}^2 + V(r), \qquad r = |\mathbf{x}|,$$

and it is transcribed to operator notation as

$$\mathfrak{H} = \frac{1}{2m}\,\mathfrak{P}^2 + \mathscr{V}(\mathfrak{R}),$$

$$\mathfrak{R} \equiv \sqrt{\mathfrak{X}^2} \equiv \underset{x,\ldots}{\mathsf{S}}\, |\mathbf{x}, \ldots\rangle \sqrt{x^2} \langle \mathbf{x}, \ldots| = \underset{\mathbf{x},\ldots}{\mathsf{S}}\, |\mathbf{x}, \ldots\rangle |\mathbf{x}| \langle \mathbf{x}, \ldots|. \tag{12.39}$$

The possibility of defining a Hermitian operator $\mathfrak{R}$ is assured by the fact that the $\mathbf{x}^2$ are real and positive so that $|\mathbf{x}| = \sqrt{\mathbf{x}^2}$, the eigenvalues of $\mathfrak{R}$, are real (and positive).

The energy eigenstates of central-force problems may simultaneously be eigenstates of angular momentum since $[\mathfrak{J}_i, \mathfrak{H}] = 0$, as we shall now show. It is because central-force problems are very important in physics and because the Hamiltonians for such problems commute with angular-momentum operators, that angular-momentum eigenstates are of great importance in quantum theory.

To show that $[\mathfrak{J}_i, \mathfrak{H}] = 0$, we first compute the commutator

$$\begin{aligned}[\mathfrak{J}_i, \mathfrak{P}^2] &= \sum_{j=1}^{3} [\mathfrak{J}_i, \mathfrak{P}_j^2] \\ &= 2i\hbar \sum_{j=1}^{3} \sum_{k=1}^{3} \epsilon_{ijk} \mathfrak{P}_j \mathfrak{P}_k = 0;\end{aligned} \tag{12.40}$$

the last equality holds because exchanging the dummy indices only changes the sign of the next-to-last member of Eqs. (12.40). This means that

$$[\mathfrak{J}_i, \mathfrak{X}^2] = 0 \tag{12.41}$$

also. Since $\mathfrak{H}$, according to Eq. (12.39), depends only upon $\mathfrak{P}^2$ and $\mathfrak{X}^2$, we have

$$[\mathfrak{J}_i, \mathfrak{H}] = 0,$$

as promised; and we see that $\mathfrak{J}_i$, $\mathfrak{J}^2$ and $\mathfrak{H}$ are simultaneously diagonable Hermitian operators.

To see how to analyse the Hamiltonian for central potentials, let us return briefly to the classical problem, and in particular let us consider the kinetic energy. The momentum squared may be written in the form

$$\mathbf{p}^2 = \left(\frac{\mathbf{p}\cdot\mathbf{r}}{r}\right)^2 + \left(\frac{\mathbf{r}\times\mathbf{p}}{r}\right)^2 = p_r^2 + \frac{\mathbf{L}^2}{r^2},$$

where p_r is the radial momentum and $\mathbf{L}$ is the angular momentum $\mathbf{r} \times \mathbf{p}$. We want to make a similar division between radial momentum and angular momentum in the quantum-mechanical Hamiltonian.

Now the standard technique for solving central potential problems is to find wave functions corresponding to eigenstates of the Hamiltonian (and therefore of energy). For this purpose, the eigenvalue equation

$$\mathfrak{H}|E, \ldots\rangle = E|E, \ldots\rangle$$

is converted into

$$\langle \mathbf{x}, \ldots|\mathfrak{H}|E, \ldots\rangle = E\langle x, \ldots|E, \ldots\rangle$$

or

$$\left[\frac{1}{2m}\mathsf{P}^2 + \mathsf{V}(r)\right]\psi_{E,\ldots}(\mathbf{x}, \ldots)$$

$$= \left[-\frac{\hbar^2}{2m}\nabla^2 + V(r)\right]\psi_{E,\ldots}(\mathbf{x}, \ldots) = E\psi_{E,\ldots}(\mathbf{x}, \ldots). \tag{12.42}$$

Thus the eigenvalue problem becomes one of solving a partial differential equation with appropriate boundary conditions.

Spherical coordinates obviously are appropriate in problems such as these, in which there is spherical symmetry. Letting θ be the colatitude and φ the azimuthal angle, we have

$$\nabla^2 = \frac{1}{r^2}\partial_r r^2 \partial_r + \frac{1}{r^2}\left[\frac{1}{\sin\theta}\partial_\theta \sin\theta\,\partial_\theta + \frac{1}{\sin^2\theta}\partial_\varphi^2\right]$$

$$= \frac{1}{r}\partial_r^2 r + \frac{1}{r^2}\left[\partial_\theta^2 + \cot\theta\,\partial_\theta + \frac{1}{\sin^2\theta}\partial_\varphi^2\right]$$

and hence

$$\mathsf{P}^2 = \left(\frac{\hbar}{i}\frac{1}{r}\partial_r r\right)^2 - \frac{1}{r^2}\hbar^2\left[\partial_\theta^2 + \cot\theta\,\partial_\theta + \frac{1}{\sin^2\theta}\partial_\varphi^2\right]. \tag{12.43}$$

Also, we have

$$\mathsf{L}_1 = (\mathbf{x} \times \mathsf{P})_1 = -\frac{\hbar}{i}(\sin\varphi\,\partial_\theta + \cos\varphi\cot\theta\,\partial_\varphi) \tag{12.44a}$$

$$\mathsf{L}_2 = (\mathbf{x} \times \mathsf{P})_2 = \frac{\hbar}{i}(\cos\varphi\,\partial_\theta - \sin\varphi\cot\theta\,\partial_\varphi) \tag{12.44b}$$

$$\mathsf{L}_3 = \frac{\hbar}{i}\partial_\varphi \tag{12.44c}$$

and hence

$$\mathsf{L}_\pm = \mathsf{L}_1 \pm i\mathsf{L}_2 = \hbar e^{\pm i\varphi}[\pm\partial_\theta + i\cot\theta\,\partial_\varphi]. \tag{12.45}$$

From the L_i it is a simple matter to obtain

$$\mathsf{L}^2 = -\hbar^2\left[\partial_\theta^2 + \cot\theta\,\partial_\theta + \frac{1}{\sin^2\theta}\partial_\varphi^2\right], \tag{12.46}$$

so that Eq. (12.43) becomes

$$\mathbf{P}^2 = \left(\frac{\hbar}{i}\frac{1}{r}\partial_r r\right)^2 + \frac{1}{r^2}\mathsf{L}^2. \tag{12.47}$$

This suggests that we define

$$\mathsf{P}_r = \frac{\hbar}{i}\frac{1}{r}\partial_r r$$

as the radial momentum; it is left for the reader to show that P_r is Hermitian whenever the wave functions are physically reasonable ones.

Inserting (12.46) into Eq. (12.42) yields

$$\left[\frac{1}{2m}\mathsf{P}_r^2 + V(r) - E + \frac{1}{2m}\frac{1}{r^2}\mathsf{L}^2\right]\psi_{E,\ldots}(r, \theta, \varphi, \ldots) = 0, \tag{12.48}$$

or, upon multiplication by r^2,

$$\left\{r^2\left[\frac{1}{2m}\mathsf{P}_r^2 + V(r) - E\right] + \frac{1}{2m}\mathsf{L}^2\right\}\psi_{E,\ldots}(r, \theta, \varphi, \ldots) = 0.$$

Here the term in brackets depends only upon r and its partial derivatives whereas the last term depends only upon the angular variables θ and φ; therefore, provided the boundary conditions are separable in the variables r, θ, φ, it is appropriate to separate variables.

Let $\hbar^2\lambda$ be the separation constant; that is, let

$$\begin{aligned}\mathsf{L}^2\psi_{E,\ldots}(r, \theta, \varphi, \ldots) &= -r^2[\mathsf{P}_r^2 + 2mV(r) - 2mE]\psi_{E,\ldots}(r, \theta, \varphi, \ldots)\\ &= \hbar^2\lambda\psi_{E,\ldots}(r, \theta, \varphi, \ldots)\end{aligned}$$

Hence $\hbar^2\lambda$ is the eigenvalue of L^2 for the solution function $\psi_{E\ldots}(r, \theta, \varphi, \ldots)$. Furthermore, let $\psi_{E,\ldots}(r, \theta, \varphi, \ldots) = R_{E,\ldots}(r, \ldots)W_\lambda(\theta, \varphi)$, so that we can write

$$(\mathsf{L}^2 - \hbar^2\lambda)W_\lambda(\theta, \varphi) = 0 \tag{12.49}$$

and

$$\left[\frac{1}{2m}\mathbf{P}_r^2 + V(r) + \frac{\hbar^2\lambda}{2mr^2}\right]R_{E,\ldots}(r, \ldots) = 0. \tag{12.50}$$

All the information about the potential now resides in Eq. (12.50); we can solve Eq. (12.49) quite generally without any consideration of the specific form of the potential function. In fact, we shall defer any further consideration of Eq. (12.50) to a later chapter.

Using Eq. (12.46) in (12.49), we get

$$\left[\partial_\theta^2 + \cot\theta\,\partial_\theta + \frac{1}{\sin^2\theta}\partial_\varphi^2 + \lambda\right]W_\lambda(\theta,\varphi) = 0$$

and it is clear that a further separation of variables is in order. Multiplying this last equation by $\sin^2\theta$, letting $W_\lambda(\theta,\varphi) = \Theta_\lambda(\theta)\Phi(\varphi)$ and dividing by the latter, we get

$$\frac{1}{W_\lambda(\theta,\varphi)}[\sin\theta\,\partial_\theta\sin\theta\,\partial_\theta + \lambda\sin^2\theta + \partial_\varphi^2]W_\lambda(\theta,\varphi)$$

$$= \frac{1}{\Theta_\lambda(\theta)}[\sin\theta\,\partial_\theta\sin\theta\,\partial_\theta + \lambda\sin^2\theta]\Theta_\lambda(\theta) + \frac{1}{\Phi(\varphi)}\partial_\varphi^2\,\Phi(\varphi) = 0,$$

provided $W_\lambda(\theta,\varphi) \neq 0$. Here let μ be the separation constant so that we have

$$\partial_\varphi^2\,\Phi(\varphi) = \mu\Phi(\varphi) \tag{12.51a}$$

$$[\sin\theta\,\partial_\theta\sin\theta\,\partial_\theta + \lambda\sin^2\theta + \mu]\Theta_\lambda(\theta) = 0. \tag{12.51b}$$

The first of Eqs. (12.51) is the easier to solve; we have

$$\Phi(\varphi) = ae^{\sqrt{\mu}\varphi} + be^{-\sqrt{\mu}\varphi}.$$

However, the condition that $\Phi(\varphi) = \Phi(\varphi + 2\pi n)$, for any integer n, must be imposed if φ is to represent the azimuthal angle (see, however, Problem 7); hence

$$a[e^{\sqrt{\mu}(\varphi+2\pi n)} - e^{\sqrt{\mu}\varphi}] + b[e^{-\sqrt{\mu}(\varphi+2\pi n)} - e^{-\sqrt{\mu}\varphi}] = 0$$

must hold for all values of a, b. This implies that

$$e^{2\pi n\sqrt{\mu}} = e^{-2\pi n\sqrt{\mu}} = 1, \qquad \text{all integers } n,$$

and consequently that $n\sqrt{\mu} = ik$, where k is an integer, for all integer values of n. This must hold for $n = 1$ in particular, and in that case, $\sqrt{\mu} = im$, m being an integer, or

$$\mu = -m^2. \tag{12.52}$$

The desired solutions then have the form

$$\Phi_m(\varphi) = \Phi_m(0)e^{im\varphi}, \qquad m \text{ an integer}, \tag{12.53}$$

where $\Phi_m(0)$ is a constant. Actually, a solution to Eq. (12.51a) taking (12.52) into account, may be a linear combination of $\Phi_m(\varphi)$ and $\Phi_{-m}(\varphi)$.

Next, we must solve Eq. (12.51b), in which we now replace μ with $-m^2$ and add a subscript m to the solution function:

$$\left[\sin\theta\frac{d}{d\theta}\sin\theta\frac{d}{d\theta} + \lambda\sin^2\theta - m^2\right]\Theta_{\lambda m}(\theta) = 0.$$

This can be converted to a more familiar form by substituting $x \equiv \cos\theta$, $\sqrt{1-x^2} = \sin\theta$, $d/d\theta = -\sqrt{1-x^2}\, d/dx$, and then multiplying by $(1-x^2)^{-1}$; the result is, when $F_{\lambda m}(x) \equiv \Theta_{\lambda m}(\theta)$,

$$\left[\frac{d}{dx}\left((1-x^2)\frac{d}{dx}\right) + \lambda - \frac{m^2}{1-x^2}\right]F_{\lambda m}(x) = 0$$

or

$$\left[(1-x^2)\frac{d^2}{dx^2} - 2x\frac{d}{dx} + \lambda - \frac{m^2}{1-x^2}\right]F_{\lambda m}(x) = 0. \tag{12.54}$$

Since m is an integer, it may take on the value zero; this yields the simpler equation

$$\left[(1-x^2)\frac{d^2}{dx^2} - 2x\frac{d}{dx} + \lambda\right]F_{\lambda 0}(x) = 0, \tag{12.55}$$

and initially we shall restrict our attention to this equation. Assumption of a solution of the form

$$F_{\lambda 0}(x) = \sum_{n=0}^{\infty} a_n x^{n+\rho}$$

yields, upon substitution into the differential equation,

$$\sum_{n=0}^{\infty} a_n[(\rho+n)(\rho+n-1)(x^{\rho+n-2} - x^{\rho+n}) - 2(\rho+n)x^{\rho+n} + \lambda x^{\rho+n}] = 0.$$

Collecting coefficients of equal power of x, we get

$$a_0\rho(\rho-1)x^{\rho-2} + a_1(\rho+1)\rho x^{\rho-1}$$
$$+ \sum_{n=0}^{\infty} \{a_{n+2}(\rho+n+2)(\rho+n+1) - a_n[(\rho+n)(\rho+n+1) - \lambda]\}x^{\rho+n} = 0.$$

The coefficients of each power of x must vanish in order that the solution be correct for all values of x. The first two terms can be made to vanish without imposing any restrictions on a_0 and a_1 if we choose $\rho = 0$; it can be shown that choosing $a_1 = 0$ and $\rho = 1$ or $a_0 = 0$ and $\rho = -1$ lead to exactly the same solutions as does choosing $\rho = 0$.

For $\rho = 0$, the coefficient of x^n becomes

$$a_{n+2}(n+2)(n+1) - a_n[n(n+1) - \lambda]$$

and its vanishing implies that

$$a_{n+2} = \frac{n(n+1) - \lambda}{(n+2)(n+1)}a_n$$

or

$$a_n = \frac{(n-1)(n-2)-\lambda}{n(n-1)} a_{n-2}.$$

Since a_0 and a_1 both are arbitrary, this result yields two independent sets of solutions, one involving even powers of x and the other involving odd powers of x. Typical coefficients in the two series are

$$a_{2n} = \frac{[(2n-1)(2n-2)-\lambda]}{2n(2n-1)} \frac{[(2n-3)(2n-4)-\lambda]}{(2n-2)(2n-3)} \cdots \frac{[3\cdot 2-\lambda]}{4\cdot 3} \frac{(-\lambda)}{2\cdot 1} a_0$$

and

$$a_{2n+1} = \frac{[2n(2n-1)-\lambda]}{(2n+1)\cdot 2n} \frac{[(2n-2)(2n-3)-\lambda]}{(2n-1)(2n-2)} \cdots \frac{4\cdot 3-\lambda}{5\cdot 4} \frac{2\cdot 1-\lambda}{3\cdot 2} a_1.$$

Hence the most general solution having the series form we have assumed here is

$$F_{\lambda 0}(x) = F_{\lambda 0}{}^{(e)}(x) + F_{\lambda 0}{}^{(0)}(x)$$

where

$$F_{\lambda 0}{}^{(e)}(x) \equiv a_0 \sum_{n=0}^{\infty} \frac{\prod_{v=0}^{n} [(2v-1)(2v-2)-\lambda]}{(2n)!(2-\lambda)} x^{2n},$$

$$F_{\lambda 0}{}^{(0)}(x) \equiv a_1 \sum_{n=0}^{\infty} \frac{\prod_{v=0}^{n} [(2v-1)2v-1]}{(2n+1)!(-\lambda)} x^{2n+1},$$

which have been normalized so that $F_{\lambda 0}{}^{(e)}(0) = a_0$, $F_{\lambda 0}{}^{(0)}(0) = a_1$.

Now the domain of the variable $x = \cos\theta$ is given by $-1 \le x \le 1$ for real values of θ, and we wish to have the solutions $F_{\lambda 0}(x)$ be convergent for all x within this domain. To test convergence of these series, beginning with $F_{\lambda 0}{}^{(e)}(x)$ (i.e. $a_1 = 0$), we use the ratio test, which yields

$$\lim_{n\to\infty} \left| \frac{\dfrac{a_0}{2-\lambda} \dfrac{\prod_{v=0}^{n} [(2v-1)(2v-2)-\lambda]}{(2n)!} x^{2n}}{\dfrac{a_0}{2-\lambda} \dfrac{\prod_{v=0}^{n-1} [(2v-1)(2v-2)-\lambda]}{(2n-2)!} x^{2n-2}} \right| = \lim_{n\to\infty} \frac{(2n-1)(2n-2)}{2n(2n-1)} |x|^2 = |x|^2.$$

For $|x| < 1$, this limit is less than unity and convergence is assured. However, since $\sum_{n=0}^{\infty} x^{2n}$ diverges for $x \ge 1$, and since in the limit as $n \to \infty$ the coefficients of this geometric series are the same as those of our series solution, the limit comparison test tells us that our series will diverge, *unless* a_n vanish for all $n \geq n_0$ for some n_0. Thus convergence is assured for all $|x| \leq 1$ only if the series has a finite number of terms, and the latter applies if and only if

$$\lambda = (2n_0 - 1)(2n_0 - 2)$$

for some n_0.

Precisely the same conclusion follows for consideration of the odd series, i.e. for $a_0 = 0$. Hence we must have

$$\lambda = l(l + 1), \qquad l \text{ an integer},$$

if we are to obtain physically reasonable solutions to the Legendre equation. Furthermore, if l is odd, none of the factors

$$(2v - 1)(2v - 2) - \lambda = (2v - 1)(2v - 2) - l(l + 1)$$

in the coefficients of the even series can vanish. If one were to vanish for some v we should have

$$l = 2v - 2, \qquad l + 1 = 2v - 1,$$

which is clearly impossible, since l is odd. Hence *if l is an odd integer, only the odd solutions $F_{\lambda 0}{}^{(0)}(x)$ exist.* Similarly, *if l is an even integer, only the even solution $F_{\lambda 0}{}^{(e)}(x)$ exists.*

With this in mind, we relabel our solutions

$$P_l(x) = \begin{cases} F_{\lambda 0}{}^{(e)}(x) & \text{if } \lambda = l(l + 1),\ l \text{ an even integer} \\ F_{\lambda 0}{}^{(0)}(x) & \text{if } \lambda = l(l + 1),\ l \text{ an odd integer.} \end{cases}$$

In addition, we normalize our solutions by choosing a_0 and a_1 such that

$$P_l(1) = 1.$$

Fortunately, a compact expression for $P_l(x)$, thus defined, exists in the form of *Rodrigue's formula,*

$$P_l(x) = \frac{(-1)^l}{(2l)!!} \frac{d^l}{dx^l}(1 - x^2)^l, \tag{12.56}$$

where the double factorial $n!!$ is defined by

$$n!! \equiv n(n - 2)(n - 4) \cdot \cdots \cdot \begin{cases} 4 \cdot 2 & \text{if } n \text{ is even} \\ 3 \cdot 1 & \text{if } n \text{ is odd;} \end{cases}$$

for $n = 2l$ an even positive integer, we have

$$(2l)!! = 2^l l!$$

Rodrigue's formula may be utilized to show that the $P_l(x)$ satisfy

$$\int_{-1}^{1} P_l(x)P_k(x)\, dx = \frac{2}{2l + 1}\delta_{l,k}. \tag{12.57}$$

Now we return to Eq. (12.54) with $m \neq 0$ allowed. Because m occurs in the differential equation involving φ whereas λ does not, the values of λ and m must be independent of each other. Hence if solutions exist for $m = 0$ only when $\lambda = l(l + 1)$, that must be the case for $m \neq 0$ also. Therefore we must solve

$$\left[\frac{d}{dx}\left((1 - x^2)\frac{d}{dx}\right) + l(l + 1) - \frac{m^2}{1 - x^2}\right]P_l^m(x) = 0, \qquad (12.58)$$

where $P_l^m(x) = F_{\lambda m}{}^{(e,o)}(x)$ for $\lambda = l(l + 1)$.

Equation (12.58) can be cast into a more convenient form through the substitution

$$P_l^m(x) = (-1)^m(1 - x^2)^{m/2}y(x),$$

for then it becomes

$$(1 - x^2)\frac{d^2 y}{dx^2} - 2(m + 1)x\frac{dy}{dx} + [(l - m)(l + m + 1)]y = 0. \qquad (12.59)$$

However, m-fold differentiation of Eq. (12.55) when $\lambda = l(l + 1)$ yields

$$\left[(1 - x^2)\frac{d^2}{dx^2} - 2(m + 1)x\frac{d}{dx} + (l - m)(l + m + 1)\right]\frac{d^m}{dx^m}P_l(x) = 0. \qquad (12.60)$$

Comparison of Eqs. (12.59) and (12.60) shows that we may set

$$y(x) = \frac{d^m}{dx^m}P_l(x)$$

or

$$P_l^m(x) = (-1)^m(1 - x^2)^{m/2}\frac{d^m}{dx^m}P_l(x)$$

$$= (-1)^{l+m}[(2l)!!]^{-1}(1 - x^2)^{m/2}\frac{d^{l+m}}{dx^{l+m}}(1 - x^2)^l \qquad (12.61)$$

It turns out that this solution to Eq. (12.58) is acceptable for negative as well as positive values of m, although it is correct in an obvious way only for positive values of m according to the manner in which we have obtained it.

Now we are in a position to combine our results. In terms of the angle θ, (12.61) becomes

$$\Theta_l{}^m(\theta) = \Theta_0\frac{(-1)^{l+m}}{(2l)!!}\sin^m\theta\frac{d^{l+m}}{d(\cos\theta)^{l+m}}\sin^{2l}\theta.$$

The product of this with $\Phi_m(\varphi)$ has the form

$$Y_l^m(\theta, \varphi) = N_{lm}\frac{(-1)^l}{(2l)!!}\sin^m\theta\frac{d^{l+m}}{d(\cos\theta)^{l+m}}\sin^{2l}\theta\, e^{im\varphi}, \qquad (12.62)$$

where N_{lm} is a normalization constant determined by the condition

$$\int_{4\pi} d\Omega\, Y_{l'}^{m'*}(\theta, \varphi) Y_l^m(\theta, \varphi) = \int_0^{2\pi} d\varphi \int_0^{\pi} d\theta \sin\theta\, Y_{l'}^{m'*}(\theta, \varphi) Y_l^m(\theta, \varphi)$$

$$= \delta_{ll'}\, \delta_{mm'}. \tag{12.63}$$

The N_{lm} satisfying this condition are

$$N_{lm} = \sqrt{\frac{2l+1}{4\pi} \frac{(l-m)!}{(l+m)!}}.$$

The $Y_l^m(\theta, \varphi)$ given in (12.62) with this value of N_{lm} are called the normalized *spherical harmonics*; thus

$$Y_l^m(\theta, \varphi) = \sqrt{\frac{2l+1}{4\pi} \frac{(l-m)!}{(l+m)!}} \frac{(-1)^l}{(2l)!!} \sin^m\theta \frac{d^{l+m}}{d(\cos\theta)^{l+m}} \sin^{2l}\theta\, e^{im\varphi}. \tag{12.64}$$

Aside from obtaining the eigenfunctions $Y_l^m(\theta, \varphi)$ of L^2 and L_3, whose eigenvalues are, respectively, $\hbar^2 l(l+1)$ and $\hbar m$, we have learned a very important fact about the quantity l, sometimes called the orbital angular-momentum quantum number. The latter may take on only integer values, unlike the abstract angular-momentum quantum number j we studied earlier, which could take on half-integer values. Hence orbital angular-momentum cannot exhaust the possible features of angular momentum we studied earlier. Either the abstract theory is richer than physics, or some other kind of angular momentum, aside from orbital, must exist. As we shall see, the latter is the case.

12.6 SPIN ANGULAR MOMENTUM

Since the eigenstates of the orbital angular-momentum operators $\mathfrak{L}^2$ and $\mathfrak{L}_3$, with $\mathfrak{L} = \mathfrak{X} \times \mathfrak{P}$, are states $|l, m_l\rangle$, with l an integer and m_l, $-l \le m_l \le l$, also an integer, these states must be orthogonal to *any* state with angular-momentum quantum number $j = (2n+1)/2$, n an integer. The latter must therefore involve a kind of angular momentum that is fundamentally different from orbital angular momentum. This new kind of angular momentum, which has no analog in classical physics, we call *spin*. The parameters defining the spin state of a particle are independent of the parameters defining the spatial state (e.g. $\mathbf{x}$ or $\mathbf{p}$; r, θ, φ or p, l, m_l; etc.), and hence we may discuss spin states in the abstract, without referring to particle position, etc. This means that the spin operators must commute with the spatial operators.

Let's consider a spin state $|s, \mu\rangle$ and spin operators $\mathfrak{S}_1$, $\mathfrak{S}_2$, $\mathfrak{S}_3$, from which we can form $\mathfrak{S}^2$ and $\mathfrak{S}_\pm = (\mathfrak{S}_1 \pm i\mathfrak{S}_2)$. The $\mathfrak{S}_i$ are to be analogous with the $\mathfrak{J}_i$ since they represent a kind of angular momentum, and they must have the same commutation relations as the $\mathfrak{J}_i$ (with an exception which we shall note later);

thus

$$[\mathfrak{S}_j, \mathfrak{S}_k] = i\hbar \sum_{l=1}^{3} \epsilon_{jkl}\mathfrak{S}_l = i\hbar\epsilon_{jkl}\mathfrak{S}_l,$$

$$[\mathfrak{S}_j, \mathfrak{S}^2] = 0.$$

The state $|s, \mu\rangle$ has the properties

$$\mathfrak{S}^2|s, \mu\rangle = \hbar^2 s(s+1)|s, \mu\rangle$$

$$\mathfrak{S}_3|s, \mu\rangle = \hbar\mu|s, \mu\rangle.$$

In the space of spin states of a particle with spin quantum number s, the set of states $\{|s, \mu\rangle\}$, with all possible values of μ, constitutes a complete set. Hence we may write

$$\mathfrak{S}_3 = \mathfrak{S}_3 \sum_{\mu} |s, \mu\rangle\langle s, \mu| = \sum_{\mu} |s, \mu\rangle\hbar\mu\langle s, \mu| .$$

Since $|s, \mu\rangle$ is orthogonal to $|l, m_l\rangle$, we have

$$\mathfrak{S}_3|l, m_l\rangle = \sum_{\mu} |s, \mu\rangle\hbar\mu\langle s, \mu|l, m_l\rangle = 0.$$

Since the choice of the 3-axis for quantizing angular momentum was quite arbitrary—we could have chosen the 1-axis or the 2-axis—we have in general

$$\mathfrak{S}_i|l, m_l\rangle = 0. \tag{12.65}$$

Similarly,

$$\mathfrak{L}_i|s, \mu\rangle = 0. \tag{12.66}$$

Now if $\mathfrak{S}$ really plays the role of angular momentum, we should be able to form, from the spin and orbital angular momenta, the total angular momentum of a particle. The example of a spinning planet, although not a correct classical analog for the quantum system, suggests how this should be done. The total angular momentum of the planet is the vector sum of its angular momentum of rotation about its own axis and its angular momentum of revolution about the sun. We assume that this classical rule for adding angular momenta as vector quantities holds in quantum theory, and write

$$\mathfrak{J} = \mathfrak{L} + \mathfrak{S}, \qquad \text{or} \qquad \mathfrak{J}_i = \mathfrak{L}_i + \mathfrak{S}_i. \tag{12.67}$$

Indeed, if $\mathfrak{S}$ really does represent angular momentum, this addition rule must hold.

Furthermore, the state of a particle with spin angular-momentum quantum numbers s, μ is representable as the tensor product of the spatial-state vector and the spin-state vector; e.g.

$$|x, \ldots; s, \mu\rangle = |x, \ldots\rangle|s, \mu\rangle,$$

or

$$|r, l, m, \ldots; s, \mu\rangle = |r, l, m, \ldots\rangle|s, \mu\rangle.$$

Because of Eqs. (12.65)–(12.66), this yields

$$\begin{aligned}\mathfrak{J}_3|l, m_l, \ldots; s, \mu\rangle &= [\mathfrak{J}_3|l, m_l, \ldots\rangle]|s, \mu\rangle + |l, m_l, \ldots\rangle\mathfrak{J}_3|s, \mu\rangle \\ &= [\mathfrak{L}_3|l, m_l, \ldots\rangle]|s, \mu\rangle + |l, m_l, \ldots\rangle\mathfrak{S}_3|s, \mu\rangle \\ &= (m_l + \mu)|l, m_l, \ldots; s, \mu\rangle,\end{aligned}$$

when we take into account the rules for operating on tensor products given in Eq. (10.35). Hence $|l, m_l, \ldots; s, \mu\rangle$ is an eigenstate of $\mathfrak{J}_3$. However, it is not in general en eigenstate of $\mathfrak{J}^2 = (\mathfrak{L}^2 + \mathfrak{S}^2 + 2\mathfrak{L}\cdot\mathfrak{S})$ because it is not an eigenstate of $\mathfrak{L}\cdot\mathfrak{S}$. This will be elaborated when we discuss the addition of angular momenta.

Earlier in this section it was mentioned that the $\mathfrak{S}_i$ do not share all the commutation rules of the $\mathfrak{J}_i$ or $\mathfrak{L}_i$. Consider, for example,

$$[\mathfrak{X}_i, \mathfrak{J}_j] = [\mathfrak{X}_i, \mathfrak{L}_j + \mathfrak{S}_j] = [\mathfrak{X}_i, \mathfrak{L}_j] + [\mathfrak{X}_i, \mathfrak{S}_j].$$

Now

$$[\mathfrak{X}_i, \mathfrak{J}_j] = [\mathfrak{X}_i, \mathfrak{L}_j] = i\hbar \sum_{k=1}^{3} \epsilon_{ijk}\mathfrak{X}_k,$$

according to Eq. (12.37) and the discussion leading up to that equation: hence

$$[\mathfrak{X}_i, \mathfrak{S}_j] = 0.$$

Similarly,

$$[\mathfrak{P}_i, \mathfrak{S}_j] = 0.$$

Therefore, the spin operators commute with all spatial operators. This is intimately connected with the fact that the space of the spin states is orthogonal to the space of the states determined by the spatial operators.

Although we were led to the notion of spin states by the existence of states with angular-momentum quantum number $j = (2n + 1)/2$, there is no need for us to restrict the spin quantum numbers to half-odd integers. The considerations of this section apply to all Hermitian operators having the commutation relations (12.9)–(12.10) and commuting with all spatial operators. We saw earlier that operators with the former property have eigenstates with eigenvalues $n/2$, n an integer; hence the maximum component of spin angular momentum may be any half-integer multiple of $\hbar$.

The spin operators have the peculiar property that they are not canonically conjugate to any observable quantity. For example, in the wave-function representation

$$[\mathsf{L}_3, \varphi] = \hbar/i$$

and L_3 is canonically conjugate to φ. On the other hand, although $\mathfrak{S}_3$ appears to generate rotations about the 3-axis, the fact that $\mathfrak{S}_3$ commutes with all spatial operators guarantees that

$$[\mathfrak{S}_3, \varphi] = 0.$$

Observable quantities without observable canonical conjugates, such as spin angular momenta, are called internal variables, and their quantum numbers, such as s, μ, are called *internal quantum numbers.* The variables that would be canonically conjugate to the observable operators, being unobservable, are called *internal* variables; symmetries under transformations of these internal variables are called *internal symmetries.* Canonical conjugacy has no meaning except for observable quantities.

12.7 COUPLING OF ANGULAR MOMENTA

Instead of the four mutually commuting operators, $\mathfrak{L}^2$, $\mathfrak{S}^2$, $\mathfrak{L}_3$, $\mathfrak{S}_3$, we could choose the four commuting operators $\mathfrak{J}^2 = (\mathfrak{L} + \mathfrak{S})^2$, $\mathfrak{J}_3$, $\mathfrak{L}^2$ and $\mathfrak{S}^2$ to define an angular-momentum state. A state defined by the former set we denote, as before,

$$|l, m_l; s, \mu\rangle,$$

whereas a state defined by the latter we denote

$$|j, m, l, s\rangle,$$

with

$$\begin{aligned}
\mathfrak{J}^2|j, m, l, s\rangle &= \hbar^2 j(j+1)|j, m, l, s\rangle,\\
\mathfrak{J}_3|j, m, l, s\rangle &= \hbar m|j, m, l, s\rangle,\\
\mathfrak{L}^2|j, m, l, s\rangle &= \hbar^2 l(l+1)|j, m, l, s\rangle,\\
\mathfrak{S}^2|j, m, l, s\rangle &= \hbar^2 s(s+1)|j, m, l, s\rangle.
\end{aligned}$$

If a system is in angular-momentum state $|l, m_l; s, \mu\rangle$ we say that l, m_l, s and μ are all *good quantum numbers* for the system; if it is in state $|j, m, l, s\rangle$, then j, m, l and s are good quantum numbers for the system. Good quantum numbers are the numbers defining the eigenstate in which a system finds itself.

Since the $|l, m_l; s, \mu\rangle$ form a complete set of angular momentum eigenstates, we can write $|j, m, l, s\rangle$ as a linear combination of them. Hence

$$|j, m, l, s\rangle = \sum_{m_l}\sum_{\mu} |l, m_l; s, \mu\rangle\langle l, m_l; s, \mu|j, m, l, s\rangle \tag{12.68}$$

and the coefficients $\langle l, m_l; s, \mu|j, m, l, s\rangle$ are called coupling coefficients, or Clebsch–Gordan coefficients. We shall not take up a general study of these coefficients, but simply note some of their most important properties. First, we have

$$\begin{aligned}\mathfrak{J}_3|j, m, l, s\rangle &= m|j, m, l, s\rangle\\ &= \sum_{m_l}\sum_{\mu}(\mathfrak{L}_3 + \mathfrak{S}_3)|l, m_l; s, \mu\rangle\langle l, m_l; s, \mu|j, m, l, s\rangle\\ &= \sum_{m_l,\mu}(m_l + \mu)|l, m_l; s, \mu\rangle\langle l, m_l; s, \mu|j, m, l, s\rangle.\end{aligned}$$

Using Eq. (12.68) again to replace $m|j, m, l, s\rangle$ with a sum over m_l and μ of states $|l, m_l; s, \mu\rangle$, we get

$$\sum_{m_l}\sum_{\mu}(m - m_l - \mu)|l, m_l; s, \mu\rangle\langle l, m_l; s, \mu|j, m, l, s\rangle = 0.$$

Since the $|l, m_l; s, \mu\rangle$ are linearly independent, their coefficients in this sum must vanish, i.e.

$$(m - m_l - \mu)\langle l, m_l; s, \mu|j, m, l, s\rangle = 0;$$

but this implies

$$\langle l, m_l; s, \mu|j, m, l, s\rangle = 0 \qquad \text{unless} \qquad m = m_l + \mu. \tag{12.69}$$

Therefore one of the sums in Eq. (12.68) can be dropped by replacing, for example, m_l with $m - \mu$:

$$|j, m, l, s\rangle = \sum_{\mu}|l, m - \mu; s, \mu\rangle\langle l, m - \mu; s, \mu|j, m, l, s\rangle. \tag{12.70}$$

Equation (12.70) can be inverted by writing

$$|l, m_l; s, \mu\rangle = \sum_{j}\sum_{\mu}|j, m, l, s\rangle\langle j, m, l, s|l, m_l; s, \mu\rangle$$

since the $|j, m, l, s\rangle$ must also form a complete set of angular-momentum states for a particle. Use of Eq. (12.69) eliminates the sum over m to yield

$$|l, m_l; s, \mu\rangle = \sum_{j}|j, m_l + \mu, l, s\rangle\langle j, m_l + \mu, l, s\rangle|l, m_l; s, \mu\rangle.$$

Furthermore, it is clear that

$$\begin{aligned}\langle j, m, l, s|j', m', l, s\rangle &= \sum_{\mu}\langle j, m, l, s|l, m - \mu; s, \mu\rangle\langle l, m' - \mu; s, \mu|j', m', l, s\rangle\\ &= \sum_{\mu}\langle j, m, l, s|l, m - \mu; s, \mu\rangle\langle j', m', l, s|l, m' - \mu; s, \mu\rangle^*\\ &= \delta_{jj'}\,\delta_{mm'}.\end{aligned} \tag{12.71}$$

However, since we are free to let

$$|l, m' - \mu; s, \mu\rangle \to e^{i\alpha}|l, m' - \mu; s, \mu\rangle$$

and

$$|j', m', l, s\rangle \to e^{i\beta}|j', m', l, s\rangle,$$

we could replace

$$\langle j', m', l, s|l, m' - \mu; s, \mu\rangle$$

with

$$e^{i(\alpha-\beta)}\langle j', m', l, s|l, m' - \mu; s, \mu\rangle$$

without changing any physics. We might as well choose α and β in such a way that $\langle j', m', l, s|l, m' - \mu; s, \mu\rangle$ is real; then Eq. (12.71) becomes

$$\sum_{\mu} \langle j, m, l, s|l, m - \mu; s, \mu\rangle\langle j', m', l, s|l, m' - \mu; s, \mu\rangle = \delta_{jj'}\,\delta_{mm'}. \quad (12.72)$$

This expresses the orthogonality of the Clebsch–Gordan (C.–G.) coefficients. In general we shall assume that phases have been chosen in such a way that the C.–G. coefficients are real.

Obviously we also have

$$\sum_{j} \langle l, m_l; s, \mu|j, m_l + \mu, l, s\rangle\langle l, m_l'; s, \mu'|j, m_l' + \mu', l, s\rangle^* = \delta_{m_l m_l'}\,\delta_{\mu\mu'}. \quad (12.73)$$

The maximum value of m available to the state $|j, m, l, s\rangle$ is evidently

$$m_{\max} = (m_l)_{\max} + \mu_{\max} = l + s;$$

hence

$$j_{\max} = l + s.$$

For this maximum value of j, and for the corresponding maximum value of m, $m_{\max} = l + s$, we have

$$|l + s, l + s, l, s\rangle = \sum_{\mu} |l, l + s - \mu; s, \mu\rangle\langle l, l + s - \mu; s, \mu|l + s, l + s, l, s\rangle.$$

However, the maximum value of m_l is l, and hence

$$l + s - \mu \le l, \qquad \text{or} \qquad s \le \mu.$$

On the other hand, $s \ge \mu$ because s is defined as the maximum allowable value of μ; hence $\mu = s$ and the sum above has only one term. Equation (12.72) then tells us that

$$\langle l, l; s, s|l + s, l + s, l, s\rangle^2 - 1,$$

and we assume the phases to be chosen such that

$$\langle l, l; s, s|l + s, l + s, l, s\rangle = 1$$

for the special case when all the quantum numbers take on their maximum values.

Although m_l and μ must both be maximal to form the state $|l + s, l + s, l, s\rangle = |l, l; s, s\rangle$, μ has two possible values in the sum in

$$|l + s, l + s - 1, l, s\rangle = \sum_{\mu = s-1}^{s} |l, l + s - 1 - \mu; s, \mu\rangle$$

$$\cdot \langle l, l + s - 1 - \mu; s, \mu | l + s, l + s - 1, l, s\rangle.$$

There is another state, linearly independent of $|l + s, l + s - 1, l, s\rangle$ that can be formed from these same two states $|l, l - 1, s, s\rangle$ and $|l, l; s, s - 1\rangle$, namely

$$|l + s - 1, l + s - 1, l, s\rangle$$

$$= \sum_{\mu = s-1}^{s} |l, l + s - 1 - \mu; s, \mu\rangle\langle l, l + s - 1 - \mu; s, \mu | l + s - 1, l + s - 1, l, s\rangle,$$

etc. (See Table 12.1.) Hence for every state $|l, m - \mu; s, \mu\rangle$ there is a state $|j, m, l, s\rangle$; but there are $(2l + 1)(2s + 1)$ of the former, since m_l can take on $(2l + 1)$ values and μ can take on $(2s + 1)$ values. Since, for each allowed value of j, there are $2j + 1$ states, it follows that

$$\sum_{j = j_{\min}}^{l+s} (2j + 1) = (2l + 1)(2s + 1).$$

But

$$\sum_{k=a}^{b} k = \tfrac{1}{2}[b(b + 1) - a(a - 1)];$$

this yields

$$(l + s)(l + s + 1) - j_{\min}(j_{\min} - 1) + l + s - j_{\min} + 1 = (2l + 1)(2s + 1)$$

or

$$j_{\min}^2 = (l - s)^2$$

or

$$j_{\min} = |l - s|,$$

since $j_{\min} \geq 0$.

Therefore, $l + s \geq j \geq |l - s|$ gives the allowed values of j, and for each allowed $j, j \geq m \geq -j$ gives the allowed values of m.

Because all allowed values of m differ from $m_{\max} = j_{\max} = l + s$ by an integer, it follows that all allowed values of j differ from $l + s$ by an integer, also; hence

$$j = l + s, l + s - 1, \ldots, |l - s| \tag{12.74}$$

Table 12.1. Correspondence between eigenstates $|jmls\rangle$ and $|lm_l;s\mu\rangle$.

There are exactly as many states in the left-hand column as in the right-hand column. Any state in a given horizontal grouping in the left-hand column is a linear combination of states in the same horizontal grouping of the right-hand column; this statement remains correct under interchange of the words "right" and "left".

$\lvert j, m, l, s\rangle$		$\lvert l, m-\mu; s, \mu\rangle$	
	$(l, s$ fixed)		
j	m	$m-\mu$	μ
$l+s$	$l+s$	l	s
$l+s$ $l+s-1$	$l+s-1$ $l+s-1$	$l-1$ l	s $s-1$
$l+s$ $l+s-1$ $l+s-2$	$l+s-2$ $l+s-2$ $l+s-2$	$l-2$ $l-1$ l	s $s-1$ $s-2$
$\vdots$	$\vdots$	$\vdots$	$\vdots$
$l+s$ $l+s-1$ $\vdots$ $\lvert l-s\rvert$	$\lvert l-s\rvert$ $\lvert l-s\rvert$ $\vdots$ $\lvert l-s\rvert$	$\lvert l-s\rvert-s$ $\lvert l-s\rvert-s+1$ $\vdots$ l	s $s-1$ $\vdots$ $\lvert l-s\rvert-l$
$\vdots$	$\vdots$	$\vdots$	$\vdots$
$l+s$ $l+s-1$	$-l-s+1$ $-l-s+1$	$-l+1$ $-l$	$-s$ $-s+1$
$l+s$	$-l-s$	$-l$	$-s$

are in fact the allowed values of j. Finally, it can be shown that

$$\langle l, m-\mu; s, \mu | j, m, l, s\rangle = 0 \qquad \text{unless} \qquad \Delta(j, l, s), \tag{12.75}$$

where by $\Delta(j, l, s)$ we mean that a triangle with sides j, l and s may be formed.

This last condition provides the justification for the so-called vector model for angular momentum used in elementary courses, in which one carries out calculations under the assumption that the angular momenta can be treated as vectors of length l, s and j, j being the length of the vector sum of l and s. The normal quantum conditions are imposed, namely that l be an integer, s a half-integer, that $\mathfrak{L}^2, \ldots$ have eigenvalues $\hbar^2 l(l+1), \ldots$, etc. The formal coupling of states through the C.–G. coefficients is ignored, and calculations are carried out much as they would be in classical physics. This technique yields many correct results and, because it is an informal approach, is often a much quicker method than the formal quantum-theoretical techniques. On the other hand, it does have some limitations.

The coefficients* $\langle l, m-\mu;\ s, \mu | j, m, ls\rangle$ can be evaluated numerically for various values of j, m, l, s and μ, and several useful tables of them exist. We shall not, however, obtain formulas for their numerical values; these formulas and their derivations are to be found in a number of readily available works on the quantum theory of angular momentum.

Later we shall see that the C.–G. coefficients discussed here may be used for coupling angular momenta in general; we are not restricted to coupling orbital angular momentum with spin angular momentum.

12.8 ANGULAR MOMENTUM AND WAVE FUNCTIONS; SPINORS

Since each spin operator commutes with all spatial operators (e.g. $\mathfrak{X}_j$, $\mathfrak{P}_j$), a complete set of compatible operators might include any three mutually commuting spatial operators (e.g. $\mathfrak{X}_1, \mathfrak{X}_2, \mathfrak{X}_3$ or $\mathfrak{R}$, Θ, Φ or $\mathfrak{P}_r$, $\mathfrak{L}^2$, $\mathfrak{L}_3$) together with two spin operators, $\mathfrak{S}^2$ and, say $\mathfrak{S}_3$. A wave function then would include among its arguments the spin quantum numbers s and μ; e.g.

$$\psi_1(r, \theta, \varphi; s, \mu) = \langle r; \theta, \varphi; s, \mu | 1\rangle \tag{12.76}$$

where

$$|r; \theta, \varphi; s, \mu\rangle \equiv |r\rangle|\theta, \varphi\rangle|s, \mu\rangle.$$

The normalization conditions on $|r\rangle$ and $|\theta, \varphi\rangle = |\cos\theta\rangle|\varphi\rangle$ are a little different from the usual conditions, because for spherical coordinates

$$\underset{r}{\mathsf{S}} = \int_0^\infty dr\, r^2, \qquad \underset{\theta}{\mathsf{S}} = \int_0^\pi d\theta \sin\theta = \int_{-1}^1 d\cos\theta, \qquad \underset{\varphi}{\mathsf{S}} = \int_0^{2\pi} d\varphi.$$

---

* Several different notations for the C.–G. coefficients are in use. For practical calculations in which these coefficients are used frequently, perhaps Rose's notation, $C(s, l, j;\ \mu, m-\mu)$ is the most efficient: the Wigner $3-j$ symbols,

$$\begin{pmatrix} s & l & j \\ \mu & m_l & m \end{pmatrix}$$

also are very useful. For definitions of these and other coupling symbols, consult A. R. Edmonds, *Angular Momentum in Quantum Mechanics*, Princeton University Press (1957), M. E. Rose, *Elementary Theory of Angular Momentum*, John Wiley and Sons (1957), or E. P. Wigner, *Group Theory*, Academic Press (1959).

The normalization conditions are

$$\langle r|r'\rangle = \frac{1}{r^2}\delta(r - r'), \qquad \langle \cos\theta|\cos\theta'\rangle = \delta(\cos\theta - \cos\theta'),$$

$$\langle \varphi|\varphi'\rangle = \delta(\varphi - \varphi'), \qquad \langle \theta, \varphi|\theta', \varphi'\rangle = \delta(\cos\theta - \cos\theta')\delta(\varphi - \varphi') \quad (12.77)$$

so that

$$\mathop{\mathsf{S}}_{r} \langle r|r'\rangle = 1, \qquad \mathop{\mathsf{S}}_{\theta,\varphi} \langle \theta, \varphi|\theta', \varphi'\rangle = 1, \text{etc.}$$

The effect of an angular-momentum operator $\mathfrak{M}$ on an arbitrary state $|1\rangle$ can be determined by the usual technique of introducing a complete set of states:

$$\mathfrak{M}|1\rangle = \mathop{\mathsf{S}}_{p,l,m_l,\ldots} \mathfrak{M}|p; l, m_l; s, \mu\rangle\langle p; l, m_l; s, \mu|1\rangle$$

$$= \mathop{\mathsf{S}}_{p,j,m,\ldots} \mathfrak{M}|p; j, m, l, s\rangle\langle p; j, m, l, s|1\rangle.$$

In particular, if $\mathfrak{M} = \mathfrak{J}^2 = \mathfrak{L}^2 + \mathfrak{S}^2 + 2\,\mathfrak{L}\cdot\mathfrak{S}$, we get

$$\mathfrak{J}^2|1\rangle = \mathop{\mathsf{S}}_{p,j,\ldots} \hbar^2 j(j+1)|p; j, m, l, s\rangle\langle p; j, m, l, s|1\rangle$$

$$= \mathop{\mathsf{S}}_{p,j,\ldots} \hbar^2[l(l+1) + s(s+1) + 2\,\mathfrak{L}\cdot\mathfrak{S}]|p; j, m, l, s\rangle\langle p; j, m, l, s|1\rangle,$$

or

$$\mathfrak{L}\cdot\mathfrak{S}|1\rangle = \frac{\hbar^2}{2}\mathop{\mathsf{S}}_{p,j,\ldots} [j(j+1) - l(l+1) - s(s+1)]|p; j, m, l, s\rangle\langle p; j, m, l, s|1\rangle.$$

This allows us to calculate such matrix elements as

$$\langle p'; j', m', l', s'|\,\mathfrak{L}\cdot\mathfrak{S}|1\rangle = \frac{\hbar^2}{2}[j(j+1) - l(l+1) - s(s+1)]\langle p'; j', m', l', s'|1\rangle.$$

Of particular interest is a technique for handling the wave functions of particles with spin. If $\mathfrak{S}$ represents any spin operator, the wave function

$$\langle r; \theta, \varphi; s, \mu|\mathfrak{S}|1\rangle = \mathop{\mathsf{S}}_{r',\theta',\ldots} \langle r; \theta, \varphi; s, \mu|\mathfrak{S}|r'; \theta', \varphi'; s', \mu'\rangle\langle r'; \theta', \varphi'; s', \mu'|1\rangle$$

$$= \mathop{\mathsf{S}}_{r',\theta',\ldots} \langle s, \mu|\mathfrak{S}|s', \mu'\rangle\frac{1}{r^2}\,\delta(r - r')\,\delta(\cos\theta - \cos\theta')\,\delta(\varphi - \varphi')$$

$$\cdot\langle r; \theta, \varphi; s', \mu'|1\rangle$$

$$= \mathop{\mathsf{S}}_{s',\mu'} \langle s, \mu|\mathfrak{S}|s', \mu'\rangle\langle r; \theta, \varphi; s', \mu'|1\rangle$$

can be found from the wave functions $\langle r; \theta, \varphi; s', \mu'|1\rangle$ through matrix techniques, as the last member suggests. Since $\mathfrak{S}^2$ commutes with all spin operators, $\mathfrak{S}|s', \mu'\rangle$

is bound to be an eigenstate of $\mathfrak{S}^2$ with eigenvalue $\hbar^2 s'(s'+1)$, and hence

$$\langle s, \mu|\mathfrak{S}|s', \mu'\rangle = \delta_{s,s'}\langle s, \mu|\mathfrak{S}|s, \mu'\rangle \equiv \delta_{s,s'}\langle \mu|\mathfrak{S}|\mu'\rangle.$$

Therefore

$$\langle r;\theta,\varphi;s,\mu|\mathfrak{S}|1\rangle = \underset{\mu'}{\mathsf{S}}\langle\mu|\mathfrak{S}|\mu'\rangle\langle r;\theta,\varphi;s,\mu'|1\rangle. \tag{12.78}$$

The simplest example occurs for $s = \frac{1}{2}$; then, replacing $\mathfrak{J}$ with $\mathfrak{S}$ and j, m with $s = \frac{1}{2}$, μ in Eqs. (12.27–29), we find

$$[\langle\mu|\mathfrak{S}^2|\mu'\rangle] = \hbar^2\frac{3}{4}\begin{bmatrix}1 & 0\\ 0 & 1\end{bmatrix} \quad \begin{matrix}\mu \downarrow \; \mu' \rightarrow: \frac{1}{2}, -\frac{1}{2}\end{matrix} \tag{12.79a}$$

$$[\langle\mu|\mathfrak{S}_3|\mu'\rangle] = \frac{\hbar}{2}\begin{bmatrix}1 & 0\\ 0 & -1\end{bmatrix} \equiv \frac{\hbar}{2}\sigma_3 \tag{12.79b}$$

$$[\langle\mu|\mathfrak{S}_+|\mu'\rangle] = \hbar\begin{bmatrix}0 & 1\\ 0 & 0\end{bmatrix} \equiv \hbar\sigma_+ \tag{12.79c}$$

$$[\langle\mu|\mathfrak{S}_-|\mu'\rangle] = \hbar\begin{bmatrix}0 & 0\\ 1 & 0\end{bmatrix} \equiv \hbar\sigma_- \tag{12.79d}$$

$$[\langle\mu|\mathfrak{S}_1|\mu'\rangle] = \frac{\hbar}{2}\begin{bmatrix}0 & 1\\ 1 & 0\end{bmatrix} \equiv \frac{\hbar}{2}\sigma_1 \tag{12.79e}$$

$$[\langle\mu|\mathfrak{S}_2|\mu'\rangle] = \frac{\hbar}{2}\begin{bmatrix}0 & -i\\ i & 0\end{bmatrix} \equiv \tfrac{1}{2}\hbar\sigma_2, \tag{12.79f}$$

as the matrices for the spin operators. The matrices σ_i are known as the Pauli spin matrices.

Similarly,

$$[\langle r;\theta,\varphi;\tfrac{1}{2},\mu'|1\rangle] = \begin{Bmatrix}\psi_1(r,\theta,\varphi;\frac{1}{2},\frac{1}{2})\\ \psi_1(r,\theta,\varphi;\frac{1}{2},-\frac{1}{2})\end{Bmatrix} \begin{matrix}\mu'\\ \frac{1}{2}\\ -\frac{1}{2}\end{matrix} \tag{12.80}$$

is a column vector representing the wave function of a particle with intrinsic spin; it is often called a spinor. As a shorthand, we may write

$$\boldsymbol{\psi}_1 = \begin{Bmatrix}\psi_{1,1/2}\\ \psi_{1,-1/2}\end{Bmatrix} \tag{12.81}$$

for a spinor.

The operator–state vector notation for spin-$\frac{1}{2}$ particles then can be replaced with the spinor notation; e.g.

$$\langle r;\theta,\varphi;s,\mu|\mathfrak{S}_3|1\rangle = \tfrac{1}{2}\hbar\sigma_3\boldsymbol{\psi}_1$$
$$= \left\{\begin{matrix} \tfrac{1}{2}\hbar\psi_1(r,\theta,\varphi;\tfrac{1}{2},\tfrac{1}{2}) \\ -\tfrac{1}{2}\hbar\psi_1(r,\theta,\varphi;\tfrac{1}{2},-\tfrac{1}{2})\end{matrix}\right\}. \quad (12.82)$$

Note that $\psi_1(r,\theta,\varphi;\frac{1}{2},\frac{1}{2})$ is the probability density amplitude for a particle in state 1 to be located in position r, θ, φ and have spin angular-momentum quantum numbers $s = \frac{1}{2}, \mu = \frac{1}{2}$. The quantum number μ is the only spin quantum number of importance here, because, since $\mathfrak{S}^2$ commutes with everything, $|1\rangle$ *must* be an eigenstate of $\mathfrak{S}^2$; thus, unless both components of the spinor are to vanish, $|1\rangle$ must have spin $s = \frac{1}{2}$.

Note that a basis for the spinors is formed by the two-element column vectors

$$[\langle\tfrac{1}{2},\lambda|\tfrac{1}{2},\tfrac{1}{2}\rangle] = [\delta_{\lambda,1/2}] = \left\{\begin{matrix}1\\0\end{matrix}\right\}\begin{matrix}\lambda\\+\frac{1}{2}\\-\frac{1}{2}\end{matrix} \quad (12.83)$$

$$[\langle\tfrac{1}{2},\lambda|\tfrac{1}{2},-\tfrac{1}{2}\rangle] = [\delta_{\lambda,-1/2}] = \left\{\begin{matrix}0\\1\end{matrix}\right\}\begin{matrix}+\frac{1}{2}\\-\frac{1}{2}\end{matrix} \quad (12.84)$$

To get the spin-up component of state $|1\rangle$, the projection operator $|\frac{1}{2},\frac{1}{2}\rangle\langle\frac{1}{2},\frac{1}{2}|$ may be used. Since $|1\rangle$ is a linear combination of tensor products of spatial states and spin states,

$$1|\rangle = \underset{p,q,r}{\mathsf{S}}\sum_{\mu}\alpha^{(1)}_{p,q,r;\mu}|p,q,r\rangle|\tfrac{1}{2},\mu\rangle,$$

where p, q, r are spatial quantum numbers, we obtain

$$|\tfrac{1}{2},\tfrac{1}{2}\rangle\langle\tfrac{1}{2},\tfrac{1}{2}|1\rangle = \underset{p,q,r}{\mathsf{S}}\sum_{\mu}\alpha^{(1)}_{p,q,r;\mu}|p,q,r\rangle|\tfrac{1}{2},\tfrac{1}{2}\rangle\langle\tfrac{1}{2},\tfrac{1}{2}|\tfrac{1}{2},\mu\rangle$$
$$= \underset{p,q,r}{\mathsf{S}}\,\alpha^{(1)}_{p,q,r;1/2}|p,q,r\rangle|\tfrac{1}{2}\tfrac{1}{2}\rangle$$
$$\equiv |1;+\tfrac{1}{2}\rangle,$$

the last member being a shorthand notation for the spin-up projection of $|1\rangle$. Similarly,

$$|\tfrac{1}{2},-\tfrac{1}{2}\rangle\langle\tfrac{1}{2},-\tfrac{1}{2}|1\rangle = \underset{p,q,r}{\mathsf{S}}\,\alpha^{(1)}_{p,q,r;-1/2}|p,q,r\rangle|\tfrac{1}{2},-\tfrac{1}{2}\rangle \equiv |1;-\tfrac{1}{2}\rangle.$$

Clearly

$$|1\rangle = |1;\tfrac{1}{2}\rangle + |1;-\tfrac{1}{2}\rangle = \sum_{\mu=-1/2}^{1/2}|1;\mu\rangle = \sum_{\mu,\lambda=-1/2}^{1/2}|1;\lambda\rangle\langle\tfrac{1}{2},\lambda|\tfrac{1}{2},\mu\rangle. \quad (12.85)$$

The *configuration-space*, or position-space parts of the state $|1\rangle$, which are given by

$$\langle \tfrac{1}{2}, \mu | 1\rangle = \mathop{\mathsf{S}}_{p,q,r} \alpha^{(1)}_{p,q,r;\mu} |p, q, r\rangle \tag{12.86}$$

may differ for different values of the spin projection μ. In fact, Eqs. (12.83) and (12.84) allow us to write (12.86) as a column vector:

$$\begin{aligned}[\langle \tfrac{1}{2}, \mu | 1\rangle] &= \sum_{\lambda=-1/2}^{1/2} [\langle \tfrac{1}{2}, \mu | \tfrac{1}{2}, \lambda\rangle] \langle \tfrac{1}{2}, \lambda | 1\rangle \\ &= \begin{Bmatrix} \langle \tfrac{1}{2}, \tfrac{1}{2} | 1\rangle \\ \langle \tfrac{1}{2}, -\tfrac{1}{2} | 1\rangle \end{Bmatrix} \qquad \begin{matrix} \mu = +\tfrac{1}{2} \\ \mu = -\tfrac{1}{2}. \end{matrix}\end{aligned} \tag{12.87}$$

Note that the components here are state vectors in configuration space, but inner products in spin space. The corresponding spinor wave functions are

$$[\langle \mathbf{x} | \langle \tfrac{1}{2}, \mu | 1\rangle] = \begin{Bmatrix} \psi_{1,1/2}(\mathbf{x}) \\ \psi_{1,-1/2}(\mathbf{x}) \end{Bmatrix}, \tag{12.88}$$

where

$$\psi_{1,\mu}(\mathbf{x}) = \mathop{\mathsf{S}}_{p,q,r} \alpha^{(1)}_{p,q,r;\mu} \langle \mathbf{x} | p, q, r\rangle = \langle \mathbf{x} | \langle \tfrac{1}{2}, \mu | 1\rangle.$$

This spinor notation allows us to replace the $\mathfrak{S}_i$ everywhere with σ_i, and thereby to obtain a concrete realization of the spin operators. Hence

$$\begin{aligned}[\langle \tfrac{1}{2}, \mu | \mathfrak{S}_i | 1\rangle] &= \left[\sum_{\lambda} \langle \tfrac{1}{2}, \mu | \mathfrak{S}_i | \tfrac{1}{2}, \lambda\rangle \langle \tfrac{1}{2}, \lambda | 1\rangle\right] \\ &= [\langle \tfrac{1}{2}, \mu | \mathfrak{S}_i | \tfrac{1}{2}, \lambda\rangle][\langle \tfrac{1}{2}, \lambda | 1\rangle] \quad \text{(summation convention)} \\ &= \tfrac{1}{2}\hbar\sigma_i \begin{Bmatrix} \langle \tfrac{1}{2}, \tfrac{1}{2} | 1\rangle \\ \langle \tfrac{1}{2}, -\tfrac{1}{2} | 1\rangle \end{Bmatrix}.\end{aligned} \tag{12.89}$$

For instance,

$$\begin{aligned}\begin{Bmatrix} \langle \tfrac{1}{2}, \tfrac{1}{2} | \mathfrak{S}_3 | 1\rangle \\ \langle \tfrac{1}{2}, -\tfrac{1}{2} | \mathfrak{S}_3 | 1\rangle \end{Bmatrix} &= \tfrac{1}{2}\hbar \begin{bmatrix} 1 & 0 \\ 0 & -1 \end{bmatrix} \begin{Bmatrix} \langle \tfrac{1}{2}, \tfrac{1}{2} | 1\rangle \\ \langle \tfrac{1}{2}, -\tfrac{1}{2} | 1\rangle \end{Bmatrix} \\ &= \tfrac{1}{2}\hbar \begin{Bmatrix} \langle \tfrac{1}{2}, \tfrac{1}{2} | 1\rangle \\ -\langle \tfrac{1}{2}, -\tfrac{1}{2} | 1\rangle \end{Bmatrix}.\end{aligned}$$

The spinors and spin matrices afford great conveniences and economies in translating problems into calculations.

One final note on spinors: the Hermitian conjugate is obtained by exchanging rows and columns, and taking the Hermitian conjugate of each component. Thus

$$\begin{Bmatrix} \langle\frac{1}{2},\frac{1}{2}|1\rangle \\ \langle\frac{1}{2},-\frac{1}{2}|1\rangle \end{Bmatrix}^{\dagger} = (\langle\tfrac{1}{2},\tfrac{1}{2}|1\rangle^{\dagger}, \langle\tfrac{1}{2},-\tfrac{1}{2}|1\rangle^{\dagger}) = (\langle 1|\tfrac{1}{2},\tfrac{1}{2}\rangle, \langle 1|\tfrac{1}{2},-\tfrac{1}{2}\rangle).$$

Since wave functions are numbers, the Hermitian conjugate for them is the same as a complex conjugate:

$$\begin{Bmatrix} \psi_{1,1/2} \\ \psi_{1,-1/2} \end{Bmatrix}^{\dagger} = (\psi^*_{1,1/2}, \psi^*_{1,-1/2}). \tag{12.90}$$

This notation can be extended to particles with higher spins. For $s = 1$, for example, the column vector or spinor has three components corresponding to $\mu = +1, 0, -1$, and the spin matrices are 3×3 matrices. In general, the spinors have $2s + 1$ components and the spin matrices have $(2s + 1)^2$ elements.

12.9 FORCES ON A CHARGED PARTICLE IN AN ELECTROMAGNETIC FIELD: THE EFFECT OF SPIN

The Hamiltonian for a particle with charge q in an electromagnetic field with scalar potential $\phi(\mathbf{x})$ and vector potential $\mathbf{A}(\mathbf{x})$ is, according to Eqs. (9.54),

$$\mathsf{H} = \frac{1}{2m}[\mathsf{P} - q\mathbf{A}(\mathbf{x})]^2 + q\phi(\mathbf{x}) + V(\mathbf{x}) \tag{12.91}$$

where $V(\mathbf{x})$ represents the potential energy from any other interactions that may be present. We shall work in the wave-function representation because of the dominant role played by $\mathbf{x}$ in our calculations.

Now according to Eq. (9.50), the four-velocity is given classically by

$$u^{\mu} = \frac{1}{m}(p^{\mu} - qA^{\mu})$$

in the presence of an electromagnetic field, and in the nonrelativistic approximation the spatial components become

$$\mathbf{v} = \frac{1}{m}(\mathbf{p} - q\mathbf{A}) \tag{12.92}$$

where $\mathbf{p}$ is the canonical momentum obtained from the Lagrangian. Hence the force on a particle is given by

$$\mathbf{F} = m\frac{d\mathbf{v}}{dt} = \frac{d}{dt}(\mathbf{p} - q\mathbf{A})$$

Transcription of this equation to quantum-mechanical form, according to the

Heisenberg equations of motion for expectation values, yields

$$\langle \mathsf{F} \rangle = \frac{d}{dt}\langle \mathbf{P} - q\mathbf{A} \rangle = \frac{1}{i\hbar}\left\langle \frac{\partial}{\partial t}(\mathbf{P} - q\mathbf{A}) \right\rangle - \frac{1}{i\hbar}\langle [\mathsf{H}, \mathbf{P} - q\mathbf{A}] \rangle$$
$$= -q\left\langle \frac{\partial \mathbf{A}}{\partial t} \right\rangle - \frac{1}{i\hbar}\langle [\mathsf{H}, \mathbf{P} - q\mathbf{A}] \rangle \tag{12.93}$$

Consider first the last two terms of $\mathfrak{H}$ as written in Eq. (12.91) in the computation of the commutator:

$$\frac{1}{i\hbar}[q\phi(\mathbf{x}) + V(\mathbf{x}), \mathbf{P} - q\mathbf{A}] = \frac{1}{i\hbar}q[\phi, \mathbf{P}] + \frac{1}{i\hbar}[V, \mathbf{P}]$$
$$= q\nabla\phi + \nabla V. \tag{12.94}$$

The remainder of the commutator calculation consists of evaluating

$$\frac{1}{i\hbar}\frac{1}{2m}[(\mathbf{P} - q\mathbf{A})^2, \mathbf{P} - q\mathbf{A}],$$

and this is done most easily by considering one vector component at a time. Thus

$$[(\mathbf{P} - q\mathbf{A})^2, \mathsf{P}_j - qA_j] = \sum_{k=1}^{3} \{(\mathsf{P}_k - qA_k), [(\mathsf{P}_k - qA_k), (\mathsf{P}_j - qA_j)]\}$$
$$= \sum_{k=1}^{3} \{(\mathsf{P}_k - eA_k), i\hbar q(\partial_k A_j - \partial_j A_k)\}$$
$$= \sum_{k=1}^{3}\sum_{l=1}^{3} i\hbar q\epsilon_{kjl}\{(\mathsf{P}_k - qA_k), (\nabla \times \mathbf{A})_l\}$$
$$= \sum_{k=1}^{3}\sum_{l=1}^{3} i\hbar qm\epsilon_{kjl}\{\mathsf{U}_k, B_l\}$$
$$= -i\hbar qm(\mathsf{U} \times \mathbf{B} - \mathbf{B} \times \mathsf{U})_j, \tag{12.95}$$

where $\mathbf{B}$ is the magnetic induction field and

$$\mathsf{U} \equiv \frac{1}{m}(\mathbf{P} - q\mathbf{A}) = \frac{1}{m}\left(\frac{\hbar}{i}\nabla - q\mathbf{A}\right) \tag{12.96}$$

represents the *velocity operator*, as can be verified by computing

$$\frac{d}{dt}\langle \mathbf{x} \rangle = -\frac{1}{i\hbar}\langle [\mathsf{H}, \mathbf{x}] \rangle.$$

Insertion of these results in the last term of Eq. (12.93) yields

$$\langle \mathbf{F} \rangle = \left\langle -q\frac{\partial \mathbf{A}}{\partial t} - q\nabla\phi - \nabla V + q\tfrac{1}{2}(\mathbf{U} \times \mathbf{B} - \mathbf{B} \times \mathbf{U}) \right\rangle$$

$$= \langle q\mathbf{E} + q\tfrac{1}{2}(\mathbf{U} \times \mathbf{B} - \mathbf{B} \times \mathbf{U}) - \nabla V \rangle, \tag{12.97}$$

where

$$\mathbf{E} = -\nabla\phi - \frac{\partial \mathbf{A}}{\partial t}$$

is the electric field vector. Now classically, $\mathbf{v} \times \mathbf{B} - \mathbf{B} \times \mathbf{v} = 2\mathbf{v} \times \mathbf{B}$ since $\mathbf{v}$ and $\mathbf{B}$ commute in classical physics, and hence

$$q\tfrac{1}{2}(\mathbf{U} \times \mathbf{B} - \mathbf{B} \times \mathbf{U}) \tag{12.98}$$

is the quantum-mechanical transcription of the Lorentz force $q\mathbf{v} \times \mathbf{B}$, whereas $q\mathbf{E}$ is the force arising from the electrostatic field. Equation (12.97) therefore corresponds nicely with the classical equation for force that would result from the electromagnetic potentials ϕ, $\mathbf{A}$ and from scalar potential function $V(\mathbf{x})$.

Expression (12.98) represents an important phenomenon in the transcription of classical results into quantum-theoretical language. Frequently the order of factors in a classical expression is irrelevant as, for example, in the interaction energy of a magnetic dipole, having moment $\mathbf{M}$, with a magnetic field $\mathbf{B}$:

$$W = -\mathbf{M} \cdot \mathbf{B} = -\mathbf{B} \cdot \mathbf{M}.$$

In quantum physics, however, the order of factors may be relevant; in particular, if the magnetic moment is proportional to orbital angular momentum $\mathbf{L}$, say

$$\mathbf{M} = \gamma_L \mathbf{L};$$

then we have

$$W = -\gamma_L \mathbf{L} \cdot \mathbf{B}.$$

In quantum theory,

$$\mathbf{L} \cdot \mathbf{B} \neq \mathbf{B} \cdot \mathbf{L}$$

and we must decide, therefore, whether the correct order is $\mathbf{L} \cdot \mathbf{B}$ or $\mathbf{B} \cdot \mathbf{L}$ in transcribing the classical expression to a quantum-mechanical form. In such cases, the correct transcription is

$$\mathbf{L} \cdot \mathbf{B} \leftrightarrow \tfrac{1}{2}(\mathbf{L} \cdot \mathbf{B} + \mathbf{B} \cdot \mathbf{L});$$

this ensures that the quantity on the right is Hermitian when $\mathbf{B}$ and $\mathbf{L}$ are. This accounts for the correspondence

$$\mathbf{v} \times \mathbf{B} \leftrightarrow \tfrac{1}{2}(\mathbf{U} \times \mathbf{B} - \mathbf{B} \times \mathbf{U}).$$

(Why this choice of signs?)

Another related example is

$$(\mathbf{p} - q\mathbf{A})^2 = \mathbf{p}^2 - 2q\mathbf{A}\cdot\mathbf{p} + q^2\mathbf{A}^2 \leftrightarrow \mathsf{P}^2 - q(\mathbf{A}\cdot\mathsf{P} + \mathsf{P}\cdot\mathbf{A}) + q^2\mathbf{A}^2.$$

In Eq. (12.97) we are interested particularly in the possibility that a particle may have spin angular momentum, and that this angular momentum may give rise to a magnetic moment, say

$$\mathsf{M}_s = \gamma_s \mathsf{S}, \tag{12.99}$$

where γ_s is called the spin gyromagnetic ratio and S is the spin operator (a spin matrix in the spinor formalism). Then there would exist an interaction between the magnetic dipole moment M_s and the magnetic field $\mathbf{B}$, with interaction energy

$$\mathsf{V}_s = -\mathsf{M}_s\cdot\mathbf{B} = -\mathbf{B}\cdot\mathsf{M}_s. \tag{12.100}$$

(Why is the order irrelevant here?) According to the equations we have just developed, this will contribute a force

$$\mathsf{F}_s = -\nabla\mathsf{V}_s = \nabla(\mathsf{M}_s\cdot\mathbf{B})$$

to the total force on a particle with spin. This, in fact provides a technique for physically separating particles with different spin components, as we shall see in the next section.

Before taking up the problem of separating spin components, which involves a nonhomogeneous magnetic field, let us consider the motion of a particle in a homogeneous field. Let $\mathbf{B}_0$ be a constant and let

$$\mathbf{A}(x) = \tfrac{1}{2}\mathbf{B}_0 \times \mathbf{x}$$

so that

$$\mathbf{B}(x) = \tfrac{1}{2}\nabla \times (\mathbf{B}_0 \times \mathbf{x}) = \tfrac{1}{2}[\mathbf{B}_0(\nabla\cdot\mathbf{x}) - \mathbf{B}_0\cdot\nabla\mathbf{x}] = \mathbf{B}_0.$$

Then we have, if we let $\mathsf{V}(\mathbf{x}) = -\mathsf{M}_s\cdot\mathbf{B}_0 + V_0(\mathbf{x})$ in Eq. (12.91),

$$\begin{aligned}\mathsf{H} &= \frac{1}{2m}\left[\left(\mathsf{P} - \frac{q}{2}\mathbf{B}_0\times\mathbf{x}\right)^2\right] + q\phi - \mathsf{M}_s\cdot\mathbf{B}_0 + V_0(\mathbf{x})\\ &= \frac{1}{2m}\left[\mathsf{P}^2 - q(\mathbf{B}_0\times\mathbf{x})\cdot\mathsf{P} + \frac{q^2}{4}(\mathbf{B}_0^2\mathbf{x}^2 - [\mathbf{B}_0\cdot\mathbf{x}]^2)\right] + q\phi - \mathsf{M}_s\cdot\mathbf{B}_0 + V_0(\mathbf{x})\\ &= \frac{1}{2m}\left[\mathsf{P}^2 + \frac{q^2}{4}(\mathbf{B}_0^2\mathbf{x}^2 - [\mathbf{B}_0\cdot\mathbf{x}]^2)\right] + q\phi - \left[\frac{q}{2m}\mathsf{L} + \mathsf{M}_s\right]\cdot\mathbf{B}_0 + V_0(\mathbf{x}).\end{aligned} \tag{12.101}$$

The next-to-last term behaves like a magnetic moment interaction term involving total magnetic moment

$$\mathbf{M} = \frac{q}{2m}\mathbf{L} + \mathbf{M}_s.$$

This suggests that perhaps

$$\mathbf{M}_s = \frac{q}{2m}\mathbf{S} \qquad (?)$$

i.e.

$$\gamma_S = \frac{q}{2m} \qquad (?)$$

We shall see, however, that in fact the correct spin gyromagnetic ratio is

$$\gamma_s = \frac{q}{m}.$$

The terms in Eq. (12.101) involving $\mathbf{B}_0^2$ are frequently negligible because the magnetic field may be small, and in such cases the magnetic moment

$$\mathbf{M} = \gamma_L\mathbf{L} + \gamma_s\mathbf{S}, \qquad \gamma_L = \frac{q}{2m}, \qquad \gamma_s = \frac{q}{m},$$

can be studied directly in terms of the eigenvalues of H—the energy levels of the particle in the magnetic field.

12.10 THE STERN-GERLACH APPARATUS

In pre-Schrödinger-equation days (1921 and 1922), O. Stern proposed and carried out with W. Gerlach, an experiment designed to analyse the magnetic-moment spectrum of an atom by physically separating atoms with different magnetic moments. A beam of atoms (silver atoms in the first instance) was sent the length of an inhomogeneous magnetic field. For sufficiently small velocities, the Lorentz force is negligible, and the beam moves the length of the field along an essentially straight line if the atoms possess no intrinsic magnetic moment. If an intrinsic magnetic moment exists, however, the force

$$\mathbf{F}_s = \nabla(\mathbf{M}_s \cdot \mathbf{B})$$

deflects the beam.

Let the z-axis be defined by $\mathbf{B}$ so that $\mathbf{B} = \hat{\mathbf{k}}B(\mathbf{x})$ and hence

$$\mathbf{F}_s = (\mathbf{M}_s)_z\nabla B(\mathbf{x})$$

that is, the force is proportional to the component of magnetic moment along the direction of $\mathbf{B}$, but is in the direction of $\nabla B(\mathbf{x})$. The apparatus of Stern and Gerlach, shown schematically in Fig. 12.4, provided a magnetic field with gradient in the z-direction, more-or-less, so that atoms with different projections of $\mathbf{M}_s$ along $\mathbf{B}$ would experience different forces and thereby eventually would be separated spatially from one another.

Classically, one would expect to observe a beam emerging from the apparatus with a uniform and continuous spread along the z-axis resulting from the magnetic-dipole interaction. This is because all projections of $\mathbf{M}_s$ along $\mathbf{B}$, ranging from $|\mathbf{M}_s|\hat{\mathbf{B}}$ to $-|\mathbf{M}_s|\hat{\mathbf{B}}$ classically are equally probable, and hence all deflections, between some maximum and minimum, are equally probable.

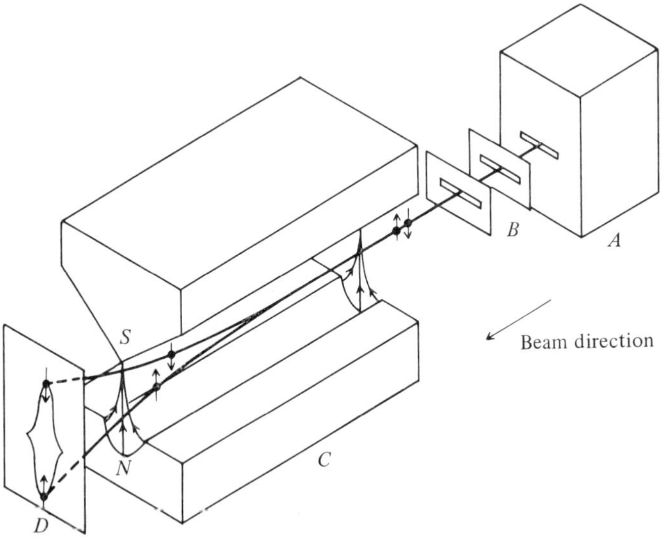

Fig. 12.4. A Stern–Gerlach apparatus for separating spin components. (A) Oven for evaporating silver. (B) Collimator for defining ribbon-like beam. (C) Magnet pole pieces producing inhomogeneous magnetic field. (D) Screen to condense silver atoms.

The silver atoms, however, produce two distinct beams, indicating that only two projections of magnetic moment along the z-axis are allowed. It turns out that the magnetic moment of a silver atom arises almost exclusively from a single electron in the atom, and hence Stern and Gerlach were actually observing the magnetic properties of an electron. Evidently the electron (and silver atom) has spin $s = \frac{1}{2}$, since the number of magnetic-moment projections was 2 and since we may assume that $\mathbf{M}_s = \gamma_s \mathbf{S}$, so that the number of projections of $\mathbf{M}_s$ along any axis would be $2s + 1$.

Assuming that the Lorentz force and electrostatic force in Eq. (12.97) are negligible, and that $V(\mathbf{x})$ is given by (12.100), a spin-$\frac{1}{2}$ particle in state $|1\rangle$ with spinor wave function $\boldsymbol{\psi}_1(\mathbf{x}, \ldots)$ experiences a mean force

$$\langle \mathbf{F} \rangle = (\boldsymbol{\psi}_1, \gamma_s \mathbf{S}_z(\nabla B)\boldsymbol{\psi}_1) = (\boldsymbol{\psi}_1, \gamma_s \tfrac{1}{2}\hbar\sigma_3(\nabla B)\boldsymbol{\psi}_1),$$

if $\mathbf{B} = \hat{\mathbf{k}}B(\mathbf{x})$, since in spinor notation the spin operator is $\frac{1}{2}\hbar\boldsymbol{\sigma} = \mathbf{S}$. Letting

$$\nabla B = \hat{\mathbf{k}}(B_0/d)$$

be a constant, d being some characteristic length, this yields

$$\langle \mathbf{F} \rangle = \hat{\mathbf{k}}\tfrac{1}{2}\hbar\gamma_s \frac{B_0}{d}[(\psi_{1,1/2}, \psi_{1,1/2}) - (\psi_{1,-1/2}, \psi_{1,-1/2})] \tag{12.102}$$

in terms of the spinor notation (12.81); here the inner products on the right-hand side are defined by

$$(\psi_{1,\mu}, \psi_{1,\mu}) = \int d^3x |\psi_{1,\mu}(\mathbf{x})|^2 \qquad \text{(no sum over } \mu\text{)}.$$

This gives the mean force on an electron in the magnetic field.

However, an electron in a definite state of angular momentum, say in a spin-up state $| + \rangle$, with wave function

$$\boldsymbol{\psi}_{+1/2} = \begin{Bmatrix} \psi_{+,1/2}(\mathbf{x}) \\ 0 \end{Bmatrix}$$

experiences a force given by the first term on the right-hand side of Eq. (12.102), whereas one in a spin-down state $|-\rangle$, with wave function

$$\boldsymbol{\psi}_{-1/2}(\mathbf{x}) = \begin{Bmatrix} 0 \\ \psi_{-,1/2}(\mathbf{x}) \end{Bmatrix}$$

experiences a force in the opposite direction, given by the second term.

From this we conclude that the Stern–Gerlach apparatus forces electrons to "choose" among possible states having definite z-components of angular momentum, those going into spin-up states being pushed in one direction by the magnetic-field gradient, and those going into spin-down states being pushed in the opposite direction. Thus a single particle does not, in general, experience a force equal to that given by Eq. (12.102), but rather experiences either a force corresponding to the first term, or one corresponding to the second term; the respective probabilities for these alternatives are, of course,

$$(\psi_{1,1/2}, \psi_{1,1/2}) \quad \text{and} \quad (\psi_{1,-1/2}, \psi_{1,-1/2}).$$

Equation (12.102) gives the mean force experienced by a particle, but each particle has a definite spin projection along the z-axis and therefore only one term contributes for that particle. If the experiment is repeated many times for electrons initially in state $|1\rangle$, the mean force experienced by the electrons will be given by Eq. (12.102).

More explicitly, the Stern–Gerlach apparatus performs two operations. First, it provides, through the presence of a magnetic field, a quantization axis for the spins of particles in the beam. A single particle in state $|1\rangle$ will have its spin either aligned or anti-aligned with this axis, with probability

$$(\psi_1(\mathbf{x}, \tfrac{1}{2}), \psi_1(\mathbf{x}, \tfrac{1}{2})) = (\psi_{1,1/2}, \psi_{1,1/2})$$

or

$$(\psi_1(\mathbf{x}, -\tfrac{1}{2}), \psi_1(\mathbf{x}, -\tfrac{1}{2})) = (\psi_{1,-1/2}, \psi_{1,-1/2})$$

respectively. Second, the particles, with spin directions thus determined, are forced either upwards or downwards by the gradient in the magnetic field, in accordance with their spin orientations.

The Stern–Gerlach experiment, as well as other later measurements, provided a value for γ_s, through measurements of the distances by which the two beams were separated. The spin gyromagnetic ratio of the electron thus was found to be

$$\gamma_s = -e/m, \tag{12.103}$$

the negative sign representing the fact that the electron charge is negative. That this differs from the orbital gyromagnetic ratio,

$$\gamma_l = -\tfrac{1}{2}e/m$$

was rather surprising and served to emphasize the fundamental difference between spin and orbital magnetic moments.

Consider now two Stern–Gerlach devices placed in tandem, according to the scheme of Fig. 12.5. The spatial separation of beams emerging from the first device allows us to select either the spin-up or spin-down component of the initial state $|1\rangle$. This component then is fed into a second apparatus having its magnetic field aligned at an angle β relative to that of the first apparatus. Now rotating the magnetic field (and hence the coordinate frame, since $\mathbf{B}$ provides the definition of the z-axis) through β about the y-axis is equivalent to rotating the particle and its spin through angle $-\beta$ about the y-axis. Hence the effect of the second device can be analysed in terms of the rotated spinors

$$\boldsymbol{\psi}'_{1,\pm 1/2} = e^{-i\beta S_y/\hbar}\boldsymbol{\psi}_{1,\pm 1/2} = e^{-i\beta\sigma_2/2}\boldsymbol{\psi}_{1,\pm 1/2},$$

where

$$\boldsymbol{\psi}_{1,+1/2} = \begin{Bmatrix} \psi_{1,1/2} \\ 0 \end{Bmatrix}, \qquad \boldsymbol{\psi}_{1,-1/2} = \begin{Bmatrix} 0 \\ \psi_{1,-1/2} \end{Bmatrix}$$

are the spinors for electrons in the two beams emerging from the first apparatus.

Now the rotation operator used here can be expanded according to the definition of the exponential:

$$e^{i\boldsymbol{\theta}\cdot\boldsymbol{\sigma}/2} = \sum_{n=0}^{\infty} \frac{1}{n!}(i\boldsymbol{\theta}\cdot\boldsymbol{\sigma}/2)^n.$$

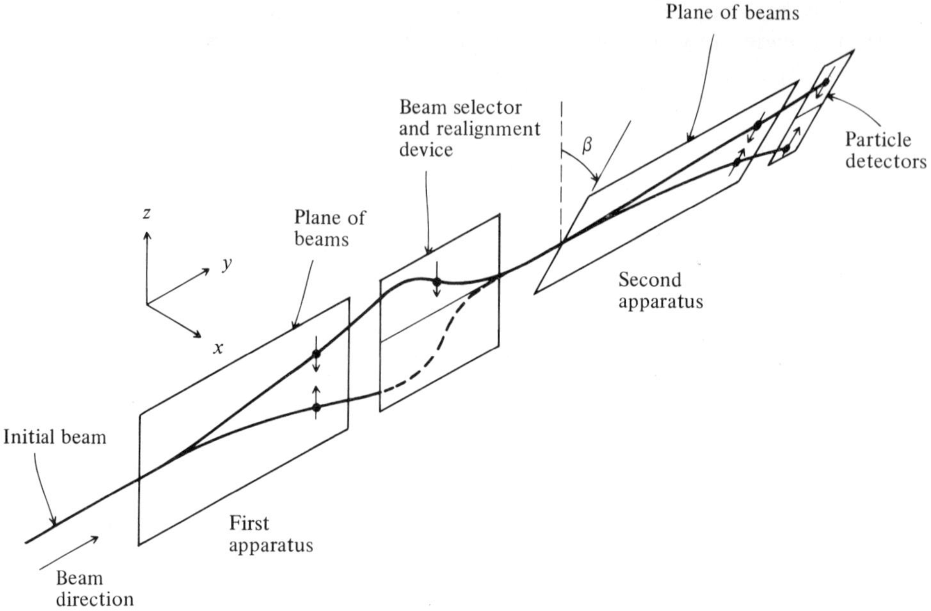

Fig. 12.5. Schematic diagram for two Stern–Gerlach-type spin separators in tandem, rotated through angle β relative to one another. Here the spin-up beam from the first device is blocked off and only the spin-down part enters the second device.

The property

$$\{\sigma_j, \sigma_k\} = 2\delta_{jk} \tag{12.104}$$

of the spin matrices allows us to simplify this considerably; thus

$$\begin{aligned}(\hat{\boldsymbol{\theta}} \cdot \boldsymbol{\sigma})^{2n} &= \left[\left(\sum_{j=1}^{3} \theta_j \sigma_j\right)^2\right]^n = \left[\sum_{j=1}^{3} \theta_j^2 \sigma_j^2 + \sum_{j \neq k = 1}^{3} \theta_j \theta_k \sigma_j \sigma_k\right]^n \\ &= \left[\sum_{j=1}^{3} \theta_j^2 + \tfrac{1}{2} \sum_{j \neq k = 1}^{3} \theta_j \theta_k \{\sigma_j, \sigma_k\}\right]^n \\ &= (\boldsymbol{\theta}^2)^n = \theta^{2n}.\end{aligned}$$

Similarly,

$$(\boldsymbol{\theta} \cdot \boldsymbol{\sigma})^{2n+1} = \hat{\boldsymbol{\theta}} \cdot \boldsymbol{\sigma} |\theta|^{2n+1}$$

and hence

$$\begin{aligned} e^{i\boldsymbol{\theta} \cdot \boldsymbol{\sigma}/2} &= \sum_{n=0}^{\infty} \frac{(-1)^n}{(2n)!} (\boldsymbol{\theta} \cdot \boldsymbol{\sigma}/2)^{2n} + i \sum_{n=0}^{\infty} \frac{(-1)^n}{(2n+1)!} (\boldsymbol{\theta} \cdot \boldsymbol{\sigma}/2)^{2n+1} \\ &= \cos \tfrac{1}{2}\theta + i \hat{\boldsymbol{\theta}} \cdot \boldsymbol{\sigma} \sin \tfrac{1}{2}\theta. \end{aligned} \tag{12.105}$$

Therefore

$$e^{-i\beta\sigma_2/2} = \cos\tfrac{1}{2}\beta - i\sigma_2 \sin\tfrac{1}{2}\beta = \begin{bmatrix} \cos\frac{1}{2}\beta & -\sin\frac{1}{2}\beta \\ \sin\frac{1}{2}\beta & \cos\frac{1}{2}\beta \end{bmatrix}.$$

Therefore the spinor corresponding to a particle in a spin-up state as it leaves the first magnet becomes

$$\boldsymbol{\psi}'_{1,+1/2} = \begin{Bmatrix} \psi_{1,1/2}\cos\frac{1}{2}\beta \\ \psi_{1,1/2}\sin\frac{1}{2}\beta \end{Bmatrix} \tag{12.106}$$

as the particle enters the second magnet, and the spinor corresponding to the spin-down state becomes

$$\boldsymbol{\psi}'_{1,-1/2} = \begin{Bmatrix} -\psi_{1,-1/2}\sin\frac{1}{2}\beta \\ \psi_{1,-1/2}\cos\frac{1}{2}\beta \end{Bmatrix}. \tag{12.107}$$

The second Stern–Gerlach device then separates the upper and lower components of these spinors into spatially distinct beams, thanks to its inhomogeneous magnetic field.

From (12.106) we see that the probability that a particle originally in a state with wave function

$$\boldsymbol{\psi}_1(\mathbf{x}) = \begin{Bmatrix} \psi_{1,1/2}(\mathbf{x}) \\ \psi_{1,-1/2}(\mathbf{x}) \end{Bmatrix}$$

has probability

$$\int d^3x|\psi_{1,1/2}(\mathbf{x})|^2\cos^2\tfrac{1}{2}\beta = (\psi_{1,1/2}, \psi_{1,1/2})\cos^2\tfrac{1}{2}\beta$$

of emerging from both magnets in spin-up states and probability

$$\int d^3x|\psi_{1,-1/2}(\mathbf{x})|^2\cos^2\tfrac{1}{2}\beta = (\psi_{1,-1/2}, \psi_{1,-1/2})\cos^2\tfrac{1}{2}\beta,$$

of emerging from both in spin-down states. The probability of emerging from the first in a spin-up state and from the second with spin down is

$$\int d^3x|\psi_{1,1/2}(\mathbf{x})|^2\sin^2\tfrac{1}{2}\beta = (\psi_{1,1/2}, \psi_{1,1/2})\sin^2\tfrac{1}{2}\beta,$$

whereas the reverse order occurs with probability

$$\int d^3x|\psi_{1,-1/2}(\mathbf{x})|^2\sin^2\tfrac{1}{2}\beta = (\psi_{1,-1/2}, \psi_{1,-1/2})\sin^2\tfrac{1}{2}\beta.$$

The fact that $\beta/2$ rather than β occurs as the argument of the circular trigonometric functions is characteristic of spin-$\frac{1}{2}$ systems and has important consequences.

12.11 PARITY

As a codicil to this chapter we consider the concept of parity, which has no useful analog in classical physics. By the parity transformation we mean the replacement of each position coordinate with its negative:

$$(x, y, z) \to (-x, -y, -z).$$

Equivalently, the parity transformation replaces a right-handed three-dimensional coordinate system with a left-handed one. This is effectively what happens when one looks in a mirror—the right hand appears to be the left, etc.; hence the frequent reference to mirrors in accounts of parity.

In classical physics parity transformations can be performed, but they have no effect at all upon the mathematical apparatus that describes a physical system; no new information is obtained by considering parity transformations in classical physics. In quantum theory some subtle effects are linked with the parity transformation—effects that play a very important role in the design and analysis of experiments.

The reason this topic belongs in this chapter is that it deals with directions—the parity transformation reverses the directions of a class of vectors (which includes the position vector) called polar vectors—and the rotations and hence angular momenta are intimately connected with directions. In particular, one may note that the differential equation for the radial factor of the wave function for a particle in a spherically symmetric potential contains no information about directions; all the directional information resides in the angular factors of such wave functions.

Now the parity operator $\mathfrak{P}$ may be defined through the equation

$$\mathfrak{P}|\mathbf{x}, \ldots\rangle = |-\mathbf{x}, \ldots\rangle;$$

thus

$$\begin{aligned}\mathfrak{P}\mathfrak{X}|\mathbf{x}, \ldots\rangle &= \mathfrak{P}\mathbf{x}|\mathbf{x}, \ldots\rangle = \mathbf{x}\mathfrak{P}|\mathbf{x}, \ldots\rangle = \mathbf{x}|-\mathbf{x}, \ldots\rangle \\ &= -\mathfrak{X}|-\mathbf{x}, \ldots\rangle = -\mathfrak{X}\mathfrak{P}|\mathbf{x}, \ldots\rangle.\end{aligned}$$

This holds for each member of the complete set of position eigenstates $|\mathbf{x}, \ldots\rangle$, and therefore we have

$$\mathfrak{P}\mathfrak{X}\mathfrak{P}^{-1} = -\mathfrak{X}. \tag{12.108a}$$

Also

$$\mathfrak{P}\mathfrak{T}\mathfrak{P}^{-1} = \mathfrak{T}, \qquad (\mathfrak{T} \text{ is the time operator}) \tag{12.108b}$$

by definition. For an eigenstate of momentum, $|\mathbf{p}, q, \ldots\rangle$, we have

$$\begin{aligned}\mathfrak{P}|\mathbf{p}, q, \ldots\rangle &= \underset{\mathbf{x}, m, \ldots}{\mathsf{S}} \mathfrak{P}|\mathbf{x}, m, \ldots\rangle\langle\mathbf{x}, m, \ldots|\mathbf{p}, q, \ldots\rangle \\ &= \underset{\mathbf{x}, m, \ldots}{\mathsf{S}} |-\mathbf{x}, m, \ldots\rangle(2\pi)^{-3/2} e^{i\mathbf{x}\cdot\mathbf{p}/\hbar}\langle m, \ldots|q, \ldots\rangle\end{aligned}$$

$$= \underset{\mathbf{x},m,\ldots}{\mathsf{S}} |\mathbf{x}, m, \ldots\rangle (2\pi)^{-3/2} e^{-i\mathbf{x}\cdot\mathbf{p}/\hbar} \langle m, \ldots | q, \ldots\rangle$$

$$= \underset{\mathbf{x},m,\ldots}{\mathsf{S}} |\mathbf{x}, m, \ldots\rangle \langle \mathbf{x}, m, \ldots | -\mathbf{p}, q, \ldots\rangle = | -\mathbf{p}, q, \ldots\rangle$$

and from this it follows that the momentum operator $\mathfrak{P}$ transforms under the parity transformation according to

$$\mathfrak{P}\mathfrak{P}\mathfrak{P}^{-1} = -\mathfrak{P}. \tag{12.109}$$

Any vector quantity $\mathbf{v}$ that, under parity, becomes $-\mathbf{v}$, is called a *polar vector*; quantum mechanically, $\mathfrak{B}$ is a polar vector if and only if

$$\mathfrak{P}\mathfrak{B}\mathfrak{P}^{-1} = -\mathfrak{B}$$

Hence $\mathfrak{X}$ and $\mathfrak{P}$ are polar vectors.

An axial vector is a vector quantity $\mathbf{a}$ that is unaffected by the reversal of all coordinate directions. A prototype for such vectors is angular momentum. Classically,

$$\mathbf{L} = \mathbf{x} \times \mathbf{p} \to (-\mathbf{x}) \times (-\mathbf{p}) = \mathbf{x} \times \mathbf{p} = \mathbf{L}$$

and hence $\mathbf{L}$ is an axial vector. In quantum mechanics we have

$$\mathfrak{P}(\mathfrak{X} \times \mathfrak{P})\mathfrak{P}^{-1} = (\mathfrak{P}\mathfrak{X}\mathfrak{P}^{-1}) \times (\mathfrak{P}\mathfrak{P}\mathfrak{P}^{-1}) = (-\mathfrak{X}) \times (-\mathfrak{P}) = \mathfrak{X} \times \mathfrak{P}$$

or

$$\mathfrak{P}\mathfrak{L}\mathfrak{P}^{-1} = \mathfrak{L}. \tag{12.110}$$

In general, then, $\mathfrak{A}$ is an axial vector operator if and only if

$$\mathfrak{P}\mathfrak{A}\mathfrak{P}^{-1} = \mathfrak{A}.$$

What about spin angular momentum and total angular momentum? By considering the effect of a rotation operator upon a position eigenstate $|\mathbf{x}, \ldots\rangle$ as we did at the beginning of this chapter, we can see that

$$\mathfrak{P}e^{i\boldsymbol{\theta}\cdot\mathfrak{J}/\hbar}|\mathbf{x}, \ldots\rangle = e^{i\boldsymbol{\theta}\cdot\mathfrak{J}/\hbar}| -\mathbf{x}, \ldots\rangle$$
$$= e^{i\boldsymbol{\theta}\cdot\mathfrak{J}/\hbar}\mathfrak{P}|\mathbf{x}, \ldots\rangle$$

for all $\boldsymbol{\theta}$ and for all $|\mathbf{x}, \ldots\rangle$, and hence we see that

$$\mathfrak{P}e^{i\boldsymbol{\theta}\cdot\mathfrak{J}/\hbar}\mathfrak{P}^{-1} = e^{i\boldsymbol{\theta}\cdot\mathfrak{J}/\hbar}.$$

It follows from this that

$$\mathfrak{P}\mathfrak{J}\mathfrak{P}^{-1} = \mathfrak{J} \tag{12.111}$$

and from Eq. (12.110) that

$$\mathfrak{P}(\mathfrak{J} - \mathfrak{L})P^{-1} = \mathfrak{P}\mathfrak{S}\mathfrak{P}^{-1} = \mathfrak{S}. \tag{12.112}$$

All angular momenta are therefore axial vectors, or *pseudo vectors*.

A scalar quantity s that is not changed by the reversal of coordinates, e.g. $\mathbf{x} \cdot \mathbf{p}$, is called a scalar, whereas a scalar quantity p that transforms according to $p \to -p$ (e.g., $\mathbf{L} \cdot \mathbf{x}$) under coordinate reversals is called a *pseudoscalar.* Thus

$$\mathfrak{P}\mathfrak{R}\mathfrak{P}^{-1} = \mathfrak{R} \qquad \text{for a scalar operator } \mathfrak{R}$$

$$\mathfrak{P}\mathfrak{Q}\mathfrak{P}^{-1} = -\mathfrak{Q} \qquad \text{for a pseudoscalar operator } \mathfrak{Q}.$$

We must be aware that the internal coordinates of intrinsic spin are inaccessible to the parity operator. Parity is a spatial operator whereas the spin is an intrinsic property of a particle that is coupled to spatial quantities through its peculiar participation in rotation transformations. Parity, however, is not a rotation, nor can it be composed from rotations, and hence there is no *a priori* reason to expect any effect on spin variables when the parity operation is carried out. That is,

$$\mathfrak{P}|s, \mu\rangle = |s, \mu\rangle,$$

i.e. $\mathfrak{P}$ always behaves like the identity operator when acting on pure *spin* states.

It is necessary in some cases to hypothesize the existence of an intrinsic parity quantum number for a particle, but the intrinsic parity is quite independent of the spin angular momentum. On the other hand, orbital angular momentum and extrinsic (or configuration-space) parity are very closely connected with each other; the extrensic parity quantum number is determined by the orbital angular momentum of a particle.

To verify this last statement, we begin by noting that $\mathfrak{P}$ is a unitary operator, for surely a physical system will behave the same way whether it is described in terms of a left-handed or right-handed coordinate system. Thus if all states $|1\rangle$ are replaced by $\mathfrak{P}|1\rangle$ and if all operators $\mathfrak{Q}$ are replaced by $\mathfrak{P}\mathfrak{Q}\mathfrak{P}^{-1}$, no inner products, and hence no predictions of experimental outcomes, will be altered. This implies that

$$\mathfrak{P}^\dagger\mathfrak{P} = \mathfrak{P}\mathfrak{P}^\dagger = 1$$

or

$$\mathfrak{P}^{-1} = \mathfrak{P}^\dagger.$$

Also it is obvious that $\mathfrak{P}^2 = 1$ since

$$\mathfrak{P}^2|\mathbf{x}, \ldots\rangle = \mathfrak{P}(\mathfrak{P}|\mathbf{x}, \ldots\rangle) = \mathfrak{P}|-\mathbf{x}, \ldots\rangle = |\mathbf{x}, \ldots\rangle.$$

This means that

$$\mathfrak{P} = \mathfrak{P}^{-1} = \mathfrak{P}^\dagger \tag{12.113}$$

and therefore $\mathfrak{P}$ is both unitary and Hermitian; its eigenvalues are of absolute value 1 and are real; that is, are either $+1$ or -1.

Let us consider the effect of parity on an eigenstate of orbital angular momentum:

$$\mathfrak{P}|l, m\rangle = \underset{\cos\theta,\varphi}{\mathsf{S}} |\cos\theta, \varphi\rangle\langle\cos\theta, \varphi|\mathfrak{P}|l, m\rangle$$

$$= \underset{\cos\theta,\varphi}{\mathsf{S}} |\cos\theta, \varphi\rangle\langle l, m|\mathfrak{P}|\cos\theta, \varphi\rangle^*$$

$$\underset{\cos\theta,\varphi}{\mathsf{S}} |\cos\theta, \varphi\rangle\langle l, m|\cos(\pi - \theta), \pi + \varphi\rangle^*$$

$$= \underset{\cos\theta,\varphi}{\mathsf{S}} |\cos\theta, \varphi\rangle\langle\cos(\pi - \theta), \varphi + \pi|l, m\rangle$$

$$= \underset{\cos\theta,\varphi}{\mathsf{S}} |\cos\theta, \varphi\rangle\, Y_l^m(\pi - \theta, \varphi + \pi).$$

The relevant part of the expression (12.64) for $Y_l^m(\theta, \varphi)$ is

$$\sin^m\theta \frac{d^{l+m}}{(d\cos\theta)^{l+m}} \sin^{2l}\theta e^{im\varphi};$$

from this we can see that

$$Y_l^m(\pi - \theta, \pi + \varphi) = (-1)^l Y_l^m(\theta, \varphi),$$

and hence that

$$\mathfrak{P}|l, m\rangle = \underset{\cos\theta,\varphi}{\mathsf{S}} |\cos\theta, \varphi\rangle(-1)^l\langle\cos\theta, \varphi|l, m\rangle = (-1)^l|l, m\rangle. \qquad (12.114)$$

Therefore the parity of an orbital-angular-momentum eigenstate is *odd*, i.e. $\mathfrak{P}$ has eigenvalue -1, whenever the orbital-angular-momentum quantum number is odd, and is *even*, i.e. $\mathfrak{P}$ has eigenvalue $+1$, whenever the orbital-angular-momentum quantum number is even.

States with well-defined intrinsic parity have either even or odd intrinsic parity also. For example, the π-meson is known to have intrinsic parity eigenvalue -1 and hence is said to have odd parity. The α-particle, on the other hand, has even intrinsic parity. Both of these particles have spin angular-momentum quantum numbers equal to zero. There is a certain ambiguity in the definition of intrinsic parity that has to do with the ability of particles to be transformed into one another under available physical interactions; some parity standard with which the parities of related particles may be compared is needed for each of several classes of particles. The intrinsic parities quoted above are based on widely adopted conventions. In any case, given the parity of one kind of particle in one of these classes of particles, and given another particle in the same class with well-defined parity, the parity of the latter must be either even or odd, since Eqs. (12.113) hold.

Later we shall see how parity affects the transition of one physical state to another. It provides selection rules that tell us when certain kinds of transitions are not allowed, according to theory, and thereby provides experimental checks upon dynamical theories. Another, related symmetry operator is the time-reversal operator, and it too provides selection rules; the latter are of importance primarily in elementary particle physics.

12.12 MOTION REVERSAL INVARIANCE

Let us define a microscopic physical system as one in which the measurable parameters are sufficiently few in number that one can give a complete mathematical description of the system in practice and need not resort to statistical considerations; it is a few-parameter, and therefore a few-particle system. A fundamental hypothesis of microscopic physics is that the behavior of the system is essentially unchanged if all the motions in the system are reversed. By "essentially unchanged" we mean that any given model for the system represents it just as well when the motions are reversed. Until recently all experimental evidence supported this hypothesis; at the present time it is not clear whether it must be discarded or not, but it is at least approximately true for most of the systems we are currently able to study experimentally. We shall assume its validity.

In order to investigate time reversal invariance, we must devise a motion-reversal operator θ, the effect of which is to reverse the motion of a system at the instant $t = 0$. Its basic properties must be

$$\theta\mathfrak{T}\theta^{-1} = -\mathfrak{T} \qquad (\mathfrak{T} \text{ is the time operator}) \tag{12.115}$$

$$\theta\mathfrak{X}\theta^{-1} = \mathfrak{X} \tag{12.116}$$

$$\theta\mathfrak{E}\theta^{-1} = \mathfrak{E} \qquad (\mathfrak{E} \text{ is the energy operator}). \tag{12.117}$$

The first of these follows from the effect we wish θ to have on a state described by a definite time, namely

$$\theta|t, \ldots\rangle = \alpha|-t, \ldots\rangle, \qquad \alpha \text{ a constant.}$$

From this it follows that

$$\begin{aligned}\theta\mathfrak{T}|t, \ldots\rangle &= \theta\mathfrak{T}\theta^{-1}\theta|t, \ldots\rangle = \theta\mathfrak{T}\theta^{-1}\alpha| - t\rangle \\ &= \theta t|t, \ldots\rangle = \alpha t| - t, \ldots\rangle = \alpha(-\mathfrak{T})| - t, \ldots\rangle,\end{aligned}$$

and comparison of the third and last members of this sequence of equations gives the desired result.

The property that θ commutes with $\mathfrak{X}$ follows from the fact that motion reversal should not be expected to affect an eigenstate of position; rather, it reverses the *order* in which positions might be occupied by a particle in motion.

The condition (12.117) that θ commutes with $\mathfrak{E}$ is imposed to guarantee that we are discussing motion reversal in the sense that the energy of the system is the same whether the system is moving "forward" or "backward". The imposition of this condition forces θ to have some peculiar properties that are quite distinct from those of the parity operator $\mathfrak{P}$, for example. In fact, it is this that prompts us to call θ a motion-reversal operator rather than a time-reversal operator. The latter might be expected to be a direct analog of the parity, or *space-reversal* operator. However, we have no way of investigating such a direct analog—it has no operational meaning since we have no way of reversing energies. (It might be said that this is why time is such a peculiar parameter when compared with position, and why the concepts of an eigenstate of time and of a time operator seem quite outlandish. Later we shall drop the notion that time has an operator representation, but for now we retain these concepts as convenient notational and manipulative devices.)

To see the strange effect produced by (12.117), consider the commutator

$$[\mathfrak{E}, \mathfrak{T}] = i\hbar.$$

If we carry out the motion reversal transformation on this commutator, we obtain

$$\theta[\mathfrak{E}, \mathfrak{T}]\theta^{-1} = [\theta\mathfrak{E}\theta^{-1}, \theta\mathfrak{T}\theta^{-1}] = [\mathfrak{E}, -\mathfrak{T}] = -i\hbar,$$

or using $[\mathfrak{E}, \mathfrak{T}] = i\hbar$,

$$\theta i\hbar\theta^{-1} = -i\hbar.$$

We would certainly be unhappy to allow $\hbar$ to fail to commute with θ since it is simply a real constant. The only alternative is to let θ have the property

$$\theta i\theta^{-1} = -i, \qquad \text{or} \qquad \{\theta, i\} = 0. \tag{12.118}$$

This is almost frightening because until now we have assumed that complex numbers commute with all operators; it seems that we are now forced to reconsider this assumption. However, our previous work was consistent with observation and "intuition" and need not be reconsidered; the motion-reversal operator simply is a very special kind of operator.

It is clear that for any state $|1(t)\rangle = \exp[-it\mathfrak{E}/h]|1(0)\rangle$ we have

$$\begin{aligned}\theta|1(t)\rangle &= \theta e^{-i\mathfrak{E}t/\hbar}|1(0)\rangle \\ &= \theta e^{-i\mathfrak{E}t/\hbar}\theta^{-1}\theta|1(0)\rangle = e^{i\mathfrak{E}t/\hbar}\theta|1(0)\rangle,\end{aligned}$$

since $\theta i\mathfrak{E}\theta^{-1} = -i\mathfrak{E}$. Now $|1(0)\rangle$ represents the system at the moment of motion reversal, i.e. at the origin of time, and hence $\theta|1(0)\rangle$ differs from $|1(0)\rangle$ at most by a constant, say β_1. Hence

$$\theta|1(0)\rangle = \beta_1|1(0)\rangle$$

and

$$\theta|1(t)\rangle = \beta_1 e^{i\mathfrak{E}t/\hbar}|1(0)\rangle = \beta_1|1(-t)\rangle.$$

The operator θ should preserve the normalization of states; therefore we expect

$$\langle 1(t)|\theta^\dagger\theta|1(t)\rangle = |\beta_1|^2\langle 1(-t)|1(-t)\rangle = \langle 1(t)|1(t)\rangle,$$

and since the operator $\exp[-i2t\mathfrak{E}/\hbar]$ that takes $|1(-t)\rangle$ into $|1(t)\rangle$ is unitary,

$$|\beta_1|^2 = 1, \qquad \text{or} \qquad \beta_1 = e^{i\alpha_1}, \qquad \alpha_1 \text{ real.}$$

This does not imply, however, that θ is unitary; in fact it is not, because it replaces complex numbers with their complex conjugates.

That θ is not a unitary operator is seen as follows: use any complete set of states $\{|m, n, \ldots\rangle\}$ to form

$$|1\rangle = \underset{m,n,\ldots}{\mathsf{S}} |m, n, \ldots\rangle\langle m, n, \ldots|1\rangle$$

and operate on both sides with θ, remembering that θ transforms all numbers into their complex conjugates; thus

$$\begin{aligned} \theta|1\rangle &= \underset{m,n,\ldots}{\mathsf{S}} \theta\langle m, n, \ldots|1\rangle|m, n, \ldots\rangle \\ &= \underset{m,n,\ldots}{\mathsf{S}} \langle m, n, \ldots|1\rangle^*\theta|m, n, \ldots\rangle \\ &= \underset{m,n,\ldots}{\mathsf{S}} \langle 1|m, n, \ldots\rangle\theta|m, n, \ldots\rangle. \end{aligned} \tag{12.119}$$

Now the set $\{\theta|m, n, \ldots\rangle\}$ of motion-reversed states also is a complete set so that

$$\underset{m,n,\ldots}{\mathsf{S}} \theta|m, n, \ldots\rangle\langle m, n, \ldots|\theta^\dagger = 1, \tag{12.120}$$

and hence we may expand $\theta|1\rangle$ in terms of this complete set:

$$\theta|1\rangle = \underset{m,n,\ldots}{\mathsf{S}} \theta|m, n, \ldots\rangle\langle m, n, \ldots|\theta^\dagger\theta|1\rangle.$$

Comparison of (12.119) with (12.120) yields

$$\langle 1|m, n, \ldots\rangle = \langle m, n, \ldots|1\rangle^* = \langle m, n, \ldots|\theta^\dagger\theta|1\rangle.$$

This is not what we would have obtained were θ unitary; θ preserves normalization, but *changes inner products to their complex conjugates.* In general,

$$\langle 2|\theta^\dagger\theta|1\rangle = \langle 1|2\rangle = \langle 2|1\rangle^*. \tag{12.121}$$

An operator with this property, that is, any operator $\mathfrak{Q}$ such that for arbitrary states $|1\rangle$ and $|2\rangle$,

$$\langle 2|\mathfrak{Q}^\dagger\mathfrak{Q}|1\rangle = \langle 1|2\rangle,$$

is called *antiunitary.* Also, any operator $\mathfrak{Q}$ with the property

$$\mathfrak{Q}[\alpha|1\rangle + \beta|2\rangle] = \alpha^*\mathfrak{Q}|1\rangle + \beta^*\mathfrak{Q}|2\rangle$$

for arbitrary $|1\rangle$ and $|2\rangle$ is called an *antilinear* operator. Hence θ is neither linear nor unitary, but rather is antilinear and antiunitary.

Let us see how θ affects some fundamental operators, e.g. $\mathfrak{P}$ and $\mathfrak{J}$. Since

$$[\mathfrak{P}_j, \mathfrak{X}_k] = (\hbar/i)\delta_{jk},$$

we have

$$\theta[\mathfrak{P}_j, \mathfrak{X}_k]\theta^{-1} = [\theta\mathfrak{P}_j\theta^{-1}, \theta\mathfrak{X}_k\theta^{-1}] = [\theta\mathfrak{P}_j\theta^{-1}, \mathfrak{X}_k]$$
$$= \theta(\hbar/i)\delta_{jk}\theta^{-1} = i\hbar\delta_{jk};$$

from this it follows that

$$\theta\mathfrak{P}\theta^{-1} = -\mathfrak{P}, \tag{12.122}$$

as indeed must be the case for motion reversal.

Recalling that

$$[\mathfrak{J}_j, \mathfrak{X}_k] = i\hbar\epsilon_{jkl}\mathfrak{X}_l,$$

we obtain

$$\theta[\mathfrak{J}_j, \mathfrak{X}_k]\theta^{-1} = [\theta\mathfrak{J}_j\theta^{-1}, \mathfrak{X}_k] = -i\hbar\epsilon_{jkl}\mathfrak{X}_l$$

or

$$\theta\mathfrak{J}\theta^{-1} = -\mathfrak{J}, \tag{12.123}$$

which, again, is what we must have if the motion-reversal operator is to perform what corresponds classically with motion reversal.

It is a standard practice to use the motion-reversal operator to define so-called *real operators* and so-called *imaginary operators.* An operator $\mathfrak{R}$ is said to be *real* if

$$[\theta, \mathfrak{R}] = 0,$$

and an operator $\mathfrak{I}$ is said to be *imaginary* if

$$\{\theta, \mathfrak{I}\} = 0;$$

operators that satisfy neither of these criteria are called *complex* operators. In this terminology, then, $\mathfrak{X}$ and $\mathfrak{E}$ are real operators whereas $\mathfrak{T}$, $\mathfrak{P}$ and $\mathfrak{J}$ are imaginary operators.

We have restricted our discussion thus far to how θ affects abstract operators and state vectors. Let us take a closer look at its behavior in terms of angular-momentum eigenstates and at how we can use what we learn to construct a practical operator for representations in which wave functions are used. Because of Eq. (12.123), we know that θ reverses the magnetic quantum number m in an angular-momentum eigenstate $|j, m\rangle$; it does not change j, but may multiply the whole state by a phase factor $\exp(i\alpha_{jm})$:

$$\theta|j, m\rangle = \exp(i\alpha_{jm})|j, m\rangle, \qquad \alpha_{jm} \text{ a real number,}$$

Therefore

$$\langle j, m'|\theta^\dagger\theta\mathfrak{J}_\pm|j, m\rangle = \langle j, m'|\theta^\dagger(\theta\mathfrak{J}_\pm\theta^{-1})\theta|j, m$$
$$= \langle j, m'|\theta^\dagger(-\mathfrak{J}_\mp)\theta|j, m\rangle = -\exp\big(i(\alpha_{jm} - \alpha_{jm'})\big)\langle j, -m'|\mathfrak{J}_\mp|j, -m\rangle$$
$$= \langle j, m|\mathfrak{J}_\mp|j, m'\rangle, \tag{12.124}$$

the last member being a consequence of the antiunitarity of θ. Now

$$\langle j, m|\mathfrak{J}_{\mp}|j, m'\rangle = \hbar\sqrt{(j \pm m')(j \mp m' + 1)}\delta_{m,m'\mp 1}$$
$$= \langle j, -m'|\mathfrak{J}_{\mp}|j, -m\rangle,$$

according to preceding results in this chapter, and use of this result in Eqs. (12.124) yields

$$-\exp[i(\alpha_{jm} - \alpha_{jm\pm 1})] = 1$$

or

$$\alpha_{jm\pm 1} = \alpha_{jm} + (2n+1)\pi, \qquad n \text{ an arbitrary integer.} \tag{12.125}$$

Now if we let β be some arbitrary complex constant, we always shall have

$$\theta|j, m\rangle\beta = \exp(i\alpha_{jm})|j, -m\rangle\beta^*;$$

hence we may write

$$\langle j, m'|\theta|j, m\rangle = \exp(i\alpha_{jm})\langle j, m'|j, -m\rangle \mathsf{K}$$
$$= \exp(i\alpha_{jm})\delta_{m',-m}\mathsf{K},$$

where K is a complex-conjugation operator, transforming everything to its right into its complex conjugate, and α_{jm} must satisfy one of the recursion relations (12.125).

It happens that there is a simple unitary operator $\mathfrak{U}$ with the property

$$\langle j, m'|\mathfrak{U}|jm\rangle = \exp(i\beta_{jm})\delta_{m,-m'},$$

namely

$$\mathfrak{U} = \exp(-i\pi\mathfrak{J}_2/\hbar)$$

with

$$\beta_{jm} = \pi(j - m).$$

Note that β_{jm} satisfies the recursion relation (12.125) with $n = -2$ for the upper sign and $n = 0$ for the lower sign. Furthermore, if we let $\alpha_{jm} = \alpha_0 + \beta_{jm} = \alpha_0 + \pi(j - m)$, the resulting α_{jm} still satisfy the recursion relations. Only the dependence of α_{jm} on j and m is of any importance, however, so we may as well take $\alpha_0 = 0$, and with this choice, we have

$$\langle j, m'|\theta|j, m\rangle = \langle j, m'|e^{-i\pi\mathfrak{J}_2/\hbar}|j, m\rangle \mathsf{K}$$
$$= d^j_{m',m}(-\pi)\mathsf{K} = \mathsf{K}d^j_{m',m}(-\pi),$$

where

$$d^j_{m',m}(\phi) \equiv \langle j, m'|e^{i\phi\mathfrak{J}_2/\hbar}|j, m\rangle.$$

An important application of this result is in the wave-function notation in which angular-momentum eigenstates are represented by one-column matrices. In this notation the matrix $d^j_{m'm}(\phi)$ represents a rotation of the spin by ϕ about the y-axis. The time-reversal operator in this notation, then, is the product of a rotation of the spin vector by an angle $-\pi$ about the y-axis, and a complex-conjugation operator, together with the instructions that (a) t be replaced by $-t$, (b) $\mathbf{x}$ be left alone. The spin-vector rotation reverses the spin quantization axis and provides an appropriate phase factor without disturbing the coordinate axes of the position vector $\mathbf{x}$ appearing in the wave function. The instructions that $t \to -t$ and $\mathbf{x} \to \mathbf{x}$ must be included since $\mathbf{x}$ and t are represented by numbers in wave functions and we have no formal operators that reverse the signs of numbers; K ensures that the operation is antiunitary and antilinear rather than unitary and linear. In the wave-function notation, the product of the operators, exclusive of the space–time instructions, is

$$\Theta = e^{-i\pi \mathsf{S}_2/\hbar}\mathsf{K}$$

where S_2 is the y-component of the spin operator. The orbital angular momentum is left out because the reversal of orbital magnetic-quantum numbers is ensured by the instruction that wave functions be complex-conjugated but $\mathbf{x}$ remain unchanged.

In this context, an interesting feature of spin-$\frac{1}{2}$ particles may be explored. For them, $\mathsf{S}_2 = \frac{1}{2}\hbar\sigma_2$, and it is easy to see that

$$e^{-i\pi\sigma_2/2} = \cos\tfrac{1}{2}\pi - i\sigma_2 \sin\tfrac{1}{2}\pi = -i\sigma_2.$$

Furthermore, $\mathsf{K}^2 = 1$ and $[\mathsf{K}, -i\sigma_2] = 0$ since the components of σ_2 are imaginary numbers. Therefore, for spin-$\frac{1}{2}$ particles,

$$\Theta^2 = (-i\sigma_2\mathsf{K})^2 = (-i\sigma_2)^2 = -1.$$

Thus two time reversals do not return the wave functions to their original values, but change their phases by π. In general, for systems with n spin-$\frac{1}{2}$ particles,

$$\Theta^2 = (-1)^n.$$

Let us neglect spin angular momentum now, and consider two special kinds of wave functions; through them we shall illustrate the effects of motion reversal on wave functions. First, consider the wave function for an eigenstate of momentum,

$$\psi_{\mathbf{k}}(\mathbf{x}) = \langle \mathbf{x}|\mathbf{k}\rangle, \qquad \mathbf{p} = \hbar\mathbf{k}.$$

Under motion reversal

$$\langle \mathbf{x}|\mathbf{k}\rangle \xrightarrow{\theta} \langle \mathbf{x}|\theta^\dagger\theta|\mathbf{k}\rangle = \langle \mathbf{k}|\mathbf{x}\rangle = \langle \mathbf{x}|\mathbf{k}\rangle^*;$$

hence

$$\psi_{\mathbf{k}}(\mathbf{x}) \xrightarrow{\theta} \psi^*_{\mathbf{k}}(\mathbf{x}).$$

However, since $\langle \mathbf{x}|\mathbf{k}\rangle = N \exp(i\mathbf{k}\cdot\mathbf{x})$, where N is a normalization constant, we have

$$\psi_{\mathbf{k}}(\mathbf{x}) \xrightarrow{\theta} \psi_{-\mathbf{k}}(\mathbf{x}).$$

Next, consider the wave function for an eigenstate of energy and orbital angular momentum

$$\psi_{E,l,m}(\mathbf{x}) = \langle \mathbf{x}|E, l, m\rangle.$$

Under motion reversal,

$$\langle \mathbf{x}|E, l, m\rangle \xrightarrow{\theta} \langle \mathbf{x}|\theta^{\dagger}\theta|E, l, m\rangle = \langle E, l, m|\mathbf{x}\rangle^{*},$$

or

$$\psi_{E,l,m}(\mathbf{x}) \xrightarrow{\theta} \psi^{*}_{E,l,m}(\mathbf{x}) = R^{*}_{E,l}(r)Y_l^{m*}(\theta, \varphi),$$

where, in the last member, variables have been separated in polar coordinates. However (see Problem 6),

$$Y_l^{m*}(\theta, \varphi) = (-1)^m Y_l^{-m}(\theta, \varphi),$$

and hence, when the radial factor is real, $R^{*}_{E,l}(r) = R_{E,l}(r)$,

$$\psi^{*}_{E,l,m}(\mathbf{x}) = (-1)^m \psi_{E,l,-m}(\mathbf{x}),$$

or

$$\psi_{E,l,m}(\mathbf{x}) \xrightarrow{\theta} (-1)^m \psi_{E,l,-m}(\mathbf{x}).$$

READING NOTES

Reference 22, Chapter 10.

Reference 24, Chapter 4.

Reference 25, Chapters 1 and 2.

Reference 38, Chapter 13.

Reference 41, Chapters 1, 2, 4.

Reference 42, Chapter 10.

Reference 43, Chapters 7, 10.

Reference 50 is substantially beyond the level of presentation given here, but presents an early unified view of group theory and its role in quantum theory. This is "cultural" reading for the advanced student.

* Works referred to in these notes will be found in the Bibliography on p. 577.

PROBLEMS

1. Verify the commutator relations (12.37) and (12.38) when $\mathfrak{J}$ is replaced by $\mathfrak{L} = \mathfrak{X} \times \mathfrak{P}$.
2. Derive Eqs. (12.44).
3. Show that the radial momentum operator

$$\mathsf{P}_r = \frac{\hbar}{i}\frac{1}{r}\frac{\partial}{\partial r}r$$

is Hermitian whenever the wave functions involved are "physically reasonable". Give a mathematical condition on the wave functions that determines in what sense "physically reasonable" is to be taken.

4. a) Show that Rodrigue's formula satisfies $P_l(1) = 1$.
   b) Use Rodrigue's formula and the self-adjointness of the differential operator to verify Eq. (12.57).
5. Verify Eq. (12.60) by differentiating (12.55) m times.
6. Show that $Y_l^{m*}(\theta, \varphi) = (-)^m Y_l^{-m}(\theta, \varphi)$. (Hint: Use L_- repeatedly on $Y_l^m(\theta, \varphi)$.)
7. The quantities

$$x_m \equiv r\sqrt{\frac{4\pi}{3}}\,Y_1^{\,m}(\theta, \varphi) = \begin{cases} -\dfrac{1}{\sqrt{2}}(x + iy), & m = 1 \\ z & m = 0 \\ \dfrac{1}{\sqrt{2}}(x - iy), & m = -1 \end{cases}$$

are the components of the so-called *spherical vectors.*

   a) Rotate the coordinate system by $\pi/2$ about the y-axis and find the $x'_m = r\sqrt{4\pi/3}\,Y_1^{\,m}(\theta', \varphi')$ in this new coordinate system in terms of the x_m.
   b) Show that $Y_{1/2}^{\pm 1/2}(\theta, \varphi) = N_\pm[\sin\theta]^{1/2}\, e^{\pm i\varphi/2}$.
   c) Show that under the rotation of part (a) the $Y_{1/2}^{\pm 1/2}(\theta', \varphi')$ are not linear combinations of the $Y_{1/2}^{\pm 1/2}(\theta', \varphi')$. This demonstrates that m and hence l cannot be $+\frac{1}{2}$ and $\frac{1}{2}$, respectively, and can be generalized to rule out half-integer values of m and l in general. This argument is superior to that given below Eq. (12.51b), because physically all that really is required of $\Phi(\varphi)$ is that $|\Phi(\varphi)|^2$ be single-valued as φ varies from 0 to 2π and this in fact does not rule out half-integer m.
8. a) Show that $\mathsf{L}_+ Y_l^m(\theta, \varphi) = \hbar[(l \mp m)(l \pm m + 1)]^{1/2} Y_l^{m\pm 1}(\theta, \varphi)$, using $\langle\theta, \varphi|\mathfrak{L}_\pm|lm\rangle = \mathsf{L}_\pm Y_l^m(\theta, \varphi)$ and your knowledge of the effect of $\mathfrak{L}_\pm$ on the state $|lm\rangle$.
   b) Use the result above to evaluate $N_{l,m}$ in Eq. (12.62).
9. Given any two vector operators, say $\mathfrak{A}, \mathfrak{B}$, find $[\mathfrak{J}_k, \mathfrak{A}\cdot\mathfrak{B}]$ for $k = 1, 2, 3$. (Hint: What happens to $\mathfrak{A}\cdot\mathfrak{B}$ under rotations?)
10. Show that for a state having well-defined z-component of angular momentum, the expectation values of the x- and y-components vanish.
11. Consider an axis z' with directional coordinates (θ, φ) in some spherical coordinate system associated in the usual way with a Cartesian coordinate system.

a) Obtain an expression for $\mathfrak{J}_{z'}$ in terms of $\mathfrak{J}_x$, $\mathfrak{J}_y$, $\mathfrak{J}_z$. (Hint: What are the direction cosines of $\mathfrak{J}_{z'}$?)
b) Show that the mean value of $\mathfrak{J}_{z'}$ for a system in the state $|j, m\rangle$ is $m \cos \theta$.

12. Given that the matrix elements $\langle jm|\mathfrak{J}_3|jm'\rangle$ and $\langle jm|\mathfrak{J}_1|jm'\rangle$ are real numbers for all m, m', show that $\langle jm|\mathfrak{J}_2|jm'\rangle$ must be imaginary. Will the latter be symmetric, antisymmetric or neither under the interchange $m \leftrightarrow m'$?
13. Let $\mathfrak{J}_i$ and $\mathfrak{J}_j$ be any two distinct components of angular momenta along orthogonal axes and let $\mathfrak{J}_k$ be the third. If $\mathfrak{Q}$ is an operator such that $[\mathfrak{Q}, \mathfrak{J}_i] = [\mathfrak{Q}, \mathfrak{J}_j] = 0$ but otherwise is arbitrary, find $[\mathfrak{Q}, \mathfrak{J}_k]$.
14. Show how the uncertainty principle is connected with the fact that the maximum eigenvalue for the component of angular momentum along a given direction is necessarily less than the square root of the eigenvalue of $\mathfrak{J}^2$.
15. Show directly through matrix calculations that

$$[\sigma_j, \sigma_k] = i2\epsilon_{jlm}\sigma_m$$

$$\{\sigma_j, \sigma_k\} = 2\delta_{jk}$$

$$\boldsymbol{\sigma} \cdot \mathbf{A}\boldsymbol{\sigma} \cdot \mathbf{B} = \mathbf{A} \cdot \mathbf{B} + i\boldsymbol{\sigma} \cdot \mathbf{A} \times \mathbf{B}$$

16. Compare the analysis of two Stern–Gerlach apparati in tandem with that of the two-slit diffraction experiment in Chapter 9.
17. Let two-spin-$\frac{1}{2}$ particles be represented by the column vectors of Eqs. (12.83) and (12.84), which we denote

$$\chi^{+1/2}(j) = \begin{Bmatrix} 1 \\ 0 \end{Bmatrix}, \qquad \chi^{-1/2}(j) = \begin{Bmatrix} 0 \\ 1 \end{Bmatrix}, \qquad j = 1, 2,$$

where the index j labels the two particles. From these eigenfunctions of

$$\mathbf{S}^2(j) = \tfrac{1}{4}\hbar^2\boldsymbol{\sigma}^2(j)$$

and

$$S_3 = \tfrac{1}{2}\hbar\sigma_3(j) \qquad j = 1, 2,$$

one can construct eigenfunctions $\xi_1{}^m$ and $\xi_0{}^0$ of the operators

$$[\mathbf{S}(1) + \mathbf{S}(2)]^2 = \mathbf{S}^2$$

and

$$S_3(1) + S_3(2) = S_3;$$

$\xi_1{}^m$ has eigenvalues $s(s + 1) = 2$ and $m = 1, 0, -1$ for $\mathbf{S}^2$ and S_3 respectively, whereas $\xi_0{}^0$ has eigenvalues $s(s + 1) = 0$ and $m = 0$ for $\mathbf{S}^2$ and S_3. These eigenfunctions are formed by constructing bilinear combinations of $\chi^\mu(1)$, $\chi^\mu(2)$:

$$\xi_1{}^m = \sum_{\mu=-1/2}^{1/2} C_{1,m;1/2,1/2,\mu}\, \chi^{m-\mu}(1)\chi^\mu(2)$$

$$\xi_0{}^0 = \sum_{\mu=-1/2}^{1/2} C_{0,0;1/2,1/2,\mu}\, \chi^{-\mu}(1)\chi^\mu(2),$$

where $C_{s,m;1/2,1/2,\mu} = \langle s, m, \frac{1}{2}, \frac{1}{2} | \frac{1}{2}, m - \mu; \frac{1}{2}, \mu \rangle$ are the Clebsch–Gordan coefficients for the addition of angular momenta.

a) Find the $C_{s,m;1/2,1/2,\mu}$; give your results in tabular form.
b) Show that the operator

$$\mathsf{P}_1 \equiv \tfrac{1}{8}[\boldsymbol{\sigma}(1) + \boldsymbol{\sigma}(2)]^2 = \frac{1}{2\hbar^2}\mathbf{S}^2$$

has the property that $\mathsf{P}_1\xi_1{}^m = \xi_1{}^m$, $\mathsf{P}_1\xi_0{}^0 = 0$, $\mathsf{P}_1{}^2 = \mathsf{P}_1$.
c) Find another operator P_0 such that $\mathsf{P}_0\xi_1{}^m = 0$, $\mathsf{P}_0\xi_0{}^0 = 1$, $\mathsf{P}_0{}^2 = \mathsf{P}_0$. The operators P_1 and P_0 are projection operators for the states $\xi_1{}^m$ and $\xi_0{}^0$, respectively. P_1 is called the triplet projection operator because $\xi_1{}^m$ has three components, and P_0 is called the singlet projection operator for analogous reasons.

18. Use the quantities $\chi^{\pm 1/2}$ defined in Problem 17 and show that for any vector $\mathbf{A}$

$$\chi^{\pm 1/2\dagger}\boldsymbol{\sigma}\cdot\mathbf{A}\chi^{\pm 1/2} = \pm Az$$

$$\chi^{\pm 1/2\dagger}\boldsymbol{\sigma}\cdot\mathbf{A}^{\mp 1/2} = Ax \mp iAy.$$

19. Use the results of Problem 18 to evaluate

$$\xi_s{}^{m\dagger}\mathsf{S}_{12}\xi_s{}^{m'} \qquad \text{for } s = 0; 1 \qquad \text{and} \qquad m', m = 0; 1, 0, -1,$$

when

$$\mathsf{S}_{12} \equiv 3\frac{\boldsymbol{\sigma}(1)\cdot\mathbf{r}\boldsymbol{\sigma}(2)\cdot\mathbf{r}}{r^2} - \boldsymbol{\sigma}(1)\cdot\boldsymbol{\sigma}(2),$$

and $\xi_s{}^m$ are defined as in Problem 17. Give your results in terms of matrices (one for $s = 0$, one for $s = 1$) labeled by m, m'. Express the matrix elements in terms of polar coordinates. Note that they involve $Y_2{}^M(\theta, \varphi)$.

20. Can linear momentum and angular momentum be completely well defined simultaneously? Does the same conclusion hold for both spin angular momentum and orbital angular momentum? Find the relevant minimum uncertainty products.

CHAPTER 13

TWO-BODY PROBLEMS; RELATIVE COORDINATES

We studied center-of-mass and barycentric coordinate frames in Section 4.3, and investigated the two-body collision problem in Section 4.4, although there we were interested primarily in how to handle such problems relativistically. Here we wish to consider the separation into center-of-mass and relative coordinates for a nonrelativistic two-body system in quantum theory. The same considerations carry over in an obvious way to relativistic problems. This chapter serves as a brief hiatus between the obviously related discussions of angular motion and radial motion in quantum theory, and is inserted here to justify studying two-body problems in terms of equations that appear to be one-body equations.

13.1 CENTER-OF-MASS AND RELATIVE COORDINATE OPERATORS

Nonrelativistically, the center of mass of a system of two particles having coordinate locations $\mathbf{x}_{(1)}$ and $\mathbf{x}_{(2)}$, and masses m_1 and m_2, respectively, is

$$\mathbf{X} = \frac{m_1\mathbf{x}_{(1)} + m_2\mathbf{x}_{(2)}}{m_1 + m_2} \tag{13.1}$$

and the total momentum of the system is

$$\mathbf{P} = \mathbf{p}_{(1)} + \mathbf{p}_{(2)} \tag{13.2}$$

if $\mathbf{p}_{(1)}$ and $\mathbf{p}_{(2)}$ are the individual momenta of the particles. Converting to quantum-mechanical operators, it is easy to check that

$$[\mathfrak{P}_j, \mathfrak{X}_k] = \frac{\hbar}{i}\delta_{jk}, \qquad j, k = 1, 2, 3, \tag{13.3}$$

using

$$[\mathfrak{P}_{(m)j}, \mathfrak{X}_{(n)k}] = \frac{\hbar}{i}\delta_{jk}\,\delta_{mn}, \tag{13.4}$$

where m, n label the two particles, $m, n = 1, 2$, and j, k label the Cartesian coordinate directions, $j, k = 1, 2, 3$. Hence $\mathfrak{P}$ and $\mathfrak{X}$ indeed represent canonically conjugate vector operators.

Now the relative position of particles 1 and 2 is given by the vector

$$\mathbf{x} = \mathbf{x}_{(2)} - \mathbf{x}_{(1)} \tag{13.5}$$

and, converting to quantum-mechanical operators, we should like to find a linear combination of $\mathfrak{P}_{(1)}$ and $\mathfrak{P}_{(2)}$ that is canonically conjugate to $\mathfrak{x}$. Thus we seek

$$\mathfrak{p} = \alpha\mathfrak{P}_{(2)} + \beta\mathfrak{P}_{(1)} \tag{13.6}$$

such that

$$[\mathfrak{p}_j, \mathfrak{x}_k] = \frac{\hbar}{i}\delta_{jk}. \tag{13.7}$$

This momentum $\mathfrak{p}$ is to represent the relative momentum of the two particles. We should like to have the $\mathfrak{p}_j$ satisfy the additional condition

$$[\mathfrak{p}_j, \mathfrak{X}_k] = 0 \tag{13.8}$$

in order that measurements of the center of the mass of the system should not interfere with, nor be interfered with by, measurements of the relative momentum. We note that the definitions of $\mathfrak{x}$ and $\mathfrak{P}$ guarantee that

$$[\mathfrak{P}_j, \mathfrak{x}_k] = 0. \tag{13.9}$$

Use of Eqs. (13.1), (13.6) and (13.4) in Eq. (13.8) yields

$$\alpha\frac{m_1}{m_1 + m_2} + \beta\frac{m_2}{m_1 + m_2} = 0$$

or

$$\beta = -\frac{m_1}{m_2}\alpha,$$

whereas Eqs. (13.5)–(13.6), when used in Eq. (13.7), give

$$\alpha - \beta = 1,$$

or

$$\alpha = \frac{m_2}{m_1 + m_2}, \qquad \beta = -\frac{m_1}{m_1 + m_2}.$$

Therefore

$$\begin{aligned}\mathfrak{p} &= \frac{m_2}{m_1 + m_2}\mathfrak{P}_{(1)} - \frac{m_1}{m_1 + m_2}\mathfrak{P}_{(2)} \\ &= \frac{m_1 m_2}{m_1 + m_2}\left(\frac{1}{m_1}\mathfrak{P}_{(1)} - \frac{1}{m_2}\mathfrak{P}_{(2)}\right)\end{aligned} \tag{13.10}$$

is the relative momentum operator for the two particles. Since

$$\mathfrak{u} \equiv \frac{1}{m_1}\mathfrak{P}_{(1)} - \frac{1}{m_2}\mathfrak{P}_{(2)}$$

evidently represents the relative velocity, we may write

$$\mathfrak{p} = \mu\mathfrak{u},$$

where

$$\mu \equiv \frac{m_1 m_2}{m_1 + m_2}$$

is the reduced mass of the two-body system. This agrees completely with the classical result for relative motion.

Finally, we should check the consistency of the definitions of relative momentum and reduced mass by evaluating the relative kinetic energy in terms of them. The kinetic energy of the motion of the center of mass is

$$\mathfrak{T}_{\text{cm}} = \frac{1}{2M}\mathfrak{P}^2, \qquad M \equiv m_1 + m_2,$$

and the total kinetic energy is

$$\mathfrak{T}_{(1)} + \mathfrak{T}_{(2)} = \frac{1}{2m_1}\mathfrak{P}_{(1)}^2 + \frac{1}{2m_2}\mathfrak{P}_{(2)}^2. \tag{13.11}$$

Hence the kinetic energy of relative motion is

$$\begin{aligned}
\mathfrak{T}_{\text{r}} &= \mathfrak{T}_{(1)} + \mathfrak{T}_{(2)} - \mathfrak{T}_{\text{cm}} \\
&= \frac{1}{2m_1}\mathfrak{P}_{(1)}^2 + \frac{1}{2m_1}\mathfrak{P}_{(2)}^2 - \frac{1}{2(m_1 + m_2)}(\mathfrak{P}_{(1)}^2 + \mathfrak{P}_{(2)}^2 - 2\mathfrak{P}_{(1)} \cdot \mathfrak{P}_{(2)}) \\
&= \frac{1}{2}\left(\frac{m_2}{m_1 M}\mathfrak{P}_{(1)}^2 + \frac{m_1}{m_2 M}\mathfrak{P}_{(2)}^2 - \frac{2}{M}\mathfrak{P}_{(1)} \cdot \mathfrak{P}_{(2)}\right) \\
&= \frac{1}{2\mu}\left(\frac{m_2}{M}\mathfrak{P}_{(1)} - \frac{m_1}{M}\mathfrak{P}_{(2)}\right)^2 = \frac{1}{2\mu}\mathfrak{p}^2.
\end{aligned} \tag{13.12}$$

Thus μ indeed behaves like the mass of a particle having momentum $\mathfrak{p}$. This is familiar from classical mechanics.

If now our two-body system is in an eigenstate $|\mathbf{P}, \ldots\rangle$ of total momentum, and if furthermore we work in the frame of reference in which the total momentum $\mathbf{P}$ vanishes, then the total kinetic-energy operator is

$$\mathfrak{T}_{\text{r}} = \frac{1}{2\mu}\mathfrak{p}^2,$$

and the center-of-mass motion will have been eliminated from the problem.

13.2 SEPARATION OF CENTER-OF-MASS AND RELATIVE-MOTION VARIABLES

We now translate these results into differential operators and show that under certain conditions the separation of variables for center-of-mass and relative motion can be accomplished.

It is easy to deduce from the definitions of Eqs. (13.1) and (13.5), together with (13.2) and (13.10), that in the wave-function representation

$$\mathsf{P}_j = \frac{\hbar}{i}\frac{\partial}{\partial X_j}, \qquad \mathsf{p}_j = \frac{\hbar}{i}\frac{\partial}{\partial x_j}. \tag{13.13}$$

In fact, one need only write

$$\mathbf{X} = \mathbf{X}(\mathbf{x}_{(1)}, \mathbf{x}_{(2)}) = \frac{m_1}{M}\mathbf{x}_{(1)} + \frac{m_2}{M}\mathbf{x}_{(2)},$$

$$\mathbf{x} = \mathbf{x}(\mathbf{x}_{(1)}, \mathbf{x}_{(2)}) = \mathbf{x}_2 - \mathbf{x}_{(1)}$$

and use

$$\frac{\partial}{\partial x_{(1)j}} = \sum_{k=1}^{3}\frac{\partial X_k}{\partial x_{(1)j}}\frac{\partial}{\partial X_k} + \sum_{k=1}^{3}\frac{\partial x_k}{\partial x_{(1)j}}\frac{\partial}{\partial x_k} = \frac{m_1}{M}\frac{\partial}{\partial X_j} - \frac{\partial}{\partial x_j},$$

$$\frac{\partial}{\partial x_{(2)j}} = \frac{m_2}{M}\frac{\partial}{\partial X_j} + \frac{\partial}{\partial x_j};$$

this yields

$$\mathsf{P}_j = \frac{\hbar}{i}\frac{\partial}{\partial x_{(1)j}} + \frac{\hbar}{i}\frac{\partial}{\partial x_{(2)j}} = \frac{\hbar}{i}\frac{\partial}{\partial X_j}$$

and

$$\mathsf{p}_j = \frac{m_1}{M}\frac{\hbar}{i}\frac{\partial}{\partial x_{(2)j}} - \frac{m_2}{M}\frac{\hbar}{i}\frac{\partial}{\partial x_{(1)j}} = \frac{\hbar}{i}\frac{\partial}{\partial x_j}.$$

Indeed, the commutation relations (13.3) and (13.7)–(13.9) guarantee that these are the correct representations. We may write

$$\mathbf{P} = \frac{\hbar}{i}\nabla_X, \qquad \mathbf{p} = \frac{\hbar}{i}\nabla_x. \tag{13.14}$$

Frequently one is interested in two-body systems in which the interactions depend only upon the mutual separation of the particles, that is, upon $\mathbf{x}_{(2)} - \mathbf{x}_{(1)} = \mathbf{x}$. The nonrelativistic Hamiltonian then is

$$\mathsf{H} = \frac{1}{2m_1}\mathbf{P}_{(1)}^2 + \frac{1}{2m_2}\mathbf{P}_{(2)}^2 + V(\mathbf{x})$$

$$= \frac{1}{2M}\mathbf{P}^2 + \frac{1}{2\mu}\mathbf{p}^2 + V(\mathbf{x}).$$

In such a situation, the solution

$$\Psi(\mathbf{X}, \mathbf{x} : t)$$

to the equation

$$\mathsf{H}\Psi(\mathbf{X}, \mathbf{x} : t) = \mathsf{E}\Psi(\mathbf{X}, \mathbf{x} : t), \qquad \mathsf{E} = i\hbar\frac{\partial}{\partial t},$$

is separable in the variables $\mathbf{X}$ and $\mathbf{x}$. Thus we may write

$$\Psi(\mathbf{X}, \mathbf{x} ; t) = \phi(\mathbf{X} ; t)\psi(\mathbf{x} ; t)$$

and utilize

$$\begin{aligned}\mathsf{E}\Psi(\mathbf{X}, \mathbf{x} : t) &= i\hbar\frac{\partial}{\partial t}\Psi(\mathbf{X}, \mathbf{x} ; t)\\ &= \phi(\mathbf{X} : t)i\hbar\frac{\partial}{\partial t}\psi(\mathbf{x} : t) + \psi(\mathbf{x} : t)i\hbar\frac{\partial}{\partial t}\phi(\mathbf{X} : t)\\ &= \phi(\mathbf{x}; t)\mathsf{E}\psi(\mathbf{x} : t) + \psi(\mathbf{x} : t)\mathsf{E}\phi(\mathbf{X} : t)\end{aligned}$$

to get

$$\frac{1}{\psi(\mathbf{x} : t)}\left[\frac{1}{2\mu}\mathbf{p}^2 + V(\mathbf{x}) - \mathsf{E}\right]\psi(\mathbf{x} : t) = \frac{1}{\phi(\mathbf{X} : t)}\left[\frac{1}{2M}\mathbf{P}^2 - \mathsf{E}\right]\phi(\mathbf{X} : t) = 0.$$

Indeed, $(1/2M)\mathbf{P}^2$ is the total Hamiltonian for the center-of-mass motion, and therefore we seek the solution $\phi(\mathbf{X} : t)$ to

$$\frac{1}{2\mu}\mathbf{P}^2\phi(\mathbf{X} ; t) = i\hbar\frac{\partial}{\partial t}\phi(\mathbf{x} ; t) \tag{13.15}$$

as the wave function for the center-of-mass motion. This dictates that the solution $\psi(\mathbf{x} : t)$ to

$$\left[\frac{1}{2\mu}\mathbf{p}^2 + V(\mathbf{x})\right]\psi(\mathbf{x} : t) = i\hbar\frac{\partial}{\partial t}\psi(\mathbf{x}; t) \tag{13.16}$$

is the appropriate solution for the relative motion. This is the Schrödinger equation for a single particle with mass μ, and hence we need solve only a one-body problem to get the wave function for relative motion of two particles whose interaction potential $V(\mathbf{x})$ depends only upon the mutual separation $\mathbf{x}$ of the particles.

Let us assume that $\Psi(\mathbf{X}, \mathbf{x} : t)$ represents an eigenstate of energy with eigenvalue E and that $\phi(\mathbf{X} ; t)$ represents an eigenstate of energy for the center-of-mass motion with eigenvalue E_{cm}. This implies that $\psi(\mathbf{x} : t)$ also represents an eigenstate of energy for the relative motion with eigenvalue

$$E_{\text{r}} = E - E_{\text{cm}}.$$

Thus we can write

$$\Psi(\mathbf{X}, \mathbf{x}: t) = \psi(\mathbf{x})\phi(\mathbf{X})e^{-i(E_r + E_{cm})t/\hbar}$$

with

$$\psi(\mathbf{x}: t) = \psi(\mathbf{x})e^{-iE_r t/\hbar}, \qquad \phi(\mathbf{X}; t) = \phi(\mathbf{X})e^{-iE_{cm}t/\hbar}.$$

Furthermore, Eq. (13.15) now can be written in the form

$$\left[-\frac{\hbar^2}{2M}\nabla_X^2 - E_{cm}\right]\phi(\mathbf{X}) = 0,$$

and if $\phi(\mathbf{X})$ is to represent a state with definite momentum $\mathbf{P}$, we have

$$\phi(\mathbf{X}) = (2\pi)^{-3/2}e^{i\mathbf{P}\cdot\mathbf{X}/\hbar}, \qquad E_{cm} = \frac{1}{2M}\mathbf{P}^2.$$

Hence, whenever the center of mass moves with definite total momentum $\mathbf{P}$ the overall wave function for the two-body system is

$$\Psi(\mathbf{X}, \mathbf{x}; t) = \psi(\mathbf{x})(2\pi)^{-3/2}e^{i\mathbf{P}\cdot\mathbf{X}/\hbar}e^{-i(E_r + \mathbf{P}^2/2M)t/\hbar}, \tag{13.17}$$

and we are left with only the differential equation

$$\left[-\frac{\hbar^2}{2\mu}\nabla^2 + V(\mathbf{x}) - E_r\right]\psi(\mathbf{x}) = 0 \tag{13.18}$$

to solve.

When the further circumstance that $V(\mathbf{x}) = V(|\mathbf{x}|) = V(r)$ prevails, Eq. (13.18) can be separated into angular and radial parts, as we have noted previously.

Thus, to solve a two-body problem under the circumstances assumed here, we may solve the Schrödinger equation (13.18) in which the reduced mass μ appears. The ordinary one-body Schrödinger equation in which the particle mass m appears, applies to a single particle with mass m moving in a fixed potential. This latter situation may be obtained from (13.18) by letting $m_2 \to \infty$ while m_1 remains finite, or

$$\mu = \frac{m_1 m_2}{m_1 + m_2} \to m_1,$$

thus the particle with finite mass m_1 moves under the influence of a fixed potential "attached" to the particle with infinite mass.

CHAPTER 14

SOLUTIONS TO THE RADIAL EQUATION

In Section 12.5 we found that a spherically symmetric potential allows for the separation of the Schrödinger equation into angular and radial equations. All of the dynamics, that is, all of the information contained in the potential function, then resides in the radial equation (12.50); hence each different potential function yields a different radial equation to be solved. In this chapter we shall consider some general features of the radial factor of the wave function for a central potential, and shall investigate solutions to the radial equations for some particularly important forms of the potential function $V(r)$.

From Eq. (12.50) and the definition for P_r given just below Eq. (12.47), we obtain the differential equation

$$\left[\frac{d^2}{dr^2} - \frac{l(l+1)}{r^2} - \frac{2m}{\hbar^2}V(r) + \frac{2m}{\hbar^2}E\right] rR_{E,l,\ldots}(r,\ldots) = 0.$$

For convenience in discussing this differential equation, we may define

$$k^2 \equiv \frac{2m}{\hbar^2}E \qquad \text{and} \qquad R_{E,l,\ldots}(r,\ldots) = \frac{f(l,k,r)}{r}; \tag{14.1}$$

these yield the equation

$$\left[\frac{d^2}{dr^2} - \frac{l(l+1)}{r^2} - \frac{2m}{\hbar^2}V(r) + k^2\right] f(l,k,r) = 0. \tag{14.2}$$

The solutions $f(l, k, r)$ to this equation are called the *radial wave functions* for the system.

Recall, from Chapter 12, that solutions to spherically symmetric problems may be written in the form

$$\begin{aligned}\psi_{E,l,m,\ldots}(r,\theta,\varphi,\ldots) &= C_{l,m}R_{E,l,\ldots}(r,\ldots)Y_l^m(\theta,\phi)\\ &= C_{l,m}\frac{f(l,k,r)}{r}Y_l^m(\theta,\varphi),\end{aligned} \tag{14.3}$$

according to the discussion preceding Eq. (12.49); here we have let $W_\lambda(\theta, \varphi) = C_{l,m}Y_l^m(\theta, \varphi)$, where $C_{l,m}$ is a constant. Since the Schrödinger equation is a homogeneous linear equation, any solution to Eq. (12.42) may be written as a linear combination of solutions (14.3) having different values of l and m:

$$\psi_{E,\ldots}(r, \theta, \varphi, \ldots) = \sum_{l,m} C_{l,m} \frac{f(l, k, r)}{r} Y_l^m(\theta, \varphi), \tag{14.4}$$

where the $f(l, k, r)$ are to be determined by the form of the potential appearing in (14.2). Therefore a complete solution of the Schrödinger equation will have been effected when we have found solutions to (14.2) satisfying appropriate boundary conditions. The values of $C_{l,m}$ appearing in (14.4) will be determined by the nature of the particular problem we may be attempting to solve.

14.1 LIMITATIONS IMPOSED BY THE FORM OF THE POTENTIAL

Equation (14.4) is known to have solutions of closed form—that is, solutions expressible in terms of standard mathematical functions—for only a limited set of potentials $V(r)$. For arbitrary l, solutions in terms of standard special functions are known to exist only for potentials of the forms

$$V(r) = \begin{cases} \alpha r^2, & \text{for all } r, \\ \beta r, & \text{for all } r, \\ \gamma/r, & \text{for all } r, \\ \delta/r^2, & \text{for all } r, \\ \begin{cases} \epsilon_1, & \text{for } r \le a, \\ \epsilon_2, & \text{for } a < r < b, \\ \epsilon_3, & \text{for } b \le r. \end{cases} \end{cases}$$

The particular case $l = 0$ yields a class of differential equations whose solutions can be given in terms of standard special functions when

$$V(r) = \begin{cases} -ae^{-r/d}, & a > 0, & \text{all } r, \\ b[e^{2(r_0 - r)/d} - 2e^{(r_0 - r)/d}], & & \text{all } r, \\ c \tanh^2 (r/d) & & \text{all } r, \end{cases}$$

and when the parameters of these potentials lie within certain limits.

In general, however, wave functions may be found by numerical integration of the differential equation together with appropriate boundary conditions. Modern high-speed computers have rendered an invaluable service in making possible the solution of many important quantum-mechanical problems, within reasonable expenditures of time and energy.

In this chapter we shall be concerned primarily with some potentials for which the solutions are obtainable in closed form. These potentials, except for the Coulomb potential, are physically pathological, but they are important precisely because they yield relatively simple solutions and because they can, under certain circumstances, be used to approximate physical situations. Before dealing with these potentials, however, we will indicate some conditions that frequently are imposed upon potentials, together with the consequences of these conditions for the radial wave functions.

14.2 ASYMPTOTIC BEHAVIOR FOR $r \to 0$

It is convenient mathematically to impose the condition

$$r^2 V(r) \underset{r\to 0}{\sim} 0 \tag{14.5}$$

upon all potentials, since then, for very small r,

$$\left[\frac{d^2}{dr^2} - \frac{l(l+1)}{r^2}\right] f(l, k, r) \underset{r\to 0}{\sim} 0; \tag{14.6}$$

that is, the asymptotic behavior of $f(l, k, r)$ near the origin of coordinates is independent of the potential $V(r)$ and of the wave number k, and is determined entirely by the value of the orbital angular-momentum quantum number l.

Many useful potentials satisfy this condition, but it is not necessarily physically reasonable. There is plenty of evidence that "hard cores" represented mathematically by δ-functions or other infinitely large positive terms are required in the representation of nuclear interactions by potential functions. In fact, it makes no sense to talk about "physically reasonable" conditions on the behavior of potentials at small values of r, because the experimental study of interactions at very small distances between particles requires very high energies, and at these energies the nonrelativistic Schrödinger equation breaks down. For example, the minimum uncertainty in the relative momentum of two particles that are no farther than Δr apart is given by

$$\Delta p \geq \hbar/2\,\Delta r$$

and for $\Delta r \simeq 1$ Fermi $\equiv 1$ F $= 10^{-13}$ cm,

$$\Delta p \geq \frac{\hbar c}{2c \cdot 1\text{ F}} = \frac{197\text{ MeV F}}{2c\text{ F}} \simeq 99\text{ MeV}/c.$$

For an electron, this corresponds to an energy uncertainty much greater than $mc^2 = 0.511$ MeV, and an electron with such an energy surely is relativistic; the same conclusion holds for nucleons (protons and neutrons), though it is less dramatically demonstrated with this choice of $\Delta r = 1$ F.

We adopt the condition (14.5), therefore, to limit the mathematical difficulties we must face in studying potentials rather than for any good physical reason.

In the approximation that (14.6) becomes an equation, the solutions to this equation are easily found to be of the form

$$a_l' r^{l+1} + b_l' r^{-l}, \tag{14.7}$$

where a' and b' are constants. For every nonnegative integer l, r^{l+1} and r^{-l} are linearly independent, and we may form two linearly independent solutions to (14.2),

$$f^{(1)}(l, k, r) \qquad \text{and} \qquad f^{(2)}(l, k, r)$$

such that

$$f^{(1)}(l, k, r) \underset{r \to 0}{\sim} \frac{k^{l+1}}{(2l+1)!!} r^{l+1} \tag{14.8}$$

and

$$f^{(2)}(l, k, r) \underset{r \to 0}{\sim} \frac{(2l-1)!!}{k^l r^l}. \tag{14.9}$$

The normalization, represented by the appearance of the constants $k^{l+1}/(2l+1)!!$ and $(2l-1)!!/k^l$, is chosen for later convenience. With this, we can write the exact solution to (14.2) in the form

$$f(l, k, r) = a_l f^{(1)}(l, k, r) + b_l f^{(2)}(l, k, r) \tag{14.10}$$

which, for small r, behaves like (14.7) if

$$a_l' = \frac{a_l k^{l+1}}{(2l+1)!!} \qquad \text{and} \qquad b_l' = \frac{b_l (2l-1)!!}{k^l}.$$

Now we can show that for such a solution to be physically acceptable, we must have $b_l = 0$; that is, only solutions with the asymptotic behavior (14.8) of $f^{(1)}(l, k, r)$ are allowed. To see this, consider the value of the probability that a particle represented by

$$\psi_E(\mathbf{r}) = \sum_{l,m} C_{l,m} \frac{f(l, k, r)}{r} Y_l^m(\theta, \varphi)$$

might be found within a very small sphere of radius ϵ centered at $r = 0$;

$$P_\epsilon = \int_{V_\epsilon} |\psi_E(\mathbf{r})|^2\, d^3r = \int_0^\epsilon dr \int_{4\pi} d\Omega\, r^2 |\psi_E(\mathbf{r})|^2$$

$$= \sum_{l',m'} \sum_{l,m} C^*_{l',m'} C_{l,m} \int_0^\epsilon f^*(l', k, r) f(l, k, r)\, dr \int_{4\pi} Y_{l'}^{m'*}(\theta, \varphi) Y_l^m(\theta, \varphi)\, d\Omega$$

$$= \sum_{l,m} |C_{l,m}|^2 \int_0^\epsilon |f(l, k, r)|^2\, dr$$

$$\underset{\epsilon \to 0}{\simeq} \sum_{l,m} |C_{l,m}|^2 \int_0^\epsilon |a_l' r^{l+1} + b_l' r^{-l}|^2\, dr$$

$$= \sum_{l,m} |C_{l,m}|^2 \left[|a_l'|^2 \frac{\epsilon^{2l+3}}{2l+3} + 2\,\mathrm{Re}\,(a_l'^* b_l') \frac{\epsilon^2}{2} + |b_l'|^2 \int_0^\epsilon r^{-2l}\, dr \right].$$

The last integral on the last line is divergent for any $\epsilon > 0$ unless $l = 0$, and hence for $l > 0$ we must have $b_l' = b_l = 0$.

The same conclusion holds even for $l = 0$, but another argument is required to demonstrate that this is so. If $l = 0$, the *centrifugal potential* term, $-l(l+1)/r^2$ vanishes, and we must solve

$$\left[\frac{d^2}{dr^2} - \frac{2m}{\hbar^2} V(r) + k^2 \right] f(0, k, r) = 0.$$

Let us assume that the leading term in the solution has the form $r^\gamma e^{\pm ikr}$; that is, let

$$f(0, k, r) \underset{r \to 0}{\sim} (\text{const.})\, r^\gamma e^{\pm ikr}.$$

Differentiation twice with respect to r yields

$$\left[\frac{d^2}{dr^2} - \frac{2m}{\hbar^2} V(r) + k^2 \right] f(0, k, r) \underset{r \to 0}{\sim} \left[\frac{\gamma(\gamma - 1)}{r^2} \pm i \frac{2k\gamma}{r} - \frac{2m}{\hbar^2} V(r) \right] f(0, k, r) \sim 0.$$

Multiplying by r^2, we obtain

$$\left[\gamma(\gamma - 1) \pm i2k\gamma r - \frac{2m^2}{\hbar^2} r^2 V(r) \right] f(0, k, r) \underset{r \to 0}{\sim} \gamma(\gamma - 1) f(0, k, r) \sim 0$$

since $\lim_{r \to 0} r^2 V(r) = 0$ is assumed. Therefore either $\gamma = 0$ or $f(0, k, r) \underset{r \to 0}{\sim} 0$. (Note that $\gamma = 1$ yields the latter alternative and hence need not be considered a separate case.) In either case, the constant b_0' in (14.7) must vanish, which is what we were to prove. Therefore we have, quite generally, $b_l = b_l' = 0$, for all nonnegative integers l, and

$$f(l, k, r) = a_l f^{(1)}(l, k, r). \tag{14.11}$$

14.3 ASYMPTOTIC BEHAVIOR AS $r \to \infty$

For large values of r, we must require the potential to vanish, since we know that particles cease to interact with one another as they get far enough apart. In fact, one frequently goes further and insists that

$$rV(r) \underset{r \to \infty}{\sim} 0, \tag{14.12}$$

which effectively rules out all "infinite-range" interactions; again this is for mathematical convenience. This last restriction simplifies the discussion of particle scattering from a potential, but it is violated by the very important Coulomb potential; indeed, the Coulomb-potential scattering problem is complicated somewhat because of this violation.

For many of our purposes the simpler assumption that $V(r)$ vanishes as r approaches infinity should be sufficient. However, there are some nonphysical potentials, such as the harmonic oscillator and infinite square-well potentials, that violate even this condition but that nevertheless are extremely useful.

Let us restrict ourselves for the moment to potentials that satisfy (14.12); for them we may write

$$\left(\frac{d^2}{dr^2} + \frac{2m}{\hbar^2}E\right) f(l, k, r) \underset{r \to \infty}{\sim} 0$$

as the asymptotic version of the radial equation. Under the assumption that this becomes an equation for sufficiently large r, we find that the asymptotic behavior of $f(l, k, r)$ is given by

$$f(l, k, r) \underset{r \to \infty}{\sim} c_{l+}e^{ikr} + c_{l-}e^{-ikr}, \tag{14.13}$$

where c_{l+} and c_{l-} are constants. Thus the asymptotic behavior of $f(l, k, r)$ is determined by $E = \hbar^2k^2/2m$ for large r, and is independent of l and, thanks to (14.12), of $V(r)$.

The two terms in (14.13) are linearly independent, and we may form two linearly independent solutions to Eq. (14.2) that behave asymptotically like these terms; thus we define $f^{(+)}(l, k, r)$ and $f^{(-)}(l, k, r)$ as the solutions to (14.2) such that

$$f^{(\pm)}(l, k, r) \underset{r \to \infty}{\sim} e^{\pm i[kr - (l+1)(\pi/2)]} \tag{14.14}$$

where the constant factors

$$e^{\mp i(l+1)\pi/2} = (\mp i)^{l+1}$$

are included for later convenience.

If $E > 0$, so that k is real, $f^{(-)}(l, k, r)$ and $f^{(+)}(l, k, r)$ represent, respectively, incoming and outgoing spherical waves, as one may easily verify by multiplying both sides of (14.14) by the time-dependent factor, $e^{-iEt/\hbar}$, of the energy eigenfunction.

If $E < 0$, we shall write

$$E = |E|e^{+i\pi},$$

so that

$$k = e^{i\pi/2}\sqrt{\frac{2m}{\hbar^2}|E|} = iK, \qquad K = \sqrt{\frac{2m}{\hbar^2}|E|}\,; \tag{14.15}$$

then

$$f^{(+)}(l, k, r) = f^{(+)}(l, iK, r)$$

decays exponentially with r in the asymptotic region, whereas

$$f^{(-)}(l, k, r) = f^{(-)}(l, iK, r)$$

grows exponentially in that region. Surely, then, $f^{(-)}(l, k, r)$ does not represent an acceptable wave function when $E < 0$, that is, when the physical system is bound. Thus $c_{l-} = 0$ when $E < 0$.

On the other hand, both $f^{(+)}(l, k, r)$ and $f^{(-)}(l, k, r)$ are quite acceptable as asymptotic solutions when $E > 0$, that is, when the physical system is unbound.

14.4 SPHERICAL FREE-PARTICLE WAVES

Let us consider solutions to Eq. (14.2) for the particular case when no potential is present, $V(r) = 0$. Then we may let $\rho \equiv kr$ and write (14.2) in the simplified form

$$\left[\frac{d^2}{d\rho^2} - \frac{l(l+1)}{\rho^2} + 1\right] f_l(\rho) = 0 \tag{14.16}$$

with

$$f_l(\rho) \equiv f(l, k, \rho/k).$$

The further substitution of $f_l(\rho) = \rho^{1/2}\mathcal{J}_l(\rho)$ leads to Bessel's equation of half-odd integer order (since l is an integer); thus

$$\frac{d}{d\rho}\rho^{1/2}\mathcal{J}_l(\rho) = \rho^{1/2}\mathcal{J}_l'(\rho) + \tfrac{1}{2}\rho^{-1/2}\mathcal{J}_l(\rho)$$

$$\frac{d^2}{d\rho^2}\rho^{1/2}\mathcal{J}_l(\rho) = \rho^{1/2}\mathcal{J}_l''(\rho) + \rho^{-1/2}\mathcal{J}_l'(\rho) - \tfrac{1}{4}\rho^{-3/2}\mathcal{J}_l(\rho),$$

and hence

$$\left\{\rho^2\frac{d^2}{d\rho^2} + \rho\frac{d}{d\rho} + [\rho^2 - (l + \tfrac{1}{2})^2]\right\}\mathcal{J}_l(\rho) = 0. \tag{14.17}$$

Solutions to Eq. (14.17) are the Bessel functions

$$J_{l+1/2}(\rho) = \sum_{n=0}^{\infty} \frac{(-1)^n}{n!\Gamma(l+n+\frac{3}{2})}\left(\frac{\rho}{2}\right)^{l+2n+1/2}$$

and the Neumann functions

$$N_{l+1/2}(\rho) = \sum_{n=0}^{\infty} \frac{(-1)^{l+1+n}}{n!\Gamma(-l+n+\frac{1}{2})}\left(\frac{\rho}{2}\right)^{-l+2n-1/2}$$
$$= (-1)^{l+1}J_{-l-1/2}(\rho).$$

These solutions are linearly independent, and any linear combination of them,

$$\mathscr{J}_l(\rho) = A_l'J_{l+1/2}(\rho) + B_l'N_{l+1/2}(\rho), \tag{14.18}$$

also is a solution to (14.17); because the Bessel and Neumann functions are linearly independent solutions to a second-order linear differential equation, any solution to (14.17) must, in fact, be expressible as some linear combination of the form (14.18).

It must be possible to write any solution to (14.16), then, in the form

$$f_l(\rho) = A_l'\rho^{1/2}J_{l+1/2}(\rho) + B_l'\rho^{1/2}N_{l+1/2}(\rho). \tag{14.19}$$

In fact, the functions $\rho^{-1/2}J_{l+1/2}(\rho)$ and $\rho^{-1/2}N_{l+1/2}(\rho)$ are so important that functions proportional to them have been given special names and symbols; thus

$$j_l(\rho) \equiv \sqrt{\frac{\pi}{2\rho}}J_{l+1/2}(\rho) \tag{14.20}$$

$$n_l(\rho) \equiv \sqrt{\frac{\pi}{2\rho}}N_{l+1/2}(\rho) \tag{14.21}$$

are called, respectively, the spherical Bessel functions and the spherical Neumann functions. If A_l' and B_l' are the constants used in Eq. (14.18), we may let $A_l = A_l'\sqrt{2/\pi}$ and $B_l = B_l'\sqrt{2/\pi}$ so that the solutions to (14.16) may be written

$$f_l(\rho) = A_l\rho j_l(\rho) + B_l\rho n_l(\rho). \tag{14.22}$$

To determine which linear combination is appropriate, we must investigate the behavior of $\rho j_l(\rho)$ and $\rho n_l(\rho)$ at small and large values of $r = \rho/k$.

A glance at the infinite series definitions for $J_{l+1/2}(\rho)$ and $N_{l+1/2}(\rho)$ shows that for $\rho = kr \to 0$ we have (see also Appendix C)

$$\rho j_l(\rho) = \sqrt{\pi/2}\,\rho^{1/2}J_{l+1/2}(\rho) \underset{\rho\to 0}{\sim} \frac{\sqrt{\pi}\rho^{l+1}}{2^{l+1}\Gamma(l+\frac{3}{2})} = \frac{\sqrt{\pi}\rho^{l+1}}{2(2l+1)!!\sqrt{\pi}}$$
$$= \frac{\rho^{l+1}}{(2l+1)!!} = \frac{k^{l+1}r^{l+1}}{(2l+1)!!} \tag{14.23}$$

and

$$\rho n_l(\rho) = \sqrt{\pi/2}\,\rho^{1/2} N_{l+1/2}(\rho) \underset{\rho\to 0}{\sim} \frac{\sqrt{\pi}(-1)^{l+1}2^{l+1/2}}{\Gamma(-l+\frac{1}{2})\rho^l\sqrt{2}} = \frac{-\sqrt{\pi}(-1)^l 2^l \Gamma(\frac{1}{2})}{\Gamma(\frac{1}{2})\rho^l \Gamma(-l+\frac{1}{2})}$$

$$= -\frac{(2l-1)!!}{\rho^l} = -\frac{(2l-1)!!}{k^l r^l}. \tag{14.24}$$

Obviously, $\rho j_l(\rho)$ and $\rho n_l(\rho)$ are, for Eq. (14.15) with $V(r) = 0$, just the solutions $f^{(1)}(l, k, r)$ and $f^{(2)}(l, k, r)$ discussed in Section 14.2; in fact, the asymptotic behavior of $\rho j_l(\rho)$ and $\rho n_l(\rho)$ is what dictated the normalization for $f^{(1,2)}(l, k, r)$ exhibited in Eqs. (14.8) and (14.9).

This asymptotic behavior of $\rho j_l(\rho)$ and $\rho n_l(\rho)$ indicates that the coefficient B_l in (14.22) must vanish if the radial wave function $f(l, k, r)$ is to be physically acceptable. Remembering that these are solutions for the free-particle equation, whose solutions we already know have the form

$$\psi_{\mathbf{k}}(\mathbf{r}) = (2\pi)^{-3/2} e^{i\mathbf{k}\cdot\mathbf{r}}$$

we may conclude from Eq. (14.3) that

$$\psi_{\mathbf{k}}(\mathbf{r}) = \sum_{l,m} C_{l,m} \frac{A_l \rho j_l(\rho)\beta}{r} Y_l^m(\theta, \varphi)$$

$$= \sum_{l,m} C_{l,m} A_l k j_l(kr) Y_l^m(\theta, \varphi)$$

$$= (2\pi)^{-3/2} e^{i\mathbf{k}\cdot\mathbf{r}}.$$

In fact, it is possible to show (see Appendix C2) that

$$C_{l,m} A_l = (2\pi)^{-3/2} \cdot 4\pi i^l Y_l^{m*}(\theta_k, \varphi_k)$$

where θ_k, φ_k define the direction of the wave vector $\mathbf{k}$. Thus

$$e^{i\mathbf{k}\cdot\mathbf{r}} = 4\pi \sum_{l,m} i^l j_l(kr) Y_l^{m*}(\theta_k, \varphi_k) Y_l^m(\theta, \varphi). \tag{14.25}$$

If k lies along the z-axis, we let $\varphi_k = 0$ along with $\theta_k = 0$; then, since $Y_l^{m*}(0, 0) = \sqrt{(2l+1)/4\pi}\,\delta_{m,0}$ and $Y_l^0(\theta, 0) = \sqrt{(2l+1)/4\pi}\,P_l(\cos\theta)$, we have

$$e^{i\mathbf{k}\cdot\mathbf{r}} = \sum_l (2l+1) i^l j_l(kr) P_l(\cos\theta), \tag{14.26}$$

where, of course, $\cos\theta = \mathbf{k}\cdot\mathbf{r}/kr$. The expansions (14.25) and (14.26) of a plane wave in terms of spherical harmonics or Legendre polynomials are known as the Rayleigh expansions for a plane wave.

So far we have considered only the small-radius behavior of the $\rho j_l(\rho)$ and $\rho n_l(\rho)$; what about their asymptotic behavior as $r \to \infty$? This is a somewhat more complicated question than that of the behavior as $r \to 0$. (See, however,

Appendix C.) It can be shown that

$$\rho j_l(\rho) \underset{\rho\to\infty}{\sim} \cos\left[\rho - \tfrac{1}{2}(l+1)\pi\right] \tag{14.27}$$

and

$$\rho n_l(\rho) \underset{\rho\to\infty}{\sim} \sin\left[\rho - \tfrac{1}{2}(l+1)\pi\right]. \tag{14.28}$$

These asymptotic forms suggest that another useful pair of functions might be

$$h_l^{(\pm)}(\rho) = j_l(\rho) \pm i n_l(\rho), \tag{14.29}$$

whose asymptotic forms are given by

$$\rho h_l^{(\pm)}(\rho) = e^{\pm i[\rho - (l+1)(\pi/2)]} = (\mp i)^{l+1} e^{\pm i\rho}. \tag{14.30}$$

Comparison of (14.30) with (14.14) shows that for $V(r) = 0$,

$$\begin{aligned} f^{(\pm)}(l, k, r) &= \rho h_l^{(\pm)}(\rho) = kr h_l^{(\pm)}(\rho) \\ &= kr[j_l(kr) \pm i n_l(kr)]. \end{aligned} \tag{14.31}$$

The $h_l^{(\pm)}$ are called *spherical Hankel functions.*

From this we see that in the case that $V(r) = 0$, we must have

$$\begin{aligned} f(l, k, r) &= \alpha_l f^{(1)}(l, k, r) = \alpha_l kr j_l(kr) \\ &= \tfrac{1}{2}\alpha_l kr[h_l^{(+)}(kr) + h_l^{(-)}(kr)] \underset{r\to\infty}{\sim} \alpha_l \cos\left[kr - \tfrac{1}{2}(l+1)\pi\right], \end{aligned} \tag{14.32}$$

where α_l is a constant to be determined by the overall normalization of the radial wave function.

While we are discussing the different varieties of spherical Bessel functions, let us consider their behavior for finite values of ρ. The solutions in Appendix C indicate that the first few orders of spherical Bessel, Neumann and Hankel functions have the forms

$$\begin{aligned} j_0(\rho) &= \frac{\sin\rho}{\rho} & n_0(\rho) &= -\frac{\cos\rho}{\rho} \\ j_1(\rho) &= \frac{\sin\rho}{\rho^2} - \frac{\cos\rho}{\rho} & n_1(\rho) &= -\frac{\cos\rho}{\rho^2} - \frac{\sin\rho}{\rho} \\ j_2(\rho) &= \left(\frac{3}{\rho^2} - \frac{1}{\rho}\right)\sin\rho - \frac{3}{\rho^2}\cos\rho & n_2(\rho) &= -\left(\frac{3}{\rho^3} - \frac{1}{\rho}\right)\cos\rho - \frac{3}{\rho^2}\sin\rho \end{aligned} \tag{14.33}$$

and

$$
\begin{aligned}
h_0^{(\pm)}(\rho) &= \pm\frac{1}{i\rho}e^{\pm i\rho} \\
h_1^{\pm}(\rho) &= -\left(\frac{1}{\rho} \pm \frac{i}{\rho^2}\right)e^{\pm i\rho} \\
h_2^{\pm}(\rho) &= \pm i\left(\frac{1}{\rho} \pm \frac{3i}{\rho^2} - \frac{3}{\rho^3}\right)e^{\pm i\rho}.
\end{aligned}
\tag{14.34}
$$

A number of useful relations among these functions and their derivatives, together with some integrals involving them, are given in Appendix C.

14.5 SQUARE-WELL POTENTIAL

The simplest kind of potential function is the so-called square-well potential of the form

$$
V(r) = \begin{cases} -V_0, & 0 \le r < a \\ 0, & r > a. \end{cases} \tag{14.35}
$$

With this potential, we actually solve two separate differential equations, namely

$$
\left[\frac{d^2}{dr^2} - \frac{l(l+1)}{r^2} + \frac{2m}{\hbar^2}V_0 + k^2\right]f_<(l, k, r) = 0, \qquad 0 \le r < a, \tag{14.36}
$$

and

$$
\left[\frac{d^2}{dr^2} - \frac{l(l+1)}{r^2} + k^2\right]f_>(l, k, r) = 0, \qquad r > a, \tag{14.37}
$$

and demand that

$$
f_<(l, k, a) = f_>(l, k, a) \tag{14.38}
$$

$$
\left.\frac{d[f_<(l, k, r)]/dr}{f_<(l, k, r)}\right|_{r=a} = \left.\frac{d[f_>(l, k, r)]/dr}{f_>(l, k, r)}\right|_{r=a} \tag{14.39}
$$

in accordance with conditions (C′) and (D′) of Section 9.4.

We have already solved these equations in the previous section. If, for $0 \le r < a$, we let

$$
\rho \equiv \kappa r, \qquad \kappa \equiv \sqrt{\frac{2m}{\hbar^2}(E + V_0)} = \sqrt{k^2 + \frac{2m}{\hbar^2}V_0} \tag{14.40}
$$

then Eq. (14.36) becomes identical with Eq. (14.16) provided we define

$$
f_l(\rho) \equiv f_<\left(l, k, \frac{\rho}{\kappa}\right), \qquad 0 \le \frac{\rho}{\kappa} < a. \tag{14.41}
$$

According to Eq. (14.22), then, the only acceptable solution to (14.36) is

$$f_<(l, k, r) = A_l \kappa r j_l(\kappa r), \tag{14.42}$$

since B_l must vanish to yield a physically acceptable solution near $r = 0$.

Since (14.37) is identical with (14.2) for $r > a$, and since the condition $r > a$ isolates the solution from the origin of the radial coordinate and hence from the condition that the radial wave function vanish as $r \to 0$, the solutions $f_>(l, k, r)$ have the form

$$f_>(l, k, r) = c_{l+} krh_l^{(+)}(kr) + c_{l-} krh_l^{(-)}(kr). \tag{14.43}$$

Here the relative values of the constants c_{l+} and c_{l-} are to be determined by the boundary conditions for very large values of r (sometimes called exterior boundary conditions).

If $E > 0$ so that the system represented by $f(l, k, r)$ is unbound, both c_{l+} and c_{l-} may be nonvanishing, and their values must be determined by the conditions under which an experiment on the system is performed. We shall consider this problem in some detail in a later chapter, when we take up the whole question of scattering and unbound states. For the moment, we shall confine our attention to bound states.

For a bound state, i.e., when $E < 0$, we use Eq. (14.15). The discussion following Eq. (14.15) indicates that the allowed solutions for $E < 0$ are

$$f_<(l, iK, r) = c_{l+} iKrh_l^{(+)}(iKr), \tag{14.44}$$

$$K = \sqrt{\frac{2m}{\hbar^2}|E|}\,.$$

Thus, for bound states, the task of finding solutions is completed when conditions (14.38) and (14.39) are applied to (14.42) and (14.44) and the radial wave functions are properly normalized. Properly normalizing consists of adjusting constants to conform to the condition

$$\int_0^\infty r^2|R_{El}(r)|^2\,dr = \int_0^\infty |f(l, k, r)|^2\,dr = 1; \tag{14.45}$$

for the bound states of the square-well potential, this means

$$|A_l|^2\kappa^2\int_0^a r^2[j_l(\kappa r)]^2\,dr + |c_{l+}|^2K^2\int_a^\infty r^2|h_l^{(+)}(iKr)|^2\,dr = 1. \tag{14.46}$$

Obviously this determines $|A_l|$ in terms of $|c_{l+}|$ and K, or $|c_{l+}|$ in terms of $|A_l|$ and K. Equations (14.38) and (14.39) then determine the remaining two constants.

In fact, Eqs. (14.42) and (14.44) inserted into (14.39) yield

$$\frac{d[rj_l(\kappa r)]/dr|_{r=a}}{j_l(\kappa a)} = \frac{d[rh_l^{(+)}(iKr)]/dr|_{r=a}}{h_l^{(+)}(iKa)} \tag{14.47}$$

in which neither A_l nore c_{l+} appears. Since κ and K are related by

$$\kappa^2 + K^2 = \frac{2m}{\hbar^2} V_0, \tag{14.48}$$

according to Eqs. (14.15) and (14.40), it is clear that K (and hence κ) is determined by (14.47). Once K and κ are thus determined, the relative magnitudes of $|A_l|$ and $|c_{l+}|$ will be established by condition (14.38), and (14.45) will determine the absolute magnitudes of the latter.

The solution of Eq. (14.47) for K is quite difficult in general and must usually be carried out numerically. The $l = 0$ case is the simplest, and even it must be completed by numerical methods; on the other hand, this case does yield to a relatively simple graphical solution, and the process of obtaining the solution is an instructive exercise.

We begin by noting that for $l = 0$, Eq. (14.2) is the same differential equation that one would have to solve for a one-dimensional system interacting through a square-well potential. The difference between the problem that we shall solve and the one-dimensional problem is that for the latter there is no requirement that the solution to (14.2) vanish at $r = 0$. Elaboration of the one-dimensional problem is left as an exercise for the reader. (See Problem 1.)

According to Eqs. (14.33) and (14.34),

$$j_0(\kappa r) = \frac{\sin \kappa r}{\kappa r}, \qquad h_0^{(+)}(iKr) = -e^{-Kr},$$

and hence, for $l = 0$, (14.47) becomes

$$\frac{r\, d(\sin \kappa r)/dr|_{r=a}}{\sin \kappa a} = \frac{r\, d(e^{-Kr})/dr|_{r=a}}{e^{-Ka}}$$

or

$$\kappa a \cot \kappa a = -Ka. \tag{14.49}$$

Letting $\xi = \kappa a$ and $\eta = Ka$ and invoking Eq. (14.48), we have the simultaneous equations

$$\xi \cot \xi = -\eta, \tag{14.50a}$$

$$\xi^2 + \eta^2 = (2m/\hbar^2) V_0 a^2, \tag{14.50b}$$

whose solutions may be found graphically.

The locus of solutions to (14.50b) such that $\xi \geq 0$, $\eta \geq 0$ is, for each value of $V_0 a^2$, the quarter circle centered at the origin of the $\xi\eta$-plane and located in the first quadrant. The locus of solutions to (14.50a) is a sequence of curves, with positive slopes, that approach asymptotes parallel to the η-axis. The simultaneous solutions to Eqs. (14.50) lie at the intersections of these loci, as shown in Fig. 14.1. It is clear that for finite values of $V_0 a^2$ there will always be a finite number of intersections, and therefore the number of bound states for a finite square-well potential is always finite; the square well has a discrete energy spectrum for its bound states, even for $l > 0$. In fact, it can be shown that quite generally the bound states form a discrete energy spectrum regardless of the shape of the potential.

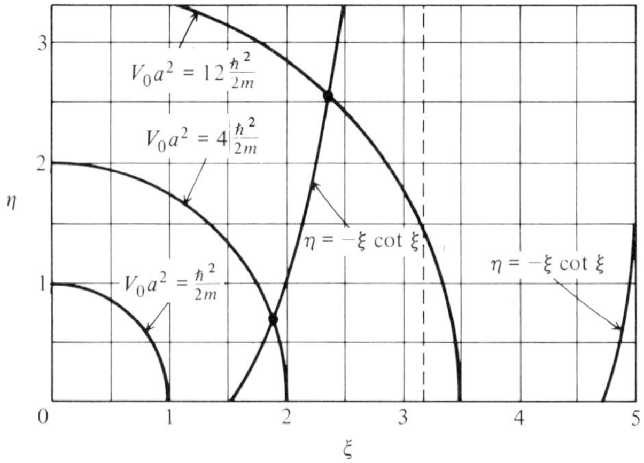

Fig. 14.1. Graphical evaluation of $\xi = \kappa a$ and $\eta = Ka$, and hence of the bound-state energy levels for the square well when $l = 0$. Three values of $V_0 a^2$ are shown. (From *Quantum Mechanics*, 3rd ed., by L. I. SCHIFF. Copyright 1968 by McGraw-Hill, Inc. Used by permission of McGraw-Hill Book Company.)

The principal application of the square well is in the study of nuclei. Rutherford's scattering experiments demonstrated quite clearly that nuclear particles do not interact with one another (except through their repulsive electromagnetic forces) unless they are within about 10^{-12} cm $= 10$ Fermi of one another. Furthermore, it is clear that the specifically nuclear forces must be very strong, since nuclear systems are bound in spite of the presence of strong repulsive electrostatic forces. The potential energies of electrostatic repulsion, for example, are of the order of

$$\frac{e^2}{4\pi\epsilon_0 r_0} \simeq \frac{1.44 \text{ MeV F}}{1 \text{ F}} \simeq 1.5 \text{ MeV}$$

even for light nuclei, where $r_0 \simeq 1$ F is approximately the nuclear radius.

One may observe that according to the above discussion the first bound state for the spherical square-well potential occurs just as cot ξ passes from positive to negative values for the first time as ξ increases through positive values, i.e. when $\xi = \pi/2$, $\eta = 0$, or where

$$(2m/\hbar^2)V_0 a^2 = \pi^2/4$$

or

$$V_0 = \frac{\pi^2(\hbar c)^2}{8mc^2a^2}.$$

Since, for a nucleon (proton or neutron), $m_N c^2 \simeq 939$ MeV $\simeq 10^3$ MeV $= 1$ GeV, and since $\hbar c = 197.32$ MeV F $\simeq 200$ MeV F, we have, upon choosing $a \simeq 10^{-13}$ cm $= 1$ F,

$$V_0 \simeq \frac{\pi^2}{2} \frac{4 \times 10^4 \text{ MeV}^2 \text{ F}^2}{4 \times 10^3 \text{ MeV F}^2} \simeq 50 \text{ MeV}$$

as the minimum potential depth required to bind a nucleon to a nucleus. Thus a very crude model for the nuclear potential can be constructed with a spherical square well having a depth greater than about 50 MeV and a radius of about 1 F.

A somewhat better estimate for the radius of the potential may be obtained under the assumption that each nucleon in a nucleus with mass number A contributes equally to the volume of the nucleus. Then we may expect the nuclear radius to be

$$r = \sqrt[3]{\frac{3}{4\pi}(\text{vol.})} \simeq \sqrt[3]{\frac{3}{4\pi}(\text{vol./nucleon})A} = r_0 A^{1/3}.$$

The coefficient $r_0 = \sqrt[3]{(3/4\pi)(\text{vol./nucleon})}$ must be determined experimentally, primarily from scattering data. Experimental values for r_0 vary from about 1.2 F to 1.5 F, depending upon the kind of experiment from which r_0 is determined. A commonly used value of r_0 is

$$r_0 \simeq 1.3 \times 10^{-13} \text{ cm} = 1.3 \text{ F}.$$

The radius of the potential for the interaction between a single nucleon and a nucleus with mass number A, then, is the distance between the "centers" of the nucleon and nucleus, or about

$$a \simeq (1.3A^{1/3} + 1) \text{ F}$$

since it turns out that a free nucleon has a radius of about 1 F.

This yields the estimate

$$V_0 \simeq 30(A^{-1/3} + 0.77)^{-1} \text{ MeV}$$

for the minimum depth of a square well in which a nucleon can be bound to a nucleus.

If the nucleon is a proton, we should include the repulsive electrostatic field in our potential also. However, for $r < a$, that is, on the inside of the nuclear potential, our use of the square well already is a crude approximation, and to add the Coulomb potential for such radii would be to complicate the mathematical problem without any real gain in physical sophistication. Hence the Coulomb potential is neglected for $r < a$. Under many circumstances it is reasonable to neglect the Coulomb potential for $r > a$ also, since often the magnitude of the Coulomb potential at the edge of the nucleus,

$$V_{\text{Coulomb}}(a) = \frac{Ze^2}{4\pi\epsilon_0 a},$$

is negligible compared with the binding energy of a nucleon. On the other hand, for problems involving the scattering of a nucleon from a nucleus rather than the bound state problem to which we have confined ourselves here, it may be quite important to include the Coulomb potential for $r > a$. The Coulomb potential may be important also if the binding energy is very small, or if the system is unstable but is maintained in a *quasibound* state for a finite time by the barrier provided by the external Coulomb potential (see Fig. 14.2); when the energy of the system lies between 0 and $Ze^2/4\pi\epsilon_0 a$, such a quasibound state may exist.

Quasibound (or quasistationary) states are considered in a later chapter on unbound systems. The investigation of such states for spherically symmetric

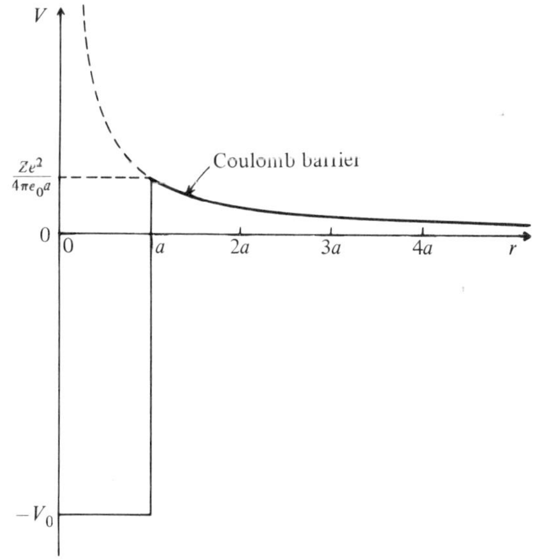

Fig. 14.2. Representation of a combined square-well and Coulomb interaction, with the interior portion of the Coulomb potential ignored. The barrier may be responsible for quasibound states.

problems is complicated by the fact that there must be a continual outward radial flow of probability current, and hence the time-dependence of the state cannot have the usual stationary time-dependence expressed through the factor $e^{-iEt/\hbar}$. Nevertheless, these states are manifestations of the purely quantum-mechanical phenomenon of *tunneling* or *barrier penetration* which has no classical analog. An introduction to this problem is obtained best through an analysis of the one-dimensional potential barrier. Although in this book we have relegated this analysis to an appendix (see Appendix D), it is important that the student understand how the phenomenon occurs, and we recommend that he consult Appendix D.

14.6 THREE-DIMENSIONAL ISOTROPIC HARMONIC OSCILLATOR

In this and the succeeding section, we consider potentials that fail to satisfy the exterior boundary conditions (14.12) but which nevertheless are important potentials in modern physics. The potential in Section 14.7 "just barely" violates (14.12), and we shall find that for bound states this presents no great problem; in fact, the Coulomb potential studied in that section corresponds to a rather realistic physical situation. However, the harmonic oscillator, which we consider in this section, violently abuses condition (14.12), and at best can represent only a very rough approximation to a physical situation. Indeed, the only possible states for a true harmonic oscillator are bound states, whereas physical systems generally possess unbound states as well.

The potential function for an isotropic (i.e. spherically symmetric) harmonic oscillator is

$$V(r) = \frac{\mu}{2}\omega_0^2 r^2 \tag{14.51}$$

where μ is the mass of the oscillating particle (or reduced mass, if we are considering a two-particle system) and ω_0 is the characteristic frequency of oscillation. Substitution of (14.51) into (14.2) yields the differential equation

$$\left[\frac{d^2}{dr^2} - \frac{l(l+1)}{r^2} - \left(\frac{\mu\omega_0}{\hbar}\right)^2 r^2 + k^2\right] f(l, k, r) = 0 \tag{14.52}$$

if we replace m in (14.2) with μ. Since condition (14.12) fails to apply here, we must look for new asymptotic solutions for $r \to \infty$.

For large r, (14.52) becomes, approximately,

$$\left[\frac{d^2}{dr^2} - \left(\frac{\mu\omega_0}{\hbar}\right)^2 r^2\right] f(l, k, r) \sim 0, \tag{14.53}$$

from which it is clear that the asymptotic behavior of $f(l, k, r)$ is independent of both k and l. The solutions to (14.53) are not obvious immediately, but evidently we may expect the differentiation operation to be equivalent to multiplication by $\pm(\mu\omega_0/\hbar)r$, at least for very large r. This suggests trying

$$f(l, k, r) \underset{r\to\infty}{\sim} \alpha e^{\pm\mu\omega_0 r^2/2\hbar},$$

since differentiation of the right-hand side twice gives

$$\alpha\left[\pm\frac{\mu\omega_0}{\hbar} + \left(\frac{\mu\omega_0}{\hbar}r\right)^2\right]e^{\pm\mu\omega_0 r^2/2\hbar} = \frac{d^2}{dr^2}\alpha e^{\pm\mu\omega_0 r^2/2\hbar};$$

thus, as $r \to \infty$,

$$\left[\frac{d^2}{dr^2} - \left(\frac{\mu\omega_0}{\hbar}\right)^2 r^2\right]\alpha e^{\pm\mu\omega_0 r^2/2\hbar} \sim 0.$$

Now we must demand $\lim_{r\to\infty} f(l, k, r) = 0$, which eliminates the positive sign in the exponent. Therefore we let

$$f(l, k, r) = \phi(l, k, r)e^{-\mu\omega_0 r^2/2\hbar} \tag{14.54}$$

and require that

$$\lim_{r\to\infty} \phi(l, k, r) = \text{constant}.$$

Now we must find $\phi(l, k, r)$, and we do so by substituting (14.54) into (14.52) and solving the resulting differential equation for $\phi(l, k, r)$. The desired equation is easily found to be

$$\left[\frac{d^2}{dr^2} - \frac{2\mu\omega_0}{\hbar}r\frac{d}{dr} - \frac{l(l+1)}{r^2} + \left(k^2 - \frac{\mu\omega_0}{\hbar}\right)\right]\phi(l, k, r) = 0. \tag{14.55}$$

At this point it is convenient to adopt the following change of variables:

$$\xi = \sqrt{\frac{\mu\omega_0}{\hbar}}\,r,$$

and to let

$$\phi_{l,k}(\xi) \equiv \phi(l, k, r). \tag{14.56}$$

Then the differential equation becomes

$$\left[\frac{d^2}{d\xi^2} - 2\xi\frac{d}{d\xi} - \frac{l(l+1)}{\xi^2} + \left(\frac{\hbar k^2}{\mu\omega_0} - 1\right)\right]\phi_{l,k}(\xi) = 0. \tag{14.57}$$

The appropriate solution to this equation may be obtained through the use of the infinite-series method of Frobenius. We assume that for some set of real coefficients $\{a_n\}$, $n = 0, 1, 2, \ldots$, we may write

$$\phi_{l,k}(\xi) = \sum_{n=0}^{\infty} a_n \xi^{p+n}, \tag{14.58}$$

where p is a constant to be determined. Insertion of (14.58) into (14.57) and application of the derivative operators term-by-term to the right-hand side of (14.58) yields

$$\sum_{n=0}^{\infty} a_n\left\{[(p+n)(p+n-1) - l(l+1)]\xi^{p+n-2} - \left[2(p+n)+1-\frac{\hbar k^2}{\mu\omega_0}\right]\xi^{p+n}\right\}$$
$$= a_0[p(p-1) - l(l+1)]\xi^{p-2} + a_1[(p+1)p - l(l+1)]\xi^{p-1}$$
$$+ \sum_{n=0}^{\infty}\left\{a_{n+2}[(p+n+2)(p+n+1) - l(l+1)] - a_n\left[2(p+n+\tfrac{1}{2}) - \frac{\hbar k^2}{\mu\omega_0}\right]\right\}\xi^{p+n}$$
$$= 0. \tag{14.59}$$

Since we must insist that $a_0 \neq 0$, and since the coefficient of each power of ξ must vanish, we must have

$$p(p-1) - l(l+1) = 0,$$

which implies that either

$$p = l+1 \qquad \text{or} \qquad p = -l.$$

For $p = -l$, however, we have

$$f(l,k,r) = \phi_{l,k}(\xi)e^{-\xi^2/2} = \phi_{l,k}\left(\sqrt{\frac{\mu\omega_0}{\hbar}}\,r\right)e^{-(\mu\omega_0/2\hbar)r^2}$$
$$\underset{r\to 0}{\sim} a_0\left(\sqrt{\frac{\mu\omega_0}{\hbar}}\,r\right)^{-l} = \text{const.}\, r^{-l};$$

hence the solution corresponding to the choice $p = -l$ is proportional to the irregular solution $f^{(2)}(l,k,r)$. This solution is not acceptable because of its behavior as $r \to 0$. Therefore we must have

$$p = l+1. \tag{14.60}$$

Next, the coefficient of $\xi^{p-1} = \xi^l$ must vanish, which means that at least one of the equations

$$p = l, \qquad p = -(l+1), \qquad a_1 = 0$$

must hold. The first two of these are inconsistent with $p = l+1$, and hence we must have

$$a_1 = 0. \tag{14.61}$$

Finally, the coefficients of ξ^{p+n} must vanish for each nonnegative integer n, and therefore the a_n's must satisfy

$$a_{n+2} = \frac{2(l+n+\frac{3}{2}) - \hbar k^2/(\mu\omega_0)}{(l+n+3)(l+n+2) - l(l+1)} a_n = \frac{2(l+n+\frac{3}{2}) - \hbar k^2/(\mu\omega_0)}{(n+2l+3)(n+2)} a_n. \quad (14.62)$$

This, together with (14.61), implies that

$$a_{2m+1} = 0, \qquad m = 0, 2, \ldots, \quad (14.63)$$

whereas repeated use of (14.62) when $n = 2m$ is an even integer implies that

$$a_{2m} = \frac{\left[2(l+2m-\frac{1}{2}) - \frac{\hbar k^2}{\mu\omega_0}\right]\left[2(l+2m-\frac{5}{2}) - \frac{\hbar k^2}{\mu\omega_0}\right]\cdots\left[2(l+\frac{3}{2}) - \frac{\hbar k^2}{\mu\omega_0}\right]}{(2l+1+2m)!!(2m)!!/(2l+1)!!} a_0$$

$$= \frac{\left[(l+2m-\frac{1}{2}) - \frac{\hbar k^2}{2\mu\omega_0}\right]\left[(l+2m-\frac{5}{2}) - \frac{\hbar k^2}{2\mu\omega_0}\right]\cdots\left[(l+\frac{3}{2}) - \frac{\hbar k^2}{\mu\omega_0}\right]}{2^m\Gamma(l+m+\frac{3}{2})\Gamma(m+1)/\Gamma(l+\frac{3}{2})} a_0 \quad (14.64)$$

where the gamma function $\Gamma(x)$ has the property $\Gamma(x+1) = x\Gamma(x)$. Therefore, the desired solution has the form

$$\phi_{l,k}(\xi) = a_0 \sum_{m=0}^{\infty} \Gamma(l+\tfrac{3}{2})\xi^{2m+l+1} \cdot \frac{\left[(l+2m-\frac{1}{2}) - \frac{\hbar k^2}{2\mu\omega_0}\right]\left[(l+2m-\frac{5}{2}) - \frac{\hbar k^2}{2\mu\omega_0}\right]\cdots\left[(l+\frac{3}{2}) - \frac{\hbar k^2}{2\mu\omega_0}\right]}{2^m\Gamma(l+m+\frac{3}{2})\Gamma(\frac{3}{2})}. \quad (14.65)$$

Now we must consider the properties of $f(l, k, r) = \phi_{l,k}(\xi)e^{-\xi^2/2}$ as $\xi \to \infty$, i.e. as $r \to \infty$. To begin with, we must be sure that the series in (14.65) converges for all positive ξ. Let us try the ratio test for this infinite series, which involves the limit, as $m \to \infty$, of

$$\left|\frac{a_{2(m+1)}}{a_{2m}}\xi^2\right| = \left|\frac{l+2m+\frac{3}{2} - \frac{\hbar k^2}{2\mu\omega_0}}{2(l+m+\frac{3}{2})(m+1)}\right|\xi^2 \underset{m\to\infty}{\sim} \frac{\xi^2}{m}. \quad (14.66)$$

According to the ratio test criterion, the series converges absolutely if the limit as $m \to \infty$ is any nonnegative number less than unity. This criterion is certainly met here, so $\phi_{l,k}(\xi)$ converges absolutely for all ξ.

For large values of ξ, it is the terms in (14.65) corresponding to rather large values of m that tend to dominate $\phi_{l,k}(\xi)$, and as ξ gets larger, the larger are the values of m for the terms that tend to dominate the series. Hence, $\phi_{l,k}(\xi)$ will behave asymptotically like the power series whose coefficients satisfy the conditions

$$a_{2m+2} = ma_{2m}, \qquad a_{2m+1} = 0, \qquad m = 0, 1, 2, \ldots$$

implied by the right-hand member of (14.66). That series is simply

$$a_0 \sum_{m=0}^{\infty} \frac{\xi^{2m+l+1}}{m!} = a_0 e^{\xi^2} \xi^{l+1}.$$

Therefore,

$$\phi_{l,k}(\xi) \underset{\xi \to \infty}{\sim} (\text{const.}) e^{\xi^2} \xi^{l+1},$$

and

$$f(l, k, r) \underset{\xi \to \infty}{\sim} (\text{const.}) e^{\xi^2} \xi^{l+1} e^{-\xi^2/2} = (\text{const.}) \xi^{l+1} e^{\xi^2/2}.$$

Hence $f(l, k, r)$ seems not only to fail to vanish for large $r = \sqrt{\hbar/\mu\omega_0}\,\xi$, but it rapidly approaches infinity. This is not physically acceptable, nor does it correspond to what one would expect for large r, because the "walls" of the harmonic oscillator potential well get increasingly high and increasingly steep as r gets larger; this should mean a decreasingly small wave function with a decreasing slope for larger and larger r values.

The fate of our series solution still is not sealed, however, for this bad behavior for larger r and ξ values will not be realized if all but a finite number of the a_n vanish; and the latter will, in fact, occur if and only if the first factor in the numerator of (14.64) vanishes for some integer $m = \lambda + 1$:

$$l + 2\lambda + \tfrac{3}{2} - \frac{\hbar k_\lambda^2}{\mu\omega_0} = 0, \tag{14.67}$$

or

$$k_\lambda = \sqrt{2\mu\omega_0(l + 2\lambda + \tfrac{3}{2})/\hbar}.$$

Remembering that $E = \hbar^2 k^2/2\mu$, we can satisfy this condition only if the energy E is restricted to the discrete set of values defined by

$$E_{\lambda,l} = (2\lambda + l + \tfrac{3}{2})\hbar\omega_0, \qquad \lambda = 0, 1, 2, \ldots \tag{14.68a}$$

or

$$E_v = (v + \tfrac{3}{2})\hbar\omega_0, \qquad v = 0, 1, 2, \ldots \tag{14.68b}$$

where in (14.68b) the orbital-angular-momentum quantum number is restricted to values of l that are less than or equal to v and that differ from v by a multiple of two.

Expressing $\hbar k^2/\mu\omega_0 = E_{\lambda l}/\hbar\omega_0$ in terms of λ, we obtain from (14.64) and (14.67) the expressions

$$a_n = \begin{cases} \dfrac{[2(m-1)-2\lambda][2(m-2)-2\lambda]\cdots[-2\lambda]}{2^m\Gamma(l+m+\frac{3}{2})\Gamma(m+1)}\Gamma(l+\tfrac{3}{2})a_0 \\ \quad = \dfrac{(-1)^m\lambda!\Gamma(l+\frac{3}{2})}{(\lambda-m)!\Gamma(l+m+\frac{3}{2})m!}a_0 \qquad \text{for} \quad n = 2m, \quad m = 0, 1, 2, \ldots, \\ 0 \qquad \text{for} \quad n = 2m+1, \quad m = 0, 1, 2, \ldots, \\ \qquad \text{and for} \quad n = 2m, \quad m = \lambda, \lambda+1, \ldots, \end{cases}$$

for the coefficients in the power series expansion of $\phi_{l,k}(\xi)/\xi^{l+1}$. Thus

$$\begin{aligned} \phi_{l,k}(\xi) &= a_0 \sum_{m=0}^{\lambda} \frac{(-1)^m\lambda!\Gamma(l+\frac{3}{2})}{(\lambda-m)!\Gamma(l+\frac{3}{2}+m)m!}\xi^{l+1+2m} \\ &= b_0 \sum_{m=0}^{\lambda} \frac{(-1)^m\xi^{l+1+2m}}{(\lambda-m)!\Gamma(l+\frac{3}{2}+m)m!}, \end{aligned} \tag{14.69}$$

where $b_0 \equiv \lambda!\Gamma(l+\frac{3}{2})a_0$.

This polynomial is proportional to a well-known polynomial that appears in mathematical physics and which has the standard form

$$L_p^q(\xi) = \sum_{k=0}^{p-q}(-1)^{k+1}\frac{\Gamma^2(p+1)\xi^k}{(p-q-k)!\Gamma(q+k+1)k!}. \tag{14.70}$$

This is called the associated Laguerre polynomial. Clearly, if we let $q = l + \frac{1}{2}$ and $p = l + \lambda + \frac{1}{2}$, we obtain

$$\phi_{l,k}(\xi) = A\xi^{l+1}L_{\lambda+l+1/2}^{l+1/2}(\xi^2),$$

where A is a constant.

In terms of r, then, our radial wave function $f(l, k, r)$ becomes

$$f(l, \sqrt{2\mu\omega_0(l+2\lambda+\tfrac{3}{2})/\hbar}, r) = B_{\lambda,l}r^{l+1}L_{\lambda+l+1/2}^{l+1/2}\left(\frac{\mu\omega_0}{\hbar}r^2\right)e^{-\mu\omega_0 r^2/2\hbar}, \tag{14.71}$$

where $B_{\lambda,l}$ is a normalization constant that is yet to be determined. The usual orthonormality condition

$$\int_0^\infty f^*(l', k_{\lambda'}, r)f(l, k_\lambda, r)\,dr = \delta_{\lambda',\lambda}\,\delta_{l',l}$$

will hold if

$$B_{\lambda,l} = \sqrt{\frac{2(\mu\omega_0/\hbar)^{l+3/2}\lambda!}{\Gamma^3(\lambda+l+\frac{3}{2})}}. \tag{14.72}$$

It should be pointed out that all states having the same value of $2\lambda + l$ are energy degenerate, according to Eq. (14.68a). Thus, for example, the states having $(\lambda, l) = (1, 1), (0, 3)$ both have the same energy, namely

$$E_{1,1} = E_{0,3} = \tfrac{9}{2}\hbar\omega_0,$$

It is clear that all values of l within a set of degenerate states will be even, or all of them will be odd, since changes of λ by an integer necessarily require a change in l by an even integer if the energy is to remain unchanged. This implies that all states in a degenerate set of states must have the same parity.

Another interesting feature of these solutions is that they are rather insensitive to the behavior of the potential for large values of r, provided the energy corresponds to one of the "low-lying" levels in the spectrum. This is quite reasonable, since for such low-energy states the likelihood of finding a particle much farther from the origin than the turning point (the value of r at which the kinetic energy vanishes),

$$\tilde{r} = \frac{1}{\mu\omega_0}\sqrt{2\mu E_{\lambda,l}} = 2\sqrt{\frac{\hbar}{\mu\omega_0}\left(\lambda + \frac{l}{2} + \frac{3}{4}\right)},$$

should be expected to be quite small. This explains, in part, the utility of the harmonic-oscillator potential in the study of nuclear structure; for if a single nucleon is considered to be moving in a potential produced by the interaction of the nucleons with all other nucleons in the nucleus, and if that nucleon is strongly bound so that it is in a low-lying energy level, the fact that the actual nuclear potential is not exactly a harmonic oscillator potential is irrelevant as long as it has the form of an oscillator potential for $r \lesssim \tilde{r} = \sqrt{2E/(\mu\omega_0^2)}$, where E is the energy of the single nucleon.

To see better how the harmonic oscillator is applied to nuclear problems, consider the following potential (Fig. 14.3):

$$V(r) = \frac{\mu\omega_0^2}{2}(r^2 - r_0^2). \tag{14.73a}$$

This potential is attractive for $r < r_0$ and repulsive for $r > r_0$; therefore its behavior for $r \gg r_0$ is distinctly nonphysical—we expect all nuclear potentials to vanish when the nucleon to which the potential is supposed to apply gets sufficiently far away from the center of the nucleus. More realistic potentials would seem to be

$$V(r) = \begin{cases} \dfrac{\mu\omega_0^2}{2}(r^2 - r_0^2), & r < r_0 \\[2ex] \dfrac{(Z-1)e^2}{4\pi\epsilon_0 r}, & r \geq r_0, \quad \text{nucleon} = \text{proton} \\[2ex] 0, & r \geq r_0, \quad \text{nucleon} = \text{neutron}. \end{cases} \tag{14.73b}$$

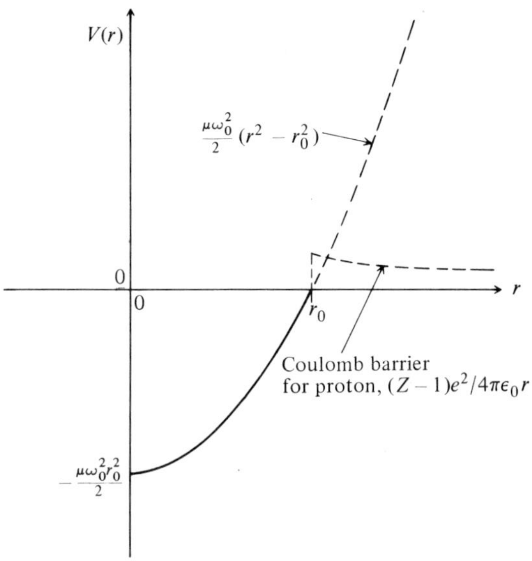

Fig. 14.3. Harmonic oscillator potential and modifications required if potential is to represent a nucleon–nucleus interaction.

These potentials with discontinuous slopes, however, are very difficult to work with, and the corresponding wave functions must be found by numerical methods. On the other hand, if one wishes to consider bound states whose binding energies are such that $\tilde{r} \ll r_0$, the wave functions will be rather insensitive to the behavior of the potential at r_0 and beyond. Hence the harmonic-oscillator wave functions provide rather good approximations to the correct wave functions for such highly bound states when the "true" potential is of the form (14.73b).

The most fruitful application of the harmonic-oscillator potential has been achieved in studies of the shell model of the nucleus, and it is in books and papers on this subject that the most extensive discussions of the potential and its various modifications are to be found.

Note that for $l = 0$ the differential equation (14.52) reduces to the Schrödinger equation for a one-dimensional harmonic oscillator if we allow r to take on negative as well as nonnegative values. Hence the $l = 0$ solutions (14.71) are also solutions for the one-dimensional harmonic oscillator, and in fact the associated Laguerre polynomials $L_{\lambda+1/2}^{1/2}(\mu\omega_0 r^2/\hbar)$, when multiplied by r, reduce to the Hermite polynomials of odd order, $H_{2\lambda+1}(\sqrt{\mu\omega_0/\hbar}\,r)$. The Hermite polynomials of even order are also solutions to the one-dimensional problem and must be included in the complete set of solutions to that problem. They are the counterparts of the solutions to (14.52) that had to be rejected in the spherically symmetric problem because they fail to vanish at the origin; since vanishing at the origin is not required in one dimension, the even-order Hermite polynomials $H_{2\lambda}(\sqrt{\mu\omega_0/\hbar}r)$ also are solutions in one dimension.

14.7 THE COULOMB POTENTIAL

Without question, the most impressive successes of quantum theory have been scored in problems dealing with essentially atomic systems, and the most important atomic problem is to obtain a theory that predicts the spectral lines—and hence energy levels—of the elements. The obvious starting point for such an effort is the hydrogen atom with its single electron moving in a Coulomb field, or any positive ion formed by stripping away all the atomic electrons except one. We shall call the study of such one-electron systems the one-electron atom problem. Here we shall confine ourselves to consideration of spinless electrons and spinless atomic nuclei.

The appropriate potential for the spinless one-electron atom problem is the pure Coulomb potential

$$V(r) = -\frac{Ze^2}{4\pi\epsilon_0 r} \tag{14.74}$$

where the minus sign indicates the attractive electron–nucleus electrostatic interaction and Z is the atomic number of the nucleus—the number of protons residing in the nucleus. This potential satisfies the condition (14.5), but it does not satisfy (14.12); in fact, it "just barely" fails to satisfy (14.12). This difficulty, however, can be overcome, as we shall see.

Differential equation (14.2), with potential (14.74), becomes (when we replace m by μ as in the previous section),

$$\left[\frac{d^2}{dr^2} - \frac{l(l+1)}{r^2} + \frac{Ze^2\mu}{2\pi\epsilon_0\hbar^2 r} + k^2\right] f(l, k, r) = 0. \tag{14.75}$$

We shall first concern ourselves with bound states, for which

$$k^2 = \frac{2\mu}{\hbar^2}E = -\frac{2\mu}{\hbar^2}|E| < 0;$$

hence we let

$$k = iK, \qquad K = \sqrt{\frac{2\mu}{\hbar^2}|E|}$$

and introduce two new variables

$$\rho = Kr, \qquad \eta = \frac{Ze^2\mu}{4\pi\epsilon_0\hbar^2 K}.$$

With these changes, (14.75) may be written

$$\left[\frac{d^2}{d\rho^2} - \frac{l(l+1)}{\rho^2} + \frac{2\eta}{\rho} - 1\right] f(l, iK, \rho/K) = 0. \tag{14.76}$$

If we look only at the dominant terms in (14.76) as $\rho \to \infty$, we obtain

$$\left[\frac{d^2}{d\rho^2} - 1\right] f(l, iK, \rho/K) \underset{\rho\to\infty}{\sim} 0, \tag{14.77}$$

the solutions to which are

$$f^{(\pm)}\left(l, iK, \frac{\rho}{K}\right) \underset{\rho\to\infty}{\sim} (\text{const.})e^{\mp\rho}.$$

(Actually this is not quite correct, but it provides a starting point for obtaining the correct solution.) The solution corresponding to the lower sign is obviously physically unacceptable, so we omit it from further consideration. Therefore we write

$$f\left(l, iK, \frac{\rho}{K}\right) = \phi_{l,K}(\rho)e^{-\rho} \tag{14.78}$$

and proceed to find a differential equation for $\phi_{l,k}(\rho)$. Thus we obtain, from direct substitution of (14.78) into (14.76),

$$\left[\frac{d^2}{d\rho^2} - 2\frac{d}{d\rho} + \frac{2\eta}{\rho} - \frac{l(l+1)}{\rho^2}\right]\phi_{l,K}(\rho) = 0. \tag{14.79}$$

Again, we shall seek a solution to this equation by assuming that $\phi_{l,k}(\rho)$ can be written as an infinite series.

Let

$$\phi_{l,K}(\rho) = \sum_{n=0}^{\infty} a_n \rho^{p+n}; \tag{14.80}$$

then

$$\frac{d}{d\rho}\phi_{l,K}(\rho) = \sum_{n=0}^{\infty} (p+n)a_n\rho^{p+n-1},$$

$$\frac{d^2}{d\rho^2}\phi_{l,K}(\rho) = \sum_{n=0}^{\infty} (p+n)(p+n-1)a_n\rho^{p+n-2},$$

so that the differential equation will be satisfied if

$$\sum_{n=0}^{\infty} a_n\{[(p+n)(p+n-1) - l(l+1)]\rho^{p+n-2} - 2(p+n-\eta)\rho^{p+n-1}\} = 0.$$

Collecting terms with equal powers of ρ yields

$$a_0[p(p-1) - l(l+1)]\rho^p + \sum_{n=1}^{\infty} \{a_{n+1}[(p+n+1)(p+n) - l(l+1)] - 2a_n(p+n-\eta)\}\rho^{p+n} = 0. \quad (14.81)$$

Since a_0 must vanish, we have either $p = l + 1$ or $p = -l$; the latter corresponds to the irregular solution, which behaves badly near $\rho = 0$, so we require that

$$p = l + 1. \quad (14.82)$$

Furthermore, the recursion relation that follows from (14.81) is

$$a_{n+1} = \frac{2(p+n-\eta)}{(p+n+1)(p+n) - l(l+1)} a_n = \frac{2(l+n+1-\eta)}{(n+1)(n+2l+2)} a_n. \quad (14.83)$$

This is enough to determine the behavior of $\phi_{l,k}(\rho)$ as $\rho \to \infty$, since the ratio-test ratio is

$$\left|\frac{\rho a_{n+1}}{a_n}\right| = \left|\frac{2(l+n+1-\eta)\rho}{(l+n+2)(l+n+1) - l(l+1)}\right| \underset{n\to\infty}{\sim} \frac{2\rho}{n}. \quad (14.84)$$

Since, for all finite ρ, this is less than unity when n is sufficiently large, we see that the series converges. However, for large ρ the series $\phi_{l,k}(\rho)$ behaves like the series having recursion relation

$$a_{n+1} = \frac{2}{n} a_n,$$

since for large ρ it is the terms with large n that determine the behavior of the series, and (14.84) implies the above recursion relation for large n. The series having this recursion relation is

$$a_0 \sum_{n=0}^{\infty} \frac{(2\rho)^n}{n!} = a_0 e^{2\rho};$$

hence

$$\phi_{l,k}(\rho) \underset{\rho\to\infty}{\sim} (\text{const.})\rho^{l+1} e^{2\rho}$$

and

$$f\left(l, iK, \frac{\rho}{K}\right) \underset{\rho\to\infty}{\sim} (\text{const.})\rho^{l+1} e^{2\rho} e^{-\rho} = (\text{const.})\rho^{l+1} e^{\rho}$$

if our infinite series solution holds and all the coefficients a_n are finite. Since this becomes infinite as $\rho \to \infty$, it is quite unacceptable.

Once again, the only way to prevent this unpleasant asymptotic behavior is to have the series terminate, and according to (14.83) this is possible only if, for some interger $n = v - 1$, $v \geq 1$,

$$l + n + 1 - \eta = l + v - \eta = 0, \qquad v = 1, 2, 3, \ldots,$$

i.e.

$$\eta = \frac{Ze^2\mu}{4\pi\epsilon_0\hbar^2 K} = v + l. \tag{14.85}$$

The only quantity on the left-hand side that is not a fixed constant is K; hence this condition restricts K, or more significantly $E = -(\hbar^2/2\mu)K^2$, to a discrete set of values defined by

$$\begin{aligned} K_{v,l} &= \frac{Ze^2\mu}{4\pi\epsilon_0\hbar^2(v+l)}, \qquad v = 1, 2, 3, \ldots; \quad l = 0, 1, 2, \ldots, \\ E_{v,l} &= -\frac{Z^2e^4\mu}{2(4\pi\epsilon_0)^2\hbar^2(v+l)^2}, \qquad v = 1, 2, 3, \ldots; \quad l = 0, 1, 2, \ldots. \end{aligned} \tag{14.86}$$

Comparison with Eq. (7.18) (with $K = 1$ as indicated in the discussion following Eq. (7.20)) shows that Eq. (14.86) is consistent with the same experimental data that gave the Bohr model its initial successes. Thus this Schrödinger-equation model for the one-electron atom does predict the elementary spectral lines of the hydrogen atom.

Now we must incorporate condition (14.85) into the series (14.80), with recursion relation (14.83), to find the one-electron atom wave functions. Equations (14.83) and (14.85) taken together yield

$$a_n = \frac{2^n\Gamma(n - v + 1)(2l + 1)!}{n!(n + 2l + 1)!\Gamma(-v + 1)}a_0,$$

and hence we have

$$\begin{aligned} \phi_{l,K_v}(\rho) &= a_0\rho^{l+1}\sum_{n=0}^{v-1}\frac{\Gamma(n - v + 1)\Gamma(2l + 2)(2\rho)^n}{\Gamma(-v + 1)\Gamma(n + 2l + 2)n!} \\ &= a_0\rho^{l+1}\sum_{n=0}^{v-1}\frac{\Gamma(v)\Gamma(2l + 2)(-2\rho)^n}{\Gamma(v - n)\Gamma(n + 2l + 2)n!}. \end{aligned} \tag{14.87}$$

Referring back to Eq. (14.70), one sees that the polynomial on the right-hand side of Eq. (14.87) is an associated Laguerre polynomial with parameters $q = 2l + 1$ and $p = 2l + v$:

$$\phi_{l,K_v}(\rho) = -a_0\frac{\Gamma(v)}{\Gamma(2l + 2)}\rho^{l+1}L_{2l+v}^{2l+1}(2\rho).$$

Therefore, the radial wave function is

$$f(l, iK_{v,l}, \rho/K_{v,l}) = A_{v,l}e^{-\rho}\rho^{l+1}L_{2l+v}^{2l+1}(2\rho), \tag{14.88}$$

where $A_{v,l}$ is a normalization constant to be determined by the orthonormality condition

$$\int_0^\infty f^*\left(l', iK_{v',l'}, \frac{\rho}{K_{v',l'}}\right) f\left(l, iK_{v,l}\frac{\rho}{K_{v,l}}\right) dr = \delta_{v,v'}\,\delta_{l,l'}. \tag{14.89}$$

It turns out that

$$A_{v,l} = 2^{l+1}\left\{K_{v,l}^3 \frac{(v-1)!}{(v+l)[(v+2l)!]^3}\right\}^{1/2}, \tag{14.90}$$

where $K_{v,l}$ is given by Eq. (14.86).

Now it is interesting to relate $K_{v,l}$ and $E_{v,l}$ to certain standard natural constants that appear in atomic physics. The fundamental constants of particular interest are

i) $\alpha = e^2/4\pi\epsilon_0\hbar c = 1/137.0360$ (fine-structure constant)
ii) $\lambda\!\!\!^{-}_{e} = \hbar/m_e c = 3.861592 \times 10^{-13}$ m
(electron reduced Compton wave length)
or
$k_e = 1/\lambda\!\!\!^{-}_{e} = 2.589606 \times 10^{12}\ \mathrm{m}^{-1}$ (electron Compton wave number)
iii) $a_1 = 1/\alpha k_e = 0.529177 \times 10^{-10}$ m (radius of first Bohr orbit);

here m_e is the mass of the electron. Equations (14.86) indicate that

$$K_{v,l} = \frac{Z}{v+l}\left(\frac{\mu}{m_e}\right)\alpha k_e = \left(\frac{\mu}{m_e}\right)\frac{Z}{v+l}\frac{1}{a_1},$$

and that

$$E_{v,l} = -\frac{\alpha^2}{2}\left(\frac{\mu}{m_e}\right)m_e c^2 \frac{Z^2}{(v+l)^2}.$$

These relationships agree with those obtained for the Bohr atom in Chapter 7.

It is common practice to replace the radial quantum number v with the *principal quantum number* $n = v + l$, so that a state is characterized by

$$n = 1, 2, 3, \ldots$$
$$l = 0, 1, 2, \ldots, n-1$$
$$m = -l, -l+1, \ldots, -1, 0, 1, \ldots, l-1, l.$$

The advantage of using $n = v + l$ is that n then defines the energy state. Note that for each value of n there are n possible different values of l, and for each l there are $2l + 1$ different possible values of m. Hence there are

$$\sum_{l=0}^{n-1} (2l + 1) = n + 2 \sum_{l=1}^{n-1} l = n^2$$

different states corresponding to the single energy level defined by n; the nth energy level is n^2-fold degenerate. These degenerate states may have different parities, in contrast with the degenerate harmonic oscillator states.

The radial wave functions for the first few one-electron atom states are, if we let $K_n \equiv K_{v,l}$ when $n = v + l$,

$$f(0, iK_1, r) = \left(\frac{\mu Z}{m_e a_1}\right)^{3/2} 2re^{-\mu Zr/m_e a_1}, \qquad n = 1, l = 0$$

$$f(0, iK_2, r) = \left(\frac{\mu Z}{2m_e a_1}\right)^{3/2} \left(2r - \frac{\mu Z}{m_e a_1} r^2\right) e^{-\mu Zr/2m_e a_1}, \qquad n = 2, l = 0$$

$$f(1, iK_2, r) = \left(\frac{\mu Z}{2m_e a_1}\right)^{3/2} \frac{\mu Z}{\sqrt{3}m_e a_1} r^2 e^{-\mu Zr/2m_e a_1}, \qquad n = 2, l = 1.$$

The last two of these correspond to degenerate states having principal quantum number $n = 2$.

These functions and the ones for $n = 3$ are plotted, along with their squares, in Figs. 14.4 and 14.5; their squares yield the probability per unit radius of finding an electron at a given value of the radius r. Figure 7.6a shows the energy spectrum of one-electron atom states.

In this discussion we have neglected the irregular solution to the radial equation, since it clearly is not allowed as a physical solution for the pure Coulomb potential. However, in problems in nuclear physics and even in atomic problems involving several electrons, the actual potential governing the interaction of a particle with a nucleus or atom will be somewhat modified from the Coulomb potential when r is small, and may behave like a true Coulomb potential only when r is rather large. In such cases, the true radial wave function is, for larger r, a linear combination of the regular and irregular solutions to the one-electron atom problem. Problems involving such situations usually require numerical solutions, and we shall not pursue the question here. It is well to recognize, however, that the regular solutions we have obtained are strictly correct only when the potential has the pure Coulomb form near the origin as well as elsewhere.

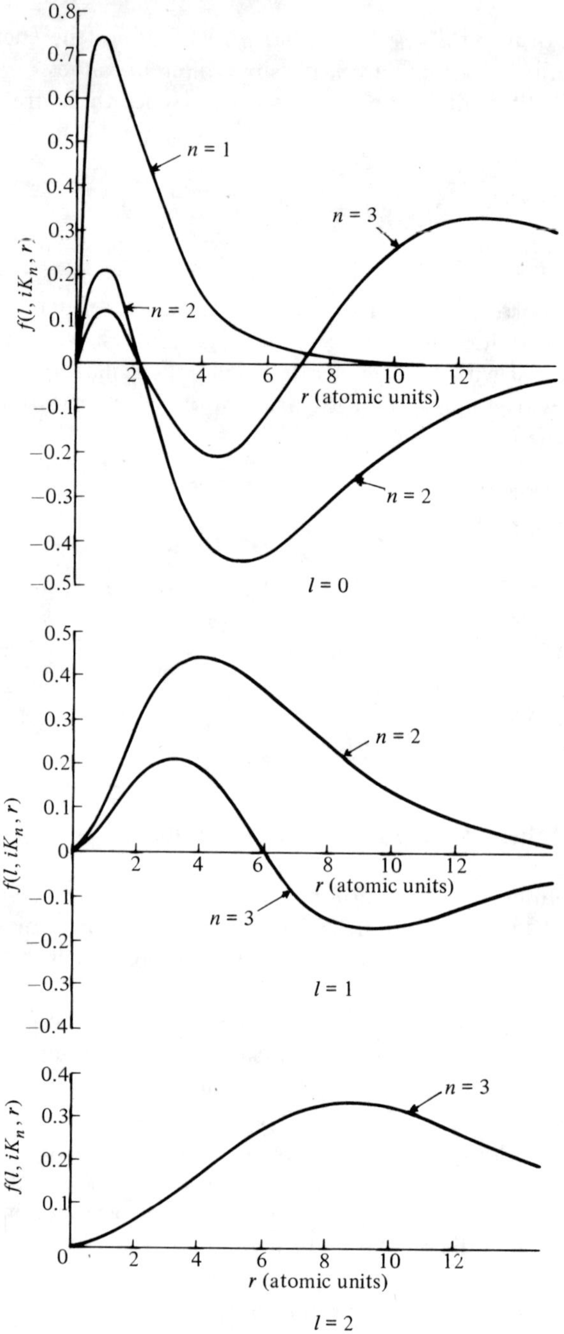

Fig. 14.4. Radial wave functions for hydrogen, with $\mu Z/m_e = 1$, for $n = 1, 2, 3$. From J. C. SLATER, *Quantum Theory of Atomic Structure*, Vol. I, McGraw-Hill Book Co., New York (1960). Reprinted through permission of the publishers.

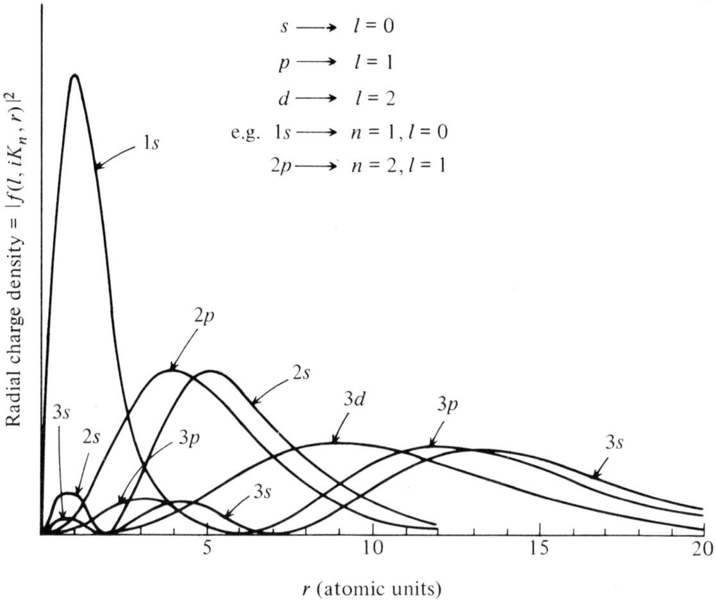

Fig. 14.5. Radial charge density for hydrogen. From J. C. SLATER, *Quantum Theory of Atomic Structure*, Vol. I, McGraw-Hill Book Co., New York (1960). Reprinted through permission of the publisher.

READING NOTES*

Reference 23, Chapter 15.

Reference 43, Chapter 4.

Reference 44, Chapter 7.

PROBLEMS

1. Equations (14.36) and (14.37) will be those for a one-dimensional square-well potential of depth V_0 and width $2a$ if we set $l = 0$, replace $0 \leq r < a$ with $0 \leq |r| < a$ and replace $r > a$ with $|r| > a$, it now being understood that r is the one-dimensional position varying from $-\infty$ to $+\infty$.

   a) Justify this statement by setting up the one-dimensional Schrödinger equation.
   b) Show that there are two sets of solutions, one set being even functions of r and the other set odd in r. (The latter coincides with the $l = 0$ solution to the radial equation in the three-dimensional spherically symmetric problem. Why?)
   c) Obtain the equation analogous to (14.49) that must be solved for the even solutions, and draw a graph similar to Fig. 14.1 that provides graphical solutions for K.

2. Solve the one-dimensional harmonic-oscillator problem using the appropriate differential equation, and give explicit solutions for the three lowest energy levels. Compare with the

---

* Works referred to in these notes will be found in the Bibliography on p. 577.

harmonic-oscillator solutions of Chapter 11. Also, respond to items (a) and (b) in Problem 1, referring them to this problem.

3. Compare the degeneracy of the one-electron atom energy levels with that of the isotropic harmonic-oscillator solutions. Comment in particular on the parities of the degenerate states in the two cases.

4. Solve the isotropic harmonic-oscillator problem in Cartesian coordinates, and compare with the results in spherical coordinates. Comment on the degeneracies of the energy levels in the two solutions.

5. The binding energy of the deuteron is about 2.23 MeV. Find the square-well depth required to yield that binding energy for a well with radius 1 F. Assume that there is only one bound state (for which $l = 0$). Do the same for several radii out to 4 F and plot the result. Estimate the mean radius $\langle r \rangle$ as a function of well radius.

6. For the one-electron atom in its ground state, calculate

   a) $\langle r \rangle$,
   b) $\langle r^2 \rangle$ and $\langle r^2 \rangle - \langle r \rangle^2$,
   c) $\langle 1/r \rangle$,
   d) $\langle 1/r^2 \rangle$,
   e) the mean kinetic energy, and
   f) the mean radial momentum.

7. Find the probability density per unit momentum interval for the electron in the one-electron atom in its $n = 1$ and $n = 2$ states. There are three quantities to be calculated.

8. Restricting your attention to the states with $m = l$, find $\langle 1/r \rangle$ for one-electron atom states in general and compare with the Bohr theory.

9. For the deuteron mentioned in Problem 5, assume the potential to be a harmonic-oscillator well and define r_0 in accordance with Fig. 14.3. Ignore the fact that potential should vanish for $r > r_0$. Find the value of ω_0 and consequently of r_0 for which the only "bound" state (with energy <0) has energy -2.23 MeV. Compare the values of r_0 and ω_0 with the results of Problem 5. Plot the square-well ground-state wave function for well radius a given by

$$V_0 a^2 = \frac{1}{r_0} \int_0^{r_0} \frac{\mu\omega_0^2}{2}(r - r_0)^2 r^2 \, dr$$

on the same axes with the harmonic oscillator ground-state wave function.

10. The term in the Hamiltonian arising from the part of the Laplacian that involves the angular variables in spherical coordinates is often called the *centrifugal barrier*. Justify this use of language and estimate the thickness of the barrier for angular momenta of $1 - 0$, 1, 2 and 3 and for kinetic energies of 1, 10, 100, 1000 MeV; give your estimates in Fermis. Comment on the significance of these estimates in relation to the study of nuclear phenomena.

CHAPTER 15

THE ELEMENTS OF PERTURBATION THEORY

Generally, physical problems are not as amenable to exact solution as the problems for which we solved the radial equations in Chapter 14. Among those we did solve, only the one involving the Coulomb potential had any clear correspondence with physical reality. However, many problems are "almost" exactly solvable in that the Hamiltonian for the system can be separated into two parts, one part yielding an exactly solvable problem and the other part being, in some sense, quite small. Such situations can be handled by approximation methods that come under the heading *perturbation theory*. We shall avoid mathematical sophistication in our treatment of perturbation theory and shall assume that throughout the discussion mathematically good behavior prevails. Our immediate motivation for considering perturbation theory is to meet the challenge of predicting certain simple atomic phenomena with the general quantum-theoretical formalism we have developed. Since realistic atomic phenomena usually do not correspond to solvable models in the form of simple Schrödinger equations, we shall need perturbation techniques to carry out our program. In the next chapter we shall see how these techniques allow us to "explain" a wide range of interesting observations quantum-theoretically.

15.1 STATIONARY-STATE PERTURBATION THEORY IN FIRST ORDER

Let the Hamiltonian operator for a given system be the sum of an exactly solvable time-independent operator $\mathfrak{H}_0$ with a discrete eigenvalue spectrum, and a time-independent operator $\mathfrak{H}_1$; that is

$$\mathfrak{H} = \mathfrak{H}_0 + \mathfrak{H}_1 \tag{15.1}$$

where in saying that $\mathfrak{H}_0$ is solvable we mean that we can readily find the eigenstates $\{|E_n, q, \ldots\rangle\}$ of $\mathfrak{H}_0$:

$$\mathfrak{H}_0|E_n, q, \ldots\rangle = E_n|E_n, q, \ldots\rangle. \tag{15.2}$$

Here the indices $q, \ldots$ distinguish linearly independent states that may have the same energy eigenvalue E_n. For convenience, we may use the single index q with the understanding that in fact there may be several indices, representing eigenvalues of several different mutually commuting operators.

In some sense $\mathfrak{H}_1$ is supposed to be small. Let us multiply $\mathfrak{H}_1$ by a parameter λ whose values are restricted to real numbers on the closed interval from 0 to 1; then we write

$$\mathfrak{H}(\lambda) = \mathfrak{H}_0 + \lambda\mathfrak{H}_1 \tag{15.3}$$

and we assume that $\mathfrak{H}(\lambda)$ has eigenvalues and eigenstates that are analytic functions of λ for all allowed values of λ. Clearly λ is a parameter that "turns on" the perturbation, for if $\lambda = 0$ we have $\mathfrak{H}(0) = \mathfrak{H}_0$, and if $\lambda = 1$, we have $\mathfrak{H}(1) = \mathfrak{H}$.

The set $\{|E_n, q\rangle\}$ of solutions to (15.2) is assumed to be complete, so that the solutions $|E_\nu(\lambda), s(\lambda)\rangle$ of the exact eigenvalue equation

$$\mathfrak{H}(\lambda)|E_\nu(\lambda), s(\lambda)\rangle = E_\nu(\lambda)|E_\nu(\lambda), s(\lambda)\rangle \tag{15.4}$$

are expressible in terms of the $\{|E_n, q\rangle\}$:

$$\begin{aligned}|E_\nu(\lambda), s(\lambda)\rangle &= \sum_{n,q}|E_n, q\rangle\langle E_n, q|E_\nu(\lambda), s(\lambda)\rangle \\ &= \sum_{n,q} a_{n,q}^{\nu,s}(\lambda)|E_n, q\rangle, \qquad (15.5)\\ a_{n,q}^{\nu,s}(\lambda) &\equiv \langle E_n, q|E_\nu(\lambda), s(\lambda)\rangle.\end{aligned}$$

Next we use our assumption that $E_\nu(\lambda)$ and $|E_\nu(\lambda), s(\lambda)\rangle$ are analytic in λ; this assumption implies that we may expand $E_\nu(\lambda)$ and $a_{n,q}^{\nu,s}(\lambda)$ in powers of λ:

$$E_\nu(\lambda) = \sum_{k=0}^{\infty} E_\nu{}^{(k)}\lambda^k, \tag{15.6a}$$

$$a_{n,q}^{\nu,s}(\lambda) = \sum_{k=0}^{\infty} a_{n,q}^{\nu,s(k)}\lambda^k. \tag{15.6b}$$

Inserting (15.3), (15.5) and (15.6) in (15.4) yields

$$(\mathfrak{H}_0 + \lambda\mathfrak{H}_1)\sum_{n,q}\sum_k a_{n,s}^{\nu,s(k)}\lambda^k|E_n, q\rangle - \sum_{n,q}\sum_k\sum_{m\le k} E_\nu{}^{(k-m)}a_{n,q}^{\nu,s(m)}\lambda^k|E_n, q\rangle = 0.$$

This must be true for all values of λ lying in the prescribed interval from 0 to 1, and hence we may set the coefficients of λ^k equal to 0 for all $k \ge 1$, thus

$$\sum_{n,q}[E_n a_{n,q}^{\nu,s(k)} + a_{n,q}^{\nu,s(k-1)}\mathfrak{H}_1 - \sum_{m\le k} E_\nu{}^{(k-m)}a_{n,q}^{\nu,s(m)}]|E_n, q\rangle = 0. \tag{15.7}$$

For $k = 0$, we may set $a_{n,q}^{\nu,s(-1)} = 0$, thereby obtaining

$$\sum_{n,q}(E_n - E_\nu{}^{(0)})a_{n,q}^{\nu,s(0)}|E_n, q\rangle = 0.$$

Hence, for each value of n, either

$$E_n = E_\nu{}^{(0)} \qquad \text{or} \qquad a_{n,q}^{\nu,s(0)} = 0.$$

This means that the only states $|E_n, q\rangle$ contributing to the exact solutions $|E(\lambda), s(\lambda)\rangle$ when $\lambda \to 0$ are those with energy eigenvalues E_n equal to $\lim_{\lambda\to 0} E_\nu(\lambda) = E_\nu{}^{(0)}$, and also that the only possible values of $E_\nu^{(0)}$ are the unperturbed energy eigenvalues E_n. Thus, to zeroth order in λ, there is no change in the energy-level structure of the system.

Suppose the system is in a certain state with unperturbed energy E_i in the limit as $\lambda \to 0$; then

$$\begin{aligned} a_{n,q}^{i,s(0)} &= \langle E_n, q | E_i(0), s(0)\rangle = \langle E_n, q | E_i^{(0)}, s(0)\rangle \\ &= \langle E_n, q | E_i, s(0)\rangle \equiv \alpha_{q,s(0)}^i \delta_{n,i}, \end{aligned} \tag{15.8}$$

and we may write

$$|E_i(0), s(0)\rangle = \sum_q \alpha_q^i |E_i, q\rangle. \tag{15.9}$$

Therefore $|E_i(0), s(0)\rangle$ is not necessarily an unperturbed eigenstate $|E_i, q\rangle$, but is a linear combination of unperturbed eigenstates with identical energies.*

Knowing that the $\alpha_{q,s(0)}^i$ may be nonvanishing for several values of q if the energy level E_i is degenerate, we proceed to seek $E_i^{(1)}$, the first-order approximation to the difference between $E_i = E_i(0)$ and the exact energy $E_i(1)$. We begin by setting $k = 1$ in (15.7) and taking the inner product of $|E_j, q'\rangle$ with the left-hand side:

$$\sum_{n,q} a_{n,q}^{i,s(0)} \langle E_j, q' | \mathfrak{H}_1 | E_n, q\rangle = \sum_{n,q} E_i^{(1)} a_{n,q}^{i,s(0)} + (E_i^{(0)} - E_n) a_{n,q}^{i,s(1)}] \delta_{n,j} \delta_{q,q'},$$

where we have set $\nu = i$ to indicate that we are seeking energy levels that are obtained from E_i by a continuous change in λ from 0 to 1. Making use of (15.8)

---

* For example, $|\mathbf{k}; \frac{1}{2}, \mu\rangle$ may represent the state of an electron with momentum $\hbar\mathbf{k}$ and with the component of spin along the direction of motion being denoted μ; this state has energy $E_i = \sqrt{m^2c^4 + \hbar^2\mathbf{k}^2c^2}$ and corresponds to a state $|E_i, q\rangle$ above. We consider the electron in a frame of reference in which $\mathbf{k}^2 \ll (\hbar/mc)^2$, and allow a weak magnetic field to be imposed in this frame, with $\mathbf{B} \perp \mathbf{k}$. For such small values of k, the problem is nonrelativistic, and we may expect the electrons to align themselves along $\mathbf{B}$. Thus the perturbed states corresponding to $|E_i(\lambda), s(\lambda)\rangle$ may be expected to be $|k; \frac{1}{2}, m\rangle$, at least in the limit as $B \to 0$, where m is the spin projection quantum number along $\hat{\mathbf{B}}$ rather than along $\hat{\mathbf{k}}$. In this case we obtain

$$\begin{aligned} |\mathbf{k}; \tfrac{1}{2}, m\rangle &= \sum_\mu |\mathbf{k}; \tfrac{1}{2}, \mu\rangle \langle \mathbf{k}; \tfrac{1}{2}, \mu | \mathbf{k}; \tfrac{1}{2}, m\rangle \\ &\underset{k\to 0}{\simeq} \sum_\mu |\mathbf{k}; \tfrac{1}{2}, \mu\rangle \langle 0; \tfrac{1}{2}, \mu | 0; \tfrac{1}{2}, m\rangle \end{aligned}$$

and $\alpha_{\mu,m}^i = \langle 0; \frac{1}{2}, \mu | 0; \frac{1}{2}, m\rangle$ may be obtained by noting that the quantization axis for the spin projections m is obtained from that for spin projections μ by a rotation of $\pi/2$ about $\hat{\mathbf{k}} \times \hat{\mathbf{B}}$.

The point is that the original choice of quantum number for spin projections along the direction of motion no longer was an appropriate, or "natural", choice after the perturbation defining a direction perpendicular to $\hat{\mathbf{k}}$ was introduced, even in the limit as $\mathbf{B} \to \mathbf{0}$. The "natural" state vectors, however, were obtainable from the original ones in this limit through the formation of linear combinations of the original states having energy E_i.

and $E_i^{(0)} = E_i$, we obtain

$$\sum_q \alpha^i_{q,s}\langle E_j, q'|\mathfrak{H}_1|E_i, q\rangle = E_i^{(1)}\alpha^i_{q',s}\delta_{i,j} + (E_i - E_j)a^{i,s(1)}_{j,q'}. \tag{15.10}$$

For the case $j = i$, this yields

$$E_i^{(1)} = \sum_q \frac{\alpha^i_{q,s}}{\alpha^i_{q',s}}\langle E_i, q'|\mathfrak{H}_1|E_i, q\rangle, \tag{15.11}$$

whereas for $j \neq i$ it gives

$$a^{i,s(1)}_{j,q'} = \sum_q \alpha^i_{q,s}\frac{\langle E_j, q'|\mathfrak{H}_1|E_i, q\rangle}{E_i - E_j}. \tag{15.12}$$

The coefficient $a^{i,s(1)}_{i,q'}$ cannot be found from (15.10), and in fact it is arbitrary except that it must provide proper normalization for $|E_i(\lambda), s(\lambda)\rangle$ through first order in λ. We choose $a^{i,s(1)}_{i,q'} = 0$, which is consistent with this last requirement. Then we get

$$|E_i(\lambda), s(\lambda)\rangle^{(1)} = \sum_{j,q'}\left[\alpha^i_{q',s}\delta_{j,i} + (1 - \delta_{j,i})\lambda\sum_q \alpha^i_{q,s}\frac{\langle E_j, q'|\mathfrak{H}_1|E_i, q\rangle}{E_i - E_j}\right]|E_j, q'\rangle \tag{15.13}$$

as the first-order approximation to the eigenstates $|E_i(\lambda), s(\lambda)\rangle$ of $\mathfrak{H}(\lambda)$.

The sums over q and q' in Eq. (15.13) are a bit awkward; however, if the states $|E_j, q\rangle$ are nondegenerate, there is only one value of q for each energy level and the q-indices become superfluous. Then we write $|E_j\rangle = |E_j, q\rangle, \alpha^i_{q,s} = 1$, and the above results become

$$E^{(1)} = \langle E_i|\mathfrak{H}_1|E_i\rangle \tag{15.11a}$$

$$a_j^{i(1)} = \frac{\langle E_j|\mathfrak{H}_1|E_i\rangle}{E_i - E_j}, \qquad j \neq i$$
$$= 0, \qquad j = i \tag{15.12a}$$

$$|E_i(\lambda)\rangle^{(1)} = |E_i\rangle + \lambda\sum_{j\neq i}\frac{\langle E_j|\mathfrak{H}_1|E_i\rangle}{E_i - E_j}|E_j\rangle. \tag{15.13a}$$

For degenerate unperturbed energy levels we must go further, as we discuss later.

In any case, Eq. (15.12) gives us a criterion for the validity of our method, that is, for the smallness of $\mathfrak{H}_1$. Since the $a^{i,s(1)}_{j,q'}$ must be much less than unity if the perturbed states are to differ little from the unperturbed ones, we must insist that the sum in (15.12) is much less than unity. This will be assured if

$$\left|\frac{\langle E_j, q'|\mathfrak{H}_1|E_i, q\rangle}{E_i - E_j}\right| \ll 1. \tag{15.14}$$

Hence $\mathfrak{H}_1$ is considered to be small if, for every pair of discrete energy levels E_i and E_j, the matrix elements of $\mathfrak{H}_1$ between all states corresponding to those levels is small compared with the magnitude of the difference between the energies of those levels.

Let us now take up the degenerate case. We may consider (15.11) a matrix equation,

$$E_i^{(1)} \begin{Bmatrix} \alpha^i_{q_1,s} \\ \alpha^i_{q_2,s} \\ \vdots \\ \alpha^i_{q_r,s} \end{Bmatrix} = \begin{bmatrix} H^i_{q_1q_1} & H^i_{q_1q_2} & \cdots & H^i_{q_1q_r} \\ H^i_{q_2q_1} & H^i_{q_2q_2} & \cdots & H^i_{q_2q_r} \\ \vdots & \vdots & & \\ H^i_{q_rq_1} & H^i_{q_rq_2} & \cdots & H^i_{q_rq_r} \end{bmatrix} \begin{Bmatrix} \alpha^i_{q_1,s} \\ \alpha^i_{q_2,s} \\ \vdots \\ \alpha^i_{q_r,s} \end{Bmatrix} \tag{15.15}$$

where

$$H^i_{q_s,q_t} \equiv \langle E_i, q_s|\mathfrak{H}_1|E_i, q_t\rangle.$$

The solutions to (15.15) are the values of $\alpha^i_{q,s}$ that diagonalize the matrix $[H^i_{q_s,q_t}]$; if we can find these values, we shall have the desired "natural" linear combinations

$$|E_i(0), s(0)\rangle = \sum_q \alpha^i_{q,s}|E_i, q\rangle$$

of the unperturbed eigenstates. These solutions exist if and only if the secular determinant for (15.15) vanishes:

$$\begin{vmatrix} H^i_{q_1q_1} - E_i^{(1)} & H^i_{q_1q_2} & \cdots & H^i_{q_1q_r} \\ H^i_{q_2q_1} & H^i_{q_2q_2} - E_i^{(1)} & \cdots & H^i_{q_2q_r} \\ \vdots & \vdots & & \vdots \\ H^i_{q_rq_1} & H^i_{q_rq_2} & \cdots & H^i_{q_rq_r} - E_i^{(1)} \end{vmatrix} = 0. \tag{15.16}$$

The r values of $E_i^{(1)}$ (some of them may be identical with one another) for which this can occur then are the actual energy shifts, to first order, arising from the perturbation $\mathfrak{H}_1$. Once the $E_i^{(1)}$ are determined, the standard methods for finding solutions to simultaneous linear homogeneous equations may be used to determine the $\alpha^i_{q,s(0)}$ up to a common multiplicative factor; the factor then is determined by the normalization condition

$$\langle E_i(0), s(0)|E_i(0), s(0)\rangle = \sum_{q,q'} \alpha^{i*}_{q',s}\alpha^i_{q,s}\langle E_i, q'|E_i, q\rangle = \sum_q |\alpha^i_{q,s}|^2 = 1.$$

Now if, in fact, $H^i_{qq'} = 0$ for all $q' \neq q$, Eq. (15.15) implies that $H^i_{qq} = E_i^{(1)}$ for all q, and the matrix $[H^i_{qq'}]$ is already diagonal. This means that we have been clever, or perhaps just lucky, in choosing the states $|E_i, q\rangle$, for they are identical with the natural states $|E(0), s(0)\rangle = |E_i, s(0)\rangle$; that is, we may let $s(0) = q$ and write $\alpha^i_{q,s} = \delta_{q,s(0)}$, which satisfies the normalization condition. Then Eqs. (15.11)–(15.13) become

$$E_i^{(1)} = \langle E_i, s(0)|\mathfrak{H}_1|E_i, s(0)\rangle \qquad \text{for all } s(0), \tag{15.11b}$$

$$a^{i,s(1)}_{j,q'} = \frac{\langle E_j, q'|\mathfrak{H}_1|E_i, s(0)\rangle}{E_i - E_j}, \tag{15.12b}$$

$$|E_i(\lambda), s(\lambda)\rangle^{(1)} = |E_i, s(0)\rangle + \lambda \sum_{\substack{j\neq i\\ q'}} \frac{\langle E_j, q'|\mathfrak{H}_1|E_i, s(0)\rangle}{E_i - E_j}|E_j, q'\rangle, \tag{15.13b}$$

which are essentially the same as Eqs. (15.11a)–(15.13a). Hence we get a trivial extension of the nondegenerate case when $\mathfrak{H}_1$ is diagonal in the degenerate states, i.e. when $\mathfrak{H}_1$ does not "connect" degenerate states having different quantum numbers q.

To see that the $|E(0), s(0)\rangle$ obtained by the prescribed diagonalization procedure are actually the solutions we desire in the more general case when $H^i_{qq'} \neq 0$ for $q \neq q'$, we reconsider Eq. (15.7) with $k = 1$ and replace $a^{i,s(0)}_{n,q}$ with $\alpha^i_{q,s}\delta_{n,i}$; then we find

$$\mathfrak{H}_1|E_i, s(0)\rangle - E_i^{(1)}|E_i, s(0)\rangle - \sum_n (E_i - E_n)\sum_q a^{i,s(1)}_{n,q}|E_n, q\rangle = 0. \tag{15.17}$$

Taking the inner product of $|E_i, s'(0)\rangle$ with the left-hand side gives us

$$\langle E_i, s'(0)|\mathfrak{H}_1|E_i, s(0)\rangle = E_i^{(1)}\delta_{s(0)s'(0)}, \tag{15.11c}$$

i.e. $\mathfrak{H}_1$ indeed is diagonalized by $\{|E_i, s(0)\rangle\}$, and this fact relieves us of the awkward sums over q in (15.11). To be complete, we should label $E_i^{(1)}$ with the auxiliary quantum number $s(0)$ of the state to which the state with energy $E_i^{(1)}(\lambda)$ reduces as $\lambda \to 0$; let $E_i^{(1)} \to E^{(1)}_{i,s(0)}$ in (15.11c).

If we carry out this diagonalization procedure for each energy level E_j, each time obtaining states $|E_j, s'(0)\rangle$, we can take the inner product of $|E_j, s'(0)\rangle$ with (15.17) for $j \neq i$ to obtain

$$\langle E_j, s'(0)|\mathfrak{H}_1|E_i, s(0)\rangle = (E_i - E_j)\sum_q a^{i,s(1)}_{j,q}\langle E_j, s'(0)|E_j, q\rangle.$$

Defining

$$a^{i,s(1)}_{j,s'} \equiv \sum_q a^{i,s(1)}_{j,q}\langle E_j, s'(0)|E_j, q\rangle \tag{15.18}$$

yields

$$a^{i,s(1)}_{j,s'} = \frac{\langle E_j, s'(0)|\mathfrak{H}_1|E_i, s(0)\rangle}{E_i - E_j} \tag{15.12c}$$

and again the awkward sum over q in Eq. (15.12) has been eliminated. Equation (15.18) corresponds to the expression

$$|E_j, s'(0)\rangle = \sum_q |E_j, q\rangle\langle E_j, q|E_j, s'(0)\rangle$$

of $|E_j, s'(0)\rangle$ as a linear combination of unperturbed states $|E_j, q\rangle$.

15.2 TWO APPLICATIONS TO ILLUSTRATE THE METHOD

a) Zeeman Effect for a Spinless Particle in a Spherically Symmetric Potential

We have seen already (Chapter 12) that in the presence of a uniform external magnetic field along the z-axis, the nonrelativistic Hamiltonian for a spinless charged particle in a potential $\mathfrak{V}(\mathfrak{X})$ is

$$\mathfrak{H} = \frac{1}{2M}\mathfrak{P}^2 + \mathfrak{V}(\mathfrak{X}) - \frac{q}{2M}\mathbf{B}\cdot\mathfrak{L} + \frac{q^2}{2M}\frac{1}{4}\mathbf{B}^2(\mathfrak{X}_1^2 + \mathfrak{X}_2^2).$$

If B is small in the sense that

$$\frac{q^2}{8M}\mathbf{B}^2\langle\mathfrak{X}_1^2 + \mathfrak{X}_2^2\rangle \ll \frac{q}{2M}B\,\langle\mathfrak{L}_3\rangle \ll \langle\mathfrak{H}_0\rangle = \left\langle\frac{1}{2M}\mathfrak{P}^2 + \mathfrak{V}(\mathfrak{X})\right\rangle$$

for all physical states of the system, then we may neglect the last term and it may be reasonable to treat the third term in $\mathfrak{H}$ as a perturbation on $\mathfrak{H}_0$. Furthermore, if $\mathfrak{V}(\mathfrak{X})$ is spherically symmetric, the eigenstates of $\mathfrak{H}_0$ may be chosen to be eigenstates of $\mathfrak{L}^2$ and $\mathfrak{L}_3$ also; thus we may label the unperturbed states $|n, l, m\rangle$, with

$$\mathfrak{H}_0|n, l, m\rangle = E_n|n, l, m\rangle$$
$$\mathfrak{L}^2|n, l, m\rangle = \hbar^2 l(l+1)|n, l, m\rangle$$
$$\mathfrak{L}_3|n, l, m\rangle = \hbar m|n, l, m\rangle.$$

Letting

$$\mathfrak{H}_1 \equiv -\frac{q}{2M}\mathbf{B}\cdot\mathfrak{L} = -\frac{q}{2M}B\mathfrak{L}_3,$$

then, we are interested in the matrix elements

$$\langle n', l', m'| -\frac{q}{2M}B\mathfrak{L}_3|n, l, m\rangle = -\frac{q\hbar}{2M}Bm\delta_{n',n}\delta_{l',l}\delta_{m',m} \tag{15.19}$$

and hence the energy of the state $|n, l, m\rangle$ is shifted from E_n to

$$E_{n,m}{}^{(1)} = E_n - \frac{q\hbar}{2M}Bm, \tag{15.20}$$

and the energy levels become dependent upon the quantum number m. This is because the magnetic field removes the spherical symmetry that existed when the particle interacted only with the potential $\mathfrak{V}(\mathfrak{X})$. The unperturbed energy levels depended only on n because of the spherical symmetry, and in first order, some of the degeneracy of these levels is removed, according to (15.20).

Because of the δ-symbols in Eq. (15.19), we see that the states $|n, l, m\rangle$ are diagonal in $\mathfrak{H} = \mathfrak{H}_0 + \mathfrak{H}_1$ as well as in $\mathfrak{H}_0$; hence Eqs. (15.12b) and (15.13b) apply and we have

$$|E_n(\lambda), l, m\rangle^{(1)} = |n, l, m\rangle.$$

To first order in perturbation theory, then, the Zeeman effect shifts energy levels but does not affect the state vectors describing the system.

We shall consider the Zeeman effect in more detail in a later section.

b) The Stark Effect for a Spinless One-electron Atom

The Stark effect results from the imposition of an external electric field on an atomic system. We consider a uniform electric field along the z-axis as a first-order perturbation on a spinless particle in a Coulomb potential; thus

$$\mathfrak{H} - \frac{1}{2M}\mathfrak{P}^2 - \frac{qq'}{4\pi\epsilon_0}\mathfrak{R}^{-1} + q\mathfrak{X}\cdot\mathbf{E},$$

or

$$\mathsf{H}(\mathbf{r}) = -\frac{\hbar^2}{2M}\nabla^2 - \frac{qq'}{4\pi\epsilon_0 r} + qEr\cos\theta$$

where

$$\langle\mathbf{r}'|\mathfrak{H}|\mathbf{r}\rangle = \mathsf{H}(\mathbf{r})\delta^3(\mathbf{r} - \mathbf{r}').$$

Here we let $\mathsf{H}(\mathbf{r}) = \mathsf{H}_0 + \mathsf{H}_1$, where

$$\mathsf{H}_0 = -\frac{\hbar^2}{2M}\nabla^2 - \frac{qq'}{4\pi\epsilon_0 r}, \qquad \mathsf{H}_1 = qEr\cos\theta$$

and we are interested in finding

$$\langle n, l', m'|\mathfrak{H}_1|n, l, m\rangle = \int d^3r\,\psi^*_{n,l',m'}(\mathbf{r})\mathsf{H}_1\psi_{n,l,m}(\mathbf{r})$$

$$= qE\int R_{n,l'}(r)R_{n,l}(r)r^3\,dr\int Y_{l'}^{m'}{}^*(\theta,\varphi)Y_l^m(\theta,\varphi)\cos\theta\,d\Omega, \quad (15.21)$$

when $R_{n,l}(r)$ are the radial factors of the wave functions for the Coulomb potential.

We can make some rather general observations at the outset that are not affected by the radial integral. First we note that the integral over azimuthal angles (φ) vanishes unless $m = m'$: hence the perturbation connects only states having identical magnetic quantum numbers m. Furthermore, the integration operation

$$\int d\Omega = \int_0^{\pi} \sin\theta\, d\theta \int_0^{2\pi} d\varphi \xrightarrow[\varphi\to\pi+\varphi]{\theta\to\pi-\theta} \int_0^{\pi} \sin\theta\, d\theta \int_0^{2\pi} d\varphi$$

is unchanged by the transformation $\mathbf{r} \to -\mathbf{r}$ (or $\theta \to \pi - \theta$, $\varphi \to \pi + \varphi$), as indicated, but the integrand $Y_{l'}^{m'*}(\theta, \varphi)\, Y_l^m(\theta, \varphi) \cos\theta$ is multiplied by $(-)^{l+l'+1}$ by this transformation. Since the integration covers all possible angles, the angular integral in (15.21) cannot be altered by this transformation (which merely rearranges the order in which angular intervals are covered in the integration process) and hence we must have

$$(-)^{l+l'+1} = 1, \qquad \text{or} \qquad l + l' + 1 = 2n, \qquad n \text{ an integer},$$

if the integral is to be nonvanishing. Therefore l and l' must differ by an odd integer, and hence $|n, l', m'\rangle$ and $|n, l, m\rangle$ must have opposite parity. Moreover, it can be shown that the angular integral vanishes unless $|l - l'| = 1$. This tells us immediately that a Stark effect can occur in first order (i.e. (15.21) may be nonvanishing) only for degenerate unperturbed energy levels that contain both even- and odd-parity states. (Recall that, in first order, $E_{i,s(0)}^{(1)}$ is calculated from states having identical unperturbed energies E_i.)

We restrict our further considerations to $n = 1, 2$ levels of one-electron (spinless) atoms. The $n = 1$ state has $l = 0$, $m = 0$ and hence is nondegenerate; therefore it has no first order Stark effect. For $n = 2$, we have $l = 0$, $m = 0$; $l = 1$, $m = -1, 0, +1$. Hence a Stark effect can exist through the nonvanishing of the matrix elements

$$\langle 2, 0, 0 | q\mathbf{X} \cdot \mathbf{E} | 2, 1, 0\rangle, \qquad \langle 2, 1, 0 | q\mathbf{X} \cdot \mathbf{E} | 2, 0, 0\rangle.$$

Since the relevant radial wave functions are

$$R_{20} = \frac{1}{\sqrt{2a^3}} e^{-r/2a}\left(1 - \frac{r}{2a}\right),$$

$$R_{21} = \frac{1}{\sqrt{6a^3}} e^{-r/2a} \frac{r}{2a},$$

we have

$$\langle 2, 1, 0 | q\mathbf{X} \cdot \mathbf{E} | 2, 0, 0\rangle = \langle 2, 0, 0 | q\mathbf{X} \cdot \mathbf{E} | 2, 1, 0\rangle$$

$$= \frac{qE}{12a^4} \int_0^{\infty} r^4 \left(1 - \frac{r}{2a}\right) e^{-r/a}\, dr = -3qEa.$$

The matrix of perturbation energies corresponding to (15.15) is then

$$\begin{array}{c} \\ 2,0,0 \\ 2,1,0 \\ 2,1,-1 \\ 2,1,1 \end{array} \begin{array}{c} \begin{array}{cccc} 2,0,0 & 2,1,0 & 2,1,-1 & 2,1,1 \end{array} \\ \begin{bmatrix} 0 & -3qEa & 0 & 0 \\ -3qEa & 0 & 0 & 0 \\ 0 & 0 & 0 & 0 \\ 0 & 0 & 0 & 0 \end{bmatrix} \end{array},$$

where we have indicated the quantum numbers labeling rows and columns in the margins. The determinant corresponding to (15.16) is thus $\Delta \equiv (E_2^{(1)})^2[(E_2^{(1)})^2 - 9q^2E^2a^2]$, and this has the four roots

$$E_2^{(1)} = -3qEa, \quad +3qEa, \quad 0, \quad 0.$$

Substituting, for example, $E_{2,1}^{(1)} = -3qEa$ into Eq. (15.15), one obtains

$$\begin{bmatrix} 3qEa & -3qEa & 0 & 0 \\ -3qEa & 3qEa & 0 & 0 \\ 0 & 0 & 3qEa & 0 \\ 0 & 0 & 0 & 3qEa \end{bmatrix} \begin{Bmatrix} \alpha^2_{1,s_1} \\ \alpha^2_{2,s_1} \\ \alpha^2_{3,s_1} \\ \alpha^2_{4,s_1} \end{Bmatrix} = 0$$

or

$$\alpha^2_{1,s_1} = \alpha^2_{2,s_1}, \qquad \alpha^2_{3,s_1} = \alpha^2_{4;s_1} = 0,$$

and hence

$$|E_2, s_1\rangle = (|2,0,0\rangle + |2,1,0\rangle)/\sqrt{2} \tag{15.22}$$

since proper normalization requires $\alpha^2_{1,s_1} = \alpha^2_{2,s_1} = 1/\sqrt{2}$. Similarly, using $E_{2,2}^{(1)} = +3qEa$ yields

$$|E_2, s_2\rangle = \frac{1}{\sqrt{2}}(|2,0,0\rangle - |2,1,0\rangle). \tag{15.23}$$

Since $E_2^{(1)} = 0$ indicates no Stark effect, the remaining two states corresponding to this value can be left unchanged:

$$E_{2,3}^{(1)} = E_{2,4}^{(1)} = 0, \qquad |E_2, s_3\rangle = |2,1,-1\rangle$$
$$|E_2, s_4\rangle = |2,1,1\rangle. \tag{15.24}$$

Now let us interpret these results physically. The operator $q\mathfrak{X}_3$ is the dipole moment operator along the z-axis, and the vanishing of $E_{2,3}^{(1)}$ and $E_{2,4}^{(1)}$ means that the states $|2,1,-1\rangle$ and $|2,1,1\rangle$ are eigenstates of $q\mathfrak{X}_3$ with eigenvalue 0; the degeneracy between these states is not removed in first-order perturbation theory. On the other hand, $|2,0,0\rangle$ and $|2,1,0\rangle$ are not eigenstates of $q\mathfrak{X}_3$; they do not possess definite dipole moments along the z-axis. Two eigenstates of $q\mathfrak{X}_3$ can be formed from these latter states if we take appropriate linear combinations of them; we found that the appropriate linear combinations are given by (15.22) and (15.23) and that these states satisfy

$$q\mathfrak{X}_3|E_2, s_1\rangle = -3qa|E_2, s_1\rangle$$

$$q\mathfrak{X}_3|E_2, s_2\rangle = +3qa|E_2, s_2\rangle.$$

Hence the natural states in terms of which to study the Stark effect are states with definite electric-dipole moments along the direction of the electric field.

15.3 DEGENERACIES PERSISTING BEYOND FIRST ORDER

It may happen that even after first-order corrections to the energy levels have been calculated and appropriate linear combinations of degenerate unperturbed states are formed, degeneracies in the energy spectrum are found to persist. This will happen when there are several states $|E_n, q_1\rangle, \ldots |E_n, q_m\rangle$ having the same unperturbed energy and for which

$$\langle E_n, q_j|\mathfrak{H}_1|E_n, q_k\rangle = 0, \qquad j \neq k$$

and

$$\langle E_n, q_j|\mathfrak{H}_1|E_n, q_j\rangle = \langle E_n, q_k|\mathfrak{H}_1|E_n, q_k\rangle \qquad \text{for all } j, k.$$

In this case we adopt the shorthand notation

$$|E_n, r\rangle = \begin{cases} |E_n, s(0)\rangle & \text{for states that are nondegenerate when first-order energy corrections are made} \\ |E_n, s(0)\rangle = |E_n, q\rangle & \text{for states that were not degenerate to begin with} \\ |E_n, q\rangle & \text{for states that remain degenerate even after first-order corrections;} \end{cases} \tag{15.25}$$

that is, $|E_n, r\rangle$ is a reasonable set of unperturbed states when the effects of first-order energy shifts are taken into account.

For example, in the Stark effect of the previous section we may write $|E_2, r\rangle$, $r = 1, 2, 3, 4$ for the four states, letting $|E_2, r\rangle = |E_2, s_r\rangle$; states $|E_2, 1\rangle$ and $|E_2, 2\rangle$ then are of the type $|E_n, s(0)\rangle$ indicated above and $|E_2, 3\rangle, |E_2, 4\rangle$ are of the type $|E_n, q\rangle$.

When degeneracy does persist through first order, it is appropriate to replace (15.11c) with

$$\langle E_i, r'|\mathfrak{H}_1|E_i, r\rangle = \delta_{r,r'}E_{i,r}{}^{(1)} \tag{15.26}$$

and (15.12c) with (see Eq. (5.12), p. 386)

$$a_{j,r'}^{i,r(1)} = \sum_{r''} \alpha_{r'',r}^{i} \frac{\langle E_j, r'|\mathfrak{H}_1|E_i, r''\rangle}{E_i - E_j}, \tag{15.27}$$

since the states that connect smoothly to $|E_i(\lambda), s(\lambda)\rangle$ may be linear combinations of the remaining degenerate states $|E_i, r\rangle, r = q$. These degeneracies may be removed by higher-order terms in the perturbation expansion. In particular, it may be possible to remove them in second order.

15.4 SECOND-ORDER STATIONARY-STATE PERTURBATION THEORY

If we wish to carry our calculations to second order in λ, Eq. (15.7) is again our key to obtaining solutions. However, we shift notation to that prescribed in Eq. (15.25) and thereby make use of the natural states obtained in first-order calculations. Therefore we replace q with r'' in Eq. (15.7) and let $k = 2$; this yields

$$\sum_{n,r''} \mathfrak{H}_1 a_{n,r''}^{i,s(1)}|E_n, r''\rangle = \sum_{n,r''} [(E_i - E_n)a_{n,r''}^{i,s(2)} + E_{i,s}{}^{(1)}a_{n,r''}^{i,s(1)} + E_{i,s}{}^{(2)}a_{n,r''}^{i,s(0)}]|E_n, r''\rangle, \tag{15.28}$$

where we assume the system to be in a state with energy E_i when $\lambda = 0$, and, when $\lambda = 1$, to be in a state with energy $E_i + E_{i,s}{}^{(1)}$ in first-order perturbation theory. Clearly, solution of the first-order problem is a prerequisite to carrying out a second-order calculation.

The coefficients $a_{n,r''}^{i,r(k)}$ are defined as before through

$$\langle E_n, r|E_i(\lambda), s(\lambda)\rangle = \sum_{k=0}^{\infty} a_{n,r}^{i,s(k)}\lambda^k,$$

and in particular we write

$$a_{n,r}^{i,s(0)} = \delta_{i,n}\alpha_{r,s(0)}^{i} \tag{15.29}$$

where

$$\alpha_{r,s(0)}^{l} = \begin{cases} \delta_{r,s(0)} & \text{if } |E_i, r\rangle \text{ is a state that is nondegenerate in first order} \\ \alpha_{q,s(0)}^{i} & \text{if } |E_i, r\rangle \text{ is a state that is degenerate through first order, and } r = q. \end{cases}$$

When this definition and Eq. (15.29) are taken into account, the inner product of $|E_j, r'\rangle$ with (15.28) gives

$$\sum_{n,r''} \langle E_j, r'|\mathfrak{H}_1|E_n, r''\rangle a_{n,r''}^{i,s(1)}$$
$$= (E_i - E_j)a_{j,r'}^{i,s(2)} + E_{i,s}{}^{(1)}a_{j,r'}^{i,s(1)} + E_{i,s}{}^{(2)}\delta_{i,j}\alpha_{r',s}^{i}. \tag{15.30}$$

First, we let $i = j$ in (15.30) and use (15.27) to get

$$\alpha^i_{r',s} E_{i,s}{}^{(2)} = \sum_{n\neq i} \sum_{r''} \langle E_j, r'|\mathfrak{H}_1|E_n, r''\rangle a^{i,s(1)}_{n,r''}$$

$$= \sum_{r'''} \alpha^i_{r''',s} \sum_{n\neq i} \sum_{r''} \frac{\langle E_j, r'|\mathfrak{H}_1|E_n, r''\rangle\langle E_n, r''|\mathfrak{H}_1|E_i, r'''\rangle}{E_i - E_n}. \tag{15.31}$$

This can be considered a matrix equation, and, in fact, an eigenvalue equation for $E_{i,s}{}^{(2)}$. Once the matrix elements

$$M_{r',r'''} \equiv \sum_{n\neq i} \sum_{r''} \frac{\langle E_i, r'|\mathfrak{H}_1|E_n, r''\rangle\langle E_n, r''|\mathfrak{H}_1|E_i, r'''\rangle}{E_i - E_n}$$

have been calculated for each r', r''', Eq. (15.31) becomes

$$\begin{bmatrix} M_{11} - E_{i,s}{}^{(2)} & M_{12} & \cdots & M_{1m} \\ M_{21} & M_{22} - E_{i,s}{}^{(2)} & \cdots & M_{2m} \\ \vdots & \vdots & & \vdots \\ M_{m1} & M_{m2} & \cdots & M_{mm} - E_{i,s}{}^{(2)} \end{bmatrix} \begin{Bmatrix} \alpha^i_{1,s} \\ \alpha^i_{2,s} \\ \vdots \\ \alpha^i_{m,s} \end{Bmatrix} = 0. \tag{15.32}$$

The eigenvalues of $E_{i,s}{}^{(2)}$ are those values of $E_{i,s}{}^{(2)}$ for which the determinant of the matrix vanishes, and for each such value there is a solution of Eq. (15.32) for the $\alpha^i_{r',s}$. Each set of α^i's then defines an appropriate eigenstate

$$|E_i, s(0)\rangle = \sum_{r'} \alpha^i_{r',s}|E_i, r'\rangle$$

to which $|E(\lambda), s(\lambda)\rangle$ reduces when $\lambda \to 0$, at least to second order.

We now assume that Eq. (15.32) has been solved, i.e. that all the $E^{(2)}_{i,s(0)}$ and $\alpha^i_{r',s}$ have been found, so that we have in hand the states $|E_i, s(0)\rangle$ in second order. Then if we replace the first-order states $|E_n, r''\rangle$, etc. with the new states $|E_n, s'(0)\rangle$, etc., Eq. (15.31) becomes

$$E^{(2)}_{i,s(0)} = \sum_{n\neq i} \sum_{s'(0)} \frac{\langle E_i, s(0)|\mathfrak{H}_1|E_n, s'(0)\rangle\langle E_n, s'(0)|\mathfrak{H}_1|E_i, s(0)\rangle}{E_i - E_n}$$

$$= \sum_{n\neq i} \sum_{s'(0)} \frac{|\langle E_i, s(0)|\mathfrak{H}_1|E_n, s'(0)\rangle|^2}{E_i - E_n}. \tag{15.33}$$

This is the second-order analog of the first-order Eq. (15.11b).

Letting $i \neq j$ in (15.30) and replacing r', r'' with $s'(0)$, $s''(0)$ gives the equation

$$(E_i - E_j)a^{i,s(2)}_{j,s'} = \sum_{n,s''} \langle E_j, s'(0)|\mathfrak{H}_1|E_n, s''(0)\rangle a^{i,s(1)}_{n,s''} - E^{(1)}_{i,s(0)} a^{i,s(1)}_{j,s'},$$

or, since $a_{i,s'(0)}^{i,s(1)}$ may be chosen equal to 0,

$$a_{j,s'}^{i,s(2)} = \sum_{\substack{n \neq i \\ s''(0)}} \frac{\langle E_j, s'(0)|\mathfrak{H}_1|E_n, s''(0)\rangle\langle E_n, s''(0)|\mathfrak{H}_1|E_i, s(0)\rangle}{(E_i - E_j)(E_i - E_n)}$$
$$- \frac{\langle E_i, s(0)|\mathfrak{H}_1|E_i, s(0)\rangle\langle E_j, s'(0)|\mathfrak{H}_1|E_i, s(0)\rangle}{(E_i - E_j)^2}. \quad (15.34)$$

This is the second-order analog of Eq. (15.13b).

The generalization to higher orders follows the same pattern. In particular, some degeneracy may remain after second-order calculations have been carried out, and one then repeats the notational adjustment outlined in the preceding section in anticipation of trying to remove in third order those degeneracies that remained even in second order. However, in third and higher orders, the calculations are sufficiently laborious to discourage the continuation of a perturbation-theoretic approach to a problem that demands corrections beyond first and second orders, unless the problem under consideration has particular theoretical value or certain properties that simplify the calculational difficulties.

A simple conceptual device is available to aid one's understanding of the energy shifts produced in various orders of perturbation theory. The first-order energy shift $E_{i,r}{}^{(1)}$ is an eigenvalue of the perturbation term in the Hamiltonian; it is the energy contributed to the system *directly* as a result of the presence of the perturbation. The form of the expression on the right-hand side of Eq. (15.33) implies that not only the state $|E_i, s(0)\rangle$, but *all* states $|E_n, s'(0)\rangle$ such that $\langle E_n, s'(0)|\mathfrak{H}_1|E_i, s(0)\rangle \neq 0$ somehow make contributions to the second-order shift in the energy of the unperturbed level E_i. A conceptual model for this is that each term in (15.33) represents a contribution to the energy level arising from a transition of the system to a state $|E_n, s'(0)\rangle$ and a subsequent return of the system to $|E_i, s(0)\rangle$; the transitions are induced by the perturbation $\mathfrak{H}_1$. These transitions, however, are unobservable; clearly they do not conserve energy. Only the "round trip" from $|E_i, s(0)\rangle$ to $|E_n, s'(0)\rangle$ and *back* again conserves energy, but the effect of the possibility of conducting such trips, thanks to the perturbation, is the second-order energy shift. The unobservable transitions are known as *virtual* transitions. We may think of third-order energy shifts as arising from three-legged round trips, from $|E_i, s(0)\rangle$ to $|E_n, s'(0)\rangle$ to $|E_m, s''(0)\rangle$ and back to $|E_i, s(0)\rangle$, for example, and fourth-order shifts are contributed by round-trip combinations of four virtual transitions, etc.

15.5 TIME-DEPENDENT PERTURBATION THEORY

We have considered an approximation technique that may be applied when $\mathfrak{H}_1$ meets a "smallness" criterion (its eigenvalues must be small compared with the separations between adjacent energy levels of the unperturbed system) and is stationary. We now develop a technique for handling the case when $\mathfrak{H}_1 = \mathfrak{H}_1(t)$ is time-dependent but still "small".

Let $\{|E, q\rangle\}$ be a complete orthonormal set of eigenstates of the unperturbed Hamiltonian $\mathfrak{H}_0$ with

$$\mathfrak{H}_0|E, q\rangle = E|E, q\rangle. \tag{15.35}$$

Assume next that at time τ a physical system with exact Hamiltonian $\mathfrak{H}(\lambda, t) = \mathfrak{H}_0 + \lambda\mathfrak{H}_1(t)$ is in state $|E, q\rangle$, which is not necessarily an eigenstate of $\mathfrak{H}(\lambda, \tau)$. At a later time $\tau + t$, the system is in state

$$|E, q(\lambda, \tau + t)\rangle,$$

which satisfies the differential equation

$$i\hbar\frac{\partial}{\partial t}|E, q(\lambda, \tau + t)\rangle = \mathfrak{H}(\lambda, \tau + t)|E, q(\lambda, \tau + t)\rangle \tag{15.36}$$

and satisfies the initial condition

$$|E, q(\lambda, \tau)\rangle = |E, q\rangle.$$

We must have

$$|E, q(\lambda, \tau)\rangle = |E, q(0, \tau)\rangle = |E, q\rangle \tag{15.37}$$

and

$$|E, q(0, \tau + t) = e^{-it\mathfrak{H}_0/\hbar}|E, q\rangle = e^{-itE/\hbar}|E, q\rangle. \tag{15.38}$$

Solving the differential equation (15.36) with its appropriate boundary condition is not a simple matter in general. Hence we shall not attack (15.36) and (15.37) directly, but shall attempt a perturbation solution, once again basing our approach on the analyticity of $|E, q(\lambda, \tau + t)\rangle$ as a function of the perturbation strength parameter λ.

We begin by expanding $|E, q(\lambda, \tau + t)\rangle$ in terms of the complete set $\{|E, q(0, \tau + t)\rangle\}$ of eigenstates of $\mathfrak{H}_0$ evaluated at time $\tau + t$:

$$\begin{aligned}|E, q(\lambda, \tau + t)\rangle &= \underset{E',q'}{\mathsf{S}}\, |E', q'(0, \tau + t)\rangle\langle E', q'(0, \tau + t)|E, q(\lambda, \tau + t)\rangle \\ &= \underset{E',q'}{\mathsf{S}}\, a_{E',q';\,E,q}(\lambda, t)e^{-iE't/\hbar}|E', q'(0, \tau)\rangle \\ &= \underset{E',q'}{\mathsf{S}}\, a_{E',q';\,E,q}(\lambda, t)e^{-iE't/\hbar}|E', q'\rangle.\end{aligned} \tag{15.39}$$

Since $|E, q(\lambda, \tau + t)\rangle$ is supposed to be analytic in λ, we may write

$$a_{E',q';\,E,q}(\lambda, t) = \sum_{k=0}^{\infty} \lambda^k a^{(k)}_{E',q';\,E,q}(t) \tag{15.40}$$

or

$$|E, q(\lambda, \tau + t)\rangle = \sum_{k=0}^{\infty} \lambda^k \mathop{\mathrm{S}}_{E',q'} a^{(k)}_{E',q';E,q}(t) e^{-iE't/\hbar}|E', q'\rangle. \tag{15.41}$$

Substitution of (15.41) into (15.36) produces the equation

$$\sum_{k=0}^{\infty} \lambda^k \mathop{\mathrm{S}}_{E',q'} \{i\hbar \dot{a}^{(k)}_{E',q';E,q}(t) + E' a^{(k)}_{E',q';E,q}(t) - a^{(k)}_{E',q';E,q}(t)\mathfrak{H}_0$$
$$- a^{(k-1)}_{E',q';E,q}(t)\mathfrak{H}_1(\tau + t)\} e^{-iE't/\hbar}|E', q'\rangle = 0.$$

Now $|E', q'\rangle$ is an eigenstate of $\mathfrak{H}_0$ so that the third term cancels the second. Since this must hold for all λ the remaining portion of the coefficient of each power of λ must vanish, and hence

$$\mathop{\mathrm{S}}_{E',q'} [i\hbar \dot{a}^{(k)}_{E',q';E,q}(t) - a^{(k-1)}_{E',q';E,q}(t)\mathfrak{H}_i(\tau + t)] e^{-iE't/\hbar}|E', q'\rangle = 0.$$

The familiar procedure of taking the inner product of $|E'', q''(0, \tau + t)\rangle = \exp(-iE''t/\hbar)|E'', q''\rangle$ with this equation yields

$$i\hbar \dot{a}^{(k)}_{E'',q'';E,q}(t) = \mathop{\mathrm{S}}_{E',q'} a^{(k-1)}_{E',q';E,q}(t) e^{i(E''-E')t/\hbar}\langle E'', q''|\mathfrak{H}_1(\tau + t)|E', q'\rangle,$$

or

$$a^{(k)}_{E'',q'';E,q}(t) - a^{(k)}_{E'',q'';E,q}(0)$$
$$= \frac{1}{i\hbar}\int_0^t dt' \mathop{\mathrm{S}}_{E',q'} a^{(k-1)}_{E',q';E,q}(t') e^{i(E''-E)t'/\hbar}\langle E'', q''|\mathfrak{H}_1(\tau + t')|E', q'\rangle.$$

This is the key equation for our approach to the perturbation problem.

For $k < 0$, we may set $a^{(k)}_{E',q';E,q}(t) = 0$; hence putting $k = 0$ in Eq. (15.42) gives $a^{(0)}_{E'',q'';E,q}(t) = a^{(0)}_{E'',q'';E,q}(0)$, for all t. Furthermore, if we set $t = 0$ in Eq. (15.41) and compare the result with Eq. (15.37), we find

$$a^{(k)}_{E'',q'';E,q}(0) = \begin{cases} 0, & \text{if } k > 0 \\ \delta(E'' - E)\delta(q'' - q), & \text{if } k = 0. \end{cases}$$

Therefore

$$a^{(0)}_{E'',q'';E,q}(t) = \delta(E'' - E)\delta(q'' - q) \tag{15.43}$$

and from Eq. (15.42),

$$a^{(k)}_{E'',q'';E,q}(t) = \frac{1}{i\hbar}\int_0^t dt' \mathop{\mathrm{S}}_{E',q'} a^{(k-1)}_{E',q';E,q}(t') e^{i\omega_{21}t'}\langle E'', q''|\mathfrak{H}_1(\tau + t')|E', q'\rangle \tag{15.44}$$

for $k \geq 1$; here $\omega_{21} = (E'' - E')/\hbar$.

Substitution of (15.43) into (15.44) when $k = 1$ gives the first-order result

$$a^{(1)}_{E',q';\,E,q}(t) = \frac{1}{i\hbar}\int_0^t dt'\, e^{i\omega_{10}t'}\langle E', q'|\mathfrak{H}_1(\tau + t')|E, q\rangle. \tag{15.45}$$

Beginning with this, it is easy to show by mathematical induction that the higher-order a-coefficients are given by

$$\begin{aligned} &a^{(k)}_{E^{(k)},q^{(k)};\,E,q}(t) \\ &= \frac{1}{(i\hbar)^k}\int_0^{t'} dt' \int_0^{t'} dt'' \cdots \int_0^{t^{(k-1)}} dt^{(k)} \underset{E^{(k-1)},q^{(k-1)}}{\mathsf{S}} \cdots \\ &\cdots \underset{E',q'}{\mathsf{S}} \exp\left(i \sum_{j=1} \omega_{j,j-1} t^{(k+1-j)}\right) \langle E^{(k)}, q^{(k)}|\mathfrak{H}_1(\tau + t')|E^{(k-1)}, q^{(k-1)}\rangle \cdots \\ &\cdots \langle E', q'|\mathfrak{H}_1(\tau + t^{(k)})|E, q\rangle, \end{aligned} \tag{15.46}$$

where $\omega_{j,j-1} = (E^{(j)} - E^{(j-1)})/\hbar$.

The object of time-dependent perturbation theory is to obtain these coefficients, since they provide a method for evaluating the probability of transition from one unperturbed state, $|E, q\rangle$, to another, $|E', q'\rangle$. To see this, we consider $a^{(1)}_{E',q';\,E,q}(t)$ whose definition, namely

$$\begin{aligned} a^{(1)}_{E',q';\,E,q}(t) &= \frac{\partial}{\partial\lambda}\langle E', q'(0, \tau + t)|E, q(\lambda, \tau + t)\rangle|_{\lambda=0} \\ &= e^{iE't/\hbar}\frac{\partial}{\partial\lambda}\langle E', q'|E, q(\lambda, \tau + t)\rangle|_{\lambda=0} \end{aligned}$$

indicates that it is a first-order approximation (in terms of the expansion parameter λ) to the probability amplitude for a system beginning in state $|E, q\rangle$ at time τ, and evolving in time according to the dynamical assumption that the Hamiltonian is

$$\mathfrak{H}(\lambda, t) = \mathfrak{H}_0 + \lambda\mathfrak{H}_1(t),$$

to be observed at a later time $\tau + t$ in the state

$$|E', q'(0, \tau + t)\rangle = e^{-iE't/\hbar}|E', q'\rangle.$$

Thus, in first order, the probability of transition from $|E, q\rangle$ to $|E', q'\rangle$ is

$$\begin{aligned} P_{10}{}^{(1)}(t) &\equiv |a^{(1)}_{E',q';\,E,q}(t)|^2 \\ &= \left|\frac{1}{i\hbar}\int_0^t dt'\, e^{i\omega_{10}t'}\langle E', q'|\mathfrak{H}_1(\tau + t')|E, q\rangle\right|^2. \end{aligned} \tag{15.47}$$

Higher approximations are obtained by replacing $a^{(1)}$ with $\sum_{k=1}^n a^{(k)}$, where n is the order of approximation.

15.6 CALCULATION OF TRANSITION RATES

We cannot make significant further progress without assuming a specific time-dependence for $\mathfrak{H}_1(t)$. The simplest time-dependence is

$$\mathfrak{H}_1(\tau + t') = \begin{cases} \mathfrak{H}_1, & 0 < t' < t \\ 0, & \text{otherwise}; \end{cases} \tag{15.48}$$

that is, the perturbation is "turned on" only during the interval of interest in our perturbation formulas. This means that it is sensible to think of the system as being in a state $|E, q\rangle$ before time τ when the perturbation is turned on, and to look for transitions to states $|E', q'\rangle$ after time $\tau + t$, for $\mathfrak{H}_0$ is the exact Hamiltonian at such times, and the eigenstates of $\mathfrak{H}_0$ are "natural" states for the system at those times.

Using Eq. (15.48), we obtain

$$\begin{aligned} a^{(1)}_{E',q';E,q}(t) &= \frac{1}{i\hbar}\langle E', q'|\mathfrak{H}_1|E, q\rangle \int_0^t dt'\, e^{i\omega_{10}t'} \\ &= \frac{1}{i\hbar}\langle E', q'|\mathfrak{H}_1|E, q\rangle e^{it\omega_{10}/2}\, \frac{\sin(\omega_{10}t/2)}{\omega_{10}/2}, \end{aligned} \tag{15.49}$$

and

$$P_{10}{}^{(1)}(t) = \frac{1}{\hbar^2}|\langle E', q'|\mathfrak{H}_1|E, q\rangle|^2 \frac{\sin^2(\omega_{10}t/2)}{(\omega_{10}/2)^2}. \tag{15.50}$$

Usually we are really interested in the transition *rates*, that is, the probability per unit time for a given transition to occur; hence we calculate

$$\frac{d}{dt}P_{10}{}^{(1)}(t) = \dot{P}_{10}{}^{(1)}(t) = \frac{2}{\hbar^2}|\langle E', q'|\mathfrak{H}_1|E, q\rangle|^2 \frac{\sin \omega_{10}t}{\omega_{10}}. \tag{15.51}$$

In general we shall be interested in this quantity when t is very long compared with a characteristic oscillation period for the transition, $t \gg 1/\omega_{10} = \hbar(E' - E)$, and therefore we should understand the behavior of $(\sin \omega_{10}t)/\omega_{10}$ in the limit of large t.

The behavior of $(\sin \omega_{10}t)/\omega_{10}$ as a function of ω_{10} is illustrated in Fig. 8.1. The general shape of the function indicates that it might be considered an approximation to the Dirac δ-function. To begin with, the central peak can be made as narrow as desired provided t is allowed to become arbitrarily large. Thus note, for example, that the first extremum away from the central peak occurs at $\omega'_{10} \simeq 4.49/t$; hence as $t \to \infty$, $\omega'_{10} \to 0$. The magnitude of this relative minimum, however, is about $0.217t$, so that as $t \to \infty$ it becomes correspondingly large. Evidently it is not fair, then, to say that (for $\omega_{10} \neq 0$)$(\sin \omega_{10}t)/\omega_{10} \to 0$ as $t \to \infty$, even though the noncentral peaks may all be made arbitrarily close to the origin; that is, the feature $\delta(\omega_{10}) = 0$ for $\omega_{10} \neq 0$—which we would like any candidate

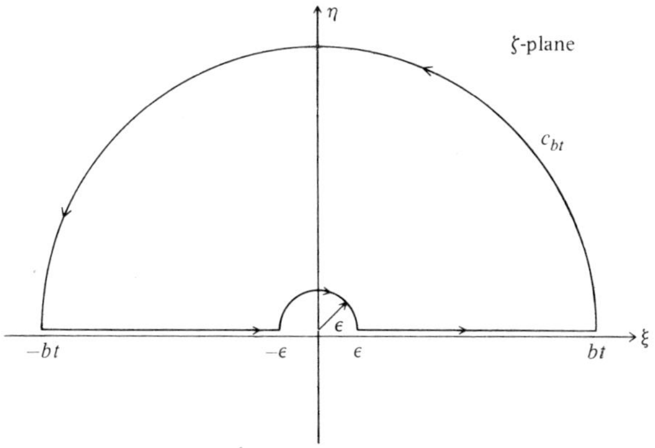

Fig. 15.1. Contour for evaluating $\int_{-b}^{0} \frac{\sin \omega_{10} t}{\omega_{10}} d\omega_{10}$.

for becoming a δ-function to have—eludes the function $(\sin \omega_{10}t)/\omega_{10}$ even in the limit of large t.

Nevertheless, for any two numbers a and b of like sign,

$$\int_a^b \frac{\sin \omega_{10} t}{\omega_{10}} d\omega_{10} = \int_{at}^{bt} \frac{\sin \xi}{\xi} d\xi \tag{15.52}$$

has the property

$$\lim_{t\to\infty} \int_a^b \frac{\sin \omega_{10} t}{\omega_{10}} d\omega_{10} = \lim_{t\to\infty} \int_{at}^{bt} \frac{\sin \xi}{\xi} d\xi = 0; \tag{15.53}$$

in fact, for any function $\varphi(\omega_{10})$ bounded on $[a, b]$,

$$\lim_{t\to\infty} \int_a^b \frac{\sin \omega_{10} t}{\omega_{10}} \varphi(\omega_{10})\, d\omega_{10} = 0.$$

Thus when it appears under the integral sign—i.e. "on the average"—$\lim_{t\to\infty} (\sin \omega_{10}t)/\omega_{10}$ does behave like 0 when $\omega_{10} \neq 0$. This is what we really are concerned about when we are studying δ-functions, and therefore $\lim_{t\to\infty} (\sin \omega_{10}t)/\omega_{10}$ remains a candidate.

To be elected, however, our candidate must have the property that when it appears under the integral sign, as in (15.52), but when $a < 0$ and $b > 0$, the integral is unity in the limit as $t \to \infty$. To simplify things a bit, we may just as well let $a = -b$; this in no way affects our result, thanks to (15.53). Then we take the integral sign to mean "principal value of the integral" and we have

$$\int_{-b}^{b} \frac{\sin \omega_{10} t}{\omega_{10}} d\omega_{10} = \text{Im} \int_{-b}^{b} \frac{e^{i\omega_{10}t}}{\omega_{10}} d\omega_{10} = \text{Im} \int_{-bt}^{bt} \frac{e^{i\xi}}{\xi} d\xi.$$

We now replace the real variable ξ with the complex variable ζ and use a contour in the ζ plane to evaluate this principal part in the limit of large t. The contour to be used is illustrated in Fig. 15.1; we shall finally let $t \to \infty$ and $\epsilon \to 0$ on this contour. Thus, since $\exp(i\zeta)/\zeta$ has no singularities inside the contour,

$$\int_{C_{bt}} \frac{e^{i\zeta}}{\zeta} d\zeta + \int_{-bt}^{-\epsilon} \frac{e^{i\zeta}}{\zeta} d\zeta + \int_{\pi}^{0} \frac{\exp(i\epsilon e^{i\varphi})}{\epsilon e^{i\varphi}} i\epsilon e^{i\varphi}\, d\varphi + \int_{\epsilon}^{bt} \frac{e^{i\zeta}}{\zeta} d\zeta = 0.$$

Along C_{bt}, ζ always has a positive imaginary part so that the integrand vanishes exponentially on this part of the contour as $t \to \infty$; the first term therefore can be ignored. As $\epsilon \to 0$, the exponent in the next-to-last term approaches unity. Hence,

$$\lim_{\substack{t\to\infty \\ \epsilon\to 0}} \left[\int_{-bt}^{-\epsilon} \frac{e^{i\zeta}}{\zeta} d\zeta + \int_{\epsilon}^{bt} \frac{e^{i\zeta}}{\zeta} d\zeta \right] = i\pi.$$

The left-hand member of this equation is the principal value whose imaginary part is the desired integral, and we therefore obtain

$$\lim_{t\to\infty} \int_{-b}^{b} \frac{\sin \omega_{10} t}{\omega_{10}} d\omega_{10} = \pi$$

and therefore

$$\lim_{t\to\infty} \frac{\sin \omega_{10} t}{\pi \omega_{10}}$$

has the δ-function properties we seek.

It is clear, then, that for $\omega_{10} t \gg \pi$, so that $t \to \infty$ provides a good approximation, Eq. (15.51) becomes

$$\dot{P}_{10}{}^{(1)} \equiv \lim_{t\to\infty} \dot{P}_{10}{}^{(1)}(t) = \frac{2\pi}{\hbar^2} |\langle E', q' | \mathfrak{H}_1 | E, q \rangle|^2 \delta(\omega_{10})$$

$$= \frac{2\pi}{\hbar} |\langle E', q' | \mathfrak{H}_1 | E, q \rangle|^2 \delta(E' - E). \tag{15.54}$$

The presence of the δ-function is comforting in that it guarantees conservation of energy; on the other hand, it is disquieting in that apparently it causes transition rates to be infinite whenever $\langle E', q' | \mathfrak{H}_1 | E, q \rangle$ is nonvanishing and $E' = E$.

The latter discomfort is overcome, however, when we realize that there may be infinitely many states available with energies lying within an energy interval of width dE' about energy E'. We let $\rho(E')\, dE'$ be the number of such states, and observe that the probability per unit time of transition to any state $|E', q'\rangle$ is given by

$$w_{10}{}^{(1)} \equiv \int_0^\infty \dot{P}_{10}{}^{(1)} \rho(E')\, dE' = \frac{2\pi}{\hbar} |\langle E, q' | \mathfrak{H}_1 | E, q \rangle|^2 \rho(E). \tag{15.55}$$

Thus, if we essentially add up the rates of transition to all possible final energy levels, taking into account the density of states, $\rho(E')$, near each energy, we obtain the result (15.55) which does not have the unpleasant feature of being necessarily either vanishing or infinite.

15.7 SCATTERING OF PARTICLES IN A BEAM TO ILLUSTRATE THE DENSITY-OF-STATES FACTOR

To understand the meaning of the density-of-states factor, $\rho(E)$, it is useful to consider the problem of particles being scattered from a beam by way of a scattering center that interacts with the particles through an interaction term $\mathfrak{H}_1$ in the Hamiltonian for the particles. The states available to the scattered beam are degenerate for two reasons:

1. There are infinitely many momentum eigenstates for each energy, different states corresponding to different directions in space.

2. If the incident particles have spin s and the target has spin S, and if the interaction is spin-independent, there are $(2s + 1)(2S + 1)$ spin states for each momentum eigenstate.

In addition, there may be arbitrarily many eigenstates within a small interval about the unperturbed energy E of an incident particle if the volume of the system is sufficiently large.

If we impose periodic boundary conditions on the system, confining our attention to a cubical box with edges of length L, then the allowed values of the magnitude of momentum are

$$|\mathbf{p}| = \hbar k = \hbar \frac{2\pi}{L}(n_1{}^2 + n_2{}^2 + n_3{}^2)^{1/2},$$

where the n_i are integers. Now each wave vector

$$\mathbf{k} = \frac{2\pi}{L}(\hat{\mathbf{i}}_1 n_1 + \hat{\mathbf{i}}_2 n_2 + \hat{\mathbf{i}}_3 n_3)$$

may be thought of as a lattice point in k-space, and, as we have seen in Chapter 5, each cube in the k-space latticework has 8 lattice points at its corners, and each corner is shared by 8 cubes; hence one point may be assigned to each cube. Thus there is one momentum eigenstate per cube of k-space, or one such state per volume $(2\pi/L)^3$ of k-space. Hence the number of momentum eigenstates in an element of volume d^3k of k-space is

$$\frac{d^3k}{(2\pi/L)^3} = \left(\frac{L}{2\pi}\right)^3 k^2\, dk\, d\Omega$$

where $d\Omega = \sin\theta\, d\theta\, d\varphi$ is an element of solid angle.

Since

$$E = \frac{\hbar^2}{2M}k^2, \qquad \frac{dE}{dk} = \frac{\hbar^2}{M}k,$$

this can be written in the form

$$\left(\frac{L}{2\pi}\right)^3 k^2 \frac{dk}{dE}\, dE\, d\Omega = \rho(E;\theta,\varphi)\, d\Omega\, dE,$$

where

$$\rho(E;\theta,\varphi) = \left(\frac{L}{2\pi}\right)^3 \frac{\sqrt{2M^3E}}{\hbar^3} = \left(\frac{L}{2\pi}\right)^3 \frac{Mk}{\hbar^2} \tag{15.56}$$

is the number of momentum states per unit energy interval per steradian. If the perturbation is spin-independent, we multiply by the spin degeneracy and get

$$\rho_{s,S}(E;\theta,\varphi) = (2s+1)(2S+1)\left(\frac{L}{2\pi}\right)^3 \frac{\sqrt{2M^3E}}{\hbar^3}$$

as the total number of final states per unit energy interval per steradian.

Integration of (15.56) over angles, ignoring spin degeneracy, yields

$$\rho(E) = \frac{L^3}{2\pi^2}\frac{\sqrt{2M^3E}}{\hbar^3}$$

as the density of states available to a final particle near energy E. Generally speaking, however, it is best not to integrate over angles until the factor $\rho_s(E;\theta,\varphi)$ is inserted into the expression for the transition rate, since $|\langle E, q'|\mathfrak{H}_1|E, q\rangle|$ might well depend upon θ or φ.

Hence the rate at which particles make transitions into momentum eigenstates with the momentum within $d\Omega$ of angles θ, φ is

$$dw_{10}{}^{(1)} = \frac{dw_{10}{}^{(1)}}{d\Omega}\, d\Omega = \frac{2\pi}{\hbar}|\langle E', q'|\mathfrak{H}_1|E, q\rangle|^2 \frac{Mk}{\hbar^2}\left(\frac{L}{2\pi}\right)^3 d\Omega \tag{15.57}$$

and

$$w_{10}{}^{(1)} = \frac{Mk}{\hbar^2}\left(\frac{L}{2\pi}\right)^3 \frac{2\pi}{\hbar}\int |\langle E, q'|\mathfrak{H}_1|E, q\rangle|^2\, d\Omega \tag{15.58}$$

is the rate at which transitions to momentum eigenstates in any direction occur.

15.8 DIMENSIONAL CONSIDERATIONS; THE SCATTERING CROSS-SECTION

Let us assume the initial-state vector to be that of an eigenstate of linear momentum with spin projection μ along the z-axis, i.e.

$$|E, q\rangle \equiv |\mathbf{k}; s, \mu\rangle$$

from which we obtain

$$\psi_{s,\mu'}(\mathbf{x}) = \langle \mathbf{x}\,; s, \mu'|\mathbf{k}\,; s, \mu\rangle = L^{-3/2}e^{i\mathbf{k}\cdot\mathbf{x}}\chi_{s,\mu'}$$

where $\chi_{s,\mu'}$ is the spin wave function with μth component $\chi^{\mu}_{s,\mu'} = \langle s, \mu'|s, \mu\rangle$ and the factor $L^{-3/2}$ is a consequence of our use of box normalization. For a local interaction, we write

$$\langle \mathbf{x}'\,; s, \mu'|\mathfrak{H}_1|\mathbf{x}\,; s, \mu\rangle = V_{\mu',\mu}(\mathbf{x})\delta^3(\mathbf{x} - \mathbf{x}')$$

where $V_{\mu',\mu}(\mathbf{x})$ represents a spin-dependent potential energy; this leads to

$$\langle \mathbf{k}'\,; s, \mu'|\mathfrak{H}_1|\mathbf{k}\,; s, \mu\rangle$$

$$= \sum_{\mu'',\mu'''}\int d^3x' \int d^3x \langle \mathbf{k}'\,; s, \mu'|\mathbf{x}'\,; s, \mu''\rangle V_{\mu'',\mu'''}(\mathbf{x})\delta^3(\mathbf{x} - \mathbf{x}')\langle \mathbf{x}\,; s, \mu''|\mathbf{k}\,; s, \mu\rangle$$

$$= \int d^3x\, L^{-3}e^{i(\mathbf{k}-\mathbf{k}')\cdot\mathbf{x}}V_{\mu',\mu}(\mathbf{x}) \equiv W_{\mu',\mu}(\mathbf{k}' - \mathbf{k})$$

$$\equiv L^{-3}U_{\mu',\mu}(\mathbf{k}' - \mathbf{k}). \tag{15.59}$$

Here $W_{\mu',\mu}(\mathbf{k}' - \mathbf{k})$ clearly has the dimensions of energy, since $V_{\mu',\mu}(\mathbf{x})$ is a potential energy function. The function $U_{\mu',\mu}(\mathbf{k}' - \mathbf{k}) = \sqrt{2\pi}FT_{\mathbf{k}'-\mathbf{k}}[V_{\mu',\mu}(\mathbf{x})]$ has dimensions (distance)^3(energy).

To obtain the first-order differential transition rate, we substitute (15.59) into Eq. (15.57) and sum over final-state spins:

$$dw_{10}{}^{(1)} = \sum_{\mu'} \frac{Mk}{(2\pi)^2\hbar^3}L^{-3}|U_{\mu',\mu}(\mathbf{k}' - \mathbf{k})|^2\, d\Omega', \tag{15.60}$$

where we have put a prime on the solid angle element to emphasize that it is the element of solid angle into which the outgoing, or final, particle emerges. The only nontrivial step remaining in the calculation of the transition rate in first order, then, is the computation of the Fourier transform of the potential function $V_{\mu',\mu}(\mathbf{x})$ in a specific problem. This approximation, in which we have assumed the outgoing particle to be in an eigenstate of linear momentum, $|\mathbf{k}'\,; s, \mu'\rangle$, is known as the first-order Born approximation; we shall discuss it in greater detail later.

Equation (15.60) still presents a problem, however; the transition rate is inversely proportional to the volume of the system although we expect this volume to be irrelevant physically. Let us check the dimensions of Eq. (15.60) to be sure that they, at least, are correct. Mass has the dimensions energy/(velocity)^2 = energy(time)^2(distance)$^{-2}$ and the wave number k has the dimensions (distance)$^{-1}$; hence we have

$$[\text{energy(time)}^2\text{(distance)}^{-2}]\frac{(\text{distance})^{-1}}{(\text{energy}\cdot\text{time})^3}(\text{distance})^{-3}$$

$$\cdot\,[(\text{distance})^3\text{energy}]^2 = (\text{time})^{-1},$$

since $d\Omega'$ is a dimensionless quantity. Hence we do have the correct dimensions for the probability of a transition per unit time. The presence of L^{-3} in Eq. (15.60) was essential in this dimensional check on our expression for the transition rate.

What we really are interested in, however, is not the transition rate given in Eq. (15.60) but rather the rate at which transitions occur, divided by the rate at which particles become available to make transitions. The latter is most easily obtained by considering the flux of incident particles, which is just the spatial density of incident particles times their velocity:

$$\mathbf{I} = \rho_i \mathbf{v} = \rho_i(\hbar/M)\mathbf{k}.$$

Our normalization of $\psi_{s,\mu'}(\mathbf{x})$ indicated that we assume each particle to occupy a volume L^3, and hence $\rho_i = L^{-3}$, or

$$\mathbf{I} = L^{-3}\frac{\hbar}{M}\mathbf{k}, \tag{15.61}$$

which has dimensions $(\text{distance})^{-2}(\text{time})^{-1}$. To obtain the actual rate at which incident particles become available for transition, we must multiply the magnitude of this flux by the area of the incident beam. However, it is better to deal with a quantity that is characteristic of the process being described and independent of such arbitrary experimental parameters as the cross-sectional area of the incident beam; therefore we simply divide (15.60) by the magnitude of $\mathbf{I}$ as given by (15.61) to obtain

$$\frac{d\sigma_{10}{}^{(1)}}{d\Omega'} = \frac{1}{|\mathbf{I}|}\frac{dw_{10}{}^{(1)}}{d\Omega'} = \sum_{\mu'} \frac{M^2}{(2\pi)^2\hbar^4}|U_{\mu',\mu}(\mathbf{k}' - \mathbf{k})|^2, \tag{15.62}$$

which is known as the differential scattering cross-section for the process to first order in perturbation theory, and has the dimensions of area. Essentially, it is the cross-sectional area, perpendicular to the path of the incoming particle, through which the incident particle must pass if it is to be scattered into solid angle $d\Omega'$ about the direction $\mathbf{k}'$ by the potential $V_{\mu',\mu}(\mathbf{x})$.

In general $d\sigma/d\Omega'$ depends on the direction of $\mathbf{k}' = (k', \theta', \varphi')$ and hence on the angular coordinates θ' and φ' of the emerging particle. An important parameter obtained from the differential cross-section is the overall cross-section,

$$\sigma_{10}{}^{(1)} \equiv \int \frac{d\sigma_{10}{}^{(1)}}{d\Omega'}\,d\Omega' = \frac{1}{|\mathbf{I}|}w_{10}{}^{(1)}.$$

The reader should compare this discussion of the cross-section with that given for the classical cross-section in Section 7.1.

15.9 HARMONIC PERTURBATIONS

An important class of perturbing time-dependent interactions is obtained by modifying (15.48) to read

$$\mathfrak{H}_1(\tau + t') = \begin{cases} \mathfrak{H}_1 \cos(\omega_0 t' + \varphi), & 0 < t' < t \\ 0, & \text{otherwise.} \end{cases} \tag{15.63}$$

Then, in place of (15.49), we have

$$a_{E',q';E,q}^{(1)}(t) = \frac{1}{i\hbar}\langle E', q'|\mathfrak{H}_1|E, q\rangle$$

$$\cdot \frac{1}{2}\Big\{\exp[i(\omega_{10} + \omega_0)t/2 + i\varphi]\frac{\sin[(\omega_{10} + \omega_0)t/2]}{(\omega_{10} + \omega_0)/2}$$

$$+ \exp[i(\omega_{10} - \omega_0)t/2 - i\varphi]\frac{\sin[(\omega_{10} - \omega_0)t/2]}{(\omega_{10} - \omega_0)/2}\Big\}.$$

In the limit of long time intervals, $t \to \infty$, the quantity in braces becomes

$$\pi\{\exp[i(\omega_{10} + \omega_0)t/2 + i\varphi]\delta(\omega_{10} + \omega_0)$$

$$+ \exp[i(\omega_{10} - \omega_0)t/2 - i\varphi]\delta(\omega_{10} - \omega_0)\},$$

so that either $\omega_{10} + \omega_0 = (E' - E + \hbar\omega_0)/\hbar = 0$ or $\omega_{10} - \omega_0 = (E' - E - \hbar)/\hbar = 0$, but not both (when $\omega_0 \neq 0$). Then

$$|a_{E',q';E,q}^{(1)}(t)|^2 = \frac{1}{\hbar^2}|\langle E', q'|\mathfrak{H}_1|E, q\rangle|^2\frac{\sin^2[(\omega_{10} \pm \omega_0)t/2]}{4[(\omega_{10} \pm \omega_0)/2]^2},$$

which differs from (15.50) only through the factor $\frac{1}{4}$ and the shift $\omega_{10} \to \omega_{10} \pm \omega_0$. Therefore (15.54) becomes

$$P_{10}^{(1)} = \frac{\pi}{2\hbar}|\langle E', q'|\mathfrak{H}_1|E, q\rangle|^2\delta(E' - E \pm \hbar\omega_0)$$

and (15.55) becomes

$$w_{10}^{(1)} = \frac{\pi}{2\hbar}|\langle E, q'|\mathfrak{H}_1|E, q\rangle|^2\rho(E \mp \hbar\omega_0).$$

The discussion of Sections 15.8 and 15.9 also follows as before with these simple modifications.

Thus the Hamiltonian (15.63) feeds energy into the system or removes it in amounts of $\hbar\omega_0$; no transition can take place to a final-state energy level that does not differ from the initial energy of the system by $\hbar\omega_0$.

READING NOTES*

Reference 22, Chapter 11.

Reference 23, Chapters 18, 19.

Reference 24, Chapter 7.

Reference 42, Chapter 9.

Reference 43, Chapter 8.

PROBLEMS

1. Show that in Eq. (15.21) the angular integral vanishes unless $|l - l'| = 1$.
2. Calculate the energy shifts resulting from the Stark effect for the isotropic harmonic oscillator when the particle is an electron
   a) in first-order perturbation theory (what is significantly different from the one-electron atom?), and
   b) in second order for the lowest possible unperturbed energy levels.
3. In the nonrelativistic approximation, the relativistic electron theory predicts the existence of the spin–orbit term
$$\frac{1}{2M_e c^2}\frac{1}{r}\left[\frac{\partial}{\partial r}V(r)\right]\mathbf{S}\cdot\mathbf{L}$$
   in the Hamiltonian for an electron in a spherical potential. For the one-electron atom, consider this as a first-order perturbation when the electron is in the $2p$ ($n = 2$, $l = 1$) states $|nlm; s\mu\rangle$, or in a linear combination of these states.
   a) Find linear combinations of the $2p$ states that are diagonal in $\mathbf{S}\cdot\mathbf{L}$ and in the unperturbed Hamiltonian.
   b) Can the spin–orbit term account for a splitting of 11,000 MHz in the spectral frequencies of these states, to first order in perturbation theory?
4. Consider the Stark effect in a one-dimensional harmonic oscillator for a particle with charge e. Find the exact solutions to the problem, and compare them with first-order perturbation solutions.
5. Let a hydrogen atom be located in a homogeneous electric field with time-dependence
$$\mathbf{E}(t) = \begin{cases} 0, & t < 0 \\ E_0\hat{\mathbf{k}}e^{-t/\tau}, & t > 0, \end{cases}$$
   (e.g. between the plates of a discharging parallel-plate capacitor). If the atom is initially in its ground state, calculate the probability that after a very long time (compared with what?) the atom will be in its (a) $2s$ state, and (b) $2p$ state.
6. Using first-order time-dependent perturbation theory, calculate the differential cross-section for the scattering of a particle by a spin-independent central potential having the

* Works referred to in these notes will be found in the Bibliography on p. 577.

form

$$V(r) = \frac{qq'}{4\pi\epsilon_0 r} e^{-\epsilon r}$$

Having calculated this, let $\epsilon \to 0$ and obtain the Rutherford differential cross-section. (Use the time dependence of Eq. (15.48) and at the appropriate point let $t \to \infty$.)

7. Discuss the effect of replacing Eq. (15.48) with

$$\mathfrak{H}_1(\tau + t') = \begin{cases} \mathfrak{H}_1, & -t < t' < t \\ 0, & \text{otherwise.} \end{cases}$$

CHAPTER 16

SEMICLASSICAL APPROACH TO RADIATION BY AN ATOMIC ELECTRON

We move now to a semiclassical discussion of the interaction of electromagnetic radiation with matter—in particular, with the electron in a one-electron atom. The discussion is semiclassical because we use the classical formulation of the electromagnetic field except that we impose a minimal quantum condition for the field by assuming Planck's radiation law. Indeed, let us begin by considering Planck's law and the bearing it has upon the rate of radiation emission and absorption by an atom.

16.1 THE EINSTEIN A- AND B-COEFFICIENTS

An atom bathed in electromagnetic radiation may be induced to shift from a state with energy E to one with energy $E' = E + \hbar\omega$ if the radiation bath contains Fourier components with frequency ω and hence can provide energy quanta of magnitude $\hbar\omega$ to the atom. This corresponds to a harmonic perturbation with frequency ω added to the unperturbed Hamiltonian for the atom. We assume that the absorbed radiation is moving toward an element of solid angle $d\Omega$ about a direction we shall label Ω, and has a polarization indicated by an index α. The rate of transition from energy level E to energy level E' will be proportional to the energy density of the ω-frequency component of the electromagnetic field. If we let $\eta_\alpha(\omega, \Omega, t)$ represent this energy density per unit energy interval, for radiation with polarization α and frequency ω moving into solid angle $d\Omega$ about direction $\Omega = (\theta, \varphi)$, when the system is in thermodynamic equilibrium at temperature T, we thus have the transition rate

$$\frac{dw_{1,0;\alpha}}{d\Omega} = b_{1,0;\alpha}\eta_\alpha(\omega, \Omega, T) \tag{16.1}$$

$b_{1,0;\alpha}$ being the coefficient of proportionality. Evidently $b_{1,0;\alpha}$ has dimensions $(\text{time})^{-1}(\text{distance})^3$.

On the other hand, the atom may *lose* energy $\hbar\omega$ by emitting radiation of frequency ω in passing from a state with energy E to one with energy $E' = E - \hbar\omega$. However, it may do so in two ways—either by being *stimulated* by the radiation bath to emit electromagnetic waves, in which case the rate of emission is again proportional to the spectral component $\eta_\alpha(\omega, \Omega, T)$ of the energy density,

$$\frac{dw_{0,1;\alpha}^{(\text{stim})}}{d\Omega} = b_{0,1;\alpha}\eta_\alpha(\omega, \Omega, T)$$

or by spontaneous emission, in which case we write

$$\frac{dw_{0,1;\alpha}^{(\text{spon})}}{d\Omega} = a_{0,1;\alpha},$$

since here we do not expect the transition rate to depend upon the energy density of the existing radiation bath. Spontaneous emission results from the interaction of an atomic electron with the internal electromagnetic field of the atom, and is predicted by the complete quantum-theoretical analysis of the interaction of the electromagnetic field with matter (quantum electrodynamics); such an analysis is beyond the scope of our work here. Thus we have

$$\frac{dw_{0,1;\alpha}}{d\Omega} = \frac{dw_{0,1;\alpha}^{(\text{stim})}}{d\Omega} + \frac{dw_{0,1;\alpha}^{(\text{spon})}}{d\Omega} = b_{0,1;\alpha}\eta_\alpha(\omega, \Omega, T) + a_{0,1;\alpha}. \tag{16.2}$$

Now the probability that the atom will be in a state with energy E when it is in thermodynamic equilibrium with its environment at temperature T is

$$P(E, T) = (\text{const.})e^{-E/\kappa T},$$

where κ is Boltzmann's constant. But if the system is in equilibrium, we expect the overall rates of emission and absorption to be equal, so that

$$\begin{aligned} 0 &= P(E, T)\frac{dw_{1,0;\alpha}}{d\Omega} - P(E', T)\frac{dw_{0,1;\alpha}}{d\Omega} \\ &= (\text{const.})\{e^{-E/\kappa T}b_{1,0;\alpha}\eta_\alpha(\omega, \Omega, T) - e^{-E'/\kappa T}[b_{0,1;\alpha}\eta_\alpha(\omega, \Omega, T) + a_{0,1;\alpha}]\}. \end{aligned}$$

Furthermore, we know that as $T \to \infty$, $\eta_\alpha(\omega, \Omega, T) \to \infty$ also, and this requires that

$$b_{1,0;\alpha} = b_{0,1;\alpha}. \tag{16.3}$$

For finite temperatures, then,

$$\begin{aligned} \eta_\alpha(\omega, \Omega, T) &= \frac{a_{0,1;\alpha}}{b_{0,1;\alpha}}[e^{(E'-E)/\kappa T} - 1]^{-1} \\ &= \frac{a_{0,1;\alpha}}{b_{0,1;\alpha}}[e^{\hbar\omega/\kappa T} - 1]^{-1}, \end{aligned} \tag{16.4}$$

since E' is taken to represent the level of greater energy.

Comparison of this with Planck's law for isotropic radiation (Eq. (6.38)—remember that here we seek the energy density per unit energy interval per steradian for a given polarization α),

$$\sum_{\alpha=1}^{2} \int \eta_\alpha(\omega, \Omega, T)\, d\Omega = 8\pi\eta_\alpha(\omega, \Omega, T) = \psi(k)\frac{dk}{d(\hbar\omega)}$$

$$= \frac{1}{\hbar c}\psi(k) = \frac{1}{\pi^2}\frac{(\omega/c)^3}{e^{\hbar\omega/\kappa T} - 1}, \tag{16.5}$$

yields

$$\frac{a_{0,1;\alpha}}{b_{0,1;\alpha}} = \left(\frac{\omega/c}{2\pi}\right)^3. \tag{16.6}$$

Hence if we can find either of the Einstein coefficients $a_{0,1;\alpha}$ or $b_{0,1;\alpha}$, we shall be able to find the other through this relation.

16.2 INTERACTION BETWEEN THE ELECTRON AND ELECTROMAGNETIC FIELD

We may estimate the coefficient $b_{1,0;\alpha} = b_{0,1;\alpha}$, and hence $a_{0,1;\alpha}$ to first-order in perturbation theory by considering the radiation field as a perturbation on the atomic electron's Hamiltonian. The electron's Hamiltonian in the presence of the field is

$$\mathfrak{H} = \frac{1}{2M}(\mathfrak{P} - q\mathfrak{A})^2 + q\Phi_0 + \mathfrak{V}(\mathfrak{X})$$

and the unperturbed Hamiltonian will be taken to be

$$\mathfrak{H}_0 = \frac{1}{2M}\mathfrak{P}^2 + q\Phi_0 + \mathfrak{V}(\mathfrak{X}),$$

where Φ_0 represents the Coulomb potential arising from the atomic nucleus and $\mathfrak{A}$ represents the vector potential of the radiation field. We choose the gauge to be such that

$$\sum_{j=1}^{3} [\mathfrak{P}_j, \mathfrak{A}_j] = 0,$$

or

$$\nabla \cdot \mathbf{A} = 0,$$

when we let $\mathfrak{P} \to \mathbf{P} = (\hbar/i)\nabla$ and $\mathfrak{A} \to \mathbf{A}$. Then

$$\mathfrak{H}_1 = -\frac{q}{M}\mathfrak{A} \cdot \mathfrak{P} + \frac{q^2}{2M}\mathfrak{A}^2.$$

We assume the field to be sufficiently weak that the $\mathfrak{A}^2$ term may be neglected.

Next, since $\mathfrak{A}$ represents a radiation field, we may write the ωth Fourier component, with wave vector $\mathbf{k} = (\omega/c)\hat{\mathbf{k}}$, in the form

$$\mathfrak{A} = \frac{1}{\omega}\mathbf{E}\cos(\omega t - \mathbf{k}\cdot\mathfrak{X})$$

where $\mathbf{E}$ is the strength of the electric field vector, and $\mathfrak{E} = \mathbf{E}\sin(\omega t - \mathbf{k}\cdot\mathfrak{X})$. This satisfies the requirements of $\mathfrak{E} = -\dot{\mathfrak{A}}$ and $\mathfrak{B} = \nabla \times \mathfrak{A}$. An electromagnetic plane wave with wave vector $\mathbf{k}$ normally might be expected to have the form

$$\frac{1}{\omega}\mathbf{E}e^{i(\omega t - \mathbf{k}\cdot\mathfrak{X})}$$

but since $\mathfrak{A}$ must be a Hermitian operator, we must take half of this plus its Hermitian conjugate as the vector potential operator.

For atomic radiations, we normally have wave numbers much less than the reciprocal of an atomic radius (wavelengths large compared with atomic sizes), i.e.

$$\mathbf{k}\cdot\langle\mathfrak{X}\rangle \ll \pi;$$

hence we may neglect the $\mathbf{k}\cdot\mathfrak{X}$ term in the cosine factor of the vector potential. This leads to a harmonic perturbation

$$\mathfrak{H}_1(t) = -\frac{q}{M\omega}\mathbf{E}\cdot\mathfrak{P}\cos\omega t. \tag{16.7}$$

Our discussion of harmonic perturbation in Chapter 15 now may be utilized directly, with

$$\mathfrak{H}_1 = -\frac{1}{2}\frac{q}{M\omega}\mathbf{E}\cdot\mathfrak{P}, \tag{16.8}$$

the $\frac{1}{2}$ occurring for the reasons discussed in our consideration of harmonic perturbation.

Our task now is to calculate $\langle E', q'|\mathfrak{H}_1|E, q\rangle$, where the $|E, q\rangle$ represent atomic electron states. For this purpose, we may use Eq. (10.5):

$$\begin{aligned}\langle E', q'|\mathfrak{P}|E, q\rangle &= M\frac{d}{dt}\langle E', q'|\mathfrak{X}|E, q\rangle \\ &= -\frac{M}{i\hbar}\langle E', q'|[\mathfrak{H}_0, \mathfrak{X}]|E, q\rangle \\ &= i\frac{M}{\hbar}(E' - E)\langle E', q'|\mathfrak{X}|E, q\rangle \\ &= iM\omega_{10}\langle E', q'|\mathfrak{X}|E, q\rangle.\end{aligned}$$

Hence

$$\langle E', q' | \mathfrak{H}_1 | E, q \rangle = i \langle E', q' | \tfrac{1}{2} q \mathbf{E} \cdot \mathfrak{X} | E, q \rangle, \tag{16.9}$$

which, aside from the irrelevant phase factor $1/i$, is the perturbation obtained from assuming an electric-dipole interaction between the dipole $q\mathfrak{X}$ and the electric field $\frac{1}{2}\mathbf{E}$. The consequence of neglecting $\mathbf{k} \cdot \langle \mathfrak{X} \rangle$, then, is to exclude all but electric-dipole interactions. This is appropriate in much of atomic physics, but in nuclear physics higher multipoles often must be considered.

16.3 CALCULATION OF THE EINSTEIN COEFFICIENTS TO FIRST ORDER IN PERTURBATION THEORY

The rate of stimulated emission of radiation for a given polarization of the electromagnetic field is

$$\begin{aligned} \int d\Omega \frac{d\omega_{1,0;\alpha}}{d\Omega} &= b_{1,0;\alpha} 4\pi \eta_\alpha(\omega, \Omega, T) \\ &= (2\pi/\hbar) |\langle E', q' | \mathfrak{H}_1 | E, q \rangle|^2 \rho(\hbar\omega) \end{aligned} \tag{16.10}$$

where $\rho(\hbar\omega)$ is the density of states available to the electromagnetic field. This final-state density factor may be obtained as follows: the energy density per unit energy interval, $8\pi\eta_\alpha(\omega, \Omega, T)$, when divided by the energy density of the electromagnetic field $\epsilon_0 \mathbf{E}^2$, gives the number of states available to the electromagnetic field per unit energy interval,

$$\rho(\hbar\omega) = \frac{8\pi\eta_\alpha(\omega, \Omega, T)}{\epsilon_0 \mathbf{E}^2} = \frac{1}{\pi^2} \frac{(\omega/c)^2}{\epsilon_0 \mathbf{E}^2} (e^{\hbar\omega/\kappa T} - 1)^{-1}. \tag{16.11}$$

Insertion of (16.11) into (16.10) produces the result

$$b_{1,0;\alpha} = \frac{4\pi}{\hbar} \frac{|\langle E', q' | \mathfrak{H}_1 | E, q \rangle|^2}{\epsilon_0 \mathbf{E}^2} = b_{0,1;\alpha}, \tag{16.12}$$

from which it follows that

$$a_{0,1;\alpha} = \frac{4\pi}{\hbar} \left(\frac{\omega}{2\pi c} \right)^3 \frac{|\langle E', q' | \mathfrak{H}_1 | E, q \rangle|^2}{\epsilon_0 \mathbf{E}^2}. \tag{16.13}$$

For the specific case $\mathfrak{H}_1 = \frac{1}{2} q \mathbf{E} \cdot \mathfrak{X}$, this yields the very useful formulas

$$\begin{aligned} b_{0,1;\alpha} &= \frac{\pi}{\hbar\epsilon_0} q^2 |\langle E', q' | \hat{\mathbf{E}} \cdot \mathfrak{X} | E, q \rangle|^2 \\ a_{0,1;\alpha} &= \frac{\pi q^2}{\hbar\epsilon_0} \left(\frac{\omega}{2\pi c} \right)^3 |\langle E', q' | \hat{\mathbf{E}} \cdot \mathfrak{X} | E, q \rangle|^2. \end{aligned} \tag{16.14}$$

The importance of the latter of these equations is that it gives the spontaneous emission rate directly in terms of the same matrix elements that determine the stimulated rate of emission; no density of states factor appears in the expression for the rate of spontaneous emission.

16.4 SELECTION RULES FOR ANGULAR MOMENTUM AND PARITY

Clearly radiative emissions can occur in the electric-dipole approximation only when the matrix elements $\langle E', q'|\hat{\mathbf{E}}\cdot\mathfrak{X}|E, q\rangle$ are nonvanishing. For spherically symmetric potentials, the states $|E, q\rangle$ will normally be eigenstates of angular momentum and parity as well as energy, and we can establish some rules ("selection rules"), or conditions on angular momenta and parity, under which these matrix elements are allowed to be nonvanishing. Here we consider spinless "electrons" bound in an atom.

First, we note that the matrix elements may be expanded in terms of position eigenstates

$$\begin{aligned}\langle E', q'|\hat{\mathbf{E}}\cdot\mathfrak{X}|E, q\rangle &= \int d^3x \int d^3x' \langle E', q'|\mathbf{x}'\rangle\langle\mathbf{x}'|\hat{\mathbf{E}}\cdot\mathfrak{X}|\mathbf{x}\rangle\langle\mathbf{x}|E, q\rangle \\ &= \int d^3x \int d^3x' \psi^*_{E',q'}(\mathbf{x}')\hat{\mathbf{E}}\cdot\mathbf{x}\,\delta^3(\mathbf{x}'-\mathbf{x})\psi_{E,q}(\mathbf{x}) \\ &= \int d^3x \psi^*_{E',q'}(\mathbf{x})\hat{\mathbf{E}}\cdot\mathbf{x}\psi_{E,q}(\mathbf{x}). \qquad (16.15)\end{aligned}$$

Next, we note that, according to Section 12.5, we may write

$$\psi_{E,q}(\mathbf{x}) = (1/r)f(l, iK, r)Y_l^m(\theta, \varphi), \qquad K = \sqrt{2M|E|} \qquad (16.16)$$

when $|E, q\rangle = |iK, l, m\rangle$ is an eigenstate of angular momentum. Furthermore

$$\hat{\mathbf{E}}\cdot\mathbf{x} = r\cos\Theta_{\mathbf{E},\mathbf{x}}$$

where $\Theta_{\mathbf{E},\mathbf{x}}$ is the angle between $\mathbf{E}$ and $\mathbf{x}$, as indicated in Fig. 16.1. Thus Eq. (16.15) becomes

$$\langle E', q'|\hat{\mathbf{E}}\cdot\mathfrak{X}|E, q\rangle = \int_0^\infty rf^*(l', iK', r)f(l, iK, r)\,dr \cdot \int Y_{l'}^{m'*}(\theta, \varphi)\cos\Theta_{\mathbf{E},\mathbf{x}}Y_l^m(\theta, \varphi)\,d\Omega.$$

The problem now is to express $\cos\Theta_{\mathbf{E},\mathbf{x}}$ in terms of the angles θ, φ and $\theta_{\mathbf{E}}$, $\varphi_{\mathbf{E}}$ so that we can evaluate the integral over $d\Omega = \sin\theta\,d\theta\,d\varphi$. This can be done using plane trigonometry, and one can show that the result may be written in the

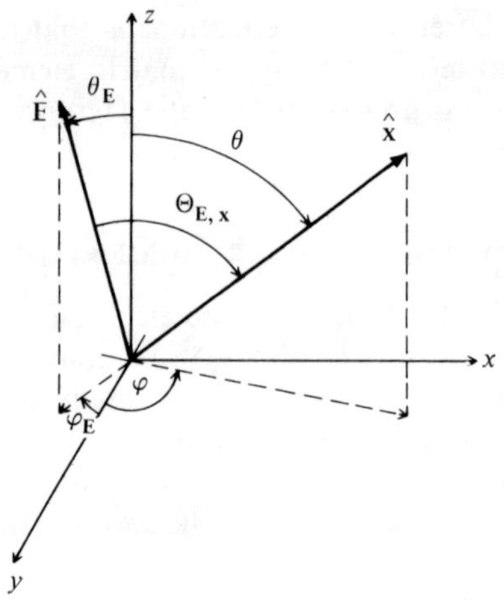

Fig. 16.1. Relevant angles for describing cos $\Theta_{\mathbf{E,x}}$.

form

$$\cos \Theta_{\mathbf{E,x}} = P_1(\cos \Theta_{\mathbf{E,x}}) = \frac{4\pi}{3} \sum_{\mu=-1}^{1} Y_1^{\mu *}(\theta_{\mathbf{E}}, \varphi_{\mathbf{E}}) Y_1^{\mu}(\theta, \varphi) \tag{16.17}$$

when the $Y_1^{\mu}(\theta, \varphi)$ are defined by Eq. (12.63). This result is a special case of a more general addition theorem for spherical harmonics:*

$$P_l(\cos \Theta_{\mathbf{E,x}}) = \frac{4\pi}{2l+1} \sum_{\mu=-l}^{l} Y_l^{\mu *}(\theta_{\mathbf{E}}, \varphi_{\mathbf{E}}) Y_l^{\mu}(\theta, \varphi).$$

Substitution of Eq. (16.17) into (16.16) gives us the key equation for evaluating our matrix elements:

$$\langle E', q' | \hat{\mathbf{E}} \cdot \mathfrak{X} | E, q \rangle = \frac{4\pi}{3} \sum_{\mu=-1}^{1} Y_1^{\mu *}(\theta_{\mathbf{E}}, \varphi_{\mathbf{E}}) \int_0^{\infty} f^*(l', iK', r) f(l, iK, r) r \, dr$$

$$\cdot \int d\Omega \, Y_{l'}^{m' *}(\theta, \varphi) Y_1^{\mu}(\theta, \varphi) Y_l^{m}(\theta, \varphi). \tag{16.18}$$

A number of general conclusions can be formulated in terms of the integral over angles without reference to the radial integral, since the former contains all the essential information about parity and angular momenta.

In the first place, $Y_1^{\mu}(\theta, \varphi)$ is odd under the parity transformation, and hence the matrix element vanishes unless l' and l refer to opposite parities, i.e. $|l' - l|$

* See, for example, M. E. Rose, *Elementary Theory of Angular Momentum*, John Wiley and Sons (1957), or A. R. Edmonds, *Angular Momentum in Quantum Mechanics*, Princeton University Press (1957), for proofs and general discussion.

must be an odd number; we write

$$\langle E', q'|\hat{\mathbf{E}}\cdot\mathbf{x}|E, q\rangle = 0 \qquad \text{unless} \qquad \Delta\Pi = -1, \text{ i.e. } \Pi'\Pi = -1 \quad (16.19)$$

where $\Delta\Pi$ represents the "difference" in parity between the states $|E, q\rangle$ and $|E', q'\rangle$, or $\Delta\Pi = \Pi'\Pi$, and Π', Π represent the parities of those states. Thus the electric-dipole operator changes the parity of the electron state.

Secondly, the integral over φ can be carried out quite easily; this integral guarantees the conservation of angular momentum along the z-axis:

$$\int_0^{2\pi} d\varphi\, e^{-i(m'-m-\mu)} = 2\pi\delta_{m',m+\mu}.$$

Thus we have a change in the electron's magnetic quantum number that equals the magnetic quantum number μ carried off by the emitted radiation:

$$\Delta m = m' - m = \mu = \pm 1, 0.$$

The restriction to values $\pm 1, 0$ is dictated by the fact that the spherical harmonic representing the dipole operator involves one unit of angular momentum.

The remaining integration over θ is somewhat more complicated in general. For $\mu = 0$, it involves evaluating

$$\int_0^{\pi} Y_{l'}^{m'*}(\theta, \varphi) \cos\theta\, Y_l^m(\theta, \varphi) \sin\theta\, d\theta$$

whereas for $\mu = \pm 1$ we must evaluate

$$\int_0^{\pi} Y_{l'}^{m'}(\theta, \varphi) \sin\theta\, Y_l^m(\theta, \varphi) \sin\theta\, d\theta.$$

These can be calculated directly for specific values of l, m, l' and m'. A more general approach involves using the Clebsch–Gordan coefficients mentioned in Section 12.7 to couple the spherical harmonic $Y_1^{\mu}(\theta, \varphi)$ with $Y_l^m(\theta, \varphi)$; thus we obtain

$$\begin{aligned} |1, \mu; l, m\rangle &= |1, \mu\rangle|l, m\rangle \\ &= \sum_{l''} |l'', m+\mu, 1, l\rangle\langle l'', m+\mu, 1, l|1, \mu; l, m\rangle \end{aligned} \qquad (16.20)$$

according to the equation following Eq. (12.69), or

$$\begin{aligned} Y_1^{\mu}(\theta, \varphi) Y_l^m(\theta, \varphi) &= \langle\theta, \varphi|1, \mu\rangle\langle\theta, \varphi|l, m\rangle \\ &= \langle\theta, \varphi|1, \mu; l, m\rangle \\ &= \sum_{l''} \langle l'', m+\mu, 1, l|1, \mu; l, m\rangle Y_{l''}^{m+\mu}(\theta, \varphi) \end{aligned} \qquad (16.20')$$

where l'', l and 1 must form a triangle. Hence l'' and l may differ by 1 or 0. Using

Eq. (16.20) in the angular integral, we get

$$\int_0^{2\pi}\int_0^{\pi} Y_{l'}^{m'*}(\theta, \varphi) Y_1^{\mu}(\theta, \varphi) Y_l^{m}(\theta, \varphi) \sin\theta \, d\theta \, d\varphi$$

$$= \sum_{l''} \langle l'', m + \mu, 1, l | 1, \mu; l, m \rangle \cdot \int_0^{2\pi}\int_0^{\pi} Y_{l'}^{m'*}(\theta, \varphi) Y_{l''}^{m+\mu}(\theta, \varphi) \sin\theta \, d\theta \, d\varphi$$

$$= \sum_{l''} \langle l'', m + \mu, 1, l | 1, \mu; l, m \rangle \delta_{l',l''} \delta_{m';m+\mu}$$

$$= \langle l', m + \mu, 1, l | 1, \mu; l, m \rangle \delta_{m',m+\mu}.$$

Hence l' must suffer the same restrictions as l'' and therefore we have $|l' - l| = \pm 1, 0$. However, the parity rule forbids $\Delta l = |l' - l| = 0$, so we have

$$\Delta l = |l' - l| = 1, \qquad \Delta m = m' - m = \pm 1, 0. \tag{16.21}$$

as the selection rules for electric-dipole radiation.

It must be remembered, however, that

1. we are considering transitions only to first order in perturbation theory, and
2. we have neglected higher multipoles than the electric dipole by neglecting $\mathbf{k} \cdot \mathfrak{X}$.

Therefore transitions may occur in which these selection rules are not obeyed, but if they do occur, they correspond to higher-order or higher-multipole transitions and therefore lead, presumably, to rather weak radiative emissions.

16.5 THE MAGNETIC QUANTUM NUMBER SELECTION RULE AND THE POLARIZATION OF EMITTED RADIATION

(a) $\Delta \mathbf{m} = \boldsymbol{\mu} = \mathbf{0}$

In this case, the matrix element is

$$\langle E', q' | \hat{\mathbf{E}} \cdot \mathfrak{X} | E, q \rangle$$

$$= \frac{4\pi}{3} Y_1^0(\theta_{\mathbf{E}}, \varphi_{\mathbf{E}}) \langle l', m, 1, l | 1, 0; l, m \rangle \cdot \delta_{m',m} \int_0^{\infty} f^*(l', iK', r) f(l, iK, r) r \, dr$$

$$= \sqrt{\frac{4\pi}{3}} \cos\theta_{\mathbf{E}} \langle l', m, 1, l | 1, 0; l, m \rangle \delta_{m',m} \int_0^{\infty} f^*(l', iK', r) f(l, iK, r) r \, dr,$$

which is proportional to $\cos\theta_{\mathbf{E}} = E_3/E$, the fraction of the electric field vector lying along the z-axis. Hence only emissions containing light with some component of the polarization along the z-axis can contribute to $\Delta m = 0$ radiation. Such radiation always has a component of the wave vector $\mathbf{k}$ perpendicular to the z-axis, and therefore no light from $\Delta m = 0$ transitions can be seen by an observer looking along the z-axis.

(b) $\mathbf{\Delta m = \mu = +1}$

The matrix element in this case is proportional to

$$Y_1^{1*}(\theta_{\mathbf{E}}, \varphi_{\mathbf{E}}) = -\sqrt{\frac{3}{4\pi}} \sin\theta_{\mathbf{E}}\, e^{-i\varphi_{\mathbf{E}}} = -\sqrt{\frac{3}{8\pi}} \frac{(E_1 - iE_2)}{E}$$

and hence only radiation having components of its polarization along the y- and/or x-axis is emitted in $\Delta m = 1$ transitions. Since this spherical harmonic is proportional to $E_1 + E_2 e^{-i\pi/2}$ the phase of the x-component of $\mathbf{E}$ leads the phase of the y-component by $\pi/2$; thus the emitted radiation is elliptically polarized, the semi-axes of the polarization ellipse being E_1 and E_2. For light emitted in the positive z-direction, $E_1 = E_2$ and we have left-circularly polarized light; light emitted in the negative z-direction is right-circularly polarized. Light emitted along the x-axis, however, is polarized in the y-direction, and that emitted along the y-axis is polarized in the x-direction.

(c) $\mathbf{\Delta m = \mu = -1}$

Here the matrix element is proportional to

$$Y_1^{-1*}(\theta_{\mathbf{E}}, \varphi_{\mathbf{E}}) = \sqrt{\frac{3}{8\pi}} \sin\theta_{\mathbf{E}}\, e^{i\varphi_{\mathbf{E}}} = \sqrt{\frac{3}{8\pi}} \frac{(E_1 + iE_2)}{E},$$

which is similar to case (b) except that the phase of E_1 now lags behind that of E_2 by $\pi/2$. The situation is just like that of the previous case, except that the polarization is reversed; radiation emitted along the positive z-axis is right-circularly polarized, and that along the negative z-axis is left-circularly polarized.

These polarization effects are actually observed in the Zeeman effect for alkali atoms. Shortly, we shall take a closer look at the normal Zeeman effect, and its cousin, the anomalous Zeeman effect. Before doing so, however, we consider how our analysis is altered by the introduction of electron spin, which we have so far ignored.

16.6 A ONE-ELECTRON ATOM IN AN EXTERNAL MAGNETIC FIELD; EFFECTS OF SPIN

In the Zeeman effect, which we take up again in the next section, an atomic electron interacts with the electromagnetic field in three distinct ways:

1. It is bound to the atom through the electrostatic potential $\Phi^{(c)}(\mathfrak{X})$ produced by the nucleus and perhaps modified by the presence of other bound electrons.

2. It interacts with an external magnetic field $\mathfrak{B}^{(m)}$ which removes some of the degeneracy present in the simplest one-electron models.

3. It radiates electromagnetic waves through its interaction with the radiation field represented by a vector potential $\mathfrak{A}^{(r)}$.

Normally, this order is the order of strength of the interactions. For compactness, we may represent these interactions by a composite four-vector potential,

$$\mathfrak{A}^{\mu} = \left(\frac{1}{c}\Phi^{(c)}, \mathfrak{A}^{(m)} + \mathfrak{A}^{(r)}\right)$$

where

$$\mathfrak{A}^{(r)} = \frac{1}{\omega}\mathbf{E}^{(r)} \cos(\omega t - \mathbf{k} \cdot \mathfrak{X})$$

and, for $\mathfrak{B}^{(m)} = \mathbf{B}^{(m)}$ uniform and parallel to the z-axis,

$$\mathfrak{A}^{(m)} = \tfrac{1}{4}\mathbf{B}^{(m)} \times \mathfrak{X}.$$

The resulting Hamiltonian can be written in the form

$$\mathfrak{H} = \mathfrak{H}_0 + \mathfrak{H}^{(m)} + \mathfrak{H}^{(r)} \tag{16.22}$$

where

$$\mathfrak{H}_0 = \frac{1}{2M}\mathfrak{P}^2 + q\Phi^{(c)}(\mathfrak{X}) + \mathfrak{V}(\mathfrak{X}), \tag{16.23a}$$

$$\begin{aligned}\mathfrak{H}^{(m)} &= -\frac{q}{M}\mathfrak{A}^{(m)} \cdot \mathfrak{P} + \frac{q^2}{2M}\mathfrak{A}^{(m)\,2} \\ &= -\frac{q}{2M}\mathbf{B}^{(m)} \cdot \mathfrak{L} + \frac{q^2}{2M} \cdot \tfrac{1}{4}\mathbf{B}^{(m)\,2}, \qquad \mathfrak{L} = \mathfrak{X} \times \mathfrak{P}\end{aligned} \tag{16.23b}$$

and

$$\mathfrak{H}^{(r)} = -\frac{q}{M}\mathfrak{A}^{(r)} \cdot \mathfrak{P} + \frac{q^2}{2M}\mathfrak{A}^{(r)\,2}. \tag{16.23c}$$

So far, however, we have continued to neglect spin angular momentum.

The two most important effects of the introduction of spin are

1. to add to (16.23b) a term $-(q/M)\mathfrak{B}^{(m)} \cdot \mathfrak{S}$, as we saw in Chapter 12, and
2. to add to (16.23a) a *spin–orbit* term,

$$\mathfrak{H}^{(\mathrm{s.o.})} = (q/M)\mathfrak{B}^{(l)} \cdot \mathfrak{S}$$

representing the interaction between the electron spin and the orbital angular momentum of the electron. The latter is actually a relativistic effect, and a non-relativistic classical calculation of $\mathfrak{B}^{(l)}$, assuming that it represents the magnetic field arising from the electron's orbital motion, is incorrect by a factor of 2; this factor is called the *Thomas precession factor* and can be explained by a relativistic calculation in which the increase in the electron's inertia as it orbits about the nucleus is taken into account. The correct spin–orbit $\mathfrak{B}^{(l)}$ is given, for a pure

Coulomb potential, by

$$\mathfrak{B}^{(l)} = -\frac{Ze}{8\pi\epsilon_0 Mc^2}\mathfrak{R}^{-3}\mathfrak{L}.$$

We must ask, however, whether the presence of electron spin significantly affects (16.23c). We restrict our considerations to the usual case in which the $\mathfrak{A}^{(r)2}$ term may be neglected, and furthermore, we restrict ourselves to the question whether the presence of spin produces an effect that is significant relative to the electric-dipole approximation we used earlier. Presumably the dominant effect of the spin will be to produce a magnetic-dipole interaction between the radiation field and the spin magnetic moment. Since, for the radiation field, $B = (1/c)E$, the magnitude of this interaction is

$$(q\hbar/M)(1/c)E,$$

whereas the magnitude of the electric-dipole interaction is

$$qE|\langle\mathfrak{X}\rangle| \simeq qEa_1,$$

where a_1 is the Bohr radius, $a_1 = \lambda_e/\alpha$ for an electron (with electronic charge $q = e$). Setting $q = e$ everywhere and taking the ratio of these two magnitudes, we obtain

$$\frac{\Delta E^{\text{spin}}}{\Delta E^{\text{el. dipole}}} \backsimeq \frac{\hbar}{Mca_1} = \alpha;$$

hence the spin contributes, to the radiation interaction, terms that are at least $\alpha < 10^{-2}$ smaller than the electric-dipole term for a one-electron atom. A first approximation should therefore neglect the spin contribution to the radiative interaction.

We shall continue to neglect terms that involve the square of electromagnetic field components. (For *very* strong magnetic fields, one must keep the $\mathfrak{B}^{(m)2}$ term and it gives rise to the quadratic Zeeman effect. We shall consistently neglect it, however.) Hence the approximate Hamiltonians of interest to us now are

$$\mathfrak{H}_0' = \frac{1}{2M}\mathfrak{P}^2 + q\Phi^{(c)}(\mathfrak{X}) + \frac{q}{M}\mathfrak{B}^{(l)}\cdot\mathfrak{S}, \tag{16.24a}$$

$$\mathfrak{H}_1{}^{(m)} = -\frac{q}{2M}(\mathfrak{L} + 2\mathfrak{S})\cdot\mathfrak{B}^{(m)}, \tag{16.24b}$$

$$\mathfrak{H}_1{}^{(r)} = -\frac{q}{M}\mathbf{E}^{(r)}\cos(\omega t)\cdot\mathfrak{P}. \tag{16.24c}$$

The first question is, can we solve (16.24a) exactly? In fact, do we require an exact solution, and if so, under what circumstances? Two cases will be distinguished, in which the relative magnitudes of $\mathfrak{B}^{(l)}$ and $\mathfrak{B}^{(m)}$ determine our course of action.

In both cases, the eigenstates, or at least the vital features of the eigenstates of the atomic electron, are obtained by considering solutions to eigenvalue equations involving $\mathfrak{H}_0' + \mathfrak{H}_1^{(m)}$. These results are then used to study the radiation occurring because of the presence of $\mathfrak{H}_1^{(r)}$ in the complete Hamiltonian.

A crucial parameter in our analysis would seem to be the magnitude of $(\hbar q/2M)\mathfrak{B}^{(l)}$ for typical electronic states; this quantity is characteristic of the atom and is independent of the external fields so that it establishes a convenient measure of the strengths of interactions. Assuming the electron to be in a state with orbital angular-momentum quantum number l and principal quantum number n, so that the approximate radius of the electron's "orbit" is $a_n = n^2 \lambda_e/\alpha$, we have

$$\left| \frac{\hbar q}{2M} \langle \mathfrak{B}^{(l)} \rangle \right| \lesssim \frac{e\hbar}{2M} \frac{Ze\alpha^3 \hbar l}{8\pi\epsilon_0 M c^2 n^6 \lambda_e^3}$$

$$= \frac{Zl}{4} \frac{\alpha^4}{n^6} Mc^2 = \alpha^2 \frac{l}{2Zn^4} E_n \tag{16.25}$$

where E_n is the binding energy of an electron in the nth level. Thus the spin–orbit interaction yields energy contributions that are of the order of α^2 smaller than the differences between low-lying electronic states, and it can properly be treated as a perturbation on the basic Hamiltonian $\mathfrak{H}_0$, provided $l/2Zn^4 \lesssim 1$.

Furthermore, we can estimate the magnetic field $\mathfrak{B}^{(l)}$ from this result. It is

$$|\langle \mathfrak{B}^{(l)} \rangle| \lesssim \frac{Zl}{n^6} \frac{\alpha^2}{2\mu_B} |E_{1\text{H}}|$$

$$= \frac{Zl}{n^6} (3.62 \times 10^{-4})\,\text{eV/bohr magneton}$$

$$= \frac{Zl}{n^6} (6.26 \times 10^4)\,\text{gauss}$$

$$= \frac{Zl}{n^6} (6.26)\,\text{weber/meter}^2, \tag{16.26}$$

where $E_{1\text{H}}$ is the hydrogen atom ground-state binding energy and where we have used the conversion factors

$$\mu_B = \frac{e\hbar}{2M_e} = 0.578\,815 \times 10^{-14}\ \text{MeV/gauss}$$

$$1\ \text{gauss} = 10^4\ \text{weber/meter}^2.$$

This provides a standard of measure for $\mathfrak{B}^{(m)}$ when we are dealing with an electron with quantum number n, l. If $|\langle \mathfrak{B}^{(m)} \rangle| \ll |\langle \mathfrak{B}^{(l)} \rangle|$, we may neglect the interaction of the external field with the electron in a first approximation; this is the *weak-field* case. Strong fields, on the other hand, are defined as those for which

$|\langle \mathfrak{B}^{(m)} \rangle| \gg |\langle \mathfrak{B}^{(l)} \rangle|$, and for them the spin–orbit interaction may be neglected in a first approximation.

16.7 THE NORMAL ZEEMAN EFFECT

We consider first the strong-field case, with the restriction that $\mathfrak{B}^{(m)}$ is not too strong, i.e.

$$|\langle (e^2/8M)\mathfrak{B}^{(m)2} \rangle| \ll |\langle (e/2M)\mathfrak{B}^{(m)} \rangle|;$$

this leads to the normal Zeeman effect. The spin–orbit term in $\mathfrak{H}_0'$ may be neglected, so that

$$\mathfrak{H}_0 + \mathfrak{H}^{(m)} = \mathfrak{H}_0 + (e/2M)(\mathfrak{L} + 2\mathfrak{S}) \cdot \mathfrak{B}^{(m)} \tag{16.27}$$

is the approximate Hamiltonian in the absence of the radiation field; we have used $q = -e$ for the electron. For a uniform magnetic field along the z-axis, the eigenstates of the system are taken to be

$$|E, l, m, \tfrac{1}{2}, \mu\rangle$$

which are eigenstates of $\mathfrak{L}^2$, $\mathfrak{L}_3$, $\mathfrak{S}^2$ and $\mathfrak{S}_3$ as well as of $\mathfrak{H}_0 + \mathfrak{H}_1{}^{(m)}$; this is appropriate, since the four angular-momentum operators all commute with the approximate Hamiltonian $\mathfrak{H}_0 + \mathfrak{H}_1{}^{(m)}$ as well as with one another.

The wave function for the electron in one of these eigenstates is

$$[\langle \mathbf{x}; \tfrac{1}{2}, \mu' | E, l, m, \tfrac{1}{2}, \mu\rangle] = \psi_{E,l,m,1/2,\mu}(\mathbf{x})$$

$$= \frac{f(l, iK, r)}{r} Y_l^m(\theta, \varphi) \cdot \begin{cases} \begin{Bmatrix} 1 \\ 0 \end{Bmatrix} & \text{if } \mu = +\frac{1}{2} \\ \begin{Bmatrix} 0 \\ 1 \end{Bmatrix} & \text{if } \mu = -\frac{1}{2}. \end{cases} \tag{16.28}$$

Here $f(l, iK, r)$ is given by Eq. (14.88) except that

$$K = \left[\frac{Z^2}{n^2}\alpha^2 \frac{Mc^2}{2} - \frac{e\hbar}{2M} B(m + 2\mu)\right]^{1/2},$$

where B is the magnitude of the magnetic-flux density of the uniform external magnetic field,

$$\mathfrak{B}^{(m)} = \mathbf{B}.$$

This follows from the fact that $\mathfrak{H}_1{}^{(m)}$ is diagonal in the basis $|E, l, m, \frac{1}{2}, \mu\rangle$ when the magnetic field is uniform along the z-axis,

$$\mathfrak{H}_1{}^{(m)}|E, l, m, \tfrac{1}{2}, \mu\rangle = \frac{e\hbar}{2M}(m + 2\mu)B|E, l, m, \tfrac{1}{2}, \mu\rangle,$$

and therefore contributes a constant to the unperturbed energy $E_n{}^{(0)} = -Z^2\alpha^2 Mc^2/2n^2$.

The column matrices in Eq. (16.28) are Pauli spinors corresponding respectively to the spin-up and spin-down states.

The main effect of the external magnetic field, then, is to break the $2(2l + 1)$-fold degeneracy that exists in the eigenstates of $\mathfrak{H}_0$; the new energy eigenvalues are

$$E_n = E_n^{(0)} + \frac{e\hbar}{2M} B(m + 2\mu). \tag{16.29}$$

Since m can take on the $2l + 1$ values $-l, \ldots, +l$, and μ can be either $-\frac{1}{2}$ or $+\frac{1}{2}$, there are $2l + 3$ different energy levels in place of the unperturbed level $E_n^{(0)}$; all but the $m = \pm l$ levels remain doubly degenerate.

The radiation emitted from the atom in the presence of a strong magnetic field will exhibit characteristic changes reflecting this removal of degeneracy by the magnetic field. To see what occurs, we must examine the matrix elements

$$\langle E_n', l', m', \tfrac{1}{2}, \mu' | \mathfrak{H}_1^{(r)} | E_n, l, m, \tfrac{1}{2}, \mu \rangle,$$

or, according to our discussion of the radiative transition rates,

$$\langle E_n', l', m', \tfrac{1}{2}, \mu' | \hat{\mathbf{E}} \cdot \boldsymbol{\mathfrak{X}} | E_n, l, m, \tfrac{1}{2}, \mu \rangle.$$

Since the operator $\hat{\mathbf{E}} \cdot \boldsymbol{\mathfrak{X}}$ does not involve any spin operators, this matrix element vanishes unless $\mu = \mu'$, and the selection rule

$$\Delta\mu = \mu' - \mu = 0 \tag{16.30}$$

must be added to our earlier selection rules; the spin orientations of the electron in its initial and final electronic states must be identical if radiation is to occur through an electric-dipole transition.

Thus electric-dipole transitions between, say, the state $|E_{n'}', l', m', \frac{1}{2}, \mu' \rangle$ and $|E_n, l, m, \frac{1}{2}, \mu \rangle$, involve the emission of light waves with frequencies

$$\begin{aligned} \omega_{10} &= \frac{E_{n'}' - E_n}{\hbar} = \frac{E_{n'}^{(0)} - E_n^{(0)}}{\hbar} + \frac{e}{2M} B(m' - m) \\ &= \frac{\Delta E^{(0)}}{\hbar} + \frac{e}{2M} B\,\Delta m = \omega_0 + \Delta\omega\,\Delta m. \end{aligned} \tag{16.31}$$

The emitted spectral lines are therefore shifted by the amounts $\Delta\omega m = eB\,\Delta m/2M$, where $\Delta m = \pm 1, 0$ is required. Furthermore, the polarization of the emitted light is determined by Δm, in accordance with our discussion in Section 16.5; see Fig. 16.2. All of this agrees with the experimental observations of the normal Zeeman effect in alkali atoms; note that the coefficient of Δm in (16.31) is the Larmor precession frequency for an electron in a magnetic field represented by B.

In much of the literature, the radiation with frequencies $\omega_0 \pm \Delta\omega$ is denoted σ since it is polarized perpendicular (German: *senkrecht*) to $\mathbf{B}$, whereas that with frequency ω_0 is denoted π since it is polarized parallel to $\mathbf{B}$.

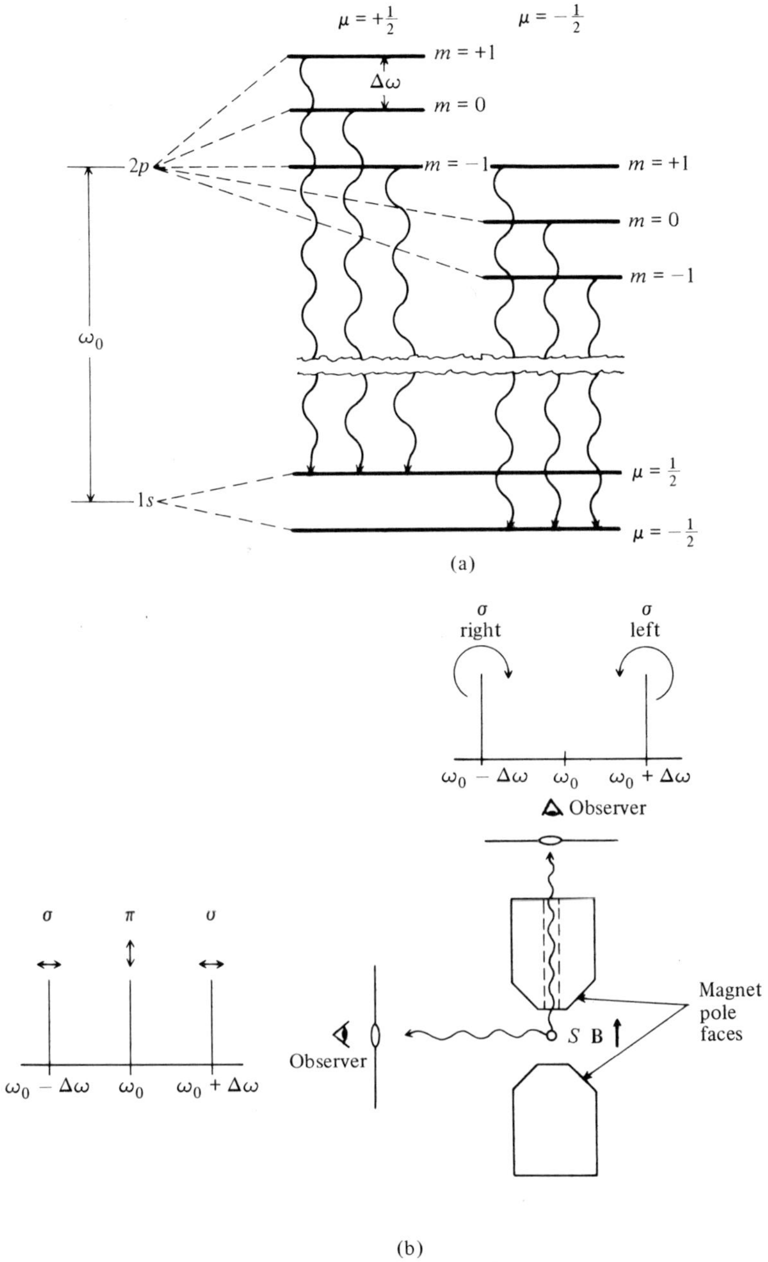

Fig. 16.2. Norman Zeeman effect. (a) Energy-level scheme and allowed radiative transitions. (b) Experimental setup with atom at S, showing polarizations of emitted radiation moving parallel and perpendicular to $\mathbf{B}$.

16.8 THE ANOMALOUS ZEEMAN EFFECT

Suppose now that $|\langle \mathfrak{B}^{(l)} \rangle| \gg |\langle \mathfrak{B}^{(m)} \rangle| = B$; then $\mathfrak{B}^{(m)}$ is considered a weak field and is treated as a perturbation on $\mathfrak{H}_0'$. The first problem in that case is to find the eigenstates of $\mathfrak{H}_0'$. We shall not actually do this, since we can learn a great deal through the much simpler expedient of finding the appropriate quantum numbers of the system aside from energy.

The eigenstates of operators $\mathfrak{L}_3$ and $\mathfrak{S}_3$ cannot be diagonalized simultaneously with $\mathfrak{H}_0'$, since they do not commute with $\mathfrak{L} \cdot \mathfrak{S}$; hence the eigenstates we used for the normal Zeeman effect will no longer work. However, the operator $\mathfrak{J} = \mathfrak{L} + \mathfrak{S}$ does commute with $\mathfrak{L} \cdot \mathfrak{S}$:

$$\begin{aligned}
[\mathfrak{J}_j, \mathfrak{L} \cdot \mathfrak{S}] &= \sum_{k=1}^{3} \{[\mathfrak{L}_j, \mathfrak{L}_k]\mathfrak{S}_k + [\mathfrak{S}_j, \mathfrak{S}_k]\mathfrak{L}_k\} \\
&= \sum_{k,l} [i\epsilon_{jkl}\mathfrak{L}_l\mathfrak{S}_k + i\epsilon_{jkl}\mathfrak{S}_l\mathfrak{L}_k] \\
&= 0
\end{aligned}$$

and hence $\mathfrak{J}^2$ and $\mathfrak{J}_3$ are appropriate diagonalizable operators to replace $\mathfrak{L}_3$ and $\mathfrak{S}_3$. The operators $\mathfrak{L}^2$ and $\mathfrak{S}^2$ commute with $\mathfrak{H}_0'$, $\mathfrak{J}^2$ and $\mathfrak{J}_3$ as well as with each other, and they continue to be useful diagonalizable operators.

From this we conclude that the appropriate eigenstates must be $|n, j, m, l, \frac{1}{2}\rangle$, where

$$\begin{aligned}
\mathfrak{H}_0'|n, j, m, l, \tfrac{1}{2}\rangle &= E_{n,j,l}|n, j, m, l, \tfrac{1}{2}\rangle \\
\mathfrak{J}^2|n, j, m, l, \tfrac{1}{2}\rangle &= \hbar^2 j(j + 1)|n, j, m, l, \tfrac{1}{2}\rangle \\
\mathfrak{J}_3|n, j, m, l, \tfrac{1}{2}\rangle &= \hbar m|n, j, m, l, \tfrac{1}{2}\rangle \\
\mathfrak{L}^2|n, j, m, l, \tfrac{1}{2}\rangle &= \hbar^2 l(l + 1)|n, j, m, l, \tfrac{1}{2}\rangle \\
\mathfrak{S}^2|n, j, m, l, \tfrac{1}{2}\rangle &= \tfrac{3}{4}\hbar^2|n, j, m, l, \tfrac{1}{2}\rangle.
\end{aligned}$$

The energy eigenvalues $E_{n,j,l}$ will now depend upon j and l as well as upon n.

Finding the effects of $\mathfrak{H}_1^{(m)}$ on the system in first-order perturbation theory is now reduced to finding the matrix elements

$$\langle n, j, m', l', \tfrac{1}{2}|\mathfrak{H}_1^{(m)}|n, j, m, l, \tfrac{1}{2}\rangle = \frac{e}{2M}B\langle n, j, m', l', \tfrac{1}{2}|\mathfrak{J}_3 + \mathfrak{S}_3|n, j, m, l, \tfrac{1}{2}\rangle$$

when $\langle \mathfrak{B}^{(m)} \rangle = \mathbf{B}$ is a uniform magnetic field along the z-axis. Clearly the $\mathfrak{J}_3$ term yields

$$\frac{e\hbar}{2M}Bm\delta_{m',m}\delta_{l',l},$$

and since the $\mathfrak{S}_3$-term does not involve orbital angular momentum, it too will vanish unless $l' = l$. However, the actual evaluation of the $\mathfrak{S}_3$-term involves some tricks.

One way to approach the problem is to note that

$$\langle j, m'|\mathfrak{S}_3\boldsymbol{\mathfrak{J}}^2|j, m\rangle = \langle j, m'|\mathfrak{S}_3|j, m\rangle \cdot \hbar^2 j(j+1)$$

which suggests that we deal with

$$\langle j, m'|\mathfrak{S}_3|j, m\rangle = \frac{1}{\hbar^2 j(j+1)}\langle j, m'|\tfrac{1}{2}(\mathfrak{S}_3\boldsymbol{\mathfrak{J}}^2 + \boldsymbol{\mathfrak{J}}^2\mathfrak{S}_3)|j, m\rangle;$$

here we have symmetrized the operator $\mathfrak{S}_3\boldsymbol{\mathfrak{J}}^2$ to obtain a Hermitian operator on the right-hand side. First we consider the product

$$\begin{aligned}\mathfrak{S}_3\boldsymbol{\mathfrak{J}}^2 &= \mathfrak{S}_3(\mathfrak{J}_1^2 + \mathfrak{J}_2^2 + \mathfrak{J}_3^2)\\ &= \mathfrak{J}_3(\mathfrak{S}_3\mathfrak{J}_3 + \mathfrak{S}_2\mathfrak{J}_2 + \mathfrak{S}_1\mathfrak{J}_1) + (\mathfrak{S}_3\mathfrak{J}_1 - \mathfrak{J}_3\mathfrak{S}_1)\mathfrak{J}_1 + (\mathfrak{S}_3\mathfrak{J}_2 - \mathfrak{J}_3\mathfrak{S}_2)\mathfrak{J}_2\\ &= \mathfrak{J}_3\boldsymbol{\mathfrak{S}}\cdot\boldsymbol{\mathfrak{J}} - (\mathfrak{J}_3\mathfrak{S}_1 - \mathfrak{S}_3\mathfrak{J}_1)\mathfrak{J}_1 + (\mathfrak{J}_2\mathfrak{S}_3 - \mathfrak{S}_2\mathfrak{J}_3)\mathfrak{J}_2\\ &\quad + ([\mathfrak{S}_3, \mathfrak{J}_2] - [\mathfrak{J}_3, \mathfrak{S}_2])\mathfrak{J}_2,\end{aligned}$$

noting that

$$[\mathfrak{S}_3, \mathfrak{J}_2] = [\mathfrak{S}_3, \mathfrak{S}_2] = [\mathfrak{J}_3, \mathfrak{S}_2],$$

which ensures the vanishing of the last term.

If we define

$$\mathfrak{K}_j \equiv \epsilon_{jkl}(\mathfrak{J}_k\mathfrak{S}_l - \mathfrak{S}_k\mathfrak{J}_l) = 2\epsilon_{jkl}\mathfrak{L}_k\mathfrak{S}_l = 2(\boldsymbol{\mathfrak{L}} \times \boldsymbol{\mathfrak{S}})_j,$$

and

$$\mathfrak{Q}_j \equiv (\boldsymbol{\mathfrak{K}} \times \boldsymbol{\mathfrak{J}})_j = 2[(\boldsymbol{\mathfrak{L}} \times \boldsymbol{\mathfrak{S}}) \times \boldsymbol{\mathfrak{J}}]_j = \epsilon_{jkl}\mathfrak{K}_k\mathfrak{J}_l$$

where sums over repeated indices are implied, we obtain

$$\mathfrak{S}_3\boldsymbol{\mathfrak{J}}^2 = \mathfrak{J}_3\boldsymbol{\mathfrak{S}}\cdot\boldsymbol{\mathfrak{J}} + \mathfrak{Q}_3.$$

Hence we have

$$\begin{aligned}\tfrac{1}{2}[\mathfrak{S}_3\boldsymbol{\mathfrak{J}}^2 + \boldsymbol{\mathfrak{J}}^2\mathfrak{S}_3] &= \tfrac{1}{2}[\mathfrak{J}_3\boldsymbol{\mathfrak{S}}\cdot\boldsymbol{\mathfrak{J}} + \boldsymbol{\mathfrak{J}}\cdot\boldsymbol{\mathfrak{S}}\mathfrak{J}_3] + \tfrac{1}{2}(\mathfrak{Q}_3 + \mathfrak{Q}_3^\dagger)\\ &= \mathfrak{J}_3\boldsymbol{\mathfrak{S}}\cdot\boldsymbol{\mathfrak{J}} + \tfrac{1}{2}(\mathfrak{Q}_3 + \mathfrak{Q}_3^\dagger),\end{aligned}$$

the latter equation following from the fact that $\mathfrak{J}_3$ commutes from $\boldsymbol{\mathfrak{S}}\cdot\boldsymbol{\mathfrak{J}} = \boldsymbol{\mathfrak{J}}\cdot\boldsymbol{\mathfrak{S}}$.

It is easy to show that, since $\mathfrak{K}$ is a Hermitian operator,

$$\mathfrak{Q}_3^\dagger = \mathfrak{Q}_3 + \epsilon_{3jk}[\mathfrak{J}_k, \mathfrak{K}_j]$$

and hence

$$\tfrac{1}{2}(\mathfrak{Q}_3 + \mathfrak{Q}_3^\dagger) = \mathfrak{Q}_3 + \tfrac{1}{2}([\mathfrak{J}_2, \mathfrak{K}_1] - [\mathfrak{J}_1, \mathfrak{K}_2]).$$

It is left as an exercise for the reader to show that furthermore $\mathfrak{J}_k, \mathfrak{K}_l$ satisfy the commutation relations,

$$[\mathfrak{J}_k, \mathfrak{R}_l] = i\hbar\epsilon_{klm}\mathfrak{R}_m, \tag{16.32}$$

reminiscent of the commutators between $\mathfrak{L}$ and $\mathfrak{X}$ obtained in Section 12.6. From this it follows that

$$\tfrac{1}{2}(\mathfrak{Q}_3 + \mathfrak{Q}_3^\dagger) = \mathfrak{Q}_3 - i\hbar\mathfrak{R}_3.$$

Thus we seek the matrix elements

$$\langle jm'|\mathfrak{S}_3|jm\rangle = \frac{1}{\hbar^2 j(j+1)}\langle jm'|\mathfrak{J}_3\mathfrak{S}\cdot\mathfrak{J}|jm\rangle + \langle jm'|\mathfrak{Q}_3 - i\hbar\mathfrak{R}_3|jm\rangle),$$

and we now show that the last matrix element on the right-hand side vanishes.

From the commutation relation (16.32), we see that the correspondence

$$\mathfrak{J} \leftrightarrow \mathfrak{L}, \qquad \mathfrak{R} \leftrightarrow \mathfrak{X}$$

yields identical commutator relations. Furthermore, just as $\mathfrak{L}\cdot\mathfrak{X} = 0$, we have

$$\mathfrak{J}\cdot\mathfrak{R} = 2\epsilon_{jkl}\mathfrak{J}_j\mathfrak{L}_k\mathfrak{S}_l = \epsilon_{jkl}(\mathfrak{L}_j\mathfrak{L}_k\mathfrak{S}_l + \mathfrak{S}_j\mathfrak{L}_k\mathfrak{S}_l) = 0,$$

which confirms the analogy implied by this correspondence.

Since

$$\langle l, m'|\mathfrak{X}|l, m\rangle = 0,$$

as parity considerations will show, we must therefore have

$$\langle j, m'|\mathfrak{R}|j, m\rangle = 0.$$

Also, it is not difficult to verify, again through parity considerations, that

$$\langle l, m'|\mathfrak{X} \times \mathfrak{L}|l, m\rangle = 0,$$

which implies that

$$\langle j, m'|\mathfrak{Q}|j, m\rangle = \langle j, m'|\mathfrak{R} \times \mathfrak{J}|j, m\rangle = 0.$$

Finally, then, we have

$$\begin{aligned}\langle j, m'|\mathfrak{S}_3|j, m\rangle &= \frac{1}{\hbar^2 j(j+1)}\langle j, m'|\mathfrak{J}_3\mathfrak{S}\cdot\mathfrak{J}|j, m\rangle \\ &= \frac{m'}{\hbar j(j+1)}\langle j, m'|\tfrac{1}{2}(\mathfrak{J}^2 + \mathfrak{S}^2 - \mathfrak{L}^2)|j, m\rangle \\ &= \frac{\hbar m}{2}\frac{j(j+1) + s(s+1) - l(l+1)}{j(j+1)}\delta_{m',m},\end{aligned} \tag{16.33}$$

from which it follows that, for $s = \frac{1}{2}$,

$$\langle n,j,m',l',\tfrac{1}{2}|\mathfrak{H}_1(m)|n,j,m,l,\tfrac{1}{2}\rangle = \frac{e\hbar}{2M}Bm\left(1 + \frac{j(j+1) + \frac{3}{4} - l(l+1)}{2j(j+1)}\right)\delta_{l,l'}\delta_{m,m'}$$

$$= g\frac{e\hbar}{2M}Bm\delta_{l,l'}\delta_{m,m'} \tag{16.34}$$

where

$$g \equiv 1 + \frac{j(j+1) + \frac{3}{4} - l(l+1)}{2j(j+1)} \tag{16.35}$$

is known as the Landé g-factor.

Hence, according to first-order perturbation theory, the energy levels are given by

$$E_{n,j,l,m} = E_{n,j,l}^{(0)} + g\hbar\omega_L m, \tag{16.36}$$

where

$$\omega_L = (e\hbar/2M)B$$

is the Larmor precession frequency and $E_{n,j,l}^{(0)}$ are the eigenvalues of $\mathfrak{H}_0'$ for the states $|n, j, m, l, \frac{1}{2}\rangle$.

The effect of the spin–orbit interaction is to make the energy levels dependent upon j and l—that is, to remove some of the $2(2l + 1)$-fold degeneracy of the eigenstates of $\mathfrak{H}_0$. Thus it splits the degenerate $2p$ states of the one-electron atom into two levels, the $2p_{1/2}$, $2p_{3/2}$ levels, where the subscripts refer to the allowed values of j. The external magnetic field then removes the remaining $(2j + 1)$-fold degeneracy, as shown in Fig. 16.3. This figure also shows the allowed electric dipole radiative transitions.

Obviously the anomalous Zeeman effect is not anomalous; we can deal quite effectively with the weak-field splitting of the spectral lines, and can even obtain correct Landé g-factor to predict the frequencies of the spectral shifts imposed by the magnetic field. The effect was named, however, in pre-quantum mechanics when the existence of the electron spin, which obviously plays a vital role in this effect, was unrecognized.

Many of the results we have obtained can be derived semiclassically without all the quantum-mechanical formalism we have utilized. It is important, however, to achieve these results through the formalism in order to see that it works quite well, and in order to gain practice in making practical calculations.

We could go on to do more along these lines. We have ignored the quadratic Zeeman effect as well as the Paschen–Beck phenomenon and other interesting effects familiar to spectroscopists. However, our aim is not to cover the field of atomic spectroscopy, but rather to show that the theory works for such phenomena. In fact, we have not even accomplished that, since we have simply accepted the

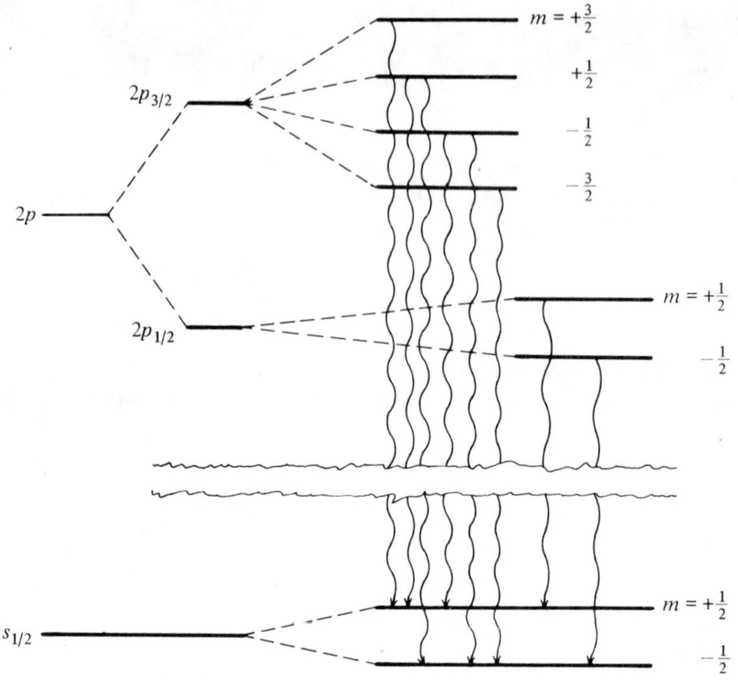

Fig. 16.3. The anomalous Zeeman effect for $2p \rightarrow 1s$ transitions of a one-electron atom.

form of the spin–orbit coupling term without being able to derive it from first principles. Furthermore, we have dealt only with one-electron problems; although we expect the theory to work for systems containing more than one electron, we have not developed any techniques for generalizing to such problems. Therefore we attempt in the next two chapters to rectify this situation. The spin–orbit coupling is a relativistic effect, so we take up the problem of finding a relativistic wave equation for the electron in the next chapter. After that, we consider the elementary problems encountered when we deal with many identical particles quantum-mechanically.

READING NOTES*

Reference 22, Chapter 14, 15.

Reference 23, Chapter 18.

Reference 24, Chapter 7.

Reference 43, Chapter 11.

---

* Works referred to in these notes will be found in the Bibliography on p. 577.

PROBLEMS

1. Using

$$\mathfrak{K}_j \equiv 2\sum_{k,l} \epsilon_{jkl}\mathfrak{L}_k\mathfrak{S}_l$$

and

$$\mathfrak{Q}_j \equiv \sum_{k,l} \epsilon_{jkl}\mathfrak{K}_k\mathfrak{J}_l,$$

show that

a) $\mathfrak{K}$ is Hermitian

b) $\mathfrak{Q}_k^\dagger = \mathfrak{Q}_k + \sum_{j,l} \epsilon_{kjl}[\mathfrak{J}_l, \mathfrak{K}_j]$

c) $[\mathfrak{J}_k, \mathfrak{K}_l] = i\hbar\epsilon_{klm}\mathfrak{K}_m$,

which are used in obtaining Eq. (16.33).

2. The Einstein coefficients $a_{0,1;\alpha}$, $b_{0,1;\alpha}$ and $b_{1,0;\alpha}$ refer to transitions between a state (represented by index 0) with energy E and a state (represented by index 1) with energy E'. Replace the index 1 with n, the latter representing an arbitrary state with energy E_n, and find the number of atoms remaining in state 0 after a time t, if initially there are N_0 in that state and none in any others. This will have the form $N_0 e^{-t/\tau}$; hence find τ in terms the Einstein coefficients and η.

3. How many quanta of the radiation field are there in the radiation field near the transition frequency when the induced emission rate equals the spontaneous emission rate? Find the temperature at which this occurs for optical frequencies, say for $\lambda = 6 \times 10^3$ Å $=$ 0.6 μm.

4. Describe the difference between the normal Zeeman effect and the anomalous Zeeman effect in terms of the effect of the magnetic field on the coupling of the orbital and spin angular momentum of the electron. Give a rough estimate of the difference in frequency resolution required for measuring the two effects. Which effect is the more demanding on the resolving power?

5. Find the mean lifetime for a $2p$ ($n = 2$, $l = 1$) electron with $m_l = 0$ against electric-dipole radiation.

CHAPTER 17

RELATIVISTIC QUANTUM THEORY

If the two central theories of modern physics, special relativity and quantum mechanics, are to provide a solid foundation for further investigation of physical laws, they must be applicable simultaneously. We cannot be satisfied unless quantum theory can be applied equally well to systems in all inertial frames of reference. In fact, in Chapter 8 we came across a relativistic equation, the Klein–Gordon equation, that seemed to be the appropriate transcription of the relativistic prequantum equation for a free particle into quantum-theoretical language. However, it is quadratic in the time derivative, and this immediately raised some difficulties.

The presence of the quadratic time derivative in the Klein–Gordon equation gives rise to three disturbing circumstances, as we shall see, namely

1. that there are both positive- and negative-energy solutions to the equation, whereas we seem to observe only positive-energy particles in nature;

2. that positive-energy solutions to the Klein–Gordon equation are not localizable in the sense that it is impossible to find a superposition of momentum eigenstates that diagonalizes the position operator $\mathfrak{X}$; and

3. the probability density cannot be made positive-definite and hence the probability interpretation of the quantum theory would seem to be inapplicable.

It can be shown, in fact, that these are inevitable consequences of the relation $p^\mu p_\mu - M^2c^2 = 0$ when it is applied to quantum theory. On the other hand, it is possible to reinterpret quantities in such a way that these consequences are no longer barriers to our acceptance of such relativistic transcriptions of quantum mechanics as rather adequate theoretical equations. To obtain "really satisfactory" theories of relativistic quantum-mechanical systems, one must go somewhat beyond the quantum descriptions developed in the preceding chapters, and must adopt the deeper point of view of quantum field theory. This, however, is not within the scope of our present discussion, and here we shall restrict ourselves to seeing what can be salvaged of ordinary relativistic quantum theory.

17.1 THE KLEIN–GORDON EQUATION

In state-vector language, the Klein–Gordon equation reads

$$(\mathfrak{P}^\mu \mathfrak{P}_\mu - M^2c^2)|M, q\rangle = 0, \tag{17.1}$$

where the index M simply indicates that the state $|M, q\rangle$ is an eigenstate of the operator $\mathfrak{P}^\mu \mathfrak{P}_\mu$, having eigenvalue M^2c^2. The operator $\mathfrak{P}^\nu$ commutes with $\mathfrak{P}^\mu \mathfrak{P}_\mu$, and hence it is possible to find solutions to the Klein–Gordon equation that are eigenstates of $\mathfrak{P}^\nu$ with eigenvalues, say $p^\nu = (E/c, + \mathbf{p})$, as we saw in Chapter 8. Let us denote such a state $|M, p; q\rangle$, where q represents any other quantum number possessed by the system. Then

$$\mathfrak{P}^\nu |M, p, q\rangle = p^\nu |M, p, q\rangle \tag{17.2}$$

and we must have

$$p_\nu p^\nu = M^2c^2 \qquad \text{or} \qquad E^2 - \mathbf{p}^2c^2 = M^2c^4$$

because of the Klein–Gordon equation. Hence

$$E = \pm\sqrt{\mathbf{p}^2c^2 + M^2c^4},$$

just as we found in Chapter 8. We may "solve" the problem of negative energies simply by ignoring the minus sign, but this merely postpones the agony.

To see how this is related to the localization problem, let us consider a state vector $|M, x_0, q\rangle$ representing a particle of mass M situated at the point $x_0{}^\nu = (ct_0, \mathbf{x}_0)$ in four-space. Let this particle be displaced to a point x^μ:

$$|M, x, q\rangle = \exp(i(x_\mu - x_{0\mu})\mathfrak{P}^\mu/\hbar)|M, x_0, q\rangle. \tag{17.3}$$

The question is, does $\langle M, x, q|M, x_0, q\rangle$ vanish when $x^\mu - x_0{}^\mu = (0, \delta\mathbf{x})$? That is, are the states representing a particle of mass M at different locations orthogonal?

To find out, we must have a completeness relation for momentum eigenstates, and this relation should be relativistically covariant. The nonrelativistic completeness relation was

$$(1/\hbar^3)\int d^3p\,|\mathbf{p}\rangle\langle\mathbf{p}| = 1,$$

but this is not obviously Lorentz-covariant, since d^3p transforms to a new inertial frame in a complicated way. Furthermore, the covariance of the expression will depend upon the normalization of the states $|\mathbf{p}\rangle$. Let us begin by normalizing our momentum eigenstates covariantly:

$$\langle M', p', q'|M, p, q\rangle = \frac{\hbar^4}{c}\delta(M' - M)\delta^3(\mathbf{p}' - \mathbf{p})\frac{p_0}{Mc}\delta_{q',q}. \tag{17.4}$$

This is covariant, since the mass M is an invariant, $\delta_{q',q}$ is independent of reference frame, and one can show that under a Lorentz transformation

$$\delta^3(\mathbf{p}' - \mathbf{p}) \to \frac{p_0}{\gamma}[p_0 - \boldsymbol{\beta}\cdot\mathbf{p}]^{-1}\delta^3(\mathbf{p}' - \mathbf{p})$$

$$p_0 \to \gamma[p_0 - \boldsymbol{\beta}\cdot\mathbf{p}].$$

More dramatically, we may write (17.4) in the form

$$\langle M', p', q'|M, p, q\rangle = \hbar^4\delta^4(p' - p)\delta_{q',q} \tag{17.5}$$

where we have used

$$\delta(M' - M) = \frac{dp_0}{dM}\delta(p_0' - p_0),$$

which follows from the basic properties of δ-functions, and

$$\frac{dp_0}{dM} = \frac{Mc^2}{p_0}$$

which is obtained from $p_0^2 - \mathbf{p}^2 - M^2c^2 = 0$ when $\mathbf{p}$ is kept fixed but p_0 and M are allowed to vary.

This normalization indicates that our completeness condition should be

$$\underset{q}{\mathsf{S}}\int c\,dM \int \frac{d^3p}{\hbar^4}\frac{Mc}{p_0}|M, p, q\rangle\langle M, p, q| = 1. \tag{17.6}$$

Usually the sum over masses $(c/\hbar)\int dM$ in (17.6) and the δ-function on masses $(\hbar/c)\delta(M' - M)$ in (17.4) are dropped, it being understood that we are dealing with a single kind of particle and hence with only one mass. Equation (17.6) can be written in a slightly more spectacular form also, when we define $p^2 \equiv p^\mu p_\mu$ and recognize that

$$\delta(p^2 - M^2c^2) = \frac{1}{2p_0}[\delta(p^0 - \sqrt{\mathbf{p}^2 + M^2c^2}) + \delta(p^0 + \sqrt{\mathbf{p}^2 + M^2c^2})].$$

If we let $\delta_+(p_2 - M^2c^2)$ represent only the first term so that p^0 is restricted to positive values, we have

$$2Mc\delta_+(p^2 - M^2c^2) = \frac{Mc}{p_0}\delta(p_0 - \sqrt{\mathbf{p}^2 + M^2c^2}).$$

Then we see that

$$\frac{1}{\hbar^4}\int d^4p\,2Mc\delta_+(p^2 - M^2c^2) = \frac{1}{\hbar^4}\int d^3p\,\frac{Mc}{p_0}.$$

Thus (17.6) can be written

$$\underset{q}{\mathsf{S}}\int dM \int \frac{d^4p}{\hbar^4}\delta_+(p^2 - M^2c^2)2Mc^2|M, p, q\rangle\langle M, p, q| = 1. \tag{17.7}$$

The beauty of this form is that it is manifestly covariant when the state vectors are covariantly normalized, and in addition it assures us that p_0 is positive and that $\mathbf{p}^2$ and M^2 are properly related to each other.

Now, returning to the question following Eq. (17.3), we may write

$$\begin{aligned}\langle M, x, q|M, x_0, q\rangle &= \mathop{\mathsf{S}}_{q'}\int \frac{d^3p}{\hbar^3}\frac{Mc}{p_0}\langle M, x_0, q|e^{-i(x_\mu - x_{0\mu})p^\mu/\hbar}|M, p, q'\rangle \\ &\qquad \langle M, p, q'|M, x_0, q\rangle \\ &= \int \frac{d^3p}{\hbar^3}\frac{Mc}{p_0}|\langle M, x_0, q|M, p, q\rangle|^2\, e^{-i(x_\mu - x_{0\mu})p^\mu/\hbar} \\ &= \int \frac{d^3p}{\hbar^3}\frac{Mc}{p_0}|\langle M, x_0, q|M, p, q\rangle|^2\, e^{+i\delta\mathbf{x}\cdot\mathbf{p}/\hbar} \end{aligned} \tag{17.8}$$

where

$$\langle M, p, q'|M, x_0, q\rangle = \langle M, p, q|M, x_0, q\rangle\delta_{q,q'}$$

since the q's represent non-space–time quantum numbers. However $\langle M, x, q|M, p, q\rangle$ is the wave function for a plane wave and therefore its absolute square cannot depend on x; furthermore, if it depends on p, it must do so through the invariant $p^2 = M^2c^2$, since the state vectors are supposed to have been normalized covariantly. Hence $|\langle M, x_0, q|M, p, q\rangle|^2 = |N|^2$ cannot depend on p and may be removed from the integral. Thus we obtain

$$\begin{aligned}\langle M, x, q|M, x_0, q\rangle &= |N|^2\frac{Mc}{h^3}\int_0^\infty \frac{\mathbf{p}^2\,dp}{\sqrt{\mathbf{p}^2 + M^2c^2}}\int d\Omega\, e^{-i\delta\mathbf{x}\cdot\mathbf{p}/\hbar} \\ &= 4\pi|N|^2\frac{Mc}{h^3}\int_0^\infty \frac{\mathbf{p}^2\,dp}{\sqrt{\mathbf{p}^2 + Mc^2}}\frac{\sin(p\,\delta x/\hbar)}{(p\,\delta x/\hbar)} \\ &= 4\pi|N|^2\frac{1}{(\delta x)^3}\int_0^\infty \frac{\xi\sin\xi\,d\xi}{\sqrt{1 + [\xi^2\hbar^2/M^2c^2(\delta x)^2]}} \\ &= 4\pi|N|^2\frac{1}{(\delta x)^3}F\left(\frac{Mc\,\delta x}{\hbar}\right). \end{aligned} \tag{17.9}$$

The integral defining $F(Mc\,\delta x/\hbar)$ is very complicated, but it can be evaluated in terms of Bessel functions [e.g. see A. ERDELYI, *et al.*, *Higher Transcendental Functions*, Vol. 2, McGraw-Hill Book Co., Inc., New York (1953), p. 24]; in fact

$$F\left(\frac{Mc\,\delta x}{\hbar}\right) = \left(\frac{Mc\,\delta x}{\hbar}\right)^2 K_1\left(\frac{Mc\,\delta x}{\hbar}\right),$$

where $K_1(y)$ is a modified Bessel function that has the following asymptotic behavior:

$$K_1(y) \underset{y\to 0}{\sim} \frac{1}{y}$$

$$K_1(y) \underset{y\to\infty}{\sim} \sqrt{\frac{\pi}{2y}}\, e^{-y};$$

see Fig. 17.1 [see also M. ABRAMOWITZ and I. A. STEGUN, Ed., *Handbook of Mathematical Functions*, National Bureau of Standards, Washington, D.C. (1964), pp. 374–378].

Hence

$$\begin{aligned}\langle M, x, q|M, x_0, q\rangle &= 4\pi|N|^2\frac{(Mc)^2}{\hbar^2\,\delta x}K_1\left(\frac{Mc\,\delta x}{\hbar}\right)\\ &\underset{\delta x\to 0}{\sim} 4\pi|N|^2\frac{Mc}{\hbar}\left(\frac{1}{\delta x}\right)^2\\ &\underset{\delta x\to\infty}{\sim} |N|^2\left(2\pi\frac{Mc}{\hbar}\frac{1}{\delta x}\right)^{3/2} e^{-Mc\delta x/\hbar}.\end{aligned} \tag{17.10}$$

This clearly does not vanish for $\delta x \neq 0$, and the "width" of the distribution in δx is approximately $\hbar/Mc$, the reduced Compton wavelength for the particle. Since $|M, x_0, q\rangle$ and $|M, x, q\rangle$ are therefore not orthogonal when $\mathbf{x} \neq \mathbf{x}_0$, they cannot be eigenstates of the Hermitian operator $\mathfrak{X}$; we cannot form, from the momentum eigenstates of a particle with mass M and positive energy, eigenstates of position. This must mean that the set $|M, p, q\rangle$ is not a complete set if we insist on having positive energies!

We cannot pursue this problem in depth here; the main point is that the positive energy states are not sufficient to provide a satisfactory theory. The alert observer will have noted that the p_0 in the denominator in Eq. (17.8)—a factor required to provide Lorentz covariance—is the culprit. That in turn arises from the requirement that quadratic kinematic quantities such as $p^2 = M^2c^2$ be the invariants in relativistic systems, and it is from this circumstance that the ambiguity in the sign of the energy occurs. Negative energies, then, are inescapable in relativistic quantum theory, and we must learn how to deal with them.

We considered in Section 9.6 the problem of finding a current density for the Klein–Gordon equation. There also, the unpleasantness that developed arose from the fact that the invariant operator of the theory is the quadratic operator $\partial^\mu\partial_\mu$, so that the probability density for Klein–Gordon particles is not positive-definite. Since $\partial_0 = (1/c)(\partial/\partial t)$ is proportional to the energy operator, and is also the source of our difficulty with the nonpositive-definiteness of the probability density, it is clear that the latter is another manifestation of our inability to do relativistic quantum theory without dealing with the negative-energy problem.

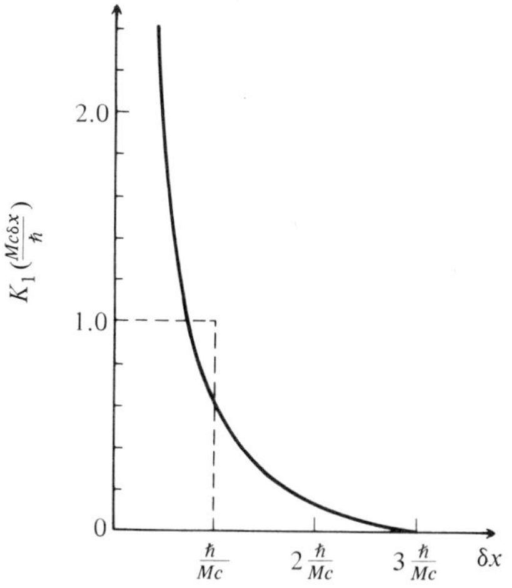

Fig. 17.1. The function $K_1(Mc\,\delta x/\hbar)$. Note that the "width" of the curve is essentially h/Mc. From M. ABRAMOWITZ and J. A. STEGUN (ed.), *Handbook of Mathematical Functions*, National Bureau of Standards, Washington, D.C. (1964). Reprinted through permission of the publisher.

A temporary way out, however, is to multiply the probability density and probability current density, given in Eqs. (9.61) and (9.62), by the charge of the particle, q_0, and reinterpret the results as the electrical charge and current densities:

$$\rho = q_0 \frac{i\hbar}{2Mc^2}\left[\psi^* \frac{\partial}{\partial t}\psi - \left(\frac{\partial}{\partial t}\psi^*\right)\psi\right]$$

$$\mathbf{j} = q_0 \frac{\hbar}{i2Mc}[\psi^* \nabla\psi - (\nabla\psi^*)\psi].$$

If ψ has been normalized in such a way that $\psi^*\psi$ is a volume density in some frame of reference, then $\psi^*\psi$ is not invariant, and therefore ρ is not the zeroth component of a four-vector since it transforms like $(1/V)(1/c)(\partial/\partial t)$ rather than like $(1/c)(\partial/\partial t)$. However, $\int d^3x\rho$ does transform like the zeroth component of a four-vector; similarly, $\int d^3x\mathbf{j}$ transforms like the spatial part of a four-vector, and we have, for the four-current of the Klein–Gordon particle,

$$J^\mu = q_0 i \frac{\hbar}{2Mc}\int d^3x[\psi^*\partial^\mu\psi - (\partial^\mu\psi^*)\psi]. \tag{17.11}$$

If, for example, ψ is an eigenfunction of $\mathfrak{P}^\mu$, we then have

$$J^\mu = \frac{q_0}{Mc}p^\mu$$

and

$$J^\mu J_\mu = q_0^2$$

as we should require. Therefore this reinterpretation seems to be an appropriate way of handling the problem. The charge *density* need not be positive-definite as long as the total charge is q_0; we have just seen that the total charge indeed is q_0.

This suggests further that we reinterpret the negative energies. If we have $q_0\rho$ representing charge density, and if it is proportional to the energy operator, then a negative charge density implies a negative energy. Perhaps, then, the eigenvalues of the *energy* operator are really equal to the energy times the sign of the charge density relative to the charge on the particle in question; that is, if the positive energy (normal) states correspond to positive charges, the "negative-energy" states are really positive-energy states for particles that are identical with the normal particles except that they have negative electric charge. Thus

$$\frac{E}{c} = \frac{q}{q_0}\sqrt{\mathbf{p}^2 + M^2c^2}$$

where q_0 is the charge on the "normal particle", and $q = \pm q_0$.

Although this view has some difficulties in the naïve form presented here, it provides a basis for the presently accepted view of relativistic particle physics, and we shall adopt it. Every particle, then, has associated with it an *antiparticle* of opposite charge; an uncharged particle may (but not necessarily will) be its own antiparticle if it has no internal quantum numbers such as spin. Although a one-particle theory for a relativistic particle may be an adequate approximation under many circumstances, a relativistic theory necessarily must be a many-particle theory. For example, if one tries to localize a particle to within $\hbar/mc$ of a point in space, the antiparticle ("negative-energy") states appear as a necessary part of the spectrum of momentum eigenstates taking part in the formation of the localized state, as we have seen.

For a satisfactory consideration of antiparticle states, the formalism of quantum field theory is required. More can be learned from our one-particle approach, however, and we now seek to find solutions to the Klein–Gordon equation in the presence of a Coulomb field; we attempt to solve the problem of a one-"Klein–Gordon electron" atom.

17.2 THE ONE-PARTICLE KLEIN–GORDON ATOM

The general problem of introducing potential functions into the Klein–Gordon equation without disturbing the Lorentz covariance of the equation is not trivial. In Chapter 9, however, we found that the interaction of a charged particle with the electromagnetic field can be accomplished rather simply in a covariant way. According to Eq. (9.50) and the discussion following it, we merely make the transcription

$$p_\mu \to p_\mu - qA_\mu$$

in order to achieve the simplest possible coupling ("minimal coupling") between the particle and the electromagnetic field. Thus the Klein–Gordon differential equation becomes

$$[(i\hbar\partial^\mu - qA^\mu)(i\hbar\partial_\mu - qA_\mu) - M^2c^2]\psi(x) = 0. \tag{17.12}$$

It is left as an exercise for the student to show that in the nonrelativistic limit this reduces to the usual Schrödinger equation with electromagnetic interaction.

Our primary concern will be to find the energy levels of such a system when the four-potential in some frame of reference is a pure Coulomb field. We must ask, then, under what circumstances Eq. (17.12) is separable in time and space coordinates. When may the solutions to (17.12) be eigenfunctions of energy? It is not difficult to see that they may be energy eigenfunctions when $A_\mu(x)$ is time independent: $A_\mu(x) = A_\mu(\mathbf{x})$. In that case we may write

$$\psi(x) = \psi_E(\mathbf{x})e^{-iEt/\hbar}$$

obtaining

$$\frac{1}{c^2}[E - q\phi(\mathbf{x})]^2\psi_E(\mathbf{x}) = \left[\left(\frac{\hbar}{i}\boldsymbol{\nabla} - q\mathbf{A}(\mathbf{x})\right)^2 + M^2c^2\right]\psi_E(\mathbf{x}),$$

since $A^\mu(x) = (\phi(x)/c, \mathbf{A}(x))$. Specializing to an attractive Coulomb field, $\mathbf{A}(x) = \mathbf{0}$, $\phi(x) = \phi(\mathbf{x}) = \phi(r) = -Ze/4\pi\epsilon_0 r$, and letting our Klein–Gordon particle have charge $q = e$, we finally get

$$\left\{-\hbar^2\boldsymbol{\nabla}^2 + M^2c^2 - \frac{1}{c^2}[E - e\phi(r)]^2\right\}\psi_E(\mathbf{x}) = 0.$$

Now this equation is separable in spherical coordinates since ϕ depends only on r; therefore we take

$$\psi_{E,l,m}(\mathbf{x}) = C_{E,l,m}\frac{g(l, \kappa, r)}{r}Y_l^m(\theta, \varphi), \tag{17.13}$$

where κ is a constant that depends directly on the energy E, and $C_{E,l,m}$ is a normalization constant. Use of the identity

$$\begin{aligned}\boldsymbol{\nabla}^2 &= \frac{1}{r}\partial_r^2 r + \frac{1}{r^2}\left[\partial_\theta^2 + \cot\theta\,\partial_\theta + \frac{1}{\sin^2\theta}\partial^2\varphi\right] \\ &= \frac{1}{r}\partial_r^2 r - \frac{1}{\hbar^2 r^2}\mathbf{L}^2\end{aligned}$$

obtained in Section 12.5 [see especially Eq. (12.46)], leads to the radial equation

$$\left\{\frac{d^2}{dr^2} - \frac{l(l+1)}{r^2} - \frac{M^2c^2}{\hbar^2} + \frac{1}{\hbar^2c^2}[E - e\phi(r)]^2\right\}g(l, \kappa, r) = 0;$$

here we have used the relationship

$$\mathbf{L}^2 Y_l^m(\theta, \varphi) = \hbar^2 l(l + 1) Y_l^m(\theta, \varphi)$$

also.

Substitution of $\phi(r) = -Ze/4\pi\epsilon_0 r$ and of

$$\xi \equiv \kappa r, \qquad \kappa \equiv \frac{Mc}{\hbar}\left(1 - \frac{E^2}{M^2c^4}\right)^{1/2} \tag{17.14}$$

into this result yields

$$\left[\frac{d^2}{d\xi^2} + 2\frac{Z\alpha E}{\hbar\kappa c\xi} - 1 - \frac{l(l + 1) - Z^2\alpha^2}{\xi^2}\right] g\left(l, \kappa, \frac{\xi}{\kappa}\right) = 0.$$

For convenience, we shall take

$$\eta = \frac{Z\alpha E}{\hbar\kappa c}, \qquad g_l(\xi) \equiv g\left(l, \kappa, \frac{\xi}{\kappa}\right). \tag{17.15}$$

Although this definition of η is different from that given above in Eq. (14.76), it is such that our new radial equation,

$$\left[\frac{d^2}{d\xi^2} - \frac{l(l + 1) - Z^2\alpha^2}{\xi^2} + 2\frac{\eta}{\xi} - 1\right] g_l(\xi) = 0 \tag{17.16}$$

resembles very closely Eq. (14.76), the only change being $l(l + 1) \rightarrow l(l + 1) - Z^2\alpha^2$. This means that in order that a_0 does not vanish in Eq. (14.81), we must have

$$p(p - 1) = l(l + 1) - Z^2\alpha^2,$$

or, instead of (14.82),

$$p = \tfrac{1}{2}[1 \pm \sqrt{(2l + 1)^2 - 4Z^2\alpha^2}].$$

The solutions corresponding to the minus sign are irregular at $r = 0$, and hence we reject them, as before.

Recalling that in order to get solutions that behave properly as $r \rightarrow \infty$ it was necessary for $p + n - \eta$ in Eq. (14.83) to vanish for some positive integer $v - 1$, we see that for the Klein–Gordon equation we must have

$$-\tfrac{1}{2} + \tfrac{1}{2}\sqrt{(2l + 1)^2 - 4Z^2\alpha^2} + v = \eta, \qquad v = 1, 2, 3, \ldots.$$

Since η is a function of E, we may solve this equation for E, using the second of Eqs. (17.14) and the first of (17.15):

$$E = Mc^2\left[1 + \left(\frac{Z\alpha}{\eta}\right)^2\right]^{-1/2} \tag{17.17}$$

Using

$$\eta = (\nu - \tfrac{1}{2}) + (l + \tfrac{1}{2})\left[1 - \left(\frac{2Z\alpha}{2l + 1}\right)^2\right]^{1/2}$$

$$\simeq (\nu + l)\left[1 - \frac{Z^2\alpha^2}{(2l + 1)(\nu + l)}\right] \quad \text{for small } Z\alpha, \tag{17.18}$$

we obtain, to fourth degree in $Z\alpha$,

$$E_{\nu,l} \simeq Mc^2\left[1 - \frac{1}{2}\left(\frac{Z\alpha}{\nu + l}\right)^2 - \frac{1}{2}\left(\frac{Z\alpha}{\nu + l}\right)^4\left(\frac{2(\nu + l)}{2l + 1} - \frac{3}{4}\right)\right]. \tag{17.19}$$

The second term agrees precisely with Eq. (14.86), but the third represents an l-dependent energy term that removes some of the degeneracy of the nonrelativistic spinless one-electron atom. Evidently the latter produces a "fine structure" to the energy levels, being of order $(Z\alpha)^2$ smaller than the separation between levels predicted by the nonrelativistic theory.

Setting $n = \nu + l$, we see that for a fixed value of the principal quantum number n, the maximum separation between fine-structure energy levels is

$$\frac{Mc^2}{2}\left(\frac{Z\alpha}{n}\right)^4\left(\frac{2n}{1} - \frac{2n}{2n - 1}\right) = Mc^2\frac{(Z\alpha)^4}{n^3}\frac{n - 1}{n - \frac{1}{2}},$$

since $0 \leq l \leq n - 1$. This, however, is much larger than the observed fine structure for the hydrogen atom. The hydrogen atom's fine-structure, then, must be attributed to causes other than the simple relativistic correction supplied by the Klein–Gordon equation, and the assumption that it arises from the existence of spin angular momentum for the electron seems to remain unchallenged by our relativistic calculation.

Indeed this raises an interesting question: can spin angular momentum be incorporated in the Klein–Gordon equation without destroying its covariant features? The answer is no; the spin operators generate rotations but not special Lorentz transformations, and the three spin operators together with the identity matrix exhaust the linearly independent 2×2 matrices. Hence, if we are to be able to generalize the notion of spin to relativistically covariant theories, we shall need six linearly independent antisymmetric matrices, and therefore shall require spin operators that cannot be represented by matrices of dimensions less than 4×4. The Klein–Gordon equation provides a possible basis for theories of spinless particles, but not for theories in which half-odd-integer spin angular momentum plays a role.

17.3 THE DIRAC EQUATION

Although the energy eigenvalues for Klein–Gordon particles in a Coulomb field may be found, they do not seem to correspond to the energy levels observed in atomic systems. Furthermore, the interpretation of the wave function is left in some doubt because of the fundamental difficulties arising from the quadratic form of the Klein–Gordon equation. P. A. M. Dirac, in 1926, attempted to circumvent these latter problems by working with a relativistically covariant equation that is linear in the four-momentum operator $\mathfrak{P}^\mu$. Although his linear equation did not avoid the difficulties just mentioned, it did lead to energy eigenvalues that agree with those of the hydrogen atom, and it provided a basis for a cogent interpretation of the negative energy levels for Fermi particles (with half-odd-integer spin quantum numbers). Most important, perhaps, is the fact that in the Dirac theory, spin appears as an integral part of the equation of motion, and such things as the spin–orbit effect and the Pauli ($\mathfrak{S} \cdot \mathfrak{B}$) term in the nonrelativistic Hamiltonian appear automatically in the nonrelativistic reduction of the Dirac equation for a particle in an electromagnetic field.

For a free particle with mass M, Dirac's equation may be written in the covariant form

$$(\Gamma_\mu \mathfrak{P}^\mu - Mc)|M, q\rangle = 0, \tag{17.20}$$

where the Γ_μ are a set of four operators that presumably form a four-vector. Since this represents a free particle, the Γ_μ evidently cannot depend on position or time variables, and hence

$$[\Gamma_\mu, \mathfrak{P}_\nu] = 0. \tag{17.21}$$

This does not mean, however, that we can simultaneously diagonalize the $\mathfrak{P}_\nu$ and Γ_μ, for we do not know the commutation relations for the Γ_μ among themselves.

If $|M, q\rangle$ is to represent a free particle, we must also have the relativistic relationship

$$(\mathfrak{P}^n \mathfrak{P}_\mu - M^2c^2)|M, q\rangle = 0$$

satisfied for all such states. Hence Eq. (17.20) is a linear equation that must be satisfied *in addition to* the Klein–Gordon equation. This leads us to suspect that we are really not avoiding the localization problems, etc. that existed for the Klein–Gordon equation, and this suspicion is well founded. On the other hand, the imposition of Eq. (17.20) may well provide a different expression for current density as well as a different nonrelativistic reduction of the equations of motion by providing a Hamiltonian for the system. The notion of a Hamiltonian was noticeably absent from our treatment of the Klein–Gordon equation.

Condition (17.20) suggests that we consider operating twice on $|M, q\rangle$ with $\Gamma_\mu \mathfrak{P}^\mu$, obtaining

$$\begin{aligned} \Gamma_\mu \mathfrak{P}^\mu \Gamma_\nu \mathfrak{P}^\nu |M, q\rangle = M^2c^2|M, q\rangle &= \mathfrak{P}_\mu \mathfrak{P}^\mu |M, q\rangle \\ &= g_{\mu\nu} \mathfrak{P}^\mu \mathfrak{P}^\nu |M, q\rangle \end{aligned}$$

or

$$(\Gamma_\mu \Gamma_\nu - g_{\mu\nu}) \mathfrak{P}^\mu \mathfrak{P}^\nu |M, q\rangle = 0. \tag{17.22}$$

It is useful here to use the identity

$$\Gamma_\mu \Gamma_\nu = \tfrac{1}{2}[\Gamma_\mu, \Gamma_\nu] + \tfrac{1}{2}\{\Gamma_\mu, \Gamma_\nu\},$$

for then we notice that

$$\begin{aligned}\tfrac{1}{2}[\Gamma_\mu, \Gamma_\nu]\mathfrak{P}^\mu\mathfrak{P}^\nu &= \tfrac{1}{2}[\Gamma_\nu, \Gamma_\mu]\mathfrak{P}^\nu\mathfrak{P}^\mu = \tfrac{1}{2}[\Gamma_\nu, \Gamma_\mu]\mathfrak{P}^\mu\mathfrak{P}^\nu \\ &= -\tfrac{1}{2}[\Gamma_\mu, \Gamma_\nu]\mathfrak{P}^\mu\mathfrak{P}^\nu = 0,\end{aligned} \tag{17.23}$$

which is obtained by interchanging dummy indices and using the fact that $[\mathfrak{P}^\mu, \mathfrak{P}^\nu] = 0$. Hence Eq. (17.22) tells us nothing about $[\Gamma_\mu, \Gamma_\nu]$, but we may use it to learn something about $\{\Gamma_\mu, \Gamma_\nu\}$. Because of Eq. (17.23) we have

$$[\{\Gamma_\mu, \Gamma_\nu\} - 2g_{\mu\nu}]\mathfrak{P}^\mu\mathfrak{P}^\nu|M, q\rangle = 0 \tag{17.24}$$

A sufficient, though not necessary, condition for (17.24) to be satisfied is

$$\{\Gamma_\mu, \Gamma_\nu\} = 2g_{\mu\nu}. \tag{17.25}$$

Let us assume that this condition holds; thus

$$\Gamma_\mu\Gamma_\nu = -\Gamma_\nu\Gamma_\mu, \qquad \mu \neq \nu \tag{17.26a}$$

$$\Gamma_0^2 = I \tag{17.26b}$$

$$\Gamma_m^2 = -I, \qquad m = 1, 2, 3. \tag{17.26c}$$

It follows that the eigenvalues of Γ_0 are ± 1 and those of Γ_m are $\pm i$; however, we cannot have all the Γ_μ diagonalized simultaneously, since they do not commute with one another. Since Γ_0 has real eigenvalues, it is Hermitian, and since the Γ_m have imaginary eigenvalues, they are antihermitian.

This immediately raises a serious question: If the Γ_m are not Hermitian, how can $\Gamma_\mu\mathfrak{P}^\mu$ have real eigenvalues Mc? Because of (17.19),

$$(\Gamma_\mu\mathfrak{P}^\mu)^\dagger = \mathfrak{P}^\mu\Gamma_\mu^\dagger = \Gamma_\mu^\dagger\mathfrak{P}^\mu;$$

but the right-hand member does not equal $\Gamma_\mu\mathfrak{P}^\mu$, and therefore $\Gamma_\mu\mathfrak{P}^\mu$ is not Hermitian. However, Eqs. (17.26) also show that

$$\begin{aligned}(\Gamma_0\Gamma_\mu\mathfrak{P}^\mu)^\dagger &= \mathfrak{P}^\mu\Gamma_\mu^\dagger\Gamma_0 = \Gamma_0(\Gamma_0\Gamma_\mu^\dagger\Gamma_0)\mathfrak{P}^\mu \\ &= \Gamma_0\Gamma_\mu\mathfrak{P}^\mu\end{aligned}$$

since

$$\Gamma_0\Gamma_0\Gamma_0 = \Gamma_0^2\Gamma_0 = \Gamma_0 = \Gamma_0^\dagger$$

$$\Gamma_0\Gamma_m\Gamma_0 = -\Gamma_0^2\Gamma_m = -\Gamma_m = \Gamma_m^\dagger.$$

From this it follows that $\Gamma_0\Gamma_\mu P^\mu$ *is* a Hermitian operator, as is $\Gamma_0 M$, and that the equation we want to solve is

$$\Gamma_0(\Gamma_\mu\mathfrak{P}^\mu - Mc)|M, q\rangle = 0 \tag{17.27}$$

rather than (17.20). An alternative approach is to stick with (17.20), but to consider the bra corresponding to $|M, q\rangle$ to be

$$\langle M, q|\Gamma_0;$$

the two approaches are equivalent, and we shall choose the latter for now.

Now (17.26) tells us right away that the Γ_μ's are not constant operators—that is, that each Γ_μ must have more than one eigenvalue—for operators with just one eigenvalue commute with all other operators. Also, they commute with the $\mathfrak{P}^\mu$ and do not involve $\mathfrak{X}^\mu$; they also commute with $\mathfrak{X}^\mu$ since otherwise the Dirac equation would not be of first degree in $\mathfrak{P}^\mu$. Hence they must be completely independent of space–time and momentum–energy variables, and must represent internal degrees of freedom for the system. The state vectors for the system then are tensor products of the space–time state vectors $|M, q\rangle$ with internal coordinate state vectors $|r\rangle$:

$$|M, q; r\rangle = |M, q\rangle|r\rangle$$

since the Γ^μ's commute with $\mathfrak{X}^\mu$'s and $\mathfrak{P}^\mu$'s.

Concentrating on the internal degrees of freedom, we consider matrix elements

$$[\gamma_\mu]_{r'r} = \langle r'|\Gamma_\mu|r\rangle. \tag{17.28}$$

In the first place, the trace of γ_μ, defined as

$$\text{Tr}\,\gamma_\mu = \sum_r \langle r|\Gamma_\mu|r\rangle$$

vanishes, since, for $\nu \neq \mu$,

$$\begin{aligned}
\text{Tr}\,\gamma_\mu &= \sum_r \langle r|\Gamma_\mu|r\rangle = \pm \sum_r \langle r|\Gamma_\mu\Gamma_\nu^2|r\rangle \\
&= \mp \sum_r \langle r|\Gamma_\nu\Gamma_\mu\Gamma_\nu|r\rangle \\
&= \mp \sum_{r,s,t} \langle r|\Gamma_\nu|s\rangle\langle s|\Gamma_\mu|t\rangle\langle t|\Gamma_\nu|r\rangle \\
&= \mp \sum_{r,s,t} \langle t|\Gamma_\nu|r\rangle\langle r|\Gamma_\mu|s\rangle\langle s|\Gamma_\nu|t\rangle \qquad (r \to t,\, t \to s,\, s \to r) \\
&= \mp \sum_{r,s,t} \langle r|\Gamma_\mu|s\rangle\langle s|\Gamma_\nu|t\rangle\langle t|\Gamma_\nu|r\rangle = \mp \sum_r \langle r|\Gamma_\mu\Gamma_\nu^2|r\rangle \\
&= -\text{Tr}\,\gamma_\mu = 0.
\end{aligned}$$

Since the eigenvalues of Γ_0 are ± 1, this means that Γ_0 must have even dimensionality; the $+1$'s on the diagonal of $[\gamma_0]_{r'r}$ when the latter is diagonal must cancel the -1's, since the trace is the sum of diagonal elements of a matrix. The same holds for the Γ_m's, which must have the same dimension as Γ_0, as Eq. (17.26a) implies.

To obtain a convenient 4 × 4 representation for the Γ_μ's, we let γ_0 be diagonal:

$$\gamma_0 = \begin{bmatrix} 1 & 0 & 0 & 0 \\ 0 & 1 & 0 & 0 \\ 0 & 0 & -1 & 0 \\ 0 & 0 & 0 & -1 \end{bmatrix};$$

the negative diagonal elements are needed to satisfy the requirement that the trace vanish. It is convenient to think of this as being composed of four 2 × 2 matrices:

$$\gamma_0 = \begin{bmatrix} 1 & 0 \\ 0 & -1 \end{bmatrix} \tag{17.29}$$

where

$$1 = \begin{bmatrix} 1 & 0 \\ 0 & 1 \end{bmatrix}, \qquad 0 = \begin{bmatrix} 0 & 0 \\ 0 & 0 \end{bmatrix}.$$

Then the γ_m have the form

$$\gamma_m = \begin{bmatrix} a_m & b_m \\ c_m & d_m \end{bmatrix}, \qquad m = 1, 2, 3.$$

Now $\gamma_m = -\gamma_m^\dagger$ implies that a_m and d_m are antihermitian 2 × 2 matrices,

$$a_m^\dagger = -a_m, \qquad d_m^\dagger = -d_m,$$

whereas

$$c_m = -b_m^\dagger.$$

The anticommutation relations (17.26a) further imply that

$$\gamma_0\gamma_m + \gamma_m\gamma_0 = 0$$

or

$$a_m = d_m = 0.$$

Therefore

$$\gamma_m = \begin{bmatrix} 0 & b_m \\ -b_m^\dagger & 0 \end{bmatrix}.$$

Use of this in Eq. (17.26c) gives

$$b_m^\dagger = b_m^{-1},$$

or in Eq. (17.26a) gives

$$b_m b_n^\dagger + b_n b_m^\dagger = 0, \qquad n \neq m.$$

These conditions are satisfied by

$$b_m = \sigma_m, \qquad m = 1, 2, 3,$$

or

$$\gamma_m = \begin{bmatrix} 0 & \sigma_m \\ -\sigma_m & 0 \end{bmatrix}. \tag{17.30}$$

Thus armed with a four-dimensional representation for the Γ_μ's, we may now attempt to see what physics might be contained in Dirac's equation. It should be emphasized that the four dimensions of our internal coordinate space represented by $|r\rangle$ are only incidentally related to the fact that the Γ_μ's are four-vectors; we found the lowest-dimensional representation for the Γ_μ's, and did not even seek higher-dimensional representations. The r-indices on the $(\gamma_\mu)_{r'r} = \langle r'|\Gamma_\mu|r\rangle$ matrices are internal-space indices, whereas the μ label coordinate axes in four-space.

Perhaps the first project in seeking a physical interpretation is to find an expression for the probability four-current density. For this purpose, we shall need the complete set of position states $|M, x; r\rangle$; we recall that

$$\langle M, x'; r|\mathfrak{P}^\mu|M, x; r\rangle = \delta^4(\mathbf{x}' - \mathbf{x})\delta_{r',r} i\hbar\partial^\mu.$$

Using this together with Eq. (17.20) and recalling that we must use $\langle\ |\Gamma_0$ in place of the usual bra $\langle\ |$ when we use Eq. (17.20), we obtain

$$\begin{aligned} 0 &= \langle M, x; r'|\Gamma_0(\Gamma_\mu \mathfrak{P}^\mu - Mc)|M, x; r\rangle \\ &= \sum_{r''} \int d^4x'' \langle M, x; r'|\Gamma_0\Gamma_\mu\mathfrak{P}^\mu - \Gamma_0 Mc|M, x''; r''\rangle\langle M, x''; r''|M, q, r\rangle \\ &= \sum_{r''} (\gamma_0\gamma_\mu)_{r',r''} i\hbar\partial^\mu \psi_{M,q,r}(x; r'') - Mc \sum_{r''} (\gamma_0)_{r',r''}\psi_{M,q,r}(x; r''), \end{aligned}$$

where

$$\psi_{M,q,r}(x; r'') \equiv \langle M, x; r''|M, q, r\rangle$$

may be thought of as a column vector with index r''; in matrix notation

$$\boldsymbol{\psi}_{M,q,r}(x) = \begin{Bmatrix} \psi_{M,q,r}(x; 1) \\ \psi_{M,q,r}(x; 2) \\ \psi_{M,q,r}(x; 3) \\ \psi_{M,q,r}(x; 4) \end{Bmatrix} = [\psi_{M,q,r}(x; r')],$$

if the four values of r' are taken to be 1, 2, 3, 4. Thus, in matrix form, our equation for $\boldsymbol{\psi}_{M,q,r}(\mathbf{x})$ becomes

$$\gamma_0(\gamma_\mu i\hbar\partial^\mu - Mc)\boldsymbol{\psi}_{M,q,r}(x) = 0. \tag{17.31}$$

The Hermitian conjugate of (17.31) is

$$\boldsymbol{\psi}^\dagger_{M,q,r}(x)(-i\hbar\overleftarrow{\partial}^\mu\gamma^\dagger_\mu - Mc)\gamma^\dagger_0 = \boldsymbol{\psi}^\dagger_{M,q,r}(x)\gamma_0(-i\hbar\overleftarrow{\partial}^\mu\gamma_\mu - Mc) = 0 \tag{17.32}$$

where the arrow over ∂^μ indicates that it operates to the left, and

$$\boldsymbol{\psi}^\dagger_{M,q,r}(x) = \{\psi^*_{M,q,r}(x;1), \psi^*_{M,q,r}(x;2)\psi^*_{M,q,r}(x;3)\psi^*_{M,q,r}(x;4)\}$$

is a row vector. Here we have used $\gamma_0^\dagger = \gamma_0$ and $\gamma^\dagger_\mu\gamma_0 = \gamma_0\gamma_\mu$ which followed from (17.26). It is common practice to define

$$\overline{\boldsymbol{\psi}}_{M,q,r}(x) \equiv \boldsymbol{\psi}^\dagger_{M,q,r}(x)\gamma_0$$

so that Eq. (17.32) becomes

$$\overline{\boldsymbol{\psi}}_{M,q,r}(i\hbar\overleftarrow{\partial}^\mu\gamma_\mu + Mc) = 0. \tag{17.33}$$

In any case, matrix multiplication of (18.31) on the left by $\boldsymbol{\psi}^\dagger$ yields

$$\overline{\boldsymbol{\psi}}_{M,q,r}(\gamma_\mu i\hbar\overrightarrow{\partial}^\mu - Mc)\boldsymbol{\psi}_{M,q,r} = 0,$$

whereas matrix multiplication of (17.33) by $\boldsymbol{\psi}$ on the right gives

$$\overline{\boldsymbol{\psi}}_{M,q,r}(i\hbar\overleftarrow{\partial}^\mu\gamma_\mu + Mc)\boldsymbol{\psi}_{M,q,r} = 0.$$

The sum of these then gives

$$\overline{\boldsymbol{\psi}}_{M,q,r}(x)\gamma_\mu(i\hbar\overleftarrow{\partial}^\mu + i\hbar\overrightarrow{\partial}^\mu)\boldsymbol{\psi}_{M,q,r}(x) = i\hbar\partial^\mu\overline{\boldsymbol{\psi}}_{M,q,r}(x)\gamma_\mu\boldsymbol{\psi}_{M,q,r}(x) = 0,$$

and this suggests that

$$j_\mu(x) = c\overline{\boldsymbol{\psi}}_{M,q,r}(x)\gamma_\mu\boldsymbol{\psi}_{M,q,r}(x) = c\boldsymbol{\psi}^\dagger_{M,q,r}(x)\gamma_0\gamma_\mu\boldsymbol{\psi}_{M,q,r}(x) \tag{17.34}$$

be considered the density of the four-current. The probability density and probability current are then obtained from

$$j_\mu(x) = (c\rho(x), -\mathbf{j}(x)),$$

that is,

$$\rho(x) = \overline{\boldsymbol{\psi}}_{M,q,r}(x)\gamma_0\boldsymbol{\psi}_{M,q,r}(x) = \boldsymbol{\psi}^\dagger_{M,q,r}(x)\boldsymbol{\psi}^\dagger_{M,q,r}(x) \tag{17.35}$$

$$\begin{aligned}\mathbf{j}(x) &= c\overline{\boldsymbol{\psi}}_{M,q,r}(x)\boldsymbol{\gamma}\boldsymbol{\psi}_{M,q,r}(x)\\ &= c\boldsymbol{\psi}^\dagger_{M,q,r}(x)\boldsymbol{\alpha}\boldsymbol{\psi}_{M,q,r}(x)\end{aligned} \tag{17.36}$$

where

$$\boldsymbol{\alpha} \equiv \gamma_0\boldsymbol{\gamma}, \qquad \boldsymbol{\gamma} = (\gamma^1, \gamma^2, \gamma^3) = (-\gamma_1, -\gamma_2, -\gamma_3)$$

Clearly $\rho(x)$ is positive-definite and hence can be interpreted as a probability density. Furthermore, we may interpret $c\boldsymbol{\alpha}$ as the velocity operator, for if we write the Dirac Equation (17.27) in the form

$$c\mathfrak{P}^0|M, q, r\rangle = \mathfrak{H}|M, q, r\rangle$$

so that

$$\mathfrak{H} = c\Gamma_0\boldsymbol{\Gamma}\cdot\mathfrak{P} + \boldsymbol{\Gamma}_0 Mc^2$$

is the Hamiltonian for the system, we have

$$\begin{aligned}
&\frac{d}{dt}\langle Mq', r'|\mathfrak{X}|M, q, r\rangle \\
&\quad = \left\langle M, q', r' \left| -\frac{1}{i\hbar}[\mathfrak{H}, \mathfrak{X}] \right| M, q, r \right\rangle \\
&\quad = \langle M, q', r'|c\Gamma_0\boldsymbol{\Gamma}|M, q, r\rangle \\
&\quad = c\gamma_0\boldsymbol{\gamma}\,\delta(q' - q) \\
&\quad = c\boldsymbol{\alpha}\,\delta(q' - q).
\end{aligned}$$

Properly, we now should prove that the γ_μ transform under Lorentz transformations like covariant four-vectors. This, however, is an involved procedure and we shall forgo it.* Assuming that the γ_μ do transform properly, we see that the transformation properties of $j_\mu(x)$ depend upon the normalization of the $\psi(x)$. If $\bar{\psi}(x)\psi(x)$ is an invariant quantity, then $j_\mu(x)$ is a four-vector.

17.4 MOMENTUM EIGENSTATES FOR THE DIRAC EQUATION

The momentum eigenfunctions $\boldsymbol{\psi}_{M,p,r}(x)$ can be obtained easily from Eq. (17.31) if we use our representation (17.29) and (17.30) for the γ_μ; thus Eq. (18.31) becomes

$$[i\hbar\partial_0 + \boldsymbol{\alpha}\cdot i\hbar\nabla - \gamma_0 Mc]\boldsymbol{\psi}_{M,p,r}(x) = 0.$$

If we let

$$\boldsymbol{\psi}_{M,pr}(x) = \begin{Bmatrix} u_+(x) \\ u_-(x) \end{Bmatrix}, \text{ where } \quad u_+(x) \equiv \begin{Bmatrix} u_1(x) \\ u_2(x) \end{Bmatrix}, \quad u_-(x) \equiv \begin{Bmatrix} u_3(x) \\ u_4(x) \end{Bmatrix},$$

and use

$$\gamma_0 = \begin{bmatrix} 1 & 0 \\ 0 & -1 \end{bmatrix}, \qquad \boldsymbol{\alpha} = \gamma_0\boldsymbol{\gamma} = \begin{bmatrix} 0 & \boldsymbol{\sigma} \\ \boldsymbol{\sigma} & 0 \end{bmatrix},$$

---

* See, for example, J. D. Bjorken and S. D. Drell, *Relativistic Quantum Mechanics*, McGraw-Hill Book Co., New York, N.Y. (1964).

then this becomes

$$\left(i\hbar\frac{\partial}{\partial t} - Mc^2\right)u_+(x) - c\boldsymbol{\sigma}\cdot\frac{\hbar}{i}\nabla u_-(x) = 0$$

$$\left(i\hbar\frac{\partial}{\partial t} + Mc^2\right)u_-(x) - c\boldsymbol{\sigma}\cdot\frac{\hbar}{i}\nabla u_+(x) = 0.$$

When $\psi_{M,p,r}(x)$ are momentum eigenfunctions, we have

$$\psi_{M,p,r}(x) = \psi_{M,p,r}(0)\exp(-ip^\mu x_\mu/\hbar), \qquad u_{+,-}(x) = u_{+,-}\exp(-ip^\mu x_\mu/\hbar),$$

where $u_{+,-} \equiv u_{+,-}(0)$, and hence

$$\begin{aligned}(E - Mc^2)u_+ - c\boldsymbol{\sigma}\cdot\mathbf{p}u_- &= 0\\ (E + Mc^2)u_- - c\boldsymbol{\sigma}\cdot\mathbf{p}u_+ &= 0.\end{aligned} \tag{17.37}$$

These equations immediately yield

$$u_+ = c^2\frac{\boldsymbol{\sigma}\cdot\mathbf{p}}{E - Mc^2}\frac{\boldsymbol{\sigma}\cdot\mathbf{p}}{E + Mc^2}u_+ = \frac{\mathbf{p}^2c^2}{E^2 - M^2c^4}u_+$$

which can be satisfied only if

$$E = \pm\sqrt{M^2c^4 + \mathbf{p}^2c^2}.$$

Thus we have the same ambiguity in the sign of E that we had for the Klein–Gordon equation. Once we have decided on the sign of E, however, we may choose u_+ or u_-, and the other is determined by Eqs. (17.37), aside from an arbitrary multiplicative constant. Thus there are four solutions, aside from the normalization constant—two for each sign of the energy.

If $E > 0$, we have

$$u_- = \frac{c\mathbf{p}\cdot\boldsymbol{\sigma}}{E + Mc^2}u_+$$

and hence $u_- \simeq (v/c)u_+$ is the smaller of the two two-component spinors. We may, for $v/c \ll 1$, let

$$u_+^{(1)} = \begin{Bmatrix}1\\0\end{Bmatrix} \quad \text{or} \quad u_+^{(2)} = \begin{Bmatrix}0\\1\end{Bmatrix},$$

which lead, respectively, to

$$\psi_{M,p,1}(x) = N \begin{Bmatrix} 1 \\ 0 \\ \dfrac{cp_3}{E + Mc^2} \\ \dfrac{cp_+}{E + Mc^2} \end{Bmatrix} \exp(-ip^\mu x_\mu/\hbar),$$

$$\psi_{M,p,2}(x) = N \begin{Bmatrix} 0 \\ 1 \\ \dfrac{cp_-}{E + Mc^2} \\ \dfrac{-cp_3}{E + Mc^2} \end{Bmatrix} \exp(-ip^\mu x_\mu/\hbar) \tag{17.38a}$$

where $p_\pm = p_1 \pm ip_2$, and N is a normalization constant. Similarly, if $E < 0$, we have

$$u_-^{(3)} = \begin{Bmatrix} 1 \\ 0 \end{Bmatrix} \quad \text{or} \quad u_-^{(4)} = \begin{Bmatrix} 0 \\ 1 \end{Bmatrix},$$

or, respectively,

$$\psi_{M,p,3}(x) = N \begin{Bmatrix} \dfrac{cp_3}{E - Mc^2} \\ \dfrac{cp_+}{E - Mc^2} \\ 1 \\ 0 \end{Bmatrix} \exp(-ip^\mu x_\mu/\hbar),$$

$$\psi_{M,p,4}(x) = N \begin{Bmatrix} \dfrac{cp_-}{E - Mc^2} \\ \dfrac{-cp_3}{E - Mc^2} \\ 0 \\ 1 \end{Bmatrix} \exp(-ip^\mu x_\mu/\hbar). \tag{17.38b}$$

It is usual to require that $\bar{\psi}_{M,p,r}(x)\psi_{M,p,r}(x) = 1$, so that

$$N = \sqrt{\frac{E + Mc^2}{2Mc^2}}. \tag{17.39}$$

An important feature of these solutions is that for $E > 0$, the lower two components are small in the nonrelativistic limit, whereas for $E < 0$, the upper two components are small nonrelativistically. Thus $r = 3, 4$ correspond to negative-energy solutions, nonrelativistically, and $r = 1, 2$ correspond to the "normal" positive-energy solutions. As $v \to c$, however, that is for relativistic particles, the positive- and negative-energy components both contribute significantly.

17.5 ELECTROMAGNETIC POTENTIALS AND THE DIRAC EQUATION

From here forward, we shall deal with the wave equation

$$(\gamma_\mu \mathsf{P}^\mu - Mc)\psi(x) = (\gamma_\mu i\hbar\partial^\mu - Mc)\psi(x) = 0 \tag{17.40}$$

or its modification when interactions are introduced. The Dirac conjugate of this equation is

$$\bar{\psi}(x)(\gamma_\mu \overleftarrow{\mathsf{P}}^\mu + Mc) = \bar{\psi}(x)(\gamma_\mu i\hbar\overleftarrow{\partial}^\mu + Mc) = 0. \tag{17.41}$$

The introduction of electromagnetic fields again is accomplished covariantly through the minimal coupling

$$\mathsf{P}^\mu \to \mathsf{P}^\mu - qA^\mu;$$

hence (17.40) becomes

$$[\gamma_\mu(\mathsf{P}^\mu - qA^\mu) - Mc]\psi(x) = [\gamma_\mu(i\hbar\partial^\mu - qA^\mu) - Mc]\psi(x) = 0. \tag{17.42}$$

In order to compare this with the Klein–Gordon equation with electromagnetic interaction, for the purpose of seeing whether it has any promise of improving upon the hydrogen-atom energy-level spectrum predicted by that equation, we multiply (17.42) on the left by $\gamma_\mu(\mathsf{P}^\mu - qA^\mu) + Mc$, obtaining

$$\begin{aligned} 0 &= [\gamma_\mu(\mathsf{P}^\mu - qA^\mu)\gamma_\nu(\mathsf{P}^\nu - qA^\nu) - M^2c^2]\psi(x) \\ &= \{(\mathsf{P}_\mu - qA_\mu)(\mathsf{P}^\mu - qA^\mu) - (q/2)[\gamma_\mu, \gamma_\nu](\mathsf{P}^\mu A^\nu + A^\mu \mathsf{P}^\nu) - M^2c^2\}\psi(x). \end{aligned} \tag{17.43}$$

We have used

$$\gamma_\mu\gamma_\nu = \tfrac{1}{2}\{\gamma_\mu, \gamma_\nu\} + \tfrac{1}{2}[\gamma_\mu, \gamma_\nu] = g_{\mu\nu} + \tfrac{1}{2}[\gamma_\mu, \gamma_\nu].$$

Now

$$\mathsf{P}^\mu A^\nu + A^\mu \mathsf{P}^\nu = i\hbar\partial^\mu A^\nu + (A^\nu \mathsf{P}^\mu + A^\mu \mathsf{P}^\nu)$$

and the quantity in parentheses is symmetric under interchange of μ, ν whereas $[\gamma_\mu, \gamma_\nu]$ is antisymmetric. Therefore

$$\begin{aligned}[\gamma_\mu, \gamma_\nu](\mathsf{P}^\mu A^\nu + A^\mu \mathsf{P}^\nu) &= [\gamma_\mu, \gamma_\nu] i\hbar\partial^\mu A^\nu \\ &= i\hbar[\gamma_\mu, \gamma_\nu]\cdot\tfrac{1}{2}(\partial^\mu A^\nu - \partial^\nu A^\mu) \\ &= \hbar(i/2)[\gamma_\mu, \gamma_\nu]F^{\mu\nu},\end{aligned}$$

according to Eq. (5.8). Let us now define

$$\sigma_{\mu\nu} = (i/2)[\gamma_\mu, \gamma_\nu] \tag{17.44}$$

so that (17.43) becomes

$$[(\mathsf{P}_\mu - qA_\mu)(\mathsf{P}^\mu - qA^\mu) - (\hbar q/2)\sigma_{\mu\nu}F^{\mu\nu} - M^2c^2]\psi(x) = 0. \tag{17.45}$$

The Klein–Gordon equation lacks the second term and therefore may be expected to have a significantly different energy-level spectrum from the Dirac equation.

Using our representation (17.29) and (17.30) for the matrices, we can find the $\sigma_{\mu\nu}$. Thus

$$\begin{aligned}\sigma_{0m} &= i\begin{bmatrix} 0 & \sigma_m \\ \sigma_m & 0 \end{bmatrix} = i\alpha_m, \qquad m = 1, 2, 3, \\ \sigma_{\mu\mu} &= 0, \\ \sigma_{mn} &= \epsilon_{mnk}\begin{bmatrix} \sigma_k & 0 \\ 0 & \sigma_k \end{bmatrix} \equiv \epsilon_{mnk}\sigma'_k, \qquad m, n = 1, 2, 3,\end{aligned} \tag{17.46}$$

where $\sigma'_k \equiv \begin{bmatrix} \sigma_k & 0 \\ 0 & \sigma_k \end{bmatrix}$; or, if we let μ, ν label rows and columns,

$$[\sigma_{\mu\nu}] = \begin{bmatrix} 0 & i\alpha_1 & i\alpha_2 & i\alpha_3 \\ -i\alpha_1 & 0 & \sigma'_3 & -\sigma'_2 \\ -i\alpha_2 & -\sigma'_3 & 0 & \sigma'_1 \\ -i\alpha_3 & \sigma'_2 & -\sigma'_1 & 0 \end{bmatrix}. \tag{17.47}$$

Recognizing that $\sigma_{\mu\nu}F^{\mu\nu}$ is the trace of the matrix product of (17.47) with the transpose of (5.10), we calculate

$$\sigma_{\mu\nu}F^{\mu\nu} = -2[(i/c)\boldsymbol{\alpha}\cdot\mathbf{E} + \boldsymbol{\sigma}'\cdot\mathbf{B}]. \tag{17.48}$$

Insertion of (17.48) into (17.45) yields

$$[(\mathsf{P}^\mu - qA^\mu)(\mathsf{P}_\mu - qA_\mu) - M^2c^2 + i(q/c)\boldsymbol{\alpha}\cdot\mathbf{E} + \hbar q\boldsymbol{\sigma}'\cdot\mathbf{B}]\psi(x) = 0. \tag{17.49}$$

This looks strikingly like a relativistic version of the Pauli equation (Schrödinger equation for a spin $-\frac{1}{2}$ particle with a term $(q\hbar/2M)\boldsymbol{\sigma}\cdot\mathbf{B}$ in the Hamiltonian). The extra term $i(q/c)\boldsymbol{\alpha}\cdot\mathbf{E}$ looks rather strange, but it can be shown that this leads to the spin–orbit effect in the nonrelativistic reduction of (17.49).

17.6 SPIN ANGULAR MOMENTUM AND THE SPIN–ORBIT EFFECT

Let us now restrict ourselves to a spherically symmetric electrostatic potential,

$$A^\mu = \left(\frac{1}{c}\phi(x), 0\right), \qquad V(r) \equiv q\phi(x) = q\phi(|\mathbf{x}|),$$

in some specified frame of reference. For such a potential, it is not difficult to see that the orbital angular momentum $\mathfrak{L} = \mathfrak{X} \times \mathfrak{P}$ is a constant of the motion for the Klein–Gordon equation. This is not the case for the Dirac equation, however. Using the technique in the discussion following Eq. (17.36), we find that with

$$\mathfrak{H} = c\Gamma_0 \boldsymbol{\Gamma} \cdot \mathfrak{P} + \Gamma_0 \mathfrak{V}(\mathfrak{R}) + \Gamma_0 Mc^2,$$

or

$$\mathsf{H} = c\boldsymbol{\alpha} \cdot \mathbf{P} + \gamma_0 V(r) + \gamma_0 Mc^2$$

in the wave-function representation,

$$\frac{d}{dt}\mathbf{L} = -\frac{1}{i\hbar}[\mathsf{H}, \mathbf{L}] = -\frac{1}{i\hbar}c[\boldsymbol{\alpha} \cdot \mathbf{P}, \mathbf{L}] = c\boldsymbol{\alpha} \times \mathbf{P}.$$

On the other hand, we also have

$$\frac{d}{dt}\boldsymbol{\sigma}' = -\frac{1}{i\hbar}[\mathsf{H}, \boldsymbol{\sigma}'] = -\frac{1}{i\hbar}[\boldsymbol{\alpha} \cdot \mathbf{P}, \boldsymbol{\sigma}'] = -\frac{2}{\hbar}\boldsymbol{\alpha} \times \mathbf{P},$$

since

$$[\alpha_m, \sigma_n'] = 2i\epsilon_{mnk}\alpha_k$$

as can be verified from direct calculation using

$$\alpha_m = \begin{bmatrix} 0 & \sigma_m \\ \sigma_m & 0 \end{bmatrix}, \qquad \sigma_n' = \begin{bmatrix} \sigma_n & 0 \\ 0 & \sigma_n \end{bmatrix}.$$

Hence we have

$$\frac{d}{dt}\left(\mathbf{L} + \frac{\hbar}{2}\boldsymbol{\sigma}'\right) = 0$$

and the quantity

$$\mathbf{J} \equiv \mathbf{L} + \mathbf{S}, \qquad \text{where} \qquad \mathbf{S} \equiv \tfrac{1}{2}\hbar\boldsymbol{\sigma}' \tag{17.50}$$

is a constant of the motion. Evidently $\mathbf{S} = \frac{1}{2}\hbar\boldsymbol{\sigma}'$ can be interpreted as a relativistic spin operator, and the spin and orbital angular momentum are, in the Dirac equation with a central potential, coupled so that their sum is a constant of the motion, though neither separately commutes with the Hamiltonian. The Dirac equation therefore has spin degrees of freedom whereas the Klein–Gordon equation could not.

This coupling of spin and orbital angular momenta must be accomplished by the spin–orbit term in the nonrelativistic approximation to the Dirac equation. Let us see explicitly how the spin–orbit term arises in the nonrelativistic reduction of the Dirac equation with a central electrostatic potential. When the potential is included in the Dirac equation, Eq. (17.37) becomes

$$[E - Mc^2 - V(r)]\psi_+(x) - c\boldsymbol{\sigma}\cdot\mathbf{P}\psi_-(x) = 0$$
$$[E + Mc^2 - V(r)]\psi_-(x) - c\boldsymbol{\sigma}\cdot\mathbf{P}\psi_+(x) = 0.$$

Assuming E to be positive, $E = Mc^2 + E'$, $E' > -Mc^2$, we see again that $\psi_-(x)$ is of order v/c smaller than $\psi_+(x)$, so we eliminate $\psi_-(x)$ in these equations in favor of $\psi_+(x)$; thus we have

$$\{E - Mc^2 - V(r) - c^2\boldsymbol{\sigma}\cdot\mathbf{P}[E + Mc^2 - V(r)]^{-1}\boldsymbol{\sigma}\cdot\mathbf{P}\}\psi_+(x) = 0. \quad (17.51)$$

We introduce the nonrelativistic approximation by noting, with $E' = E - mc^2$,

$$[E + Mc^2 - V(r)]^{-1} = [2Mc^2 + E' - V(r)]^{-1} \simeq \frac{1}{2Mc^2}\left[1 - \frac{E' - V(r)}{2Mc^2}\right]$$

since $E' - V(r) \ll Mc^2$ is to be assumed. Equation (17.51) then becomes

$$\left\{E' - V(r) - \frac{1}{2M}\left[1 - \frac{E' - V(r)}{2Mc^2}\right](\boldsymbol{\sigma}\cdot\mathbf{P})^2 - \frac{\hbar}{i4M^2c^2}[\boldsymbol{\sigma}\cdot\nabla V(r)]\boldsymbol{\sigma}\cdot\mathbf{P}\right\}\psi_+(x) = 0$$

in this approximation. Now

$$\boldsymbol{\sigma}\cdot\mathbf{A}\,\boldsymbol{\sigma}\cdot\mathbf{B} = \sum_{j,k} A_jB_k\sigma_j\sigma_k = \sum_{j,k} A_jB_k(\tfrac{1}{2}\{\sigma_j\sigma_k\} + \tfrac{1}{2}[\sigma_j, \sigma_k])$$
$$= \mathbf{A}\cdot\mathbf{B} + i\boldsymbol{\sigma}\cdot(\mathbf{A}\times\mathbf{B}),$$

$$\nabla V(r) = \mathbf{r}(1/r)\frac{\partial}{\partial r}V(r),$$

and, to a first approximation, we may take

$$\frac{E' - V(r)}{2Mc^2} \approx \frac{1}{2Mc^2}\frac{1}{2M}\mathbf{P}^2.$$

From this we obtain

$$\frac{1}{2M}\left[1 - \frac{E' - V(r)}{2Mc^2}\right](\boldsymbol{\sigma}\cdot\mathbf{P})^2 \simeq \frac{1}{2M}\mathbf{P}^2 - \frac{1}{8M^3c^2}(\mathbf{P}^2)^2$$

and

$$[\boldsymbol{\sigma}\cdot\nabla V(r)]\boldsymbol{\sigma}\cdot\mathbf{P} = \frac{\hbar}{i}\frac{\partial V(r)}{\partial r}\frac{\partial}{\partial r} + i\boldsymbol{\sigma}\cdot\frac{1}{r}\frac{\partial V(r)}{\partial r}\mathbf{r}\times\mathbf{P}$$

$$= \frac{\hbar}{i}\frac{\partial V(r)}{\partial r}\frac{\partial}{\partial r} + \frac{2i}{\hbar}\frac{1}{r}\frac{\partial V(r)}{\partial r}\mathbf{S}\cdot\mathbf{L};$$

hence

$$E'\psi_+(x) \simeq \left[\frac{1}{2M}\mathbf{P}^2 - \frac{1}{8M^3c^2}(\mathbf{P}^2)^2 + V(r) - \frac{\hbar^2}{4M^2c^2}\frac{\partial V(r)}{\partial r}\frac{\partial}{\partial r} + \frac{1}{2M^2c^2}\frac{1}{r}\frac{\partial V(r)}{\partial r}\mathbf{S}\cdot\mathbf{L}\right]\psi_+(x) = 0. \tag{17.52}$$

The right-hand side of this equation corresponds to the Hamiltonian in the nonrelativistic reduction. The first and third terms are the ones usually appearing in the Schrödinger equation, and the second is the first correction in the expansion of $c\sqrt{\mathbf{p}^2 + m^2c^2} - mc^2$. The last term is the desired spin–orbit term, and includes the relativistic Thomas precession factor as we should expect. There is no classical analog for the fourth term, and we shall not need to deal further with it.*

The spin–orbit term should be compared with the equations following (16.23c) for the case

$$V(r) = \frac{Zeq}{4\pi\epsilon_0}\frac{1}{r}.$$

17.7 THE DIRAC ONE-ELECTRON ATOM

The crucial test for the Dirac equation is whether it does better than the Klein–Gordon equation at predicting energy levels for the one-electron atom. By now we have high hopes that it does, thanks to the interaction term $-\frac{1}{2}\hbar q\sigma_{\mu\nu}F^{\mu\nu}$ in Eq. (17.45) and to the presence of the spin–orbit term in Eq. (17.52).

Our first job is to separate variables between angular and radial coordinates in the equation

$$[\gamma_0 E - c\boldsymbol{\gamma}\cdot\mathbf{P} - \gamma_0 V(r) - Mc^2]\psi(x) = 0.$$

We shall assume that $\psi_E(\mathbf{x})$ is an eigenfunction of energy with energy eigenvalue E, and for the sake of simplicity, we multiply from the left by γ_0, obtaining

$$E\psi_E(x) = [c\boldsymbol{\alpha}\cdot\mathbf{P} + V(r) - \gamma_0 Mc^2]\psi_E(\mathbf{x}). \tag{17.53}$$

To separate variables, we wish to have $\boldsymbol{\alpha}\cdot\mathbf{P}$ decomposed into a term not depending on $\mathbf{x}\times\mathbf{P}$ and one depending only multiplicatively on r. Useful

---

* See E. U. CONDON and G. H. SHORTLEY, *The Theory of Atomic Spectra*, Cambridge University Press, London (1935), p. 130.

identities to have available in seeking such a decomposition are

$$\boldsymbol{\alpha}\cdot\mathbf{A}\boldsymbol{\alpha}\cdot\mathbf{B} = \boldsymbol{\sigma}'\cdot\mathbf{A}\boldsymbol{\sigma}'\cdot\mathbf{B} = \mathbf{A}\cdot\mathbf{B} + i\boldsymbol{\sigma}'\cdot(\mathbf{A}\times\mathbf{B})$$

$$\boldsymbol{\alpha}\cdot\mathbf{A}\boldsymbol{\sigma}'\cdot\mathbf{B} = \boldsymbol{\sigma}'\cdot\mathbf{A}\boldsymbol{\alpha}\cdot\mathbf{B} = \begin{bmatrix} 0 & \mathbf{A}\cdot\mathbf{B} \\ \mathbf{A}\cdot\mathbf{B} & 0 \end{bmatrix} + i\boldsymbol{\alpha}\cdot(\mathbf{A}\times\mathbf{B}). \quad (17.54)$$

From the latter of these, we see that if $\mathbf{A} = \mathbf{x}$, and $\mathbf{B} = \mathbf{L} = \mathbf{x}\times\mathbf{P}$, we get

$$\boldsymbol{\alpha}\cdot[\mathbf{r}(\mathbf{r}\cdot\mathbf{P}) - r^2\mathbf{P}] = \frac{1}{i}\boldsymbol{\alpha}\cdot\mathbf{r}\boldsymbol{\sigma}'\cdot\mathbf{L}$$

or

$$\boldsymbol{\alpha}\cdot\mathbf{P} = \frac{1}{r^2}(\boldsymbol{\alpha}\cdot\mathbf{r})(\mathbf{r}\cdot\mathbf{P}) + i\frac{1}{r^2}(\boldsymbol{\alpha}\cdot\mathbf{r})(\boldsymbol{\sigma}'\cdot\mathbf{L}),$$

which appears to have the desired form, the first term being the radial term and the second being the angular term. Experience shows, however, that a slight modification simplifies the solution of the problem; thus we define the Hermitian operators

$$\mathsf{P}_r \equiv \frac{1}{r}(\mathbf{r}\cdot\mathbf{P} - i\hbar)$$

$$\mathsf{K} \equiv \gamma_0\frac{1}{\hbar}(\boldsymbol{\sigma}'\cdot\mathbf{L} + \hbar) \quad (17.55)$$

$$\alpha_r \equiv \frac{1}{r}\boldsymbol{\alpha}\cdot\mathbf{r}$$

so that

$$\boldsymbol{\alpha}\cdot\mathbf{P} = \alpha_r\mathsf{P}_r + (i\hbar/r)\alpha_r\gamma_0\mathsf{K}. \quad (17.56)$$

Now $\mathbf{L}\times\mathbf{L} = i\hbar\mathbf{L}$, and hence

$$\begin{aligned}\mathsf{K}^2 &= \hbar^{-2}[(\boldsymbol{\sigma}'\cdot\mathbf{L})^2 + 2\hbar\boldsymbol{\sigma}\cdot\mathbf{L} + \hbar^2] \\ &= \hbar^{-2}[\mathbf{L}^2 + \hbar\sigma'\cdot\mathbf{L} + \hbar^2] = \hbar^{-2}(\mathbf{L} + \tfrac{1}{2}\hbar\boldsymbol{\sigma}')^2 + \tfrac{1}{4}. \\ &= \hbar^{-2}\mathbf{J}^2 + \tfrac{1}{4}. \end{aligned} \quad (17.57)$$

This means that the eigenvalues of K^2 are $\frac{1}{4}$ greater than those of $\hbar^{-2}\mathbf{J}^2$, but the latter are simply $j(j+1)$, where j is a positive half-odd integer. Therefore the eigenvalues k of K must be $\pm(j+\frac{1}{2}) = \pm 1, \pm 2, \pm 3, \ldots$.

Since K commutes with $\mathsf{H} = c\boldsymbol{\alpha}\cdot\mathbf{P} + V(r) - \gamma_0 Mc^2$ as is easily verified, both K and H may be diagonalized simultaneously. Furthermore, there are only two 4×4 Dirac matrices appearing in the Hamiltonian now, namely γ_0 and α_r, and any two convenient matrices satisfying

$$\alpha_r^2 = \gamma_0^2 = 1, \qquad \{\alpha_r, \gamma_0\} = 0$$

may be chosen to represent them. If we continue to keep γ_0 diagonal as in Eq. (17.29), we may let

$$\alpha_r \equiv \begin{bmatrix} 0 & -iI \\ iI & 0 \end{bmatrix} = \begin{bmatrix} 0 & 0 & -i & 0 \\ 0 & 0 & 0 & -i \\ i & 0 & 0 & 0 \\ 0 & i & 0 & 0 \end{bmatrix}. \tag{17.58}$$

Inserting (17.56), with definition (17.55), into Eq. (17.53) gives us the obviously separable equation

$$E\psi_E(\mathbf{x}) = \left[c\alpha_r \mathsf{P}_r + V(r) + \frac{i\hbar c}{r}\alpha_r\gamma_0 \mathsf{K} - \gamma_0 M_c^2\right]\psi_E(\mathbf{x}). \tag{17.59}$$

Separating variables and assuming $\psi_E(x)$ to be diagonal in K as well as E,

$$\psi_{E,k}(\mathbf{x}) = \frac{1}{r}\phi_{E,k}(r)W_k(\theta, \varphi)$$

with

$$\mathsf{K}\psi_{E,k}(\mathbf{x}) = k\psi_{E,k}(\mathbf{x}),$$

we obtain the radial equation

$$E\phi_{E,k}(r) = r\left[c\alpha_r \mathsf{P}_r + V(r) + i\frac{\hbar ck}{r}\alpha_r\gamma_0 - \gamma_0 M_c^2\right]\frac{1}{r}\phi_{E,k}(r).$$

Now $\phi_{E,k}(r)$ has two distinct components in the representation we have chosen for α_r, γ_0, and we may write

$$\phi_{E,k}(r) = \begin{Bmatrix} f_{E,k}(r) \\ g_{E,k}(r) \end{Bmatrix}$$

and use

$$\mathsf{P}_r = \frac{\hbar}{i}\left(\frac{\partial}{\partial r} + \frac{1}{r}\right), \qquad r\mathsf{P}_r\frac{1}{r} = \frac{\hbar}{i}\frac{\partial}{\partial r}$$

together with (17.29) and (17.58) to cast this equation into the form

$$[E + Mc^2 - V(r)]f_{E,k}(r) = -\hbar c\left[\frac{d}{dr} + \frac{k}{r}\right]g_{E,k}(r)$$

$$[E - Mc^2 - V(r)]g_{E,k}(r) = \hbar c\left[\frac{d}{dr} - \frac{k}{r}\right]f_{E,k}(r).$$

We shall concern ourselves only with finding the bound state energy levels of the system when it has positive energies. Hence we shall have $0 < E < Mc^2$, and we may simplify the above equations by defining the positive-real quantities

$$\kappa_1 \equiv \frac{Mc^2 + E}{\hbar c}, \qquad \kappa_2 \equiv \frac{Mc^2 - E}{\hbar c},$$

$$\kappa = \sqrt{\kappa_1 \kappa_2} = \sqrt{\frac{M^2c^4 - E^2}{\hbar^2 c^2}},$$

$$\rho \equiv \kappa r. \tag{17.60}$$

From this we get the coupled differential equations

$$\begin{aligned} \left(\frac{d}{d\rho} + \frac{k}{\rho}\right) g_k(\rho) + \left(\frac{\kappa_1}{\kappa} - \frac{V(r)}{\hbar c \kappa}\right) f_k(\rho) &= 0, \\ \left(\frac{d}{d\rho} - \frac{k}{\rho}\right) f_k(\rho) + \left(\frac{\kappa_2}{\kappa} + \frac{V(r)}{\hbar c \kappa}\right) g_k(\rho) &= 0, \end{aligned} \tag{17.61}$$

where

$$f_k(\rho) \equiv f_{E,k}(r), \qquad g_k(\rho) \equiv g_{E,k}(r).$$

Now we specialize to the Coulomb potential,

$$V(r) = q\frac{Ze}{4\pi\epsilon_0 r} = q\frac{Ze\kappa}{4\pi\epsilon_0 \rho} = -\frac{Z\alpha\hbar c\kappa}{\rho} \tag{17.62}$$

when

$$q = -e, \qquad \alpha = e^2/4\pi\epsilon_0\hbar c.$$

From our solutions of the nonrelativistic spinless one-electron problem, we may infer that at large values of ρ,

$$f_k(\rho) \underset{\rho\to\infty}{\sim} \rho^m e^{-\rho},$$

$$g_k(\rho) \underset{\rho\to\infty}{\sim} \rho^m e^{-\rho},$$

where m is some real number presumably determined by k and κ. Hence we write

$$f_k(\rho) = p_k(\rho)e^{-\rho},$$

$$g_k(\rho) = q_k(\rho)e^{-\rho},$$

and with this and (17.62), Eqs. (17.61) become

$$\begin{aligned} p_k'(\rho) - p_k(\rho) + \frac{k}{\rho}p_k(\rho) + \left(\frac{\kappa_1}{\kappa} + \frac{Z\alpha}{\rho}\right) q_k(\rho) &= 0 \\ q_k'(\rho) - q_k(\rho) - \frac{k}{\rho}q_k(\rho) + \left(\frac{\kappa_2}{\kappa} - \frac{Z\alpha}{\rho}\right) p_k(\rho) &= 0. \end{aligned} \tag{17.63}$$

We now seek infinite series solutions

$$
\begin{aligned}
p_k(\rho) &= \rho^m \sum_{n=0}^{\infty} a_n \rho^n \\
q_k(\rho) &= \rho^m \sum_{n=0}^{\infty} b_n \rho^n
\end{aligned}
\tag{17.64}
$$

to these equations. Again we may expect that these series will have to terminate in order to preserve the desired asymptotic behavior as $\rho \to \infty$; we shall see that this is indeed so by considering the coupled recursion relations for the a_n and b_n.

Putting the series (17.64) into Eqs. (17.63) gives the equations

$$
\rho^{m-1}[(m+k)a_0 + Z\alpha b_0] + \sum_{n=0}^{\infty}\left[(m+n+k+1)a_{n+1} + Z\alpha b_{n+1} - a_n - \frac{\kappa_1}{\kappa} b_n\right]\rho^{m+n} = 0,
$$

$$
\rho^{m-1}[(m-k)b_0 - Z\alpha a_0] + \sum_{n=0}^{\infty}\left[(m+n-k+1)b_{n+1} - Z\alpha a_{n+1} - b_n - \frac{\kappa_2}{\kappa} a_n\right]\rho^{m+n} = 0,
$$

and since these must be true identically, we have

$$
(m+k)a_0 + Z\alpha b_0 = (m-k)b_0 - Z\alpha a_0 = 0 \tag{17.65}
$$

and

$$
\begin{aligned}
(m+n+k+1)a_{n+1} + Z\alpha b_{n+1} - a_n - \frac{\kappa_1}{\kappa} b_n &= 0, \\
(m+n-k+1)b_{n+1} - Z\alpha a_{n+1} - b_n - \frac{\kappa_2}{\kappa} a_n &= 0.
\end{aligned}
\tag{17.66}
$$

Equations (17.65) imply that

$$
m = \pm\sqrt{k^2 - (Z\alpha)^2} \tag{17.67}
$$

since a_0, b_0 must not vanish; hence m depends only on k and not on κ. Furthermore, both a_n and b_n can be eliminated from Eqs. (17.66) by taking the difference between κ times the first equation and κ_1 times the second:

$$
[\kappa(m+n+k+1) + Z\alpha\kappa_1]a_{n+1} = [\kappa_1(m+n-k+1) - Z\alpha\kappa]b_{n+1}
$$

or

$$
b_n = \frac{\kappa(m+n+k) + Z\alpha\kappa_1}{\kappa_1(m+n-k) - Z\alpha\kappa} a_n. \tag{17.68}
$$

The asymptotic behavior of p_k and q_k for large ρ will be determined by the behavior of a_n and b_n for large n, and from (17.68) and (17.66), we see that these in turn are governed by

$$b_n \underset{n\to\infty}{\sim} \frac{\kappa}{\kappa_1} a_n, \qquad (n+1)a_{n+1} \underset{n\to\infty}{\sim} 2a_n, \qquad (n+1)b_{n+1} \underset{n\to\infty}{\sim} 2b_n.$$

But this implies that for large n, the terms $a_n\rho^n$ and $b_n\rho^n$ are essentially the same as those in the expansion

$$\sum_{n=0}^{\infty} \frac{(2\rho)^n}{n!}$$

of $e^{2\rho}$. Hence the a_n and b_n must actually vanish for large n. This means that for some nonnegative integer ν, $a_{\nu+1} = b_{\nu+1} = 0$. Equations (17.66) then imply that

$$a_\nu = -\sqrt{\frac{\kappa_1}{\kappa_2}}\, b_\nu$$

and this, inserted in Eq. (17.68), yields

$$2\kappa(m+\nu) + Z\alpha(\kappa_1 - \kappa_2) = 2\kappa(m+\nu) + Z\alpha \cdot \frac{2E}{\hbar c} = 0.$$

Upon squaring $\kappa(m+\nu)$, we obtain from this

$$\kappa^2(m+\nu)^2 = \frac{M^2c^4 - E^2}{\hbar^2c^2}(m+\nu)^2 = Z^2\alpha^2\frac{E^2}{\hbar^2c^2}$$

or, upon solving $E = E_{k,\nu}$,

$$E_{k,\nu} = Mc^2\left[1 + \left(\frac{Z\alpha}{m+\nu}\right)^2\right]^{-1/2}, \qquad m = \sqrt{k^2 - Z^2\alpha^2}, \tag{17.69}$$

where we have used the positive sign in Eq. (17.67) in order to obtain good behavior from p_k, q_k as $\rho \to 0$.

It is clear that to order $(Z\alpha)^2$, Eq. (17.69) gives the same energy levels as the nonrelativistic Schrödinger one-electron atom when we set $n = |k| + \nu = 1, 2, 3, \ldots$;

$$E_n \simeq Mc^2 - \frac{(Z\alpha)^2}{n^2}\frac{Mc^2}{2},$$

the second term giving the electron's binding energy. To order $(Z\alpha)^4$, we have

$$E_{n,k} = Mc^2 - \left(\frac{Z\alpha}{n}\right)^2\frac{Mc^2}{2} - \left(\frac{Z\alpha}{n}\right)^4\left(\frac{n}{|k|} - \frac{3}{4}\right)\frac{Mc^2}{2};$$

the last term gives the fine structure. Now, since $|k|$ can take on values from 1 to n, the maximum separation between levels with a given n is

$$\left(\frac{Z\alpha}{n}\right)^4 (n-1)\frac{Mc^2}{2}.$$

This is just $(n - \frac{1}{2})/(2n)$ as large as the fine-structure splitting obtained at the end of Section 17.3 for the Klein–Gordon atom, and it agrees well with experiment.

Notice that the spin and orbital angular momenta are not separately involved in determining the energy levels. The quantity j, determined by the eigenvalues $\hbar^2 j(j+1)$ of $\mathbf{J}^2 = (\mathbf{L} + \mathbf{S})^2$, however, is involved through the eigenvalues $k = \pm(j + \frac{1}{2})$ of K. Thus the spin and orbital angular momenta are inextricably coupled in the Dirac hydrogen atom. For each value of j, however, there are exactly two possible values of l that may couple with the spin $s = \frac{1}{2}$ to yield that value of j, namely $l = j \pm \frac{1}{2}$. We may therefore think of a given energy eigenstate as being a superposition of different eigenstates of l, and the definition (17.55) of K can be used to clarify which values of k, j and l belong to a given energy level determined by $n = v + |k|$.

Since

$$\begin{aligned} \mathsf{K} &= \gamma_0 \frac{1}{\hbar}\left[\frac{2}{\hbar}\mathbf{S}\cdot\mathbf{L} + \hbar\right] \\ &= \gamma_0\left[\frac{1}{\hbar^2}(\mathbf{J}^2 - \mathbf{L}^2 - \mathbf{S}^2) + 1\right], \end{aligned}$$

we have, when K operates on an eigenstate of $\mathbf{J}^2$, $\mathbf{L}^2$,$\mathbf{S}^2$ with eigenvalues determined by $j, l, s = \frac{1}{2}$,

$$\begin{aligned} \mathsf{K} &\to \gamma_0[j(j+1) - l(l+1) - s(s+1) + 1] \\ &= \gamma_0[j(j+1) - l(l+1) + \tfrac{1}{4}]. \end{aligned}$$

In a nonrelativistic approximation, only the upper components of ψ are large, and hence we may let $\gamma_0 \to 1$ in this approximation; then

$$k = \begin{cases} -l, & j = l - \frac{1}{2} \\ l+1, & j = l + \frac{1}{2}. \end{cases}$$

From this we can construct a table of values of $n = v + |k|, k, l, j$ that indicates the composition of the energy levels. Remember that k must not vanish. In Table 17.1 we have listed the values of these quantum numbers, together with the spectroscopic notation for each state, for $n = 1, 2, 3$. The first state listed under each value of n, however, does not exist. Each of these states has $v = 0$ and $k < 0$, which is impossible; we see this as follows: Let $n = v = 0$ in Eqs. (17.66) remembering that $a_{v+1} = b_{v+1} = 0$, so that

$$\frac{b_0}{a_0} = -\sqrt{\frac{\kappa_2}{\kappa_1}},$$

Table 17.1. Nonrelativistic Decomposition of Dirac Hydrogen Atom Energy Levels

n	v	k	l	j	notation
*1	0	-1	1	$\frac{1}{2}$	
1	0	$+1$	0	$\frac{1}{2}$	${}^2S_{1/2}$
*2	0	-2	2	$\frac{3}{2}$	
2	0	$+2$	1	$\frac{3}{2}$	${}^2P_{3/2}$
2	1	-1	1	$\frac{1}{2}$	${}^2P_{1/2}$
2	1	$+1$	0	$\frac{1}{2}$	${}^2S_{1/2}$
*3	0	-3	3	$\frac{5}{2}$	
3	0	$+3$	2	$\frac{5}{2}$	${}^2D_{5/2}$
3	1	-2	2	$\frac{3}{2}$	${}^2D_{3/2}$
3	1	$+2$	1	$\frac{3}{2}$	${}^2P_{3/2}$
3	2	-1	1	$\frac{1}{2}$	${}^2P_{1/2}$
3	2	-1	0	$\frac{1}{2}$	${}^2S_{1/2}$

* These rows define states that do not exist.

and compare this with Eq. (17.65),

$$\frac{b_0}{a_0} = \frac{Z\alpha}{m - k}.$$

When $k < 0$, the second equation is positive whereas the first is necessarily negative, since $m > 0$ and κ_1, κ_2 are real. Therefore we must exclude the cases $v = 0, k < 0$, since the recursion relations cannot be satisfied for these values.

Here we have obtained only the general expression for the energies of bound states of the Dirac hydrogen-like atom. More detailed analysis of the Dirac model for hydrogen shows that it provides an excellent model for this simplest atomic system, although it has some important imperfections that can be removed only by invoking quantum field theory. These imperfections are intimately connected with the negative-energy states that haunt all relativistic theories. The most useful conceptual device for dealing with negative energy states for the Dirac equation is Dirac's *hole theory*, in which all the negative-energy states are filled and one positive-energy state is filled when we are dealing with a *one-electron system*. The absence of an electron in a negative-energy state then is equivalent to the presence of a positively charged particle of the same mass, spin and magnitude of electric charge. We have not yet dealt formally with the Pauli principle, which states that no two electrons (or other identical particles with half-odd-integer spins) may occupy the same physical state, and it is this principle that provides a basis for the hole theory.

READING NOTES*

Reference 24, Chapter 11.
Reference 42, Chapter 11.
Reference 43, Chapter 13.
Reference 44, Chapter 23.

PROBLEMS

1. Show that in the nonrelativistic limit Eq. (17.12) reduces to the usual Schrödinger equation with appropriate electromagnetic interaction terms.
2. Derive Eq. (17.39).
3. Prove the identities (17.54).
4. Show that K defined in Eqs. (17.55) commutes with H as given by the operator on the right-hand side of Eq. (17.53).
5. Prove the following identities for the γ-matrices:

$$\gamma_0\gamma_\mu\gamma_0 = 2g_{\mu 0}\gamma_0 - \gamma_\mu$$
$$\gamma_0\gamma_\mu\gamma_\nu\gamma_0 = 2g_{\nu 0}\gamma_0\gamma_\mu - 2g_{\mu 0}\gamma_0\gamma_\nu + \gamma_\mu\gamma_\nu$$
$$\gamma_\mu\gamma^\mu = 4$$
$$\gamma_\mu\gamma_\nu\gamma^\mu = -2\gamma_\nu$$
$$\gamma_\mu\gamma_\nu\gamma_\lambda\gamma^\mu = 4g_{\nu\lambda}.$$

6. Show that for arbitrary 4-vectors A^μ and B^μ,

$$\gamma_\mu A^\mu \gamma_\nu B^\nu = A^\mu B_\mu - i\sigma^{\mu\nu}A_\mu B_\nu.$$

7. a) If ψ is a solution to the Dirac equation for a free particle,

$$(\gamma_\mu \mathsf{P}^\mu - Mc)\psi(x) = (i\hbar\gamma_\mu\partial^\mu - Mc)\psi(x)$$

(i.e. γ_0 times Eq. (17.32)), prove that

$$\bar{\psi}\gamma^\mu\psi = \frac{1}{2M}[\bar{\psi}\mathsf{P}^\mu\psi - (\mathsf{P}^\mu\bar{\psi})\psi] - \frac{i}{2M}\mathsf{P}_\nu\bar{\psi}\sigma^{\mu\nu}\psi.$$

b) Use this result to show that

$$\bar{\psi}\gamma_0\psi = \frac{M}{E} = \frac{1}{\gamma}, \qquad \gamma = \frac{1}{\sqrt{1-\beta^2}}.$$

8. Find the energies of the $2\,{}^2P_{3/2}$, $2\,{}^2P_{1/2}$ and $2\,{}^2S_{1/2}$ Dirac hydrogen-atom energy levels. Calculate the differences between them in frequency (MHz). Compare this with observed splittings. Calculate the splitting of $n = 2$ levels for the Klein–Gordon atom and compare.

---

* Works referred to in these notes will be found in the Bibliography on p. 577.

CHAPTER 18

MANY-PARTICLE SYSTEMS; IDENTICAL PARTICLES

We have concentrated heretofore on systems consisting of one or two particles, and even when dealing with two particles, we have transformed the coordinates of the system to center-of-mass and relative coordinates in order to convert the two-body problem into two one-body problems. Furthermore, we have assumed that the particles involved are in some way distinguishable from one another.

Now even in the two-body problem it is not always possible to separate center-of-mass and relative coordinates in the equations of motion for the system, and in many-body systems such simple transformations of coordinates can rarely be used to obtain exact solutions for state vectors or wave functions. Moreover, the assumption that particles that are basically similar (i.e. have the same masses, electric charges, spin, etc.) can be distinguished from one another is not justified in quantum theory, even in many situations in which it would be justified classically. In fact, the indistinguishability of individual particles of the same type leads to a number of surprising effects that are experimentally observable but that cannot be accounted for in classical theory.

In classical physics, one may keep track of individual particles by following their positions in space; the particles can be arbitrarily well-localized at all times and their locations may be used as labels for the particles. In quantum theory, however, a particle cannot remain well-localized even if its position is well-established within some small volume of space at a given time, for the phenomenon of wave-packet spreading eventually destroys the localization of the particle. This is illustrated schematically in Fig. 18.1. The uncertainty principle thus forces us to give up the notion that we can "keep our eye on" a particle and distinguish it from all similar particles for an arbitrarily long time; eventually the wave functions for two similar particles will overlap and then it will be impossible to determine which particle is being observed when a particle is observed in the region of space where the wave functions both have finite absolute values.

Our discussion of the many-particle problem will be formulated nonrelativistically in terms of the Hamiltonian operator. However, most of the work can be generalized by simply replacing the Schrödinger equation with appropriate equations of motion. The really appropriate language and formalism for the many-body-problem are those of quantum field theory, in which a *number-of-particles operator* plays a central role. Nevertheless, we can deal with the basic

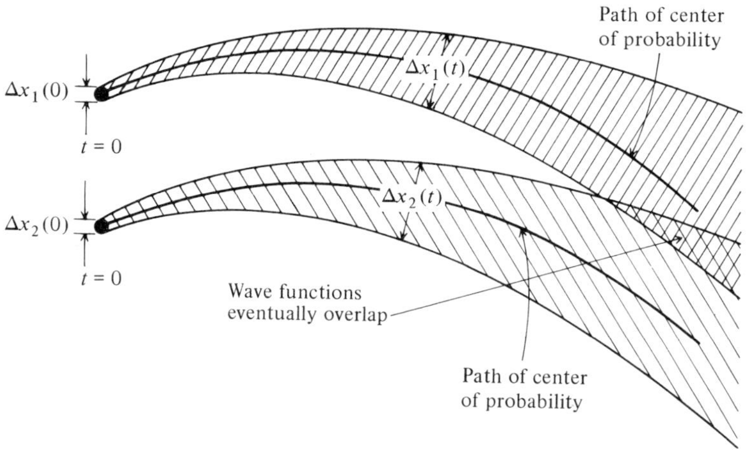

Fig. 18.1 The spreading of wave functions with time and its effect on our ability to distinguish between identical particles. (After Blokhintsev, Ref. 22.)

questions and can uncover some interesting phenomena through the medium of the nonrelativistic version of quantum mechanics.

If we can construct a many-particle Hamiltonian that governs the time-evolution of a system, say

$$\mathfrak{H} = \mathfrak{H}(\mathfrak{P}_1, \mathfrak{X}_1, \mathfrak{S}_1, \ldots; \mathfrak{P}_2, \mathfrak{X}_2, \mathfrak{S}_2, \ldots; \ldots; \mathfrak{P}_N, \mathfrak{X}_N, \mathfrak{S}_N, \ldots)$$

where the indices $i = 1, 2, \ldots, N$ correspond to the N particles in the system, then the Schrödinger state vectors for the system must satisfy

$$i\hbar\frac{\partial}{\partial t}|q, r, \ldots, (t)\rangle = \mathfrak{H}|q, r, \ldots (t)\rangle$$

where $q, r, \ldots$ are eigenvalues of a set of mutually commuting operators that are constants of the motion. In particular, one usually deals with eigenstates of $\mathfrak{H}$, namely

$$|E, q, r, \ldots, (t)\rangle = e^{-iE(t-t_0)/\hbar}|E, q, r, \ldots\rangle$$

such that

$$\mathfrak{H}|E, q, r, \ldots\rangle = E|E, q, r, \ldots\rangle;$$

any other state vector can be constructed as a linear combination of these energy eigenstates.

Now by "identical particles" we shall mean particles i, j such that the interchange of operators $\mathfrak{P}_i$, $\mathfrak{X}_i$, $\mathfrak{S}_i, \ldots$ with $\mathfrak{P}_j$, $\mathfrak{X}_j$, $\mathfrak{S}_j, \ldots$ does not affect any boundary or initial conditions on the system, nor does it affect $\mathfrak{H}$:

$$\mathfrak{H}(\mathfrak{P}_1, \ldots; \mathfrak{P}_2, \ldots; \ldots; \mathfrak{P}_i, \ldots; \ldots; \mathfrak{P}_j, \ldots; \ldots; \mathfrak{P}_N, \ldots)$$
$$= \mathfrak{H}(\mathfrak{P}_1, \ldots; \mathfrak{P}_2, \ldots; \mathfrak{P}_j, \ldots; \ldots; \mathfrak{P}_i, \ldots; \ldots; \mathfrak{P}_N, \ldots)$$

where it is understood that the indices on $\mathfrak{X}, \mathfrak{S}, \ldots$ undergo the interchange $i \leftrightarrow j$ also. For example, consider two particles having identical masses and identical intrinsic properties of all kinds (e.g. electric charge, spin angular momentum, etc.) so that they react identically to external forces and cannot be distinguished from one another at the boundaries of the experiment; let them interact with one another through the central potential $\mathfrak{V}'(|\mathfrak{X}_2 - \mathfrak{X}_1|)$. When the external force on one of the particles is represented by a potential $\mathfrak{V}(\mathfrak{X}_i)$, the Hamiltonian for the pair of particles is

$$\mathfrak{H}(1, 2) = \frac{1}{2M}\mathfrak{P}_1{}^2 + \frac{1}{2M}\mathfrak{P}_2{}^2 + \mathfrak{V}(\mathfrak{X}_1) + \mathfrak{V}(\mathfrak{X}_2) + \cdots + \mathfrak{V}'(|\mathfrak{X}_2 - \mathfrak{X}_1|)$$
$$= \mathfrak{H}(2, 1)$$

and the particles are therefore considered identical.

18.1 THE PARTICLE-INTERCHANGE OPERATOR AND ITS EIGENSTATES

An operator $\mathfrak{P}_{ij}$ may be introduced to effect the exchange of a pair of particles i and j in a state vector. In order to develop the concept of exchange operators, we shall use eigenstates of single-particle operators to describe many-particle systems. Thus if $\mathfrak{Q}_i, \mathfrak{R}_i, \ldots$ are one-particle operators for the ith particle in the system, $i = 1, 2, \ldots, N$, an appropriate eigenstate of these operators is

$$|q_1, r_1, \ldots; q_2, r_2, \ldots; \ldots; q_i, r_i, \ldots; \ldots; q_N, r_N, \ldots\rangle$$
$$= |q_1, r_1, \ldots\rangle|q_2, r_2, \ldots\rangle \cdots |q_i, r_i, \ldots\rangle \cdots |q_N, r_N, \ldots\rangle.$$

A many-particle system can exist in such an eigenstate only if the particles do not interact with one another, but it is always possible to express the true state of the system as a linear combination of those eigenstates. As a shorthand, we may write

$$|q_1, \ldots, q_N\rangle \qquad \text{or} \qquad |1, \ldots, N\rangle,$$

allowing a single quantum number q_i to stand for a whole set of one-particle quantum numbers $(q_i, r_i, \ldots)$, or simply letting the label i stand for the set of quantum numbers describing the state of that particle. In any case, the ordering of the quantum numbers is important; the eigenvalues in the ith position are those belonging to the ith particle.

The particle-exchange operator may be defined through its effect on such state vectors:

$$\mathfrak{P}_{ij}|q_1, \ldots, q_i, \ldots, q_j, \ldots, q_N\rangle = |q_1, \ldots, q_j, \ldots, q_i, \ldots, q_N\rangle$$

or

$$\mathfrak{P}_{ij}|1, \ldots, i, \ldots, j, \ldots, N\rangle = |1, \ldots, j, \ldots, i, \ldots, N\rangle \tag{18.1}$$

That is, $\mathfrak{P}_{ij}$ puts particle i in the state previously occupied by particle j and vice versa. Clearly $\mathfrak{P}_{ij}^{-1} = \mathfrak{P}_{ij}$, the order of indices on the operator being irrelevant, for

$$\begin{aligned} \mathfrak{P}_{ij}\mathfrak{P}_{ij}|1, \ldots, i, \ldots, j, \ldots, N\rangle &= \mathfrak{P}_{ij}|1, \ldots, j, \ldots, i, \ldots, N\rangle \\ &= |1, \ldots, i, \ldots, j, \ldots, N\rangle \end{aligned}$$

i.e.

$$\mathfrak{P}_{ij}^2 = 1.$$

Furthermore, $\mathfrak{P}_{ij}$ must be unitary because it must preserve inner products; thus

$$\begin{aligned} &\langle q_1', \ldots, q_j', \ldots, q_i', \ldots, q_N'|q_1, \ldots, q_j, \ldots, q_i, \ldots, q_N\rangle \\ &\quad = \langle q_1', \ldots, q_i', \ldots, q_j', \ldots, q_N'|\mathfrak{P}_{ij}^{\dagger}\mathfrak{P}_{ij}|q_1, \ldots, q_i, \ldots, q_j, \ldots, q_N\rangle \\ &\quad = \delta(q_1' - q_1) \cdots \delta(q_j' - q_j) \cdots \delta(q_i' - q_i) \cdots \delta(q_N' - q_N) \\ &\quad = \langle q_1', \ldots, q_i', \ldots, q_j', \ldots, q_N'|q_1, \ldots, q_i, \ldots, q_j, \ldots, q_N\rangle, \end{aligned}$$

from which it follows that

$$\mathfrak{P}_{ij}^{\dagger}\,\mathfrak{P}_{ij} = 1$$

and, in the usual way, that

$$\mathfrak{P}_{ij}\mathfrak{P}_{ij}^{\dagger} = 1$$

or

$$\mathfrak{P}_{ij}^{\dagger} - \mathfrak{P}_{ij}^{-1} = \mathfrak{P}_{ij} \tag{18.2}$$

The particle-exchange operator therefore is at once Hermitian and unitary. It follows that it must have eigenvalues that are real and lie on the unit circle; i.e. it has eigenvalues ± 1.

Now if particles i and j are identical, we have

$$\begin{aligned} \mathfrak{P}_{ij}\mathfrak{H}(1, \ldots, i, \ldots, j, \ldots, N)\mathfrak{P}_{ij}^{-1} &= \mathfrak{H}(1, \ldots, j, \ldots, i, \ldots, N) \\ &= \mathfrak{H}(1, \ldots, i, \ldots, j, \ldots, N), \end{aligned}$$

or

$$[\mathfrak{P}_{ij}, \mathfrak{H}(1, \ldots, i, \ldots, j, \ldots, N)] = 0. \tag{18.3}$$

As a consequence, when some state $|E; 1, \ldots, i, \ldots, j, \ldots, N\rangle$ is an eigenstate of $\mathfrak{H}$ with eigenvalue E, then so is

$$\mathfrak{P}_{ij}|E; 1, \ldots, i, \ldots, j, \ldots, N\rangle = |E; 1, \ldots, j, \ldots, i, \ldots, N\rangle,$$

for

$$\mathfrak{P}_{ij}\mathfrak{H}|E;1,\ldots,i,\ldots,j,\ldots,N\rangle = E\mathfrak{P}_{ij}|E;1,\ldots,i,\ldots,j,\ldots,N\rangle$$
$$= \mathfrak{H}\mathfrak{P}_{ij}|E;1,\ldots,i,\ldots,j,\ldots,N\rangle.$$

Here, by $|E;1,\ldots,N\rangle$ we mean an eigenstate of $\mathfrak{H}(1,\ldots,N)$, which is not necessarily a product of single-particle state vectors $|q_1,\ldots,q_N\rangle$; it may be written as a linear combination of products of single-particle state vectors, however:

$$|E;1,\ldots,N\rangle = \sum_{q_1,\ldots,q_N} a^E_{q_1,\ldots,q_N}|q_1,\ldots,q_N\rangle.$$

The question now arises whether $|E;1,\ldots,i,\ldots,j,\ldots,N\rangle$ and $\mathfrak{P}_{ij}|E;1,\ldots,i,\ldots,j,\ldots,N\rangle = |E;1,\ldots,j,\ldots,i,\ldots,N\rangle$ represent the same physical state of the system. There is no strong theoretical ground upon which to base a decision, although the existence of a complete set of physical observables, which we assumed, does imply in a subtle way that these state vectors represent the same state. The assumption that a complete set of observables exists, however, can be modified without doing violence to the structure of quantum theory, in such a way that the two state vectors represent different states. Therefore we must refer to experiments to find an answer to this question.

We shall not pursue the details, either experimental or mathematical, of the investigation of this question, but it has been found experimentally that for all known particles that can be subjected to relevant experimental tests, the state vectors $|E;1,\ldots,i,\ldots,j,\ldots,N\rangle$ and $\mathfrak{P}_{ij}|E;1,\ldots,i,\ldots,j,\ldots,N\rangle$ represent the same physical state; such particles include electrons, protons, neutrons and photons. Hence we shall assume in general that these two state vectors for a system of identical particles (e.g. electrons) represent the same physical state and therefore differ from each other only by a phase factor. As we have seen, that phase factor must be ± 1, one of the eigenvalues of $\mathfrak{P}_{ij}$.

This assumption is often called the *principle of identity of particles*: the only allowed states of a system of identical particles are those that are eigenstates of $\mathfrak{P}_{ij}$ for all possible pairs of indices i and j of the identical particles. It is not difficult to show that if i, j, k and l label identical particles and if the eigenvalue of $\mathfrak{P}_{ij}$ is $+1$ for the state vector representing the system, then the eigenvalue of $\mathfrak{P}_{kl}$ is $+1$ also. An analogous result holds if $\mathfrak{P}_{ij}$ has eigenvalue -1; then $\mathfrak{P}_{kl}$ has eigenvalue -1 also.

Thus there are two distinct classes of identical particles. Those that obey

$$\mathfrak{P}_{ij}|E;1,\ldots,i,\ldots,j,\ldots,N\rangle = |E;1,\ldots,i,\ldots,j,\ldots,N\rangle \tag{18.4}$$

are said to obey *Bose–Einstein statistics* and are called *bosons*, whereas those that obey

$$\mathfrak{P}_{ij}|E;1,\ldots,i,\ldots,j,\ldots,N\rangle = -|E;1,\ldots,i,\ldots,j,\ldots,N\rangle \tag{18.5}$$

are said to obey *Fermi–Dirac* statistics and are called *fermions*. The state vectors of the former are *symmetric* under interchange of any single pair of particles, and those of the latter are *antisymmetric*. Furthermore, since

$$\frac{\partial}{\partial t}\mathfrak{P}_{ij} = 0, \qquad [\mathfrak{H}(1, \ldots, N), \mathfrak{P}_{ij}] = 0$$

it is clear that the eigenvalues of $\mathfrak{P}_{ij}$ are conserved quantities. Therefore a state vector that is symmetric under particle interchange at some instant of time remains so for all time, and an antisymmetric state vector is perpetually antisymmetric. Once a boson, always a boson, and once a fermion, always a fermion.

Another important experimental fact is that particles having integer spin quantum number, $s = 0, 1, 2, \ldots$, are always bosons, and those having half-odd-integer spin quantum number, $s = \frac{1}{2}, \frac{3}{2}, \frac{5}{2}, \ldots$, are always fermions. This is in agreement with theoretical predictions concerning the connection between spin and statistics based upon relativistic quantum field theory. Because of the quantum conditions on total angular momentum, this means that any aggregate of bosons that has a definite total angular momentum is itself a boson, whereas an aggregate of fermions that forms an eigenstate of angular momentum is (a) a boson if there is an even number of fermions in the aggregate, or (b) a fermion if the number of fermions is odd. Thus the proton, with spin $s = \frac{1}{2}$, is a fermion, as is the neutron, but the deuteron, which is a bound state of a proton plus a neutron, is a boson and has, according to experiment, a spin angular momentum $s = 1$. On the other hand, H^3 and He^3 nuclei are both fermions (with spin angular momentum as $s = \frac{1}{2}$). The α-particle, or He^4 nucleus, is a boson with $s = 0$. The photon has $s = 1$ and is a boson also.

18.2 THE CONSTRUCTION OF STATE VECTORS AND WAVE FUNCTIONS FOR BOSONS AND FERMIONS

Consider a pair of noninteracting identical particles, the state vector for one being denoted $|q_1, r_1, \ldots\rangle_1$, and for the other $|q_2, r_2, \ldots\rangle_2$. If we ignore the identity of the particles we may expect the state vector for the pair to be the tensor product

$$|q_1, r_1, \ldots; q_2, r_2, \ldots\rangle \equiv |q_1, r_1, \ldots\rangle_1 |q_2, r_2, \ldots\rangle_2,$$

where the set $(q_1, r_1, \ldots)$ specifies the state of particle 1 and $(q_2, r_2, \ldots)$ specifies the state of particle 2; the subscripts on the state vectors label the particles. Now assume these particles to be identical bosons; the product state vector above is no longer adequate since it is not necessarily symmetric under the interchange of particles. The correct boson state vector must be

$$\begin{aligned} |q_1, r_1, \ldots; q_2, r_2, \ldots\rangle_+ &= B_2(1 + \mathfrak{P}_{12})|q_1, r_1, \ldots; q_2, r_2, \ldots\rangle \\ &= B_2(|q_1, r_1, \ldots; q_2, r_2, \ldots\rangle + |q_2, r_2, \ldots; q_1, r_1, \ldots\rangle) \\ &= B_2(|q_1, r_1, \ldots\rangle_1 |q_2, r_2, \ldots\rangle_2 + |q_2, r_2, \ldots\rangle_1 |q_1, r_1, \ldots\rangle_2) \qquad (18.6) \end{aligned}$$

where B_2 is a normalization constant. Clearly

$$\mathfrak{P}_{12}|q_1, r_1, \ldots; q_2, r_2, \ldots\rangle_+ = |q_1, r_1, \ldots; q_2, r_2, \ldots\rangle_+,$$

as required. To find B_2, we impose the requirement

$$\begin{aligned} {}_+\langle q_1', r_1', \ldots; q_2', r_2', \ldots | q_1, r_1, \ldots; q_2, r_2, \ldots\rangle_+ \\ = \delta(q_1' - q_1)\delta(r_1' - r_1)\cdots\delta(q_2' - q_2)\delta(r_2' - r_2)\cdots, \end{aligned}$$

from which it follows that

$$B_2 = \begin{cases} \frac{1}{2} & \text{if } q_1 = q_2, r_1 = r_2, \ldots \\ 1/\sqrt{2} & \text{otherwise;} \end{cases}$$

the reason for the two different values is that $|q_1, r_1, \ldots; q_2, r_2, \ldots\rangle$ is automatically symmetric when $q_1 = q_2, r_1 = r_2, \ldots$ and the symmetrization procedure is not required when the states of the two particles are identical. If the two states are not identical and symmetrization is required, the cross-terms in the inner product, e.g. $\langle q_1', r_1', \ldots; q_2', r_2', \ldots | q_2, r_2, \ldots; q_1, r_1, \ldots\rangle$ must vanish and this leads to $B_2 = 1/\sqrt{2}$.

It is implicit in this discussion that $|q_i, r_i, \ldots\rangle_j$ is an eigenstate of the operators $\mathfrak{Q}_j, \mathfrak{R}_j, \ldots$ with eigenvalues $q_i, r_i, \ldots$. For example,

$$\begin{aligned} \mathfrak{Q}_1|q_1, r_1, \ldots\rangle_1 &= q_1|q_1, \ldots\rangle_1, \qquad \mathfrak{Q}_1|q_2, r_2, \ldots\rangle_1 = q_2|q_2, r_2, \ldots\rangle_1 \\ \mathfrak{Q}_2|q_1, r_1, \ldots\rangle_1 &= \mathfrak{Q}_2|q_2, r_2, \ldots\rangle_1 = 0 \\ \mathfrak{Q}_1|q_1, r_1, \ldots\rangle_2 &= \mathfrak{Q}_1|q_2, r_2, \ldots\rangle_2 = 0 \\ \mathfrak{Q}_2|q_1, r_1, \ldots\rangle_2 &= q_1|q_1, r_1, \ldots\rangle_2, \qquad Q_2|q_2, r_2, \ldots\rangle_2 = q_2|q_2, r_2 \ldots\rangle_2 \\ \mathfrak{Q}_1|q_1, r_1, \ldots; q_2, r_2, \ldots\rangle &= \mathfrak{Q}_1|q_1, r_1, \ldots\rangle_1|q_2, r_2, \ldots\rangle_2 \\ &= (\mathfrak{Q}_1|q_1, r_1, \ldots\rangle_1)|q_2, r_2, \ldots\rangle_2 \\ &\quad + |q_1, r_1, \ldots\rangle_1(\mathfrak{Q}_1|q_2, r_2, \ldots\rangle_2) \\ &= q_1|q_1, r_1, \ldots; q_2, r_2, \ldots\rangle, \end{aligned}$$

etc., in accordance with the rules outlined in Section 10.4. Note that $|q_1, r_1, \ldots; q_2, r_2, \ldots\rangle_+$ is *not* an eigenstate of $\mathfrak{Q}_1$ unless $q_1 = q_2$, for

$$\begin{aligned} \mathfrak{Q}_1|q_1, r_1, \ldots; q_2, r_2, \ldots\rangle_+ = B_2(q_1|q_1, r_1, \ldots; q_2, r_2, \ldots\rangle \\ + q_2|q_2, r_2, \ldots; q_1, r_1, \ldots\rangle). \end{aligned}$$

On the other hand, $|q_1, r_1, \ldots; q_2, r_2, \ldots\rangle_+$ *is* an eigenstate of $(\mathfrak{Q}_1 + \mathfrak{Q}_2)$ with eigenvalue $q_1 + q_2$, and of $\mathfrak{Q}_1\mathfrak{Q}_2$ with eigenvalue $q_1 q_2$.

The generalization to larger numbers of particles is straightforward. For N identical bosons, we must form the eigenstate of $\mathfrak{P}_{ij}$ for all $i \neq j$ such that the eigenvalue of $\mathfrak{P}_{ij}$ is $+1$. This is

$$|q_1, r_1, \ldots; \ldots; q_N, r_N, \ldots\rangle_+ \equiv B_N \sum_P \mathfrak{P}|q_1, r_1, \ldots; \ldots; q_N, r_N \ldots\rangle, \quad (18.7)$$

where $\sum_P \mathfrak{P}$ means the summation over all $N!$ permutations of the indices $1, \ldots, N$, and B_N is a normalization constant. In fact, if the N particles are in $n \leq N$ different states, N_1 being the number in the first of these states, $\ldots$, N_n being the number in the nth state, then $B_N = (N!N_1!N_2!\ldots N_n!)^{-1/2}$ where $\sum_{i=1}^n N_i = N$. As before, we use the order of listing eigenvalues in the state vector to indicate the particle to which a given eigenvalue is assigned; thus if $\mathfrak{Q}_3$ is an operator for particle 3, we have

$$\mathfrak{Q}_3|q_1, q_2, q_3, \ldots, q_N\rangle = q_3|q_1, q_2, q_3, \ldots, q_N\rangle$$

and

$$\mathfrak{Q}_3|q_2, q_3, q_1, \ldots, q_N\rangle = q_1|q_2, q_3, q_1, \ldots, q_N\rangle.$$

For three particles, for example,

$$\begin{aligned}|q_1, q_2, q_3\rangle_+ = B_3(&|q_1, q_2, q_3\rangle + |q_3, q_1, q_2\rangle \\ &+ |q_2, q_3, q_1\rangle + |q_3, q_2, q_1\rangle \\ &+ |q_1, q_3, q_2\rangle + |q_2, q_1, q_3\rangle),\end{aligned}$$

where

$$B_3 = \begin{cases} 1/\sqrt{6} & \text{if } q_1 \neq q_2 \neq q_3 \quad (N_1 = N_2 = N_3 = 1) \\ \frac{1}{6} & \text{if } q_1 = q_2 = q_3 \quad (N_1 = 3, N_2 = N_3 = 0) \\ 1/2\sqrt{3} & \text{otherwise} \quad (N_1 = 2, N_2 = 1, N_3 = 0). \end{cases}$$

We reiterate that this symmetrized state vector is inadequate when we deal with particles that interact with one another, since the unsymmetrized state vectors for the system then cannot be written as simple tensor products of single-particle state vectors. However, the tensor-product state vectors do form a complete set of state vectors for the system, and hence an arbitrary unsymmetrized state vector may be expanded in terms of such products:

$$\begin{aligned}|1\,(N\text{ particles})\rangle &= \mathop{\mathsf{S}}_{q_1,\ldots,q_N} |q_1, \ldots, q_N\rangle\langle q_1, \ldots, q_N|1\,(N\text{ particles})\rangle \\ &= \mathop{\mathsf{S}}_{q_1,\ldots,q_N} a^{(1)}_{q_1,\ldots,q_N}|q_1, \ldots, q_N\rangle.\end{aligned}$$

We can then form a completely symmetric state vector for the system by constructing

$$|1\,(N\text{ particles})\rangle = \mathop{\mathsf{S}}_{q_1,\ldots,q_N} a^{(1)}_{q_1,\ldots,q_N}|q_1, \ldots, q_N\rangle_+.$$

Actually, $q_1, \ldots, q_n$ are no longer the appropriate quantities over which to sum since there may be several particles, say N_i of them, in the same state, say the state denoted by q_i; then the state of the system is defined by the numbers $N_1, \ldots, N_n$, if there are n states available to each particle in the system. The number N_i is called the *occupation number* for the one-particle state defined by q_i, and obviously $\sum_{i=1}^{n} N_i = N$. Therefore we may write

$$|1(N \text{ particles})\rangle = \underset{N_1,\ldots,N_n}{\mathsf{S}} C^{(1)}(N_1, \ldots, N_n)|N_1, \ldots, N_n\rangle_+$$

where

$$|N_1, \ldots, N_n\rangle_+ \equiv |\underbrace{q_1, q_1, q_1, \ldots, q_1}_{N_1}, \ldots, \underbrace{q_n, q_n, \ldots, q_n}_{N_n}\rangle_+$$

and

$$C^{(1)}(N_1, \ldots, N_n) = N_1!N_2! \cdots N_n! a^{(1)}_{q_1, q_2, \ldots, q_n}$$

The symmetrized state vectors $|N_1, \ldots, N_n\rangle_+$ thus form a complete orthonormal set of state vectors for a boson system. It is a particularly convenient set when the interparticle interactions are weak and can be considered as perturbations; then $|N_1, \ldots, N_n\rangle_+$ may be used as a zeroth approximation to the state vector for the system and perturbation theory may be used to evaluate the $C^{(1)}(N_1, \ldots, N_n)$.

In practical calculations it is frequently convenient to use wave functions, and these must be appropriately symmetrized. It is customary to regard the spatial locations $\mathbf{x}_1, \ldots, \mathbf{x}_N$ of the particles as labels distinguishing the particles, and hence $|\mathbf{x}_1, \ldots; \mathbf{x}_2, \ldots; \ldots; \mathbf{x}_N, \ldots\rangle$ is not symmetrized in forming wave functions, for to do so would be to destroy the labeling function of spatial coordinates. Thus

$$\psi_{q_1,\ldots,q_N+}(\mathbf{x}_1, \ldots, \mathbf{x}_N) \equiv \langle \mathbf{x}_1, \ldots, \mathbf{x}_N | q_1, \ldots, q_N\rangle_+ \tag{18.8}$$

is the probability amplitude that a boson found at $\mathbf{x}_1$ (and therefore called particle 1) is in one of the states $q_1, \ldots, q_N$, that a boson found at $\mathbf{x}_2$ (particle 2) is in one of the remaining states among $q_1, \ldots, q_N$, etc. It follows that

$$\psi_{q_1,\ldots,q_N+}(\mathbf{x}_1, \ldots, \mathbf{x}_N) = B_N \sum_P \mathsf{P}\psi_{q_1}(\mathbf{x}_1)\psi_{q_2}(\mathbf{x}_2) \cdots \psi_{q_N}(\mathbf{x}_N), \tag{18.9}$$

where $\sum_P \mathsf{P}$ is the sum over the $N!$ permutations of the labels $\mathbf{x}_1, \ldots, \mathbf{x}_N$.

Essentially the same procedures are followed in forming state vectors and wave functions for fermions, except that $\sum_P \mathfrak{P}$ and $\sum_P \mathsf{P}$ are replaced by $\sum_P (-)^P \mathfrak{P}$ and $\sum_P (-)^P \mathsf{P}$, where

$$(-)^P = \begin{cases} +1 & \text{if } P \text{ is an even permutation} \\ -1 & \text{if } P \text{ is an odd permutation.} \end{cases}$$

For example, the two-fermion state vector is

$$
\begin{aligned}
|q_1, q_2\rangle_- &\equiv (1/\sqrt{2})(1 - \mathfrak{P}_{12})|q_1, q_2\rangle \\
&= (1/\sqrt{2})(|q_1, q_2\rangle - |q_2, q_1\rangle) \\
&= (1/\sqrt{2})(|q_1\rangle_1|q_2\rangle_2 - |q_2\rangle_1|q_1\rangle_2)
\end{aligned} \tag{18.10}
$$

and for three fermions we have

$$
\begin{aligned}
|q_1, q_2, q_3\rangle_- \equiv (1/\sqrt{6})(&|q_1, q_2, q_3\rangle + |q_3, q_1, q_2\rangle \\
&+ |q_2, q_3, q_1\rangle - |q_1, q_3, q_2\rangle \\
&- |q_3, q_2, q_1\rangle - |q_2, q_1, q_3\rangle)
\end{aligned}
$$

as the antisymmetric state vector. It is easy to see, for example, that each of these is an eigenstate of $\mathfrak{P}_{12}$ with eigenvalue -1. The normalization for these states is unique. Why? (What are the possible occupation numbers for the states?) In general,

$$|q_1, \ldots, q_N\rangle_- = \frac{1}{\sqrt{N!}} \sum_P (-)^P \mathfrak{P} |q_1, \ldots, q_N\rangle \tag{18.11}$$

and

$$\psi_{q_1, \ldots, q_N -}(\mathbf{x}_1, \ldots, \mathbf{x}_N) = \frac{1}{\sqrt{N!}} \sum_P (-)^P \mathrm{P} \psi_{q_1}(\mathbf{x}_1), \ldots, \psi_{q_N}(\mathbf{x}_N). \tag{18.12}$$

A convenient technique for summarizing the antisymmetric nature of these fermion wave functions is provided by the properties of determinants. In fact we have

$$\psi_{q_1, \ldots, q_N -}(\mathbf{x}_1, \ldots, \mathbf{x}_N) = \frac{1}{\sqrt{N!}} \begin{vmatrix} \psi_{q_1}(\mathbf{x}_1) & \psi_{q_2}(\mathbf{x}_1) & \cdots & \psi_{q_N}(\mathbf{x}_1) \\ \psi_{q_1}(\mathbf{x}_2) & \psi_{q_2}(\mathbf{x}_2) & \cdots & \psi_{q_N}(\mathbf{x}_2) \\ \vdots & \vdots & & \vdots \\ \psi_{q_1}(\mathbf{x}_N) & \psi_{q_2}(\mathbf{x}_N) & \cdots & \psi_{q_N}(\mathbf{x}_N) \end{vmatrix}. \tag{18.13}$$

This is known as the Slater determinant.

Again, $\psi_{q_1, \ldots, q_N -}(\mathbf{x}_1, \ldots, \mathbf{x}_N)$ or $|q_1, \ldots, q_N\rangle_-$ will not serve to describe systems of interacting particles, but they do form complete antisymmetric orthonormal bases in terms of which arbitrary states of an N-fermion system may be expanded.

In fact, it is important to note that $|q_1, \ldots, q_N\rangle_\pm$ are linear combinations of the permutations of $|q_1, \ldots, q_N\rangle$. This can be interpreted as meaning that there is an interaction among identical bosons that renders $|q_1, \ldots, q_N\rangle$ an inappropriate state vector and forces symmetrization of the single-particle state vectors. Similarly, there is an interaction among identical fermions that forces antisymmetrization of their single-particle state vectors. These interactions are obviously very strong. They cannot be formulated precisely in terms of potential functions, but are taken into account through the appropriate symmetrization or antisymmetrization of state vectors or wave functions.

18.3 THE PAULI EXCLUSION PRINCIPLE FOR FERMIONS

An extremely important consequence of the rule that the state vector representing identical fermions must be antisymmetric under the exchange of any pair of the fermions is that no two identical fermions may possess the same quantum numbers; no two of them can occupy the same physical state. This follows from

$$\mathfrak{P}_{ij}|q_1, \ldots, q_i, \ldots, q_j, \ldots, q_N\rangle_- = -|q_1, \ldots, q_j, \ldots, q_i, \ldots, q_N\rangle_-$$

when $q_i = q_j$, $r_i = r_j$, etc.; for then if $q, r, \ldots$ represent the common values of q_i and q_j, r_i and r_j, etc., we have

$$\begin{aligned}\mathfrak{P}_{ij}|q_1, \ldots, q, \ldots, q, \ldots, q_N\rangle_- &= |q_1, \ldots, q, \ldots, q, \ldots, q_N\rangle_- \\ &= -|q_1, \ldots, q, \ldots, q, \ldots, q_N\rangle_- = 0.\end{aligned}$$

A state in which two particles have identical quantum numbers for all physical variables is necessarily symmetric under exchange of those particles, but if the particles are fermions the state must also be antisymmetric under particle exchange and therefore vanishes.

The application of this result to atomic electrons is known as the *Pauli exclusion principle.* Actually, Pauli enunciated his exclusion principle in 1925, before there were wave functions or state vectors to antisymmetrize. At that time physicists were feverishly attempting to find a completely general set of principles from which all observed atomic spectral phenomena could be derived and which at the same time would "predict" the periodic table of elements. The most advanced physical models available to them were elaborations of the Bohr model of the hydrogen atom, and Pauli was attempting to reconcile the observed anomalous Zeeman effect with the periodic table of the elements through such models. He found eventually that such a reconciliation could be effected through the hypothesis that no two electrons can occupy the same state in an atom. This rule, in fact, made Bohr-like atoms seem plausible at last, for without it one could not satisfactorily explain why atomic electrons do not all occupy the lowest Bohr orbit.

With this exclusion principle, the fundamental scheme of "atom-building" finally seemed to be comprehensible, at least for the lightest elements. Later in this chapter, we shall see how the exclusion principle, which we shall now think of not as a separate hypothesis, but rather as a consequence of the antisymmetry of many-electron (electrons have spin quantum number $s = \frac{1}{2}$ and are fermions) state vectors or wave functions, provides a key to understanding atomic structure.

It was Pauli, too, who in 1940 first gave a theoretical justification for the connection between spin and statistics, that is, between the spin quantum number of a species of particles and their possession of Bose–Einstein or Fermi–Dirac statistics. His justification requires the use of quantum field theory.

18.4 THE HELIUM ATOM

The first challenge one must face in building up a theory of many-electron atoms is that of solving the simplest two-electron atom problem. We shall not obtain a complete solution to the problem, but shall restrict ourselves to calculations that indicate the central features of such a solution. Even the two-electron atom is sufficiently complicated for numerical methods and rather sophisticated approximation techniques to be required in serious attempts at solution, but the new phenomena arising from the identity of the electrons can be seen in rather straightforward approaches to the problem. Electron spin will be seen to play a vital role even though we shall not be concerned with magnetic fields; this underlies the connection, between the Zeeman effect and the periodic table, that was perceived by Pauli and led to his discovery of the exclusion principle.

The Hamiltonian for the helium atom may be written in the form

$$\mathsf{H} = -\frac{1}{2\mu}\nabla_1^2 - \frac{1}{2\mu}\nabla_2^2 - \frac{2e^2}{4\pi\epsilon_0 r_1} - \frac{2e^2}{4\pi\epsilon_0 r_2} + \frac{e^2}{4\pi\epsilon_0 r_{12}} + \mathsf{W}(\mathbf{r}_1, \mathbf{r}_2, \nabla_1, \nabla_2, \mathbf{S}_1, \mathbf{S}_2) \tag{18.14}$$

in wave-function notation, where $r_{12} = |\mathbf{x}_2 - \mathbf{x}_1|$ and $\mathbf{S}_1, \mathbf{S}_2$ represent the spins of the two electrons. Here W represents "complications" to the problem, including the spin–orbit effect, spin–spin interactions between the electrons, other relativistic corrections, etc. These effects will all be assumed to be very small (as indeed they are), and we shall neglect W completely in our approximation. Then we may approximate H by

$$\mathsf{H} \cong \mathsf{H}_{01} + \mathsf{H}_{02} + V(r_{12}) \tag{18.15}$$

where

$$\mathsf{H}_{0i} = -\frac{1}{2\mu}\nabla_i^2 - \frac{2\alpha\hbar c}{r_i}, \qquad V(r_{12}) = \frac{\alpha\hbar c}{r_{12}}, \qquad \alpha = \frac{e^2}{4\pi\epsilon_0\hbar c}. \tag{18.16}$$

Obviously $-2\alpha\hbar c/r_i$ represents the attraction between the He nucleus (located at the origin in the relative coordinates of the nucleus and the ith electron) and the ith electron, and $V(r_{12})$ represents the repulsion between the two electrons. The reduced mass of the electron–He nucleus system is $\mu \equiv m_e m_{He}/m_e + m_{He}$.

Equation (18.15) is not separable in the coordinates $\mathbf{x}_1$ and $\mathbf{x}_2$ and hence the solution is not of the form $\psi_{q_1, q_2; \ldots}(\mathbf{x}_1, \mathbf{x}_2; \ldots)$ but rather is a linear combination of such wave functions. Therefore we consider $V(r_{12})$ as a perturbation on $\mathsf{H}_0 \equiv$

$\mathsf{H}_{01} + \mathsf{H}_{02}$ and begin our analysis by finding solutions to the problem having H_0 as Hamiltonian. This procedure is not well justified since $\langle V(r_{12})\rangle$ cannot be expected to be very much smaller than $\langle 2\alpha hc/r_1\rangle$ and hence the separation in unperturbed energy levels is not likely to be very much bigger than the expectation values of the perturbation. Nevertheless, we can obtain rough estimates from such calculations and, more importantly, qualitative features of the exact solution should become evident as we proceed.

Now let the solutions to the single-particle problems be eigenstates of orbital angular momentum and spin angular momentum,

$$\begin{gathered}\psi_i(\mathbf{x}_j)\chi_i(s_j) \equiv \psi_{n_i,l_i,m_i}(\mathbf{x}_j)\chi_{\mu_i}(s_j);\\ \mathsf{H}_{0i}\psi_{n_i,l_i,m_i}(\mathbf{x}_i)\chi_{\mu_i}(s_i) = E_{n_i}\psi_{n_i,l_i,m_i}(\mathbf{x}_i)\chi_{\mu_i}(s_i).\end{gathered} \tag{18.17}$$

These are the simple one-electron hydrogen atom wave functions when spin effects are neglected in the Hamiltonian. It is the neglect of spin effects, in particular the spin–orbit coupling, that allows us to use a wave function that is a product of a spatial factor and a spinor; more generally, we should have had to use a linear combination of such products.

The unsymmetrized two-particle unperturbed wave function then has the form

$$\Psi_{12}(\mathbf{x}_1, \mathbf{x}_2; s_1, s_2) = \Psi_{12}(\mathbf{x}_1, \mathbf{x}_2)\chi_{12}(s_1, s_2),$$

with

$$\psi_{12}(\mathbf{x}_1, \mathbf{x}_2) = \psi_1(\mathbf{x}_1)\psi_2(\mathbf{x}_2)$$

$$\chi_{12}(s_1, s_2) = \chi(s_1)\chi(s_2),$$

where we have separated space and spin variables for the two-particle functions also. The exchange operator P_{12} can, in this case, be factored into a space-exchange part and a spin-exchange part, respectively $\mathsf{P}_{12}{}^{(\mathrm{M})}$ and $\mathsf{P}_{12}{}^{(\mathrm{B})}$, M referring to Majorana and B referring to Bartlett (since Majorana and Bartlett, respectively, introduced these operators into nuclear physics in a significant way):

$$\mathsf{P}_{12} = \mathsf{P}_{12}{}^{(\mathrm{M})}\mathsf{P}_{12}{}^{(\mathrm{B})} \tag{18.18}$$

where

$$\mathsf{P}_{12}{}^{(\mathrm{M})}\Psi_{12}(\mathbf{x}_1, \mathbf{x}_2; s_1, s_2) = \Psi_{12}(\mathbf{x}_2, \mathbf{x}_1; s_1, s_2) \tag{18.19a}$$

$$\mathsf{P}_{12}{}^{(\mathrm{B})}\Psi_{12}(\mathbf{x}_1, \mathbf{x}_2; s_1, s_2) = \Psi_{12}(\mathbf{x}_1, \mathbf{x}_2; s_2, s_1) \tag{18.19b}$$

$$\begin{aligned}\mathsf{P}_{12}\Psi_{12}(\mathbf{x}_1, \mathbf{x}_2; s_1, s_2) &= \mathsf{P}_{12}{}^{(\mathrm{M})}\mathsf{P}_{12}{}^{(\mathrm{B})}\Psi(\mathbf{x}_1, \mathbf{x}_2; s_1, s_2)\\ &= \Psi_{12}(\mathbf{x}_2, \mathbf{x}_1; s_2, s_1).\end{aligned} \tag{18.19c}$$

Our perturbation is symmetric under P_{12}, as it must be; more importantly, it involves only $|\mathbf{x}_2 - \mathbf{x}_1|$ and hence is symmetric under $\mathsf{P}_{12}{}^{(\mathrm{M})}$ and also under $\mathsf{P}_{12}{}^{(\mathrm{B})}$. We conclude that physically acceptable states of the system must be eigenstates of $\mathsf{P}_{12}{}^{(M)}$ *and* of $\mathsf{P}_{12}{}^{(B)}$, since the two electrons are supposed to be identical. Furthermore, since the electrons are fermions, the allowed eigenvalue of P_{12} is -1; this means that when $\mathsf{P}_{12}{}^{(M)}$ has eigenvalue $+1$, $\mathsf{P}_{12}{}^{(B)}$ must have eigenvalue -1 and vice versa.

Let us consider first the possible spin states $\chi_{12}(s_1, s_2)$ and the linear combinations of them that are eigenstates of $\mathsf{P}_{12}{}^{(B)}$. To begin with, the possible magnetic quantum numbers μ_1 and μ_2 for the electrons are $\pm\frac{1}{2}$; hence $\mu' = \mu_1 + \mu_2$ can take on one of the three values $+1, 0, -1$. When $\mu' = +1$, it is clear that $\mu_1 = \mu_2 = +\frac{1}{2}$, and the spin wave function is

$$\chi_{1/2}(s_1)\chi_{1/2}(s_2) = \chi_{1/2}(s_2)\chi_{1/2}(s_1)$$

which is clearly symmetric in the particle spin labels s_1 and s_2. Similarly, if $\mu' = -1$, then $\mu_1 = \mu_2 = -\frac{1}{2}$ and the spin function is

$$\chi_{-1/2}(s_1)\chi_{-1/2}(s_2) = \chi_{-1/2}(s_2)\chi_{-1/2}(s_1)$$

which again is symmetric in s_1 and s_2.

However, if $\mu = 0$, we have two alternatives, $\mu_1 = \frac{1}{2}, \mu_2 = -\frac{1}{2}$, and $\mu_1 = -\frac{1}{2}$, $\mu_2 = +\frac{1}{2}$, and neither of the product functions is either symmetric or antisymmetric. Therefore we form the symmetric and antisymmetric combinations

$$(1/\sqrt{2})(1 + \mathsf{P}_{12}{}^{(B)})\chi_{1/2}(s_1)\chi_{1/2}(s_2) = (1/\sqrt{2})[\chi_{1/2}(s_1)\chi_{-1/2}(s_2) + \chi_{1/2}(s_2)\chi_{-1/2}(s_1)]$$
$$(18.20)$$
$$(1/\sqrt{2})(1 - \mathsf{P}_{12}{}^{(B)})\chi_{1/2}(s_1)\chi_{1/2}(s_2) = (1/\sqrt{2})[\chi_{1/2}(s_1)\chi_{-1/2}(s_2) - \chi_{1/2}(s_2)\chi_{-1/2}(s_1)],$$

respectively.

In this way we obtain three symmetric spin functions with $\mu' = +1, 0, -1$ and one antisymmetric spin function with $\mu' = 0$. This suggests strongly that the symmetric functions correspond to the three spin orientations of a spin-1 system and that the lone antisymmetric function is the function for a spin-0 system. That indeed is the case, and we write

$$\begin{aligned}
\chi_{1,+1}(s_1, s_2) &\equiv \chi_{1/2}(s_1)\chi_{1/2}(s_2) \\
\chi_{1,0}(s_1, s_2) &\equiv (1/\sqrt{2})[\chi_{1/2}(s_1)\chi_{-1/2}(s_2) + \chi_{1/2}(s_2)\chi_{-1/2}(s_1)] \\
\chi_{1,-1}(s_1, s_2) &\equiv \chi_{-1/2}(s_1)\chi_{-1/2}(s_2) \\
\chi_{0,0}(s_1, s_2) &\equiv (1/\sqrt{2})[\chi_{1/2}(s_1)\chi_{-1/2}(s_2) - \chi_{1/2}(s_2)\chi_{-1/2}(s_1)].
\end{aligned} \qquad (18.21)$$

This notation indicates that the coefficients of the product spinors on the right-hand side are the Clebsch–Gordan coefficients $\langle s, \mu', s', s''|s', \mu_1; s'', \mu_2\rangle$ when $s' = s'' = \frac{1}{2}$ and $s = 1$ or 0.

Now we turn to the space wave functions $\psi_{12}(\mathbf{x}_1, \mathbf{x}_2)$. Since $\chi_{1,\mu'}(s_1, s_2)$ is symmetric under particle interchange, the space function multiplying it must be antisymmetric in order to yield a totally antisymmetric wave function for the system. Since the total spin s contributed by the spinors is one unit, $s = 1$, we call the corresponding wave function the "triplet" $(2s + 1 = 3)$ wave function, writing

$$\Psi_{1,2;t}(\mathbf{x}_1, \mathbf{x}_2; s_1, s_2) = \psi_{1,2;t}(\mathbf{x}_1, \mathbf{x}_2)\chi_{1,\mu'}(s_1, s_2), \tag{18.22a}$$

where

$$\begin{aligned}\psi_{1,2;t}(\mathbf{x}_1, \mathbf{x}_2) &\equiv \frac{1}{\sqrt{2}}(1 - \mathsf{P}_{12}{}^{(\mathrm{M})})\psi_1(\mathbf{x}_1)\psi_2(\mathbf{x}_2) \\ &= \frac{1}{\sqrt{2}}[\psi_1(\mathbf{x}_1)\psi_2(\mathbf{x}_2) - \psi_1(\mathbf{x}_2)\psi_2(\mathbf{x}_1)].\end{aligned} \tag{18.22b}$$

On the other hand, the singlet state, characterized by $\chi_{0,0}(s_1, s_2)$ $(2s + 1 = 1$ for $s = 0)$ must be multiplied by a symmetric function of space variables $\mathbf{x}_1, \mathbf{x}_2$ and we write

$$\Psi_{1,2;s}(\mathbf{x}_1, \mathbf{x}_2; s_1, s_2) = \psi_{1,2;s}(\mathbf{x}_1, \mathbf{x}_2)\chi_{0,0}(s_1, s_2) \tag{18.23a}$$

where

$$\begin{aligned}\psi_{1,2;s}(\mathbf{x}_1, \mathbf{x}_2) &\equiv \frac{1}{\sqrt{2}}(1 + \mathsf{P}_{12}{}^{(\mathrm{M})})\psi_1(\mathbf{x}_1)\psi_2(\mathbf{x}_2) \\ &= \frac{1}{\sqrt{2}}[\psi_1(\mathbf{x}_1)\psi_2(\mathbf{x}_2) + \psi_1(\mathbf{x}_2)\psi_2(\mathbf{x}_1)].\end{aligned} \tag{18.23b}$$

The appropriate zeroth-order wave functions for the system, then, are $\Psi_{1,2;t}$ and $\Psi_{1,2;s}$, and they can be characterized by the total spin quantum number or by t $(s = 1)$ or s $(s = 0)$.

The triplet state, which corresponds in the vector model to a parallel alignment of the spins of the two electrons, is often referred to as *orthohelium*, and the singlet state is called *parahelium*, corresponding to an antiparallel alignment of spins.

A general rule in quantum physics is that states with fewer nodes or nodal surfaces (surfaces on which the wave-function vanishes) have less energy than states with more nodes. This is intuitively evident since the kinetic energy, being determined in the wave-function representation by second-order spatial derivatives, is proportional to the curvature of the wave function, and the more nodes there are, the more curvature there must be and hence the more kinetic energy must be represented by that wave function. In the six-dimensional space of $(\mathbf{x}_1, \mathbf{x}_2)$, $\psi_{1,2;s}(\mathbf{x}_1, \mathbf{x}_2)$ can have nodes only where $\psi_1(\mathbf{x}_1)$ or $\psi_2(\mathbf{x}_2)$ have nodes, but $\psi_{1,2;t}(\mathbf{x}_1, \mathbf{x}_2)$ has all those nodes together with the nodal hyperplane $\mathbf{x}_1 = \mathbf{x}_2$. Therefore $\psi_{1,2;t}(\mathbf{x}_1, \mathbf{x}_2)$ must have lower energy than $\psi_{1,2;s}(\mathbf{x}_1, \mathbf{x}_2)$. However, when $\psi_1 = \psi_2$, only the symmetric state represented by $\psi_{1,2;s}(\mathbf{x}_1, \mathbf{x}_2)$ exists, and in particular when ψ_1 is the hydrogen-atom ground-state wave function, there is only a singlet, or parahelium, state available to the two-electron system. Since the perturbation is assumed to be too small to interchange any of the unperturbed energy levels,

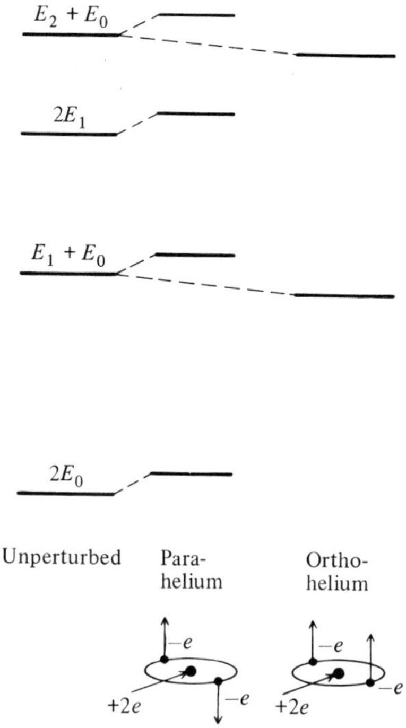

Fig. 18.2. Schematic illustration of energy-levels split by $V(r_{12})$ in the helium atom.

the ground state of the two-electron system must be a parahelium state; but the next level will be orthohelium, followed closely by a corresponding parahelium state. See Fig. 18.2.

18.5 SOLUTIONS TO THE PERTURBATION PROBLEM; THE EXCHANGE INTEGRAL

We can extend our insights into the nature of the two-electron atom by proceeding with the purturbation calculation. The calculation is not very different from the Stark-effect calculation we did earlier. Since the perturbation does not involve spin operators, we shall simply solve for the spatial wave function, beginning with the unperturbed product wave function $\psi_{12}(\mathbf{x}_1, \mathbf{x}_2) = \psi_1(\mathbf{x}_1)\psi_2(\mathbf{x}_2)$. If E_{n_1} is the eigenvalue of H_{01} for $\psi_1(\mathbf{x}_1)$ and E_{n_2} is the eigenvalue of H_{02} for $\psi_2(\mathbf{x}_2)$, we have

$$\mathsf{H}_0\psi_{12}(\mathbf{x}_1, \mathbf{x}_2) = (E_{n_1} + E_{n_2})\psi_{12}(\mathbf{x}_1, \mathbf{x}_2). \tag{18.24}$$

But $\psi_{21}(\mathbf{x}_1, \mathbf{x}_2) = \psi_2(\mathbf{x}_1)\psi_1(\mathbf{x}_2)$ has exactly the same unperturbed energy eigenvalue, so that $\psi_{12}(\mathbf{x}_1, \mathbf{x}_2)$ and $\psi_{21}(\mathbf{x}_1, \mathbf{x}_2)$ are degenerate wave functions. Assuming that no accidental degeneracies occur because of fortuitous peculiarities in the energy spectrum for one electron, we may write

$$\psi(\mathbf{x}_1, \mathbf{x}_2) = a\psi_{12}(\mathbf{x}_1, \mathbf{x}_2) + b\psi_{21}(\mathbf{x}_1, \mathbf{x}_2)$$

as the most general wave function for a state having energy $E_{n_1} + E_{n_2}$.

Now we require that $\psi(\mathbf{x}_1, \mathbf{x}_2)$ be an energy eigenfunction with energy E when H is the Hamiltonian operator:

$$\begin{aligned} \mathsf{H}\psi(\mathbf{x}_1, \mathbf{x}_2) &= \mathsf{H}_0\psi(\mathbf{x}_1, \mathbf{x}_2) + V(r_{12})\psi(\mathbf{x}_1, \mathbf{x}_2) \\ &= E\psi(\mathbf{x}_1, \mathbf{x}_2). \end{aligned} \tag{18.25}$$

Multiplication of this on the left by $\psi_{12}^*(\mathbf{x}_1, \mathbf{x}_2)$ and integration over d^3x_1 and d^3x_2 yields

$$(E_{n_1 n_2} + V_{11} - E)a + V_{12}b = 0, \tag{18.26}$$

where

$$\begin{aligned} E_{n_1 n_2} &= E_{n_1} + E_{n_2}, \\ V_{11} &\equiv \int \psi_{12}^*(\mathbf{x}_1, \mathbf{x}_2)V(r_{12})\psi_{12}(\mathbf{x}_1, \mathbf{x}_2)\, d^3x_1\, d^3x_2, \\ V_{12} &\equiv \int \psi_{12}^*(\mathbf{x}_1, \mathbf{x}_2)V(r_{12})\psi_{21}(\mathbf{x}_1, \mathbf{x}_2)\, d^3x_1\, d^3x_2, \end{aligned} \tag{18.27}$$

and multiplication by $\psi_{21}^*(\mathbf{x}_1, \mathbf{x}_2)$ and subsequent integration over d^3x_1 and d^3x_2 gives

$$(E_{n_1 n_2} + V_{22} - E)b + V_{21}a = 0, \tag{18.28}$$

where

$$\begin{aligned} V_{22} &= \int \psi_{21}^*(\mathbf{x}_1, \mathbf{x}_2)V(r_{12})\psi_{21}(\mathbf{x}_1, \mathbf{x}_2)\, d^3x_1\, d^3x_2 \\ V_{21} &= \int \psi_{21}^*(\mathbf{x}_1, \mathbf{x}_2)V(r_{12})\psi_{12}(\mathbf{x}_1, \mathbf{x}_2)\, d^3x_1\, d^3x_2. \end{aligned} \tag{18.29}$$

Since $\mathbf{x}_1$ and $\mathbf{x}_2$ are dummy variables in the integrals, it is clear that $V_{11} = V_{22}$ and $V_{12} = V_{21}^* = V_{21} = V_{12}^*$, with

$$\begin{aligned} K \equiv V_{11} = V_{22} &= \frac{e^2}{4\pi\epsilon_0} \int |\psi_1(\mathbf{x}_1)|^2 |\psi_2(\mathbf{x}_2)|^2 \frac{1}{r_{12}}\, d^3x_1\, d^3x_2 \\ A \equiv V_{12} = V_{21} &= \frac{e^2}{4\pi\epsilon_0} \int \psi_1^*(\mathbf{x}_1)\psi_2^*(\mathbf{x}_2)\frac{1}{r_{12}}\psi_2(\mathbf{x}_1)\psi_1(\mathbf{x}_2)\, d^3x_1\, d^3x_2. \end{aligned} \tag{18.30}$$

Thus we must solve the coupled linear equations

$$\begin{aligned} (E_{n_1 n_2} + K - E)a + Ab &= 0 \\ Aa + (E_{n_1 n_2} + K - E)b &= 0. \end{aligned} \tag{18.31}$$

These equations possess a solution if and only if the secular determinant vanishes:

$$\begin{vmatrix} (E_{n_1n_2} - E + K) & A \\ A & (E_{n_1n_2} - E + K) \end{vmatrix} = (E_{n_1n_2} - E + K)^2 - A^2 = 0;$$

that is, the possible eigenvalues E are

$$\begin{aligned} E_{\pm} &= E_{n_1n_2} + K \pm A \\ &= E_{n_1} + E_{n_2} + K \pm A. \end{aligned} \tag{18.32}$$

Substituting $E = E_+$ into Eqs. (18.31) yields $b = a$ and substituting $E = E_-$ gives $b = -a$.

Therefore there are two spatial solutions to the wave equation, one symmetric under particle interchange,

$$\psi_+(\mathbf{x}_1, \mathbf{x}_2) = \begin{cases} \dfrac{1}{\sqrt{2}}[\psi_1(\mathbf{x}_1)\psi_2(\mathbf{x}_2) + \psi_2(\mathbf{x}_1)\psi_1(\mathbf{x}_2)] & \text{if } \psi_1(\mathbf{x}) \neq \psi_2(\mathbf{x}) \\ & ; \\ \psi_1(\mathbf{x}_1)\psi_2(\mathbf{x}_2) & \text{if } \psi_1(\mathbf{x}) = \psi_2(\mathbf{x}), \end{cases}$$

$$E_+ = E_{n_1} + E_{n_2} + K + A, \tag{18.33}$$

and one antisymmetric under particle interchange

$$\psi_-(\mathbf{x}_1, \mathbf{x}_2) = \frac{1}{\sqrt{2}}[\psi_1(\mathbf{x}_1)\psi_2(\mathbf{x}_2) - \psi_2(\mathbf{x}_1)\psi_1(\mathbf{x}_2)],$$

$$E_- = E_{n_1} + E_{n_2} + K - A. \tag{18.34}$$

In order that the overall wave function, including spins, be antisymmetric, the first of these must be a singlet state (antisymmetric in spin variables) and the second must be a triplet state (symmetric in spin variables) as we saw in the last section. We see quite clearly that the triplet state indeed does have the lower energy E_- when the single-particle spatial wave functions are distinct. The difference between singlet and triplet energy levels E_+ and E_- is $2A$; let us seek an interpretation for the quantity A and also for K. Note that when the single-particle states are identical, $K = A$.

The matrix of the perturbation energy is

$$\begin{bmatrix} K & A \\ A & K \end{bmatrix},$$

and above we diagonalized this matrix. The integral K, being on the diagonal of the original perturbation matrix, is called the *direct energy* and can be thought of as the mean energy contributed by the direct Coulomb interaction of the electric charges of the two electrons in their single-particle states. If we define

$$\rho_{ij}(\mathbf{x}) \equiv \psi_i^*(\mathbf{x})\psi_j(\mathbf{x}) \tag{18.35}$$

we obtain

$$K = \frac{1}{4\pi\epsilon_0} \int \frac{e\rho_{11}(\mathbf{x}_1) e\rho_{22}(\mathbf{x}_2)}{|\mathbf{x}_2 - \mathbf{x}_1|} d^3x_1\, d^3x_2, \tag{18.36}$$

which is the sum of all the elementary coulomb interactions between elements of charge $e\rho_{11}(\mathbf{x}_i) d^3x_1$ and $e\rho_{22}(\mathbf{x}_2)\, d^3x_2$.

The quantity A is similar, except that for it we deal not with the charge density $\rho_{ii}(\mathbf{x})$ for a particle in a single state, but rather with *exchange densities* $\rho_{ij}(\mathbf{x})$ with $i \neq j$; the appearance of such *densities* is evidently a consequence of the identity of the two particles and with our inability to determine which particle is in which state. Thus we have

$$A = \frac{1}{4\pi\epsilon_0} \int \frac{e\rho_{12}(\mathbf{x}_1) e\rho_{12}^*(\mathbf{x}_2)}{|\mathbf{x}_2 - \mathbf{x}_1|} d^3x_1\, d^3x_2, \tag{18.37}$$

where $\rho_{12}^*(\mathbf{x}) = \rho_{21}(\mathbf{x})$. This *exchange energy* is responsible for the splitting of the singlet and triplet (parahelium and orthohelium) states in the helium atom.

This discussion has centered around the qualitative features of the helium atom problem, and we have chosen an approach that is not particularly accurate in the lowest orders of perturbation theory. Also, we have neglected the degeneracy resulting from the existence of several m values for each orbital angular momentum quantum number l; in general, the perturbation matrix may be much larger than the 2×2 matrix we considered. Careful application of this approach in higher orders of perturbation theory gives good agreement with experimental observation of helium energy levels, and some other approaches give good agreement while requiring less calculation. The importance of good quantitative agreement between experiment and the perturbation calculation is that in the perturbation calculation there are no adjustable parameters; this gives us considerable confidence that the theory is correct—that we understand the helium atom in terms of the hydrogen atom. The basis of that understanding is to be found in the direct and exchange integrals K and A that we obtained simply from first-order degenerate perturbation theory.

18.6 THE PERIODIC TABLE

Our success with the helium atom suggests that we might be able to understand, at least qualitatively, the periodic table in terms of the central features of one- and two-electron atoms. Since atoms are electrically neutral, the number of electrons in an atom must be equal to the number of protons in the atomic nucleus, this number being denoted Z. The mass of the atom is determined almost entirely by the nucleus, which contains Z protons and $A - Z$ neutrons, or A nucleons; A is called the atomic-mass number, and Z is the atomic number of an atom. The chemical behavior of an atomic species is determined by Z, and hence when there are several nuclei having the same Z but different A (isotopes) they have essentially the same chemical behavior. The stable nuclear species are clustered around the line $A = 2Z$ on an A–Z plot, but tend to stay above this line for large Z; i.e. heavy nuclides tend to have more neutrons than protons.

The periodic table is based upon chemical behavior and hence we shall be concerned only with the atomic number Z. Now the Pauli principle requires that there be no more than one electron in each one-electron state. Furthermore, we should expect the low-lying energy levels to be filled first, and as Z gets larger, levels of higher and higher energies will be filled. To start with, we assume the energy levels of the electrons are essentially those of the one-electron atom, considering interparticle interactions as perturbations that affect the locations of energy levels but not their ordering. Eventually, however, as the number of electrons increases, it may be expected that the energy contributed by interelectron coulomb interactions will compete with the energy contributed by increasing the orbital angular momenta of the highest-lying electrons, and when this happens, the ordering of energy levels will be altered somewhat.

For a one-electron atom, the relevant quantum numbers are n, l, m and μ, the last being the electron spin magnetic quantum number. There are two values of μ for each fixed set of values for n, l, m; there are $2l + 1$ values of m for each l and there are n possible values of l. Hence for a fixed n we have

$$2 \sum_{l=0}^{n-1} (2l + 1) = 2n^2$$

available one-electron states. Table 18.1 shows the number of states available for each n and for all values of n below some fixed value n_0. In practice, the useful part of this table is limited by the maximum atomic number Z of existing atomic nuclei, which happens to be slightly over 100. It is customary to call a set of states with fixed n a *shell*; here we have listed the first five shells.

Table 18.1

n_0	States with $n = n_0$	States with $n \leq n_0$
1	2	2
2	8	10
3	18	28
4	32	60
5	50	110
⋮	⋮	⋮

When a shell is filled, there are two electrons in each orbital angular momentum state, and they have opposite spin magnetic quantum numbers. Also, for each electron with orbital magnetic quantum number m there is one with magnetic quantum number $-m$. This suggests that the total angular momentum of the electrons in a closed shell should be 0 and hence that the charge distribution for electrons in the shell should be spherically symmetric and that the magnetic moment for the shell should vanish. If this were the case, atoms with $Z = 2$, 10, 28, 60 and 110, according to our simple picture, would have closed-shell electron structures and should be particularly resistant to reaction with other atoms. This is because, according to Gauss's theorem, the spherically symmetric charge distribution outside the small, positive nucleus leads to a vanishing electric field outside the atom.

Comparison of this prediction with the periodic table is quite satisfying when we note that atoms with $Z = 2$ and $Z = 10$ are helium and neon, two so-called inert gases. However, $Z = 28$ is nickel which is not inert and which lies about midway between the inert gases argon ($Z = 18$) and krypton ($Z = 36$). Our "magic numbers" fail after $Z = 10$. However, it is encouraging that argon, the next inert gas following neon in the periodic table, has 8 more electrons than neon; perhaps this is because the $l = 2$ levels in the $n = 3$ shell are particularly hard to fill because of the interelectronic repulsion among the electrons. After all, the $n = 3$, $l = 0$ and $n = 3$, $l = 1$ *subshells* individually should be spherically symmetric and hence might provide quite nonreactive electron structures.

This suggests that we consider the numbers of electrons in atoms with closed subshells rather than with closed shells, and that we see which of these correspond to inert gases. We then may seek to discover why the other closed subshells, and particularly closed shells, do not correspond to inert gases. It should be mentioned here that the reason that we naïvely expect closed shells to be particularly important is that all the electrons in a given shell have the same mean distance from the nucleus if interelectronic forces and other perturbations are neglected. In the zeroth approximation, this means that the addition of a new electron through interaction with another atom would be quite unlikely because the new electron would have small probability of penetrating the charge cloud of the closed shell, the radius of which is somewhat smaller than that of the shell in which the additional electron would reside. Furthermore, the removal of an electron is improbable because the electrons from another atom would be unlikely to penetrate the closed-shell charge cloud in order to expose that bound electron to the positive nuclear charge of the other atom. That closed shells do not always yield inert gases simply means that the mean radius depends on other quantum numbers besides n when interelectronic forces are taken into account.

The closed subshells correspond to $Z = 2, 4, 10, 12, 18, 28, 30, 36, 46, 60, 62, 68, 78, 92, 110, \ldots$. Of these, the inert gases correspond to $Z = 2$, 10, 18 and 36; above $Z = 36$, things seem to get out of hand in terms of our simple picture, which is hardly surprising. It turns out that this just reflects quite drastic rearrangements of the energy levels as the number of electrons increases. Rather sophisticated analysis and enormously complicated computations are required to carry the theory even qualitatively beyond the first two shells, but it is quite clear that we are on the right track.

Table 18.2. Order of filling Single-particle States in the Periodic Table
Upper right: scheme becomes apparent when n and l are added

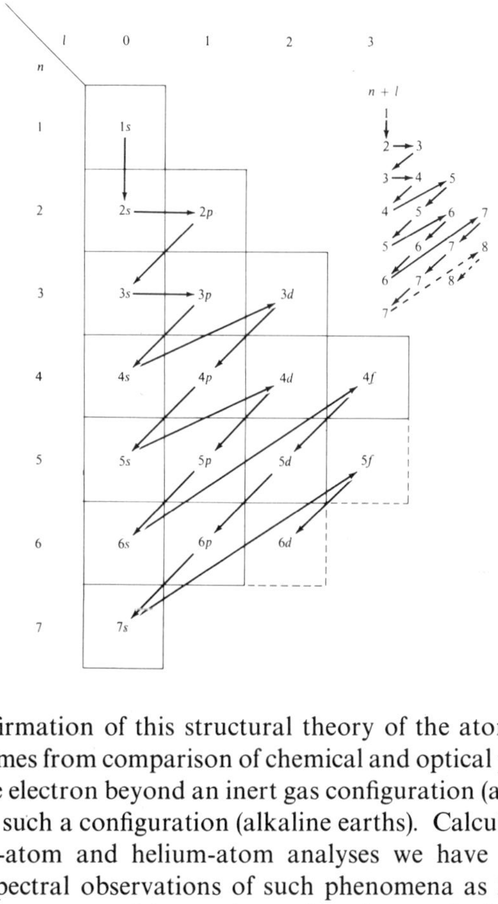

Further confirmation of this structural theory of the atom in terms of one-electron states comes from comparison of chemical and optical properties of atoms that have just one electron beyond an inert gas configuration (alkali atoms) or two electrons beyond such a configuration (alkaline earths). Calculations very similar to the hydrogen-atom and helium-atom analyses we have given before yield predictions for spectral observations of such phenomena as Zeeman and Stark effects, and the agreement with experiment is quite good. Also, the qualitative features of the effects are quite similar from one inert-gas configuration to another. Corresponding tests of the theory for one electron less than an inert gas, or two less, have been conducted, and again the theory stands up quite well.

A useful mnemonic device for following the order of filling of subshells is given in Table 18.2; Table 18.3 gives a detailed summary of the filling of shells, which should be compared with the periodic table, Table 18.4.

18.7 MANY-ELECTRON ATOMS; THE HARTREE–FOCK MODEL

Although the perturbation-theoretic approach may be exploited in analysing the two-electron atom, it is not readily applied to three and more electrons. The most fruitful method for obtaining reasonably precise theoretical predictions for many-electron atoms is the variational technique, which utilizes the fact that the expectation value of the ground-state energy of a system is minimal when the exact wave function (or state vector) for the system is used. This leads to the Hartree–Fock self-consistent model for many-electron atoms. It is not a strictly theoretical model, and does not use only first principles in its formulation; rather it is semi-empirical in that it utilizes observed features of the periodic table to determine which single-particle states in an atom will be filled in a given atom in its ground state.

This is not the appropriate place for a full discussion of the Ritz variational method upon which the Hartree–Fock approach is based, but we give instead a brief plausibility argument. Let Q be a Hermitian operator, and let $\psi(\mathbf{x})$ be a wave function for a system; we assume $\psi(\mathbf{x})$ to be properly normalized,

$$\int_V \psi^*(\mathbf{x})\psi(\mathbf{x})\, d^3x = 1. \tag{18.38}$$

Now consider the integral

$$\int_V \psi^*(\mathbf{x})\mathsf{Q}\psi(\mathbf{x})\, d^3x$$

when, for convenience, V is a finite volume; we wish to minimize the value of this integral subject to the condition (18.38). Therefore, we must have

$$\delta\left[\int_V \psi^*(\mathbf{x})\mathsf{Q}\psi(\mathbf{x})\, d^3x - \lambda \int_V \psi^*(\mathbf{x})\psi(\mathbf{x})\, d^3x\right] = 0$$

where λ is for now an undetermined constant that plays the role of a Lagrangian multiplier. That is, we want any small change in $\psi(\mathbf{x})$ or $\psi^*(\mathbf{x})$ to leave the integral in brackets unchanged to first order in $\delta\psi(\mathbf{x})$ or $\delta\psi^*(\mathbf{x})$.

The variation symbol can be moved inside the integrals since the limits on the integrals are to remain unchanged; only the form of ψ or ψ^* is to change. Hence, we require that

$$\int \{\delta\psi^*(\mathbf{x})[\mathsf{Q} - \lambda]\psi(\mathbf{x}) + \psi^*(\mathbf{x})[\mathsf{Q} - \lambda]\delta\psi(\mathbf{x})\}\, d^3x = 0.$$

Since the variations $\delta\psi(\mathbf{x})$ and $\delta\psi^*(\mathbf{x})$ are to be considered independent and

Table 18.3. The Distribution of Electrons in Atoms*

Element	*K*	*L*		*M*			*N*		Ground term	Ionization potential, eV
	1,0 1*s*	2,0 2*s*	2,1 2*p*	3,0 3*s*	3,1 3*p*	3,2 3*d*	4,0 4*s*	4,1 4*p*		
H 1	1	—	—	—	—	—	—	—	$^2S_{1/2}$	13.539
He 2	2	—	—	—	—	—	—	—	$^1S_0$	24.45
Li 3	2	1	—	—	—	—	—	—	$^2S_{1/2}$	5.37
Be 4	2	2	—	—	—	—	—	—	$^1S_0$	9.48
B 5	2	2	1	—	—	—	—	—	$^2P_{1/2}$	8.4
C 6	2	2	2	—	—	—	—	—	$^3P_0$	11.217
N 7	2	2	3	—	—	—	—	—	$^4S_{3/2}$	14.47
O 8	2	2	4	—	—	—	—	—	$^3P_2$	13.56
F 9	2	2	5	—	—	—	—	—	$^2P_{3/2}$	18.6
Ne 10	2	2	6	—	—	—	—	—	$^1S_0$	21.5
Na 11				1	—	—	—	—	$^2S_{1/2}$	5.12
Mg 12				2	—	—	—	—	$^1S_0$	7.61
Al 13				2	1	—	—	—	$^2P_{1/2}$	5.96
Si 14		Neon configuration		2	2	—	—	—	$^3P_0$	7.39
P 15				2	3	—	—	—	$^4S_{3/2}$	10.3
S 16				2	4	—	—	—	$^3P_2$	10.31
Cl 17				2	5	—	—	—	$^2P_{3/2}$	12.96
Ar 18				2	6	—	—	—	$^1S_0$	15.69
K 19						—	1	—	$^2S_{1/2}$	4.32
Ca 20						—	2	—	$^1S_0$	6.09
Sc 21						1	2	—	$^2D_{3/2}$	6.57
Ti 22				Argon configuration		2	2	—	$^3F_2$	6.80
V 23						3	2	—	$^4F_{3/2}$	6.76
Cr 24						5	1	—	$^7F_3$	6.74
Mn 25						5	2	—	$^6S_{5/2}$	7.40
Fe 26						6	2	—	$^5D_4$	7.83
Co 27						7	2	—	$^4F_{9/2}$	7.81
Ni 28						8	2	—	$^3F_4$	7.606
Cu 29						10	1	—	$^2S_{1/2}$	7.69
Zn 30						10	2	—	$^1S_0$	9.35
Ga 31						10	2	1	$^2P_{1/2}$	5.97
Ge 32						10	2	2	$^3P_0$	7.85
As 33						10	2	3	$^4S_{3/2}$	9.4
Se 34						10	2	4	$^3P_2$	
Br 35						10	2	5	$^2P_{3/2}$	11.80
Kr 36						10	2	6	$^1S_0$	13.940

* From D. I. Blokhintsev, *Principles of Quantum Mechanics*, Allyn and Bacon, Inc., Boston (1964). Reprinted through permission of the publisher.

Table 18.3 (continued)

Element	Configuration of internal layers	N		O			P	Ground term	Ionization potential, eV
		4,2 4*d*	4,3 4*f*	5,0 5*s*	5,1 5*p*	5,2 5*d*	6,0 6*s*		
Rb 37	Krypton configuration	—	—	1	—	—	—	$^2S_{1/2}$	4.16
Sr 38		—	—	2	—	—	—	$^1S_0$	5.67
Y 39		1	—	2	—	—	—	$^2D_{3/2}$	6.5
Zr 40		2	—	2	—	—	—	$^3F_2$	
Nb 41		4	—	1	—	—	—	$^6D_{1/2}$	
Mo 42		5	—	1	—	—	—	$^7S_3$	7.35
Tc 43		6	—	1	—	—	—	$^6S_{5/2}$	
Ru 44		7	—	1	—	—	—	$^5F_5$	7.7
Rh 45		8	—	1	—	—	—	$^4F_{9/2}$	7.7
Pd 46		10	—	—	—	—	—	$^1S_0$	8.5
Ag 47	Palladium configuration		—	1	—	—	—	$^2S_{1/2}$	7.54
Cd 48			—	2	—	—	—	$^1S_0$	8.95
In 49			—	2	1	—	—	$^2P_{1/2}$	5.76
Sn 50			—	2	2	—	—	$^3P_0$	7.37
Sb 51			—	2	3	—	—	$^4S_{3/2}$	8.5
Te 52			—	2	4	—	—	$^3P_2$	
J 53			—	2	5	—	—	$^2P_{3/2}$	10.44
Xe 54			—	2	6	—	—	$^1S_0$	12.078
Cs 55	46 electrons between 1*s* and 4*d*		—	8 electrons between 5*s* and 5*p*		—	1	$^2S_{1/2}$	3.88
Ba 56			—			—	2	$^1S_0$	5.19
La 57			—			1	2	$^2D_{3/2}$	
Ce 58			2			—	2	^3H	
Pr 59			3			—	2	^4I	
Nd 60			4			—	2	^5I	
Pm 61			5			—	2	^6H	
Sm 62			6			—	2	^7F	
Eu 63			7			—	2	^8S	
Gd 64			7			1	2	^9D	
Tb 65			9			—	2	^6H	
Dy 66			10			—	2	^5I	
Ho 67			11			—	2	^4I	
Er 68			12			—	2	^3H	
Tu 69			13			—	2	^2F	
Yo 70			14			—	2	^1S	
Lu 71			14			1	2	$^3D_{3/2}$	

Table 18.3 (continued)

Element	Configuration of internal layers	*O* 5,2 5*d*	*O* 5,3 5*f*	*P* 6,0 6*s*	*P* 6,1 6*p*	*P* 6,2 6*d*	*Q* 7,0 7*s*	Ground term	Ionization potential, eV
Nf 72	68 electrons between 1*s* and 5*p*	2	—	2	—	—	—	$^3F_2$	
Ta 73		3	—	2	—	—	—	$^4F_{3/2}$	
W 74		4	—	2	—	—	—	$^5D_0$	
Re 75		5	—	2	—	—	—	$^6S_{5/2}$	
Os 76		6	—	2	—	—	—	$^5D_4$	
Ir 77		7	—	2	—	—	—	$^4F_{9/2}$	
Pt 78		8	—	2	—	—	—	$^3D_3$	
Au 79	78 electrons between 1*s* and 5*d*		—	1	—	—	—	$^2S_{1/2}$	9.20
Hg 80			—	2	—	—	—	$^1S_0$	10.39
Tl 81			—	2	1	—	—	$^2P_{1/2}$	6.08
Pb 82			—	2	2	—	—	$^3P_0$	7.39
Bi 83			—	2	3	—	—	$^4S_{3/2}$	8.0
Po 84			—	2	4	—	—	$^3P_2$	
At 85			—	2	5	—	—	$^2P_{3/2}$	
Rn 86			—	2	6	—	—	$^1S_0$	10.689
Fr 87	78 electrons between 1*s* and 5*d*		—	2	6	—	1	$^2S_{1/2}$	
Ra 88			—	2	6	—	2	$^1S_0$	
Ac 89			—	2	6	1	2	$^2D_{3/2}$	
Th 90			—	—	—	2	2	^3F	
Pa 91			2	2	6	1	2	^4K	
U 92			3	2	6	1	2	^5L	
Np 93			4	2	6	—	2	^6M	
Pu 94			6	2	6	—	2	^7F	
Am 95			7	2	6	—	2	^8S	
Cm 96			7	2	6	1	2	^9D	
Bk 97			8	2	6	1	2		
Cf 98			10	2	6	—	2		
Es 99			11	2	6	—	2		
Fm 100			12	2	6	—	2		
Md 101			13	2	6	—	2		
102			14	2	6	—	2		
103			14	2	6	1	2		

Table 18.4. The Periodic Table of Elements*

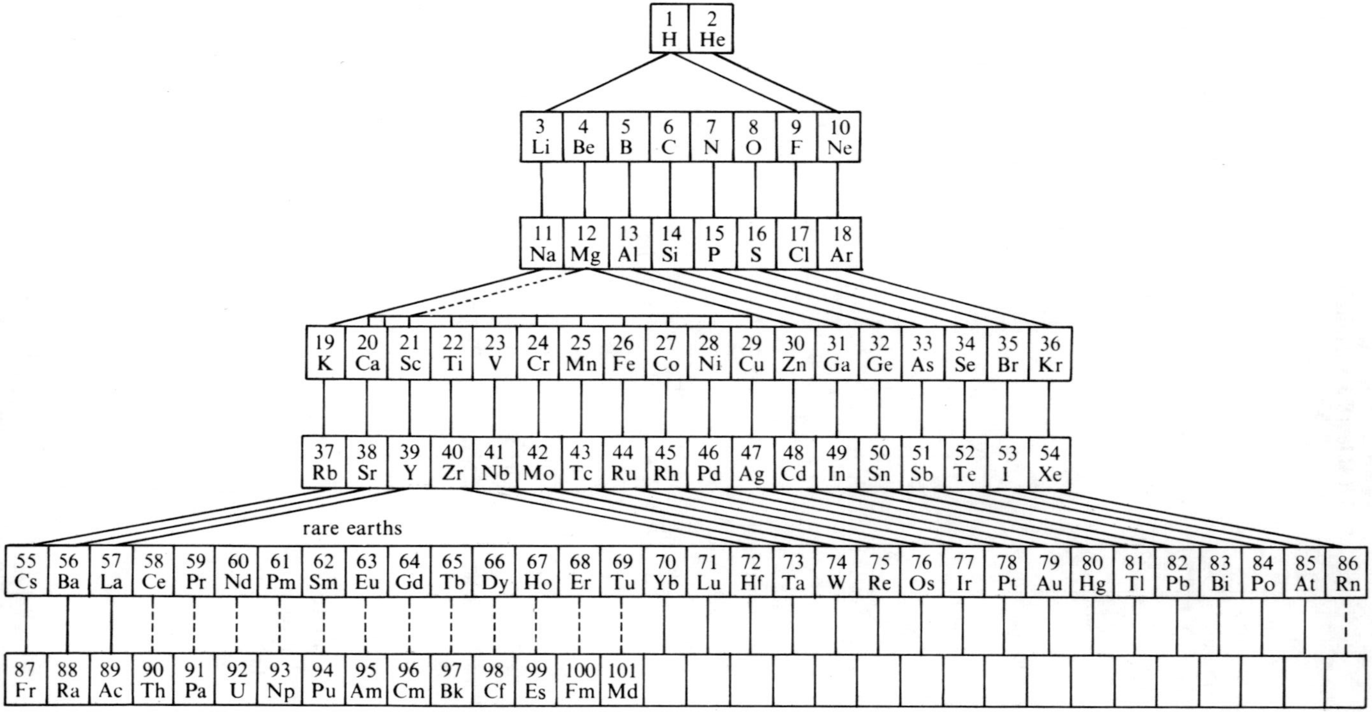

* From D. I. BLOKHINTSEV, *Principles of Quantum Mechanics*, Allyn and Bacon, Inc., Boston (1964). Reprinted through permission of the publisher.

arbitrary, their coefficients must vanish; this means that

$$\mathsf{Q}\psi(\mathbf{x}) = \lambda\psi(\mathbf{x}) \tag{18.39a}$$

and, since Q is Hermitian,

$$\psi^*(\mathbf{x})\mathsf{Q}^\dagger = \lambda\psi^*(\mathbf{x}) \tag{18.39b}$$

when $\psi(\mathbf{x})$ is the wave function that minimizes the integral. From this we conclude that the $\psi(\mathbf{x})$ that minimizes $\int_V \psi^*\mathsf{Q}\psi\, d^3x$ is an eigenfunction of Q and that the appropriate value of λ is an eigenvalue of Q. In particular, the minimized integral will be smallest when λ is the smallest possible eigenvalue λ_0 of Q; hence in general

$$\int_V \varphi^*(\mathbf{x})\mathsf{Q}\varphi(\mathbf{x})\, d^3x \geq \lambda_0. \tag{18.40}$$

We have argued here that a necessary condition for $\int_V \psi^*\mathsf{Q}\psi\, d^3x$ to be minimal is that ψ be an eigenfunction of Q and best of all, it should be the eigenfunction with the smallest eigenvalue λ_0. We have avoided the strictly mathematical questions of whether the integrals or their minima exist; we must assume that they do.

To apply this to the many-electron atom, we let Q be the many-particle Hamiltonian

$$\begin{aligned} \mathsf{H} &= \sum_{i=1}^{Z} \frac{1}{2M}\mathsf{P}_i^2 - \sum_{i=1}^{Z} \frac{Ze^2}{4\pi\epsilon_0 r_i} + \sum_{i<j} \frac{e^2}{4\pi\epsilon_0 r_{ij}} \\ &= \mathsf{F} + \mathsf{G} \end{aligned} \tag{18.41}$$

where

$$\mathsf{F} = \sum_{i=1}^{Z} \mathsf{F}_i, \qquad \mathsf{F}_i = \frac{1}{2M}\mathsf{P}_i^2 - \frac{Ze^2}{4\pi\epsilon_0 r_i} \tag{18.42a}$$

$$\mathsf{G} = \sum_{i<j} \mathsf{G}_{ij}, \qquad \mathsf{G}_{ij} = \frac{e^2}{4\pi\epsilon_0}\frac{1}{r_{ij}}. \tag{18.42b}$$

We then write the wave function for the electrons in the form

$$\Psi_{1,\ldots,Z^-}(\mathbf{x}_1, \ldots, \mathbf{x}_Z) = \frac{1}{\sqrt{Z!}} \sum_P (-)^P \mathsf{P} \prod_{i=1}^{Z} \psi_i(\mathbf{x}_i)$$

where the $\psi_i(\mathbf{x}_i) = R_{n_i l_i}(r_i) Y_{l_i}^{m_i}(\theta_i, \varphi_i)\chi_{\mu_i}(s_i)$ are single-particle wave functions in a spherically symmetric potential. The procedure will be to vary the radial factors in these single-particle functions. This will yield coupled wave equations for the single-particle wave functions. Unfortunately, these coupled equations require knowledge of matrix elements of the potential before they can be solved; for the purpose of calculating these matrix elements, we use the hydrogen atom wave functions for the single-particle states that are known to be filled in the ground state of the Z-electron atom. The coupled equations are then solved to yield new single-particle wave functions for the Z electrons. These wave functions then replace the hydrogen-atom wave functions in the matrix element; new coupled equations are obtained; and new, presumably more accurate, single-particle wave functions are obtained. This procedure continues until one is happy with the precision of his estimated wave functions.

We now outline the mathematical method. First, consider

$$(\Psi_{1,\ldots,Z-}(\mathbf{x}_1,\ldots,\mathbf{x}_Z), \mathsf{F}\Psi_{i,\ldots,Z-}(\mathbf{x}_1,\ldots,\mathbf{x}_Z))$$

$$= \sum_{i=1}^{Z} (\Psi_{1,\ldots,Z-}, \mathsf{F}_i\Psi_{1,\ldots,Z-});$$

but if $i \neq j$, $(\Psi_{1,\ldots,Z-}, \mathsf{F}_i\Psi_{1,\ldots,Z-}) = (\Psi_{1,\ldots,Z-}, \mathsf{F}_j\Psi_{1,\ldots,Z-})$ because replacing i with j simply amounts to interchanging the roles of the ith and jth particles, i.e. to replacing $\Psi_{1,\ldots,Z-}$ with $-\Psi_{1,\ldots,Z-}$. Hence

$$(\Psi_{1,\ldots,Z-}(\mathbf{x}_1,\ldots,\mathbf{x}_Z), \mathsf{F}\Psi_{1,\ldots,Z-}(\mathbf{x}_1,\ldots,\mathbf{x}_Z)) = Z(\Psi_{1,\ldots,Z-}, \mathsf{F}_1\Psi_{1,\ldots,Z-}).$$

However,

$$(\Psi_{1,\ldots,Z-}, \mathsf{F}_1\Psi_{1,\ldots,Z-})$$

$$= \frac{1}{Z!}\int d^3x_1 \ldots \int d^3x_Z \sum_P (-)^P \mathsf{P}\psi_1(\mathbf{x}_1)\ldots\psi_Z(\mathbf{x}_Z)$$

$$\cdot\, \mathsf{F}_1 \sum_{P'} (-)^{P'}\, \mathsf{P}\psi_1(\mathbf{x}_1)\ldots\psi_Z(\mathbf{x}_Z)$$

$$= \frac{1}{Z!}\sum_{i=1}^{Z} (\psi_i(\mathbf{x}_1), \mathsf{F}_1\psi_i(\mathbf{x}_1))(Z-1)!$$

$$= \frac{1}{Z}\sum_{i=1}^{Z} (\psi_i, \mathsf{F}_1\psi_i)$$

where we have used

$$(\psi_j, \psi_k) = \int d^3x\, \psi_j^*(\mathbf{x})\psi_k(\mathbf{x}) = \delta_{j,k};$$

hence

$$(\Psi_{1,\ldots,Z-}, \mathsf{F}\Psi_{1,\ldots,Z-}) = \sum_{i=1}^{Z} (\psi_i, \mathsf{F}_1\psi_i).$$

Careful study of the matrix element $(\Psi_{1,\ldots,Z-}, \mathsf{G}\Psi_{1,\ldots,Z-})$ shows, in a similar way, that

$$(\Psi_{1,\ldots,Z-}, \mathsf{G}\Psi_{1,\ldots,Z-}) = \sum_{i<j} \big[(\psi_i(\mathbf{x}_1)\psi_j(\mathbf{x}_2), \mathsf{G}_{12}\psi_i(\mathbf{x}_1)\psi_j(\mathbf{x}_2)) - (\psi_i(\mathbf{x}_1)\psi_j(\mathbf{x}_2), \mathsf{G}_{12}\psi_i(\mathbf{x}_2)\,\psi_j(\mathbf{x}_1))\big].$$

Here the first term is the direct integral and second is the exchange integral for the two-body operator $\mathsf{G}_{12} = e^2/4\pi\epsilon_0 r_{12}$.

Now we vary the integral

$$\int \Psi^*_{1,\ldots,Z-}(\mathbf{x}_1, \ldots, \mathbf{x}_Z)\mathsf{H}\Psi_{1,\ldots,Z-}(\mathbf{x}_1, \ldots, \mathbf{x}_Z)\, d^3x_1 \cdots d^3x_Z$$

$$-E \int \Psi^*_{1,\ldots,Z-}(\mathbf{x}_1, \ldots, \mathbf{x}_Z)\Psi_{1\,\ldots,Z-}(\mathbf{x}_1, \ldots, \mathbf{x}_Z)\, d^3x_1 \cdots d^3x_Z$$

with respect to each $\psi_i(\mathbf{x})$, using in H the result for the matrix elements of F and G obtained above. The resulting coupled equations are

$$\left[-\frac{h^2}{2M}\nabla_i^2 - \frac{Ze^2}{4\pi\epsilon_0 r_i} + \sum_{j\neq i}\left(\psi_j, \frac{e^2}{4\pi\epsilon_0 r_{ij}}\psi_j\right) - E_i\right]\psi_i(\mathbf{x}_i) = \sum_{j\neq i}\left(\psi_j, \frac{e^2}{4\pi\epsilon_0 r_{ij}}\psi_i\right)\psi_j(\mathbf{x}_i). \tag{18.43}$$

The term on the right-hand side of these Hartree–Fock equations arises from the exchange integral. We have suppressed spin variables in writing out this expression, but nevertheless spins are involved in the calculation. Of particular interest is the fact that the exchange integral must vanish unless the spins of state i and state j are parallel, since the Coulomb potential does not involve spin operators. Therefore, as far as atomic electrons in this model are concerned, the identity of particles produces exchange effects only when the electrons are in the same spin states; this is as we should expect.

The computational procedure now would be to calculate $(\psi_j, (1/r_{ij})\psi_i)$ for all i, j, using the hydrogen atom radial wave functions, and with these estimates for the potential matrix elements, to solve the set of equations for $\psi_i(\mathbf{x})$. Inserting these results in the matrix elements followed by recalculation of $\psi_i(\mathbf{x})$ should improve the estimate of $\psi_i(\mathbf{x})$ (as well as of the eigenvalues E_i of the single-particle energies), etc. This self-consistent procedure has been very important in confirming quantum theory as the appropriate foundation for the study of atomic structure. Also, from it one obtains the best available single-particle wave functions for the study of many-electron atoms.

18.8 ELEMENTS OF QUANTUM STATISTICAL MECHANICS

The preceding discussion of atomic systems was intended as a brief though rather incomplete illustration of the application of the principle of identity of particles to an actual physical problem. Let us now take a more general view of things, focusing on some questions that might be raised by a person more interested in statistical mechanics than in atomic theory. For example, what is the probability that n particles will be distributed, in particular numbers, to particular ones among z available physical states?

The question, as stated, is incomplete; we get different answers depending on what additional assumptions we make. First consider n distinguishable particles that can be put into any of the z states with equal probability; what happens to one particle does not affect what happens to any other. Now there are z states into which particle 1 may be put, z states into which particle 2 may be put, ..., and z states into which particle n may be put. Hence there are z^n different ways of distributing the n particles, each being equally probable; therefore the probability of any particular arrangement is $1/z^n$. Since we don't care *which* particles went into which states, but rather care only that certain states be occupied by certain numbers of particles, we aren't really interested in the ordering of the particles $1, \ldots, n$; hence there are $n!$ different arrangements that have n_1 particles in state 1, n_2 particles in state 2, ..., n_z particles in state z, with $\sum_{i=1}^{z} n_i = n$. Therefore the probability of having a given partition $n_1, n_2, \ldots, n_z$ of the n particles is simply

$$p_{\mathrm{MB}} = \frac{n!}{z^n}$$

when the particles are distinguishable and no restriction is placed upon the number of particles in a given state. This is the probability for Maxwell–Boltzmann statistics.

Now let us drop the assumption that the particles are distinguishable, retaining, however, the assumption that the fate of one particle in no way affects that of another. This is the behavior of identical bosons, there being no exclusion principle to control the number of them occupying any given state. We may think of the z states in terms of $z + 1$ walls or partitions separating the states. The n particles and $z + 1$ walls now may be thought of as lying in a row, the number of particles in a given state being the number lying between the walls separating that state from the others; there must be a wall fixed at each end of the row. Now there are $(n + z - 1)!$ ways of permuting the n particles and $z - 1$ movable walls, but the $(z - 1)!$ permutations of walls must not be counted since the ordering of walls serves to label the z states. Therefore there are $(n + z - 1)!/(z - 1)!$ different ways of arranging the n particles; but since the particles are indistinguishable, the $n!$ permutations of particles leads to equivalent distributions of particles. Finally, then, there are

$$\frac{(n + z - 1)!}{n!(z - 1)!} = \binom{n + z - 1}{n}$$

different and equally probable ways of distributing the n identical particles among z states when no exclusion principle is operating, and the probability of any one arrangement is

$$p_{\text{BE}} = \frac{1}{\binom{n + z - 1}{n}} = \frac{n!(z - 1)!}{(n + z - 1)!}.$$

This is the probability for Bose–Einstein statistics.

If we have identical particles for which the exclusion principle operates, i.e. fermions, we must have $z \geq n$ since only one particle is allowed in each state. There are z ways of assigning the first particle, $z - 1$ ways of assigning the second, $\ldots$, and $z - n + 1$ ways of assigning the nth particle to a state. However, the $n!$ different orderings of this procedure are equivalent since the particles are indistinguishable. There are, then,

$$\frac{z(z - 1)\cdots(z - n + 1)!}{n!} = \frac{z!}{n!(z - n)!} = \binom{z}{n}$$

different equally probable ways of distributing n fermions among z states, the probability of any one such distribution, therefore, being

$$p_{\text{FD}} = \frac{1}{\binom{z}{n}} = \frac{n!(z - n)!}{z!}.$$

This is the probability for Fermi–Dirac statistics.

From a physical point of view, we are interested primarily in how particles are distributed among states with various energies. Let us assume that there are Z_i states with energy $E_i, i = 1, 2, \ldots$, and that there are n_i particles distributed among these Z_i states. The probability of having n_1 particles with energy $E_1, \ldots, n_k$ particles with energy E_k, etc. then is

$$P = \prod_{i=1}^{\infty} p_i,$$

where

$$p_i = \begin{cases} p_{\mathrm{MB}i} = n_i!/Z_i^{n_i} & \text{for distinguishable particles} \\ p_{\mathrm{BE}i} = 1\Big/\dbinom{n_i + Z_i - 1}{n_i} & \text{for identical bosons} \\ p_{\mathrm{FD}i} = 1\Big/\dbinom{Z_i}{n_i} & \text{for identical fermions.} \end{cases}$$

If the total number of particles is N, we have the constraint $\sum_i n_i = N$; also, the total energy is $E = \sum_i n_i E_i$.

What set of values $(n_1, n_2, \ldots, n_k, \ldots)$ will maximize P? That is, what is the most probable distribution of N particles among the energy levels? To find out, we use the method of Lagrangian multipliers and vary

$$Q \equiv \ln P + \alpha N + \beta E$$
$$= \sum_i (\ln p_i + \alpha n_i + \beta n_i E_i)$$

with respect to n_k, $k = 1, 2, \ldots$. Maximizing $\ln P$ corresponds to maximizing P itself, and use of $\ln P$ allows us to bring together the constraints and the quantity to be maximized in a convenient way. Differentiation of Q with respect to n_k gives

$$\frac{dQ}{dn_k} = \frac{d}{dn_k} \ln p_k + \alpha + \beta E_k$$

and the appropriate value of n_k is that which makes this quantity vanish; evidently it depends upon E_k, Z_k, α and β, the latter being as yet undetermined constants.

To find $\ln p_k$ and $d(\ln p_k)/dn_k$ we must resort to the expressions for p_k given above for the three particular cases. Each of these expressions involves $n_k!$, however, and the derivative of $\ln (n_k!)$ is not an elementary function. To make matters relatively simple, we assume that n_k and $Z_k - n_k$ are very large integers so that we may use the approximation

$$n! \underset{n \to \infty}{\sim} \sqrt{2\pi}\, n^{n+1/2} e^{-n},$$

known as Stirling's formula. With this we obtain

$$\frac{d}{dn_k} \ln (n_k!) \simeq \ln n_k + \frac{1}{2n_k} \simeq \ln n_k$$

$$\frac{d}{dn_k} \ln (Z_k + n_k - 1)! \simeq \ln (Z_k + n_k - 1) + \frac{1}{2(Z_k + n_k - 1)} \simeq \ln (Z_k + n_k - 1)$$

$$\frac{d}{dn_k} \ln (Z_k - n_k)! \simeq -\ln (Z_k - n_k) - \frac{1}{2(Z_k - n_k)} \simeq -\ln (Z_k - n_k)$$

and the three maximization criteria

$$\ln n_k - \ln Z_k + \alpha + \beta E_k = 0 \qquad \text{(MB)}$$

$$\ln n_k - \ln (Z_k + n_k - 1) + \alpha + \beta E_k = 0 \qquad \text{(BE)}$$

$$\ln n_k - \ln (Z_k - n_k) + \alpha + \beta E_k = 0. \qquad \text{(FD)}$$

Thus

$$n_k = \begin{cases} Z_k e^{-\alpha - \beta E_k} & \text{(MB)} \quad (18.44a) \\ \dfrac{Z_k}{e^{\alpha + \beta E_k} - 1} & \text{(BE)} \quad (18.44b) \\ \dfrac{Z_k}{e^{\alpha + \beta E_k} + 1} & \text{(FD)}, \quad (18.44c) \end{cases}$$

where, for the Bose–Einstein case, we make use of the assumption that $Z_k \gg 1$.

The Maxwell–Boltzmann case is what we expect from classical statistical mechanics, and this allows us to identify β with $1/\kappa T$ since $\exp(-E_k/\kappa T)$ is proportional to the classical probability that a particle will have energy E_k in a system in thermodynamic equilibrium at temperature T. We also have the result that the BE and FD statistics approach MB or classical statistics when $E_k \gg \kappa T$, in accordance with observation. It is convenient to write $\alpha = \alpha_{\text{M,B,F}} - \beta E_{\text{M,B,F}}$, the indices M, B and F corresponding to MB, BE or FD statistics, respectively. We can evaluate α through the behavior of n_k at low temperatures. As $T \to 0$, we expect n_k to vanish for all but the lowest single-particle (ground state) energy level for the system when MB or BE statistics apply; hence

$$\alpha_{\text{M}} = \ln \frac{Z_0}{N}, \qquad E_{\text{M}} = E_0$$

for MB statistics, and

$$\alpha_{\text{B}} = \ln \frac{N + Z_0}{N}, \qquad E_{\text{B}} = E_0$$

where Z_0 is the degeneracy of the lowest energy level and E_0 is its energy. These choices guarantee that all N particles will be in the lowest energy level when $T = 0$.

For fermions the situation is different; only one particle can be in each available state, and hence only Z_0 particles can be in the lowest energy level. Hence $\alpha_{\text{F}} = 0$, E_{F} = energy of states j_{F} such that

$$\sum_{i=1}^{j_{\text{F}}-1} Z_i < N \leq \sum_{i=1}^{j_{\text{F}}} Z_i.$$

Actually, for the states j_{F}, $n_{j_{\text{F}}} = N - \sum_{i=1}^{j_{\text{F}}-1} Z_i$; thus when $N = \sum_{i=1}^{j_{\text{F}}} Z_i$, this gives $n_{j_F} = Z_{j_F}$, as Eq. (18.44b) predicts, but otherwise n_{j_F} is given by the equation just quoted rather than by Eq. (18.44b). The quantity E_F is called the Fermi energy; all states with energies below it are filled when $T = 0$, and there is at least one state with energy E_F occupied at $T = 0$, but all states with energies exceeding E_F are empty at absolute zero.

The function

$$f(E, T) \equiv \frac{1}{e^{(E-E_F)/\kappa T} + 1}$$

is illustrated in Fig. 18.3. It is called the Fermi function and gives the fraction of fermion states with energy E that are filled when the temperature is T. The effect of the exclusion principle is clearly visible here; it is to force particles to occupy states with nonvanishing energies, even at very low temperatures, whenever the number of fermions exceeds the number of states with energy $E = 0$. This is a strictly nonclassical result.

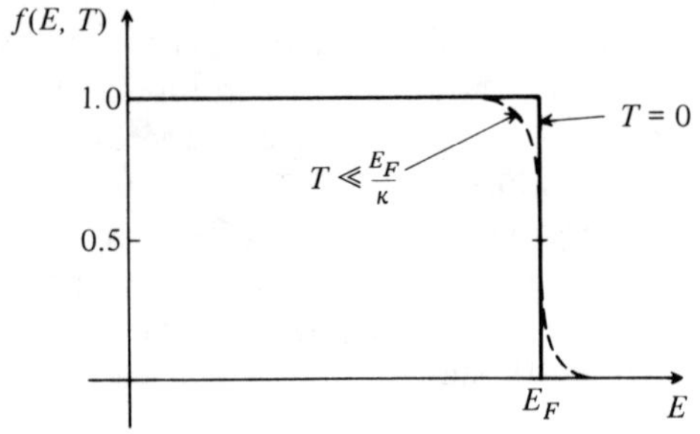

Fig. 18.3. The Fermi function.

Note that

$$b(E_k, T) = \frac{1}{e^{\alpha+\beta E_k} - 1} = \frac{n_k}{Z_k}$$

is the Planck's law probability that a photon has energy E_k in a black body in thermal equilibrium at temperature $T = 1/\kappa\beta$ if we set $\alpha = 0$. Here we have important evidence that the photon is a boson.

18.9 THE FERMI GAS

If at $T = 0$ we have N identical fermions, with spin $\frac{1}{2}$ and mass M, that do not interact with one another, and if we impose periodic boundary conditions on their wave functions, then

$$E_F = \frac{p_F^2}{2M}$$

where p_F is the momentum of the fastest fermions. Also

$$N = 2\int d\Omega \int_0^{k_F} \frac{L^3 k^2}{(2\pi)^3}\, dk = \frac{L^3}{\pi^2}\int_0^{k_F} k^2\, dk$$
$$= \frac{L^3}{3\pi^2} k_F^3, \tag{18.45}$$

where $k_F = p_F/\hbar$, $L/2\pi$ is the period of the boundary conditions (L is the length of an edge of the cubic normalization box) and the factor 2 comes from the fact that there are two spin orientations for each particle. We have simply added together all the states having energy less than $E_F = p_F^2/2M$ in a gas of noninteracting spin-$\frac{1}{2}$ fermions. The density of fermions is $\rho = N/L^3 = N/V$, and hence

$$k_F = (3\pi^2\rho)^{1/3}, \qquad p_F = \hbar(3\pi^2\rho)^{1/3}. \tag{18.46}$$

This leads to

$$E_F = \frac{\hbar^2}{2M}(3\pi^2\rho)^{2/3}. \tag{18.47}$$

The total kinetic energy of the fermion gas is given by

$$E_T = 2\int_0^{k_F} \frac{\hbar^2 k^2}{2M}\frac{L^3}{(2\pi)^3}\, d^3k = \frac{\hbar^2 L^3}{2M\pi^2}\int_0^{k_F} k^4\, dk$$
$$= \frac{\hbar^2 L^3}{10M\pi^2} k_F^5, \tag{18.48}$$

or

$$E_T = \tfrac{3}{5} N E_F. \tag{18.49}$$

Each fermion, then, contributes a mean energy of $\frac{3}{5}E_F$ to the total kinetic energy of the gas.

18.10 THE THOMAS–FERMI ATOMIC MODEL

A rather crude but quite useful model for the atom utilizes the Fermi gas as a starting point; this is the Thomas–Fermi atomic model. Here each atomic electron is assumed to move in a central potential $V(\mathbf{x})$ that arises from the nucleus and the combined effect of all other electrons. This potential is assumed to vary less rapidly than the charge density of the electron cloud $\rho(\mathbf{x})$ so that locally, in the neighborhood of each point $\mathbf{x}$, the electrons have the kinetic energy of Fermi gas electrons. These assumptions lead to the expression

$$\mathscr{E}(\mathbf{x}) = \mathscr{E}_K(\mathbf{x}) + \mathscr{E}_V(\mathbf{x})$$

for the energy density of an electron, where

$$\mathscr{E}_K(\mathbf{x}) = \rho(\mathbf{x}) \cdot \tfrac{3}{5} E_F(\mathbf{x})$$
$$= \frac{\hbar^2}{10\pi^2 M} [3\pi^2 \rho(\mathbf{x})]^{5/3} = \lambda \rho^{5/3}(\mathbf{x}) \tag{18.50}$$

and

$$\mathscr{E}_V(\mathbf{x}) = -e\rho(\mathbf{x}) V(\mathbf{x}) \tag{18.51}$$

are the kinetic and potential energy densities, respectively, for an electron.

The potential $V(\mathbf{x})$ is estimated to be

$$V(\mathbf{x}) = \frac{Ze}{4\pi\epsilon_0 |\mathbf{x}|} - \frac{e}{4\pi\epsilon_0} \int \frac{\rho(\mathbf{x}')}{|\mathbf{x} - \mathbf{x}'|} d^3x',$$

the first term representing the nuclear Coulomb potential and the second potential arising from the combined repulsive effects of other electrons. The total energy of the system then is

$$E = \int d^3x \left[\lambda \rho^{5/3}(\mathbf{x}) - \frac{e^2}{4\pi\epsilon_0} \rho(\mathbf{x}) \frac{Z}{|\mathbf{x}|} + \frac{e^2}{4\pi\epsilon_0} \rho(\mathbf{x}) \int d^3x' \frac{\rho(\mathbf{x}')}{|\mathbf{x} - \mathbf{x}'|} \right]. \tag{18.52}$$

This may be minimized with respect to $\rho(\mathbf{x})$ subject to the constraint

$$\int \rho(\mathbf{x})\, d^3x = Z.$$

Thus if ϵ/Z is the Lagrangian multiplier multiplying $Z = \int \rho(\mathbf{x})\, d^3x$ for this minimization, we must have, upon variation of $\rho(\mathbf{x})$,

$$\delta(E - \epsilon) = \int d^3x\, \delta\rho(\mathbf{x}) \left[\tfrac{5}{3} \lambda \rho^{2/3}(\mathbf{x}) - \frac{e^2}{4\pi\epsilon_0} \frac{Z}{|\mathbf{x}|} + \frac{e^2}{4\pi\epsilon_0} \int d^3x' \frac{\rho(\mathbf{x}')}{|x - x'|} - \epsilon \right]$$
$$= 0,$$

or, since $\delta\rho(\mathbf{x})$ is an arbitrary variation in the charge density,

$$\tfrac{5}{3} \lambda \rho^{2/3}(\mathbf{x}) - \frac{Ze^2}{4\pi\epsilon_0 |\mathbf{x}|} + \frac{e^2}{4\pi\epsilon_0} \int d^3x' \frac{\rho(\mathbf{x}')}{|\mathbf{x} - \mathbf{x}'|} - \epsilon = 0. \tag{18.53}$$

Let us define $V_0 = \epsilon/e$; then the minimization condition becomes

$$\rho(\mathbf{x}) = \left\{ \frac{3e}{5\lambda} [V(x) + V_0] \right\}^{3/2}.$$

Since $\rho(\mathbf{x}) \to 0$ as $|\mathbf{x}| \to \infty$ as well as $V(\mathbf{x}) \to 0$ as $|\mathbf{x}| \to \infty$, we must have $V_0 = \epsilon = 0$. Hence the charge density that leads to minimal energy under our assumptions is

$$\rho(\mathbf{x}) = \left[\frac{3e}{5\lambda} V(\mathbf{x}) \right]^{3/2}, \qquad \lambda = \frac{\hbar^2}{10\pi^2 M} (3\pi^2)^{5/3}. \tag{18.54}$$

Now we recall what one of Maxwell's equations tells us about the electric potential $V(\mathbf{x})$; it says

$$\nabla^2 V(\mathbf{x}) = -\frac{\eta(\mathbf{x})}{\epsilon_0} \tag{18.55}$$

where $\eta(\mathbf{x})$ is the charge density; in our case,

$$\eta(\mathbf{x}) = -e\rho(\mathbf{x}) + Ze\delta(\mathbf{x}) \tag{18.56}$$

and hence

$$\nabla^2 V(\mathbf{x}) = \frac{e}{\epsilon_0}\rho(\mathbf{x}) - \frac{Ze}{\epsilon_0}\delta(\mathbf{x})$$

or, using the connection between $\rho(\mathbf{x})$ and $V(\mathbf{x})$ obtained from the minimization condition,

$$\nabla^2 V(\mathbf{x}) = \frac{e}{\epsilon_0}\left(\frac{2Me}{\hbar^2}\right)^{3/2}\frac{1}{3\pi^2}V^{3/2}(\mathbf{x}) - \frac{Ze}{\epsilon_0}\delta(\mathbf{x}). \tag{18.57}$$

Since we are assuming that $V(\mathbf{x}) = V(|\mathbf{x}|)$, i.e. that the potential is spherically symmetric, we have, letting $|\mathbf{x}| = r$,

$$\frac{1}{r}\frac{d^2}{dr^2}rV(r) = \frac{e}{3\pi^2\epsilon_0}\left(\frac{2Me}{\hbar^2}\right)^{3/2}V^{3/2}(r) - \frac{Ze}{4\pi\epsilon_0}\frac{\delta(r)}{r^2}$$

since $\delta(\mathbf{x}) = (1/4\pi)\,\delta(r)/r^2$ in radial coordinates. The substitution $\xi = \alpha r$ and $rV(r) = \phi(\xi)$ lead to

$$\alpha^2\phi''(\xi) = \frac{e}{3\pi^2\epsilon_0}\left(\frac{2Me}{\hbar^2}\right)^{2/2}\alpha^{1/2}\xi^{-1/2}\phi^{3/2}(\xi) - \frac{Ze}{4\pi\epsilon_0}\alpha^2\frac{\delta(\xi)}{\xi},$$

and we may set $\alpha = 2Me/h^2(e/(3\pi^2\epsilon_0))^{2/3}$, converting this to the differential equation

$$\phi''(\xi) = \xi^{-1/2}\phi^{3/2}(\xi) - \frac{Ze}{4\pi\epsilon_0}\frac{\delta(\xi)}{\xi} \tag{18.58}$$

subject to the boundary conditions

$$\phi(0) = -\frac{Ze}{4\pi\epsilon_0}, \qquad \phi(\infty) = 0. \tag{18.59}$$

In fact, the first boundary condition and the $\delta(\xi)/\xi$ term serve the same purpose; the presence of the $\delta(\xi)/\xi$ term simply ensures satisfaction of the first boundary condition, and we can eliminate it if we retain the boundary condition. Hence we may solve

$$\phi''(\xi) = \xi^{-1/2}\phi^{3/2}(\xi), \tag{18.60}$$

known as the Thomas–Fermi equation, subject to the above boundary conditions,

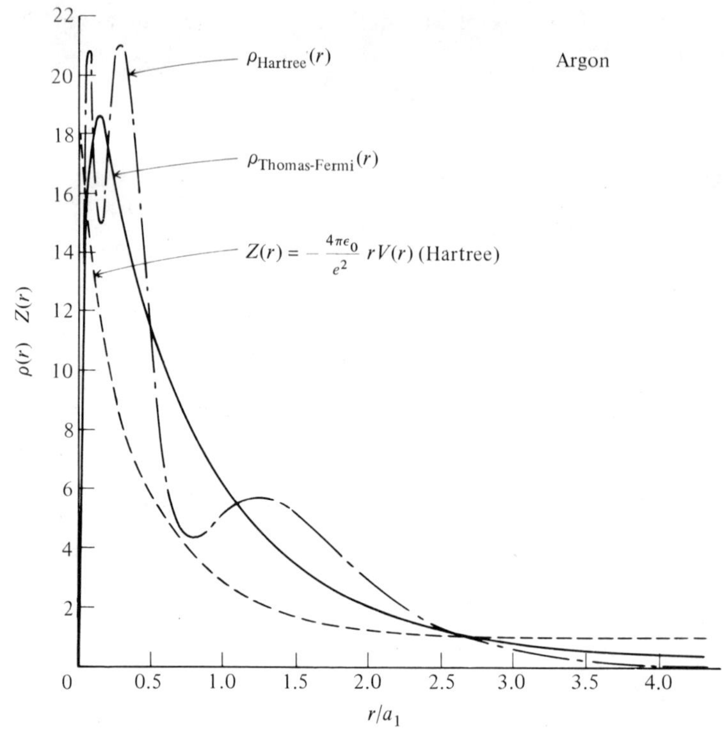

Fig. 18.4. Atomic charge densities $\rho(r)$, as calculated from the Thomas–Fermi model and from the Hartree model, for argon. From R. M. EISBERG, *Fundamentals of Modern Physics*, John Wiley and Sons, New York (1961). Reprinted through permission of the publisher.

to obtain an expression for the atomic potential $V(r)$ and also an expression for the electron density $\rho(r)$. The Thomas–Fermi equation has been solved numerically for various atomic systems. In Fig. 18.4, the results for argon are shown and are compared with similar results obtained from the Hartree model.

READING NOTES*

Reference 22, Chapters 17, 19.

Reference 24, Chapter 10.

Reference 42, Chapter 8.

Reference 43, Chapter 10.

Reference 44, Chapter 12.

---

* Works referred to in these notes will be found in the Bibliography on p. 577.

PROBLEMS

1. a) For three electrons, write down the spin parts of the wave functions in terms of the one-electron spinors.
   b) Bring together those with a given total spin angular momentum, for each possible total spin angular momentum, and identify the magnetic quantum numbers of each eigenfunction of total spin.
2. Reconsider the helium-atom problem of Sections 18.4 and 18.5 assuming that electrons are bosons rather than fermions. Compare the results with the fermion case discussed in the text.
3. Find the mean momentum and the mean energy of N particles in a Fermi gas.
4. A graph of the solution to Eq. (18.60) is given in Fig. 18.5. Use this to estimate, according to the Thomas–Fermi model,
   a) the mean radius of electrons in an atom.
   b) the mean magnitude of electron momenta.
   c) the mean electron–electron Coulomb interaction energy.
   d) the mean kinetic energy of an atomic electron.

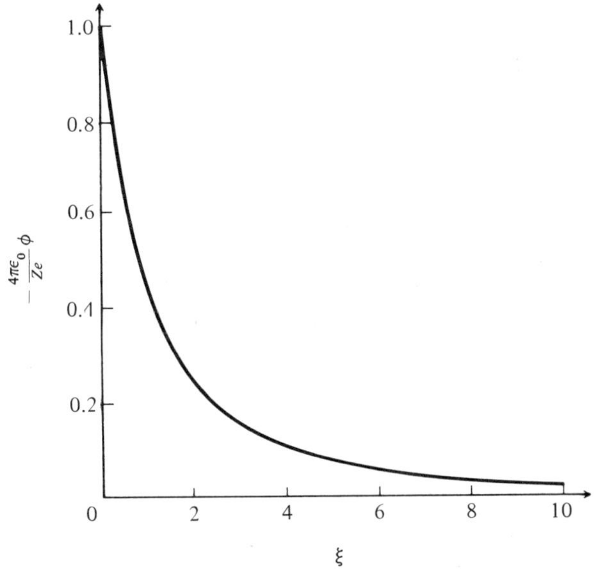

Fig. 18.5. The solution function for the Thomas–Fermi equation. Data from L. D. LANDAU and E. M. LIFSHITZ, *Quantum Mechanics*, Addison-Wesley (1958).

CHAPTER 19

THE PHYSICS OF UNBOUND SYSTEMS; SCATTERING THEORY

There are essentially two ways in which we learn about microscopic physical systems; either we let them voluntarily tell us about themselves through, for example, their spontaneous emission of electromagnetic waves or their spontaneous breakup into smaller (less massive) parts, or we probe them for information by bombarding them with available particles of our choosing, and observe the behavior of the particles or fragments emanating from the collision. Much of our knowledge of atomic structure comes from the essentially voluntary statements of atoms that we decode in our analysis of atomic spectra. Of course, the situations in which the atoms found themselves were often carefully controlled by experimenters through the imposition of electric or magnetic fields, and the willingness of the atoms to "speak" was guaranteed through heating the system containing the atoms to a sufficiently high temperature. Our early understanding of nuclear systems similarly was obtained primarily through the analysis of decay processes.

Actually, scattering processes are very similar to decay processes. In fact, scattering consists of the formation of an agglomerate of particles through the collision of two less massive agglomerates followed by the decay of the system thus formed. For example, a deuteron, which is a bound neutron plus proton state, may collide with a Be^8 nucleus, forming an excited, unbound ten-nucleon system; this system then may decay into Be^7 plus a triton, or H^3 nucleus (proton bound with two neutrons):

$$H^2 + Be^8 \rightarrow H^3 + Be^7.$$

If this happens sufficiently rapidly, we are not tempted to think of the ten-nucleon system as a separate entity. However, there are various ways in which any slowness in the interaction may manifest itself in the behavior of $H^3 + Be^7$ in the final state, and when this happens we see quite clearly that the scattering process consists of a formation stage and a decay stage.

This is an example of an inelastic scattering; another inelastic process is

$$H^2 + Be^8 \rightarrow H^2 + Be^{8*}$$

where by Be^{8*} we mean an excited state of the Be^8 nucleus rather than the ground state Be^8. Since Be^{8*} is more energetic than Be^8, the relative H^2–Be^{8*} kinetic energy must be less than the incident H^2–Be^8 kinetic energy. The excited state will eventually lose its extra energy, perhaps through the emission of a photon:

$$H^2 + Be^8 \rightarrow H^2 + Be^{8*}$$
$$\qquad\qquad\qquad\qquad\quad \llcorner\!\!\longrightarrow \gamma + Be^8.$$

We may look upon this as an inelastic scattering process with a three-body final state, $H^2 + \nu + Be^8$, or as an inelastic process with a two-body final state, $H^2 + Be^{8*}$, if the observer fails to witness the emerging photon, γ.

The elastic version of this scattering is

$$H^2 + Be^8 \rightarrow H^2 + Be^8$$

where the incident and emerging particles both retain their original masses and identities. Even such elastic scattering, however, may exhibit evidence of the formation of an intermediate system that subsequently decays into the *elastic channel* $H^2 + Be^8$.

In any case, all of our knowledge about the structures of microscopic physical systems comes from the analysis of unbound states. Scattering processes are characterized by:

1. the identities or characteristics of the bombarding and target particles and their relative energies,
2. the identities of the emerging particles,
3. the scattering cross-section for the observed process at various incident relative kinetic energies,
4. the angular distributions of the emerging particles relative to the direction of incidence of the bombarding particle at each incident relative kinetic energy.

Decay processes are characterized by:

1. the identity of the decaying particle,
2. the identities of the emerging particles and their kinetic energies,
3. the transition rates, or decay rates (at various kinetic energies if there are more than two decay products),
4. angular distributions of decay products in terms of the relative motion of two products when there are more than two decay products.

The role of quantum theory in studying such unbound systems is to predict all else when items 1 and 2 are given for decay processes or for scattering processes.

This requires having a quantum-mechanical model for the interaction (e.g. the form of the potential in nonrelativistic elastic scattering) responsible for the decay or scattering process. Only for rather simple nonrelativistic problems have we been able to develop models that seem quite satisfactory, and even these break down when minor complexities are introduced. Hence we have no universally applicable theory for unbound systems, any more than we have for bound systems; atomic and nuclear scattering and decay problems, for example, involve very different interactions, and even different approaches. However, we can formulate such problems in rather general terms, thereby providing a meeting ground upon which experimentalist and theorist can compare data with the predictions of various models.

In this chapter we shall discuss the general formulation of the scattering problem, restricting ourselves to nonrelativistic physics. Decay problems are handled quite similarly, with obvious modifications; in fact, we investigated electromagnetic decays of atoms in an earlier chapter. Except for some early remarks on inelastic scattering, we shall restrict ourselves to elastic scattering processes, also. Finally, as a related but separate topic, we discuss the penetration of potential barriers in three dimensions; this has obvious applications, in decay problems such as alpha-particle decay.

19.1 FORMAL THEORY OF SCATTERING

We begin by taking a formal operator approach to the scattering of particle A by particle B in the elastic channel:

$$A + B \to A + B.$$

To begin with, particles A and B are assumed to be sufficiently far from each other that they do not interact; their interaction range is finite and much smaller than their initial separation. Furthermore, each particle is well localized so that it makes sense to speak of the particles as being well separated from each other, and each has rather well-defined momentum, $\mathbf{p}_A$ and $\mathbf{p}_B$ respectively. Of course the momentum and position cannot be arbitrarily well defined simultaneously; however, if $\Delta p_A \ll p_A$ we may say that the momentum of particle A is rather well defined, and at the same time we may have $\Delta x_A \ll |\mathbf{x}_A - \mathbf{x}_B|$ and

$$\Delta p_A \Delta x_A \simeq \hbar.$$

This is assumed to be the case.

In time, the wave packets must spread, and we may question whether these conditions persist. All that is required is that they persist essentially until the collision takes place. Later we shall formulate the condition this imposes upon the wave packets, but for now we simply assume that such a condition is satisfied.

Let us denote the two-particle state vector representing the initial state of this system by

$$|\mathbf{p}_A, a; \mathbf{p}_B, b\rangle = |\mathbf{p}_A, a\rangle|\mathbf{p}_B, b\rangle \tag{19.1}$$

where a and b denote internal quantum numbers characterizing the two particles. These are not, in fact, eigenstates of $\mathfrak{P}_A$ and $\mathfrak{P}_B$, as noted above, but we use this notation for the sake of brevity.

Now the Hamiltonian for this system is

$$\mathfrak{H} = \mathfrak{H}_A + \mathfrak{H}_B + \mathfrak{V}_{AB}, \qquad \mathfrak{H}_{\substack{A\\B}} = \frac{1}{2M_{\substack{A\\B}}}\mathfrak{P}^2_{\substack{A\\B}} \tag{19.2}$$

if no external forces are present and if $\mathfrak{V}_{AB}$ is the part of the Hamiltonian representing mutual interaction of the particles. Let us assume that $\mathfrak{V}_{AB}$ depends only upon the relative coordinates and momenta of the two particles so that the Hamiltonian is separable in center-of-mass coordinates:

$$\begin{aligned}
\mathfrak{H} &= \frac{1}{2M}\mathfrak{P}^2 + \mathfrak{h} & \mathfrak{h} &= \frac{1}{2\mu}\mathfrak{p}^2 + \mathfrak{V}_{AB}, \\
M &= M_A + M_B, & \mu &= \frac{M_A M_B}{M_A + M_B}, \\
\mathfrak{P} &= \mathfrak{P}_A + \mathfrak{P}_B, & \mathfrak{p} &= \mu\left(\frac{\mathfrak{p}_B}{M_B} - \frac{\mathfrak{p}_A}{M_A}\right), \\
\mathfrak{X} &= \frac{1}{M}(M_A\mathfrak{X}_A + M_B\mathfrak{X}_B), & \mathfrak{x} &= \mathfrak{X}_B - \mathfrak{X}_A.
\end{aligned} \tag{19.3}$$

Then we may write

$$|\mathbf{p}_A, a; \mathbf{p}_B, b\rangle = |\mathbf{P}; \mathbf{p}, a, b\rangle \equiv |\mathbf{P}\rangle|\mathbf{p}, a, b\rangle, \tag{19.4}$$

where

$$\mathfrak{P}|\mathbf{P}\rangle = \mathbf{P}|\mathbf{P}\rangle \tag{19.5}$$

and $|\mathbf{P}\rangle$ involves no internal coordinates. Since $|\mathbf{P}\rangle$ represents the center-of-mass motion and nothing else, we may ignore it, dealing entirely with $|\mathbf{p}, a, b\rangle$. However $|\mathbf{p}, a, b\rangle$ is not an eigenstate of $\mathfrak{p}$; there is associated with it some Δp arising from Δp_A and Δp_B. It is "approximately" an eigenstate of $\mathfrak{p}$ in the same sense that $|\mathbf{p}_A, a\rangle$ is "approximately" an eigenstate of $\mathfrak{P}_A$.

Although $|\mathbf{p}, a, b\rangle$ is not even "approximately" an eigenstate of $\mathfrak{h}$, we may make it so at certain times, by equipping $\mathfrak{V}_{AB}$ with a time dependence that "turns off" the potential for large negative times. Let us take $\mathfrak{V}_{AB}(t) = \mathfrak{V}_{AB}e^{\epsilon t}$, where ϵ is chosen to be large enough that $\mathfrak{V}_{AB}(t)$ essentially vanishes when particles A and B are sufficiently far apart at the beginning of the experiment. That is, if the collision occurs (the particles make their closest approach) at $t = 0$, and the scattering experiment begins at $t = -T$, we must have $\epsilon \gg 1/T$. Then at $t = -T$, $|\mathbf{p}, a, b\rangle$ is approximately an eigenstate of

$$\mathfrak{h}(-T) = \frac{1}{2\mu}\mathfrak{P}^2 + \mathfrak{V}_{AB}(-T) \simeq \frac{1}{2\mu}\mathfrak{P}^2 \equiv \mathfrak{h}_0. \tag{19.6}$$

In general, the Schrödinger state vector representing the time evolution of relative motion of these particles is

$$|\mathbf{p}, a, b(t)\rangle_S \equiv e^{-i\mathfrak{h}(t+T)/\hbar}|\mathbf{p}, a, b\rangle \tag{19.7}$$

since $|\mathbf{p}, a, b\rangle$ is the Heisenberg state vector representing the system at $t = -T$. Presumably, however, $\mathfrak{h}$ is sufficiently complicated (i.e. the interaction $\mathfrak{V}_{AB}$ is nontrivial) that we cannot obtain $|\mathbf{p}, a, b(t)\rangle_S$ easily through this transformation. An alternative to the Heisenberg and Schrödinger pictures, which correspond, respectively, to all or none of the dynamical time dependence residing in the operators, is the interaction picture, in which the operators evolve as if the interparticle interaction $\mathfrak{V}_{AB}$ were absent:

$$\mathfrak{Q}_I(t) \equiv e^{i\mathfrak{h}_0 t/\hbar}\mathfrak{Q}_S e^{-i\mathfrak{h}_0 t/\hbar}, \tag{19.8}$$

where $\mathfrak{Q}_S$ is a Schrödinger operator representing dynamical variable $\mathfrak{Q}$. For example, we define

$$\mathfrak{h}_I(t) = e^{i\mathfrak{h}_0 t/\hbar}\mathfrak{V}(t)e^{-i\mathfrak{h}_0 t/\hbar}, \tag{19.9a}$$

where

$$\mathfrak{V}(t) = \mathfrak{V}_{AB}e^{\epsilon t}; \tag{19.9b}$$

thus for $t = 0$,

$$\mathfrak{h}_I(0) = \mathfrak{V}(0) = \mathfrak{V}_{AB}.$$

The operator $\mathfrak{h}_I$ is the interaction Hamiltonian in the interaction representation.

The state vectors in the interaction representation are given by

$$\begin{aligned} |1(t)\rangle_I &\equiv e^{i\mathfrak{h}_0 t/\hbar}|1(t)\rangle_S \\ &= e^{i\mathfrak{h}_0 t/\hbar}e^{-i\mathfrak{h}t/\hbar}|1(0)\rangle_S \\ &= e^{i\mathfrak{h}_0 t + T/\hbar}\, e^{-i\mathfrak{h}(t+T)/\hbar}|1\rangle_H \end{aligned} \tag{19.10}$$

where $|1\rangle_H$ is the Heisenberg state vector representing the condition of the system at time $t = -T$ when the experiment begins.

It is important to remember that the exponentials here are not necessarily additive since $\mathfrak{h}$ and $\mathfrak{h}_0$ do not necessarily commute. The transformation from a Heisenberg state vector to an interaction-representation (IR) state vector is thus rather complicated, being the composite of two transformations that generally do not commute with one another. For this reason, the IR state vectors are usually expressed in terms of a single transformation $e^{i\mathfrak{h}_0 t/\hbar}$ on Schrödinger state vectors. We use the interaction representation in attacking the scattering problem simply because it leads rather easily to an appropriate integral equation that we can solve formally. There are also deeper reasons for using this representation; they will become apparent as we progress.

The equation of motion obeyed by the IR state vectors is

$$\begin{aligned} i\hbar\frac{\partial}{\partial t}|1(t)\rangle_{\mathrm{I}} &= e^{i\mathfrak{h}_0 t/\hbar}\left[-\mathfrak{h}_0|1(t)\rangle_{\mathrm{S}} + i\hbar\frac{\partial}{\partial t}|1(t)\rangle_{\mathrm{S}}\right] \\ &= e^{i\mathfrak{h}_0 t/\hbar}(-\mathfrak{h}_0 + \mathfrak{h})|(t)\rangle_{\mathrm{S}} \\ &= e^{i\mathfrak{h}_0 t/\hbar}\mathfrak{V}(t)e^{-i\mathfrak{h}_0 t/\hbar}|1(t)\rangle_{\mathrm{I}} \\ &= \mathfrak{h}_{\mathrm{I}}(t)|1(t)\rangle_{\mathrm{I}}. \end{aligned} \tag{19.11}$$

This differential equation cannot be solved by simple exponentiation (see Problem 1). However, we can form from it an integral equation, namely

$$|1(t)\rangle_{\mathrm{I}} = |1\rangle_{\mathrm{H}} + \frac{1}{i\hbar}\int_{-T}^{t}\mathfrak{h}_{\mathrm{I}}(t')|1(t')\rangle_{\mathrm{I}}\,dt' \tag{19.12}$$

where $|1\rangle_{\mathrm{H}} = |1(-T)\rangle_{\mathrm{I}}$; in the scattering problem we have outlined above, $|1\rangle_{\mathrm{H}} = |\mathbf{p}, a, b\rangle$.

We now assume that $|1\rangle_{\mathrm{H}}$ is an eigenstate of $\mathfrak{h}_0$ with eigenvalue E, and show how this implies that

$$[\mathfrak{h}(t) - E]|1(t)\rangle_{\mathrm{I}} = 0, \tag{19.13}$$

at least approximately, for all finite values of t. Thus

$$\begin{aligned} i\hbar\frac{\partial}{\partial t}[\mathfrak{h}(t) - E]|1(t)\rangle_{\mathrm{I}} &= \left[i\hbar\frac{\partial}{\partial t}\mathfrak{h}_0 + i\hbar\frac{\partial}{\partial t}\mathfrak{h}_{\mathrm{I}}(t)\right]|1(t)\rangle_{\mathrm{I}} + [\mathfrak{h}_0 + \mathfrak{h}_{\mathrm{I}}(t) - E]i\hbar\frac{\partial}{\partial t}|1(t)\rangle_{\mathrm{I}} \\ &= \left\{i\hbar\frac{\partial}{\partial t}\mathfrak{h}_{\mathrm{I}}(t) + [\mathfrak{h}_0 + \mathfrak{h}_{\mathrm{I}}(t) - E]\mathfrak{h}_{\mathrm{I}}(t)\right\}|1(t)\rangle_{\mathrm{I}}. \end{aligned}$$

Now

$$\begin{aligned} i\hbar\frac{\partial}{\partial t}\mathfrak{h}_{\mathrm{I}}(t) &= i\hbar\frac{\partial}{\partial t}[e^{i\mathfrak{h}_0 t/\hbar}\mathfrak{V}(t)e^{-i\mathfrak{h}_0 t/\hbar}] \\ &= -[\mathfrak{h}_0\mathfrak{h}_{\mathrm{I}}(t) - \mathfrak{h}_{\mathrm{I}}(t)\mathfrak{h}_0 - i\epsilon\hbar\mathfrak{h}_{\mathrm{I}}(t)] \end{aligned}$$

and therefore

$$\begin{aligned} i\hbar\frac{\partial}{\partial t}[\mathfrak{h}(t) - E]|1(t)\rangle_{\mathrm{I}} &= \mathfrak{h}_{\mathrm{I}}(t)[\mathfrak{h}_0 + \mathfrak{h}_{\mathrm{I}}(t) - E + i\hbar\epsilon]|1(t)\rangle_{\mathrm{I}} \\ &= \mathfrak{h}_{\mathrm{I}}(t)[\mathfrak{h}(t) - E + i\hbar\epsilon]|1(t)\rangle_{\mathrm{I}}. \end{aligned} \tag{19.14}$$

Hence, in the limit of small ϵ, we see that $[\mathfrak{h}(t) - E]|1(t)\rangle_{\text{I}}$ satisfies the same differential equation that $|1(t)\rangle_{\text{I}}$ satisfies. However, at $t = -T$, we have $\mathfrak{h}(-T) = \mathfrak{h}_0$ and $|1(-T)\rangle_{\text{I}} = |1\rangle_{\text{H}}$, so that

$$[\mathfrak{h}(-T) - E]|1(-T)\rangle_{\text{I}} = [\mathfrak{h}_0 - E]|1\rangle_{\text{H}} = 0 \tag{19.15}$$

because of our assumption that $|1\rangle_{\text{H}}$ is an eigenstate of $\mathfrak{h}_0$ with eigenvalue E. This means that $[\mathfrak{h}(t) - E]|1(t)\rangle_{\text{I}}$ (at least for $\epsilon \to 0$) satisfies a linear homogeneous differential equation in t with initial condition $[\mathfrak{h}(t) - E]|1(t)\rangle_{\text{I}} = 0$, $t = -T$. From this it follows that $[\mathfrak{h}(t) - E]|1(t)\rangle_{\text{I}} = 0$ for all times. In picturesque terms, the "slope" of $[\mathfrak{h}(t) - E]|1(t)\rangle_{\text{I}}$ as a function of time vanishes at $t = -T$, and this forces the slope to vanish for all other values of t.

The acute observer will have noted that we cheated a little here. In the first place, we used the fact that ϵ is finite and positive to make $\mathfrak{h}_1(t)$ vanish at $t = -T$, but later had to assume ϵ to be vanishingly small to get $[\mathfrak{h}(t) - E]|1(t)\rangle_{\text{I}}$ to satisfy the desired differential equation. Earlier we noted that $\epsilon \gg 1/T$ was required, so $\epsilon \to 0$ implies $T \to \infty$, and hence we must allow T to be essentially infinite if our arguments are to be justifiable. This would seem to give us trouble with the spreading of wave packets, for over an infinite length of time, what initially was a well-localized wave-packet would be spread over all space. Our formal arguments therefore cannot represent rigorously the physical situation; rather we mean by $T = \infty$ that T is long compared with all other characteristic times involved in the experiment except the time intervals over which the spread in wave packets become significant. Then ϵ may become very small in the sense that $\hbar\epsilon$ is negligible compared with any energy resolution available in the experiment and therefore may be neglected except as a damping constant for the interaction potential.

Thus we have found that $|1(t)\rangle_{\text{I}}$ is, for all practical purposes, an eigenstate of $\mathfrak{h}(t)$ with energy eigenvalue E if $|1\rangle_{\text{H}}$ is an eigenstate of $\mathfrak{h}_0$ with eigenvalue E. This illustrates an important practical advantage of using the interaction representation; another is illustrated in the following paragraph.

As a consequence of the preceding result, the state vector

$$|1(t)\,\text{in}\rangle \equiv e^{i(E-\mathfrak{h}_0)t/\hbar}|1(t)\rangle_{\text{I}}$$

is independent of time, for

$$\begin{aligned} i\hbar\frac{\partial}{\partial t}|1(t)\,\text{in}\rangle &= e^{i(E-\mathfrak{h}_0)t/\hbar}\left[-(\text{E} - \mathfrak{h}_0) + i\hbar\frac{\partial}{\partial t}\right]|1(t)\rangle_{\text{I}} \\ &= e^{i(E-\mathfrak{h}_0)t/\hbar}[-E + \mathfrak{h}_0 + \mathfrak{h}_1(t)]|1(t)\rangle_{\text{I}} \\ &= 0. \end{aligned} \tag{19.16}$$

Hence we drop the parameter t in these new states, writing $|1\,\text{in}\rangle \equiv |1(t)\,\text{in}\rangle = e^{i(E-\mathfrak{h}_0)t/\hbar}|1(t)\rangle_{\text{I}}$. Of course, this time independence of the "in" states is approximate for the reasons indicated above; it is accurate to the extent that we can neglect the spreading in wave packets. The label "in" refers to the fact that the interaction is "turned off" at $t \geq -T \simeq \infty$ so that the initial state of the system consists of "incoming" free particles.

Now we arrive at a central result. Since $|1\text{ in}\rangle$ is time-independent, we may choose

$$|1\text{ in}\rangle = |1(0)\rangle_{\text{I}}.$$

According to Eq. (19.12) then, we have

$$\begin{aligned}
|1\text{ in}\rangle &= |1\rangle_{\text{H}} + \frac{1}{i\hbar}\int_{-T}^{0} e^{i\mathfrak{h}_0 t'/\hbar}\mathfrak{V}e^{\epsilon t'}e^{-i\mathfrak{h}_0 t'/\hbar}e^{-i(E-\mathfrak{h}_0)t'/\hbar}|1\text{ in}\rangle \\
&= |1\rangle_{\text{H}} + \frac{1}{i\hbar}\int_{-T}^{0} e^{i(\mathfrak{h}_0 - E - i\epsilon\hbar)t'/\hbar}\mathfrak{V}|1\text{ in}\rangle \\
&= |1\rangle_{\text{H}} - (\mathfrak{h}_0 - E - i\epsilon\hbar)^{-1}e^{i(\mathfrak{h}_0 - E - i\epsilon\hbar)t'/\hbar}\Big|_{-T}^{0}\,\mathfrak{V}|1\text{ in}\rangle \\
&\xrightarrow[T\to\infty]{} |1\rangle_{\text{H}} - (\mathfrak{h}_0 - E - i\epsilon\hbar)^{-1}\mathfrak{V}|1\text{ in}\rangle.
\end{aligned}$$

It is common practice to write this result, which comes from purely formal integration over t, in the form

$$|1\text{ in}\rangle = |1\rangle_{\text{H}} - \frac{1}{\mathfrak{h}_0 - E - i\epsilon}\mathfrak{V}|1\text{ in}\rangle \tag{19.12'}$$

where we have replaced $\epsilon\hbar$ with simply ϵ to save writing. The operator

$$\frac{1}{\mathfrak{h}_0 - E - i\epsilon}$$

is simply the operator inverse to $\mathfrak{h}_0 - E$, with the boundary condition $\mathfrak{V}(t)\xrightarrow[t\to-\infty]{}0$ included implicitly through the $i\epsilon$-term; we assume that such an inverse exists. Equation (19.12′) then, is simply an operator expression for the integral equation from which it has been obtained formally.

If we impose the condition $\mathfrak{V}(t)\xrightarrow[t\to+\infty]{}0$ so that we have free particles as $t \to +\infty$ (outgoing free particles), then we must take

$$\mathfrak{V}(t) = \mathfrak{V}e^{-\epsilon t}.$$

We can carry through the analysis as above, but replacing the states $|1\text{ in}\rangle$ with states $|1\text{ out}\rangle$; then we shall have

$$\left|1\begin{matrix}\text{in}\\\text{out}\end{matrix}\right\rangle = |1\rangle_{\text{H}} - \frac{1}{\mathfrak{h}_0 - E \mp i\epsilon}\mathfrak{V}\left|1\begin{matrix}\text{in}\\\text{out}\end{matrix}\right\rangle, \tag{19.17}$$

where for "out" states $|1\rangle_{\text{H}}$ is the state of the system at $t = +T \simeq \infty$, i.e. at the *end* of the experiment.

To get a more concrete view of things, let us form wave functions from these state vectors:

$$\left\langle \mathbf{x} \middle| 1 \begin{matrix}\text{in}\\ \text{out}\end{matrix} \right\rangle = \langle \mathbf{x}|1\rangle_{\mathrm{H}} - \left\langle \mathbf{x} \middle| \frac{1}{\mathfrak{h}_0 - E \mp i\epsilon} \mathfrak{V} \middle| 1 \begin{matrix}\text{in}\\ \text{out}\end{matrix} \right\rangle$$

$$= \langle \mathbf{x}|1\rangle_{\mathrm{H}} \int \frac{d^3p}{\hbar^3} \langle \mathbf{x}|\mathbf{p}, a, b\rangle \left\langle \mathbf{p}, a, b \middle| \frac{1}{(1/2\mu)\mathbf{p}^2 - E \mp i\epsilon} \mathfrak{V} \middle| 1 \begin{matrix}\text{in}\\ \text{out}\end{matrix} \right\rangle .$$

Now $\mathbf{p}^2$ is Hermitian and may operate on $|\mathbf{p}, a, b\rangle$ even though it resides in the denominator (i.e. is part of an inverse operator); this gives

$$\int \frac{d^3p}{\hbar^3} \langle \mathbf{x}|\mathbf{p}, a, b\rangle \frac{1}{(1/2\mu)\mathbf{p}^2 - E \mp i\epsilon} \left\langle \mathbf{p}, a, b \middle| \mathfrak{V} \middle| 1 \begin{matrix}\text{in}\\ \text{out}\end{matrix} \right\rangle$$

for the last term. Inserting complete sets $\int d^3\mathbf{x}'|\mathbf{x}'\rangle\langle\mathbf{x}'|$ here and assuming $\mathfrak{V}$ to be diagonal in $\mathbf{x}$,

$$\langle \mathbf{x}'|\mathfrak{V}|\mathbf{x}''\rangle = V(\mathbf{x}')\delta^3(\mathbf{x}' - \mathbf{x}'')$$

yields

$$\int \frac{d^3p}{\hbar^3} \langle \mathbf{x}|\mathbf{p}, a, b\rangle \frac{1}{(1/2\mu)\mathbf{p}^2 - E \mp i\epsilon} \int d^3x' \langle \mathbf{p}, a, b|\mathbf{x}'\rangle V(\mathbf{x}') \left\langle \mathbf{x}' \middle| 1 \begin{matrix}\text{in}\\ \text{out}\end{matrix} \right\rangle .$$

It is understood that the states denoted $|\mathbf{x}\rangle$ include internal quantum numbers corresponding to the quantum numbers a, b. Upon interchanging orders of integration and writing

$$\psi_{\mathbf{p}}(\mathbf{x}) = \langle \mathbf{x}|\mathbf{p}, a, b\rangle = e^{i\mathbf{p}\cdot\mathbf{x}/\hbar}(2\pi)^{-3/2}$$
$$\psi_{1}{}^{\binom{\text{in}}{\text{out}}}(\mathbf{x}) = \left\langle \mathbf{x} \middle| 1 \begin{matrix}\text{in}\\ \text{out}\end{matrix} \right\rangle \tag{19.18}$$

and

$$G^{(\pm)}(\mathbf{x} - \mathbf{x}') = -\frac{1}{(2\pi\hbar)^3} \int d^3p\, e^{i\mathbf{p}\cdot(\mathbf{x}-\mathbf{x}')/\hbar} \frac{1}{(1/2\mu)\mathbf{p}^2 - E \mp i\epsilon}$$
$$= -\frac{1}{(2\pi)^3} \int d^3k\, e^{i\mathbf{k}\cdot(\mathbf{x}-\mathbf{x}')} \frac{1}{(\hbar^2/2\mu)\mathbf{k}^2 - E \mp i\epsilon}, \tag{19.19}$$

we obtain

$$\psi_{1}{}^{\binom{\text{in}}{\text{out}}}(\mathbf{x}) = \phi_1(\mathbf{x}) + \int d^3x' G^{(\pm)}(\mathbf{x} - \mathbf{x}') V(\mathbf{x}') \psi_{1}{}^{\binom{\text{in}}{\text{out}}}(\mathbf{x}'), \tag{19.20a}$$

where

$$\phi_1(\mathbf{x}) \equiv \langle \mathbf{x}|1\rangle_{\mathrm{H}}. \tag{19.20b}$$

This is the integral equation for $\psi_1^{\binom{\text{in}}{\text{out}}}(\mathbf{x})$, and $G^{(\pm)}(\mathbf{x}-\mathbf{x}')$ are the Green's functions for the differential equation

$$\left(\frac{1}{2\mu}\mathbf{P}^2 - E\right)\psi(\mathbf{x}) = \left(-\frac{\hbar^2}{2\mu}\nabla^2 - E\right)\psi(\mathbf{x}) = 0$$

corresponding to the boundary condition that the wave functions $\psi_1^{\binom{\text{in}}{\text{out}}}(\mathbf{x})$ behave asymptotically as free-particle functions as the particles approach (+) or recede (−) from one another. It is left to the reader to show that

$$\lim_{\epsilon\to 0}\left(-\frac{\hbar^2}{2\mu}\nabla^2 - E\right)G^{(\pm)}(\mathbf{x}-\mathbf{x}') = \delta^3(\mathbf{x}-\mathbf{x}').$$

Notice that the time does not appear anywhere in these equations; we eliminated it through formal integration of the integral equation for $\left|1\,{\text{in} \atop \text{out}}\right\rangle$. The only residuum of time dependence is the presence of $\mp i\epsilon$ in the Green's functions.

We shall investigate the Green's functions in more detail in Section 19.4. Before doing so, let us work out a few more formal details. We may iterate Eq. (19.17) obtaining the infinite series

$$\begin{aligned}
\left|1\,{\text{in} \atop \text{out}}\right\rangle &= |1\rangle_{\mathrm{H}} - \frac{1}{\mathfrak{h}_0 - E \mp i\epsilon}\mathfrak{B}|1\rangle_{\mathrm{H}} \\
&\quad + \frac{1}{\mathfrak{h}_0 - E \mp i\epsilon}\mathfrak{B}\frac{1}{\mathfrak{h}_0 - E \mp i\epsilon}\mathfrak{B}|1\rangle_{\mathrm{H}} - \cdots \\
&= \left[1 - \frac{1}{\mathfrak{h}_0 - E \mp i\epsilon}\mathfrak{B} + \frac{1}{\mathfrak{h}_0 - E \mp i\epsilon}\mathfrak{B}\frac{1}{\mathfrak{h}_0 - E \mp i\epsilon}\mathfrak{B} - \cdots\right]|1\rangle_{\mathrm{H}} \\
&= \sum_{n=0}^{\infty}\left[-\frac{1}{\mathfrak{h}_0 - E \mp i\epsilon}\mathfrak{B}\right]^n|1\rangle_{\mathrm{H}}.
\end{aligned}$$

The terms in this series correspond to the various orders in perturbation theory, the number of times $\mathfrak{B}$ appears in a term indicating the order of perturbation theory. Since

$$\sum_{n=0}^{\infty}(-x)^n = \frac{1}{1+x},$$

we may expect, at least formally, that

$$\begin{aligned}
\left|1\,{\text{in} \atop \text{out}}\right\rangle &= \frac{1}{1 + (\mathfrak{h}_0 - E \mp i\epsilon)^{-1}\mathfrak{B}}|1\rangle_{\mathrm{H}} \\
&= [1 + (\mathfrak{h}_0 - E \mp i\epsilon)^{-1}\mathfrak{B}]^{-1}|1\rangle_{\mathrm{H}} \\
&= \{(\mathfrak{h}_0 - E \mp i\epsilon)^{-1}[(\mathfrak{h}_0 - E \mp i\epsilon) + \mathfrak{B}]\}^{-1}|1\rangle_{\mathrm{H}}
\end{aligned}$$

$$
\begin{aligned}
&= (\mathfrak{h}_0 + \mathfrak{V} - E \mp i\epsilon)^{-1}(\mathfrak{h}_0 - E \mp i\epsilon)|1\rangle_{\mathrm{H}} \\
&= (\mathfrak{h} - E \mp i\epsilon)^{-1}(\mathfrak{h} - E \mp i\epsilon - \mathfrak{V})|1\rangle_{\mathrm{H}} \\
&= \left(1 - \frac{1}{\mathfrak{h} - E \mp i\epsilon}\mathfrak{V}\right)|1\rangle_{\mathrm{H}}.
\end{aligned} \tag{19.21}
$$

It is easy to verify, without using the infinite series, that this is true. The operators

$$\Omega^{(\pm)} \equiv 1 - \frac{1}{\mathfrak{h} - E \mp i\epsilon}\mathfrak{V} \tag{19.22}$$

such that

$$\left|1 \begin{matrix}\text{in}\\ \text{out}\end{matrix}\right\rangle = \Omega^{(\pm)}|1\rangle_{\mathrm{H}} \tag{19.23}$$

are called the scattering operators; obviously they satisfy (see Eq. 19.17)

$$\Omega^{(\pm)} = 1 + \mathfrak{G}_0^{(\pm)}\mathfrak{V}\Omega^{(\pm)}, \tag{19.24}$$

where

$$\mathfrak{G}_0^{(\pm)} \equiv -\frac{1}{\mathfrak{h}_0 - E \mp i\epsilon} \tag{19.25}$$

may be called Green's operators.

The essential operators of scattering theory are formed from $\Omega^{(\pm)}$, $\mathfrak{G}_0^{(\pm)}$ and $\mathfrak{V}$. The so-called scattering matrix is simply

$$
\begin{aligned}
S_{fi} &\equiv \langle \mathbf{p}_f, q_f \text{ out}|\mathbf{p}_i, q_i \text{ in}\rangle = {}_{\mathrm{H}}\langle \mathbf{p}_f, q_f|\Omega^{(-)\dagger}\,\Omega^{(+)}|\mathbf{p}_i, q_i\rangle_{\mathrm{H}} \\
&\equiv \langle \mathbf{p}_f, q_f|\Omega^{(-)\dagger}\,\Omega^{(+)}|\mathbf{p}_i, q_i\rangle
\end{aligned} \tag{19.26}
$$

(henceforward we shall write $|\mathbf{p}, q\rangle \equiv |\mathbf{p}, q\rangle_{\mathrm{H}}$ for Heisenberg vectors). However,

$$\Omega^{(-)} = \Omega^{(+)} + \mathfrak{G}^{(-)}\mathfrak{V} - \mathfrak{G}^{(+)}\mathfrak{V}, \tag{19.27}$$

where

$$\mathfrak{G}^{(\pm)} \equiv -\frac{1}{\mathfrak{h} - E \mp i\epsilon}, \tag{19.28}$$

so that

$$\Omega^{(-)\dagger} = \Omega^{(+)\dagger} + \mathfrak{V}(\mathfrak{G}^{(+)} - \mathfrak{G}^{(-)}), \tag{19.29}$$

since $\mathfrak{V}$ is Hermitian and $\mathfrak{G}^{(\pm)\dagger} = \mathfrak{G}^{(\mp)}$. Therefore

$$
\begin{aligned}
S_{fi} &= \langle \mathbf{p}_f, q_f|\,[\Omega^{(+)\dagger} + \mathfrak{V}(\mathfrak{G}^{(+)} - \mathfrak{G}^{(-)})]\Omega^{(+)}|\mathbf{p}_i, q_i\rangle \\
&= \langle \mathbf{p}_f, q_f \text{ in}|\mathbf{p}_i, q_i \text{ in}\rangle - \langle \mathbf{p}_f, q_f|\mathfrak{V}\left(\frac{1}{\mathfrak{h} - E_f - i\epsilon} - \frac{1}{\mathfrak{h} - E_f + i\epsilon}\right)|\mathbf{p}_i, q_i \text{ in}\rangle \\
&= \langle \mathbf{p}_f, q_f \text{ in}|\mathbf{p}_i, q_i \text{ in}\rangle - \frac{i2\epsilon}{(E_i - E_f)^2 + \epsilon^2}\langle \mathbf{p}_f, q_f|\mathfrak{V}|\mathbf{p}_i, q_i \text{ in}\rangle,
\end{aligned}
$$

where $E_i = \mathbf{p}_i^2/2M$ and we recall that $|\mathbf{p}_i, q_i \text{ in}\rangle$ is an eigenstate of $\mathfrak{h}$ with eigenvalue $\mathbf{p}_i^2/2M = E_i$.

The matrix element S_{fi} is the amplitude for the probability that the system, initially in state $|\mathbf{p}_i, q_i \text{ in}\rangle$ makes a transition to $|\mathbf{p}_f, q_f \text{ out}\rangle$.

Since $\langle \mathbf{p}_f, q_f \text{ in}|\mathbf{p}_i, q_i \text{ in}\rangle$ is constant in time, but the states $|\mathbf{p}, q \text{ in}\rangle$ arise from $|\mathbf{p}, q\rangle_{\mathrm{H}}$, which are orthonormal, we might suspect that

$$\langle \mathbf{p}_f, q_f \text{ in}|\mathbf{p}_i, q_i \text{ in}\rangle = \delta^3(\mathbf{k}_f - \mathbf{k}_i)\delta(q_f - q_i);$$

that is, the orthonormality of the states $|\mathbf{p}, q\rangle_{\mathrm{H}}$ is carried over to the states $|\mathbf{p}, q \text{ in}\rangle$, even though the operator $\Omega^{(+)}$ that transforms the former into the latter is not unitary in general, as we see in the next section. The proof of this statement is beyond our present scope, but a proof exists.

Also, we note that if E is a continuous variable,

$$\lim_{\epsilon \to 0} \frac{2\epsilon}{(E_i - E)^2 + \epsilon^2} = 2\pi\delta(E_i - E)$$

since

$$\int_{-\infty}^{\infty} \frac{2\epsilon\, dE}{(E_i - E)^2 + \epsilon^2} = \int_{-\pi/2}^{\pi/2} d(\tan\phi)\frac{2}{\tan^2\phi + 1} = 2\int_{-\pi/2}^{\pi/2} d\phi = 2\pi,$$

but for $E_i \neq E$,

$$\lim_{\epsilon \to 0} \frac{2\epsilon}{(E_i - E)^2 + \epsilon^2} = 0.$$

Therefore we have

$$\begin{aligned} S_{fi} &= \delta^3(\mathbf{k}_f - \mathbf{k}_i)\delta(q_f - q_i) - i2\pi\delta(E_f - E_i)\langle \mathbf{p}_f, q_f|\mathfrak{V}|\mathbf{p}_i, q_i \text{ in}\rangle \\ &= \delta^3(\mathbf{k}_f - \mathbf{k}_i)\delta(q_f - q_i) - i2\pi\delta(E_f - E_i)\langle \mathbf{p}_f, q_f|\mathfrak{V} - \mathfrak{V}\frac{1}{\mathfrak{h} - E_i \mp i\epsilon}\mathfrak{V}|\mathbf{p}_i, q_i\rangle. \end{aligned} \tag{19.30}$$

We now define the operator

$$\begin{aligned} \mathfrak{T} \equiv \mathfrak{V}\Omega^{(+)} &= \mathfrak{V} + \mathfrak{V}\mathfrak{G}^{(+)}\mathfrak{V} \\ &= \mathfrak{V} - \mathfrak{V}\frac{1}{\mathfrak{h} - E - i\epsilon}\mathfrak{V} \end{aligned} \tag{19.31}$$

and its matrix elements

$${}_{\mathrm{H}}\langle \mathbf{p}_f, q_f|\mathfrak{T}|\mathbf{p}_i, q_i\rangle_{\mathrm{H}} \equiv T_{fi}, \tag{19.32}$$

so that the S-matrix elements may be written in the form

$$S_{fi} = \delta^3(\mathbf{k}_f - \mathbf{k}_i)\delta(q_f - q_i) - i2\pi\delta(E_f - E_i)T_{fi}. \tag{19.33}$$

Evidently no transitions are possible between states having different energies E_i and E_f; thus energy conservation is ensured by the scattering matrix. Since we are working in relative coordinates, conservation of momentum is irrelevant; in the center-of-mass frame, the total momentum is already zero. The *first term* in the expression (19.33) for S_{fi}, however, vanishes unless the relative momentum is completely unchanged by the collision; the particles must recede from one another in the same direction along which they approached one another. Also all other quantum numbers are preserved in the first term. We are usually not concerned with such situations; they represent no scattering at all—the system behaves as if no interaction were present, as far as this term is concerned. Therefore it is the T-matrix, T_{fi}, from which we expect to learn about the scattering produced by the interaction $\mathfrak{V}$ in the Hamiltonian.

19.2 UNITARITY OF THE S-MATRIX

One of the principal reasons that the S-matrix is so important in scattering theory is that it is unitary. This cannot be said for the operators $\Omega^{(\pm)}$. Evidently

$$\Omega^{(\pm)} = \mathop{\mathsf{S}}_{\mathbf{p}',q'} \left| \mathbf{p}', q' \begin{matrix}\text{in}\\ \text{out}\end{matrix} \right\rangle \langle \mathbf{p}', q'|, \tag{19.34}$$

since then

$$\begin{aligned}\Omega^{(\pm)}|\mathbf{p}, q\rangle &= \mathop{\mathsf{S}}_{\mathbf{p},q} \left| \mathbf{p}', q' \begin{matrix}\text{in}\\ \text{out}\end{matrix} \right\rangle \langle \mathbf{p}', q'|\mathbf{p}, q\rangle \\ &= \left| \mathbf{p}, q \begin{matrix}\text{in}\\ \text{out}\end{matrix} \right\rangle\end{aligned} \tag{19.35}$$

as desired. Then

$$\begin{aligned}\Omega^{(\pm)\dagger}\,\Omega^{(\pm)} &= \mathop{\mathsf{S}}_{\mathbf{p}',q'} \mathop{\mathsf{S}}_{\mathbf{p},q} |\mathbf{p}', q'\rangle \left\langle \mathbf{p}', q' \begin{matrix}\text{in}\\ \text{out}\end{matrix} \middle| \mathbf{p}, q \begin{matrix}\text{in}\\ \text{out}\end{matrix} \right\rangle \langle \mathbf{p}, q| \\ &= \mathop{\mathsf{S}}_{\mathbf{p},q} |\mathbf{p}, q\rangle\langle \mathbf{p}, q| = 1\end{aligned}$$

since the $\{|\mathbf{p}, q\rangle\}$ form a complete set. However,

$$\begin{aligned}\Omega^{(\pm)}\Omega^{(\pm)\dagger} &= \mathop{\mathsf{S}}_{\mathbf{p}',q'} \mathop{\mathsf{S}}_{\mathbf{p},q} \left| \mathbf{p}', q' \begin{matrix}\text{in}\\ \text{out}\end{matrix} \right\rangle \langle \mathbf{p}', q'|\mathbf{p}, q\rangle \left\langle \mathbf{p}, q \begin{matrix}\text{in}\\ \text{out}\end{matrix} \right| \\ &= \mathop{\mathsf{S}}_{\mathbf{p},q} \left| \mathbf{p}, q \begin{matrix}\text{in}\\ \text{out}\end{matrix} \right\rangle \left\langle \mathbf{p}, q \begin{matrix}\text{in}\\ \text{out}\end{matrix} \right| \neq 1\end{aligned}$$

since the $\left\{ \left| \mathbf{p}, q \begin{matrix}\text{in}\\ \text{out}\end{matrix} \right\rangle \right\}$ do not form a complete set if the interaction is strong enough to form bound states of the two-particle system. The in and out states obviously cannot be bound, and hence are orthogonal to any bound states, which must have lower energies.

The operator $\mathfrak{S}$, defined by

$$\mathfrak{S} = \Omega^{(-)\dagger}\Omega^{(+)} = \underset{\mathbf{p}',q'}{\mathsf{S}}\ \underset{\mathbf{p},q}{\mathsf{S}}\ |\mathbf{p}', q'\rangle\langle\mathbf{p}', q' \text{ out}|\mathbf{p}, q \text{ in}\rangle\langle\mathbf{p}, q| \tag{19.36}$$

so that

$$S_{fi} = \langle\mathbf{p}_f, q_f|\mathfrak{S}|\mathbf{p}_i, q_i\rangle, \tag{19.37}$$

is unitary, however. Thus

$$\begin{aligned}
\mathfrak{S}^\dagger\mathfrak{S} &= \Omega^{(+)\dagger}\Omega^{(-)}\Omega^{(-)\dagger}\Omega^{(+)} \\
&= \underset{\mathbf{p}''',q'''}{\mathsf{S}}\ \underset{\mathbf{p}'',q''}{\mathsf{S}}\ \underset{\mathbf{p}',q'}{\mathsf{S}}\ \underset{\mathbf{p},q}{\mathsf{S}}\ |\mathbf{p}''', q'''\rangle\langle\mathbf{p}''', q''' \text{ in}|\mathbf{p}'', q'' \text{ out}\rangle \\
&\quad\cdot \langle\mathbf{p}'', q''|\mathbf{p}', q'\rangle\langle\mathbf{p}', q' \text{ out}|\mathbf{p}, q \text{ in}\rangle\langle\mathbf{p}, q| \\
&= \underset{\mathbf{p}''',q'''}{\mathsf{S}}\ \underset{\mathbf{p}',q'}{\mathsf{S}}\ \underset{\mathbf{p},q}{\mathsf{S}}\ |\mathbf{p}''', q'''\rangle\langle\mathbf{p}''', q''' \text{ in}|\mathbf{p}', q' \text{ out}\rangle \\
&\quad\cdot \langle\mathbf{p}', q' \text{ out}|\mathbf{p}, q \text{ in}\rangle\langle\mathbf{p}, q|.
\end{aligned} \tag{19.38}$$

But we can form a complete set from the $\{|\mathbf{p}, q \text{ out}\rangle\}$ by adding to this set of states all the bound states $\{|E, q\rangle\}$ of the system:

$$\underset{\mathbf{p}',q'}{\mathsf{S}}\ |\mathbf{p}', q' \text{ out}\rangle\langle\mathbf{p}', q' \text{ out}| + \underset{\substack{\text{bound}\\\text{states}}}{\mathsf{S}}\ |E', q'\rangle\langle E', q'| = 1;$$

then

$$\begin{aligned}
\langle\mathbf{p}''', q''' \text{ in}|\mathbf{p}, q \text{ in}\rangle &= \underset{\mathbf{p}',q'}{\mathsf{S}}\ \langle\mathbf{p}''', q''' \text{ in}|\mathbf{p}', q' \text{ out}\rangle\langle\mathbf{p}', q' \text{ out}|\mathbf{p}, q \text{ in}\rangle \\
&\quad + \underset{\substack{\text{bound}\\\text{states}}}{\mathsf{S}}\ \langle\mathbf{p}''', q''' \text{ in}|E', q'\rangle\langle E', q'|\mathbf{p}, q \text{ in}\rangle \\
&= \underset{\mathbf{p}',q'}{\mathsf{S}}\ \langle\mathbf{p}''', q''' \text{ in}|\mathbf{p}', q' \text{ out}\rangle\langle\mathbf{p}', q' \text{ out}|\mathbf{p}, q \text{ in}\rangle
\end{aligned} \tag{19.39}$$

since $\langle E', q'|\mathbf{p}, q \text{ in}\rangle = 0$, the bound states being orthogonal to the in states. Using Eq. (19.39) in Eq. (19.38), we get

$$\begin{aligned}
\mathfrak{S}^\dagger\mathfrak{S} &= \underset{\mathbf{p}''',q'''}{\mathsf{S}}\ \underset{\mathbf{p},q}{\mathsf{S}}\ |\mathbf{p}''', q'''\rangle\langle\mathbf{p}''', q''' \text{ in}|\mathbf{p}, q \text{ in}\rangle\langle\mathbf{p}, q \text{ in}| \\
&= \underset{\mathbf{p},q}{\mathsf{S}}\ |\mathbf{p}, q\rangle\langle\mathbf{p}, q| = 1.
\end{aligned} \tag{19.40}$$

Similarly, we may use the orthogonality of the bound states with the out states, $\langle E', q'|\mathbf{p}, q \text{ out}\rangle = 0$, to obtain $\mathfrak{S}\mathfrak{S}^\dagger = 1$.

This is extremely important; it means that $\mathfrak{S}$ is a normalization-preserving operator in quantum theory, transforming one momentum eigenstate into another. Yet $\mathfrak{S}$ is constructed from the interactions responsible for scattering, and hence contains all the information about scattering contained in a given model for those interactions. Equations (19.36) and (19.24) show explicitly the relations between $\mathfrak{S}$ and the abstract operators $\mathfrak{h}_0$ and $\mathfrak{V}$. For this formalism to be useful, we must be able to calculate matrix elements S_{fi} or, equivalently, T_{fi}, and convert these into expressions for transition rates and cross-sections. Calculation of the T_{fi} usually involves approximation procedures, usually perturbation-theoretical calculations. Many interactions of interest, however, cannot be considered perturbations, and more sophisticated techniques are required for them.

19.3 TRANSITION RATES AND CROSS-SECTIONS

For the moment, let us assume that the matrix elements T_{fi} are known. How do we calculate transition rates and cross-sections from them?

To start with, we consider the meaning of S_{fi}. It is the amplitude for the probability that the initial state (i) (which asymptotically is approximately an eigenstate of momentum, say $|\mathbf{p}_i, q_i\rangle = |\mathbf{p}_i, a, b\rangle$) is transformed by the interaction $\mathfrak{V} = \mathfrak{V}_{AB}$ into a final state (f) which asymptotically is approximately an eigenstate of momentum, say $|\mathbf{p}_f, q_f\rangle = |\mathbf{p}_f, a, b\rangle$. Hence $|S_{fi}|^2$ is the probability of this transformation. We are interested in the cases $f \neq i$, which reveal the effect of the interaction, and hence we shall deal with

$$|S_{fi}|^2 = |2\pi\delta(E_f - E_i)T_{fi}|^2, \qquad f \neq i.$$

The presence of the δ-function ensures energy conservation, but the presence of its square is a nuisance; it makes $|S_{fi}|^2$ appear to be either infinite or zero even when we take into account the fact that E_f is a continuous variable for scattering processes.

However, we note that

$$\begin{aligned} 2\pi\delta(E_f - E_i) &= \lim_{\tau/2 \to \infty} \int_{-\tau/2}^{\tau/2} e^{i(E_f - E_i)t/\hbar} \frac{dt}{\hbar} \\ &= \begin{cases} \lim\limits_{\tau \to \infty} \dfrac{\tau}{\hbar} & \text{if } E_f = E_i \\ 0 & \text{if } E_f \neq E_i, \end{cases} \end{aligned}$$

where τ is the time interval during which the scattering takes place. Presumably τ is sufficiently large that E_f is quite well defined; in fact, the δ-function arose from letting $\epsilon \to 0$ below Eq. (19.29), and this was allowed only if $T = \tau/2 \gg 1/\epsilon$, where T is the time between the beginning of the experiment and the collision. Hence we may write, dropping the limit as $\tau \to \infty$ but assuming τ very large,

$$|S_{fi}|^2 = (\tau/\hbar)2\pi\delta(E_f - E_i)|T_{fi}|^2 = P_{fi}.$$

The rate of transition, then, is

$$\dot{P}_{fi} = \frac{d}{d\tau}|S_{fi}|^2 = \frac{2\pi}{\hbar}\delta(E_f - E_i)|T_{fi}|^2. \tag{19.41}$$

Compare this with Eq. (15.54), which is for the transition rate in first order of perturbation theory; evidently $\langle E', q'|\mathfrak{H}_1|E, q\rangle$, which in our present notation would read $\langle \mathbf{p}_f, q_f|\mathfrak{V}|\mathbf{p}_i, q_i\rangle$, is the first-order approximation to T_{fi}.

Now we may eliminate the remaining delta function just as we did in Section 15.6. If the number of states within an energy interval of width dE_f about E_f is $\rho(E_f)dE_f$, the rate of transition to any state with the quantum numbers of $|\mathbf{p}_f, q_f\rangle$ but with arbitrary energy is

$$w_{fi} = \int_0^\infty \dot{P}_{fi}\rho(E_f)\,dE_f = \frac{2\pi}{\hbar}|T_{fi}|^2\rho(E_f). \tag{19.42}$$

The work in Sections 15.7 and 15.8 may now be applied to this result. The only difference between w_{fi} obtained here and $w_{10}{}^{(1)}$ obtained in Section 15.6 and used in the subsequent sections is that here T_{fi} is exact (perturbation theory has not been invoked) whereas $\langle E', q'|\mathfrak{H}_1|E, q\rangle$ appearing in Chapter 15 was obtained through first-order perturbation theory.

Therefore we have

$$\frac{dw_{fi}}{d\Omega} = \frac{2\pi\mu k_f}{\hbar^3}\left(\frac{L}{2\pi}\right)^3 |T_{fi}|^2 \tag{19.43}$$

as the counterpart of Eq. (15.57), and

$$w_{fi} = \frac{2\pi\mu k_f}{\hbar^3}\left(\frac{L}{2\pi}\right)^3 \int |T_{fi}|^2\,d\Omega \tag{19.44}$$

as the exact version of Eq. (15.58). Similarly, Eq. (15.62) becomes

$$\frac{d\sigma_{fi}}{d\Omega} = \frac{\mu^2 L^6}{(2\pi\hbar^2)^2}|T_{fi}|^2 \tag{19.45}$$

(sec Eq. (15.59)), and of course

$$\sigma_{fi} = \frac{\mu^2 L^6}{(2\pi\hbar^2)^2}\int |T_{fi}|^2\,d\Omega. \tag{19.46}$$

Here L is the length of an edge of the cube to which periodic boundary conditions are applied; it is canceled in these equations by a factor L^{-3} that must appear in T_{fi} when periodic normalization is used in the state vectors in terms of which T_{fi} is calculated. (See Section 15.8.) Usually we choose $L = 1$.

19.4 WAVE-FUNCTION FORMULATION; GREEN'S FUNCTIONS

For purposes of calculation, it is useful in nonrelativistic quantum theory to express our formalism in terms of wave functions. We began this task earlier, obtaining in (19.20) the integral equation

$$\psi_1{}^{(\text{in}/\text{out})}(\mathbf{x}) = \phi_1(\mathbf{x}) + \int d^3x' G^{(\pm)}(\mathbf{x} - \mathbf{x}')V(\mathbf{x}')\psi_1{}^{(\text{in}/\text{out})}(\mathbf{x}').$$

The keys to even approximate solution of this equation are the Green's functions

$$G^{(\pm)}(\mathbf{x}-\mathbf{x}') = -\frac{2\mu}{\hbar^2}\frac{1}{(2\pi)^3}\int d^3k e^{i\mathbf{k}\cdot(\mathbf{x}-\mathbf{x}')}\frac{1}{\mathbf{k}^2 - K^2 \mp i\epsilon'},$$

where

$$K \equiv \frac{\sqrt{2\mu E}}{\hbar}, \qquad \epsilon' \equiv \frac{2M}{\hbar^2}\epsilon.$$

We can evaluate the integral over $\mathbf{k}$ explicitly, using spherical coordinates; thus

$$\begin{aligned} G^{(\pm)}(\mathbf{x}-\mathbf{x}') &= -\frac{2\mu}{\hbar^2}\frac{1}{(2\pi)^3}\int_0^\infty dk\, k^2 \int d\Omega \frac{e^{ik|\mathbf{x}-\mathbf{x}'|\cos\theta}}{k^2 - K^2 \mp i\epsilon'} \\ &= -\frac{2\mu}{\hbar^2(2\pi)^2}\int_0^\infty \frac{k^2\,dk}{k^2 - K^2 \mp i\epsilon'}\int_{-1}^{1} d\xi\, e^{ik|\mathbf{x}-\mathbf{x}'|\xi} \\ &= -\frac{2\mu}{\hbar^2(2\pi)^2}\int_0^\infty \frac{k}{k^2 - K^2 \mp i\epsilon'}\,\frac{e^{ik|\mathbf{x}-\mathbf{x}'|} - e^{-ik|\mathbf{x}-\mathbf{x}'|}}{i|\mathbf{x}-\mathbf{x}'|}\,dk, \end{aligned}$$

where θ is the angle between $\mathbf{k}$ and $(\mathbf{x}-\mathbf{x}')$ and $\xi \equiv \cos\theta$. The second term in the integrand may be written

$$\int_{-\infty}^{0} dk \frac{k}{k^2 - K^2 \mp i\epsilon'}\frac{e^{ik|\mathbf{x}-\mathbf{x}'|}}{i|\mathbf{x}-\mathbf{x}'|},$$

so that we have

$$G^{(\pm)}(\mathbf{x}-\mathbf{x}') = i\frac{2\mu}{(2\pi\hbar)^2}\frac{1}{|\mathbf{x}-\mathbf{x}'|}\int_{-\infty}^{\infty}\frac{ke^{ik|\mathbf{x}-\mathbf{x}'|}}{k^2 - K^2 \mp i\epsilon'}\,dk.$$

This can be evaluated using a contour integral in the complex k-plane.

Consider

$$\int_{C_R}\frac{ke^{ik\eta}}{k^2 - K^2 \mp i\epsilon}\,dk$$

when C_R is the contour indicated in Fig. 19.1 and η is a positive real number. If the radius of the arc becomes infinite, the integrand vanishes on the arc and the contour integral becomes equal to the integral along the infinite real axis:

$$\lim_{R\to\infty}\int_{C_R}\frac{ke^{ik\eta}\,dk}{k^2 - K^2 \mp i\epsilon'} = \int_{-\infty}^{\infty}\frac{ke^{ik\eta}\,dk}{k^2 - K^2 \mp i\epsilon}.$$

On the other hand, the Cauchy residue theorem tells us that the contour is equal to $2\pi i$ times the sum of coefficients of linear factors in the denominator that vanish inside the contour; each coefficient is evaluated at the point where the correspond-

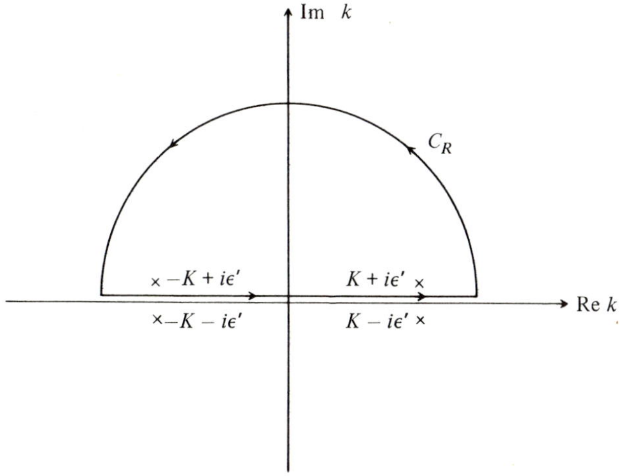

Fig. 19.1. Contour for evaluating $G^{(\pm)}(\mathbf{x} - \mathbf{x}')$.

ing factor vanishes. Since the denominator is

$$(k + \sqrt{K^2 \pm i\epsilon'})(k - \sqrt{K^2 \pm i\epsilon'}),$$

these coefficients are

$$\left.\frac{ke^{ik\eta}}{k - \sqrt{K^2 \pm i\epsilon}}\right|_{k=-\sqrt{K^2 \pm i\epsilon'}}, \qquad \left.\frac{ke^{ik\eta}}{k + \sqrt{K^2 \mp i\epsilon'}}\right|_{k=+\sqrt{K^2 \mp i\epsilon'}}.$$

The first of these applies when the lower sign in front of $i\epsilon'$ is used, for only then will $\sqrt{K^2 \pm i\epsilon'} = -K\sqrt{1 \pm i\epsilon'/K} \simeq -K \mp i\epsilon'/2$ lie within the contour (above the real axis); the second applies when the upper sign appears, for only then does $+\sqrt{K^2 \pm i\epsilon'} \simeq K \pm i\epsilon'/2$ lie inside the contour. Hence the values of the contour integral are, when $\epsilon' \to 0$,

$$i2\pi \cdot \tfrac{1}{2}\, e^{\pm iK\eta} = i\pi e^{\pm iK\eta},$$

and therefore

$$G^{(\pm)}(\mathbf{x} - \mathbf{x}') = -\frac{\mu}{2\pi\hbar^2}\frac{e^{\pm iK|\mathbf{x}-\mathbf{x}'|}}{|\mathbf{x} - \mathbf{x}'|}. \tag{19.47}$$

Our integral equation, with this explicit expression for $G^{(\pm)}(\mathbf{x} - \mathbf{x}')$, now has the form

$$\psi_1^{\binom{\text{in}}{\text{out}}}(\mathbf{x}) = \phi_1(\mathbf{x}) - \frac{\mu}{2\pi\hbar^2}\int d^3x' \frac{e^{\pm iK|\mathbf{x}-\mathbf{x}'|}}{|\mathbf{x} - \mathbf{x}'|} V(\mathbf{x}')\psi_1^{\binom{\text{in}}{\text{out}}}(\mathbf{x}'). \tag{19.48}$$

Now $\mathbf{x}$ is the location at which we wish to make an observation of the system, say the location of a particle detector in the relative coordinate system. It is far from the location of the scattering, which is the region of space near the origin within which $V(\mathbf{x}')$ is nonvanishing; hence the integration over $\mathbf{x}'$ is assumed to be over a limited region of space such that $|\mathbf{x}'| \ll |\mathbf{x}|$. Hence

$$\begin{aligned} |\mathbf{x} - \mathbf{x}'| &= \sqrt{\mathbf{x}^2 - 2\mathbf{x}\cdot\mathbf{x}' + \mathbf{x}'^2} \\ &= |\mathbf{x}|\sqrt{1 - 2\frac{\mathbf{x}\cdot\mathbf{x}'}{|\mathbf{x}|\,|\mathbf{x}'|} + \left|\frac{\mathbf{x}'}{\mathbf{x}}\right|^2} \\ &\simeq |\mathbf{x}| - \frac{\mathbf{x}\cdot\mathbf{x}'}{|\mathbf{x}|} + \text{terms of order}\left|\frac{\mathbf{x}'}{\mathbf{x}}\right|^2. \end{aligned}$$

We keep both these terms when approximating $|\mathbf{x} - \mathbf{x}'|$ in the exponent, but keep only the first term in approximating the denominator, thus obtaining

$$\psi_1^{\binom{\text{in}}{\text{out}}}(\mathbf{x}) \underset{r\to\infty}{\simeq} \phi_1(\mathbf{x}) - \frac{\mu}{2\pi\hbar^2}\frac{e^{\pm iKr}}{r}\int d^3x'\, e^{\mp \mathbf{K}'\cdot\mathbf{x}'}V(\mathbf{x}')\psi_1^{\binom{\text{in}}{\text{out}}}(\mathbf{x}'), \tag{19.49}$$

where $r \equiv |\mathbf{x}|$ and $\mathbf{K}' \equiv (\mathbf{x}/r)K$ represents the momentum of a particle entering our detector, the detector being located at position $\mathbf{x}$ and the "point of scattering" (actually a region of scattering) being located at the origin.

It is common practice to define

$$f_{K\pm} \equiv -\frac{\mu}{2\pi\hbar^2}\int d^3x'\, e^{\mp i\mathbf{K}'\cdot\mathbf{x}'}V(\mathbf{x}')\psi_1^{\binom{\text{in}}{\text{out}}}(\mathbf{x}'), \tag{19.50}$$

so that

$$\psi_1^{\binom{\text{in}}{\text{out}}}(\mathbf{x}) \underset{r\to\infty}{\simeq} \phi_1(\mathbf{x}) + \frac{e^{\pm iKr}}{r}f_{K\pm}; \tag{19.51}$$

$f_{K\pm}$ is called the *scattering amplitude*. If the initial wave is asymptotically a momentum eigenfunction with momentum $\mathbf{p}_i = \hbar\mathbf{K}$, i.e.

$$\phi_1(\mathbf{x}) = (2\pi)^{-3/2}\, e^{i\mathbf{K}\cdot\mathbf{x}}$$

or simply

$$\phi_1(\mathbf{x}) = e^{i\mathbf{K}\cdot\mathbf{x}}$$

for unit box normalization, then $f_{K\pm}$ is often written as a function of the angle θ between $\mathbf{K}$ and $\mathbf{K}'$ or between $\mathbf{p}_i$ and $\mathbf{p}_f = \hbar\mathbf{K}'$, $f_{K\pm}(\theta)$, since θ is the angle through which the incident particle is then scattered. Note that $\mathbf{K}\cdot\mathbf{K}' = K^2\cos\theta$. If the potential $V(\mathbf{x})$ is not symmetric about the direction $\hat{\mathbf{K}}$, $f_{K\pm}$ may also depend on the azimuthal angle φ measured about the axis $\hat{\mathbf{K}}$; then we write $f_{K\pm}(\theta, \varphi)$.

The scattering amplitude is connected intimately with the T-matrix. Since $\psi_{\text{sc}}(r) \equiv (e^{+iKr}/r)f_{K+}$ represents the scattered portion of the wave function $\psi_1^{(\text{in})}(\mathbf{x})$, the current density

$$\mathbf{j}_{\mathrm{sc}} \equiv \frac{\hbar}{i2\mu}[\psi_{\mathrm{sc}}{}^*\nabla\psi_{\mathrm{sc}} - (\nabla\psi_{\mathrm{sc}}{}^*)\psi_{\mathrm{sc}}]$$

$$= \frac{\hbar}{i2\mu}2iK\frac{\mathbf{x}}{r^3}|f_{K+}(\theta)|^2 = \frac{\hbar\mathbf{K}_f}{Mr^2}|f_{K+}(\theta)|^2 = \frac{\mathbf{v}_f}{r^2}|f_{K+}(\theta)|^2$$

represents the emerging current of scattered waves. If our detector allows particles to enter it through a surface $dS = \hat{\mathbf{r}}r^2\,d\Omega$ at angle θ from the incident momentum, the fraction of probability per unit time that a scattered particle will enter it is

$$d\mathbf{S}\cdot j_{\mathrm{sc}} = v|f_{K+}(\theta)|^2\,d\Omega$$

where $v = |\mathbf{v}_f| = |\mathbf{v}_i|$. The differential cross-section is obtained when this is divided by the incident intensity; the latter is just $I_{\mathrm{inc}} = v_i = v$ when we normalize to one particle per unit volume (or $|I_{\mathrm{inc}}| = L^3$ for box normalization, $(2\pi)^3$ for δ-function normalization). (See Section 15.8.) Hence

$$\frac{d\sigma}{d\Omega} = |f_{K+}(\theta)|^2$$

or (19.52)

$$L^{-3}|f_{K+}(\theta)|^2,\ (2\pi)^{-3}|f_{K+}(\theta)|^2$$

for box, δ-function normalizations, and it follows that

$$|f_{K+}(\theta)|^2 = \frac{\mu^2}{(2\pi)^2\hbar^4}|T_{fi}|^2 \tag{19.53}$$

for spinless particles when we normalize to unit volume. Therefore, evaluation of $f_{K+}(\theta)$ is equivalent to evaluation of T_{fi} (to within an irrelevant phase factor).

It is sometimes possible to arrange the wave function in such a way that $f_{K+}(\theta)$ can be identified from the asymptotic form of the wave function. On the other hand, sometimes it is not necessary to know $\psi_i^{(\mathrm{in})}$ to evaluate $f_{K+}(\theta)$ reasonably accurately; we have seen this already in the time-dependent perturbation calculations of Chapter 15. In fact we may use Eq. (19.50) to get the same results without using the time-dependent theory. To get the first-order Born approximation, for example, we simply replace $\psi^{(\mathrm{in})}(\mathbf{x})$ in Eq. (19.50) with

$$e^{\pm i\mathbf{K}\cdot\mathbf{x}'} = \phi_{\pm\mathbf{K}}(\mathbf{x}'),$$

using unit box normalization; this yields

$$f_{K\pm} = -\frac{\mu}{2\pi\hbar^2}\int d^3x'\,e^{\pm i(\mathbf{K}-\mathbf{K}')\cdot\mathbf{x}'}V(\mathbf{x}')$$

$$= -\frac{\mu}{2\pi\hbar^2}U(\mp[\mathbf{K}-\mathbf{K}']) \tag{19.54}$$

in the notation of Section 15.8. This gives exactly the same cross-section as the first-order time-dependent perturbation calculation.

19.5 WAVE PACKETS

As was stressed earlier in this chapter, the wave function $\phi_1(\mathbf{x})$ representing the pair of particles in relative coordinates at the beginning of the experiment cannot be an exact momentum eigenfunction because the particles must be sufficiently well localized at large relative separations that they do not interact with one another. The function $\phi_1(\mathbf{x})$ must therefore be a wave packet, say of the form

$$\phi_1(\mathbf{x}) = e^{i\mathbf{k}(\mathbf{x}-\mathbf{x}_0)}\pi(\mathbf{x} - \mathbf{x}_0) \tag{19.55}$$

where $\mathbf{x}_0$ is the separation of the centers of the particles at the initiation of the experiment. Here we use box normalization, the box being a unit cube. The function $\pi(\boldsymbol{\xi})$ has its maximum at $\boldsymbol{\xi} = \mathbf{0}$ and vanishes for $|\boldsymbol{\xi}| \gtrsim \Delta x$; furthermore, we may assume that $\pi(\boldsymbol{\xi}) = \pi(-\boldsymbol{\xi})$, i.e. that the wave packet is symmetric about $\mathbf{x}_0$ at the start of the experiment. The constant phase factor $e^{i\mathbf{k}\cdot\mathbf{x}_0}$ is included for later convenience.

In this form, $\phi_1(\mathbf{x})$ is not an eigenfunction of P, but it should have mean momentum $\hbar\mathbf{k}$. The reader may verify that this will be ensured if $\pi(\mathbf{x} - \mathbf{x}_0)$ is real to within a constant phase factor. We may expand $\phi_1(\mathbf{x})$ in momentum eigenfunctions,

$$\begin{aligned}\phi_1(\mathbf{x}) &= \int d^3k'\; e^{i\mathbf{k}'\cdot(\mathbf{x}-\mathbf{x}_0)}\varphi(\mathbf{k}' - \mathbf{k}) \\ &= e^{i\mathbf{k}\cdot(\mathbf{x}-\mathbf{x}_0)}\int d^3q\; e^{i\mathbf{q}\cdot(\mathbf{x}-\mathbf{x}_0)}\varphi(\mathbf{q}).\end{aligned} \tag{19.56}$$

The realness of

$$\pi(\boldsymbol{\xi}) = \int d^3q\; e^{i\mathbf{q}\cdot\boldsymbol{\xi}}\varphi(\mathbf{q}) \tag{19.57}$$

then implies that $\varphi^*(\mathbf{q}) = \varphi(-\mathbf{q})$, and the invariance of $\pi(\boldsymbol{\xi})$ under $\boldsymbol{\xi} \to -\boldsymbol{\xi}$ implies that $\varphi(\mathbf{q})$ is real.

As the experiment proceeds, $\phi_1(\mathbf{x})$ becomes a function of time, $\phi_1(\mathbf{x}, t)$. What we actually are given is $\phi_1(\mathbf{x}, -T)$. Hence, for times near $t = -T$, when $\mathfrak{h} \simeq \mathfrak{h}_0$, or if the interaction $\mathfrak{V}$ vanishes, we have (since $\mathsf{h}_0 = -(\hbar^2/2\mu)\nabla^2$)

$$\begin{aligned}\phi_1(\mathbf{x}, t) &= e^{i(t+T)h_0/\hbar}\phi_1(\mathbf{x}) \\ &= e^{-i(t+T)(\hbar/2\mu)\nabla^2]}\phi_1(\mathbf{x}) \\ &= e^{i\mathbf{k}\cdot(\mathbf{x}-\mathbf{x}_0)}\int d^3q\; e^{i\mathbf{q}(\mathbf{x}-\mathbf{x}_0)}\; e^{i\hbar(\mathbf{k}^2+\mathbf{q}^2)(t+T)2\mu}\varphi(\mathbf{q})\end{aligned}$$

$$= e^{i\mathbf{k}\cdot(\mathbf{x}-\mathbf{x}_0) - i(\hbar \mathbf{k}^2/2\mu)(t+T)} \int d^3q\, e^{i\mathbf{q}\cdot(\mathbf{x}-\mathbf{x}_0) - i(\hbar \mathbf{k}\cdot\mathbf{q}/\mu)(t+T) - i(\hbar \mathbf{q}^2/2\mu)(t+T)} \varphi(\mathbf{q})$$

$$= e^{i\mathbf{k}\cdot(\mathbf{x}-\mathbf{x}_0) - i(\hbar \mathbf{k}^2/2\mu)(t+T)} \int d^3q\, e^{i\mathbf{q}\cdot[(\mathbf{x}-\mathbf{x}_0) - (\hbar \mathbf{k}/\mu)(t+T)] - i(\hbar \mathbf{q}^2/2\mu)(t+T)} \varphi(\mathbf{q}).$$

Now as we saw in discussing the minimum uncertainty product and the spread of wave packets, it is the last term in the exponential within the integral that is responsible for spreading the wave packet in time. In order that the spreading be negligible, even at $t = T$, we must have

$$\hbar \frac{\mathbf{q}^2}{\mu} T \ll 1. \tag{19.58}$$

The momentum packet $\varphi(\mathbf{q})$ has a width $\Delta k = \Delta p/\hbar$ so that $\hbar q \lesssim \Delta p$; hence we require that

$$\frac{(\Delta \mathbf{p})^2}{\mu} \frac{T}{\hbar} \ll 1. \tag{19.59a}$$

If our wave packet is minimal, we have $\Delta p\, \Delta x \simeq \hbar$ and this becomes

$$\frac{\Delta p\, T}{\mu\, \Delta x} \ll 1, \tag{19.59b}$$

a constraint on the three parameters Δp, Δx and T. Another way of writing this becomes evident when we note that $vT = (p/\mu)T \simeq |\mathbf{x}_0|$; hence we have

$$\left(\frac{\Delta p}{p}\right) \Big/ \left(\frac{\Delta \mathbf{x}}{|\mathbf{x}_0|}\right) \ll 1 \tag{19.59c}$$

which says that the relative uncertainty in momentum must be far less than the relative uncertainty in position of the particles.

If these conditions are met, the time-dependent wave function $\phi_1(\mathbf{x}, t)$ becomes

$$\phi_1(\mathbf{x}, t) = e^{i\mathbf{k}\cdot(\mathbf{x}-\mathbf{x}_0) - i(\hbar^2 \mathbf{k}^2/2\mu)(t+T)} \pi(\mathbf{x} - \mathbf{x}_0 - \mathbf{v}(t+T)) \tag{19.60}$$

where $\mathbf{v} = \hbar \mathbf{k}/\mu$. The wave packet at time t, then, is centered at $\mathbf{x}_0 + \mathbf{v}(t + T)$ in accordance with our classical expectations, but has not changed its shape nor its width.

We have ignored interactions here. How can we include them, and what happens when they are included? They can be introduced by replacing $e^{i\mathbf{k}'\cdot(\mathbf{x}-\mathbf{x}_0)}$ in Eq. (19.56) with

$$\psi_{\mathbf{k}'}{}^{\binom{\text{in}}{\text{out}}}(\mathbf{x});$$

that is, plane waves are replaced by

$$\psi_{\mathbf{k}'}{}^{\binom{\text{in}}{\text{out}}}(\mathbf{x}) = e^{i\mathbf{k}'\cdot\mathbf{x}} - \frac{\mu}{2\pi\hbar^2} \int d^3x' \frac{e^{\pm ik'|\mathbf{x}'-\mathbf{x}|}}{|\mathbf{x}' - \mathbf{x}|} V(\mathbf{x}') \psi_{\mathbf{k}'}{}^{\binom{\text{in}}{\text{out}}}(\mathbf{x}').$$

The first term in $\psi_{\mathbf{k}'}{}^{\left(\substack{\text{in}\\ \text{out}}\right)}(\mathbf{x})$ leads to exactly the results we obtained above for $\phi_1(\mathbf{x})$, except that the wave packet now is $\pi(\mathbf{x})$ and hence is centered at the origin; this latter is because $\psi_{\mathbf{k}'}{}^{\left(\substack{\text{in}\\ \text{out}}\right)}(\mathbf{x})$ is the wave function at time $t = 0$, when the collision is taking place, and hence is represented by wave packets centered at the origin. The second contains all the scattering information. Thus we have

$$\phi_1{}^{\left(\substack{\text{in}\\ \text{out}}\right)}(\mathbf{x}) = \int d^3k' \, \varphi(\mathbf{k}' - \mathbf{k}) \psi_{\mathbf{k}'}{}^{\left(\substack{\text{in}\\ \text{out}}\right)}(\mathbf{x}). \tag{19.61}$$

Recall that $\psi_{\mathbf{k}'}{}^{\left(\substack{\text{in}\\ \text{out}}\right)}(\mathbf{x})$ is an eigenfunction of $\mathsf{h} = \mathsf{h}_0 + V(\mathbf{x})$ with eigenvalue $E' = \hbar^2\mathbf{k}'^2/2\mu$ (simply because $\left|\mathbf{p}' \substack{\text{in}\\ \text{out}}\right\rangle$ is an eigenstate of $\mathfrak{h}$ with eigenvalue $\mathbf{p}'^2/2\mu$); therefore we included the time by writing

$$\phi_1{}^{\left(\substack{\text{in}\\ \text{out}}\right)}(\mathbf{x}, t) = \int d^3k' \varphi(\mathbf{k}' - \mathbf{k}) e^{-i(\hbar^2/2\mu)\mathbf{k}'^2 t} \psi_{\mathbf{k}'}{}^{\left(\substack{\text{in}\\ \text{out}}\right)}(\mathbf{x}) \tag{19.62}$$

since $\phi_1{}^{\left(\substack{\text{in}\\ \text{out}}\right)}(\mathbf{x})$ is the wave function at $t = 0$ (not $t = -T$). When $|\mathbf{x}|$ is large, the asymptotic form (19.51) may be used for $\psi_{\mathbf{k}'}{}^{\left(\substack{\text{in}\\ \text{out}}\right)}(\mathbf{x})$; then

$$\begin{aligned}\phi_1{}^{\left(\substack{\text{in}\\ \text{out}}\right)}(\mathbf{x}, t) &= e^{i\mathbf{k}\cdot\mathbf{x}} e^{-i(\hbar/2\mu)\mathbf{k}^2 t} \pi(\mathbf{x} - \mathbf{v}t) \\ &\quad + \frac{1}{r} \int d^3k' \varphi(\mathbf{k}' - \mathbf{k}) e^{i(\hbar^2\mathbf{k}'^2/2\mu)t} e^{\pm ik'r} f_{k'\pm}(\theta).\end{aligned} \tag{19.63}$$

If we assume that $f_{k'\pm}(\theta)$ is essentially independent of k' for all $\mathbf{k}'$ within the momentum packet, we may let $f_{k'\pm}(\theta) \to f_{k\pm}(\theta)$ and take it outside the integral. Furthermore, we note that $\varphi(\mathbf{k}' - \mathbf{k}) = 0$ unless $\mathbf{k}'$ is parallel to $\mathbf{k}$, by assumption. Then, by letting $\mathbf{k}' = \mathbf{k} + \mathbf{q}$ as before and applying condition (19.59), we find that

$$\phi_1{}^{\left(\substack{\text{in}\\ \text{out}}\right)}(\mathbf{x}, t) = e^{i\mathbf{k}\cdot\mathbf{x} - i(\hbar/2\mu)\mathbf{k}^2 t} \pi(\mathbf{x} - \mathbf{v}t) + \frac{1}{r} e^{-i(\hbar\mathbf{k}^2/2\mu)t \pm ikr} \pi(-\mathbf{v}t + r\hat{\mathbf{k}}) f_{k\pm}(\theta). \tag{19.64}$$

The last term vanishes unless

$$r \simeq \pm vt;$$

for "in" states this means that scattering effects may be observed on a sphere of radius $r = vt$, which is what we should expect classically. For "out" states, we have spherical waves converging on the origin along with plane waves at negative times. At large positive times, we have only plane waves since, for positive times, $-\mathbf{v}t - r\hat{\mathbf{k}} = -\hat{\mathbf{k}}(vt + r)$ cannot vanish at any value of r, and hence $\pi(-\mathbf{v}t - r\hat{\mathbf{k}})$ is vanishingly small.

Figure 19.2 illustrates the situation for "in" waves. At $t < 0$ there is a packet of plane waves moving toward the origin, and at $t > 0$ there is an additional packet of spherical waves, represented by the second term on the right-hand side of Eq. (19.64), going outward from the origin with velocity v. For "out" waves, we have Fig. 19.3. Here for $t < 0$ there is a packet of plane waves moving toward the origin plus a packet of spherical waves converging on the origin; for $t > 0$ the spherical waves have vanished and only a plane wave moving away from the origin remains.

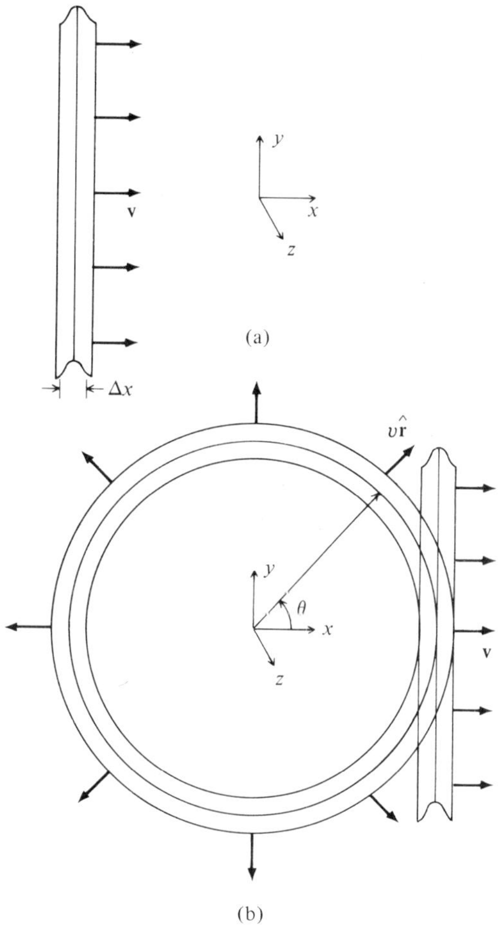

Fig. 19.2. (a) Wave packet with width Δx when $t < 0$; scattering has not yet begun. (b) Incident packet plus spherical packet of scattered waves for $t > 0$. The angular variation in intensity of scattered waves is determined by $f_{k+}(\theta)$. This is the situation represented by the "in" states.

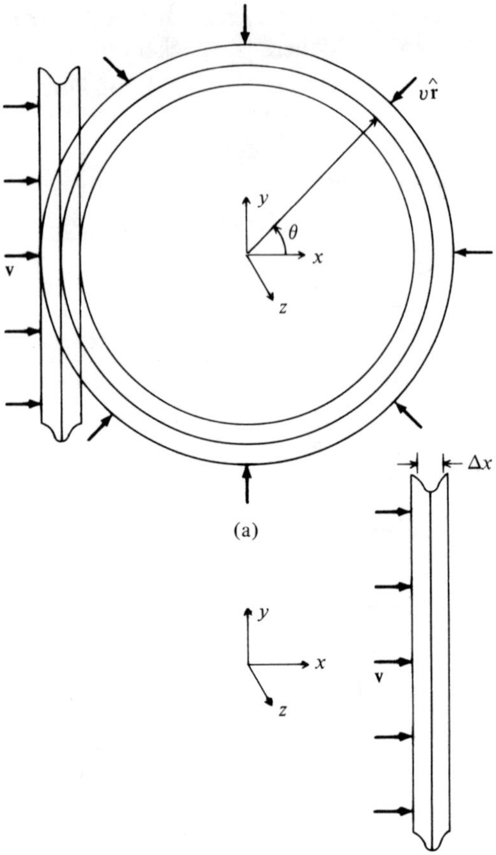

Fig. 19.3. (a) Incoming spherical waves plus plane wave packet for $t < 0$. The angular variation in intensity of incoming spherical waves is given by $f_{k-}(\theta)$. (b) Receding plane wave packet of width Δx for $t > 0$. This is the situation represented by the "out" states.

The "in" waves correspond to the situation seen by the observer who sets up the scattering experiment and initially observes plane waves. "Out" waves correspond to the situation seen by the one at the detector, who is observing final plane waves. We may think of the "in" waves as formation waves, which form the intermediate state of the system. The system then decays from this intermediate state when an outgoing plane wave is observed, and hence the "out" waves can be thought of as decay waves. Although the plane wave in Fig. 19.3 is pictured as moving along the x-axis, in an experiment it is observed moving in the direction of the detecting apparatus.

Thus

$$S_{fi} = \int d^3x\, \psi_{\mathbf{k}_f}{}^{(\text{out})}(\mathbf{x})\psi_{\mathbf{k}_i}{}^{(\text{in})}(\mathbf{x})$$

gives the amplitude for the probability of observing an outgoing plane-wave packet with momentum $\mathbf{k}_f$, given that the intermediate state of the system is formed by ingoing plane waves with momentum $\mathbf{k}_i$ and that there is an interaction represented by $\mathfrak{V}$ or $V(\mathbf{x})$.

Note that we need not actually calculate S_{fi}, nor even $\psi_{\mathbf{k}_f}{}^{(\text{out})}(\mathbf{x})$ to get the scattering cross-section. As we have seen $\psi_{\mathbf{k}_i}{}^{(\text{in})}(\mathbf{x})$ gives sufficient information through the scattering amplitude $f_{k_i+}(\theta)$

This temporal analysis of the behavior of wave packets in the scattering process serves two purposes:

1. to justify our cavalier attitude toward wave packets in Section 19.1 and provide criteria for the nonspreading of the wave packets through condition (19.59), and

2. to provide a mental picture for the scattering process. We shall now return to the time-independent view of things, neglecting wave packets and using plane waves for $\phi_1(\mathbf{x})$.

19.6 CONNECTION BETWEEN "IN" AND "OUT" FUNCTIONS

Consideration of the results of the preceding section or of the original definitions of the "in" and "out" states leads to the conclusion that the "out" states may be obtained from the "in" states through the motion-reversal operation. Therefore, as we see from Section 12.12,

$$\psi_{\mathbf{k}}{}^{(\text{out})}(\mathbf{x}) = \psi_{\mathbf{k}}{}^{(\text{in})*}(\mathbf{x}). \tag{19.65}$$

This in turn leads to the conclusion that

$$f_{k-}(\theta) = f_{k+}^*(\theta). \tag{19.66}$$

Knowing this, we may confine our attention to the "in" functions, converting to "out" functions at will via this transformation.

19.7 PARTIAL WAVE ANALYSIS OF SCATTERING

The interaction $\mathfrak{V}$, or $V(\mathbf{x})$, is frequently spherically symmetric in practical problems, or at least approximately so. In such cases it is natural to consider intermediate states of the system in terms of eigenstates of angular momentum. Indeed, the interaction may be much stronger for some angular-momentum eigenstates than for others and good approximations may be obtained by considering only those terms in the cross-section that arise from angular-momentum eigenstates for which the interaction gives large scattering amplitudes. Therefore we shall expand $\psi_{\mathbf{k}}^{(\text{in})}(\mathbf{x})$ in orbital angular-momentum eigenfunctions, noting first that

$$e^{i\mathbf{k}\cdot\mathbf{x}} = \sum_{l=0}^{\infty} i^l(2l+1)j_l(kr)P_l(\cos\theta) \tag{19.67}$$

where θ is the angle between $\mathbf{k}$ and $\mathbf{x}$.

Now $\psi_{\mathbf{k}}^{(\text{in})}(\mathbf{x})$ contains outgoing spherical waves in addition to the plane waves $e^{i\mathbf{k}\cdot\mathbf{x}}$, and a complete set of outgoing spherical waves is provided by the set $\{h_l^{(+)}(kr)\}$ of outgoing spherical Hankel functions; $h_l^{(+)}(kr)$ is a solution to the same radial wave equation that has $j_l(kr)$ as a solution. Therefore, outside the range of the potential $V(\mathbf{x})$, where $h_l^{(+)}(kr)$ is a solution to the radial equation, we must have

$$\psi_{\mathbf{k}}^{(\text{in})}(\mathbf{x}) = \sum_{l=0}^{\infty} i^l(2l+1)u^{(+)}(l,k,r)P_l(\cos\theta), \tag{19.68}$$

where

$$u^{(+)}(l,k,r) = j_l(kr) + \Lambda_l(k)h_l^{(+)}(kr) \tag{19.69}$$

for r greater than the range of the potential $V(\mathbf{x})$, $\Lambda_l(k)$ being a coefficient related to the scattering amplitude.

It is convenient to rewrite Eq. (19.69) using the asymptotic behaviors of $j_l(kr)$ and $h_l^{(+)}(kr)$ to suggest a more convenient constant than $\Lambda_l(k)$. Thus we have

$$\begin{aligned} u^{(+)}(l,k,r) &\underset{r\to\infty}{\sim} \frac{\cos\left(kr - \frac{1}{2}(l+1)\pi\right)}{kr} + \Lambda_l(k)\frac{e^{i(kr-\frac{1}{2}(l+1)\pi)}}{kr} \\ &= \frac{1}{kr}\left[\frac{[1+2\Lambda_l(k)]e^{i(kr-\frac{1}{2}(l+1)\pi)} + e^{-i(kr-\frac{1}{2}(l+1)\pi)}}{2}\right], \end{aligned} \tag{19.70}$$

which, if we let

$$1 + 2\Lambda_l(k) = e^{i2\delta_l(k)}, \tag{19.71}$$

becomes

$$u^{(+)}(l,k,r) \underset{r\to\infty}{\sim} \frac{e^{i\delta_l(k)}}{kr}\cos\left(kr - \frac{l+1}{2}\pi + \delta_l\right). \tag{19.72}$$

Therefore we let

$$\Lambda_l(k) = \tfrac{1}{2}(e^{i2\delta_l(k)} - 1) = ie^{i\delta_l(k)}\sin\delta_l(k) \tag{19.73}$$

so that Eq. (19.69) can be written

$$u^{(+)}(l,k,r) = j_l(kr) + ie^{i\delta_l(k)}\sin\delta_l(k)h_l^{(+)}(kr) \tag{19.74}$$

and Eq. (19.68) becomes

$$\psi_{\mathbf{k}}^{(\text{in})}(\mathbf{x}) = \sum_{l=0}^{\infty} i^l(2l + 1)[j_l(kr) + ie^{i\delta_l(k)} \sin \delta_l(k)\, h_l^{(+)}(kr)]P_l(\cos \theta)$$

$$= e^{i\mathbf{k}\cdot\mathbf{x}} + \sum_{l=0}^{\infty} i^{l+1}(2l + 1)e^{i\delta_l} \sin \delta_l\, h_l^{(+)}(kr)P_l(\cos \theta)$$

$$\underset{r\to\infty}{\sim} e^{i\mathbf{k}\cdot\mathbf{x}} + \frac{e^{ikr}}{r}\frac{1}{k}\sum_{l=0}^{\infty} (2l + 1)e^{i\delta_l} \sin \delta_l\, P_l(\cos \theta), \tag{19.75}$$

where $\delta_l \equiv \delta_l(k)$. It follows immediately that

$$f_{k+}(\theta) = \frac{1}{k}\sum_{l=0}^{\infty} (2l + 1)e^{i\delta_l} \sin \delta_l\, P_l(\cos \theta), \tag{19.76}$$

where all the k-dependence is contained in $\delta_l = \delta_l(k)$.

The quantity δ_l is known as the *partial-wave phase shift* since, according to (19.72), the interaction shifts the phase of the "in" wave function by δ_l relative to the phase it would have in the absence of interaction. In the sense that $kr - \frac{1}{2}(l + 1)\pi$ is a phase, being the argument of the cosine function in (19.72) when no interaction is present, it is shifted by δ_l when the interaction is turned on.

The quantity $A_l(k)/k = ie^{i\delta_l} \sin \delta_l/k$ is called the *partial-wave scattering amplitude*.

If there are no absorption processes available to the system through the interaction $\mathfrak{B}$, that is, if only elastic scattering is possible, δ_l is a real number. We shall restrict ourselves to such situations.

Now let us investigate the connection between δ_l and the S-matrix. To do this, we must obtain $\psi_{\mathbf{k}'}^{(\text{out})}(\mathbf{x})$ in terms of δ_l; but

$$\psi_{\mathbf{k}'}^{(\text{out})}(\mathbf{x}) = \psi_{\mathbf{k}'}^{(\text{in})*}(\mathbf{x}) = \sum_{l=0}^{\infty} (-i)^l(2l + 1)u^{(+)*}(l, k', r)P_l(-\cos \theta)$$

$$= \sum_{l=0}^{\infty} i^l(2l + 1)u^{(+)*}(l, k', r)\frac{4\pi}{2l + 1}\sum_{m=-l}^{l} Y_l^{m*}(\theta_{\mathbf{k}'}, \varphi_{\mathbf{k}'}) \cdot Y_l^m(\theta_{\mathbf{x}}, \varphi_{\mathbf{x}}),$$

where $P_l(-x) = (-)^l P_l(x)$ and the equation below (16.17) have been used. Now

$$u^{(+)*}(l, k', r) = j_l(k'r) - ie^{-i\delta_l} \sin \delta_l h_l^{(-)}(k'r)$$

$$= \tfrac{1}{2}[h_l^{(+)}(k'r) + (1 - 2ie^{-i\delta_l} \sin \delta_l)h_l^{(-)}(k'r)]$$

$$= e^{-i2\delta_l}\tfrac{1}{2}[e^{i2\delta_l}h_l^{(+)}(k'r) + h_l^{(-)}(k'r)]$$

$$= e^{-i2\delta_l}u_l^{(+)}(l, k', r)$$

since

$$u^{(+)}(l, k', r) = \tfrac{1}{2}\{[1 + 2A_l(k')]h_l^{(+)}(k'r) + h_l^{(-)}(k'r)\}$$
$$= \tfrac{1}{2}[e^{i2\delta_l}h_l^{(+)}(k'r) + h_l^{(-)}(k'r)].$$

Therefore

$$\psi_{\mathbf{k}'}^{(\text{out})}(\mathbf{x}) = 4\pi \sum_{l=0}^{\infty} i^l e^{-i2\delta_l} u^{(+)}(l, k', r) \cdot \sum_{m=-l}^{l} Y_l^{m*}(\theta_{\mathbf{x}'}, \varphi_{\mathbf{k}'}) Y_l^m(\theta_{\mathbf{x}}, \varphi_{\mathbf{x}})$$

and $\psi_{\mathbf{k}'}^{(\text{in})}(\mathbf{x})$ is exactly the same except that the $e^{-i2\delta_l}$ is missing.

From these facts we conclude that

$$S_{fi} = \int d^3x\, \psi_{\mathbf{k}'}^{(\text{out})*}(\mathbf{x})\psi_{\mathbf{k}}^{(\text{in})}(\mathbf{x})$$
$$= \sum_{l,l'} \sum_{m,m'} e^{i2\delta_l}(4\pi)^2 i^{(l-l')} Y_{l'}^{m'}(\theta_{\mathbf{k}'}, \varphi_{\mathbf{k}'}) Y_l^{m*}(\theta_{\mathbf{k}}, \varphi_{\mathbf{k}})$$
$$\cdot \int dr\, r^2 u^{(+)*}(l', k', r) u^{(+)}(l, k, r) \int d\Omega\, Y_{l'}^{m'*}(\theta_{\mathbf{x}}, \varphi_{\mathbf{x}}) Y_l^m(\theta_{\mathbf{x}}, \varphi_{\mathbf{x}}).$$

But the integral over $d\Omega$ yields $\delta_{l',l}\delta_{m',m}$. Also, the $u^{(+)}(l, k, r)$ form an orthonormal set of radial wave functions, since they are the radial factors of the "in" wave functions; hence (see Problem 3, Appendix C) the radial integral yields $(\pi/2k^2)\delta(k' - k)$, and

$$S_{fi} = (4\pi)^2 \sum_{l,m} e^{i2\delta_l} Y_l^{m*}(\theta_{\mathbf{k}'}, \varphi_{\mathbf{k}'}) Y_l^m(\theta_{\mathbf{k}}, \varphi_{\mathbf{k}}) \frac{\pi}{2k^2} \delta(k' - k)$$
$$= \frac{2\pi^2}{k^2} \sum_l e^{i2\delta_l}(2l + 1)P_l(\cos\theta_{\mathbf{k},\mathbf{k}'})\delta(k' - k). \tag{19.77}$$

To check this, let $\delta_l = 0$; then, since

$$\sum_{l=0}^{\infty} (2l + 1)P_l(\cos\theta_{\mathbf{k},\mathbf{k}'}) = 4\pi\delta(\widehat{\mathbf{k} - \mathbf{k}'})$$
$$\equiv 4\pi\delta(\cos\theta_{\mathbf{k}} - \cos\theta_{\mathbf{k}'})\delta(\varphi_{\mathbf{k}} - \varphi_{\mathbf{k}'}),$$

we get

$$S_{fi} = \frac{8\pi^3}{k^2}\delta(k - k')\delta(\widehat{\mathbf{k} - \mathbf{k}'}) = (2\pi)^3\delta^3(\mathbf{k} - \mathbf{k}').$$

This differs from Eq. (19.33) with $T_{fi} = 0$ by the factor $(2\pi)^3$, which comes from the fact that here we used the box normalization for the wave functions, with $L = 1$, but integrated d^3x over all space, whereas in Eq. (19.33) we assumed

δ-function normalization; it is the difference between using $e^{i\mathbf{k}\cdot\mathbf{x}}$ and $(2\pi)^{-3/2}e^{i\mathbf{k}\cdot\mathbf{x}}$ for plane waves.

Therefore we have

$$S_{fi} = \frac{2\pi^2}{k^2}\delta(k' - k)\sum_{l=0}^{\infty} e^{i2\delta_l}(2l + 1)P_l(\hat{\mathbf{k}}\cdot\hat{\mathbf{k}}')$$

for unit box normalization, or

$$S_{fi} = \frac{1}{4\pi k^2}\delta(k' - k)\sum_{l=0}^{\infty} e^{i2\delta_l}(2l + 1)P_l(\hat{\mathbf{k}}\cdot\hat{\mathbf{k}}') \tag{19.78}$$

for δ-function normalization. In either case, we write

$$S_l \equiv e^{i2\delta_l}. \tag{19.79}$$

and call this the partial-wave S-matrix element.

Using the former normalization and replacing $\delta^3(\mathbf{k}_f - \mathbf{k}_i) = \delta^3(\mathbf{k}' - \mathbf{k})$ in Eq. (19.33) with $(2\pi)^3\delta^3(\mathbf{k}' - \mathbf{k})$, we obtain

$$\begin{aligned} S_{fi} &= (2\pi)^3\delta^3(\mathbf{k}' - \mathbf{k}) - i2\pi\delta(E' - E)T_{fi} \\ &= (2\pi)^3\delta^3(\mathbf{k}' - \mathbf{k}) - i2\pi\frac{\mu}{\hbar^2 k}\delta(k' - k)T_{fi} \\ &= \frac{2\pi^2}{k^2}\delta(k' - k)\sum_{l=0}^{\infty}[(e^{i2\delta_l} - 1) + 1](2l + 1)P_l(\hat{\mathbf{k}}\cdot\hat{\mathbf{k}}') \\ &= (2\pi)^3\delta^3(\mathbf{k} - \mathbf{k}') + \frac{2\pi^2}{k^2}\delta(k' - k)\sum_{l=0}^{\infty}(e^{i2\delta_l} - 1)(2l + 1)P_l(\hat{\mathbf{k}}\cdot\hat{\mathbf{k}}') \end{aligned}$$

The identity

$$\delta(E' - E) = \delta\left(\frac{\hbar^2}{2\mu}(k'^2 - k^2)\right) = \frac{\mu}{\hbar^2 k}\delta(k' - k)$$

has been invoked here. Hence

$$\begin{aligned} T_{fi} &= -\frac{2\pi\hbar^2}{\mu k}\sum_{l=0}^{\infty} e^{i\delta_l}\sin\delta_l\,(2l + 1)P_l(\hat{\mathbf{k}}'\cdot\hat{\mathbf{k}}) \\ &= -\frac{2\pi\hbar^2}{\mu}f_{k+}(\theta), \end{aligned} \tag{19.80}$$

which confirms Eq. (19.53); the latter was obtained using the same normalization we have used here.

We note in passing that if a bound state of the two-particle system exists at energy $E_0 < 0$, then according to Eq. (14.15), it is given wave number $k_0 = iK_0$,

$K_0 = +\sqrt{2\mu|E_0|}/\hbar$. Hence, only the $h_l^{(+)}(k_0 r)$ terms in

$$\psi_{\mathbf{k}_0}^{(\text{in})}(\mathbf{x}) = \sum_{l=0}^{\infty} i^l(2l+1)\tfrac{1}{2}[e^{i2\delta_l}h_l^{(+)}(k_0 r) + h_l^{(-)}(k_0 r)]P_l(\cos\theta)$$

can survive, because the $h_l^{(-)}(k_0 r)$ terms diverge for large r when $k_0 = iK_0$, $K_0 > 0$. The only way we can attain this state of affairs is to let $S_l = e^{i2\delta_l}$ become infinite so that the $(+)$ terms swamp the $(-)$ terms. This makes $\psi_{\mathbf{k}_0}^{(\text{in})}(\mathbf{x})$ infinite, but that's not surprising because $\psi_{\mathbf{k}_0}^{(\text{in})}(\mathbf{x})$ is not a physically meaningful wave function for a bound state; it must be replaced by $\psi_{E_0}(\mathbf{x})$, which differs from $\psi_{\mathbf{k}_0}^{(\text{in})}(\mathbf{x})$ by a normalization factor that cancels the infinity. The important point here is that the ratio of the $h_l^{(+)}(k_0 r)$ to the $h_l^{(-)}(k_0 r)$ term, which is given by $S_l = e^{i2\delta_l}$, must be infinite when a bound state with orbital angular momentum l occurs at energy E_0, and therefore S_l must have a singularity (a pole) at $E = E_0$. Bound states, then, correspond to poles in the S-matrix as a function of energy, these poles lying on the negative-energy axis. We do not pursue this subject further here, but simply note that an important area of study in modern scattering theory is the investigation of singularities in the S-matrix.

19.8 THE MEANING OF PHASE SHIFTS

The scattering amplitude is not a directly observable quantity, but its absolute square is; thus

$$\frac{d\sigma}{d\Omega}(k,\theta) = |f_{k+}(\theta)|^2 = \frac{1}{k^2}\left|\sum_{l=0}^{\infty}(2l+1)e^{i\delta_l}\sin\delta_l P_l(\cos\theta)\right|^2. \tag{19.81}$$

The cross-section integrated over angles is

$$\int d\Omega \frac{d\sigma}{d\Omega} = \sigma = \frac{4\pi}{k^2}\sum_{l=0}^{\infty}(2l+1)\sin^2\delta_l = \sum_{l=0}^{\infty}\sigma_l, \tag{19.82}$$

since

$$\int d\Omega\left|\sum_{l=0}^{\infty}a_l P_l(\cos\theta)\right|^2 = \sum_{l,l'}a_{l'}^* a_l 2\pi\int_{-1}^{1}P_{l'}(x)P_l(x)\,dx$$

$$= \sum_l \frac{4\pi|a_l|^2}{2l+1}$$

and we have $a_l = (2l+1)e^{i\delta_l}\sin\delta_l$. The cross-sections are thus expressible directly in terms of phase shifts, and hence the phase shifts constitute an important meeting ground for experiment and theory. Experimenters can measure the δ_l by observing angular distributions of scattered particles, and theorists can use models for the interparticle interactions to predict phase shifts.

It should be emphasized that we have been dealing with $\psi_{\mathbf{k}}^{(\text{in})}(\mathbf{x})$ only for values of $r = |\mathbf{x}|$ where $V(\mathbf{x})$ vanishes or nearly vanishes. Let us assume that the potential is spherically symmetric and let r_0 be its range, $V(\mathbf{x}) = V(r) = 0$ for $r > r_0$.

According to Eq. (14.2), if we let $f(l, k, r) = ru^{(+)}(l, k, r)$ so that $u^{(+)}(l, k, r)$ is the exact radial wave function appearing in the expansion (19.68),

$$\frac{d^2}{dr^2} ru^{(+)}(l\ k, r) = \left[\frac{l(l+1)}{r^2} + 2\frac{\mu}{\hbar^2}V(r) - k^2\right] ru^{(+)}(l, k, r). \tag{19.83}$$

Now if $V(r)$ is negative, and therefore attractive, for $r < r_0$, $d^2[ru^{(+)}(l, k, r)]/dr^2$ will be more negative for $r < r_0$ than it would have been if the potential were absent. Since the second derivative is proportional to the curvature of a function and since a more negative second derivative means sharper bending of the function toward the r-axis, we see that an attractive potential tends to make the wave function oscillate more rapidly inside the range of the potential. This means that at $r = r_0$ the wave function will have gone through more oscillations than it would have in the absence of the potential, and hence we expect the asymptotic behavior of $u^{(+)}(l, k, r)$ to reflect this by having $\delta_l > 0$. That is, since the exterior wave function

$$\tfrac{1}{2}[e^{i2\delta_l}h_l^{(+)}(r) + h_l^{(-)}(kr)] = u^{(+)}(l, k, r), \qquad r > r_0,$$

must, as $r \to r_0$, fit smoothly onto a function that has gone through more oscillations than a free wave would have experienced, its asymptotic behavior

$$\sim \frac{1}{2}\frac{e^{i\delta_l}}{r}[e^{i[kr - \frac{1}{2}(l+1)\pi + \delta_l]} + e^{-i[kr - \frac{1}{2}(l+1)\pi + \delta_l]}]$$

must show this by having

$$kr - \frac{l+1}{2}\pi + \delta_l > kr - \frac{l+1}{2}\pi,$$

that is $\delta_l > 0$. Therefore an attractive potential leads to positive phase-shifts. Similarly a repulsive potential leads to negative phase-shifts. See Fig. 19.4.

In Chapter 14, we saw that the exterior wave function $u_e^{(+)}(l, k, r)$ in Eq. (19.69), and its logarithmic derivative, must equal the interior wave function $u_i^{(+)}(l, k, r)$ and its logarithmic derivative at $r = r_0$; $u_i^{(+)}(l, k, r)$ here is the solution to Eq. (19.83) for $r < r_0$. Furthermore, $u_i^{(+)}(l, k, r)$ must be the regular solution to Eq. (19.83), behaving like Ar^l for small values of r. The matching of the wave functions themselves at $r = r_0$ serves to evaluate the normalization constant A, and the matching of logarithmic derivatives serves to evaluate δ_l.

To see how the latter is accomplished, we define

$$b_l \equiv \left.\frac{d[u_i^{(+)}(l, k, r)]/dr}{u_i^{(+)}(l, k, r)}\right|_{r=r_0} \tag{19.84}$$

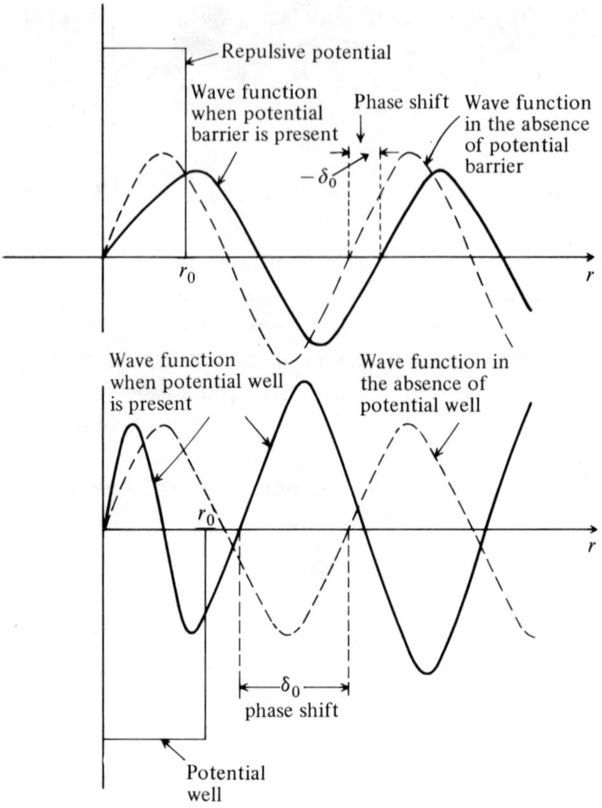

Fig. 19.4. Schematic representation of waves and phase-shifts for repulsive and attractive potentials.

and note that

$$\begin{aligned}\tfrac{1}{2}[e^{i2\delta_l}h_l^{(+)}(kr) + h_l^{(-)}(kr)] &= \tfrac{1}{2}[(e^{i2\delta_l} + 1)j_l(kr) + i(e^{i2\delta_l} - 1)n_l(kr)] \\ &= e^{i\delta_l}[(\cos\delta_l)j_l(kr) - (\sin\delta_l)n_l(kr)] \\ &= u_e^{(+)}(l, k, r). \end{aligned} \tag{19.85}$$

Therefore matching logarithmic derivatives at $r = r_0$ yields

$$k\frac{[(\cos\delta_l)j_l'(\rho_0) - (\sin\delta_l)n_l'(\rho_0)]}{(\cos\delta_l)j_l(\rho_0) - (\sin\delta_l)n_l(\rho_0)} = b_l, \qquad \rho_0 = kr_0. \tag{19.86}$$

This yields

$$\tan\delta_l = \frac{kj_l'(\rho_0) - b_l j_l(\rho_0)}{kn_l'(\rho_0) - b_l n_l(\rho_0)}. \tag{19.87}$$

When the b_l have been calculated (often numerical methods are required) for a given potential, $\tan \delta_l$ is obtained through the use of tables of spherical Bessel functions.

It follows from Eq. (19.87) that when the denominator does not vanish and b_l is essentially constant for values of k such that $(kr_0)^2 = \rho_0^2 \ll 4l - 2$, i.e. for values of k such that we may use the asymptotic forms (14.23) and (14.24) in Eq. (19.87), then

$$\tan \delta_l \propto (kr_0)^{2l+1}, \qquad k \ll \frac{4l - 2}{r_0}. \tag{19.88}$$

Hence $\tan \delta_l$ rapidly diminishes as l increases, and only the lower partial waves need be considered. The higher partial waves have a large repulsive *centrifugal potential* (represented by the term $l(l + 1)/r^2$ in the radial equation) to overcome if they are to get near the region where $V(r)$ does not vanish. Classically, a particle with momentum $p = \hbar k$ approaching a scattering center with angular momentum $\hbar l$ has a closest distance of approach to the scattering center of

$$a = \frac{\hbar l}{p} = \frac{l}{k};$$

hence if $a > r_0$, it does not interact with the scattering center and passes on by. In quantum theory the uncertainty principle requires a modification of this condition $l > kr_0$, but clearly if l is sufficiently large no scattering will occur. This indicates why partial wave analysis is a very useful tool for potentials having a definite finite range.

The conditions that b_l be essentially constant in k, and that the denominator does not vanish are frequently met in practice. The vanishing of the denominator corresponds to a resonance in the cross-section; the phase shift passes through $\pi/2$ (or an odd multiple of it) at energies such that $kn_l'(\rho_0) = b_l n_l(\rho_0)$.

19.9 SCATTERING BY A SQUARE WELL

For a square-well potential, with $V(r) = -V_0$ for $r < r_0$ and $V(r) = 0$ for $r > r_0$, the interior wave functions are just

$$u_i^{(+)}(l, k, r) = A_l j_l(Kr), \qquad K = \sqrt{\frac{2\mu}{\hbar^2} V_0 + k^2}.$$

Therefore

$$b_l = \frac{K j_l'(Kr_0)}{j_l(Kr_0)}. \tag{19.89}$$

It is easy to see that if $k^2 \ll 2\mu V_0/\hbar^2$, the b_l are not very sensitive to changes in k, and we should expect our estimate $\tan \delta_l \propto k^{2l+1}$ to be a good one. For small

$\rho_0 = kr_0$, we have

$$\tan \delta_l \simeq \frac{(l - b_l r_0)(kr_0)^{2l+1}}{(l + 1 + b_l r_0)(2l + 1)!!(2l - 1)!!};$$

in particular

$$\tan \delta_0 \simeq \frac{-b_0 r_0}{1 + b_0 r_0} kr_0, \qquad \tan \delta_1 \simeq \frac{1 - b_1 r_0}{3(2 + b_1 r_0)}(kr_0)^3.$$

These phase shifts may be made arbitrarily small by choosing k to be small, provided the denominators stay finite; hence for sufficiently small k we have

$$\delta_0 \simeq \sin \delta_0 \simeq \tan \delta_0 \simeq -\frac{b_0 r_0}{1 + b_0 r_0} kr_0, \qquad \sin \delta_1 \simeq \frac{1 - b_1 r_0}{3(2 + b_1 r_0)}(kr_0)^3.$$

The $l = 1$ (p-wave) scattering amplitude is likely to be negligible compared with the $l = 0$ (s-wave) scattering amplitude; in any case, the p-wave amplitude vanishes at $k = 0$, whereas the s-wave amplitude is

$$\frac{e^{i\delta_0} \sin \delta_0}{k} \simeq -e^{i\delta_0} \frac{b_0 r_0^2}{1 + b_0 r_0},$$

and therefore

$$\sigma(k = 0) = \sigma_0 = 4\pi \sin^2 \delta_0 = \frac{4\pi b_0^2 r_0^2}{(1 + b_0 r_0)^2}$$

according to Eq. (19.82). Explicit evaluation of (19.89) for $l = 0$ gives

$$b_0 = K \cot K r_0 - 1/r_0$$

and hence

$$\sigma(k = 0) = 4\pi r_0^2 \left(1 - \frac{\tan K r_0}{K r_0}\right)^2.$$

The difficulty of calculating the δ_l and hence the partial-wave amplitude obviously increases rapidly as the number of partial waves required increases, and hence as k increases. Nevertheless, the method is clear.

The square well provides a convenient medium for discussing the phenomenon of resonance scattering. For such cases the variation of b_l with k is very important. We continue to restrict our attention to small k, keeping only the lowest-degree terms in a Taylor-series expansion of b_l as a function of $K r_0$ about $K_0 r_0 = \sqrt{2\mu V_0} r_0/\hbar$:

$$b_l r_0 \simeq b_l^{(0)} r_0 - b_l^{(1)}(K - K_0) r_0^2.$$

But

$$K \simeq K_0\left[1 + \frac{1}{2}\left(\frac{k}{K_0}\right)^2\right] \qquad \text{if } k \ll K_0,$$

so that

$$b_l r_0 \simeq b_l^{(0)} r_0 - \frac{1}{2K_0} b_l^{(1)}(kr_0)^2.$$

Inserting this into Eq. (19.87) yields

$$\tan \delta_l = \frac{\left[l - b_l^{(0)} r_0 + \dfrac{1}{2K_0} b_l^{(1)}(kr_0)^2\right]}{\left[l + 1 + b_l^{(0)} r_0 - \dfrac{1}{2K_0} b_l^{(1)}(kr_0)^2\right]} \frac{(kr_0)^{2l+1}}{(2l-1)!!(2l+1)!!}.$$

The denominator vanishes when

$$k^2 = k_r^2 \equiv \frac{2K_0}{b_l^{(1)} r_0^2}(l + 1 + b_l^{(0)} r_0)$$

and a resonance occurs; of course this is possible only if $b_l^{(1)} > 0$. When this happens for only one l-value at a given energy, the scattering at the energy is likely to arise almost entirely from waves having that value of l, and the angular distribution (differential cross-section) at that energy will be what one would expect from waves with that orbital angular momentum. The energy dependence of the cross-section in the immediate neighborhood of a resonance normally has the form

$$\sigma_l(E) = \frac{\text{const.}}{(E - E_r)^2 + \Gamma^2/4}$$

where $E_r = (\hbar^2/2\mu)k_r^2$ and Γ^2 is a (possibly energy-dependent) quantity that characterizes the width of the resonance peak.

19.10 COULOMB SCATTERING

In one of the exercises at the end of Chapter 15, you are asked to show that the scattering from a Coulomb potential, $V(\mathbf{x}) = qq'/4\pi\epsilon_0 r$, has a cross-section equal in first-order Born approximation to the Rutherford cross-section. It turns out, however, that the second-order Born approximation to the scattering amplitude is infinite. Therefore, we should try to do an exact analysis of scattering from the Coulomb potential to see whether the quantum-mechanical cross-section is the same as the classical Rutherford cross-section. We continue to assume that the particles involved are spinless and non-identical; this corresponds to α-particle scattering from a nucleus other than He^4.

Our approach will be to solve the radial wave equation (14.75), but with Ze^2 replaced by $-ZZ'e^2$ since we wish to deal with charges (Ze) and $(Z'e)$ of like sign scattering through a repulsive Coulomb interaction; in this solution we shall keep $E = \hbar^2k^2/2\mu > 0$, since we are interested in the scattering of particles, and shall try to find $f_{k+}(\theta)$ by inspecting the asymptotic form of the wave function.

Now the solutions to

$$\left[\frac{d^2}{d\rho^2} - \frac{l(l+1)}{\rho^2} - \frac{2\eta}{\rho} + 1\right] f\left(l, k, \frac{\rho}{k}\right) = 0, \tag{19.90}$$

where

$$\rho = kr, \qquad \eta = \frac{ZZ'e^2\mu}{4\pi\epsilon_0\hbar^2k}, \tag{19.91}$$

which particularly interest us are those that behave like $e^{i\rho}$ as $\rho \to \infty$ and like ρ^{l+1} and $\rho \to 0$. For this reason, we let

$$f(l, k, \rho/k) = \rho^{l+1}e^{i\rho}g(l, k, \rho) \tag{19.92}$$

and seek a differential equation for $g(l, k, \rho)$. Straightforward differentiation of this expression for $f(l, k, \rho/k)$ shows that $g(l, k, \rho)$ must satisfy

$$\left[\frac{d^2}{d\rho^2} + 2\left(i + \frac{l+1}{\rho}\right)\frac{d}{d\rho} + 2\frac{i(l+1) - \eta}{\rho}\right] g(l, k, \rho) = 0. \tag{19.93}$$

Let

$$c \equiv -2i, \qquad b \equiv 2(l+1), \qquad a = (l+1) + i\eta; \tag{19.94}$$

then this has the form

$$\left[\frac{d^2}{d\rho^2} + \left(\frac{b}{\rho} - c\right)\frac{d}{d\rho} - \frac{ac}{\rho}\right] g(l, k, \rho) = 0. \tag{19.95}$$

Assuming that this has an infinite series solution, we write

$$g(l, k, \rho) = \sum_{m=0}^{\infty} \alpha_m \rho^m,$$

thereby obtaining the recursion relations

$$\alpha_m[m(m + b - 1)] = \alpha_{m-1}c(m + a - 1)$$

through direct differentiation. Hence

$$\alpha_m = c\frac{m + a - 1}{m(m + b - 1)}\alpha_{m-1} = c^2\frac{(m + a - 1)(m + a - 2)}{m(m-1)(m + b - 1)(m + b - 2)}\alpha_{m-2}$$

$$= \cdots = c^m\frac{(m + a - 1)(m + a - 2)\cdots a}{m(m-1)\cdots 1 \cdot (m + b - 1)(m + b - 2)\cdots b}\alpha_0$$

$$= c^m \frac{\Gamma(m + a)\Gamma(b)}{m!\Gamma(m + b)\Gamma(a)}\alpha_0$$

and we have the solution

$$g(l, k, \rho) = \alpha_1 F_1(a; b; c\rho), \tag{19.96}$$

where

$${}_1F_1(a; b; c\rho) = \sum_{m=0}^{\infty} \frac{\Gamma(a + m)\Gamma(b)}{\Gamma(a)\Gamma(b + m)} \frac{(c\rho)^m}{m!} \tag{19.97}$$

is a standard function known as the confluent hypergeometric function.

As an aid to investigating the behavior of ${}_1F_1(a; b; c\rho)$ for large values of ρ, we recall the definitions of the gamma function $\Gamma(x)$ and the beta function $B(x, y)$. The former is defined by

$$\Gamma(x) \equiv \int_0^{\infty} t^{x-1}e^{-t}\, dt \tag{19.98}$$

and has the properties

$$\Gamma(x) = (x - 1)\Gamma(x - 1), \qquad \Gamma(x)\Gamma(1 - x) = \frac{\pi}{\sin \pi x}. \tag{19.99}$$

If x is an integer,

$$x! = \Gamma(x + 1),$$

On the other hand,

$$B(x, y) = \int_0^1 (1 - t)^{y-1}t^{x-1}\, dt \tag{19.100}$$

and it can be shown rather easily that

$$B(x, y) = \frac{\Gamma(x)\Gamma(y)}{\Gamma(x + y)}. \tag{19.101}$$

Indeed, this relationship can be used to prove the second of Eqs. (19.99).

It is important to note that the integral defining $\Gamma(x)$ remains valid when the integral is taken along any semiinfinite line tilted at angle ϕ, $-\pi/2 \le \phi \le \pi/2$ relative to the real axis in the complex plane. This is because the integral in $\Gamma(x)$ can be converted into a contour integral, the contour passing from infinity to the origin just above the real axis, encircling the origin, and passing back to infinity just below the real axis (see Fig. 19.5). Since the integrand of the contour integral, which is $[i/(2 \sin \pi x)](-t)^{x-1}e^{-t}$, has no singularities in the t-plane other than at the origin and at $t = -\infty$, the contour can be rotated through any angle about the origin, as long as $\text{Re}\, t > 0$ along the contour. In particular, the change of

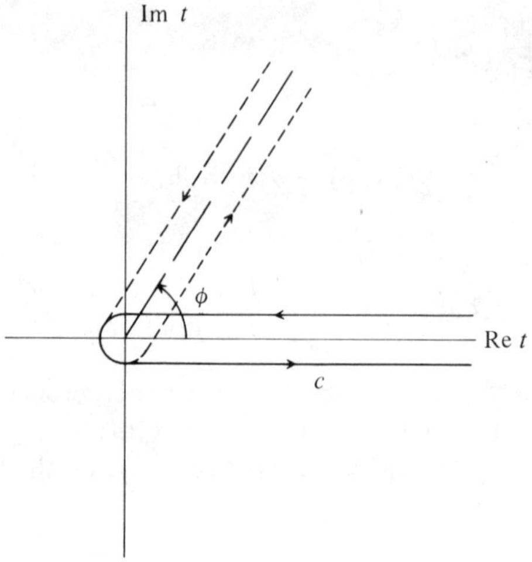

Fig. 19.5. Contour for evaluating

$$\Gamma(x) = \frac{i}{2 \sin \pi x} \int_C (-t)^{x-1} e^{-t}\, dt$$

(Hankel's contour integral form). Rotation of the contour through angle ϕ, $-\pi/2 \leq \phi \leq \pi/2$, does not affect the integral.

contour is allowed for the limiting cases $\phi = \pm\pi/2$ when $\Gamma(x)$ can be defined in terms of the integrals along the positive and negative imaginary axes:

$$\Gamma(x) = \int_0^{+i\infty} t^{x-1} e^{-t}\, dt = \int_0^{-i\infty} t^{x-1} e^{-t}\, dt. \tag{19.102}$$

We use the relation between the Γ and B functions to write the confluent hypergeometric function in the form

$$\begin{aligned} {}_1F_1(a; b; c\rho) &= \sum_{m=0}^{\infty} \frac{\Gamma(b)}{\Gamma(a)\Gamma(b-a)} \frac{\Gamma(a+m)\Gamma(b-a)}{\Gamma(b+m)} \frac{(c\rho)^m}{m!} \\ &= \frac{1}{B(a, b-a)} \sum_{m=0}^{\infty} B(a+m, b-a) \frac{(c\rho)^m}{m!}. \end{aligned}$$

Inserting the definition for $B(x, y)$ inside the summation sign and interchanging the sum and integral, we find

$$\begin{aligned} {}_1F_1(a; b; c\rho) &= \frac{1}{B(a, b-a)} \int_0^1 (1-t)^{b-a-1} \sum_{m=0}^{\infty} \frac{(c\rho t)^m}{m!} t^{a-1}\, dt \\ &= \frac{1}{B(a, b-a)} \int_0^1 (1-t)^{b-a-1} t^{a-1} e^{c\rho t}\, dt. \end{aligned} \tag{19.103}$$

In order to evaluate this integral, let us think in terms of the complex t-plane. From the definitions of a and b we see that $\operatorname{Re}(a-1) \geq 0$ and $\operatorname{Re}(b-a-1) \geq 0$; therefore the first two factors in the integrand contribute no singularities in the finite t-plane. Since $c\rho = -i2kr$ and kr is always positive, the integrand will vanish for t lying on the arc of a large semicircle in the lower half plane, and the radius of the semicircle becomes infinite. This tells us that the integral around the contour C_R in Fig. 19.6 is equal to the integral from $-\infty$ to $+\infty$ just below the real t-axis, and furthermore that this integral vanishes, according to the Cauchy residue theorem. Therefore

$$0 = \int_{-\infty}^{\infty} (1-t)^{b-a-1} t^{a-1} e^{c\rho t}\, dt$$

$$= \left(\int_{-\infty}^{0} + \int_{0}^{1} + \int_{1}^{\infty}\right)(1-t)^{b-a-1} t^{a-1} e^{c\rho t}\, dt,$$

or

$$\int_0^1 (1-t)^{b-a-1} t^{a-1} e^{c\rho t}\, dt$$

$$= -\int_{-\infty}^{0} (1-t)^{b-a-1} t^{a-1} e^{c\rho t}\, dt - e^{c\rho} \int_0^{\infty} (u+1)^{a-1}(-u)^{b-a-1} e^{c\rho u}\, du,$$

where we used $t = u + 1$ in the integral from 1 to ∞ to get the last integral.

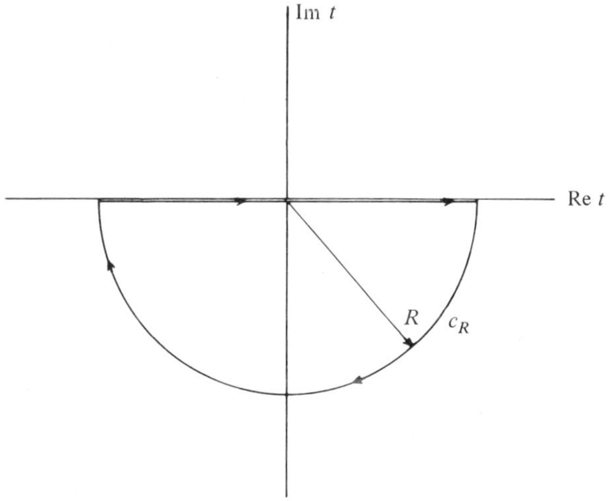

Fig. 19.6. Contour for evaluating ${}_1F_1(a; b\ c\rho)$.

Recognizing that $c\rho = -i2kr$, let us define $z = 2kr = ic\rho$ and also let us take $t' = +izt$ and $u' = +izu$ on the right-hand side. Then

$$\int_0^1 (1-t)^{b-a-1} t^{a-1} e^{c\rho t}\, dt$$

$$= -\int_{-i\infty}^{0} \left(1 + i\frac{t'}{z}\right)^{b-a-1} \left(\frac{t'}{iz}\right)^{a-1} e^{-t'} \frac{dt'}{iz}$$

$$-e^{-iz} \int_0^{+i\infty} \left(1 - i\frac{u'}{z}\right)^{a-1} \left(i\frac{u'}{z}\right)^{b-a-1} e^{-u'} \frac{du'}{iz}$$

$$\underset{z\to\infty}{\sim} -(iz)^{-a} \int_{-i\infty}^{0} t'^{a-1} e^{-t'}\, dt' + (-iz)^{-b+a} e^{-iz} \int_0^{i\infty} u'^{b-a-1} e^{-u'}\, du'$$

$$= (iz)^{-a}\Gamma(a) + (-iz)^{a-b} e^{-iz} \Gamma(b-a)$$

$$= (-c\rho)^{-a}\Gamma(a) + (c\rho)^{a-b} e^{c\rho} \Gamma(b-a).$$

Therefore

$${}_1F_1(a; b; c\rho) \underset{\rho\to\infty}{\sim} \frac{\Gamma(a)}{B(a, b-a)}(-c\rho)^{-a} + \frac{\Gamma(b-a)}{B(a, b-a)}(c\rho)^{a-b} e^{c\rho},$$

and finally, using Eqs. (19.92), (19.94), (19.96) and (19.101),

$$f(l, k, \rho/k)$$

$$\underset{\rho\to\infty}{\sim} \alpha_0 \rho^{l+1} e^{i\rho} \left[\frac{\Gamma(2l+1)}{\Gamma(l+1-i\eta)} (i2\rho)^{-l-1-i\eta} + \frac{\Gamma(2l+1)}{\Gamma(l+1+i\eta)} (-i2\rho)^{-l-1+i\eta} e^{-i2\rho} \right]$$

$$= \alpha_0 2^{-l-1} \frac{\Gamma(2l+1)}{\Gamma(l+1+i\eta)} \left[e^{-i[\rho - \frac{1}{2}(l+1)\pi - \eta \ln(2\rho/i)]} \right.$$

$$\left. + \frac{\Gamma(l+1+i\eta)}{\Gamma(l+1-i\eta)} e^{i[\rho - \frac{1}{2}(l+1)\pi - \eta \ln i2\rho]} \right]. \tag{19.104}$$

This last result can be refined a little through the following transformations: Let

$$e^{i2\xi_l} \equiv \frac{\Gamma(l+1+i\eta)}{\Gamma(l+1-i\eta)}, \tag{19.105}$$

which is justified because $\Gamma(l+1-i\eta) = \Gamma^*(l+1+i\eta)$. Also use

$$\eta \ln(\mp i2\rho) = \eta \ln 2\rho + \eta \ln(\mp i) = \eta \ln 2\rho \mp i\eta\pi/2.$$

Then we have

$$f(l, k, r) \underset{\rho\to\infty}{\sim} C_l \frac{e^{\eta\pi/2}}{\Gamma(l+1+i\eta)} e^{i\xi_l} \cdot \sin(kr - \tfrac{1}{2}l\pi - \eta \ln 2kr + \xi_l). \tag{19.106}$$

Our solutions $u_e^{(+)}(l, k, r)$ in the phase-shift analysis are essentially $f(l, k, r)/r$, and we saw that at very large distances they behave like $(1/r)\sin(kr - \frac{1}{2}l\pi + \delta_l)$. The Coulomb phase shifts are then $\xi_l - \eta \ln 2kr$; they are radius-dependent, which indicates that the asymptotic form of the $f(l, k, r)/r$ is not the same as that of $u_e^{(+)}(l, k, r)$. This is caused by the fact that the Coulomb potential has an infinite range—it has no characteristic length parameter indicating at what distance it can be neglected. Since the phase shifts $-\eta \ln 2kr$ do not vary from one l-wave to another, however, we can ignore them in calculating the cross-section; they simply correct the asymptotic form of the wave function and do not affect the cross-section.

Evidently the scattering amplitude is determined by the phase shifts ξ_l; in fact

$$f_{k+}(\theta) = \frac{1}{k}\sum_{l=0}^{\infty}(2l + 1)e^{i\xi_l}\sin\xi_l P_l(\cos\theta)$$

$$= \frac{1}{2ik}\sum_{l=0}^{\infty}(2l + 1)\left[\frac{\Gamma(l + 1 + i\eta)}{\Gamma(l + 1 - i\eta)} - 1\right]P_l(\cos\theta). \tag{19.107}$$

Here we come upon another consequence of the "infinite rangedness" of the Coulomb potential; because there is no "cut-off" radius for it, we cannot get an accurate evaluation of $f_{k+}(\theta)$ by using just the lowest few values of l in the summation. We are faced with the task of evaluating the sum over l exactly. We use several stratagems to accomplish this.

First, we know that the Rutherford cross-section is proportional to $\operatorname{cosec}^4(\theta/2) = [\sin^2(\theta/2)]^{-2} = 4(1 - \cos\theta)^{-2}$, and therefore we suspect that multiplying $f_{k+}(\theta)$ by $(1 - \cos\theta)$ will remove most of the angular dependence. Hence we try to calculate

$$(1 - \cos\theta)f_{k+}(\theta)$$
$$= \frac{1}{2ik}\sum_{l=0}^{\infty}(2l + 1)\left[\frac{\Gamma(l + 1 + i\eta)}{\Gamma(l + 1 - i\eta)} - 1\right](1 - \cos\theta)P_l(\cos\theta)$$

A recurrence relation* for the $P_l(x)$ now may be utilized. We have

$$xP_l(x) = \frac{l + 1}{2l + 1}P_{l+1}(x) + \frac{l}{2l + 1}P_{l-1}(x),$$

and hence

$$\sum_{l=0}^{\infty}(2l + 1)\left[\frac{\Gamma(l + 1 + i\eta)}{\Gamma(l + 1 - i\eta)} - 1\right](1 - x)P_l(x)$$

* See, for example, T. M. MacRobert, *Spherical Harmonics*, Dover Publications, Inc., New York, N.Y. (1948).

$$= \sum_{l=0}^{\infty} \left\{(2l + 1)\left[\frac{\Gamma(l + 1 + i\eta)}{\Gamma(l + 1 - i\eta)} - 1\right] - l\left[\frac{\Gamma(l + i\eta)}{\Gamma(l - i\eta)} - 1\right]\right.$$

$$\left. -(l + 1)\left[\frac{\Gamma(l + 2 + i\eta)}{\Gamma(l + 2 - i\eta)} - 1\right]\right\} P_l(x)$$

$$= \sum_{l=0}^{\infty} \frac{\Gamma(l + in)}{\Gamma(l + 2 - i\eta)}[(2l + 1)(l + i\eta)(l + 1 - i\eta) - l(l + 1 - i\eta)(l - i\eta)$$

$$-(l + 1)(l + 1 + i\eta)(l + i\eta)]P_l(\cos\theta)$$

$$= \sum_{l=0}^{\infty} \frac{\Gamma(l + i\eta)}{\Gamma(l + 2 - i\eta)}(2l + 1)2\eta^2 P_l(\cos\theta)$$

$$= 2ik(1 - \cos\theta) f_{k+}(\theta).$$

Evaluation of the last sum depends on finding a gimmick for handling the ratio $\Gamma(l + i\eta)/\Gamma(l + 2 - i\eta)$; the B function

$$B(l + 1, 1 - i\eta) = \Gamma(l + 1)\Gamma(1 - i\eta)/\Gamma(l + 2 - i\eta) = \int_0^1 (1 - t)^{-i\eta} t^l \, dt$$

turns out to be helpful if we use the change of variables $t = \frac{1}{2}(1 + y)$:

$$B(l + 1, 1 - i\eta) = \int_{-1}^{1} \left(\frac{1 + y}{2}\right)^l \left(\frac{1 - y}{2}\right)^{-i\eta} \frac{dy}{2}$$

$$= \int_{-1}^{1} \left(\frac{1 - y^2}{2}\right)^l \frac{(1 - y)^{-l-i\eta}}{2^{-i\eta+1}} \, dy.$$

The reason this is helpful is that $[\frac{1}{2}(1 - y^2)]^l$ appears in Rodrigue's formula

$$P_l(y) = \frac{(-)^l}{2^l l!} \frac{d^l}{dy^l}(1 - y^2)^l,$$

and since

$$\frac{d^l}{dy^l}(1 - y)^{-i\eta} = (-i\eta)(-i\eta - 1)\cdots(-i\eta - l + 1)(-)^l(1 - y)^{-l-i\eta}$$

$$= \frac{\Gamma(l + i\eta)}{\Gamma(i\eta)}(1 - y)^{-l-i\eta},$$

we may use integration by parts to convert the integrand to one involving $P_l(y)$:

$$B(l + 1, 1 - i\eta) = l!\,2^{i\eta-1} \frac{\Gamma(i\eta)}{\Gamma(l + i\eta)} \int_{-1}^{1} P_l(y)(1 - y)^{-i\eta} \, dy$$

$$= \Gamma(l + 1)\frac{\Gamma(i\eta)}{\Gamma(l + i\eta)} 2^{i\eta-1} \int_{-1}^{1} P_l(y)(1 - y)^{-i\eta} \, dy.$$

Therefore

$$2ik(1 - \cos\theta) f_{k+}(\theta)$$

$$= 2\eta^2 \sum_{l=0}^{\infty} \frac{\Gamma(l + i\eta)}{\Gamma(l+1)\Gamma(1 - i\eta)} B(l + 1, 1 - i\eta)(2l + 1) P_l(\cos\theta)$$

$$= 2^{i\eta}\eta^2 \frac{\Gamma(i\eta)}{\Gamma(1 - i\eta)} \int_{-1}^{1} (1 - y)^{-i\eta} \sum_{l=0}^{\infty} P_l(y)(2l + 1) P_l(\cos\theta)\, dy.$$

However,

$$\sum_{l=0}^{\infty} (2l + 1) P_l(\cos\theta) P_l(y) = 2\delta(y - \cos\theta),$$

so that we have

$$f_{k+}(\theta) = \frac{2^{i\eta}\eta^2}{ik} \frac{\Gamma(i\eta)}{\Gamma(1 - i\eta)} (1 - \cos\theta)^{-1-i\eta}$$

$$= \frac{2^{i\eta}\eta}{k} \frac{\Gamma(i\eta)}{\Gamma(-i\eta)} (1 - \cos\theta)^{-1-i\eta}$$

$$= \frac{\eta}{2k} e^{i\xi_{-1}} \left(\sin^2 \frac{\theta}{2}\right)^{-1-i\eta}$$

$$= \frac{\eta}{2k} e^{i[\xi_{-1} - \eta 2 \ln(\sin\theta/2)]} \operatorname{cosec}^2 \frac{\theta}{2}, \qquad (19.108)$$

where $\xi_{-1} = \arg[\Gamma(i\eta)/\Gamma(-i\eta)]$. Therefore the Coulomb scattering cross-section is just

$$\frac{d\sigma}{d\Omega} = \frac{\eta^2}{4k^2} \operatorname{cosec}^4 \frac{\theta}{2} = \left(\frac{\mu Z Z' e^2}{2\hbar^2 k^2}\right)^2 \operatorname{cosec}^4 \frac{\theta}{2}. \qquad (19.109)$$

From this very lengthy argument we conclude that the first-order Born approximation just happens to give the exact Coulomb scattering cross-section.

The mathematical details involved in obtaining this result, however, should not be allowed to obscure the central point, namely that the phase shifts can be evaluated by comparing the coefficients of e^{ikr} and e^{-ikr} in the exact wave function. An inevitable but much less important lesson is that there is ample opportunity for exercising mathematical ingenuity in finding exact solutions and evaluating infinite sums.

19.11 TRANSMISSION THROUGH A SPHERICAL POTENTIAL BARRIER

Another kind of unbound system in quantum theory is that in which a particle is trapped inside a potential barrier of finite height and width, as shown in Fig. 19.7. By "trapped" we mean that the particle has positive energy $E = \hbar^2 k^2/2\mu$, but

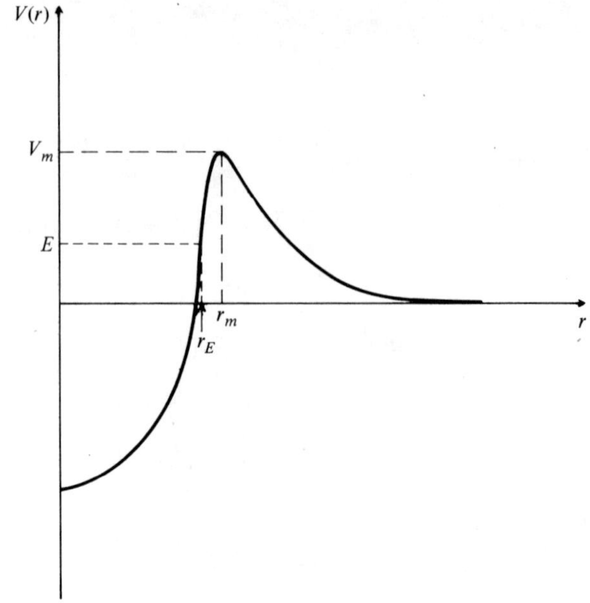

Fig. 19.7. Spherical potential with a barrier of height V_m.

$E < V_m$, where V_m is the maximum height of the barrier, and at some time the particle has probability 1 of being found within the region defined by $r < r_m$, where r_m is the solution to $V(r_m) = V_m$. Classically, the particle would be confined to the interior region defined by $r < r_E$, where r_E is the smaller solution of $V(r_E) = E$. Because of the wave nature of quantum systems and the resulting uncertainty principle, however, the particle actually has a finite probability of being found outside the interior region as time progresses. If it is observed in the exterior region, it must have positive momentum directed away from the origin when it is so observed; hence the wave function for the system must reduce, as $r \to \infty$, to a superposition of outgoing spherical waves.

Let us consider solutions

$$\psi(\mathbf{x}, t) = \sum_{l,m} a_{l,m} u_l(r, t) Y_l^m(\theta, \varphi) \tag{19.110}$$

to the Schrödinger equation

$$\left[-\frac{\hbar^2}{2\mu} \nabla^2 + V(r) \right] \psi(\mathbf{x}, t) = i\hbar \frac{\partial}{\partial t} \psi(\mathbf{x}, t).$$

Evidently $f_l(r, t) \equiv r u_l(r, t)$ satisfies

$$\left[\frac{d^2}{dr^2} - \frac{l(l+1)}{r^2} - \frac{2\mu}{\hbar^2}V(r)\right] f_l(r, t) = i\hbar\frac{\partial}{\partial t} f_l(r, t). \tag{19.111}$$

Separating variables in the usual way, we may write

$$f_l(r, t) = T(t) f(l, K', r) \tag{19.112a}$$

where

$$T(t) = e^{-iE't/\hbar}, \qquad K' = \sqrt{2\mu E'/\hbar^2}, \tag{19.112b}$$

and $f(l, K', r)$ satisfies

$$\left[\frac{d^2}{dr^2} - \frac{l(l+1)}{r^2} - \frac{2\mu}{\hbar^2}V(r) + K'^2\right] f(l, K', r) = 0. \tag{19.113}$$

We shall impose the boundary condition

$$f(l, K', r) \underset{r\to\infty}{\sim} B_l e^{ikr} \tag{19.114a}$$

or

$$u_l(r, t) \underset{r\to\infty}{\sim} B_l \frac{e^{ikr}}{r} e^{-iE't/\hbar}. \tag{19.114b}$$

This prevents H from being a Hermitian operator; therefore E' need not be real and we shall have to find an interpretation for the imaginary part of E'. To see this, we note that the radial component of the current is given by

$$\begin{aligned}\mathbf{j}_r &= \hat{\mathbf{r}} \sum_{l,m} \sum_{l',m'} a^*_{l'm'} a_{lm} Y_{l'}^{m'*}(\theta, \varphi) Y_l^m(\theta, \varphi) \\ &\quad \cdot \frac{\hbar}{i2\mu}\left\{u^*_{l'}(r, t)\frac{\partial}{\partial r}u_l(r, t) - \left[\frac{\partial}{\partial r}u^*_{l'}(r, t)\right]u_l(r, t)\right\} \\ &= \hat{\mathbf{r}} \sum_{l,m} \sum_{l',m'} a^*_{l'm'} B^*_{l'} a_{lm} B_l Y_{l'}^{m'*}(\theta, \varphi) Y_l^m(\theta, \varphi)\frac{\hbar k}{\mu} e^{i(E'^* - E')t/\hbar}.\end{aligned}$$

Hence if conservation of probability is applied to a very large sphere V_R of radius R, we get

$$\frac{d}{dt}\int_{V_R} |\psi(\mathbf{r}, t)|^2 \, d^3r + \int_{V_R} \nabla \cdot \mathbf{j} \, d^3r = 0$$

or, applying Gauss's law to the last integral and using the orthonormality of spherical harmonics,

$$\sum_{l,m} |a_{lm}|^2 \left[\frac{d}{dt}\int_0^R |u_l(r, t)|^2 r^2 \, dr + |B_l|^2 \frac{\hbar k}{\mu} e^{-i(E' - E'^*)t/\hbar}\right] = 0.$$

This can be satisfied only if

$$\frac{d}{dt}\int_0^R |u_l(r,t)|^2 r^2\,dr = -|B_l|^2\frac{\hbar k}{\mu}e^{-i(E'-E'^*)t/\hbar},$$

or

$$-\frac{i}{\hbar}(E' - E'^*)\int_0^R |f(l,K',r)|^2\,dr = -|B_l|^2\frac{\hbar k}{\mu}.$$

However,

$$-i(E' - E'^*) = 2\,\mathrm{Im}\,E';$$

hence we conclude that

$$2\,\mathrm{Im}\,E' = -\frac{\hbar^2 k}{\mu}\frac{|B_l|^2}{\int_0^\infty |f(l,K',r)|^2\,dr}$$

where we have let $R \to \infty$. Let us write

$$E' = \epsilon - i(\Gamma_l/2), \qquad \Gamma > 0. \tag{19.115}$$

Since $P_{(l)}(t) \equiv \int_0^R |u_l(r,t)|^2 r^2\,dr$ is the probability of finding the particle within a sphere of radius R, and since we now know that

$$\frac{d}{dt}P_{(l)}(t) = -\frac{\Gamma_l}{\hbar}P_{(l)}(t),$$

or

$$P_{(l)}(t) = e^{-\Gamma_l t/\hbar}P_{(l)}(0), \tag{19.116}$$

we can identify $\Gamma_l/\hbar$ as the rate of disintegration of the system in which the particle is trapped. We shall find another interpretation of Γ_l that is related to this one through the uncertainty principle.

Let us turn our attention now to the solution of the radial equation (19.113) subject to the boundary condition (19.114a) and the condition that $f(l, K', r)$ be finite at $r = 0$. In particular, we shall assume that $V(r)$ is a simple "square" barrier, with a depressed interior:

$$V(r) = \begin{cases} -V_0, & r < r_1 \\ V_m, & r_1 < r < r_2 \\ 0, & r < r_2. \end{cases}$$

The solutions to Eq. (19.113) in the three regions of space where the potential has different values are

$$\begin{aligned} f_{\mathrm{I}} &= Kr[a_{\mathrm{I}}h_l^{(+)}(Kr) + b_{\mathrm{I}}h_l^{(-)}(Kr)], && r < r_1 \\ f_{\mathrm{II}} &= Qr[a_{\mathrm{II}}h_l^{(+)}(iQr) + b_{\mathrm{II}}h_l^{(-)}(iQr)], && r_1 < r < r_2 \\ f_{\mathrm{III}} &= K'r[a_{\mathrm{III}}h_l^{(+)}(K'r) + b_{\mathrm{III}}h_l^{(-)}(K'r)], && r > r_2, \end{aligned} \tag{19.117}$$

where

$$K^2 \equiv K'^2 + \frac{2\mu}{\hbar^2}V_0, \qquad Q^2 \equiv \frac{2\mu}{\hbar^2}V_m - K'^2. \tag{19.118}$$

These functions and their logarithmic derivatives must be equal at the appropriate boundaries r_1 and r_2. The resulting transcendental equations then determine the values of four of the a's and b's and the condition that must be satisfied for such solutions for the a's and b's to exist determines permissible values of K'. The remaining two a's and b's are determined by the boundary condition (19.114) for $r \to \infty$, which means that

$$b_{\text{III}} = 0, \qquad a_{\text{III}} = B_l, \tag{19.119}$$

and the condition that f_{I} be finite at $r = 0$, which requires that

$$a_{\text{I}} = b_{\text{I}} = \tfrac{1}{2}A_l. \tag{19.120}$$

The conditions at r_1 and r_2 in general are rather complicated, simply because the $h_l^{(\pm)}(x)$ are rather complicated functions in general. For $l = 0$, however, they are quite simple; we restrict ourselves to this case in the remaining discussion. Dropping the subscripts on A_l, B_l, a_{II}, and b_{II}, we have

$$A \sin Kr_1 = ae^{-Qr_1} - be^{Qr_1}$$

$$K \cot Kr_1 = -Q\frac{ae^{-Qr_1} + be^{Qr_1}}{a^{-Qr_1} - be^{Qr_1}}$$

$$ae^{-Qr_2} - be^{Qr_2} = Be^{iK'r_2}$$

$$-Q\frac{ae^{-Qr_2} + be^{Qr_2}}{ae^{-Qr_2} - be^{Qr_2}} = iK'. \tag{19.121}$$

The second and fourth of these can be solved for b/a, and if b/a is eliminated between them, they give a condition for the solutions to be consistent, and this may be regarded as an equation for K'. This equation for K' may be written in the form

$$e^{2Q\lambda}\frac{Q - iK'}{Q + iK'}\,\frac{Q + K \cot Kr_1}{Q - K \cot Kr_1} = 1, \tag{19.122}$$

where $\lambda \equiv r_2 - r_1$.

This is an enormously difficult equation to solve for K'; recall that Q and K both are functions of K'. However, if we let $\lambda \to \infty$, we see that in this limit

$$Q + K \cot Kr_1 = 0, \qquad \text{or} \qquad \tan Kr_1 = -\frac{K}{Q}; \tag{19.123}$$

this corresponds to having a square well with depth $V_0 + V_m$, and $V_m - E'$ is the binding energy of the particle bound in this well. In this particular case, the solutions K' are such that $E' = \hbar^2 K'^2/2\mu$ is real and $E' - V_m = -Q^2\hbar^2/2\mu$ coincides with the energy levels E of the square-well potential with depth $V_0 + V_m$; Q itself will be real and the wave function $f(l, K', r)$ will behave like e^{-Qr} at very large values of r, so that in this case $k = iQ$.

For finite values of λ, however, the boundary condition at large distances forces E' to have an imaginary part. However, if λ is sufficiently large, i.e. if

$$Q\lambda = \sqrt{\frac{2\mu}{\hbar^2}V_m - K'^2}\,\lambda \gg 1,$$

we may expect that (19.123) is still approximately satisfied. Let K'_0 be a solution to Eq. (19.123) and let K' be the corresponding exact solution to Eq. (19.122); then we may approximate Eq. (19.122) by expanding everything in powers of $\Delta K' = K' - K'_0$, throwing away second and higher powers in $\Delta K'$. This yields

$$e^{2Q_0\lambda}\frac{Q_0 - iK'_0}{Q_0 + iK'_0}\frac{(K_0^2 + Q_0^2)(1 + Q_0 r_1)}{2Q_0^2 K_0^2}K'_0\Delta K' \simeq 1 \tag{19.124a}$$

or

$$\Delta K' \simeq -\frac{2Q_0^2 K_0^2(Q_0 + iK'_0)^2 e^{-2Q_0\lambda}}{(Q_0^2 + K_0'^2)(Q_0^2 + K_0^2)(1 + Q_0 r_1)K'_0}, \tag{19.124b}$$

where $Q_0 \equiv Q(K'_0)$ and $K_0 \equiv K(K'_0)$. Now since

$$\begin{aligned} K' &= \sqrt{2\mu(\epsilon - i\Gamma/2)}/\hbar \\ &= \sqrt{K_0'^2 + (2\mu/\hbar^2)(\epsilon - E - i\Gamma/2)} \end{aligned}$$

when $K_0'^2 = (2\mu/\hbar^2)E$, we have

$$K' \simeq K'_0 - \frac{\mu}{\hbar^2 K'_0}(E - \epsilon + i\Gamma/2). \tag{19.125}$$

We neglect the correction $\mu/(\hbar^2 K'_0)(\epsilon - E)$ to the real part of K'_0, assuming it to be small compared with K'_0.

The imaginary part of K' represents a qualitative change in K', even when it is small; therefore we cannot neglect it. Letting

$$\operatorname{Im}\Delta K' = -\frac{\mu}{\hbar^2 K'_0}\frac{\Gamma}{2} \tag{19.126}$$

we easily obtain

$$\Gamma \simeq \frac{\hbar K_0}{\mu}\frac{8\hbar K_0 K'_0 Q_0^3 e^{-2Q_0\lambda}}{(Q_0^2 + K_0'^2)(Q_0^2 + K_0^2)(1 + Q_0 r_1)} = \hbar\frac{\hbar K_0 K'_0}{\mu}T. \tag{19.127}$$

We noted earlier that $\Gamma/\hbar$ is a disintegration rate and had the dimension $(\text{time})^{-1}$. Also, since K_0' is a solution to Eq. (19.123), K_0' is of the order $1/r_1$. These facts provide a means of interpreting Eq. (19.127), for since $\hbar K_0/\mu$ is the velocity v_0 of the particle inside the well,

$$\frac{\hbar K_0}{2\mu} K_0' \simeq \frac{v_0}{2r_1} \tag{19.128}$$

is essentially the rate at which the particle collides with the walls of the well; hence

$$T = \frac{16 K_0 Q_0^3 e^{-2Q_0\lambda}}{(Q_0^2 + K_0'^2)(Q_0^2 + K_0^2)(1 + Q_0 r_1)} \tag{19.129}$$

is essentially the probability that the particle escapes upon colliding with the well. If, as often is the case, $K_0'^2 \ll (2\mu/\hbar^2)V_0$ and $K_0'^2 \ll (2\mu/\hbar^2)V_m$, we have

$$T \simeq \frac{16\sqrt{V_0 V_m}}{V_0 + V_m} \frac{e^{-(2/\hbar)\sqrt{2\mu V_m}\lambda}}{1 + (1/\hbar)\sqrt{2\mu V_m} r_1}.$$

When this is small, the disintegration of the system is strongly inhibited, since $\Gamma/\hbar$ represents the decay rate. This is precisely what occurs in α-particle decay, where the barrier is provided by a repulsive Coulomb force, and it explains the long α-lifetimes of certain nuclei, as Gamov has shown.

The actual transmission coefficient is the ratio $|2B/A|^2$, which may be found from Eqs. (19.121). This is the absolute square of the ratio of the amplitude of the outgoing waves outside the barrier to the amplitude of the outgoing waves inside the barrier.

The approach we have taken to this problem raises a number of apparent difficulties. In the first place, it is incorrect to think of a state as having a complex energy. The term $-i\Gamma/2$ in the energy was simply a convenient formal device to deal with the fact that a "trapped" state, often called a quasi-stationary state, could not be a truly stationary state of the system. In fact, we must have

$$\psi(\mathbf{x}, t) = \int_0^\infty a(E)\psi_E(\mathbf{x})e^{-iEt/\hbar}\, dE$$

where the ψ_E, $(\mathbf{x})$ are true energy eigenfunctions with real eigenvalues E; $\psi(\mathbf{x}, t)$ is a superposition of energy eigenstates. If we are given that the system is represented by $\psi(\mathbf{x}, 0)$ at time $t = 0$, the amplitude for the probability that it can be represented by the same function $\psi(\mathbf{x}, 0)$ (i.e. that it is in the same state) at time t is just

$$b(t) = \int d^3x\, \psi^*(\mathbf{x}, 0)\psi(\mathbf{x}, t),$$

or in our case,

$$b(t) = \int d^3x \int_0^\infty dE'\, a^*(E') \int_0^\infty dE''\, a(E'')e^{-iE''t/\hbar}\psi^*_{E'}(\mathbf{x})\psi_{E''}(\mathbf{x})$$

$$= \int_0^\infty dE'\, |a(E')|^2 e^{-iE't/\hbar}.$$

Here $|a(E')|^2$ is the probability that the system can be found in a state with energy E. However, we know that $b(t)$ here is given by

$$b(t) = \int d^3x\, \psi^*(\mathbf{x}, 0)\psi(\mathbf{x}, 0)e^{-i(E - i\Gamma/2)t/\hbar}$$

$$= e^{-i(E - i\Gamma/2)t/\hbar}$$

for $t > 0$, in the approximation that $\psi(\mathbf{x}, t)$ has not substantially changed its spatial distribution from $\psi(\mathbf{x}, 0)$ in time t. Now let us assume that the system was formed in the same way that it decayed, that is, that it grew exponentially,

$$b(t) = e^{-i(E + i\Gamma/2)t/\hbar}, \qquad t < 0,$$

with the same growth rate as decay rate; this simply assumes time-reversal invariance for the process. Then we have

$$|a(E')|^2 = \frac{1}{2\pi\hbar} \int_{-\infty}^{\infty} e^{iE't/\hbar} b(t)\, dt$$

$$= \frac{1}{2\pi\hbar}\left[\int_{-\infty}^{0} e^{i(E' - E - i\Gamma/2)t/\hbar}\, dt + \int_0^\infty e^{i(E' - E + i\Gamma/2)t/\hbar}\, dt\right]$$

$$= \frac{1}{2\pi\hbar}\left[\frac{\hbar}{i(E' - E) + \Gamma/2} - \frac{\hbar}{i(E' - E) - \Gamma/2}\right]$$

$$= \frac{1}{2\pi} \frac{\Gamma}{(E' - E)^2 + \Gamma^2/4} \tag{19.130}$$

Here we have ignored the limitation that t be less than the time required for $\psi(\mathbf{x}, t)$ to disintegrate appreciably, and hence our result is an approximation. Nevertheless, it shows the form exhibited generally for inhibited disintegrations.

The energy distribution is peaked at $E' = E$ and has a half-width of $\Gamma/2$ since

$$|a(E \pm \Gamma/2)|^2 = \tfrac{1}{2}|a(E)|^2.$$

Actually, this energy distribution is not restricted to the potential-barrier problem; it holds also when the decay is inhibited by other phenomena. For example, two particles, A and B, may collide and form a conglomerate system in which A and B lose their identities (e.g. when A and B are nuclei, becoming an excited state of a larger nucleus upon colliding). A subsequent decay of this conglomerate may be inhibited by the fact that the collision energy may be spread over many of the constituent particles in such a way that no one configuration of them can easily gain sufficient energy to decay into separate particles, say A and B, again. The distributing of energy over many particles creates an inhibition that gives rise to the same kind of quasi-stationary behavior as does the potential barrier. The elastic cross-section then displays a resonance with an energy distribution similar to Eq. (19.130)

Another difficulty, closely related to the above discussion, is that if we take $E' = E - i\Gamma/2$ seriously, so that $K' = (\sqrt{2\mu E}/\hbar) - i\sqrt{\mu/2E}(\Gamma/4\hbar)$ is to be taken seriously as a momentum, we have

$$f_{\text{III}} = Be^{iK_0'r}e^{\sqrt{\mu/2E}(\Gamma/4\hbar)r}$$

and the wave function grows exponentially for large r. This does not occur in fact, and it corresponds to allowing the system to decay for an infinite time so that eventually "all of the probability" has drifted out to the infinite sphere. Again, the term $-i\Gamma/2$ in the energy must be thought of as a formality, and is not to be interpreted literally as a part of the energy.

READING NOTES*

Reference 23, Chapter 21.

Reference 24, Chapter 8.

Reference 29 is an excellent sophisticated text on quantum theory. The approach is quite different from that of Schiff, for example. Chapter III provides a neat discussion of several aspects of scattering theory. This book is highly recommended for advanced students.

Reference 42, Chapter 11.

Reference 42, Chapters 5, 9.

PROBLEMS

1. In connection with Eq. (19.9), and also with Eq. (15.36), discuss the conditions under which

$$|1(t)\rangle = e^{-i\mathfrak{H}t/\hbar}|1(0)\rangle$$

is a correct solution to

$$i\hbar\frac{\partial}{\partial t}|1(t)\rangle = \mathfrak{H}|1(t)\rangle.$$

Why, then, is $-i\int_0^t \mathfrak{H}(t')\,dt'/\hbar$ not, in general, an appropriate replacement for $-i\mathfrak{H}t/\hbar$ in the exponential above, when $\mathfrak{H}$ depends on time?

---

* Works referred to in these notes will be found in the Bibliography on p. 577.

2. Show that for $G^{\pm}(\mathbf{x} - \mathbf{x}')$ given by Eq. (19.19),

$$\lim_{\epsilon \to 0}\left(-\frac{\hbar^2}{2\mu}\nabla^2 - E\right)G^{(\pm)}(\mathbf{x} - \mathbf{x}') = \delta^3(\mathbf{x} - \mathbf{x}').$$

3. Obtain Eq. (19.21) from Eq. (19.12′) without using the formal expedient of the geometric series sum above Eq. (19.21).

4. Consider the one-dimensional equation

$$\left[\frac{d^2}{dx^2} + k^2\right]\phi_k(x) = 0.$$

a) Using the Fourier-transform equation

$$G_k^{\pm}(x, x') = \frac{1}{2\pi}\int dq\, g_k^{(\pm)}(q)e^{iq(x-x')}$$

find the Green's function for this equation, such that

$$\left[\frac{d^2}{dx^2} + k^2\right]G_k^{(\pm)}(x, x') = \delta(x - x')$$

and the $+(-)$ sign refers to whether G_k is generating waves with x moving away from (toward) x'.

b) If a potential function is introduced so that

$$\left[\frac{d^2}{dx^2} - U(x) + k^2\right]\psi(k, x) = 0,$$

show that then if $\phi_k(x) = e^{ikx}/\sqrt{2\pi}$, the solutions corresponding to "in" or "out" solutions (for waves initially moving in the positive x-direction) are

$$\psi^{(\pm)}(k, x) = \phi_k + \int dx'\, G_k^{(\pm)}(x, x')U(x')\psi^{(\pm)}(k, x').$$

c) Let

$$h_+(k, x) \equiv -\frac{i}{2k}\int_{-\infty}^{x} e^{-ikx'}U(x')\psi^{(+)}(k, x')\,dx'$$

$$h_-(k, x) \equiv -\frac{i}{2k}\int_{x}^{\infty} e^{ikx'}U(x')\psi^{(+)}(k, x')\,dx'$$

and

$$h_{\pm}(k, x) = |h_{\pm}(k, x)|e^{i\eta_{\pm}(k,x)}.$$

Then, for a wave packet with Fourier amplitude $f(k)$ whose phase is constant, use the method of stationary phase to "follow" the motion of the wave packet.

d) Find the time at which a particle arrives at point x. Does the presence of the potential introduce a delay in the reflection of waves?

5. Referring to Eq. (19.55), show that the wave function $\phi_1(x)$ given in that equation has mean momentum $\hbar k$ if $\pi(\mathbf{x} - \mathbf{x}_0)$ has a constant phase.

6. In Figs. 19.2 and 19.3, the emerging (incident) plane waves must have amplitudes that are diminished relative to the incident (emerging) plane waves, if the law of conservation of probability is to hold. A consequence of this is the so-called *optical theorem*,

$$\sigma = (4\pi/k)\,\mathrm{Im}\, f_{k+}(0),$$

stating that the total scattering cross-section is equal to $4\pi/k$ times the imaginary part of the forward scattering amplitude.

a) Prove this theorem. (Hint: see Eqs. (19.76) and (19.82).)
b) Since the conservation of probability in scattering processes is expressed through the unitarity of $\mathfrak{S}$, use

$$\langle \mathbf{p}_j, q_j|\mathfrak{S}^\dagger\mathfrak{S}|\mathbf{p}_i, q_i\rangle = \underset{\mathbf{p}_f, q_f}{\mathrm{S}} S^*_{fj}S_{fi} = \delta^3(\mathbf{k}_j - \mathbf{k}_i)\delta(q_j - q_i)$$

to prove the theorem in a different way. (Hint: use Eqs. (19.33) and (19.46).)

7. Let two identical scattering centers be located at points $\mathbf{x}_1$ and $\mathbf{x}_2$, and show that in first-order Born approximation (Eq. (19.45)), the differential cross-section for scattering from them is

$$\frac{d\sigma}{d\Omega} = \left(\frac{d\sigma}{d\Omega}\right)_0 |e^{-i\mathbf{q}\cdot\mathbf{x}_1} + e^{-i\mathbf{q}\cdot\mathbf{x}_2}|^2, \qquad \mathbf{q} = \mathbf{k}' - \mathbf{k},$$

where $(d\sigma/d\Omega)_0$ is the cross-section from a single scatterer.

8. If the potential for each scatterer in Problem 7 is $V_0e^{-r/a}$ and if the particles are a distance l apart, their line of separation being at angle $\pi/4$ with the axis of the incident beam, find the resulting cross-section.

9. If an electron in a Coulomb field is to be bound, the following conditions must be met:

a) The potential must be attractive.
b) The total energy must be negative, and
c) No wave proportional to $e^{-ikr} = e^{\kappa r}$ may be present in the solution.

Referring to the end of Section 19.7, show that this means that $\Gamma(l + 1 + i\eta)$ must be singular, so that $l + 1 + i\eta = -m$, m an integer, and hence derive the energy-level formula for the nonrelativistic one-electron atom.

APPENDIX A

OPTICAL INTERFERENCE FRINGES

A1 INTERFERENCE FRINGES FROM A COHERENT DOUBLE SOURCE

Here we analyse, in terms of the electric-field vector, interference from a coherent double source. The source is a pair of infinite parallel lines that emit electromagnetic radiation in such a way that emissions from both lines are in phase with each other; the standard example of such an arrangement is a double slit through which light is allowed to pass. Figure A1 illustrates the geometry in a plane perpendicular to the parallel lines.

The separation of lines is given by distance d, and the plane of the sources is a distance D from a plane screen erected parallel to the plane of sources. The sources are labeled 1 and 2, and variables with these numbers as subscripts refer to the respective sources; clearly,

$$r_1 = \sqrt{D^2 + (y - d/2)^2}$$

$$r_2 = \sqrt{D^2 + (y + d/2)^2},$$

$$\mathbf{d}^2 = (\mathbf{r}_1 - \mathbf{r}_2)^2 = r_1^2 + r_2^2 - 2r_1r_2 \cos \alpha.$$

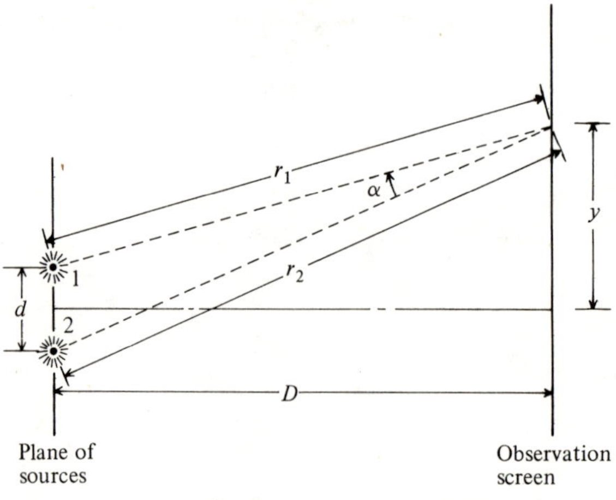

Fig. A1. Geometry of interference experiment with two coherent sources (double slit); the sources are labeled 1 and 2.

Thus,

$$\cos\alpha = \frac{r_1^2 + r_2^2 - d^2}{2r_1 r_2} = \frac{D^2 + y^2 - 3d^2/4}{[(D^2 + y^2 + d^2/4)^2 - d^2y^2]^{1/2}},$$

$$\sin\alpha = \frac{dD}{[(D^2 + y^2 + d^2/4)^2 - d^2y^2]^{1/2}}$$

Now, for a single source at 1, the electric field is

$$\mathbf{E} = E_0\hat{\mathbf{e}}_1\frac{\sin(kr_1 - \omega t)}{kr_1},$$

where k is the wave number and $\omega = ck$ the frequency of the radiation; $\hat{\mathbf{e}}_1$ is a unit vector perpendicular to $\hat{\mathbf{r}}_1$. For two coherent sources, one at 1 and the other at 2,

$$\mathbf{E} = E_0\left[\hat{\mathbf{e}}_1\frac{\sin(kr_1 - \omega t)}{kr_1} + \hat{\mathbf{e}}_2\frac{\sin(kr_2 - \omega t)}{kr_2}\right]$$

$$= E_0\left\{\hat{\mathbf{e}}_1\left[\frac{\sin(kr_1 - \omega t)}{kr_1} + \cos\alpha\frac{\sin(kr_2 - \omega t)}{kr_2}\right] + \hat{\mathbf{r}}_1\sin\alpha\frac{\sin(kr_2 - \omega t)}{kr_2}\right\}.$$

If $d \ll D$, $\alpha \to 0$ and $\sin\alpha \sim \alpha \to 0$, $\cos\alpha \to 1$; also,

$$r_1 \simeq \sqrt{D^2 + y^2}\left[1 + \frac{d^2 - 4yd}{8(D^2 + y^2)}\right] = r\left[1 + \frac{d^2 - 4yd}{8r^2}\right],$$

$$r_2 \simeq \sqrt{D^2 + y^2}\left[1 + \frac{d^2 + 4yd}{8(D^2 + y^2)}\right] = r\left[1 + \frac{d^2 + 4yd}{8r^2}\right],$$

where

$$r = \sqrt{D^2 + y^2}.$$

Hence

$$r_j \simeq r(1 + \Delta_j), \qquad j = 1, 2,$$

$$\Delta_1 = \frac{d}{r}\frac{d - 4y}{8r} \simeq \frac{d}{D}\frac{d - 4y}{8D}, \qquad \Delta_2 = \frac{d}{r}\frac{d + 4y}{8r} \simeq \frac{d}{D}\frac{d + 4y}{8D},$$

and

$$\Delta_2 - \Delta_1 = \frac{yd}{r^2} \simeq \frac{yd}{D^2}.$$

Therefore if $4y/d \ll 1$, Δ_1 and Δ_2 are "second-order" small, and in this case, to first order,

$$\frac{1}{r_1} \simeq \frac{1}{r_2} \simeq \frac{1}{r},$$

and

$$\mathbf{E} \simeq \frac{E_0}{kr}\hat{\mathbf{e}}_1[\sin(kr_1 - \omega t) + \sin(kr_2 - \omega t)]$$

$$= \frac{2E_0}{kr}\hat{\mathbf{e}}_1 \sin\left(k\frac{r_1 + r_2}{2} - \omega t\right) \cos k\frac{r_2 - r_1}{2}$$

$$= \frac{2E_0}{kr}\hat{\mathbf{e}}_1 \sin\left[k\left(r_1 + \frac{yd}{2D}\right) - \omega t\right] \cos k\frac{yd}{2D}$$

$$\simeq \frac{2E_0}{kr}\hat{\mathbf{e}}_1 \sin(kr - \omega t) \cos k\frac{yd}{2D}.$$

Thus the electric field at a point with vertical coordinate y on the screen is approximately what it would have been for a single source of intensity E_0 and wave number k, except that it is phase-shifted by about $kyd/2D$ and multiplied by an amplitude factor $2\cos(yd/2D)$.

The intensity I of radiation at the point in question is proportional to $\mathbf{E}^2$, or

$$I \propto \frac{4E_0^2}{k^2r^2}\cos^2(kyd/2D)\sin^2(kr - \omega t) \simeq 4I_0\cos^2(kyd/2D),$$

where I_0 is the single-source intensity. Clearly, when $kyd/2D = n\pi$, the intensity is maximized, and when $kyd/2D = (n + \frac{1}{2})\pi$, the intensity will be minimal (vanishing). This produces a fringe pattern with maxima and minima respectively at the following values of y:

$$y = n(D/d)\lambda \qquad \text{(bright fringes)}$$

$$y = (n + \tfrac{1}{2})(D/d)\lambda. \qquad \text{(dark fringes)}$$

A2 CIRCULAR FRINGES

Let us consider now circular fringes of the kind observed in the Michelson–Morley experiment. We consider an extended source of radiation whose emissions are reflected by two parallel mirrors separated from one another by a distance d. The source is a distance r_1 from the point at which its observed radiations are reflected from the closer mirror, and the observation point is a distance r_2 from the point of reflection from the closer mirror (see Fig. A2); we let $r \equiv r_1 + r_2$. The closer mirror is half-silvered so that radiation can pass to the farther mirror and be reflected. The normal angle of incidence on both mirrors is θ.

It is easy to see that at the observation point

$$\mathbf{E} \propto \hat{\mathbf{e}}\{\sin(kr - \omega t) + \sin[k(r + 2d\cos\theta) - \omega t]\}$$

$$= \hat{\mathbf{e}}2\sin[k(r + d\cos\theta) - \omega t]\cos(kd\cos\theta),$$

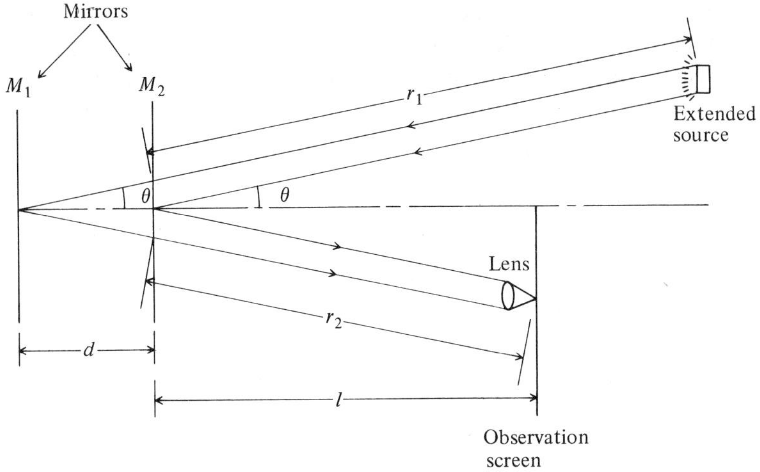

Fig. A2. Geometry of circular fringe interference experiment.

where $\hat{\mathbf{e}}$ is a unit vector normal to the propagation direction through the observation point. Here the phase-shift is $kd\cos\theta$ and the amplitude factor is $2\cos(kd\cos\theta)$; hence the intensity is proportional to $\cos^2(kd\cos\theta)$ and maxima appear at $kd\cos\theta = n\pi$, minima at $kd\cos\theta = (n+\frac{1}{2})\pi$. Keeping n and k (or λ) fixed, $\cos\theta$ is a function of the number (or order) of fringes as counted from the center ($\theta = 0$):

$$\cos\theta = n\lambda/2d \qquad \text{(bright fringes)}$$

$$\cos\theta = (n+\tfrac{1}{2})\lambda/2d \qquad \text{(dark fringes).}$$

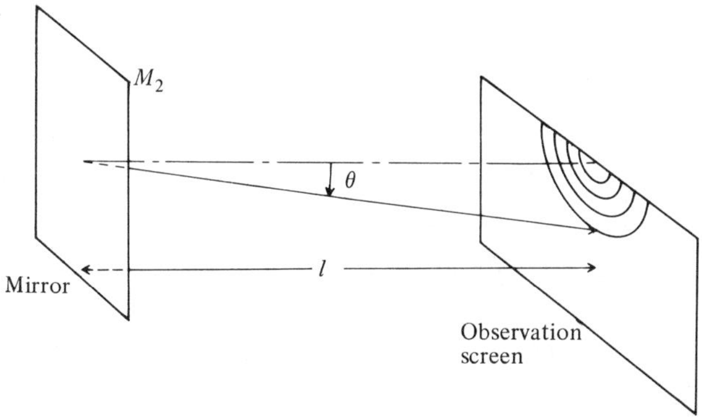

Fig. A3. Circular fringes from the experiment of Fig. A2.

This is convenient because in the experimental apparatus θ and d can be fixed, and then any changes in the number of fringes contained within an angle θ (or within radius $\rho = l \tan \theta$ on the observation screen; see Fig. A3) corresponds to a change in λ. In another important application, λ and θ are fixed and the number of fringes within angle θ is a function of d.

APPENDIX B

THE PHASE OF A WAVE WITH WELL-DETERMINED MOMENTUM

Since

$$\psi_{\mathbf{p},E}(\mathbf{x}, t) = Ae^{ih(\mathbf{x},t)},$$

we have

$$\psi_{\mathbf{r},E}(\mathbf{x} + \Delta\mathbf{x}, t + \Delta t) = Ae^{ih(\mathbf{x}+\Delta\mathbf{x},t+\Delta t)}$$

and this must be periodic in $\mathbf{x}, t$. The wave vector $\mathbf{k}$ can be decomposed into components along $\mathbf{k}$ and perpendicular to $\mathbf{k}$; this decomposition will provide a convenient Cartesian coordinate system. Let

$$\mathbf{k} = \hat{\mathbf{n}}_0 k + \hat{\mathbf{n}}_1 0 + \hat{\mathbf{n}}_2 0,$$

where $\hat{\mathbf{n}}_0$ is parallel to $\mathbf{k}$, and $\hat{\mathbf{n}}_1$ and $\hat{\mathbf{n}}_2$ are perpendicular to each other as well as to $\mathbf{k}$. Thus the wave numbers in directions perpendicular to $\mathbf{k}$, say $\mathbf{k}_1$ and $\mathbf{k}_2$ vanish, and that means that there is no variation in the wave function if $\mathbf{x}$ is displaced by $\Delta\mathbf{x}$ perpendicular to $\mathbf{k}$. Thus, if we let $\mathbf{x} = \hat{\mathbf{n}}_0\xi_0 + \hat{\mathbf{n}}_1\xi_1 + \hat{\mathbf{n}}_2\xi_2$ and $\Delta\mathbf{x} = \hat{\mathbf{n}}_0\,\Delta\xi_0 + \hat{\mathbf{n}}_1\,\Delta\xi_1 + \hat{\mathbf{n}}_2\,\Delta\xi_2$, we must have $h(\mathbf{x} + \hat{\mathbf{n}}_1\Delta\xi_1 + \hat{\mathbf{n}}_2\Delta\xi_2, t) = h(x, t)$ for any values of $\Delta\xi_1, \Delta\xi_2$,

To apply this result to finding the form of $h(\mathbf{x}, t)$, we perform a 3-dimensional Taylor expansion of $h(\mathbf{x} + \Delta\mathbf{x}, t)$ about $\mathbf{x}$:

$$\begin{aligned} h(\mathbf{x} + \Delta\mathbf{x}, t) &= \sum_{n=0}^{\infty} \frac{1}{n!}(\Delta\mathbf{x}\cdot\nabla)^n h(\mathbf{x}, t) \\ &= \sum_{n=0}^{\infty} \frac{1}{n!} \sum_{i_1,\ldots,i_n=0}^{2} \Delta\xi_{i_n}\Delta\xi_{i_2}\cdots\Delta\xi_{i_n} \frac{\partial^n}{\partial\xi_{i_1}\,\partial\xi_{i_2}\cdots\partial\xi_{im}} h(\mathbf{x}, t). \end{aligned}$$

The only way this can be independent of $\Delta\xi_1$ and $\Delta\xi_2$ is for every coefficient of $\Delta\xi_1$ or $\Delta\xi_2$ to vanish. This will imply

$$\frac{\partial^m}{\partial\xi_1^m} h(\mathbf{x}, t) = \frac{\partial^m}{\partial\xi_2^m} h(\mathbf{x}, t) = 0$$

for all $m \geq 1$, so that the expansion becomes

$$h(\mathbf{x} + \Delta\mathbf{x}, t) = \sum_{n=0}^{\infty} \frac{1}{n!}(\Delta\xi_0)^n \frac{\partial^n}{\partial\xi_0^n} h(\mathbf{x}, t)$$

$$= h(\mathbf{x}, t) + \sum_{n=1}^{\infty} \frac{1}{n!}(\Delta\xi_0)^n \frac{\partial^n}{\partial\xi_0^n} h(\mathbf{x}, t).$$

On the other hand, whenever $\Delta\xi_0 = j\lambda$ where j is an integer, $h(\mathbf{x} + \hat{\mathbf{k}}\Delta\xi_0, t) = h(\mathbf{x} + \hat{\mathbf{k}}j\lambda, t) = h(\mathbf{x}, t) + 2\pi l$, where l is an integer, if we are to have periodicity in space along $\hat{\mathbf{k}}$. Thus

$$2\pi l = \sum_{n=1}^{\infty} \frac{1}{n!}(j\lambda)^n \frac{\partial^n}{\partial\xi^n} h(\mathbf{x}, t)$$

for any integer j and for some integer l. Also, since l must change by one unit each time j does so, we have $l = l_0 \pm j$, where l_0 is some integer, for all j. Substituting this in the above equation and equating powers of j (since j is arbitrary), we get

$$l_0 = 0, \qquad \frac{\partial^n}{\partial\xi_0^n} h(\mathbf{x}, t) = \begin{cases} 0 & \text{if } n > 1 \\ \pm 2\pi/\lambda & \text{if } n = 1 \end{cases}.$$

Hence

$$h(\mathbf{x}, t) = \pm\frac{2\pi}{\lambda}\xi_0 + h_0(t) = \pm k\xi_0 + h_0(t) = \pm\mathbf{k}\cdot\mathbf{x} + h_0(t).$$

A parallel argument for variations in t yields

$$h(\mathbf{x}, t) = \pm(\mathbf{k}\cdot\mathbf{x} - \omega t).$$

The negative sign in front of ω comes from the necessity that the phase velocity be in the direction of propagation of the phase, and conventionally one chooses the upper sign for the whole expression. Hence

$$\psi_{\mathbf{p},E}(\mathbf{x}, t) = Ae^{i(\mathbf{k}\cdot\mathbf{x} - \omega t)}, \qquad \mathbf{p} = \hbar\mathbf{k},$$

is the only possible wave function for a wave with well-defined momentum $\mathbf{p}$.

APPENDIX C

ON SPHERICAL BESSEL FUNCTIONS

C1 SPHERICAL BESSEL FUNCTIONS

In Section 14.4 we discussed the solutions to the equation

$$\left[\frac{d^2}{d\rho^2} - \frac{l(l+1)}{\rho^2} + 1\right] f_l(\rho) = 0 \tag{C.1}$$

by converting this equation to Bessel's equation of half-odd-integer orders, Eq. (14.17). The familiar cylindrical Bessel functions were then utilized to obtain appropriate solutions $f_l(\rho)$ to (C.1). However, it is a rather simple matter to obtain solutions to (C.1) directly, and it is instructive to do so.

First, we let $\rho \to \infty$ and get the asymptotic behavior of $f_l(\rho)$:

$$\left(\frac{d^2}{d\rho^2} + 1\right) f_l(\rho) \underset{\rho\to\infty}{\sim} 0$$

and hence

$$f_l(\rho) \underset{\rho\to\infty}{\sim} (\text{const.})e^{\pm i\rho}.$$

Define

$$\rho h_l^{(\pm)}(\rho) = \phi_l^{(\pm)}(\rho) e^{\pm i\rho}; \tag{C.2}$$

then the $\phi_l^{(\pm)}(\rho)$ satisfy

$$\left[\frac{d^2}{d\rho^2} \pm i2\frac{d}{d\rho} - \frac{l(l+1)}{\rho^2}\right] \phi_l^{(\pm)}(\rho) = 0,$$

and it is clear that $\phi_l^{(-)}(\rho) = \phi_l^{(+)*}(\rho^*)$ up to a constant factor.

We now assume

$$\phi_l^{(+)}(\rho) = \sum_{n=0}^{\infty} a_n \rho^{p+n} \tag{C.3}$$

and obtain

$$a_0[p(p-1) - l(l+1)]\rho^{p-2}$$
$$+ \sum_{n=0}^{\infty} \{a_{n+1}[(n+p)(n+p+1) - l(l+1)] + a_n 2i(n+p)\}\rho^{p+n-1} = 0.$$

Since a_0 may not vanish, this gives the usual values for p,

$$p = -l, l + 1$$

and the recursion relation obtained from the higher powers of ρ gives

$$a_{n+1} = \frac{2}{i}\frac{p+n}{(2p+n)(n+1)}a_n$$

or

$$a_n = \left(\frac{2}{i}\right)^n \frac{\Gamma(p+n)\Gamma(2p)}{\Gamma(p)\Gamma(2p+n)n!}a_0 \qquad \text{if } p \geq 0$$

$$= \left(\frac{2}{i}\right)^n \frac{\Gamma(-2p-n+1)\Gamma(-p+1)}{\Gamma(-2p+1)\Gamma(-p-n+1)n!}a_0 \qquad \text{if } p \leq 0.$$

If $p = l + 1$, the solution $\phi_l^{(+)}(\rho)$ is given by

$$a_0\rho^{l+1}\sum_{n=0}^{\infty}\frac{(l+n)!(2l+1)!}{l!(2l+n+1)!n!}\left(\frac{2\rho}{i}\right)^n,$$

and it is clear that this power series fails to terminate. Asymptotically it behaves like the power series with recursion formula

$$a_{n+1} = \frac{2}{i(n+1)}a_n$$

namely,

$$a_0 e^{-i2\rho}\rho^{l+1} = a_0\sum_{n=0}^{\infty}\frac{(2\rho/i)^n}{n!}\rho^{l+1}.$$

We may reject this solution, since it yields

$$\phi_l^{(+)}(\rho)e^{i\rho} \underset{\rho\to\infty}{\sim} (\text{const.})e^{-i\rho}\rho^{l+1}$$

contrary to our hypothesis about the asymptotic behavior of $\phi_l^{(+)}(\rho)$.

The solution of interest to us, then, corresponds to $p = -l$, or

$$\phi_l^{(+)}(\rho) = a_0\rho^{-l}\sum_{n=0}^{\infty}\frac{\Gamma(2l-n+1)\Gamma(l+1)}{\Gamma(2l+1)\Gamma(l-n+1)n!}\left(\frac{2\rho}{i}\right)^n. \tag{C.4}$$

Now this is a polynomial since $1/\Gamma(l - n + 1)$ vanishes for integers $n \geq l + 1$; hence the upper limit on the sum may be replaced with l. As $\rho \to \infty$, we have

$$\phi_l^{(+)}(\rho) \underset{\rho\to\infty}{\sim} a_0\frac{l!}{(2l)!}\left(\frac{2}{i}\right)^l = a_0\frac{e^{-il\pi/2}}{(2l-1)!!}.$$

Therefore

$$\rho h_l^{(+)}(\rho) = \phi_l^{(+)}(\rho)e^{i\rho} \underset{\rho\to\infty}{\sim} \frac{a_0}{(2l-1)!!}e^{i(\rho - l\pi/2)}. \tag{C.5}$$

Imposing the condition that

$$\rho h_l^{(+)}(\rho) \underset{\rho\to\infty}{\sim} e^{i[\rho-(l+1)\pi/2]}$$

yields

$$a_0 = (2l-1)!!/i;$$

thus we obtain

$$\begin{aligned}
\phi_l^{(+)}(\rho) &= \sum_{n=0}^{l} \frac{(2l-n)!\,i^{-n-1}}{(l-n)!\,n!}(2\rho)^{n-l} \\
&= \sum_{n=0}^{\infty} \frac{(2l-2n-1)!\,(-1)^{n+1}(2\rho)^{2n-l+1}}{(l-2n-1)!\,(2n+1)!} \\
&\quad + i\sum_{n=0}^{\infty} \frac{(2l-2n)!\,(-1)^{n+1}(2\rho)^{2n-l}}{(l-2n)!\,(2n)!} \\
&= \phi_l^{(1)}(\rho) + i\phi_l^{(2)}(\rho).
\end{aligned} \tag{C.6}$$

Here we have defined

$$\phi_l^{(1)}(\rho) \equiv \sum_{n=0}^{\infty} \frac{(2l-2n-1)!\,(-1)^{n+1}(2\rho)^{2n+1-l}}{(l-2n-1)!\,(2n+1)!} \tag{C.7}$$

$$\phi_l^{(2)}(\rho) \equiv \sum_{n=0}^{\infty} \frac{(2l-2n)!\,(-1)^{n+1}(2\rho)^{2n-l}}{(l-2n)!\,(2n)!}. \tag{C.8}$$

The function $\phi_l^{(-)}(\rho)$ is obtained easily from $\phi_l^{(+)}(\rho)$; in fact, $\phi_l^{(-)}(\rho) = \phi_l^{(+)*}(\rho^*)$, or, for real ρ,

$$\begin{aligned}
\phi_l^{(-)}(\rho) &= \sum_{n=0}^{\infty} \frac{(2l-n)!\,i^{n+1}}{(l-n)!\,n!}(2\rho)^{n-l} \\
&= \phi_l^{(1)}(\rho) - i\phi_l^{(2)}(\rho).
\end{aligned} \tag{C.9}$$

Furthermore, by defining

$$\rho j_l(\rho) \equiv \mathrm{Re}\,[\rho h_l^{(+)}(\rho)] = \mathrm{Re}\,[\phi_l^{(+)}(\rho)e^{i\rho}]$$

and

$$\rho n_l(\rho) \equiv \mathrm{Im}\,[\rho h_l^{(+)}(\rho)] = \mathrm{Im}\,[\phi_l^{(+)}(\rho)e^{i\rho}]$$

when $\rho = \rho^*$ is real, we obtain

$$\rho j_l(\rho) = \phi_l^{(1)}(\rho)\cos\rho - \phi_l^{(2)}(\rho)\sin\rho \tag{C.10}$$

and

$$\rho n_l(\rho) = \phi_l^{(1)}(\rho)\sin\rho + \phi_l^{(2)}(\rho)\cos\rho. \tag{C.11}$$

The reader should compare these expressions $\rho h_l^{(\pm)}(\rho)$, $\rho j_l(\rho)$ and $\rho n_l(\rho)$ with those given in Section 14.4, at least for the first few values of l.

C2 THE EXPANSION OF PLANE WAVES IN LEGENDRE POLYNOMIALS

Since the position-dependent exponential function

$$e^{i\mathbf{k}\cdot\mathbf{r}} = e^{ikr\cos\theta} = e^{ikrx}, \qquad x = \cos\theta = \hat{\mathbf{k}}\cdot\hat{\mathbf{r}},$$

which characterizes plane waves, is a function of $x = \cos\theta$, it can be expanded in terms of the complete set of orthogonal Legendre polynomials $\{P_l(x)\}$. Thus

$$e^{i\mathbf{k}\cdot\mathbf{r}} = \sum_{l=0}^{\infty} B_l(kr)P_l(x). \tag{C.12}$$

Recalling the orthogonality relation

$$\int_{-1}^{1} P_{l'}(x)P_l(x)\,dx = \frac{2}{2l+1}\delta_{l,l'},$$

we obtain

$$B_l(kr) = (l+\tfrac{1}{2})\int_{-1}^{1} e^{ikrx}P_l(x)\,dx = (l+\tfrac{1}{2})\int_{-1}^{1} e^{i\rho x}P_l(x)\,dx, \qquad \rho \equiv kr.$$

In order to evaluate this integral, we shall utilize Rodrigue's formula for the Legendre polynomial:

$$\begin{aligned} B_l(\rho) &= (l+\tfrac{1}{2})\frac{(-)^l}{2^l l!}\int_{-1}^{1} e^{i\rho x}\frac{d^l}{dx^l}(1-x^2)^l\,dx \\ &= (l+\tfrac{1}{2})\frac{(i\rho)^l}{2^l l!}\int_{-1}^{1} e^{i\rho x}(1-x^2)^l\,dx, \end{aligned}$$

where the last equality is obtained through repeated integration by parts. Now we show that $\rho B_l(\rho)$ satisfies Eq. (C.1); we merely differentiate twice with respect to ρ:

$$\begin{aligned} \frac{d^2}{d\rho^2}\rho B_l(\rho) &= (2l+1)\frac{i^l}{2^{l+1}l!}\left[\frac{l(l+1)}{\rho^2} - 1\right]\rho^{l+1}\int_{-1}^{1} e^{i\rho x}(1-x^2)^l\,dx \\ &= \left[\frac{l(l+1)}{\rho^2} - 1\right]B_l(\rho). \end{aligned}$$

Hence, $B_l(\rho)$ is proportional to $j_l(\rho)$.

To find the constant of proportionality, we consider* the limit of

$$\int_{-1}^{1} e^{i\rho x}(1 - x^2)^l\, dx \qquad \text{as } \rho \to 0:$$

$$\int_{-1}^{1} e^{i\rho x}(1 - x^2)^l\, dx \underset{\rho\to 0}{\to} \int_{-1}^{1} (1 - x^2)^l\, dx = 2\int_{0}^{1} x^{2(\frac{1}{2}-1)}(1 - x^2)^{l+1-1} x\, dx$$

$$= \int_0^1 \xi^{\frac{1}{2}-1}(1 - \xi)^{l+1-1}\, d\xi = B(\tfrac{1}{2}, l+1)$$

$$= \frac{\Gamma(\frac{1}{2})\Gamma(l+1)}{\Gamma(l+\frac{3}{2})} = \frac{2^{l+1} l!}{(2l+1)!!}.$$

Hence

$$B_l(\rho) \underset{\rho\to 0}{\sim} \frac{(2l+1)i^l}{(2l+1)!!}\rho^l \sim (2l+1)i^l j_l(\rho),$$

and since $B_l(\rho)$ in general is proportional to $j_l(\rho)$, we have

$$B_l(\rho) = (2l+1)i^l j_l(\rho).$$

Substitution into Eq. (C.12) gives

$$e^{i\mathbf{k}\cdot\mathbf{r}} = \sum_{l=0}^{\infty} (2l+1)i^l j_l(kr)P_l(\cos\theta).$$

C3 SUMMARY OF USEFUL RELATIONSHIPS AMONG SPHERICAL BESSEL FUNCTIONS

In this section we provide a brief summary of relationships among solutions to the spherical Bessel equations of different orders, and the derivatives of those solutions. Also, some problems related to them are given. The formulas given below are valid for spherical Bessel, Neumann or Hankel functions, represented here by the generic symbol $g_l(\rho)$.

For $l > 0$ it can be shown that

$$\frac{2l+1}{\rho} g_l(\rho) = g_{l-1}(\rho) + g_{l+1}(\rho)$$

and that

$$\frac{d}{d\rho} g_l(\rho) = g_{l-1}(\rho) - \frac{l+1}{\rho} g_l(\rho) = \frac{l}{\rho} g_l(\rho) - g_{l+1}(\rho),$$

---

* See, for example, P. FRANKLIN, *Methods of Advanced Calculus*, McGraw-Hill Book Company, New York (1944), Section 98; or consult a table of integrals.

the last equation holding also for $l = 0$. These can be combined to yield

$$\frac{d}{d\rho}g_l(\rho) = \frac{1}{2l+1}[lg_{l-1}(\rho) - (l+1)g_{l+1}(\rho)]$$

for $l \geq 0$. A useful indefinite integral formula is

$$\int g_l^2(\rho)\rho^2\, d\rho = \tfrac{1}{2}\rho^2[g_l^2(\rho) - g_{l-1}(\rho)g_{l+1}(\rho)]$$

for $l > 0$.

PROBLEMS

1. Let $f_l(\rho)$ and $g_l(\rho)$ be linearly independent solutions to the spherical Bessel equation of order l.

   a) Using the differential equations satisfied by these functions, show that the Wronskian of $f_l(\rho)$ and $g_l(\rho)$ defined by

   $$\mathrm{W}[f_l(\rho), g_l(\rho)] = f_l(\rho)\frac{d}{d\rho}g_l(\rho) - \left[\frac{d}{d\rho}f_l(\rho)\right]g_l(\rho)$$

   is given by

   $$\mathrm{W}[f_l(\rho), g_l(\rho)] = a/\rho^2,$$

   where a is a constant.

   b) Show that if $f_l(\rho) = j_l(\rho)$, $g_l(\rho) = n_l(\rho)$, then $a = 1$, and that if $f_l(\rho) = h_l^{(+)}(\rho)$, $g_l(\rho) = h_l^{(-)}(\rho)$, then $a = 2/i$. (Hint: use the asymptotic behavior of the functions.)

   c) Show that this also yields

   $$f_l(\rho)g_{l-1}(\rho) - g_l(\rho)f_{l-1}(\rho) = a/\rho^2.$$

2. Derive, using the formulas given in the text of this section,

   $$\frac{d}{d\rho}\rho^{l+1}g_l(\rho) = \rho^{l+1}g_{l-1}(\rho),$$

   $$\frac{d}{d\rho}\rho^{l}g_l(\rho) = \rho^{-1}g_{l+1}(\rho).$$

3. a) Using the differential equations for $f_l(\alpha\rho)$, $g_l(\beta\rho)$, show that

   $$\int_0^R f_l(\alpha\rho)g_l(\beta\rho)\rho^2\, d\rho = \frac{R^2}{\alpha^2 - \beta^2}[\beta f_l(\alpha R)g_l'(\beta R) - \alpha g_l(\beta R)f_l'(\alpha R)]$$

   where $f_l'(\xi) \equiv df_l(\xi)/d\xi$ and similarly for $g_l'(\xi)$.

   b) Show that the right-hand side of the above equation can also be written

   $$\frac{R^2}{\alpha^2 - \beta^2}[\beta f_l(\alpha R)g_{l-1}(\beta R) - \alpha f_{l-1}(\alpha R)g_l(\beta R)].$$

c) Show that

$$\lim_{R\to\infty}\int_0^R j_l(\alpha\rho)j_l(\beta\rho)\rho^2\,d\rho = \lim_{R\to\infty}\frac{1}{2\alpha\beta}\left\{\frac{(-1)^l}{\alpha+\beta}\sin\left[(\alpha+\beta)R\right] + \frac{1}{\alpha-\beta}\sin\left[(\alpha-\beta)R\right]\right\}.$$

d) The first term on the right-hand side oscillates rapidly and therefore averages to zero. The second term, however, is, in the limit, $\pi\delta(\alpha-\beta)$. Demonstrate this by showing that

$$\lim_{K\to\infty}\int_a^b \frac{\sin K\xi}{\xi}\,d\xi = \begin{cases} 0 & \text{if } a, b \text{ have the same sign} \\ \pi & \text{otherwise.} \end{cases}$$

Use

$$\int_{-\infty}^{\infty}\frac{\sin K\xi}{\xi}\,d\xi$$

and evaluate this by contour integration.

e) Finally, show that

$$\int_0^\infty j_l(kr)j_l(k'r)r^2\,dr = \frac{\pi}{2k^2}\delta(k-k').$$

APPENDIX D

ONE-DIMENSIONAL BARRIER PENETRATION

Let us consider the simple potential barrier illustrated in Fig. D1. The corresponding one-dimensional Schrödinger equation for a particle with mass μ is

$$\left[-\frac{\hbar^2}{2\mu}\frac{d^2}{dx^2} + V(x) - E\right]\psi(x) = 0 \tag{D.1}$$

with

$$V(x) = \begin{cases} 0, & x > |a| \\ V_0, & x \le |a|. \end{cases}$$

It is convenient to multiply Eq. (D.1) by $-2\mu/\hbar^2$ and define

$$k \equiv \sqrt{2\mu E}/\hbar, \qquad W(x) \equiv (2\mu/\hbar^2)V(x), \qquad W_0 \equiv (2\mu/\hbar^2)V_0,$$

whence we have

$$\left[\frac{d^2}{dx^2} - W(x) + k^2\right]\psi(x) = 0. \tag{D.2}$$

Now we divide the x-axis into three regions labeled I (for $x < -a$), II (for $|x| \le a$) and III (for $x > a$), and assume that a simple wave with fixed wave number k is incident from the left. Physical considerations indicate that in region I we should have the incident wave plus a reflected wave,

$$\psi_{\underset{\sim}{k}\,\mathrm{I}}(x) = \alpha(e^{ikx} + Re^{-ikx}), \qquad x < -a, \tag{D.3}$$

whereas in region III there should be only a transmitted wave,

$$\psi_{\underset{\sim}{k}\,\mathrm{III}}(x) = \alpha T e^{ikx}, \qquad x > a. \tag{D.4}$$

In region II the differential equation is

$$\left[\frac{d^2}{dx^2} - (W_0 - k^2)\right]\psi_{\underset{\sim}{k}\,\mathrm{II}}(x) = 0, \qquad |x| \le a$$

which has solutions of the form

$$\psi_{\underset{\sim}{k}\,\mathrm{II}}(x) = \alpha A(e^{-\kappa x} + \beta e^{\kappa x}), \qquad \kappa \equiv \sqrt{W_0 - k^2}, \qquad |x| \le a, \tag{D.5}$$

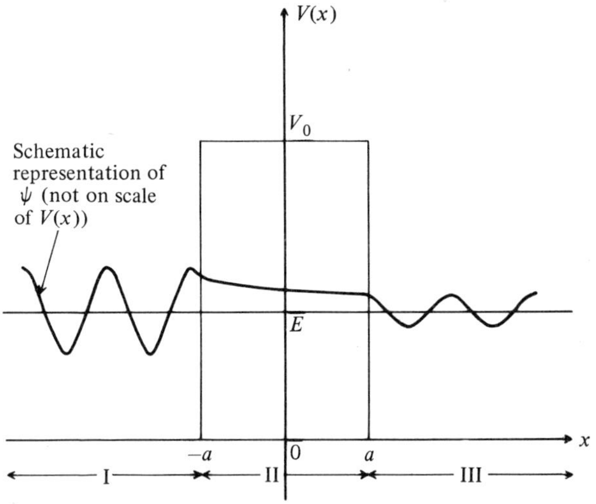

Fig. D1. "Square" potential barrier of width $2a$ and height V_0. The wave function is shown schematically to indicate transmission through the barrier.

where κ is imaginary if $E > V_0$ and real if $E < V_0$. The problem now is to use boundary conditions at $|x| = a$ to fix the values of A, β, R and T; overall normalization will determine α.

The boundary conditions at $x = \pm a$ are

$$\psi_{\underline{k}\,\mathrm{I}}(-a) = \psi_{\underline{k}\,\mathrm{II}}(-a) \qquad \psi_{\underline{k}\,\mathrm{II}}(a) = \psi_{\underline{k}\,\mathrm{III}}(a)$$

$$\left.\frac{d}{dx}\ln\psi_{\underline{k}\,\mathrm{I}}(x)\right|_{-a} = \left.\frac{d}{dx}\ln\psi_{\underline{k}\,\mathrm{II}}(x)\right|_{-a} \qquad \left.\frac{d}{dx}\ln\psi_{\underline{k}\,\mathrm{II}}(x)\right|_{a} = \left.\frac{d}{dx}\ln\psi_{\underline{k}\,\mathrm{III}}(x)\right|_{a},$$

which, upon substitution of Eqs. (D.3–D.5), become

$$e^{-ika} + Re^{ika} = A(e^{\kappa a} + \beta e^{-\kappa a}) \tag{D.6a}$$

$$Te^{ika} = A(e^{-\kappa a} + \beta e^{\kappa a}) \tag{D.6b}$$

$$ik\frac{1 - Re^{i2ka}}{1 + Re^{i2ka}} = -\kappa\frac{1 - \beta e^{-2\kappa a}}{1 + \beta e^{-2\kappa a}} \tag{D.6c}$$

$$ik = -\kappa\frac{1 - \beta e^{2\kappa a}}{1 + \beta e^{2\kappa a}} \tag{D.6d}$$

Obviously, the last of these is easy to solve for β:

$$\beta = e^{-2\kappa a}\frac{1 + ik/\kappa}{1 - ik/\kappa}. \tag{D.7}$$

Substitution of Eq. (D.7) into Eq. (D.6c) then gives

$$\frac{k}{i\kappa}\frac{1 - Re^{i2ka}}{1 + Re^{i2ka}} = \frac{1 - ik\lambda/\kappa}{\lambda - ik/\mu} \equiv \eta, \qquad \lambda \equiv \frac{1 + e^{-4\kappa a}}{1 - e^{-4\kappa a}}, \tag{D.8}$$

the right-hand side of the first equality serving to define η. This equation can be solved for R in terms of η, yielding

$$R = -e^{-i2ka}\frac{\kappa\eta + ik}{\kappa\eta - ik},$$

or, when the definition of η is used,

$$\begin{aligned} R &= -e^{-i2ka}\frac{\kappa^2 + k^2}{(\kappa^2 - k^2) - i2k\kappa\lambda} \\ &= -e^{-i2ka}(\kappa^2 + k^2)\left[(\kappa^2 - k^2) - i2k\kappa\frac{1 + e^{-4\kappa a}}{1 - e^{-4\kappa a}}\right]^{-1}. \end{aligned} \tag{D.9}$$

Now the coefficient A may be obtained directly from Eq. (D.6a),

$$A = e^{-i(k - i\kappa)a}\frac{1 + Re^{i2ka}}{1 + \beta e^{-2\kappa a}}$$

and Eq. (D.6b) may be used to obtain T:

$$\begin{aligned} T &= e^{-2(ik + \kappa)a}\frac{1 + \beta e^{2\kappa a}}{1 + \beta e^{-2\kappa a}}(1 + Re^{i2ka}) \\ &= -i4e^{-2(ik + \kappa)a}\frac{k\kappa(1 - e^{-4\kappa a})^{-1}}{(\kappa^2 - k^2) - i2\lambda k\kappa} \\ &= -i4e^{-2(ik + \kappa)a}k\kappa(1 - e^{-4\kappa a})^{-1}\left[(\kappa^2 - k^2) - i2k\kappa\frac{1 + e^{-4\kappa a}}{1 - e^{-4\kappa a}}\right]^{-1}. \end{aligned} \tag{D.10}$$

It is not actually these quantities that we wish to investigate, but rather it is the currents or intensities associated with the waves in regions I and III; they happen to be intimately related to R and T. One can see quite easily that the current in region III is $J_T = |\alpha|^2|T|^2\hbar k/\mu$, whereas in region I it is $J_0 - J_R = |\alpha|^2(1 - |R|^2)\hbar k/\mu$, where J_0 is the incident current and $J_R = |\alpha|^2|R|^2\hbar k/\mu$ is the reflected current. The ratio of transmitted current to incident current is therefore $J_T/J_0 = |T|^2$ and $|T|^2$ is called the *transmission coefficient*. Similarly, the ratio of reflected to incident current is $J_R/J_0 = |R|^2$, and $|R|^2$ is the reflection coefficient. With a little algebra, it is easy to show that $|R|^2 + |T|^2 = 1$, which is required by the conservation of probability.

Let us look carefully at the coefficients of transmission and reflection, for κ real, i.e., $V_0 > E$:

$$|T|^2 = 16e^{-4\kappa a}(1 - e^{-2\kappa a})^{-2}\frac{k^2\kappa^2}{(\kappa^2 - k^2)^2 + 4\lambda^2 k^2\kappa^2}$$

$$|R|^2 = \frac{(\kappa^2 + k^2)^2}{(\kappa^2 - k^2)^2 + 4\lambda^2 k^2 \kappa^2}.$$

Notice that $|R|^2$ and the ratio in $|T|^2$ are symmetric in k^2 and κ^2. This is useful when we consider the opposite extremes $V_0 \gg E$ and $E \gg V_0 - E$. In the former case, we have $\kappa^2 \gg k^2$, and in this limit,

$$|T|^2 \underset{k/\kappa \to 0}{\sim} \frac{4}{\sinh^2 2\kappa a}\frac{k^2}{\kappa^2}$$

$$|R|^2 \underset{k/\kappa \to 0}{\sim} 1 - \frac{4}{\sinh^2 2\kappa a}\frac{k^2}{\kappa^2}.$$

The remarkable fact is that even though $|T|^2$ may be small, it does not vanish. This quite contradicts classical prediction, and is often referred to as the *tunneling effect* of quantum mechanics.

In the case $V_0 > E \gg V_0 - E$, i.e. $k^2 \gg \kappa^2$,

$$|T|^2 \underset{\kappa/k \to 0}{\sim} \frac{4}{\sinh^2 2\kappa a}\frac{\kappa^2}{k^2}$$

and the complementary expression for $|R|^2$.

If, furthermore, $\kappa a \ll 1$ this becomes

$$|T|^2 \underset{\substack{\kappa/k \to 0 \\ \kappa a \to 0}}{\sim} \frac{1}{k^2 a^2}.$$

Thus if the energy nearly equals the height of a barrier of sufficiently small width, the penetration varies inversely with the energy (and hence with the height of the barrier). On the other hand, if the barrier is much above the energy, the penetration is proportional to the energy, as we found above.

An interesting special case occurs when we relate the barrier height to the barrier width through the demand

$$\int_{-\infty}^{\infty} W(x)\, dx = \gamma, \qquad \gamma \text{ a constant.}$$

Thus $W = \gamma/2a$. In the limit $a \to 0$, this potential yields $W(x) = \gamma\delta(x)$ so that the barrier is infinitely high with infinitesimally small width. In this limit, $\kappa \simeq \sqrt{\gamma/a}/\hbar$ for any finite energy, and one readily can find that in this limit

$$R = -\frac{i\gamma}{2k(1 + i\gamma/2k)}, \qquad T = 1 + R = \frac{2k}{2k + i\gamma}.$$

The wave function becomes

$$\psi_{\underset{\sim}{k}}(x) = \alpha[(e^{ikx} + Re^{-ikx})\theta(-x) + Te^{ikx}\theta(x)],$$

where

$$\theta(x) = \begin{cases} 1, & x > 0 \\ \frac{1}{2}, & x = 0 \\ 0, & x < 0. \end{cases}$$

It is left to the reader to show directly that this wave function satisfies

$$\left[\frac{d^2}{dx^2} - \gamma\delta(x) + k^2\right]\psi_{\underset{\sim}{k}}(x) = 0$$

and that $\alpha = (2\pi)^{-1/2}$ guarantees

$$\int_{-\infty}^{\infty} \psi^*_{\underset{\sim}{k}'}(x)\psi_{\underset{\sim}{k}}(x)\, dx = \delta(k' - k).$$

The former can be accomplished by showing that $(d/dx)\theta(x) = \delta(x)$.

BIBLIOGRAPHY

History and Criticism

1. M. JAMMER, *The Conceptual Development of Quantum Mechanics*, McGraw-Hill (1966).
2. P. A. SCHILPP, *Albert Einstein: Philosopher-Scientist*, Vol. I, II, Harper Brothers (1951).

Electromagnetism

3. J. D. JACKSON, *Classical Electrodynamics*, Wiley (1962).
4. J. B. MARION, *Classical Electromagnetic Radiation*, Academic Press (1965).
5. J. C. SLATER and N. H. FRANK, *Electromagnetism*, McGraw-Hill (1947).

Electromagnetism and Relativity

6. R. BECKER and F. SAUTER, *Electromagnetic Fields and Interactions*, Vol. I, Blaisdell (1964).
7. L. D. LANDAU and E. M. LIFSHITZ, *The Classical Theory of Fields*, Revised 2nd Ed., Addison-Wesley (1962).
8. W. K. H. PANOFSKY and M. PHILLIPS, *Classical Electricity and Magnetism*, 2nd Ed., Addison-Wesley (1962).

Relativity

9. A. EINSTEIN, *The Meaning of Relativity*, Princeton University Press (1953).
10. A. EINSTEIN *et al.*, *The Principle of Relativity*, Dover (1923).
11. C. MØLLER, *The Theory of Relativity*, Oxford University Press (1952).
12. W. PAULI, *Theory of Relativity*, Pergamon Press (1958).
13. W. RINDLER, *Special Relativity*, Interscience-Wiley (1960).

Modern Physics

14. R. M. EISBERG, *Fundamentals of Modern Physics*, Wiley (1961).
15. R. B. LEIGHTON, *Principles of Modern Physics*, McGraw-Hill (1959).
16. F. K. RICHTMEYER, E. H. KENNARD and T. LAURITSEN, *Introduction to Modern Physics*, 5th Ed., McGraw-Hill (1953).

Fourier Analysis

17. M. J. LIGHTHILL, *Fourier Analysis and Generalized Functions*, Cambridge University Press (1960).
18. I. N. SNEDDON, *Fourier Transforms*, McGraw-Hill (1951).

Quantum Theory

19. D. B. BEARD, *Quantum Mechanics*, Allyn and Bacon (1963).
20. H. A. BETHE and R. JACKIW, *Intermediate Quantum Mechanics*, 2nd Ed., W. A. Benjamin (1968).
21. H. A. BETHE and E. E. SALPETER, *Quantum Mechanics of One- and Two-Electron Atoms*, Springer-Verlag (1957).
22. D. I. BLOKHINTSEV, *Principles of Quantum Mechanics*, Allyn and Bacon (1963).
23. D. BOHM, *Quantum Theory*, Prentice-Hall (1951).
24. P. A. M. DIRAC, *Principles of Quantum Mechanics*, 4th Ed., Oxford University Press (1958).

25. A. R. EDMONDS, *Angular Momentum in Quantum Mechanics*, Princeton University Press (1957).
26. E. FERMI, *Notes on Quantum Mechanics*, University of Chicago Press (1961).
27. R. P. FEYNMAN, *et al., The Feynman Lectures on Physics*, Vol. III, Addison-Wesley (1964).
28. R. P. FEYNMAN and A. HIBBS, *Quantum Mechanics and Path Integrals*, McGraw-Hill (1966).
29. K. GOTTFRIED, *Quantum Mechanics*, Vol. I, W. A. Benjamin (1966).
30. E. IKENBERRY, *Quantum Mechanics for Mathematicians and Physicists*, Oxford University Press (1962).
31. J. M. JAUCH, *Foundations of Quantum Mechanics*, Addison-Wesley, 1967.
32. F. A. KAEMPFFER, *Concepts in Quantum Mechanics*, Academic Press (1965).
33. A. KATZ, *Classical Mechanics, Quantum Mechanics and Field Theory*, Academic Press (1965).
34. E. C. KEMBLE, *Fundamentals of Quantum Mechanics*, Dover (1937).
35. L. D. LANDAU and E. M. LIFSHITZ, *Quantum Mechanics*, Addison-Wesley (1958).
36. P. T. MATTHEWS, *Introduction to Quantum Mechanics*, McGraw-Hill (1963).
37. E. MERTZBACHER, *Quantum Mechanics*, Wiley (1961).
38. A. MESSIAH, *Quantum Mechanics*, Vols. I and II, Wiley (1962).
39. L. C. PAULING and B. E. WILSON, Jr., *Introduction to Quantum Mechanics with Applications to Chemistry*, McGraw-Hill (1935).
40. E. PERSICO. *Fundamentals of Quantum Mechanics,* Prentice-Hall (1950).
41. M. E. ROSE, *Elementary Theory of Angular Momentum*, Wiley (1957)
42. D. SAXON, *Elementary Quantum Mechanics*, Holden-Day (1968).
43. L. I. SCHIFF, *Quantum Mechanics*, 3rd Ed., McGraw-Hill (1968).
44. J. C. SLATER, *Quantum Theory of Atomic Structure*, Vol. I, McGraw-Hill (1960).
45. P. STEHLE, *Quantum Mechanics*, Holden-Day (1966).
46. G. L. TRIGG, *Quantum Mechanics*, Van Nostrand (1964).
47. J. VON NEUMANN, *Mathematical Foundations of Quantum Mechanics*, Princeton University Press (1955).
48. H. WEYL, *The Theory of Groups and Quantum Mechanics*, Dover (1931).
49. R. L. WHITE, *Basic Quantum Mechanics*, McGraw-Hill (1966).
50. E. P. WIGNER, *Group Theory and its Applications to the Quantum Mechanics of Atomic Spectra*, Academic Press (1959).

INDEX